Texts in Computational Science and Engineering

19

Editors

More information about this series at http://www.springer.com/series/5151

John A. Trangenstein

Scientific Computing

Vol. II – Eigenvalues and Optimization

Springer

John A. Trangenstein
Professor Emeritus of Mathematics
Department of Mathematics
Duke University
Durham
North Carolina, USA

Additional material to this book can be downloaded from http://extras.springer.com.

ISSN 1611-0994 ISSN 2197-179X (electronic)
Texts in Computational Science and Engineering
ISBN 978-3-030-09871-1 ISBN 978-3-319-69107-7 (eBook)
https://doi.org/10.1007/978-3-319-69107-7

Mathematics Subject Classification (2010): 15, 65

Printed on acid-free paper

This Springer imprint is published by the registered company Springer International Publishing AG part of Springer Nature.
The registered company address is: Gewerbestrasse 11, 6330 Cham, Switzerland

To my daughter Pamela

Preface

This is the second volume in a three-volume book about scientific computing. The primary goal in these volumes is to present many of the important computational topics and algorithms used in applications such as engineering and physics, together with the theory needed to understand their proper operation. However, a secondary goal in the design of this book is to allow readers to experiment with a number of interactive programs *within the book*, so that readers can improve their understanding of the problems and algorithms. This interactivity is available in the HTML form of the book, through JavaScript programs.

The intended audience for this book are upper level undergraduate students and beginning graduate students. Due to the self-contained and comprehensive treatment of the topics, this book should also serve as a useful reference for practicing numerical scientists. Instructors could use this book for multisemester courses on numerical methods. They could also use individual chapters for specialized courses such as numerical linear algebra, constrained optimization, or numerical solution of ordinary differential equations.

In order to read all volumes of this book, readers should have a basic understanding of both linear algebra and multivariable calculus. However, for this volume it will suffice to be familiar with linear algebra and single variable calculus. Some of the basic ideas for both of these prerequisites are reviewed in this text, but at a level that would be very hard to follow without prior familiarity with those topics. Some experience with computer programming would also be helpful, but not essential. Students should understand the purpose of a computer program, and roughly how it operates on computer memory to generate output.

Many of the computer programming examples will describe the use of a Linux operating system. This is the only publicly available option in our mathematics department, and it is freely available to all. Students who are using proprietary operating systems, such as Microsoft and Apple systems, will need to replace statements specific to Linux with the corresponding statements that are appropriate to their environment.

This book also references a large number of programs available in several programming languages, such as C, C^{++}, Fortran and JavaScript, as well as

MATLAB modules. These programs should provide examples that can train readers to develop their own programs, from existing software whenever possible or from scratch whenever necessary.

Chapters begin with an overview of topics and goals, followed by recommended books and relevant software. Some chapters also contain a case study, in which the techniques of the chapter are used to solve an important scientific computing problem in depth.

Chapter 1 presents numerical methods for finding eigenvalues and eigenvectors. These mathematical tools are important in such diverse problem areas as fundamental modes of vibration in engineering, or growth and decay rates in economics. Later in this chapter, we develop the singular value decomposition more carefully. At the end of the chapter, we also show how to use eigenvalues and eigenvectors to solve linear recurrences, or to compute functions of matrices, such as exponentials or square roots. This chapter depends on the material in Chap. 6 of Volume I.

Chapter 2 examines some important methods for solving large and typically sparse systems of linear equations. These numerical methods are especially useful for implementing implicit methods for partial differential equations (which are discussed in two other books by this author [172, 173]). This chapter depends on the material in Chap. 1.

Chapter 3 discusses the numerical solution of systems of nonlinear equations. Here the mathematical analysis depends strongly on multivariable calculus. There are some useful generalizations of methods from Chap. 5 of Volume I to the solution of problems in this chapter. There are also some useful generalizations of iterative methods for linear systems that extend to nonlinear systems. The chapter ends with a case study involving the Lennard-Jones potential from chemistry. This chapter depends on the material in Chap. 5 of Volume I and Chap. 2.

Chapter 4 examines constrained optimization methods. This subject is usually left to books that are devoted solely to this topic. The chapter begins with linear programming problems, advances to quadratic programming problems, and ends with general nonlinear programming problems. This chapter depends on the material in Chap. 3 and Chap. 6 of Volume I.

In summary, this volume covers mathematical and numerical analysis, algorithm selection, and software development. The goal is to prepare readers to build programs for solving important problems in their chosen discipline. Furthermore, they should develop enough mathematical sophistication to know the limitations of the pieces of their algorithm and to recognize when numerical features are due to programming bugs rather than the correct response of their problem.

I am indebted to many teachers and colleagues who have shaped my professional experience. I thank Jim Douglas Jr. for introducing me to numerical analysis as an undergrad. (Indeed, I could also thank a class in category theory for motivating me to look for an alternative field of mathematical study.) John Dennis, James Bunch, and Jorge Moré all provided a firm foundation for my training in numerical analysis, while Todd Dupont, Jim Bramble, and Al Schatz gave me important training in finite element analysis for my PhD thesis. But I did not really learn to program until I met Bill Gragg, who also emphasized the importance of classical analysis in the development of fundamental algorithms. I also learned from my students, particularly Randy LeVeque, who was in the first numerical analysis class I ever taught. Finally, I want to thank Bill Allard for many conversations about the deficiencies in numerical analysis texts. I hope that this book moves the field a bit in the direction that Bill envisions.

Most of all, I want to thank my family for their love and support.

Durham, NC, USA John A. Trangenstein
July 7, 2017

Contents

Contents for Volume 1

Contents for Volume 3

Chapter 1
Eigenvalues and Eigenvectors

The solution of the algebraic eigenvalue problem has for long had a particular fascination for me because it illustrates so well the difference between what might be termed classical mathematics and practical numerical analysis. The eigenvalue problem has a deceptively simple formulation and the background theory has been known for many years; yet the determination of accurate solutions presents a wide variety of challenging problems.

James H. Wilkinson, ***The Algebraic Eigenvalue Problem***, *p. v (1965)*

Abstract This chapter begins with the basic theory of eigenvalues and eigenvectors of matrices. Essential concepts such as characteristic polynomials, the Fundamental Theorem of Algebra, the Gerschgorin circle theorem, invariant subspaces, change of basis, spectral radius and the distance between subspaces are developed. Hermitian matrices are analyzed through the spectral theorem, and a perturbation analysis of their eigenvalue problem is performed. This chapter presents and examines algorithms for finding eigenvalues of Hermitian tridiagonal matrices, such as bisection, the power method, QL, QR, implicit QR, divide and conquer and dqds. Reduction of general Hermitian matrices to tridigonal form, and the Lanczos process are also discussed. Next, the eigenvalue problem for general matrices is examined. Theory for the Schur decomposition and the Jordan form are presented. Perturbation theory and conditions numbers lead to *a posteriori* estimates for general eigenvalue problems. Numerical methods for upper Hessenberg matricces are discussed, followed by general techniques for orthogonal similarity transformation to upper Hessenberg form. Then the chapter turns to the singular value decomposition, with theory discussing its existence, pseudo-inverses and the minimax theorem. Methods for reducing general matrices to bidiagonal form, and techniques for finding singular value decompositions of bidiagonal matrices follow next. The chapter ends with discussions of linear recurrences and functions of matrices.

Additional Material: The details of the computer programs referred in the text are available in the Springer website (http://extras.springer.com/2018/978-3-319-69107-7) for authorized users.

J.A. Trangenstein, *Scientific Computing*, Texts in Computational Science and Engineering 19, https://doi.org/10.1007/978-3-319-69107-7_1

1.1 Overview

Our goal in this chapter is to develop numerical methods to find directions that are left unchanged by a linear transformation, and the scaling factor the linear transformation applies to such a direction. Mathematically speaking, the problem takes the following simple form. Given a square matrix **A**, we want to find a nonzero vector **eigenvector** **x** and a scalar **eigenvalue** λ so that

$$\mathbf{A}\mathbf{x} = \mathbf{x}\lambda \ .$$

These invariant directions **x** can have important physical utility, such as the principal axes of stress and strain in solid mechanics (see, for example, Fung [76, p. 72ff], Malvern [121, p. 85ff] or Truesdell [174, p. 17f]). Similar ideas are used in combinatorics to solve linear recurrences (see, for example, Roberts [150, p. 210ff] or Tucker [175, p. 297ff]). Eigenvalues are also used to analyze the stability of critical points in systems of ordinary differential equations (see, for example, Drazin [56, p. 151ff] or Kelley and Peterson [109, p. 52ff]). In economics, eigenvectors are used to find positive production vectors that keep steady demand (see Szyld et al. [169]).

We will begin by developing some fundamental concepts for eigenvalues and eigenvectors in Sect. 1.2. For example, we will see that eigenvalues are zeros of the **characteristic polynomial** of the associated matrix. Thus, our work in this chapter might draw on our experience with solving nonlinear equations in Chap. 5 of Volume I. However, in order to find eigenvectors we will find it useful to employ unitary matrices to transform a given matrix with desired eigenvectors into a simpler form. These transformations will use ideas we developed for solving least squares problems in Chap. 6 of Volume I. We will find that the perturbation theory for eigenvalues and eigenvectors is even more difficult than for the least squares problem. In cases where several linearly independent directions share a single eigenvalue, it is improper to discuss the perturbation of an eigenvector; rather, we will discuss perturbations of **invariant subspaces**.

We will develop separate theory and algorithms for Hermitian and non-Hermitian matrices in Sect. 1.3. The very important **Spectral Theorem** 1.3.1 will show that Hermitian matrices have real eigenvalues and mutually orthogonal eigenvectors. This result will allow us to develop very fast and accurate numerical methods for finding eigenvalues and eigenvectors of Hermitian matrices. On the other hand, non-Hermitian matrices may possess **eigenvector deficiencies** that are very difficult to determine numerically. As a result, we will have to develop more complicated theory and algorithms to treat these matrices in Sect. 1.4.

We will re-examine the **singular value decomposition** in Sect. 1.5. In particular, we will use eigenvalues and eigenvectors to prove the existence of the singular value decomposition, and to develop an efficient algorithm for its computation. Linear recurrences will be treated in Sect. 1.6. We will end the chapter with a discussion of functions of matrices in Sect. 1.7.

For additional reading on the material in this chapter, we recommend the books by Bai et al. [9], Demmel [48], Parlett [137], Stewart [166], Wilkinson [186], Wilkinson and Reinsch [187].

For eigenproblem software, we recommend LAPACK (written in Fortran), CLAPACK (translated from Fortran to C), numpy.linalg (written in Python), and GNU Scientific Library (GSL) Eigensystems (written in C). We also recommend MATLAB commands eig, condeig, hess, schur, svd, norm, mpower, expm and sqrtm. Scilab provides several functions to compute eigenvalues and singular values.

1.2 Fundamental Concepts

Suppose that $\mathbf{A}$ is an $n \times n$ matrix. If we can find a nonzero n-vector $\mathbf{x}$ and a scalar λ so that $\mathbf{A}\mathbf{x} = \mathbf{x}\lambda$, then $\mathbf{A}$ leaves the direction of $\mathbf{x}$ unchanged. Along that direction, the linear transformation corresponds to a simple rescaling. This suggests the following definition.

Definition 1.2.1 If $\mathbf{A}$ is an $n \times n$ matrix and $\mathbf{x}$ is a nonzero n-vector, then $\mathbf{x}$ is an **eigenvector** of $\mathbf{A}$ with **eigenvalue** λ if and only if $\mathbf{A}\mathbf{x} = \mathbf{x}\lambda$.
In this section, we will present some important theory about eigenvectors and eigenvalues. We will see in Sect. 1.2.2 that eigenvalues are zeros of the characteristic polynomial of the matrix. Afterward, we will determine how eigenvalues and eigenvectors are affected by certain matrix operations, such as the inverse and Hermitian. In Sect. 1.2.4, we will use the fundamental theorem of algebra to show that matrices have a full set of eigenvalues. We will learn how to estimate eigenvalues by the Gerschgorin Circle theorem in Sect. 1.2.5. Eigenvalues will be associated with certain matrix invariants, especially the trace and determinant, in Sect. 1.2.6. We will discuss the effect of a change of basis on the eigenvalues and eigenvectors of a matrix in Sects. 1.2.8 and 1.2.9. Finally, we will develop a method to measure the distance between subspaces in Sect. 1.2.11; this measurement will be useful in studying the effect of matrix perturbations on eigenvectors. However, we will begin by finding eigenvalues and eigenvectors of some matrices that have geometric or algebraic utility.

1.2.1 Examples

Example 1.2.1 Consider the linear transformation given by reflection about some nonzero 2-vector u. Obviously, the reflection of u is u. If $v \perp u$, denoting that $\mathbf{v}$ and $\mathbf{u}$ are **orthogonal vectors** , then the reflection of v is $-v$. Thus a reflection in two-dimensional space has at least two eigenvalues, namely 1 and -1. An illustration of reflection is shown in Fig. 1.1.

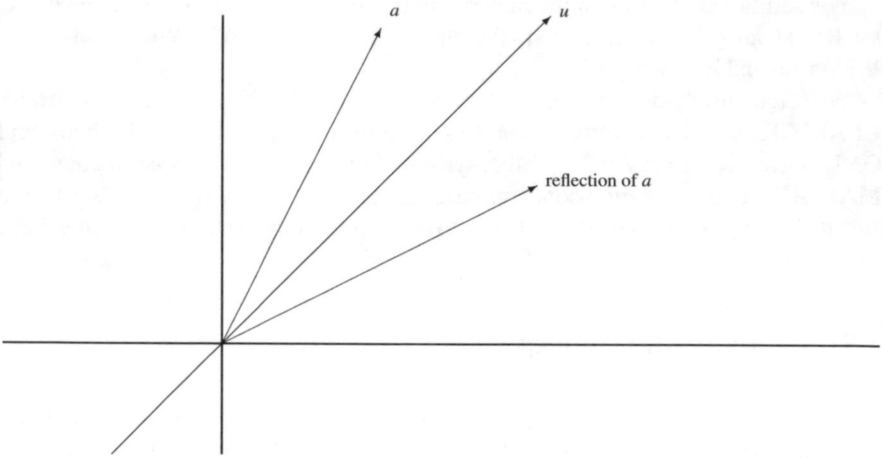

Fig. 1.1 Reflection of vector **a** about line containing vector **u**

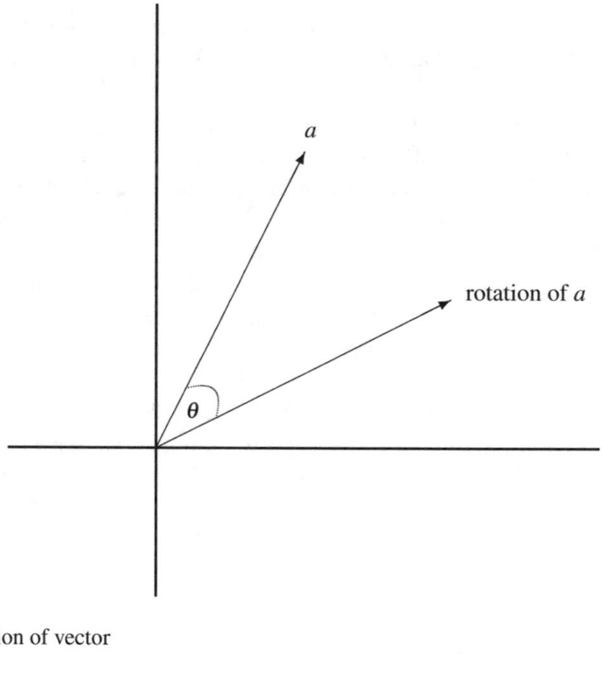

Fig. 1.2 Rotation of vector

Example 1.2.2 Consider the linear transformation given by rotation counterclockwise through some angle $\theta \neq \pm\pi$ in the plane. Figure 1.2 shows that such a linear transformation has no real invariant directions, and therefore no real eigenvalues.

Our next example shows that for some special matrices, it is easy to find eigenvalues and eigenvectors.

Example 1.2.3 If $\mathbf{D}$ is an $n \times n$ diagonal matrix with diagonal entries δ_i, then

$$\text{for } 1 \leq i \leq n \text{ , } \mathbf{D}\mathbf{e}_i = \mathbf{e}_i \delta_i \text{ .}$$

In other words, the axis vectors $\mathbf{e}_i$ are eigenvectors of a diagonal matrix $\mathbf{D}$, and the diagonal entries of $\mathbf{D}$ are the eigenvalues.

1.2.2 Characteristic Polynomial

In general, if $\mathbf{x}$ is an eigenvector of $\mathbf{A}$ with eigenvalue λ, then

$$(\mathbf{A} - \mathbf{I}\lambda)\mathbf{x} = \mathbf{0} \text{ ,}$$

where $\mathbf{I}$ represents the identity matrix. Since $\mathbf{x} \neq \mathbf{0}$, Lemma 3.2.9 in Chap. 3 of Volume I shows that $\mathbf{A} - \mathbf{I}\lambda$ is singular. Then Lemma 3.2.19 in Chap. 3 of Volume I completes the proof of the following lemma.

Lemma 1.2.1 *Suppose that* $\mathbf{A}$ *is an* $n \times n$ *matrix. Then* λ *is an eigenvalue of* $\mathbf{A}$ *if and only if* $\det(\mathbf{A} - \mathbf{I}\lambda) = 0$. *Furthermore, if* λ *is an eigenvalue of* $\mathbf{A}$ *then* $\mathbf{x} \neq \mathbf{0}$ *is a corresponding eigenvector of* $\mathbf{A}$ *if and only if* $(\mathbf{A} - \mathbf{I}\lambda)\mathbf{x} = \mathbf{0}$.
The first statement in this lemma gives us a polynomial equation to solve for an eigenvalue λ. If we can find a solution of this equation, then the second result in this lemma gives us an equation to solve for the eigenvector $\mathbf{x}$.

Lemma 1.2.1 suggests the following definition.

Definition 1.2.2 If $\mathbf{A}$ is a square matrix, then $\det(\mathbf{A} - \mathbf{I}\lambda)$ is called the **characteristic polynomial** of $\mathbf{A}$.
The eigenvalues of a matrix are the zeros of its characteristic polynomial.

Example 1.2.4 Let us find the eigenvalues and eigenvectors of

$$\mathbf{A} = \begin{bmatrix} 1 & -1 \\ 2 & 4 \end{bmatrix} .$$

Lemma 1.2.1 shows that the eigenvalues satisfy

$$0 = \det(\mathbf{A} - \mathbf{I}\lambda) = \det \begin{bmatrix} 1-\lambda & -1 \\ 2 & 4-\lambda \end{bmatrix} = (1-\lambda)(4-\lambda) + 2 = (\lambda - 3)(\lambda - 2) .$$

Thus the eigenvalues are either 2 or 3.

Let us find an eigenvector for $\lambda = 2$. Let $\mathscr{N}(\mathbf{A})$ denote the **nullspace** of the matrix $\mathbf{A}$. Since $\mathbf{x} \in \mathscr{N}(\mathbf{A} - \mathbf{I}\lambda)$, we solve

$$\mathbf{0} = (\mathbf{A} - \mathbf{I}2)\mathbf{x} = \begin{bmatrix} -1 & -1 \\ 2 & 2 \end{bmatrix} \begin{bmatrix} \xi_1 \\ \xi_2 \end{bmatrix} .$$

It is easy to see that we can choose

$$\mathbf{x} = \begin{bmatrix} 1 \\ -1 \end{bmatrix} ,$$

or the eigenvector could be chosen to be any nonzero scalar multiple of this vector.

Next, let us find an eigenvector for $\lambda = 3$. We solve

$$\mathbf{0} = (\mathbf{A} - \mathbf{I}3)\mathbf{x} = \begin{bmatrix} -2 & -1 \\ 2 & 1 \end{bmatrix} \begin{bmatrix} \xi_1 \\ \xi_2 \end{bmatrix} .$$

In this case, we can take

$$\mathbf{x} = \begin{bmatrix} 1 \\ -2 \end{bmatrix}$$

or any nonzero scalar multiple of this vector.

At this point, we can form a matrix $\mathbf{X}$ with columns given by the eigenvectors we have found, and create a diagonal matrix $\boldsymbol{\Lambda}$ from the eigenvalues. Then we have

$$\mathbf{AX} \equiv \begin{bmatrix} 1 & -1 \\ 2 & 4 \end{bmatrix} \begin{bmatrix} 1 & 1 \\ -1 & -2 \end{bmatrix} = \begin{bmatrix} 1 & 1 \\ -1 & -2 \end{bmatrix} \begin{bmatrix} 2 & 0 \\ 0 & 3 \end{bmatrix} \equiv \mathbf{X}\boldsymbol{\Lambda} .$$

Example 1.2.5 Let us find the eigenvalues of the rotation

$$\mathbf{A} = \begin{bmatrix} \cos\theta & -\sin\theta \\ \sin\theta & \cos\theta \end{bmatrix} .$$

Lemma 1.2.1 shows that the eigenvalues satisfy

$$\begin{aligned} 0 = \det(\mathbf{A} - \mathbf{I}\lambda) &= \det \begin{bmatrix} \cos\theta - \lambda & -\sin\theta \\ \sin\theta & \cos\theta - \lambda \end{bmatrix} \\ &= (\cos\theta - \lambda)^2 + \sin^2\theta = (\lambda - \cos\theta - \mathrm{i}\sin\theta)(\lambda - \cos\theta + \mathrm{i}\sin\theta) . \end{aligned}$$

Thus the eigenvalues are

$$\lambda = \cos\theta \pm i\sin\theta = \mathrm{e}^{\pm i\theta} .$$

These computations show that unless $\sin\theta = 0$ (i.e., $\theta = \pm\pi$), the eigenvalues of a 2×2 rotation are complex.

Example 1.2.6 Suppose that $\mathbf{P}$ is a projector. Then Lemma 3.2.11 in Chap. 3 of Volume I implies that $\mathbf{P}^2 = \mathbf{P}$. If $\mathbf{P}$ has an eigenvector $\mathbf{x}$ with eigenvalue λ, then

$$\mathbf{x}\lambda = \mathbf{P}\mathbf{x} = \mathbf{P}\mathbf{P}\mathbf{x} = \mathbf{P}\mathbf{x}\lambda = \mathbf{x}\lambda^2 .$$

Since eigenvectors are nonzero, we conclude that $\lambda^2 = \lambda$, or $0 = \lambda(\lambda - 1)$. Thus the eigenvalues of a projector are either 0 or 1.

Example 1.2.7 Suppose that $\mathbf{R}$ is an $n \times n$ right-triangular matrix. Then $\mathbf{R} - \mathbf{I}\lambda$ is also right-triangular, and the discussion in Sect. 3.11 in Chap. 3 of Volume I shows that

$$\det(\mathbf{R} - \mathbf{I}\lambda) = \prod_{i=1}^{n} (\varrho_{ii} - \lambda) .$$

Thus the eigenvalues of a right-triangular matrix are its diagonal entries. Similarly, the eigenvalues of a left-triangular matrix are its diagonal entries.

Example 1.2.7 is a special case of the following more general result.

Lemma 1.2.2 *If* $\mathbf{A}$ *is an* $n \times n$ *matrix, then* $\det(\mathbf{A} - \mathbf{I}\lambda)$ *is a polynomial of degree at most* n *in* λ.

Proof This can be demonstrated by using the Laplace expansion in Theorem 3.2.4 in Chap. 3 of Volume I.

The following lemma will be useful when we discuss deflation in the proof of the Spectral Theorem 1.3.1.

Lemma 1.2.3 *Suppose that* $\mathbf{A}$ *is a square matrix, and we can partition*

$$\mathbf{A} = \begin{bmatrix} \mathbf{A}_{11} & \mathbf{A}_{12} \\ \mathbf{0} & \mathbf{A}_{22} \end{bmatrix} .$$

Then

$$\det \mathbf{A} = \det \mathbf{A}_{11} \cdot \det \mathbf{A}_{22} ,$$

and the eigenvalues of $\mathbf{A}$ *are either eigenvalues of* $\mathbf{A}_{11}$ *or eigenvalues of* $\mathbf{A}_{22}$.

Proof Suppose that $\mathbf{A}$ is $n \times n$ and $\mathbf{A}_{11}$ is $k \times k$. Let π denote a permutation of $\{1, \ldots, n\}$, π_1 denote a permutation of $\{1, \ldots, k\}$ and π_2 denote a permutation of $\{k+1, \ldots, n\}$. The Laplace expansion (3.7) in Chap. 3 of Volume I gives us

$$\det \mathbf{A} = \sum_{\text{permutations } \pi} (-1)^{\mathbf{N}(\pi)} \alpha_{\pi(1),1} \ldots \alpha_{\pi(n),n}$$

then the zero block in $\mathbf{A}$ implies that

$$= \sum_{\text{permutations } (\pi_1,\pi_2)} (-1)^{\mathbf{N}(\pi_1)+\mathbf{N}(\pi_2)} \alpha_{\pi_1(1),1} \ldots \alpha_{\pi_1(k),k} \alpha_{\pi_2(k+1),k+1} \ldots \alpha_{\pi_2(n),n}$$

then distributing the products gives us

$$
\begin{aligned}
&= \left(\sum_{\text{permutations } \pi_1} (-1)^{\mathbf{N}(\pi_1)} \alpha_{\pi_1(1),1} \ldots \alpha_{\pi_1(k),k} \right) \\
&\times \left(\sum_{\text{permutations } \pi_2} (-1)^{\mathbf{N}(\pi_2)} \alpha_{\pi_2(k+1),k+1} \ldots \alpha_{\pi_2(n),n} \right) \\
&= \det \mathbf{A}_{11} \cdot \det \mathbf{A}_{22} \ .
\end{aligned}
$$

We now see that the characteristic polynomial for $\mathbf{A}$ can be factored as

$$\det(\mathbf{A} - \mathbf{I}\lambda) = \det(\mathbf{A}_{11} - \mathbf{I}\lambda)\det(\mathbf{A}_{22} - \mathbf{I}\lambda) \ .$$

The final claim in the lemma follows easily from this factorization.

Exercise 1.2.1 Find the eigenvalues and eigenvectors of $\begin{bmatrix} 3 & 4 \\ 4 & -3 \end{bmatrix}$.

Exercise 1.2.2 Find the eigenvalues and eigenvectors of $\begin{bmatrix} \alpha & \beta \\ \beta & \alpha \end{bmatrix}$ in terms of α and β.

Exercise 1.2.3 Find the matrix $\mathbf{X}$ of eigenvectors and matrix $\boldsymbol{\Lambda}$ of eigenvalues for

$$\mathbf{A} = \begin{bmatrix} 1 & 1/4 & 0 \\ 0 & 1/2 & 0 \\ 0 & 1/4 & 1 \end{bmatrix}$$

and verify that $\mathbf{AX} = \mathbf{X}\boldsymbol{\Lambda}$.

Exercise 1.2.4 Find the matrix $\mathbf{X}$ of eigenvectors and matrix $\boldsymbol{\Lambda}$ of eigenvalues for

$$\mathbf{A} = \begin{bmatrix} 1 & 1/2 & 0 \\ 0 & 1/2 & 1 \\ 0 & 0 & 1 \end{bmatrix}$$

and verify that $\mathbf{AX} = \mathbf{X}\boldsymbol{\Lambda}$.

Exercise 1.2.5 Let $\mathbf{J} = \begin{bmatrix} 1 & 1 \\ 0 & 1 \end{bmatrix}$.

1. Show that both the identity matrix $\mathbf{I}$ and $\mathbf{J}$ have the same characteristic polynomial.
2. Find all linearly independent eigenvectors of $\mathbf{J}$.
3. Find an eigenvector of $\mathbf{I}$ that is not an eigenvector of $\mathbf{J}$.

Exercise 1.2.6 If

$$\mathbf{A} = \begin{bmatrix} 0 & 0 & 1 \\ 0 & 1 & 0 \\ 1 & 0 & 0 \end{bmatrix} ,$$

find the eigenvalues and eigenvectors of **A**.

Exercise 1.2.7 Find the matrix **X** of eigenvalues and matrix $\mathbf{\Lambda}$ of eigenvectors for

$$\mathbf{R} = \begin{bmatrix} 2 & -3 & 5 \\ & -1 & 5 \\ & & 4 \end{bmatrix} ,$$

and verify that $\mathbf{R}\mathbf{X} = \mathbf{X}\mathbf{\Lambda}$.

Exercise 1.2.8 If $\mathbf{I}_{ij}$ is an interchange matrix with $i \neq j$, find the eigenvalues and eigenvectors of $\mathbf{I}_{ij}$.

Exercise 1.2.9 Develop an algorithm to find the eigenvalues of a 2×2 matrix with complex entries. When you are done, compare your algorithm to the computations beginning with line 590 of LAPACK routine zlaqr0.

Exercise 1.2.10 Consider the matrix

$$\mathbf{A} = \begin{bmatrix} 3 & 2 & 2 & -4 \\ 2 & 3 & 2 & -1 \\ 1 & 1 & 2 & -1 \\ 2 & 2 & 2 & -1 \end{bmatrix} .$$

1. Find the characteristic polynomial for **A**.
2. Show that $\lambda = 1$ is a zero of this characteristic polynomial, and find the dimension of $\mathscr{N}(\mathbf{A} - \mathbf{I})$.
3. Show that $\lambda = 2$ is a zero of this characteristic polynomial, and find the dimension of $\mathscr{N}(\mathbf{A} - \mathbf{I}2)$.
4. Show that $\lambda = 3$ is a zero of this characteristic polynomial, and find the dimension of $\mathscr{N}(\mathbf{A} - \mathbf{I}3)$.

1.2.3 Inverses and Hermitians

Next, we would like to discuss how the inverse and Hermitian operators affect eigenvectors and eigenvalues.

Lemma 1.2.4 *If* $\mathbf{A}$ *is a square matrix, then* $\mathbf{A}$ *is nonsingular if and only if none of its eigenvalues are zero. If* $\mathbf{A}$ *is nonsingular and* $\mathbf{A}$ *has an eigenvalue* λ *with eigenvector* $\mathbf{x}$*, then* $\mathbf{x}$ *is an eigenvector of* $\mathbf{A}^{-1}$ *with eigenvalue* $1/\lambda$*. Here* $\mathbf{A}^{-1}$ *denotes the* **inverse** *of the matrix* $\mathbf{A}$.

Proof To prove the first claim, it is equivalent to show that $\mathbf{A}$ is singular if and only if it has a zero eigenvalue. If $\mathbf{A}$ has a zero eigenvalue, then there is a nonzero vector $\mathbf{x}$ so that $\mathbf{A}\mathbf{x} = \mathbf{0}$, so $\mathbf{A}$ is singular. Conversely, if $\mathbf{A}$ is singular, then there is a nonzero vector $\mathbf{z}$ so that $\mathbf{A}\mathbf{z} = \mathbf{0}$, so $\mathbf{z}$ is an eigenvector of $\mathbf{A}$ with eigenvalue zero.

If $\mathbf{A}$ is nonsingular and $\mathbf{A}\mathbf{x} = \mathbf{x}\lambda$, then $\lambda \neq 0$. We can multiply by $\mathbf{A}^{-1}$ and divide by λ to get $\mathbf{x}/\lambda = \mathbf{A}^{-1}\mathbf{x}$. This shows that $\mathbf{x}$ is an eigenvector of $\mathbf{A}^{-1}$ with eigenvalue $1/\lambda$.

The following lemma is a simple consequence of our study of characteristic polynomials.

Lemma 1.2.5 *If* λ *is an eigenvalue of the square matrix* $\mathbf{A}$*, then* $\overline{\lambda}$ *is an eigenvalue of* $\mathbf{A}^H$*. Here* $\mathbf{A}^H$ *represents the complex conjugate transpose* (a.k.a. **Hermitian**) *of the matrix* $\mathbf{A}$.

Proof Lemma 1.2.1 shows that

$$0 = \det(\mathbf{A} - \mathbf{I}\lambda)$$

then Lemma 3.2.18 in Chap. 3 of Volume Igives us

$$= \det(\mathbf{A}^H - \mathbf{I}\overline{\lambda}) \; .$$

Finally, Lemma 1.2.1 proves that $\overline{\lambda}$ is an eigenvalue of $\mathbf{A}^H$.

Exercise 1.2.11 If $\mathbf{A} = \mathbf{A}^H$, prove that $\det(\mathbf{A})$ is real.

Exercise 1.2.12 If $\mathbf{u}$ and $\mathbf{v}$ are n-vectors, find the eigenvalues and eigenvectors of $\mathbf{A} = \mathbf{u}\mathbf{v}^H$.

Exercise 1.2.13 If u and v are n-vectors, use eigenvalues to show that $\det(\mathbf{I} + \mathbf{v}\mathbf{u}^H) = 1 + v^H\mathbf{u}$.

Exercise 1.2.14 If $\mathbf{R}$ is right triangular with distinct diagonal entries, how would you find the eigenvalues of $\mathbf{R}$, and how would you find the eigenvectors?

1.2.4 Fundamental Theorem of Algebra

The following very important theorem tells us where to look for eigenvalues, and how many to expect.

Theorem 1.2.1 (Fundamental Theorem of Algebra) *Every polynomial of degree $n \geq 1$ with complex coefficients has n zeros in the complex plane (counting multiplicity)*

Proof Let $p(\zeta)$ be a polynomial of degree n. First, we will show that $p(\zeta)$ has at least one zero. As $|\zeta| \to \infty$ we must have that $|p(\zeta)| \to \infty$, so the minimum of $|p(\zeta)|$ occurs at some finite value ζ_0. Suppose that $|p(\zeta_0)| > 0$; we will find a contradiction. Since ζ_0 is a zero of $p(\zeta) - p(\zeta_0)$, there must be an integer $m \leq n$ so that

$$p(\zeta) = p(\zeta_0) + (\zeta - \zeta_0)^m q(\zeta)$$

for some polynomial $q(\zeta)$ of degree $n - m$ such that $q(\zeta_0) \neq 0$. Let

$$p(\zeta_0) = \varrho e^{i\psi} \ , \ q(\zeta_0) = \omega e^{i\phi}$$

be the polar representations of these two nonzero complex numbers. Since $p(\zeta_0)$ and $q(\zeta_0)$ are nonzero, ϱ and ω are both positive. Let $\theta = (\psi + \pi - \phi)/m$ and choose ε so that $0 < \varepsilon < (\varrho/\omega)^{1/m}$. For this range of values of ε, we can find some $\alpha > 0$ so that

$$\|\zeta - \zeta_0| \leq \varepsilon \Rightarrow |q(\zeta) - z(\zeta_0)| \leq \alpha|\zeta - \zeta_0| \ .$$

If necessary, we can make ε to be even smaller by requiring that $\varepsilon < \omega/\alpha$. Define $\zeta = \zeta_0 + \varepsilon e^{i\theta}$. Then

$$\begin{aligned} p(\zeta) &= \varrho e^{i\psi} + \varepsilon^m e^{im\theta} [\omega e^{i\phi} + q(\zeta) - q(\zeta_0)] \\ &= (\varrho - \varepsilon^m \omega) e^{i\psi} + [q(\zeta) - q(\zeta_0)] \varepsilon^m \omega e^{i\theta} \end{aligned}$$

has modulus satisfying

$$\begin{aligned} |p(\zeta)| &\leq |\varrho - \varepsilon^m \omega| + \alpha\omega\varepsilon^{m+1} \\ &= \varrho - \varepsilon^m [\omega - \alpha\varepsilon] < \varrho \ . \end{aligned}$$

This shows that we can choose ζ so that $|p(\zeta)|$ is smaller than $|p(\zeta_0)|$. Since $|p(\zeta_0)|$ is the minimum, our assumption that $|p(\zeta_0)| > 0$ leads to a contradiction. As a result, p must have a zero.

Next, we show that a polynomial of degree n must have n zeros, counting multiplicity. The previous part of the proof shows that every polynomial of degree one has one zero. Inductively, assume that all polynomials of degree at most $n - 1$ have $n - 1$ zeros, counting multiplicity. If ζ_0 is a zero of multiplicity m of an nth degree polynomial $p(\zeta)$, then we can find a polynomial $q(\zeta)$ of degree at most $n - m$ so that

$$p(\zeta) = (\zeta - \zeta_0)^m q(\zeta)$$

If $n > m$, then our inductive hypothesis shows that q has $n - m$ zeros counting multiplicity. These, together with ζ_0, give us n zeros of $p(\zeta)$.

Since the eigenvalues of an $n \times n$ matrix $\mathbf{A}$ are zeros of the characteristic polynomial $\det(\mathbf{A} - \mathbf{I}\lambda) = 0$, the Fundamental Theorem of Algebra implies the following:

Corollary 1.2.1 *Every $n \times n$ matrix has n complex eigenvalues, counting multiplicity.*

Exercise 1.2.15 What are the real and imaginary parts of the three solutions to $\zeta^3 = 1$ for a complex scalar ζ? (Hint: use the polar form for complex numbers.)

Exercise 1.2.16 Find all complex numbers ξ that satisfy the equation $e^{i\xi} = -1$.

Exercise 1.2.17 Find all complex numbers θ that satisfy the equation $e^{i\theta} = i$.

Exercise 1.2.18 Where are the complex numbers ζ that satisfy $\overline{\zeta} = 1/\zeta$?

Exercise 1.2.19 Plot the curve given by the complex number $e^{(1+i)\theta}$ as θ increases from 0 to 2π.

Exercise 1.2.20 Find the lengths and inner product $\mathbf{z}_1 \cdot \mathbf{z}_2$ for $\mathbf{z}_1 = [2 - 4i, 4i]^\top$ and $\mathbf{z}_2 = [2 + 4i, 4]^\top$.

1.2.5 Gerschgorin Circle Theorem

The fundamental theorem of algebra does not imply that every $n \times n$ matrix has n linearly independent eigenvectors, nor does it mean that eigenvalues are easy to compute. In fact, the easy examples of Sect. 1.2.2 give us a false sense of how difficult it is to find eigenvalues of general matrices. Fortunately, the following theorem can give us an estimate of the location of the eigenvalues.

Theorem 1.2.2 (Gerschgorin Circle) *If λ is an eigenvalue of an $n \times n$ matrix $\mathbf{A}$ with components α_{ij}, then there is an integer $i \in [1, n]$ so that*

$$|\lambda - \alpha_{ii}| \leq \sum_{j \neq i} |\alpha_{ij}| \ .$$

Proof Suppose that $\mathbf{x}$ is an eigenvector of $\mathbf{A}$ corresponding to λ, and that $\mathbf{x}$ has components ξ_j. We can assume that $\mathbf{x}$ has been scaled so that $1 = \|\mathbf{x}\|_\infty = |\xi_i|$ for some index i. Then $|\xi_j| \leq 1$ for all j, and

$$\mathbf{e}_i{}^H \mathbf{A}\mathbf{x} = \sum_{j=1}^{n} \alpha_{ij}\xi_j = \mathbf{e}_i{}^H \mathbf{x}\lambda = \xi_i \lambda \ .$$

This can be rewritten in the form

$$(\lambda - \alpha_{ii})\xi_i = \sum_{j \neq i} \alpha_{ij}\xi_j \, .$$

We can take absolute values of both sides to see that

$$|\lambda - \alpha_{ii}| = |\lambda - \alpha_{ii}|\,|\xi_i| = \left|\sum_{j \neq i} \alpha_{ij}\xi_j\right| \leq \sum_{j \neq i} |\alpha_{ij}\xi_j| \leq \sum_{j \neq i} |\alpha_{ij}| \; .$$

Geometrically, This theorem says that the eigenvalues of $\mathbf{A}$ are contained in circles with centers given by the diagonal entries of $\mathbf{A}$, and radii given by the sum of the corresponding off-diagonal entries of $\mathbf{A}$.

Exercise 1.2.21 Use the Gerschgorin Circle Theorem to locate the eigenvalues of the following matrices:

1.

$$\begin{bmatrix} 1 & 0 & -1 \\ 1 & 2 & 1 \\ 2 & 2 & 3 \end{bmatrix}$$

2.

$$\begin{bmatrix} 1 & 2 & 2 \\ 0 & 2 & 1 \\ -1 & 2 & 2 \end{bmatrix}$$

3.

$$\begin{bmatrix} -2 & -8 & -12 \\ 1 & 4 & 4 \\ 0 & 0 & 1 \end{bmatrix}$$

4.

$$\begin{bmatrix} 2-\mathrm{i} & 0 & \mathrm{i} \\ 0 & 1+\mathrm{i} & 0 \\ \mathrm{i} & 0 & 2-\mathrm{i} \end{bmatrix}$$

5.

$$\begin{bmatrix} 3 & 2 & 2 & -4 \\ 2 & 3 & 2 & -1 \\ 1 & 1 & 2 & -1 \\ 2 & 2 & 2 & -1 \end{bmatrix}$$

Exercise 1.2.22 Suppose that $\mathbf{A}$ is Hermitian, has positive diagonal entries, and is **strictly diagonally dominant**, meaning that

$$|\mathbf{A}_{ii}| > \sum_{j \neq i} \left|\mathbf{A}_{ij}\right| .$$

Show that $\mathbf{A}$ is positive, as in Definition 3.13.1 in Chap. 3 of Volume I.

1.2.6 Trace and Determinant

In constructing models for physical systems, it is often essential that the model be **invariant** under certain changes in the frame of reference (see Marsden and Hughes [124, p. 8] or Truesdell [174, p. 22]). Practically speaking, physical laws should not change with the viewer's coordinate system. When the physical law involves a linear transformation, it is natural to search for the invariants of that transformation, and express the physical law in terms of those invariants.

To begin our discussion of invariants, we will examine some small but general characteristic polynomials, and try to discover a pattern. This discussion will lead us to the trace and determinant of a matrix. We will relate these two quantities to the eigenvalues of a matrix in this section. Later, in Lemma 1.2.13 we will show that a change of basis preserves eigenvalues, and therefore preserves the trace and determinant.

Example 1.2.8 The characteristic polynomial of a general 2×2 matrix

$$\mathbf{A} = \begin{bmatrix} \alpha_{11} & \alpha_{12} \\ \alpha_{21} & \alpha_{22} \end{bmatrix}$$

is

$$\det(\mathbf{A} - \mathbf{I}\lambda) = \det \begin{bmatrix} \alpha_{11} - \lambda & \alpha_{12} \\ \alpha_{21} & \alpha_{22} - \lambda \end{bmatrix} = (\alpha_{11} - \lambda)(\alpha_{22} - \lambda) - \alpha_{21}\alpha_{12}$$
$$= \lambda^2 - (\alpha_{11} + \alpha_{22})\lambda + \alpha_{11}\alpha_{22} - \alpha_{21}\alpha_{12} .$$

If λ_1 and λ_2 are the zeros of this characteristic polynomial, then we must have

$$0 = (\lambda_1 - \lambda)(\lambda_2 - \lambda) = \lambda^2 - (\lambda_1 + \lambda_2)\lambda + \lambda_1\lambda_2 .$$

We conclude that

$$\lambda_1 + \lambda_2 = \alpha_{11} + \alpha_{22} \text{ and } \lambda_1\lambda_2 = \alpha_{11}\alpha_{22} - \alpha_{21}\alpha_{12} = \det \mathbf{A} .$$

Example 1.2.9 The characteristic polynomial of a general 3 × 3 matrix

$$\mathbf{A} = \begin{bmatrix} \alpha_{11} & \alpha_{12} & \alpha_{13} \\ \alpha_{21} & \alpha_{22} & \alpha_{23} \\ \alpha_{31} & \alpha_{32} & \alpha_{33} \end{bmatrix}$$

is

$$\det(\mathbf{A} - \mathbf{I}\lambda) = \det \begin{bmatrix} \alpha_{11} - \lambda & \alpha_{12} & \alpha_{13} \\ \alpha_{21} & \alpha_{22} - \lambda & \alpha_{23} \\ \alpha_{31} & \alpha_{32} & \alpha_{33} - \lambda \end{bmatrix} =$$

$$(\alpha_{11} - \lambda) \det \begin{bmatrix} \alpha_{22} - \lambda & \alpha_{23} \\ \alpha_{32} & \alpha_{33} - \lambda \end{bmatrix} - \alpha_{12} \det \begin{bmatrix} \alpha_{21} & \alpha_{23} \\ \alpha_{31} & \alpha_{33} - \lambda \end{bmatrix} + \alpha_{13} \det \begin{bmatrix} \alpha_{21} & \alpha_{22} - \lambda \\ \alpha_{31} & \alpha_{32} \end{bmatrix}$$

$$\begin{aligned} &= -\lambda^3 + (\alpha_{11} + \alpha_{22} + \alpha_{33})\lambda^2 \\ &- (\alpha_{11}\alpha_{22} + \alpha_{22}\alpha_{33} + \alpha_{33}\alpha_{11} - \alpha_{12}\alpha_{21} - \alpha_{23}\alpha_{32} - \alpha_{31}\alpha_{13})\lambda \\ &+ (\alpha_{11}\alpha_{22}\alpha_{33} + \alpha_{12}\alpha_{23}\alpha_{31} + \alpha_{13}\alpha_{21}\alpha_{32} - \alpha_{11}\alpha_{32}\alpha_{23} - \alpha_{22}\alpha_{31}\alpha_{13} - \alpha_{33}\alpha_{21}\alpha_{12}) \,. \end{aligned}$$

If λ_1, λ_2 and λ_3 are the zeros of this characteristic polynomial, then we must have

$$0=(\lambda_1-\lambda)(\lambda_2\lambda)(\lambda_3-\lambda)=-\lambda^3+(\lambda_1+\lambda_2+\lambda_3)\lambda^2-(\lambda_1\lambda_2+\lambda_2\lambda_3+\lambda_3\lambda_1)\lambda+\lambda_1\lambda_2\lambda_3.$$

We conclude that

$$\begin{aligned} &\lambda_1 + \lambda_2 + \lambda_3 = \alpha_{11} + \alpha_{22} + \alpha_{33} \,, \\ &\lambda_1\lambda_2+\lambda_2\lambda_3 + \lambda_3\lambda_1 = \alpha_{11}\alpha_{22} + \alpha_{22}\alpha_{33} + \alpha_{33}\alpha_{11} - \alpha_{12}\alpha_{21} - \alpha_{23}\alpha_{32} - \alpha_{31}\alpha_{13} \text{ and} \\ &\lambda_1\lambda_2\lambda_3 = \alpha_{11}\alpha_{22}\alpha_{33} + \alpha_{12}\alpha_{23}\alpha_{31} + \alpha_{13}\alpha_{21}\alpha_{32} - \alpha_{11}\alpha_{32}\alpha_{23} - \alpha_{22}\alpha_{31}\alpha_{13} - \alpha_{33}\alpha_{21}\alpha_{12} \\ &\qquad = \det \mathbf{A} \,. \end{aligned}$$

Careful readers will have noticed a pattern that suggests the following definition.

Definition 1.2.3 The **trace** of a square matrix is the sum of its diagonal entries, and is written

$$\mathrm{tr}(\mathbf{A}) = \sum_{i=1}^{n} \alpha_{ii} \,.$$

The previous definition allows us to state the following results.

Lemma 1.2.6 *If* $\mathbf{A}$ *and* $\mathbf{b}$ *are* $n \times n$ *matrices, then*

$$tr(\mathbf{AB}) = tr(\mathbf{BA}) \,.$$

Also, if $\mathbf{u}$ *and* $\mathbf{w}$ *are n-vectors, then*

$$tr(\mathbf{u}\mathbf{w}^H) = \mathbf{w} \cdot \mathbf{u} \; .$$

Here $\mathbf{w} \cdot \mathbf{u}$ *represents the* **inner product** *of the vectors* $\mathbf{x}$ *and* $\mathbf{y}$.

Proof It is easy to see that

$$\mathrm{tr}(\mathbf{AB}) = \sum_{i=1}^{n} (\mathbf{AB})_{ii} = \sum_{i=1}^{n}\sum_{j=1}^{n} \mathbf{A}_{ij}\mathbf{B}_{ji} = \sum_{j=1}^{n}\sum_{i=1}^{n} \mathbf{B}_{ji}\mathbf{A}_{ij} = \sum_{j=1}^{n} (\mathbf{BA})_{jj}\ \mathrm{tr}(\mathbf{BA})$$

and that

$$\mathrm{tr}(\mathbf{u}\mathbf{w}^H) = \sum_{i=1}^{n} \mathbf{u}_i \overline{\mathbf{w}_i} = \sum_{i=1}^{n} \overline{\mathbf{w}_i}\mathbf{u}_i = \mathbf{w} \cdot \mathbf{u} \; .$$

Lemma 1.2.7 *If* $\mathbf{A}$ *is an* $n \times n$ *matrix, then*

$$\det(\mathbf{A} - \mathbf{I}\lambda) = (-\lambda)^n + (-\lambda)^{n-1}\, tr(\mathbf{A}) \ldots + \det(\mathbf{A}) \; .$$

Proof By setting $\lambda = 0$, we see that the constant term in the polynomial on the right must be $\det(\mathbf{A})$. We can use expansion by minors to show that the trace is the coefficient of $(-\lambda)^{n-1}$.

Lemma 1.2.8 *Let* $\mathbf{A}$ *be an* $n \times n$ *matrix. Then the trace of* $\mathbf{A}$ *is the sum of its eigenvalues (counting multiplicity), and the determinant of* $\mathbf{A}$ *is the product of the eigenvalues.*

Proof Since the Fundamental Theorem of Algebra 1.2.1 shows that $\mathbf{A}$ has n eigenvalues (counting multiplicity), we can factor

$$\det(\mathbf{A} - \mathbf{I}\lambda) = \prod_{i=1}^{n} (\lambda_i - \lambda)$$

and expand the product to get

$$= (-\lambda)^n + (-\lambda)^{n-1} \left(\sum_{i=1}^{n} \lambda_i \right) + \ldots + \prod_{i=1}^{n} \lambda_i \; .$$

Example 1.2.10 We saw that the eigenvalues of

$$\mathbf{A} = \begin{bmatrix} \cos\theta & -\sin\theta \\ \sin\theta & \cos\theta \end{bmatrix}$$

are $\lambda_1 = \cos\theta + \mathrm{i}\sin\theta$ and $\lambda_2 = \cos\theta - \mathrm{i}\sin\theta$. Note that

$$\mathrm{tr}(\mathbf{A}) = 2\cos\theta = \lambda_1 + \lambda_2 \text{ and}$$

$$\det(\mathbf{A}) = \cos^2\theta + \sin^2\theta = \lambda_1\lambda_2 .$$

Exercise 1.2.23 Find the coefficients of the characteristic polynomial for

$$\mathbf{A} = \begin{bmatrix} 2 & -1 & 1 \\ -1 & 0 & 1 \\ 1 & 1 & 2 \end{bmatrix} .$$

Then find the eigenvalues of **A**, and relate the eigenvalues of **A** to the coefficients of its characteristic polynomial.

Exercise 1.2.24 Suppose that **S** is a real symmetric 3×3 matrix (possibly a stress tensor in solid mechanics), and let $\sigma = \mathrm{tr}(\mathbf{S})/3$ (possibly the mean stress in solid mechanics). What are the coefficients of the characteristic polynomial for $\mathbf{S} - \mathbf{I}\sigma$ (which is called the deviatoric stress in solid mechanics)?

Exercise 1.2.25 For each of the following matrices, find the eigenvalues and then verify that the trace is the sum of the eigenvalues and the determinant is the product.

1.

$$\begin{bmatrix} 3 & 4 & 2 \\ 0 & 1 & 2 \\ 0 & 0 & 0 \end{bmatrix}$$

2.

$$\begin{bmatrix} 0 & 0 & 2 \\ 0 & 2 & 0 \\ 2 & 0 & 0 \end{bmatrix}$$

1.2.7 Invariant Subspace

It is possible for the characteristic polynomial of a matrix to have a root of multiplicity greater than one. When this happens, there may be more than one linearly independent eigenvector for the given eigenvalue. As a result, we need to generalize the notion of eigenvectors, by introducing the following definition.

Definition 1.2.4 Let **A** be an $n \times n$ matrix. Then a subspace $\mathscr{V}$ of n-vectors is **invariant** under **A** if and only if for all $\mathbf{v} \in \mathscr{V}$ we have $\mathbf{A}\mathbf{v} \in \mathscr{V}$.

For example, the nullspace of a matrix $\mathbf{A}$ is invariant under $\mathbf{A}$. Also, if λ is an eigenvalue of $\mathbf{A}$ then the span of all eigenvectors for λ is an invariant subspace under $\mathbf{A}$.

The next lemma will help to characterize the circumstances under which an invariant subspace exists.

Lemma 1.2.9 *Suppose that* $\mathbf{A}$ *is a square* $n \times n$ *matrix and* $\mathscr{V}$ *is an invariant subspace under* $\mathbf{A}$ *with dimension* $k < n$. *Then there is a unitary matrix* $\mathbf{U}$ *and a block right-triangular matrix* $\mathbf{C}$ *so that* $\mathbf{U}^H\mathbf{A}\mathbf{U} = \mathbf{C}$. *Furthermore, we can partition*

$$\mathbf{C} = \begin{bmatrix} \mathbf{C}_{11} & \mathbf{C}_{12} \\ & \mathbf{C}_{22} \end{bmatrix} ,$$

where $\mathbf{C}_{11}$ *is* $k \times k$.

Proof Let $\mathbf{u}_1, \ldots, \mathbf{u}_k$ be an orthonormal basis for $\mathscr{V}$, and use Corollary 3.2.1 in Chap. 3 of Volume I to extend this to an orthonormal basis $\{\mathbf{u}_j\}_{j=1}^n$ for all n-vectors. Since $\mathscr{V}$ is invariant under $\mathbf{A}$, for all $1 \le i \le k$ there is a set of scalars $\{\gamma_{ij}\}_{j=1}^k$ so that

$$\mathbf{A}\mathbf{u}_i = \sum_{j=1}^k \mathbf{u}_j \gamma_{ij} .$$

Since $\{\mathbf{u}_j\}_{j=1}^n$ is a basis for all n-vectors, for all $k < i \le n$ there is a set of scalars $\{\gamma_{ij}\}_{j=1}^n$ so that

$$\mathbf{A}\mathbf{u}_i = \sum_{j=1}^n \mathbf{u}_j \gamma_{ij} .$$

This suggests that we define $\gamma_{ij} = 0$ for $1 \le i \le k$ and $k < j \le n$, and let γ_{ij} be the entries of an $n \times n$ matrix $\mathbf{C}$. Then we have shown that $\mathbf{A}\mathbf{U} = \mathbf{U}\mathbf{C}$ where

$$\mathbf{C} = \begin{bmatrix} \mathbf{C}_{11} & \mathbf{C}_{12} \\ & \mathbf{C}_{22} \end{bmatrix} .$$

1.2.8 Change of Basis

We can relate linear transformations to matrices by a choice of basis. Distinct choices of bases lead to distinct matrix representations of a single linear transformation. Eventually, we would like to know what aspects of a linear transformation are independent of the choice of basis.

We will begin by representing a single point in space with respect to two different bases. Suppose that the columns of the two $m \times m$ matrices $\mathbf{B}$ and $\widetilde{\mathbf{B}}$ each form a basis for the set of all m-vectors. In other words, either of these two sets of vectors is a representation, with respect to the axis vectors, of a basis for all m-vectors. Definition 3.2.17 in Chap. 3 of Volume I shows that both $\mathbf{B}$ and $\widetilde{\mathbf{B}}$ must be nonsingular.

Suppose that we are given an m-vector $\mathbf{c}$, which is the representation of a point in space with respect to the axis vectors. To find the representation of this point with respect to the basis given by the columns of $\mathbf{B}$, we find the vector $\mathbf{x}$ that solves

$$\mathbf{B}\mathbf{x} = \mathbf{c} .$$

Similarly, the representation of $\mathbf{c}$ with respect to the second basis is $\widetilde{\mathbf{x}}$ where

$$\widetilde{\mathbf{B}}\widetilde{\mathbf{x}} = \mathbf{c} .$$

To change from the representation in the first basis to the representation in the second basis, we use the equation

$$\widetilde{\mathbf{x}} = \widetilde{\mathbf{B}}^{-1}\mathbf{B}\mathbf{x} .$$

This observation leads to the following useful lemma.

Lemma 1.2.10 *Suppose that* $\mathbf{B}$ *and* $\widetilde{\mathbf{B}}$ *are two nonsingular* $m \times m$ *matrices. Let* $\mathscr{A}$ *be a* **linear transformation** *on* m*-vectors, meaning that*

$$\mathscr{A}(\mathbf{c}\gamma) = \mathscr{A}(\mathbf{c})\gamma \text{ for all } m\text{-vectors } \mathbf{c} \text{ and all scalars } \gamma \text{ , and}$$
$$\mathscr{A}(\mathbf{c} + \widetilde{\mathbf{c}}) = \mathscr{A}(\mathbf{c}) + \mathscr{A}(\widetilde{\mathbf{c}}) \text{ for all } m\text{-vectors } \mathbf{c} \text{ and } \widetilde{\mathbf{c}} .$$

Let $\mathbf{c}$ *be an* m*-vector. Suppose that with respect to the basis given by the columns of* $\mathbf{B}$, $\mathbf{c}$ *has the representation* $\mathbf{x}$ *and* $\mathscr{A}(\mathbf{c})$ *has the representation* $\mathbf{y}$*; in other words,*

$$\mathbf{c} = \mathbf{B}\mathbf{x} \text{ and } \mathscr{A}(\mathbf{c}) = \mathbf{B}\mathbf{y} .$$

Also suppose that with respect to the basis given by the columns of $\widetilde{\mathbf{B}}$, $\mathbf{c}$ *has the representation* $\widetilde{\mathbf{x}}$ *and* $\mathscr{A}(\mathbf{c})$ *has the representation* $\widetilde{\mathbf{y}}$*. Then the matrix representation* $\widetilde{\mathbf{A}}$ *of* $\mathscr{A}$ *with respect to the second basis is related to the matrix representation* $\mathbf{A}$ *with respect to the first basis by*

$$\widetilde{\mathbf{B}}\widetilde{\mathbf{A}}\widetilde{\mathbf{B}}^{-1} = \mathbf{B}\mathbf{A}\mathbf{B}^{-1} .$$

Proof For any m-vector $\mathbf{c}$, we have

$$\widetilde{\mathbf{B}}\widetilde{\mathbf{A}}\widetilde{\mathbf{B}}^{-1}\mathbf{c} = \widetilde{\mathbf{B}}\widetilde{\mathbf{A}}\widetilde{\mathbf{x}} = \widetilde{\mathbf{B}}\widetilde{\mathbf{y}} = \mathscr{A}(\mathbf{c}) = \mathbf{B}\mathbf{y} = \mathbf{B}\mathbf{A}\mathbf{x} = \mathbf{B}\mathbf{A}\mathbf{B}^{-1}\mathbf{c} .$$

Here is a related result that will be helpful in discussing the distance between subspaces in Sect. 1.2.11.

Lemma 1.2.11 *The orthogonal projector onto a subspace is unique.*

Proof Suppose that the columns of **A** form a basis for the given subspace. Any other basis for the subspace would be a linear combination of these vectors, and could be written as the columns of **AB** for some nonsingular matrix **B**. Lemma 6.4.3 in Chap. 6 of Volume I shows that

$$\mathbf{A}\mathbf{A}^{\dagger} = \mathbf{A}\left(\mathbf{A}^H\mathbf{A}\right)^{-1}\mathbf{A}^H$$

is an orthogonal projector onto the given subspace. Since

$$\{\mathbf{AB}\}\left(\{\mathbf{AB}\}^H\{\mathbf{AB}\}\right)^{-1}\{\mathbf{AB}\}^H = \{\mathbf{AB}\}\,\mathbf{B}^{-1}\left(\mathbf{A}^H\mathbf{A}\right)^{-1}\mathbf{B}^{-H}\mathbf{B}^H\mathbf{A}^H$$
$$= \mathbf{A}\left(\mathbf{A}^H\mathbf{A}\right)^{-1}\mathbf{A}^H\,,$$

the two bases for the subspace produce the same orthogonal projector.

1.2.9 Similarity Transformations

Let us return to our initial discussion about invariants of linear transformations. Suppose that a linear transformation $\mathscr{A}$ mapping n-vectors to n-vectors has $n \times n$ matrix representation **A** with respect to some basis given by the columns of **B**. If we change to a new basis given by the columns of $\widetilde{\mathbf{B}}$ as in Lemma 1.2.10, then the matrix representation of $\mathscr{A}$ with respect to the new basis is

$$\widetilde{\mathbf{B}\mathbf{A}\mathbf{B}^{-1}} = \mathbf{B}\mathbf{A}\mathbf{B}^{-1}\,.$$

Let λ be an eigenvector of **A**:

$$\mathbf{A}\mathbf{x} = \mathbf{x}\lambda \text{ and } \mathbf{x} \neq 0\,.$$

Then

$$\mathbf{B}\mathbf{x}\lambda = (\mathbf{B}\mathbf{A}\mathbf{B}^{-1})\mathbf{B}\mathbf{x} = (\widetilde{\mathbf{B}\mathbf{A}\mathbf{B}^{-1}})\mathbf{B}\mathbf{x} \Rightarrow (\widetilde{\mathbf{B}}^{-1}\mathbf{B}\mathbf{x})\lambda = \widetilde{\mathbf{A}}(\widetilde{\mathbf{B}}^{-1}\mathbf{B}\mathbf{x})\,.$$

Thus, λ is an eigenvector of $\widetilde{\mathbf{A}}$ with eigenvector $\widetilde{\mathbf{B}}^{-1}\mathbf{B}\mathbf{x}$. This proves the following lemma.

Lemma 1.2.12 *The eigenvalues of a linear transformation are invariant under a change of basis.*

The form of the change of basis equation suggests the following definition.

Definition 1.2.5 Suppose that $\mathbf{A}$ is an $n \times n$ matrix and $\mathbf{B}$ is a nonsingular $n \times n$ matrix. Then the matrix $\mathbf{BAB}^{-1}$ is said to be **similar** to $\mathbf{A}$, and the mapping

$$\mathscr{S}_{\mathbf{B}}(\mathbf{A}) = \mathbf{BAB}^{-1}$$

is called a **similarity transformation**.

This definition leads to the following easy result.

Lemma 1.2.13 *Suppose that* $\mathbf{A}$ *is an* $n \times n$ *matrix,* $\mathbf{B}$ *is a nonsingular* $n \times n$ *matrix, and* $\mathbf{C} = \mathbf{BAB}^{-1}$ *is similar to* $\mathbf{A}$*. If* $\mathbf{x}$ *is an eigenvector of* $\mathbf{A}$ *with eigenvalue* λ*, then* $\mathbf{Bx}$ *is an eigenvector of* $\mathbf{C}$ *with eigenvalue* λ*.*

Proof The claim holds because

$$\mathbf{CBx} = \left(\mathbf{BAB}^{-1}\right)\mathbf{Bx} = \mathbf{BAx} = \mathbf{Bx}\lambda \,.$$

Here are some other useful results about eigenvalues.

Lemma 1.2.14 *If* λ *is an eigenvalue of* $\mathbf{A}$ *and* σ *is a scalar, then* $\lambda - \sigma$ *is an eigenvalue of* $\mathbf{A} - \mathbf{I}\sigma$*, and* $\lambda\sigma$ *is an eigenvalue of* $\mathbf{A}\sigma$*.*

Proof If $\mathbf{Ax} = \mathbf{x}\lambda$, then $(\mathbf{A} - \mathbf{I}\sigma)\mathbf{x} = \mathbf{x}(\lambda - \sigma)$. Also, if $\mathbf{Ax} = \mathbf{x}\lambda$, then $(\mathbf{A}\sigma)\mathbf{x} = \mathbf{Ax}\sigma = \mathbf{x}\lambda\sigma$.

1.2.10 Spectral Radius

Since the eigenvalues of a matrix are independent of a change of basis, we can use them to measure the linear transformation that corresponds to the matrix. This suggests the following useful definition.

Definition 1.2.6 If $\mathbf{A}$ is a square matrix, then

$$\varrho(\mathbf{A}) \equiv \max\{|\lambda| : \lambda \text{ is an eigenvalue of } \mathbf{A}\} \tag{1.1}$$

is called the **spectral radius** of $\mathbf{A}$.

The following lemma will be used in the proof of Lemma 1.2.23.

Lemma 1.2.15 *If* $\|\cdot\|$ *is a norm on square matrices that is consistent with some vector norm, then for all square matrices* $\mathbf{A}$

$$\varrho(\mathbf{A}) \le \|\mathbf{A}\| \,. \tag{1.2}$$

Proof If λ is an eigenvalue of $\mathbf{A}$ with eigenvector $\mathbf{x}$, then

$$\|\mathbf{x}\| \, |\lambda| = \|\mathbf{x}\lambda\| = \|\mathbf{Ax}\| \le \|\mathbf{A}\| \, \|\mathbf{x}\| \,.$$

Here the inequality holds because of the Definition 3.5.4 in Chap. 3 of Volume I of a consistent matrix norm. Since $\mathbf{x}$ is nonzero, we conclude that $|\lambda| \leq \|\mathbf{A}\|$ for all eigenvalues of $\mathbf{A}$.

1.2.11 Distance Between Subspaces

In order to discuss the effect of matrix perturbations on eigenvectors and more generally on invariant subspaces of matrices, we will need to develop some theory about the distances between subspaces. There are several reasons for developing this notion. For example, matrices may have multiple linearly independent eigenvectors for a single eigenvalue. If floating point computations lead to a perturbed matrix with several nearly equal eigenvalues, we will want to measure how close the relevant invariant subspaces of the two matrices might be. In other cases, we might develop a numerical method for finding eigenvalues and eigenvectors for which the convergence rate depends on the separation between the eigenvalues. We could shift the matrix by the identity matrix times a scalar close to the average of the cluster of eigenvalues to increase the separation of the eigenvalues, and examine how well the associated invariant subspaces are separated.

The results in this section will be rather technical and abstract. We will try to describe the results geometrically, and to relate them to their applications in Sects. 1.3.2 and 1.4.2. Suffice it to say that the perturbation analysis of eigenvalues and eigenvectors is quite a bit more difficult than the perturbation analysis of least squares problems in Sect. 6.5 in Chap. 6 of Volume I, which was in turn significantly more complicated than the perturbation analysis for systems of linear equations in Sect. 3.6 in Chap. 3 of Volume I.

The following definition will introduce two new ideas. The first idea appears in Stewart [163, p. 733], who adopted it from Kato [106]. The second idea appears in a paper by Stewart [165, p. 651], and is given its formal definition in Golub and van Loan [84, p. 82].

Definition 1.2.7 Let $\mathscr{V}_1$ and $\mathscr{V}_2$ be two subspaces of the set of all n-vectors. Then the **gap** between the two subspaces is

$$\gamma(\mathscr{V}_1, \mathscr{V}_2) = \max \left\{ \sup_{\substack{\mathbf{v}_1 \in \mathscr{V}_1 \\ \|\mathbf{v}_1\|_2 = 1}} \inf_{\mathbf{v}_2 \in \mathscr{V}_2} \|\mathbf{v}_1 - \mathbf{v}_2\|_2 \,,\; \sup_{\substack{\mathbf{v}_2 \in \mathscr{V}_2 \\ \|\mathbf{v}_2\|_2 = 1}} \inf_{\mathbf{v}_1 \in \mathscr{V}_1} \|\mathbf{v}_1 - \mathbf{v}_2\|_2 \right\} . \tag{1.3}$$

Next, let $\mathbf{P}_i$ be the orthogonal$i = 1, 2$. Then the **distance** between the two subspaces is

$$\operatorname{dist}(\mathscr{V}_1, \mathscr{V}_2) = \|\mathbf{P}_1 - \mathbf{P}_2\|_2 . \tag{1.4}$$

Basically, the gap between two subspaces measures the least upper bound on the distance from a unit vector in one subspace to any vector in the other subspace. If $\mathscr{V}_1$ is contained in $\mathscr{V}_2$, then the least upper bound on the distance from any unit vector in $\mathscr{V}_1$ to a vector in $\mathscr{V}_2$ would be zero. However, the least upper bound on the distance from a unit vector in $\mathscr{V}_2$ to a vector in $\mathscr{V}_1$ would be greater than zero. To obtain a measure of the distance between two distinct subspaces, the gap is taken to be the larger of these least upper bounds.

At this point, the distance between two subspaces is harder to describe geometrically. In Lemma 1.2.20, we will show that under certain circumstances, the distance between two subspaces is the maximum of the sine of a principal angle between the two subspaces. Principal angles between subspaces will be discussed in Lemma 1.2.19.

The Definition 1.2.7 of the gap between two subspaces leads to the following easy observation.

Lemma 1.2.16 *If $\mathscr{V}_1$ and $\mathscr{V}_2$ are two subspaces of n-vectors and $\gamma(\mathscr{V}_1, \mathscr{V}_2) < 1$ then* $\dim \mathscr{V}_1 = \dim \mathscr{V}_2$.

Proof We will prove the contrapositive of this claim. Suppose that $\dim \mathscr{V}_1 > \dim \mathscr{V}_2$. We begin by using the Gram-Schmidt process in Sect. 6.8 in Chap. 6 of Volume I to find $\mathbf{v}_1 \in \mathscr{V}_1$ so that $\mathbf{v}_1 \perp \mathscr{V}_2$ and $\|\mathbf{v}_1\|_2 = 1$. Here $\mathscr{U} \perp \mathscr{W}$ denotes that the two sets of vectors $\mathscr{U}$ and $\mathscr{W}$ are **orthogonal**. Then

$$\inf_{\mathbf{v}_2 \in \mathscr{V}_2} \|\mathbf{v}_1 - \mathbf{v}_2\|_2 = \inf_{\mathbf{v}_2 \in \mathscr{V}_2} \sqrt{\|\mathbf{v}_1\|_2^2 + \|\mathbf{v}_2\|_2^2} = \|\mathbf{v}_1\|_2 = 1 \; .$$

The definition of the gap now implies that $\gamma(\mathscr{V}_1, \mathscr{V}_2) \geq 1$.

The definition (1.4) of the distance between subspaces leads to the following result, which can be found in Golub and van Loan [84, p. 82].

Lemma 1.2.17 *Suppose that* $\mathbf{U} = \begin{bmatrix} \mathbf{U}_1 & \mathbf{U}_2 \end{bmatrix}$ *and* $\mathbf{W} = \begin{bmatrix} \mathbf{W}_1 & \mathbf{W}_2 \end{bmatrix}$ *are two unitary matrices with* $\mathbf{U}_1$ *and* $\mathbf{W}_1$ *having the same dimensions. Then*

$$dist(\mathscr{R}(\mathbf{U}_1), \mathscr{R}(\mathbf{W}_1)) = \left\|{\mathbf{W}_1}^H \mathbf{U}_2\right\|_2 = \left\|{\mathbf{U}_1}^H \mathbf{W}_2\right\|_2 \; . \tag{1.5}$$

Here $\mathscr{R}(\mathbf{A})$ represents the **range** *of the matrix* **A**.

Proof Let

$${\mathbf{U}_1}^H \mathbf{W}_1 = \mathbf{Q} \boldsymbol{\Sigma} \mathbf{V}^H$$

be the singular value decomposition of ${\mathbf{U}_1}^H \mathbf{W}_1$. Then

$$\min_{\|\mathbf{x}\|_2 = 1} \left\|{\mathbf{U}_1}^H \mathbf{W}_1 \mathbf{x}\right\|_2 = \sigma_r \; ,$$

where σ_r is the smallest diagonal entry of $\boldsymbol{\Sigma}$. Since the singular value decomposition of $\mathbf{W}_1{}^H\mathbf{U}_1$ is $\mathbf{U}\boldsymbol{\Sigma}\mathbf{Q}^H$, we also have that

$$\min_{\|\mathbf{x}\|_2=1} \left\|\mathbf{W}_1{}^H\mathbf{U}_1\mathbf{x}\right\|_2 = \sigma_r \min_{\|\mathbf{x}\|_2=1} \left\|\mathbf{U}_1{}^H\mathbf{W}_1\mathbf{x}\right\|_2 .$$

Suppose that $\|\mathbf{x}\|_2 = 1$. Since $\mathbf{U}^H\mathbf{W}$ is unitary,

$$1 = \left\|\mathbf{U}^H\mathbf{W}\begin{bmatrix}\mathbf{x}\\ \mathbf{0}\end{bmatrix}\right\|_2^2 = \left\|\mathbf{U}_1{}^H\mathbf{W}_1\mathbf{x}\right\|_2^2 + \left\|\mathbf{U}_2{}^H\mathbf{W}_1\mathbf{x}\right\|_2^2 .$$

It follows that

$$\left\|\mathbf{U}_2{}^H\mathbf{W}_1\right\|_2^2 = \max_{\|\mathbf{x}\|_2=1} \left\|\mathbf{U}_2{}^H\mathbf{W}_1\mathbf{x}\right\|_2^2 = 1 - \min_{\|\mathbf{x}\|_2=1} \left\|\mathbf{U}_1{}^H\mathbf{W}_1\mathbf{x}\right\|_2^2 .$$

We can replace $\mathbf{U}^H\mathbf{W}$ with $\mathbf{W}^H\mathbf{U}$ to get

$$\left\|\mathbf{W}_2{}^H\mathbf{U}_1\right\|_2^2 = 1 - \min_{\|\mathbf{x}\|_2=1} \left\|\mathbf{W}_1{}^H\mathbf{U}_1\mathbf{x}\right\|_2^2 = 1 - \min_{\|\mathbf{x}\|_2=1} \left\|\mathbf{U}_1{}^H\mathbf{W}_1\mathbf{x}\right\|_2^2 = \left\|\mathbf{U}_2{}^H\mathbf{W}_1\right\|_2^2 .$$

Finally, we observe that

$$\operatorname{dist}\left(\mathscr{R}\left(\mathbf{U}_1\right), \mathscr{R}\left(\mathbf{W}_1\right)\right)^2 \equiv \left\|\mathbf{U}_1\mathbf{U}_1{}^H - \mathbf{W}_1\mathbf{W}_1{}^H\right\|_2^2$$

and since $\mathbf{U}$ and $\mathbf{W}$ are unitary, we have

$$= \left\|\begin{bmatrix}\mathbf{U}_1{}^H\\ \mathbf{U}_2{}^H\end{bmatrix}\left(\mathbf{U}_1\mathbf{U}_1{}^H - \mathbf{W}_1\mathbf{W}_1{}^H\right)\begin{bmatrix}\mathbf{W}_1 & \mathbf{W}_2\end{bmatrix}\right\|_2^2 = \left\|\begin{bmatrix}\mathbf{0} & \mathbf{U}_1{}^H\mathbf{W}_2\\ -\mathbf{U}_2{}^H\mathbf{W}_1 & \mathbf{0}\end{bmatrix}\right\|_2^2$$

$$= \max_{\begin{bmatrix}\mathbf{x}_1\\ \mathbf{x}_2\end{bmatrix}\neq\mathbf{0}} \frac{\left\|\begin{bmatrix}\mathbf{U}_1{}^H\mathbf{W}_2\mathbf{x}_2\\ -\mathbf{U}_2{}^H\mathbf{W}_1\mathbf{x}_1\end{bmatrix}\right\|_2^2}{\left\|\begin{bmatrix}\mathbf{x}_1\\ \mathbf{x}_2\end{bmatrix}\right\|_2^2} = \max_{\begin{bmatrix}\mathbf{x}_1\\ \mathbf{x}_2\end{bmatrix}\neq\mathbf{0}} \frac{\|\mathbf{U}_1{}^H\mathbf{W}_2\|_2^2\left(\|\mathbf{x}_2\|_2^2 + \|\mathbf{x}_1\|_2^2\right)}{\left\|\begin{bmatrix}\mathbf{x}_1\\ \mathbf{x}_2\end{bmatrix}\right\|_2^2}$$

$$= \left\|\mathbf{U}_1{}^H\mathbf{W}_2\right\|_2^2 .$$

Lemma 1.2.20 will characterize the distance between subspaces in terms of the **principal angles between subspaces**. These angles are given through the cosines or sines that are the diagonal values in the arrays $\mathbf{C}$ or $\mathbf{S}$ in the next lemma. Our presentation of this lemma will follow that in Björck and Golub [20, p. 582].

Lemma 1.2.18 *Let* $\mathbf{Q}_\mathbf{A}$ *be an* $m \times n_\mathbf{A}$ *matrix such that* $\mathbf{Q}_\mathbf{A}{}^H\mathbf{Q}_\mathbf{A} = \mathbf{I}$, *and let* $\mathbf{Q}_\mathbf{B}$ *be an* $m \times n_\mathbf{B}$ *matrix such that* $\mathbf{Q}_\mathbf{B}{}^H\mathbf{Q}_\mathbf{B} = \mathbf{I}$. *Then there is an integer*

$$n = \min\{n_\mathbf{A} , n_\mathbf{B}\} ,$$

an $m \times n$ *matrix* $\mathbf{U_A}$ *such that* $\mathbf{U_A}^H\mathbf{U_A} = \mathbf{I}$*, an* $n_\mathbf{B} \times n$ *matrix* $\mathbf{Y_B}$ *such that* $\mathbf{Y_B}^H\mathbf{Y_B} = \mathbf{I}$*, and an* $n \times n$ *diagonal matrix* $\mathbf{C}$ *with diagonal entries in* $(0, 1]$ *so that*

$$\mathbf{Q_A}\mathbf{Q_A}^H\mathbf{Q_B} = \mathbf{U_A}\mathbf{C}\mathbf{Y_B}^H \; . \tag{1.6}$$

Furthermore, there is an $m \times n$ *matrix* $\mathbf{W_A}$ *and an* $n \times n$ *diagonal matrix* $\mathbf{S}$ *with diagonal entries in* $[0, 1)$ *so that*

$$\left(\mathbf{I} - \mathbf{Q_A}\mathbf{Q_A}^H\right)\mathbf{Q_B} = \mathbf{W_A}\mathbf{S}\mathbf{Y_B}^H \; . \tag{1.7}$$

Finally,

$$\mathbf{C}^2 + \mathbf{S}^2 = \mathbf{I} \; .$$

Proof First, we note that for all $n_\mathbf{B}$-vectors $\mathbf{x}$ we have

$$\left\|\mathbf{Q_A}^H\mathbf{Q_B}\mathbf{x}\right\|_2^2 = \mathbf{x}^H\mathbf{Q_B}^H\mathbf{Q_A}\mathbf{Q_A}^H\mathbf{Q_B}\mathbf{x} \le \left\|\mathbf{Q_A}\mathbf{Q_A}^H\right\|_2 \|\mathbf{Q_B}\mathbf{x}\|_2^2$$

and since $\mathbf{Q_A}\mathbf{Q_A}^H$ is an orthogonal projector, Lemma 3.5.6 in Chap. 3 of Volume I gives us

$$\le \|\mathbf{Q_B}\mathbf{x}\|_2^2 = \|\mathbf{x}\|_2^2 \; .$$

Next, let the singular value decomposition of $\mathbf{Q_A}^H\mathbf{Q_B}$ be

$$\mathbf{Q_A}^H\mathbf{Q_B} = \mathbf{Y_A}\mathbf{C}\mathbf{Y_B}^H \; ,$$

in which zero singular values are discarded. Here $\mathbf{Y_A}$ is $n_\mathbf{A} \times n$ and $\mathbf{Y_B}$ is $n_\mathbf{B} \times n$ where

$$n = \text{rank}\left(\mathbf{Q_A}^H\mathbf{Q_B}\right) \le \min\{\text{rank}\left(\mathbf{Q_A}\right) \; , \; \text{rank}\left(\mathbf{Q_B}\right)\} = \min\{n_\mathbf{A} \; , \; n_\mathbf{B}\} \; .$$

We denote by $\text{rank}\left(\mathbf{A}\right)$ the **rank** of the matrix $\mathbf{A}$. Since $\mathbf{C}$ comes from a such a singular value decomposition, its diagonal entries are positive. Also, for any $n_\mathbf{B}$-vectors $\mathbf{x}$ we have

$$\left\|\mathbf{C}\mathbf{Y_B}^H\mathbf{x}\right\|_2^2 = \left\|\mathbf{Y_A}\mathbf{C}\mathbf{Y_B}^H\mathbf{x}\right\|_2^2 = \left\|\mathbf{Q_A}^H\mathbf{Q_B}\mathbf{x}\right\|_2^2 \le \|\mathbf{x}\|_2^2 = \left\|\mathbf{Y_B}^H\mathbf{x}\right\|_2^2 \; .$$

This proves that the diagonal entries of $\mathbf{C}$ are at most one. Claim (1.6) follows easily from the singular value decomposition, by choosing $\mathbf{U_A} = \mathbf{Q_A}\mathbf{Y_A}$.

Lemma 1.5.1 will prove the existence of the singular value decomposition. This lemma will show that the columns of $\mathbf{Y_A}$ are eigenvectors of $\left(\mathbf{Q_A}\mathbf{Q_B}^H\right)\left(\mathbf{Q_A}\mathbf{Q_B}^H\right)^H$, and that the columns of $\mathbf{Y_B}$ are eigenvectors of $\left(\mathbf{Q_A}\mathbf{Q_B}^H\right)^H\left(\mathbf{Q_A}\mathbf{Q_B}^H\right)$. Since

$$\begin{aligned}&\left[\left(\mathbf{I}-\mathbf{Q_A}\mathbf{Q_A}^H\right)\mathbf{Q_B}\right]^H\left[\left(\mathbf{I}-\mathbf{Q_A}\mathbf{Q_A}^H\right)\mathbf{Q_B}\right]=\mathbf{Q_B}^H\left(\mathbf{I}-\mathbf{Q_A}\mathbf{Q_A}^H\right)\mathbf{Q_B}\\&=\mathbf{I}-\left(\mathbf{Q_A}^H\mathbf{Q_B}\right)^H\left(\mathbf{Q_A}^H\mathbf{Q_B}\right)=\mathbf{I}-\mathbf{Y_B}\mathbf{C}\mathbf{Y_A}^H\mathbf{Y_A}\mathbf{C}\mathbf{Y_B}^H\\&=\mathbf{Y_B}\left(\mathbf{I}-\mathbf{C}^2\right)\mathbf{Y_B}^H\,,\end{aligned}$$

we see that the columns of $\mathbf{Y_B}$ are eigenvectors of $\left(\mathbf{I}-\mathbf{Q_A}\mathbf{Q_A}^H\right)\mathbf{Q_B}$ pre-multiplied by its transpose. Thus a singular value decomposition of this matrix can be written in the form

$$\left(\mathbf{I}-\mathbf{Q_A}\mathbf{Q_A}^H\right)\mathbf{Q_B}=\mathbf{W_A}\mathbf{S}\mathbf{Y_B}^H\,,$$

where $\mathbf{W_A}^H\mathbf{W_A}=\mathbf{I}$, and $\mathbf{S}$ is a diagonal matrix with nonnegative diagonal entries. In order to preserve the matrix $\mathbf{Y_B}$ in this singular value decomposition, *we do not discard zero singular values in* $\mathbf{S}$. This proves (1.7).

Finally, we note that for all $n_\mathbf{B}$-vectors $\mathbf{x}$, we have

$$\begin{aligned}&\left(\mathbf{Y_B}^H\mathbf{x}\right)^H\mathbf{Y_B}^H\mathbf{x}=\|\mathbf{x}\|_2^2=\|\mathbf{Q_B}\mathbf{x}\|_2^2\\&=\left\|\mathbf{Q_A}\mathbf{Q_A}^H\mathbf{Q_B}\mathbf{x}\right\|_2+\left\|\left(\mathbf{I}-\mathbf{Q_A}\mathbf{Q_A}^H\right)\mathbf{Q_B}\mathbf{x}\right\|_2^2\\&=\left\|\mathbf{U_A}\mathbf{C}\mathbf{Y_B}^H\mathbf{x}\right\|_2^2+\left\|\mathbf{W_A}\mathbf{S}\mathbf{Y_B}^H\mathbf{x}\right\|_2^2=\left\|\mathbf{C}\mathbf{Y_B}^H\mathbf{x}\right\|_2^2+\left\|\mathbf{S}\mathbf{Y_B}^H\mathbf{x}\right\|_2^2\\&=\left(\mathbf{Y_B}^H\mathbf{x}\right)^H\left(\mathbf{C}^2+\mathbf{S}^2\right)\mathbf{Y_B}^H\mathbf{x}\,.\end{aligned}$$

Since $\mathbf{x}$ is arbitrary and $\mathbf{Y_B}$ is nonsingular, we conclude that $\mathbf{C}^2+\mathbf{S}^2=\mathbf{I}$. Since the diagonal entries of $\mathbf{C}$ lie in $(0,1]$, it follows that the diagonal entries of $\mathbf{S}$ lie in $[0,1)$.

Lemma 1.2.19 *Suppose that* $\mathbf{A}$ *is an* $m\times n_\mathbf{A}$ *matrix with rank* $(\mathbf{A})=n_\mathbf{A}$, *and* $\mathbf{B}$ *is an* $m\times n_\mathbf{B}$ *matrix with rank* $(\mathbf{B})=n_\mathbf{B}\le n_\mathbf{A}$. *Assume that* $2n_\mathbf{B}\le m$. *Then there is a* $n_\mathbf{B}\times n_\mathbf{B}$ *diagonal matrix* $\mathbf{C}$ *with diagonal entries in* $[0,1]$, *a diagonal matrix* $\mathbf{S}$ *with diagonal entries in* $[0,1]$ *so that* $\mathbf{C}^2+\mathbf{S}^2=\mathbf{I}$, *and a unitary matrix*

$$\mathbf{U}=\left[\mathbf{U}_1\ \mathbf{U}_2\ \mathbf{U}_3\right]$$

with $\mathbf{U}_1$ *and* $\mathbf{U}_2$ *both* $n\times n_\mathbf{B}$ *matrices, such that* $\mathscr{R}(\mathbf{U}_1\mathbf{C}+\mathbf{U}_2\mathbf{S})=\mathscr{R}(\mathbf{A})$. *Furthermore,*

$$\mathbf{P}=\left[\mathbf{U}_1\ \mathbf{U}_2\ \mathbf{U}_3\right]\begin{bmatrix}\mathbf{C}&-\mathbf{S}&\mathbf{0}\\\mathbf{S}&\mathbf{C}&\mathbf{0}\\\mathbf{0}&\mathbf{0}&\mathbf{I}\end{bmatrix}\begin{bmatrix}\mathbf{U}_1^H\\\mathbf{U}_2^H\\\mathbf{U}_3^H\end{bmatrix}\tag{1.8}$$

is unitary, and such that for all m-vectors $\mathbf{y}\in\mathscr{R}(\mathbf{B})$ *we have* $\mathbf{P}\mathbf{y}\in\mathscr{R}(\mathbf{A})$.

Proof Using the techniques in Sects. 6.7.1.3 or 6.8.1.1 in Chap. 6 of Volume I, we can factor

$$\mathbf{A} = \mathbf{Q_A}\mathbf{R_A} \text{ and } \mathbf{B} = \mathbf{Q_B}\mathbf{R_B}$$

where $\mathbf{R_A}$ is $n_\mathbf{A} \times n_\mathbf{A}$ right-triangular and nonsingular, $\mathbf{R_B}$ is $n_\mathbf{B} \times n_\mathbf{B}$ right-triangular and nonsingular,

$$\mathbf{Q_A}^H\mathbf{Q_A} = \mathbf{I} \text{ and } \mathbf{Q_B}^H\mathbf{Q_B} = \mathbf{I} .$$

First, we will prove the claims of this lemma in the case when $\mathscr{N}\left(\mathbf{A}^H\mathbf{B}\right) = \{\mathbf{0}\}$. Since $\mathbf{R_A}$ and $\mathbf{R_B}$ are nonsingular, it follows that

$$\mathscr{N}\left(\mathbf{Q_A}^H\mathbf{Q_B}\right) = \{\mathbf{0}\} .$$

The Fundamental Theorem of Linear Algebra 3.2.3 in Chap. 3 of Volume I implies that $\mathscr{R}\left(\mathbf{Q_B}^H\mathbf{Q_A}\right)$ is the set of all $n_\mathbf{B}$ vectors, and that

$$n_\mathbf{B} \le n_\mathbf{A} .$$

It is now easy to see that

$$\mathscr{R}\left(\mathbf{Q_B}\mathbf{Q_B}^H\mathbf{Q_A}\right) = \mathscr{R}\left(\mathbf{Q_B}\right) .$$

Indeed, for any $n_\mathbf{B}$-vector $\mathbf{y}$ we can find an $n_\mathbf{A}$-vector $\mathbf{x}$ so that

$$\mathbf{Q_B}\mathbf{Q_B}^H\mathbf{Q_A}\mathbf{x} = \mathbf{Q_B}\mathbf{y}$$

by solving $\mathbf{Q_B}^H\mathbf{Q_A}\mathbf{x} = \mathbf{y}$.

Lemma 1.2.18 guarantees that there is an integer

$$n \le \min\{n_\mathbf{A} \ , \ n_\mathbf{B}\} = n_\mathbf{B} ,$$

an $m \times n$ matrix $\mathbf{U}_1$ such that $\mathbf{U}_1{}^H\mathbf{U}_1 = \mathbf{I}$, an $n_\mathbf{A} \times n$ matrix $\mathbf{Y_A}$ such that $\mathbf{Y_A}^H\mathbf{Y_A} = \mathbf{I}$ and an $n \times n$ diagonal matrix $\mathbf{C}$ with diagonal entries in $(0, 1]$ so that

$$\mathbf{Q_B}\mathbf{Q_B}^H\mathbf{Q_A} = \mathbf{U}_1\mathbf{C}\mathbf{Y_A}^H .$$

Since this is a singular value decomposition, Lemma 1.5.4 will show that

$$\mathscr{R}\left(\mathbf{U}_1\right) = \mathscr{R}\left(\mathbf{Q_B}\mathbf{Q_B}^H\mathbf{Q_A}\right) = \mathscr{R}\left(\mathbf{Q_B}\right) .$$

We conclude that

$$n = \dim \mathscr{R}\left(\mathbf{U}_1\right) = \dim \mathscr{R}\left(\mathbf{Q_B}\right) = n_\mathbf{B} .$$

Lemma 1.2.18 also guarantees that there is an $m \times n_{\mathbf{B}}$ matrix $\mathbf{U}_2$ such that $\mathbf{U}_2{}^H\mathbf{U}_2 = \mathbf{I}$ and a diagonal matrix $\mathbf{S}$ with diagonal entries in $[0, 1)$ such that

$$\left(\mathbf{I} - \mathbf{Q}_{\mathbf{B}}\mathbf{Q}_{\mathbf{B}}{}^H\right)\mathbf{Q}_{\mathbf{A}} = \mathbf{U}_2\mathbf{S}\mathbf{Y}_{\mathbf{A}}{}^H \; .$$

This is also a singular value decomposition, so Lemma 1.5.4 will show that

$$\mathscr{R}\left(\left(\mathbf{I} - \mathbf{Q}_{\mathbf{B}}\mathbf{Q}_{\mathbf{B}}{}^H\right)\mathbf{Q}_{\mathbf{A}}\right) \subset \mathscr{R}\left(\mathbf{U}_2\right) \; .$$

This relationship is not an equality, because $\mathbf{S}$ could have zero diagonal entries.

Without loss of generality, we may assume that the diagonal entries of $\mathbf{C}$ and $\mathbf{S}$ have been arranged so that any zero diagonal entries of $\mathbf{S}$ are ordered last. We can then partition

$$\left(\mathbf{I} - \mathbf{Q}_{\mathbf{B}}\mathbf{Q}_{\mathbf{B}}{}^H\right)\mathbf{Q}_{\mathbf{A}} = \begin{bmatrix}\mathbf{U}_{2,+} & \mathbf{U}_{2,0}\end{bmatrix}\begin{bmatrix}\mathbf{S}_{+} & \mathbf{0} \\ \mathbf{0} & \mathbf{0}\end{bmatrix}\begin{bmatrix}\mathbf{Y}_{\mathbf{A},+}{}^H \\ \mathbf{Y}_{\mathbf{A},0}{}^H\end{bmatrix} ,$$

where $\mathbf{U}_{2,+}$ is $m \times k$ and $k \leq n_{\mathbf{B}}$. With this partitioning, we have

$$\mathscr{R}\left(\mathbf{U}_{2,+}\right) = \mathscr{R}\left(\left(\mathbf{I} - \mathbf{Q}_{\mathbf{B}}\mathbf{Q}_{\mathbf{B}}{}^H\right)\mathbf{Q}_{\mathbf{A}}\right) \subset \mathscr{R}\left(\mathbf{Q}_{\mathbf{B}}\right)^{\perp} = \mathscr{R}\left(\mathbf{U}_1\right)^{\perp} \; .$$

It follows that

$$k = \dim \mathscr{R}\left(\mathbf{U}_{2,+}\right) \leq \dim \mathscr{R}\left(\mathbf{U}_1\right)^{\perp} = m - n_{\mathbf{B}} \; .$$

We conclude that $k \leq \min\{n_{\mathbf{B}} \, , \; m - n_{\mathbf{B}}\}$.

Since $n_{\mathbf{B}} \leq m - n_{\mathbf{B}}$, we can assume that

$$\mathscr{R}\left(\mathbf{U}_{2,0}\right) \subset \mathscr{R}\left(\mathbf{U}_1\right)^{\perp} \; .$$

This is because the columns of $\mathbf{U}_{2,0}$ are multiplied by zero singular values in the corresponding singular value decomposition. Then

$$\mathscr{R}\left(\mathbf{U}_2\right) = \mathscr{R}\left(\begin{bmatrix}\mathbf{U}_{2,+} & \mathbf{U}_{2,0}\end{bmatrix}\right) \subset \mathscr{R}\left(\mathbf{U}_1\right)^{\perp} \; .$$

If necessary, we can use Corollary 3.2.1 in Chap. 3 of Volume I to extend the columns of $\mathbf{U}_2$ to an orthonormal basis for $\mathscr{R}\left(\mathbf{U}_1\right)^{\perp}$, given by the columns of the matrix $\begin{bmatrix}\mathbf{U}_2 & \mathbf{U}_3\end{bmatrix}$. Here $\mathbf{U}_3$ is $m \times (m - 2n_{\mathbf{B}})$.

At this point, it is useful to note that

$$\begin{aligned}\mathbf{Q}_{\mathbf{A}}\mathbf{Y}_{\mathbf{A}} &= \left[\mathbf{Q}_{\mathbf{B}}\mathbf{Q}_{\mathbf{B}}{}^H\mathbf{Q}_{\mathbf{A}} + \left(\mathbf{I} - \mathbf{Q}_{\mathbf{B}}\mathbf{Q}_{\mathbf{B}}{}^H\right)\mathbf{Q}_{\mathbf{A}}\right]\mathbf{Y}_{\mathbf{A}} \\ &= \left[\mathbf{U}_1\mathbf{C}\mathbf{Y}_{\mathbf{A}}{}^H + \mathbf{U}_2\mathbf{S}\mathbf{Y}_{\mathbf{A}}{}^H\right]\mathbf{Y}_{\mathbf{A}} = \mathbf{U}_1\mathbf{C} + \mathbf{U}_2\mathbf{S} \; .\end{aligned}$$

Since Lemma 1.2.18 guarantees that

$$\mathbf{C}^2 + \mathbf{S}^2 = \mathbf{I}$$

it is easy to see that $\mathbf{P}$, given by (1.8), is unitary. Furthermore, if $\mathbf{y} \in \mathscr{R}(\mathbf{B}) = \mathscr{R}(\mathbf{Q_B}) = \mathscr{R}(\mathbf{U}_1)$, then there exists an $n_\mathbf{B}$-vector $\mathbf{x}$ so that $\mathbf{y} = \mathbf{U}_1\mathbf{x}$. As a result,

$$\begin{aligned}\mathbf{P}\mathbf{y} &= \begin{bmatrix}\mathbf{U}_1 & \mathbf{U}_2 & \mathbf{U}_3\end{bmatrix}\begin{bmatrix}\mathbf{C} & -\mathbf{S} & \mathbf{0}\\ \mathbf{S} & \mathbf{C} & \mathbf{0}\\ \mathbf{0} & \mathbf{0} & \mathbf{I}\end{bmatrix}\begin{bmatrix}\mathbf{U}_1{}^H\mathbf{U}_1\\ \mathbf{U}_2{}^H\mathbf{U}_1\\ \mathbf{U}_3{}^H\mathbf{U}_1\end{bmatrix}\mathbf{x}\\ &= \begin{bmatrix}\mathbf{U}_1 & \mathbf{U}_2 & \mathbf{U}_3\end{bmatrix}\begin{bmatrix}\mathbf{C} & -\mathbf{S} & \mathbf{0}\\ \mathbf{S} & \mathbf{C} & \mathbf{0}\\ \mathbf{0} & \mathbf{0} & \mathbf{I}\end{bmatrix}\begin{bmatrix}\mathbf{I}\\ \mathbf{0}\\ \mathbf{0}\end{bmatrix}\mathbf{x} = \begin{bmatrix}\mathbf{U}_1 & \mathbf{U}_2 & \mathbf{U}_3\end{bmatrix}\begin{bmatrix}\mathbf{C}\\ \mathbf{S}\\ \mathbf{0}\end{bmatrix}\mathbf{x}\\ &= (\mathbf{U}_1\mathbf{C} + \mathbf{U}_2\mathbf{S})\,\mathbf{x} = \mathbf{Q_A}\mathbf{Y_A}\mathbf{x}\end{aligned}$$

is in the range of $\mathbf{A}$. This proves the lemma when $\mathscr{N}(\mathbf{A}^H\mathbf{B}) = \{\mathbf{0}\}$.

If $\mathscr{N}(\mathbf{A}^H\mathbf{B}) \neq \{\mathbf{0}\}$, then

$$\mathscr{R}\left(\mathbf{Q_B}\mathbf{Q_B}{}^H\mathbf{Q_A}\right) \subset \mathscr{R}(\mathbf{Q_B}) .$$

We can adjust the singular value decomposition

$$\mathbf{Q_B}\mathbf{Q_B}{}^H\mathbf{Q_A} = \mathbf{U}_1\mathbf{C}\mathbf{Y_A}{}^H$$

to allow an $n_\mathbf{B} \times n_\mathbf{B}$ matrix $\mathbf{C}$ to have one or more zero diagonal entries, and $\mathbf{Y_A}$ remains $n_\mathbf{A} \times n_\mathbf{B}$. The columns of $\mathbf{U}_1$ corresponding to nonzero diagonal entries of $\mathbf{C}$ would form an orthonormal basis for $\mathscr{R}\left(\mathbf{Q_B}\mathbf{Q_B}{}^H\mathbf{Q_A}\right)$, and then Corollary 3.2.1 in Chap. 3 of Volume I could be used to extend this to an orthonormal basis for $\mathscr{R}(\mathbf{Q_B})$, in a matrix denoted by $\mathbf{U}_1$. The remainder of the proof would proceed as before.

The matrix $\mathbf{C}$ in Lemma 1.2.19 corresponds to the cosines of the principal angles between the subspaces given by the ranges of $\mathbf{A}$ and $\mathbf{B}$; similarly $\mathbf{S}$ is the array of sines of the principal angles.

The next lemma characterizes the gap between subspaces in terms of the sines of the principal angles between the subspaces.

Lemma 1.2.20 *Let n and m be integers, where $1 \leq n$ and $2n \leq m$. Suppose that $\mathscr{V}_1$ and $\mathscr{V}_2$ are two n-dimensional subspaces of the set of all m-vectors, and define $\mathscr{P}_1$ and $\mathscr{P}_2$ to be the orthogonal projectors onto $\mathscr{V}_1$ and $\mathscr{V}_2$, respectively. Then*

$$\gamma(\mathscr{V}_1, \mathscr{V}_2) = \|\mathscr{P}_1 - \mathscr{P}_2\|_2 = \max_i \sigma_i ,$$

where σ_i is a sine of a principal angle between the two subspaces.

Proof First, we note that for all $\mathbf{v}_1 \in \mathscr{V}_1$

$$\inf_{\mathbf{v}_2 \in \mathscr{V}_2} \|\mathbf{v}_1 - \mathbf{v}_2\|_2 = \|\mathbf{v}_1 - \mathscr{P}_2 \mathbf{v}_1\|_2 = \|\mathscr{P}_1 \mathbf{v}_1 - \mathscr{P}_2 \mathbf{v}_1\|_2 \ .$$

It follows that

$$\sup_{\substack{\mathbf{v}_1 \in \mathscr{V}_1 \\ \|\mathbf{v}_1\|_2 = 1}} \inf_{\mathbf{v}_2 \in \mathscr{V}_2} \|\mathbf{v}_1 - \mathbf{v}_2\|_2 = \sup_{\substack{\mathbf{v}_1 \in \mathscr{V}_1 \\ \|\mathbf{v}_1\|_2 = 1}} \|(\mathscr{P}_1 - \mathscr{P}_2)\, \mathbf{v}_1\|_2 = \|\mathscr{P}_1 - \mathscr{P}_2\|_2 \ ,$$

and

$$\gamma(\mathscr{V}_1, \mathscr{V}_2) = \|\mathscr{P}_1 - \mathscr{P}_2\|_2 \ .$$

Next, let the columns of $\mathbf{B}$ form a basis for $\mathscr{V}_1$, and the columns of $\mathbf{A}$ form a basis for $\mathscr{V}_2$. Then it is easy to see from the proof of Lemma 1.2.19 that

$$\mathscr{P}_1 = \mathbf{U}_1 {\mathbf{U}_1}^H = \begin{bmatrix} \mathbf{U}_1 & \mathbf{U}_2 & \mathbf{U}_3 \end{bmatrix} \begin{bmatrix} \mathbf{I} & \mathbf{0} & \mathbf{0} \\ \mathbf{0} & \mathbf{0} & \mathbf{0} \\ \mathbf{0} & \mathbf{0} & \mathbf{0} \end{bmatrix} \begin{bmatrix} \mathbf{U}_1 \\ \mathbf{U}_2 \\ \mathbf{U}_3 \end{bmatrix} \text{ and}$$

$$\mathscr{P}_2 = \mathbf{Q}_\mathbf{A} {\mathbf{Q}_\mathbf{A}}^H = (\mathbf{Q}_\mathbf{A} \mathbf{Y}_\mathbf{A})\,(\mathbf{Q}_\mathbf{A} \mathbf{Y}_\mathbf{A})^H = (\mathbf{U}_1 \mathbf{C} + \mathbf{U}_2 \mathbf{S})\,(\mathbf{U}_1 \mathbf{C} + \mathbf{U}_2 \mathbf{S})^H$$
$$= \begin{bmatrix} \mathbf{U}_1 & \mathbf{U}_2 & \mathbf{U}_3 \end{bmatrix} \begin{bmatrix} \mathbf{C}^2 & \mathbf{CS} & \mathbf{0} \\ \mathbf{SC} & \mathbf{S}^2 & \mathbf{0} \\ \mathbf{0} & \mathbf{0} & \mathbf{0} \end{bmatrix} \begin{bmatrix} {\mathbf{U}_1}^H \\ {\mathbf{U}_2}^H \\ {\mathbf{U}_3}^H \end{bmatrix} \ .$$

These imply that

$$\mathscr{P}_1 - \mathscr{P}_2 = \begin{bmatrix} \mathbf{U}_1 & \mathbf{U}_2 & \mathbf{U}_3 \end{bmatrix} \begin{bmatrix} \mathbf{I} - \mathbf{C}^2 & -\mathbf{CS} & \mathbf{0} \\ -\mathbf{SC} & -\mathbf{S}^2 & \mathbf{0} \\ \mathbf{0} & \mathbf{0} & \mathbf{0} \end{bmatrix} \begin{bmatrix} {\mathbf{U}_1}^H \\ {\mathbf{U}_2}^H \\ {\mathbf{U}_3}^H \end{bmatrix}$$
$$= \begin{bmatrix} \mathbf{U}_1 & \mathbf{U}_2 & \mathbf{U}_3 \end{bmatrix} \begin{bmatrix} \mathbf{S}^2 & -\mathbf{CS} & \mathbf{0} \\ -\mathbf{SC} & -\mathbf{S}^2 & \mathbf{0} \\ \mathbf{0} & \mathbf{0} & \mathbf{0} \end{bmatrix} \begin{bmatrix} {\mathbf{U}_1}^H \\ {\mathbf{U}_2}^H \\ {\mathbf{U}_3}^H \end{bmatrix} \ .$$

Since

$$\begin{bmatrix} \sigma^2 & -\gamma\sigma \\ -\sigma\gamma & -\sigma^2 \end{bmatrix} \begin{bmatrix} 1+\sigma & \gamma \\ -\gamma & 1+\sigma \end{bmatrix} = \begin{bmatrix} 1+\sigma & \gamma \\ -\gamma & 1+\sigma \end{bmatrix} \begin{bmatrix} \sigma & \\ & -\sigma \end{bmatrix} ,$$

the eigenvectors and eigenvalues associated with a cosine-sine pair are easy to determine. Since Lemma 1.3.3 will show that $\|\mathscr{P}_1 - \mathscr{P}_2\|_2$ is the largest modulus of one its eigenvalues, the claimed results have been proved.

As noted by Stewart [163, p. 734], the assumption that $2n \le m$ is for convenience only; otherwise, we may work with the orthogonal complements of the two subspaces in Lemma 1.2.20.

Our next lemma guarantees the existence of a solution to a system of quadratic equations of a particular form. This lemma will be used to prove Lemma 1.2.22, and is similar to a result due to Stewart [163, p. 737].

Lemma 1.2.21 *Suppose that* $\mathbf{g}$ *is an n-vector,* $\mathscr{T}$ *is an invertible linear transformation on n-vectors, and* ϕ *is a function mapping n-vectors to n-vectors. Assume that there is a constant* η *so that for all n-vectors* $\mathbf{x}$ *and* $\mathbf{y}$*,*

$$\|\phi(\mathbf{x})\|_2 \le \eta \|\mathbf{x}\|_2^2 \text{ and}$$
$$\|\phi(\mathbf{x}) - \phi(\mathbf{y})\|_2 \le 2\eta \max\{\|\mathbf{x}\|_2, \|\mathbf{y}\|_2\} \|\mathbf{x} - \mathbf{y}\|_2 \,.$$

Define the sequence $\{\mathbf{x}_i\}_{i=0}^{\infty}$ *of n-vectors by*

$$\mathbf{x}_0 = \mathbf{0} \text{ and } \mathbf{x}_{i+1} = \mathscr{T}^{-1}\left[\mathbf{g} - \phi(\mathbf{x}_i)\right] \text{ for all } i \ge 0 \,.$$

If

$$4\eta \|\mathbf{g}\|_2 \left\|\mathscr{T}^{-1}\right\|_2^2 < 1 \,, \tag{1.9}$$

then $\mathbf{x}_i \to \mathbf{x}$ *where* $\|\mathbf{x}\|_2 < 2\left\|\mathscr{T}^{-1}\mathbf{g}\right\|_2$*. This lemma is also true if we replace the 2-norm of* $\mathscr{T}^{-1}$ *with the Frobenius norm.*

Proof Note that

$$\|\mathbf{x}_1\|_2 = \left\|\mathscr{T}^{-1}\mathbf{g}\right\|_2 \le \left\|\mathscr{T}^{-1}\right\|_2 \|\mathbf{g}\|_2 \,,$$

and that

$$\begin{aligned} \|\mathbf{x}_2\|_2 &= \left\|\mathscr{T}^{-1}\mathbf{g} - \mathscr{T}^{-1}\phi(\mathbf{x}_1)\right\|_2 \le \|\mathbf{x}_1\|_2 + \|\mathscr{T}^{-1}\|_2 \eta \|\mathbf{x}_1\|_2^2 \\ &= \|\mathbf{x}_1\|_2 \left(1 + \eta \left\|\mathscr{T}^{-1}\right\|_2^2 \|\mathbf{g}\|_2\right) \,. \end{aligned}$$

Let

$$\kappa_2 = \eta \|\mathbf{g}\|_2 \left\|\mathscr{T}^{-1}\right\|_2^2 \,,$$

and assume inductively that for some $i \ge 2$ we have

$$\|\mathbf{x}_i\|_2 \le \|\mathbf{x}_1\|_2 \left(1 + \kappa_i\right) \,.$$

Then

$$\|\mathbf{x}_{i+1}\|_2 = \left\|\mathscr{T}^{-1}\mathbf{g} - \mathscr{T}^{-1}\phi(\mathbf{x}_i)\right\|_2 \le \|\mathbf{x}_1\|_2 + \left\|\mathscr{T}^{-1}\right\|_2 \eta\|\mathbf{x}_i\|_2^2$$
$$\le \|\mathbf{x}_1\|_2 \left[1 + \eta \left\|\mathscr{T}^{-1}\right\|_2 \|\mathbf{x}_i\|_2^2/\|\mathbf{x}_1\|_2\right] \le \|\mathbf{x}_1\|_2 \left[1 + \eta \left\|\mathscr{T}^{-1}\right\|_2 \|\mathbf{x}_1\|_2(1+\kappa_i)^2\right]$$
$$\le \|\mathbf{x}_1\|_2 \left[1 + \kappa_2(1+\kappa_i)^2\right] .$$

Thus for $i \ge 2$ we can take

$$\kappa_{i+1} = \kappa_2(1+\kappa_i)^2 .$$

Note that assumption (1.9) implies that $\kappa_2 < 1/4$. Also note that if $i \ge 2$ and $\kappa_i < 1$, then $\kappa_{i+1} < 4\kappa_2 < 1$. If $\kappa \in (0,1)$, then

$$|f(\kappa) - f(\kappa')| = |\kappa - \kappa'|\,\kappa_2(2 + \kappa + \kappa') < |\kappa - \kappa'| ,$$

and the function

$$f(\kappa) = \kappa_2(1+\kappa)^2$$

is contractive. Then the contractive mapping Theorem 5.4.3 in Chap. 5 of Volume I implies that $\kappa_i \in [0,1]$ for all i, and that $\kappa_i \to \kappa$ where κ is the positive root of

$$\kappa = \kappa_2(1+\kappa)^2 .$$

It is easy to see that this root is

$$\kappa = \frac{2\kappa_2}{1 - 2\kappa_2 + \sqrt{1 - 4\kappa_2}} .$$

Now we can see that

$$\|\mathbf{x}_i\|_2 \le \|\mathbf{x}_1\|_2(1+\kappa_i) \le 2\|\mathbf{x}_1\|_2$$

for all i. Also, the function

$$F(\mathbf{x}) = \mathscr{T}^{-1}\left[\mathbf{g} - \phi(\mathbf{x})\right]$$

is contractive on the set of n-vectors with norm at most $2\|\mathbf{x}_1\|_2$, because

$$\|F(\mathbf{x}) - F(\mathbf{y})\|_2 = \|\mathscr{T}^{-1}\left[\phi(\mathbf{x}) - \phi(\mathbf{y})\right]\|_2 \le \|\mathscr{T}^{-1}\|_2 \eta \max\{\|\mathbf{x}\|_2, \|\mathbf{y}\|_2\}\|\mathbf{x} - \mathbf{y}\|_2$$
$$\le 4\eta\|\mathscr{T}^{-1}\|_2\|\mathbf{x}_1\|_2\|\mathbf{x} - \mathbf{y}\|_2 \le 4\eta\|\mathscr{T}^{-1}\|_2^2\|\mathbf{g}\|_2\|\mathbf{x} - \mathbf{y}\|_2 = 4\kappa_2\|\mathbf{x} - \mathbf{y}\|_2 .$$

Then the Contractive Mapping Theorem 3.2.2 will imply that $\mathbf{x}_i \to \mathbf{x}$ where

$$\|\mathbf{x}\|_2 \leq 2\|\mathbf{x}_1\|_2 \,.$$

Next, we will prove a result that allows us to determine an orthonormal basis for an invariant subspace of a matrix. This lemma will be used to prove the Stewart Perturbation Theorem 1.2.3. Our discussion of the next lemma will follow that in Stewart [163, p. 739].

Lemma 1.2.22 *Let* $\mathbf{A}$ *be an* $m \times m$ *matrix and let* $\mathbf{U}$ *be an* $m \times m$ *unitary matrix. Partition*

$$\mathbf{U} = \begin{bmatrix} \mathbf{U}_1 & \mathbf{U}_2 \end{bmatrix}$$

where $\mathbf{U}_1$ *is* $m \times n$*, and let*

$$\mathbf{U}^H \mathbf{A} \mathbf{U} = \begin{bmatrix} {\mathbf{U}_1}^H \\ {\mathbf{U}_2}^H \end{bmatrix} \mathbf{A} \begin{bmatrix} \mathbf{U}_1 & \mathbf{U}_2 \end{bmatrix} \equiv \begin{bmatrix} \mathbf{B}_{11} & \mathbf{B}_{12} \\ \mathbf{B}_{21} & \mathbf{B}_{22} \end{bmatrix} .$$

Define the operator $\mathscr{T}$ *mapping* $(m-n) \times n$ *matrices to* $(m-n) \times n$ *matrices by*

$$\mathscr{T}\mathbf{P} = \mathbf{P}\mathbf{B}_{11} - \mathbf{B}_{22}\mathbf{P} \,. \tag{1.10}$$

Assume that $\mathscr{T}$ *is invertible, and that*

$$\|\mathbf{B}_{12}\|_2 \|\mathbf{B}_{21}\|_2 \left\|\mathscr{T}^{-1}\right\|_2^2 < 1/4 \,. \tag{1.11}$$

Then there is a $(m-n) \times n$ *matrix* $\mathbf{P}$ *such that*

$$\|\mathbf{P}\|_2 < 2\|\mathbf{B}_{21}\|_2 \left\|\mathscr{T}^{-1}\right\|_2$$

and such that the columns of

$$\mathbf{W}_1 \equiv (\mathbf{U}_1 + \mathbf{U}_2\mathbf{P}) \left(\mathbf{I} + \mathbf{P}^H\mathbf{P}\right)^{-1/2}$$

span an invariant subspace of $\mathbf{A}$*. This lemma is also true if we replace the 2-norm of* $\mathscr{T}^{-1}$ *with the Frobenius norm.*

Proof Note that for any $(m-n) \times n$ matrix $\mathbf{P}$, the matrix

$$\mathbf{W} \equiv \begin{bmatrix} \mathbf{U}_1 & \mathbf{U}_2 \end{bmatrix} \begin{bmatrix} \mathbf{I} & -\mathbf{P}^H \\ \mathbf{P} & \mathbf{I} \end{bmatrix} \begin{bmatrix} \left(\mathbf{I} + \mathbf{P}^H\mathbf{P}\right)^{-1/2} & \\ & \left(\mathbf{I} + \mathbf{P}\mathbf{P}^H\right)^{-1/2} \end{bmatrix}$$

is unitary. Furthermore,

$$\mathbf{W}^H\mathbf{A}\mathbf{W} = \begin{bmatrix} (\mathbf{I}+\mathbf{P}^H\mathbf{P})^{-1/2} & \\ & (\mathbf{I}+\mathbf{P}\mathbf{P}^H)^{-1/2} \end{bmatrix} \begin{bmatrix} \mathbf{I} & \mathbf{P}^H \\ -\mathbf{P} & \mathbf{I} \end{bmatrix} \begin{bmatrix} \mathbf{B}_{11} & \mathbf{B}_{12} \\ \mathbf{B}_{21} & \mathbf{B}_{22} \end{bmatrix}$$
$$\times \begin{bmatrix} \mathbf{I} & -\mathbf{P}^H \\ \mathbf{P} & \mathbf{I} \end{bmatrix} \begin{bmatrix} (\mathbf{I}+\mathbf{P}^H\mathbf{P})^{-1/2} & \\ & (\mathbf{I}+\mathbf{P}\mathbf{P}^H)^{-1/2} \end{bmatrix}$$
$$= \begin{bmatrix} (\mathbf{I}+\mathbf{P}^H\mathbf{P})^{-1/2} & \\ & (\mathbf{I}+\mathbf{P}\mathbf{P}^H)^{-1/2} \end{bmatrix}$$
$$\times \begin{bmatrix} \mathbf{B}_{11}+\mathbf{B}_{12}\mathbf{P}+\mathbf{P}^H\mathbf{B}_{21}+\mathbf{P}^H\mathbf{B}_{22}\mathbf{P} & \mathbf{B}_{12}-\mathbf{B}_{11}\mathbf{P}^H+\mathbf{P}^H\mathbf{B}_{22}-\mathbf{P}^H\mathbf{B}_{21}\mathbf{P}^H \\ -\mathbf{P}\mathbf{B}_{11}-\mathbf{P}\mathbf{B}_{12}\mathbf{P}+\mathbf{B}_{21}+\mathbf{B}_{22}\mathbf{P} & \mathbf{B}_{22}-\mathbf{B}_{21}\mathbf{P}^H-\mathbf{P}\mathbf{B}_{12}+\mathbf{P}\mathbf{B}_{11}\mathbf{P}^H \end{bmatrix}$$
$$\times \begin{bmatrix} (\mathbf{I}+\mathbf{P}^H\mathbf{P})^{-1/2} & \\ & (\mathbf{I}+\mathbf{P}\mathbf{P}^H)^{-1/2} \end{bmatrix} .$$

In order for the first n columns of $\mathbf{W}$ to span an invariant subspace of $\mathbf{W}^H\mathbf{A}\mathbf{W}$, we require

$$-\mathbf{P}\mathbf{B}_{11}-\mathbf{P}\mathbf{B}_{12}\mathbf{P}+\mathbf{B}_{21}+\mathbf{B}_{22}\mathbf{P}=\mathbf{0} ,$$

which is equivalent to the equation

$$\mathscr{T}(\mathbf{P}) = \mathbf{B}_{21}-\mathbf{P}\mathbf{B}_{12}\mathbf{P} , \tag{1.12}$$

where $\mathscr{T}$ was defined by (1.10). Since

$$\|\mathbf{P}\mathbf{B}_{12}\mathbf{P}\|_2 \le \|\mathbf{B}_{12}\|_2\|\mathbf{P}\|_2^2$$

and

$$\begin{aligned} \|\mathbf{P}\mathbf{B}_{12}\mathbf{P}-\mathbf{P}'\mathbf{B}_{12}\mathbf{P}'\|_2 &= \|\mathbf{P}\mathbf{B}_{12}(\mathbf{P}-\mathbf{P}')+(\mathbf{P}-\mathbf{P}')\mathbf{B}_{12}\mathbf{P}'\|_2 \\ &\le \|\mathbf{P}\|_2\|\mathbf{B}_{12}\|_2\|\mathbf{P}-\mathbf{P}'\|_2+\|\mathbf{P}-\mathbf{P}'\|_2\|\mathbf{B}_{12}\|_2\|\mathbf{P}'\|_2 \\ &\le 2\max\{\|\mathbf{P}\|_2, \|\mathbf{P}'\|_2\}\|\mathbf{B}_{12}\|_2\|\mathbf{P}-\mathbf{P}'\|_2 \end{aligned}$$

assumption (1.11) shows that the hypotheses of Lemma 1.2.21 are satisfied. This lemma guarantees that Eq. (1.12) has a solution $\mathbf{P}$ with

$$\|\mathbf{P}\|_2 < 2\|\mathbf{B}_{21}\|_2\|\mathscr{T}^{-1}\|_2 .$$

We can now see that

$$
\begin{aligned}
\mathbf{A}\mathbf{W}_1 &= \mathbf{A}(\mathbf{U}_1 + \mathbf{U}_2\mathbf{P})\left(\mathbf{I} + \mathbf{P}^H\mathbf{P}\right)^{-1/2} \\
&= \left[\mathbf{U}_1\ \mathbf{U}_2\right]\begin{bmatrix}\mathbf{B}_{11} & \mathbf{B}_{12} \\ \mathbf{B}_{21} & \mathbf{B}_{22}\end{bmatrix}\begin{bmatrix}\mathbf{I} \\ \mathbf{P}\end{bmatrix}\left(\mathbf{I} + \mathbf{P}^H\mathbf{P}\right)^{-1/2} \\
&= \left[\mathbf{U}_1\ \mathbf{U}_2\right]\begin{bmatrix}\mathbf{B}_{11} + \mathbf{B}_{12}\mathbf{P} \\ \mathbf{B}_{21} + \mathbf{B}_{22}\mathbf{P}\end{bmatrix}\left(\mathbf{I} + \mathbf{P}^H\mathbf{P}\right)^{-1/2} \\
&= \left[\mathbf{U}_1\ \mathbf{U}_2\right]\begin{bmatrix}\mathbf{B}_{11} + \mathbf{B}_{12}\mathbf{P} \\ \mathbf{P}(\mathbf{B}_{11} + \mathbf{B}_{12}\mathbf{P}\end{bmatrix}\left(\mathbf{I} + \mathbf{P}^H\mathbf{P}\right)^{-1/2} \\
&= \left[\mathbf{U}_1\ \mathbf{U}_2\right]\begin{bmatrix}\mathbf{I} \\ \mathbf{P}\end{bmatrix}(\mathbf{B}_{11} + \mathbf{B}_{12}\mathbf{P})\left(\mathbf{I} + \mathbf{P}^H\mathbf{P}\right)^{-1/2} \\
&= \mathbf{W}_1\left(\mathbf{I} + \mathbf{P}^H\mathbf{P}\right)^{1/2}(\mathbf{B}_{11} + \mathbf{B}_{12}\mathbf{P})\left(\mathbf{I} + \mathbf{P}^H\mathbf{P}\right)^{-1/2} \; .
\end{aligned}
$$

We remark that the operator $\mathscr{T}$, defined by (1.10), is invertible if and only if $\mathbf{B}_{11}$ and $\mathbf{B}_{22}$ have no common eigenvalues. In fact, we have the following stronger result.

Lemma 1.2.23 *Let* **A** *and* **B** *be square matrices and define*

$$
\mathscr{T}\mathbf{P} = \mathbf{P}\mathbf{A} - \mathbf{B}\mathbf{P} \; .
$$

Then the eigenvalues of $\mathscr{T}$ *are the differences between the eigenvalues of* **A** *and* **B**. *Consequently,*

$$
\|\mathscr{T}\|_2 \ge \max_{\substack{\lambda_{\mathbf{A}} \text{ eigenvalue of } \mathbf{A} \\ \lambda_{\mathbf{B}} \text{ eigenvalue of } \mathbf{B}}} |\lambda_{\mathbf{A}} - \lambda_{\mathbf{B}}| \text{ and } \frac{1}{\|\mathscr{T}^{-1}\|_2} \le \min_{\substack{\lambda_{\mathbf{A}} \text{ eigenvalue of } \mathbf{A} \\ \lambda_{\mathbf{B}} \text{ eigenvalue of } \mathbf{B}}} |\lambda_{\mathbf{A}} - \lambda_{\mathbf{B}}| \; .
$$

Proof Let $\mathbf{A}\mathbf{X}_{\mathbf{A}} = \mathbf{X}_{\mathbf{A}}\mathbf{J}_{\mathbf{A}}$ be the Jordan canonical form for **A**, described in Sect. 1.4.1.5. Then Theorem 1.4.6 shows that $\mathbf{X}_{\mathbf{A}}$ is nonsingular, and for all column indices j either $\mathbf{J}_{\mathbf{A}}\mathbf{e}_j = \mathbf{e}_j\lambda_{\mathbf{A}}$ or $\mathbf{J}_{\mathbf{A}}\mathbf{e}_j = \mathbf{e}_{j-1} + \mathbf{e}_j\lambda_{\mathbf{A}}$.

Next, suppose that $\mathscr{T}\mathbf{P} = \mathbf{P}\lambda$, and $\mathbf{P} \neq \mathbf{0}$. Since $\mathbf{X}_{\mathbf{A}}$ is nonsingular, there is a first index j for which $\mathbf{P}\mathbf{X}_{\mathbf{A}}\mathbf{e}_j \neq \mathbf{0}$. There are two cases for the value of j. First, if $\mathbf{J}_{\mathbf{A}}\mathbf{e}_j = \mathbf{e}_j\lambda_{\mathbf{A}}$, then

$$
\mathbf{P}\mathbf{X}_{\mathbf{A}}\mathbf{e}_j\lambda = (\mathbf{P}\mathbf{A} - \mathbf{B}\mathbf{P})\mathbf{X}_{\mathbf{A}}\mathbf{e}_j = \mathbf{P}\mathbf{X}_{\mathbf{A}}\mathbf{J}_{\mathbf{A}}\mathbf{e}_j - \mathbf{B}\mathbf{P}\mathbf{X}_{\mathbf{A}}\mathbf{e}_j = \mathbf{P}\mathbf{X}_{\mathbf{A}}\mathbf{e}_j\lambda_{\mathbf{A}} - \mathbf{B}\mathbf{P}\mathbf{X}_{\mathbf{A}}\mathbf{e}_j \; ,
$$

so

$$
[\mathbf{B} - \mathbf{I}(\lambda - \lambda_{\mathbf{A}})]\,\mathbf{P}\mathbf{X}_{\mathbf{A}}\mathbf{e}_j = \mathbf{0} \; .
$$

This shows that $\lambda - \lambda_{\mathbf{A}}$ is an eigenvalue of **B**.

Otherwise, we have $\mathbf{J_A e}_j = \mathbf{e}_{j-1} + \mathbf{e}_j \lambda_{\mathbf{A}}$. In this case,

$$\begin{aligned}\mathbf{PX_A e}_j \lambda &= (\mathbf{PA} - \mathbf{BP})\mathbf{X_A e}_j = \mathbf{PX_A J_A e}_j - \mathbf{BPX_A e}_j \\ &= \mathbf{PX_A}(\mathbf{e}_{j-1} + \mathbf{e}_j \lambda_{\mathbf{A}}) - \mathbf{BPX_A e}_j = \mathbf{PX_A e}_j \lambda_{\mathbf{A}} - \mathbf{BPX_A e}_j \;,\end{aligned}$$

where the last equation holds because the assumption that j is the first index so that $\mathbf{PX_A e}_j \neq \mathbf{0}$ implies that $\mathbf{PX_A e}_{j-1} = \mathbf{0}$. The last equation can be rewritten as

$$[\mathbf{B} - \mathbf{I}(\lambda - \lambda_{\mathbf{A}})]\,\mathbf{PX_A e}_j = \mathbf{0}\;.$$

This shows that $\lambda - \lambda_{\mathbf{A}}$ is an eigenvalue of $\mathbf{B}$.

Lemma 1.2.15 now shows that

$$\|\mathscr{T}\|_2 \geq \max_{\substack{\lambda_{\mathbf{A}} \text{ an eigenvalue of } \mathbf{A} \\ \lambda_{\mathbf{B}} \text{ an eigenvalue of } \mathbf{B}}} |\lambda_{\mathbf{A}} - \lambda_{\mathbf{B}}|\;.$$

Also, Lemma 1.2.4 shows that the eigenvalues of $\mathscr{T}^{-1}$ are of the form $1/(\lambda_{\mathbf{A}} - \lambda_{\mathbf{B}})$, and so Lemma 1.2.15 proves that

$$\left\|\mathscr{T}^{-1}\right\|_2 \geq \max_{\substack{\lambda_{\mathbf{A}} \text{ an eigenvalue of } \mathbf{A} \\ \lambda_{\mathbf{B}} \text{ an eigenvalue of } \mathbf{B}}} \frac{1}{|\lambda_{\mathbf{A}} - \lambda_{\mathbf{B}}|} \geq \frac{1}{\min_{\substack{\lambda_{\mathbf{A}} \text{ an eigenvalue of } \mathbf{A} \\ \lambda_{\mathbf{B}} \text{ an eigenvalue of } \mathbf{B}}} |\lambda_{\mathbf{A}} - \lambda_{\mathbf{B}}|}\;.$$

This proves the final claim of the lemma.

We conclude this section with the following theorem, which will be useful in estimating the distance between invariant subspaces for a matrix and its perturbation. This theorem is due to Stewart [163, p. 745], and will be used to prove Theorem 1.4.10.

Theorem 1.2.3 (Stewart Perturbation) *Suppose that* $\mathbf{A}$ *and* $\widetilde{\mathbf{A}}$ *are* $m \times m$ *matrices. Assume that*

$$\mathbf{U} = \begin{bmatrix}\mathbf{U}_1 & \mathbf{U}_2\end{bmatrix}$$

is an $m \times m$ *unitary matrix, and that the range of the* $m \times n$ *matrix* $\mathbf{U}_1$ *is an invariant subspace under* $\mathbf{A}$. *Define*

$$\begin{bmatrix}\mathbf{B}_{11} & \mathbf{B}_{12} \\ \mathbf{0} & \mathbf{B}_{22}\end{bmatrix} = \begin{bmatrix}\mathbf{U}_1{}^H \\ \mathbf{U}_2{}^H\end{bmatrix} \mathbf{A} \begin{bmatrix}\mathbf{U}_1 & \mathbf{U}_2\end{bmatrix} \text{ and } \begin{bmatrix}\widetilde{\mathbf{B}}_{11} & \widetilde{\mathbf{B}}_{12} \\ \widetilde{\mathbf{B}}_{21} & \widetilde{\mathbf{B}}_{22}\end{bmatrix} = \begin{bmatrix}\mathbf{U}_1{}^H \\ \mathbf{U}_2{}^H\end{bmatrix} \widetilde{\mathbf{A}} \begin{bmatrix}\mathbf{U}_1 & \mathbf{U}_2\end{bmatrix}\;.$$

Suppose that

$$\delta \equiv \inf_{\|\mathbf{P}\|_2 = 1} \|\mathbf{PB}_{11} - \mathbf{B}_{22}\mathbf{P}\|_2 - \left\|\widetilde{\mathbf{B}}_{11} - \mathbf{B}_{11}\right\|_2 - \left\|\widetilde{\mathbf{B}}_{22} - \mathbf{B}_{22}\right\|_2 > 0\;,$$

and assume that

$$4 \left\|\widetilde{\mathbf{B}}_{21}\right\|_2 \left(\|\mathbf{B}_{12}\|_2 + \left\|\widetilde{\mathbf{B}}_{12} - \mathbf{B}_{12}\right\|_2\right) \le \delta^2 \,. \tag{1.13}$$

Then there is an $(m-n) \times n$ *matrix* $\mathbf{P}$ *satisfying*

$$\|\mathbf{P}\|_2 \le 2 \frac{\left\|\widetilde{\mathbf{B}}_{21}\right\|_2}{\delta} \,,$$

and such that the columns of the matrix $\mathbf{W}_1 = (\mathbf{U}_1 + \mathbf{U}_2\mathbf{P})\left(\mathbf{I} + \mathbf{P}^H\mathbf{P}\right)^{-1/2}$ *are an orthonormal basis for a subspace invariant under* $\widetilde{\mathbf{A}}$. *This theorem is also true if we replace the matrix 2-norms with Frobenius norms.*

Proof Define the linear transformation $\widetilde{\mathscr{T}}$ mapping $(m-n)\times n$ matrices to $(m-n)\times n$ matrices be defined by

$$\widetilde{\mathscr{T}}\,\mathbf{P} = \mathbf{P}\widetilde{\mathbf{B}}_{11} - \widetilde{\mathbf{B}}_{22}\mathbf{P} \,.$$

Note that

$$\begin{aligned}
&\inf_{\|\mathbf{P}\|_2=1} \left\|\widetilde{\mathscr{T}}\,\mathbf{P}\right\|_2 = \inf_{\|\mathbf{P}\|_2=1} \|\mathbf{P}\widetilde{\mathbf{B}}_{11} - \widetilde{\mathbf{B}}_{22}\mathbf{P}\|_2 \\
&= \inf_{\|\mathbf{P}\|_2=1} \left\|\mathbf{P}\mathbf{B}_{11} - \mathbf{B}_{22}\mathbf{P} + \mathbf{P}(\widetilde{\mathbf{B}}_{11} - \mathbf{B}_{11}) - (\widetilde{\mathbf{B}}_{22} - \mathbf{B}_{22})\mathbf{P}\right\|_2 \\
&\ge \inf_{\|\mathbf{P}\|_2=1} \|\mathbf{P}\mathbf{B}_{11} - \mathbf{B}_{22}\mathbf{P}\|_2 - \|\widetilde{\mathbf{B}}_{11} - \mathbf{B}_{11}\|_2 - \|\widetilde{\mathbf{B}}_{22} - \mathbf{B}_{22}\|_2 = \delta > 0 \,,
\end{aligned}$$

so $\widetilde{\mathscr{T}}$ is invertible. It follows that for all $\mathbf{P}$ we have

$$\left\|\widetilde{\mathscr{T}}^{-1}\right\|_2 = \sup_{\widetilde{\mathscr{T}}\mathbf{P}\neq\mathbf{0}} \frac{\|\mathbf{P}\|_2}{\left\|\widetilde{\mathscr{T}}\mathbf{P}\right\|_2} = \frac{1}{\inf_{\mathbf{P}\neq\mathbf{0}} \left\|\widetilde{\mathscr{T}}\mathbf{P}\right\|_2 / \|\mathbf{P}\|_2} \,.$$

Assumption (1.13) implies that

$$\begin{aligned}
&\|\widetilde{\mathbf{B}}_{21}\|_2 \|\widetilde{\mathbf{B}}_{12}\|_2 \left\|\widetilde{\mathscr{T}}^{-1}\right\|_2^2 = \|\widetilde{\mathbf{B}}_{21} - \mathbf{B}_{21}\|_2 \|\widetilde{\mathbf{B}}_{12}\|_2 \left\|\widetilde{\mathscr{T}}^{-1}\right\|_2^2 \\
&\le \|\widetilde{\mathbf{B}}_{21} - \mathbf{B}_{21}\|_2 (\|\mathbf{B}_{12}\|_2 + \|\widetilde{\mathbf{B}}_{12} - \mathbf{B}_{12}\|_2) \frac{1}{\delta^2} < \frac{1}{4} \,,
\end{aligned}$$

so we can apply Lemma 1.2.22 to find $\mathbf{P}$ as claimed.

Exercise 1.2.26 Consider the matrix

$$\mathbf{A} = \begin{bmatrix} 3 & 2 & 2 & -4 \\ 2 & 3 & 2 & -1 \\ 1 & 1 & 2 & -1 \\ 2 & 2 & 2 & -1 \end{bmatrix} .$$

By examining the characteristic polynomial of **A**, it is easy to see that the eigenvalues of **A** are 1, 2 and 3. Find the distance between the invariant subspace

$$\mathscr{V}_1 = \{\mathbf{x} : \mathbf{A}\mathbf{x} = \mathbf{x}\}$$

and the invariant subspace

$$\mathscr{V}_2 = \{\mathbf{x} : \mathbf{A}\mathbf{x} = \mathbf{x}2\} \ .$$

1.3 Hermitian Matrices

At this point in this chapter, we have developed various fundamental concepts regarding eigenvalues and eigenvectors. The Fundamental Theorem of Algebra 1.2.1 was used to show that every square matrix of size n with complex entries has n eigenvalues in the complex plane. However, we said very little about the existence of eigenvectors. For general square matrices, such a discussion is very difficult.

However, for Hermitian matrices we can make very strong statements about eigenvectors, as well as their eigenvalues. Readers should pay particular close attention to the Spectral Theorem 1.3.1, because it guarantees that eigenvalues of Hermitian matrices are real, and that their eigenvectors can be chosen to be orthonormal. This fact will lead to very accurate and efficient algorithms for computing eigenvalues and eigenvectors of Hermitian matrices.

Our computational approach to the Hermitian eigenproblem will be similar to our strategy in solving systems of linear equations or least squares problems. Initially, we will identify a simple problem for which the computation is easy, and then find a way to reduce all problems to the simple form. For linear equations, the simple form was a triangular or trapezoidal matrix; for least squares problems, the simple form was a matrix with mutually orthogonal columns. For Hermitian eigenproblems, the simple form will be a real symmetric tridiagonal matrix.

1.3.1 Theory

1.3.1.1 Eigenvalues and Eigenvectors

Let us begin by proving a general result about eigenvalues and eigenvectors of Hermitian matrices.

Lemma 1.3.1 *Suppose that* **A** *is an Hermitian matrix. Then the eigenvalues of* **A** *are real, and its eigenvectors corresponding to distinct eigenvalues are orthogonal.*

Proof For all vectors **x**, the scalar

$$\left(\mathbf{x}^H\mathbf{A}\mathbf{x}\right)^H = \mathbf{x}^H\mathbf{A}^H\mathbf{x} = \mathbf{x}^H\mathbf{A}\mathbf{x} \ ,$$

is real. If $\mathbf{x}$ is an eigenvector of $\mathbf{A}$ with eigenvalue λ, then

$$\|\mathbf{x}\|^2\lambda = \mathbf{x}^H\mathbf{x}\lambda = \mathbf{x}^H\mathbf{A}\mathbf{x} = \left(\mathbf{x}^H\mathbf{A}\mathbf{x}\right)^H = \left(\mathbf{x}^H\mathbf{x}\lambda\right)^H = \|\mathbf{x}\|^2\overline{\lambda} .$$

Since $\mathbf{x} \neq 0$, we conclude that $\lambda = \overline{\lambda}$, which proves that the eigenvalue is real.

Next, suppose that $\mathbf{A}\mathbf{x}_1 = \mathbf{x}_1\lambda_1$ and $\mathbf{A}\mathbf{x}_2 = \mathbf{x}_2\lambda_2$ with $\lambda_1 \neq \lambda_2$. Then

$$\mathbf{x}_2 \cdot \mathbf{x}_1\lambda_1 = \mathbf{x}_2{}^H\mathbf{A}\mathbf{x}_1 = \left(\mathbf{x}_1{}^H\mathbf{A}\mathbf{x}_2\right)^H = (\mathbf{x}_1 \cdot \mathbf{x}_2\lambda_2)^H = (\mathbf{x}_1 \cdot \mathbf{x}_2)^H\lambda_2 = \mathbf{x}_2 \cdot \mathbf{x}_1\lambda_2 .$$

Since $\lambda_1 \neq \lambda_2$, we conclude that $\mathbf{x}_2 \cdot \mathbf{x}_1 = 0$.

Recall that Lemma 1.2.12 showed that a change of basis preserves eigenvalues of a matrix. The next lemma will give us a more useful similarity transformation to apply to Hermitian matrices.

Lemma 1.3.2 *If* $\mathbf{A}$ *is an* $n \times n$ *Hermitian matrix and* $\mathbf{Q}$ *is an* $n \times n$ *unitary matrix, then* $\mathbf{A}$ *and* $\mathbf{Q}^H\mathbf{A}\mathbf{Q}$ *have the same eigenvalues.*

Proof If $\mathbf{A}\mathbf{x} = \mathbf{x}\lambda$, then

$$\left(\mathbf{Q}^H\mathbf{A}\mathbf{Q}\right)\left(\mathbf{Q}^H\mathbf{x}\right) = \left(\mathbf{Q}^H\mathbf{x}\right)\lambda .$$

Since $\|\mathbf{x}\|_2 = \|\mathbf{Q}^H\mathbf{x}\|_2$, $\mathbf{x}$ is nonzero whenever $\mathbf{Q}^H\mathbf{x}$ is nonzero.

1.3.1.2 Spectral Theorem

The next theorem shows that every Hermitian matrix has an orthonormal basis of eigenvectors, and corresponding real eigenvalues.

Theorem 1.3.1 (Spectral) *If* $\mathbf{A}$ *is an* $n \times n$ *Hermitian matrix, then there is an* $n \times n$ *unitary matrix* $\mathbf{U}$ *and an* $n \times n$ *real diagonal matrix* $\boldsymbol{\Lambda}$ *so that* $\mathbf{A}\mathbf{U} = \mathbf{U}\boldsymbol{\Lambda}$. *In particular, if* $\mathbf{A}$ *is a real symmetric matrix, then* $\mathbf{U}$ *is real and orthogonal.*

Proof We will find the eigenvalues and eigenvectors of $\mathbf{A}$ onc at a time. During such a process, we will use the values of a new eigenvalue and eigenvector to reduce the size of the problem for the remaining eigenvalues. This process is called **deflation**.

The fundamental theorem of algebra shows that the characteristic polynomial $\det(\mathbf{A} - \mathbf{I}\lambda)$ has n zeros, counting multiplicity, and Lemma 1.3.1 shows that these zeros are real. Suppose that $\mathbf{x}$ is an n-vector such that $\|\mathbf{x}\|_2 = 1$ and $\mathbf{A}\mathbf{x} = \mathbf{x}\lambda$. We can find a Householder reflector $\mathbf{H}$ so that $\mathbf{H}^H\mathbf{x} = \mathbf{e}_1$. Since $\mathbf{H}\mathbf{H}^H = \mathbf{I}$, it follows that $\mathbf{H}\mathbf{e}_1 = \mathbf{x}$. Thus the first column of $\mathbf{H}$ is $\mathbf{x}$, and we can partition

$$\mathbf{H} = \left[\mathbf{x}\,,\,\mathbf{Q}\right] .$$

Then

$$\mathbf{H}^H\mathbf{A}\mathbf{H} = \begin{bmatrix}\mathbf{x}^H\\ \mathbf{Q}^H\end{bmatrix}\mathbf{A}\left[\mathbf{x}\;\mathbf{Q}\right] = \begin{bmatrix}\lambda & \mathbf{0}\\ \mathbf{0} & \mathbf{Q}^H\mathbf{A}\mathbf{Q}\end{bmatrix} .$$

Lemma 1.2.3 implies that the remaining eigenvalues of $\mathbf{A}$ are eigenvalues of the $(n-1) \times (n-1)$ matrix $\mathbf{Q}^H\mathbf{A}\mathbf{Q}$.

Inductively, suppose that for some $k \geq 1$ we have found a unitary matrix $\mathbf{U}$ and a $k \times k$ diagonal matrix $\boldsymbol{\Lambda}$ so that

$$\mathbf{U}^H\mathbf{A}\mathbf{u} = \begin{bmatrix} \mathbf{U}_1{}^H \\ \mathbf{U}_2{}^H \end{bmatrix} \mathbf{A} \begin{bmatrix} \mathbf{U}_1 & \mathbf{U}_2 \end{bmatrix} = \begin{bmatrix} \boldsymbol{\Lambda} & \\ & \mathbf{U}_2{}^H\mathbf{A}\mathbf{U}_2 \end{bmatrix} .$$

As in the beginning of this proof, we can find a vector $\mathbf{y}$ and a real scalar λ so that

$$\mathbf{U}_2{}^H\mathbf{A}\mathbf{U}_2\mathbf{y} = \mathbf{y}\lambda \text{ where } \|\mathbf{y}\|_2 = 1 .$$

Next, we find a Householder reflector so that $\mathbf{H}\mathbf{y} = \mathbf{e}_1$. Afterward, we can partition

$$\mathbf{H} = \begin{bmatrix} \mathbf{y} & \mathbf{Q} \end{bmatrix} .$$

Then

$$\begin{bmatrix} \mathbf{I} & \\ & \mathbf{H} \end{bmatrix} \begin{bmatrix} \mathbf{U}_1{}^H \\ \mathbf{U}_2{}^H \end{bmatrix} \mathbf{A} \begin{bmatrix} \mathbf{U}_1 & \mathbf{U}_2 \end{bmatrix} \begin{bmatrix} \mathbf{I} & \\ & \mathbf{H} \end{bmatrix} = \begin{bmatrix} \boldsymbol{\Lambda} & & \\ & \lambda & \\ & & \mathbf{Q}^H\mathbf{U}_2{}^H\mathbf{A}\mathbf{U}_2\mathbf{Q} \end{bmatrix} ,$$

so the deflation process has been extended to $k+1$. Also note that

$$\begin{bmatrix} \mathbf{U}_1 & \mathbf{U}_2 \end{bmatrix} \begin{bmatrix} \mathbf{I} & \\ & \mathbf{H} \end{bmatrix} = \begin{bmatrix} \mathbf{U}_1 & \mathbf{U}_2\mathbf{H} \end{bmatrix} = \begin{bmatrix} \mathbf{U}_1 & \mathbf{U}_2\mathbf{y} & \mathbf{U}_2\mathbf{Q} \end{bmatrix}$$

is a unitary matrix whose first $k+1$ columns are eigenvectors of $\mathbf{A}$. We can continue in this way until we arrive at the spectral decomposition $\mathbf{U}^H\mathbf{A}\mathbf{U} = \boldsymbol{\Lambda}$.

1.3.1.3 Norms

In Lemma 3.5.4 in Chap. 3 of Volume I, we defined three important subordinate matrix norms, and showed how to evaluate the norms directly. The matrix 2-norm is harder to determine than the matrix 1-norm or ∞-norm, as the next lemma will show. This lemma also provides a different way to evaluate the Frobenius norm of a matrix, which was originally defined in Eq. (3.21) in Chap. 3 of Volume I.

Lemma 1.3.3 *If* $\mathbf{A}$ *is an Hermitian matrix with eigenvalues* $\{\lambda_i\}_{i=1}^n$*, then*

$$\|\mathbf{A}\|_2 = \max_{1 \leq i \leq n} |\lambda_i| \text{ and } \|\mathbf{A}\|_F = \sqrt{\sum_{i=1}^n \lambda_i^2} . \tag{1.14}$$

Here $\|\mathbf{A}\|_F$ *denotes the* **Frobenius matrix norm** *of* $\mathbf{A}$.

Proof The spectral theorem shows that $\mathbf{A} = \mathbf{U}\boldsymbol{\Lambda}\mathbf{U}^H$ where $\mathbf{U}$ is unitary and $\boldsymbol{\Lambda}$ is real and diagonal. Then

$$\|\mathbf{A}\|_2^2 = \max_{\mathbf{x}\neq\mathbf{0}} \frac{\|\mathbf{A}\mathbf{x}\|_2^2}{\|\mathbf{x}\|_2^2} = \max_{\mathbf{x}\neq\mathbf{0}} \frac{\mathbf{x}^H\mathbf{A}^H\mathbf{A}\mathbf{x}}{\mathbf{x}\cdot\mathbf{x}} = \max_{\mathbf{x}\neq\mathbf{0}} \frac{\mathbf{x}^H\mathbf{U}\boldsymbol{\Lambda}^2\mathbf{U}^H\mathbf{x}}{(\mathbf{U}^H\mathbf{x})\cdot(\mathbf{U}^H\mathbf{x})} = \max_{\mathbf{y}\neq\mathbf{0}} \frac{\mathbf{y}^H\boldsymbol{\Lambda}^2\mathbf{y}}{\mathbf{y}\cdot\mathbf{y}}$$
$$= \max_{1\leq i\leq n} |\lambda_i|^2 \ .$$

This proves the first claim. Since Lemma 3.5.9 in Chap. 3 of Volume I shows that $\|\mathbf{X}\|_F = \|\mathbf{X}^H\|_F$ for all matrices $\mathbf{X}$,

$$\|\mathbf{A}\|_F^2 = \sum_{j=1}^n \|\mathbf{A}\mathbf{e}_j\|_2^2 = \sum_{j=1}^n \|\mathbf{U}\boldsymbol{\Lambda}\mathbf{U}^H\mathbf{e}_j\|_2^2 = \sum_{j=1}^n \|\boldsymbol{\Lambda}\mathbf{U}^H\mathbf{e}_j\|_2^2 = \left\|\boldsymbol{\Lambda}\mathbf{U}^H\right\|_F^2$$
$$= \|\mathbf{U}\boldsymbol{\Lambda}\|_F^2 = \sum_{j=1}^n \|\mathbf{U}\boldsymbol{\Lambda}\mathbf{e}_j\|_2^2 = \sum_{j=1}^n \|\boldsymbol{\Lambda}\mathbf{e}_j\|_2^2 = \|\boldsymbol{\Lambda}\|_F^2 = \sum_{i=1}^n \lambda_i^2 \ .$$

1.3.1.4 Minimax Theorem

The next lemma develops the notion of a **Rayleigh quotient**, and shows how it relates to the eigenvalues of an Hermitian matrix.

Lemma 1.3.4 *If the eigenvalues of an $n \times n$ Hermitian matrix $\mathbf{A}$ are*

$$\lambda_1 \leq \lambda_2 \leq \ldots \leq \lambda_n \ ,$$

then for any nonzero n-vector $\mathbf{x}$

$$\lambda_1 \leq \frac{\mathbf{x}^H\mathbf{A}\mathbf{x}}{\mathbf{x}\cdot\mathbf{x}} \leq \lambda_n \ .$$

Proof The Spectral Theorem 1.3.1 guarantees that there is an orthogonal matrix $\mathbf{Q}$ such that $\mathbf{A}\mathbf{Q} = \mathbf{Q}\boldsymbol{\Lambda}$, where $\boldsymbol{\Lambda}$ is diagonal. Let $\mathbf{z} = \mathbf{Q}^H\mathbf{x}$ have entries ζ_i. Then

$$\frac{\mathbf{x}^H\mathbf{A}\mathbf{x}}{\mathbf{x}\cdot\mathbf{x}} = \frac{\mathbf{z}^H\boldsymbol{\Lambda}\mathbf{z}}{\mathbf{z}\cdot\mathbf{z}} = \frac{\sum_{i=1}^n \lambda_i\zeta_i^2}{\sum_{i=1}^n \zeta_i^2}$$

is a weighted average of the eigenvalues of $\mathbf{A}$. The claimed result now follows easily.

The previous lemma is a special case of the following much stronger and more famous result, often attributed to Courant and Fischer; see, for example, Courant [41, pp. 31–34]. We will use the following theorem to prove the eigenvalue perturbation Lemma 1.3.5.

Theorem 1.3.2 (Minimax) *Let $\mathscr{S}$ denote a subspace of the linear space of all n-vectors. If the eigenvalues of an $n \times n$ Hermitian matrix $\mathbf{A}$ are ordered*

$$\lambda_1 \leq \lambda_2 \leq \ldots \leq \lambda_n ,$$

then

$$\lambda_i = \min_{\dim \mathscr{S}=i} \max_{\substack{\mathbf{x}\in\mathscr{S} \\ \mathbf{x}\neq\mathbf{0}}} \frac{\mathbf{x}^H\mathbf{A}\mathbf{x}}{\mathbf{x}\cdot\mathbf{x}} = \max_{\dim \mathscr{S}=n-i+1} \min_{\substack{\mathbf{x}\in\mathscr{S} \\ \mathbf{x}\neq\mathbf{0}}} \frac{\mathbf{x}^H\mathbf{A}\mathbf{x}}{\mathbf{x}\cdot\mathbf{x}} . \tag{1.15}$$

Proof The Rayleigh quotient Lemma 1.3.4 shows that the first equation in (1.15) is valid for $i = 1$. Assume inductively that it is valid for $i - 1 \geq 1$. The Spectral Theorem 1.3.1 says that $\mathbf{A} = \mathbf{U}\boldsymbol{\Lambda}\mathbf{U}^H$ where $\mathbf{U}$ is unitary and $\boldsymbol{\Lambda}$ is diagonal with diagonal entries $\lambda_1, \cdots, \lambda_n$. Partition

$$\mathbf{U} = \begin{bmatrix}\mathbf{U}_{i-1} & \mathbf{U}_{n-i+1}\end{bmatrix} \text{ and } \boldsymbol{\Lambda} = \begin{bmatrix}\boldsymbol{\Lambda}_{i-1} & \\ & \boldsymbol{\Lambda}_{n-i+1}\end{bmatrix}$$

where $\mathbf{U}_{i-1}$ is $n \times (i-1)$ and $\boldsymbol{\Lambda}_{i-1}$ is $(i-1) \times (i-1)$. Given a subspace $\mathscr{S}$ of dimension i, let the columns of the $n \times i$ matrix $\mathbf{S}$ form a basis for $\mathscr{S}$, and note that $\mathbf{B} = {\mathbf{U}_{i-1}}^H\mathbf{S}$ is an $(i-1) \times i$ matrix. The contrapositive of Lemma 3.2.6 in Chap. 3 of Volume I shows that $\mathscr{N}(\mathbf{B}) \neq \{\mathbf{0}\}$, so there exists a nonzero vector $\mathbf{z}$ such that $\mathbf{B}\mathbf{z} = \mathbf{0}$. Let $\mathbf{s} = \mathbf{S}\mathbf{z}$. Then

$${\mathbf{U}_{i-1}}^H\mathbf{s} = {\mathbf{U}_{i-1}}^H\mathbf{S}\mathbf{z} = \mathbf{B}\mathbf{z} = \mathbf{0} .$$

Let $\mathbf{t} = {\mathbf{U}_{n-i+1}}^H\mathbf{s}$. Then

$$\frac{\mathbf{s}^H\mathbf{A}\mathbf{s}}{\mathbf{s}\cdot\mathbf{s}} = \frac{\mathbf{s}^H\begin{bmatrix}\mathbf{U}_{i-1} & \mathbf{U}_{n-i+1}\end{bmatrix}\begin{bmatrix}\boldsymbol{\Lambda}_{i-1} & \\ & \boldsymbol{\Lambda}_{n-i+1}\end{bmatrix}\begin{bmatrix}{\mathbf{U}_{i-1}}^H \\ {\mathbf{U}_{n-i+1}}^H\end{bmatrix}\mathbf{s}}{\mathbf{s}^H\begin{bmatrix}\mathbf{U}_{i-1} & \mathbf{U}_{n-i+1}\end{bmatrix}\begin{bmatrix}{\mathbf{U}_{i-1}}^H \\ {\mathbf{U}_{n-i+1}}^H\end{bmatrix}\mathbf{s}}$$

$$= \frac{\begin{bmatrix}\mathbf{0} & \mathbf{t}^H\end{bmatrix}\begin{bmatrix}\boldsymbol{\Lambda}_{i-1} & \\ & \boldsymbol{\Lambda}_{n-i+1}\end{bmatrix}\begin{bmatrix}\mathbf{0} \\ \mathbf{t}\end{bmatrix}}{\begin{bmatrix}\mathbf{0} & \mathbf{t}^H\end{bmatrix}\begin{bmatrix}\mathbf{0} \\ \mathbf{t}\end{bmatrix}} = \frac{\mathbf{t}^H\boldsymbol{\Lambda}_{n-i+1}\mathbf{t}}{\mathbf{t}\cdot\mathbf{t}} \geq \lambda_i .$$

Since $\mathscr{S}$ of dimension i was arbitrary, this shows that

$$\min_{\dim \mathscr{S}=i} \max_{\substack{\mathbf{x}\in\mathscr{S} \\ \mathbf{x}\neq\mathbf{0}}} \frac{\mathbf{x}^H\mathbf{A}\mathbf{x}}{\mathbf{x}\cdot\mathbf{x}} \geq \lambda_i . \tag{1.16}$$

Next, we can repartition

$$\mathbf{U} = \begin{bmatrix} \mathbf{U}_i & \mathbf{U}_{n-i} \end{bmatrix} \text{ and } \boldsymbol{\Lambda} = \begin{bmatrix} \boldsymbol{\Lambda}_i & \\ & \boldsymbol{\Lambda}_{n-i} \end{bmatrix}$$

where $\mathbf{U}_i$ is $n \times i$ and $\boldsymbol{\Lambda}_i$ is $i \times i$. Then

$$\max_{\substack{\mathbf{x}=\mathbf{U}_i\mathbf{t} \\ \mathbf{x}\neq\mathbf{0}}} \frac{\mathbf{x}^H\mathbf{A}\mathbf{x}}{\mathbf{x}\cdot\mathbf{x}} = \max_{\mathbf{t}\neq\mathbf{0}} \frac{\mathbf{t}^H{\mathbf{U}_i}^H \begin{bmatrix} \mathbf{U}_{i-1} & \mathbf{U}_{n-i+1} \end{bmatrix} \begin{bmatrix} \boldsymbol{\Lambda}_{i-1} & \\ & \boldsymbol{\Lambda}_{n-i+1} \end{bmatrix} \begin{bmatrix} {\mathbf{U}_{i-1}}^H \\ {\mathbf{U}_{n-i+1}}^H \end{bmatrix} \mathbf{U}_i\mathbf{t}}{\mathbf{t}^H{\mathbf{U}_i}^H\mathbf{U}_i\mathbf{t}}$$

$$= \max_{\mathbf{t}\neq\mathbf{0}} \frac{\mathbf{t}^H\boldsymbol{\Lambda}_i\mathbf{t}}{\mathbf{t}\cdot\mathbf{t}} = \lambda_i\ .$$

This proves that

$$\min_{\dim\mathscr{S}=i} \max_{\substack{\mathbf{x}\in\mathscr{S} \\ \mathbf{x}\neq\mathbf{0}}} \frac{\mathbf{x}^H\mathbf{A}\mathbf{x}}{\mathbf{x}\cdot\mathbf{x}} \leq \lambda_i\ . \tag{1.17}$$

Equations (1.16) and (1.17) imply the first equation in (1.15). The proof of the second equation in the claim is similar.

Exercise 1.3.1 If $\mathbf{u}$ is an m-vector, what are the eigenvalues and eigenvectors of $\mathbf{u}\mathbf{u}^H$?

Exercise 1.3.2 What are the eigenvalues and eigenvectors of an Hermitian elementary reflector, given by Eq. (6.8) in Chap. 6 of Volume I?

Exercise 1.3.3 Suppose that we partition the Hermitian matrix $\mathbf{A}$ as follows:

$$\mathbf{A} = \begin{bmatrix} \alpha & \mathbf{a}^H \\ \mathbf{a} & \mathbf{B} \end{bmatrix}.$$

Show that there is an eigenvalue of $\mathbf{A}$ in the closed interval $[\alpha - \|\mathbf{a}\|_2, \alpha + \|\mathbf{a}\|_2]$. Note that this interval is different from a Gerschgorin circle in Lemma 1.2.2.

Exercise 1.3.4 Find the orthogonal matrix $\mathbf{U}$ and diagonal matrix $\boldsymbol{\Lambda}$ guaranteed by the Spectral Theorem 1.3.1 for the matrix

$$\mathbf{A} = \begin{bmatrix} 0 & 1 \\ 1 & 0 \end{bmatrix}.$$

Exercise 1.3.5 If $\mathbf{A}$ is Hermitian and positive (see Definition 3.13.1 in Chap. 3 of Volume I), show that both $\mathbf{A}^2$ and $\mathbf{A}^{-1}$ are Hermitian and positive.

Exercise 1.3.6 Find the minimum value of

$$\phi(\mathbf{x}) = \frac{\mathbf{x}_1^2 - \mathbf{x}_1\mathbf{x}_2 + \mathbf{x}_2^2}{\mathbf{x}_1^2 + \mathbf{x}_2^2}$$

over all nonzero values of $\mathbf{x}$.

Exercise 1.3.7 Suppose that $\mathbf{A}$ and $\mathbf{B}$ are $m \times m$ Hermitian matrices, and that $\mathbf{B}$ is positive. Show that the smallest eigenvalue of $\mathbf{A} + \mathbf{B}$ is larger than the smallest eigenvalue of $\mathbf{A}$.

Exercise 1.3.8 (Polar Decomposition) Show that for every $m \times n$ matrix $\mathbf{A}$ with rank $(\mathbf{A}) = n$, there is a unitary matrix $\mathbf{U}$ and an Hermitian positive matrix $\mathbf{P}$ so that $\mathbf{A} = \mathbf{U}\mathbf{P}$. (Hint: apply the spectral theorem to $\mathbf{A}^H\mathbf{A}$.)

1.3.2 Perturbation Analysis

Before developing algorithms to compute eigenvectors and eigenvalues, we would like to analyze the sensitivity of the eigenvalue problem to perturbations. We will begin by developing approximations for the perturbations to an eigenvalue and its corresponding eigenvector, due to some perturbation in the matrix. These initial approximations will be instructive, but will not provide error bounds.

Afterward, we will estimate the perturbations in an eigenvalue due to a perturbed matrix in Lemma 1.3.5, and then examine the perturbations of some corresponding invariant subspace in Lemma 1.3.6. In both of these lemmas, the perturbations to the matrix are presumably of the sort that might arise in some algorithm to perform an eigenvalue-revealing matrix factorization. However, the algorithms we will use to find eigenvalues and eigenvectors will be iterative, so we will also need error estimates to determine when to stop the iteration. For such purposes, we will present some error estimates in Lemmas 1.3.7 and 1.3.7.

1.3.2.1 Perturbation Approximations

Suppose that $\mathbf{A}$ is an Hermitian $n \times n$ matrix. Let the scalar λ and the n-vector $\mathbf{x}$ satisfy $\mathbf{A}\mathbf{x} = \mathbf{x}\lambda$ and $\|\mathbf{x}\| = 1$. Also let $\mathbf{E}$ be an $n \times n$ Hermitian matrix with small entries. We would like to find an n-vector $\widetilde{\mathbf{x}}$ and a scalar $\widetilde{\lambda}$ so that $(\mathbf{A} + \mathbf{E})\widetilde{\mathbf{x}} = \widetilde{\mathbf{x}}\widetilde{\lambda}$, in order to estimate $\|\mathbf{x} - \widetilde{\mathbf{x}}\|_2$ and $|\lambda - \widetilde{\lambda}|$.

As in the proof by deflation of the Spectral Theorem 1.3.1, we can find an Hermitian unitary matrix $\mathbf{H}$ whose first column is $\mathbf{x}$, and partition

$$\mathbf{H} = \begin{bmatrix} \mathbf{x} & \mathbf{Q} \end{bmatrix} .$$

Assuming that $\widetilde{\mathbf{x}}$ is not orthogonal to $\mathbf{x}$, we can scale $\widetilde{\mathbf{x}}$ so that

$$\mathbf{H}\widetilde{\mathbf{x}} = \begin{bmatrix} 1 \\ \mathbf{p} \end{bmatrix} .$$

Then

$$\mathbf{H}\mathbf{A}\mathbf{H} = \begin{bmatrix} \mathbf{x}^H \\ \mathbf{Q}^H \end{bmatrix} \mathbf{A} \begin{bmatrix} \mathbf{x} \ \mathbf{Q} \end{bmatrix} = \begin{bmatrix} \lambda & \\ & \mathbf{Q}^H\mathbf{A}\mathbf{Q} \end{bmatrix} ,$$

and

$$\mathbf{H}(\mathbf{A}+\mathbf{E})\mathbf{H} = \begin{bmatrix} \mathbf{x}^H \\ \mathbf{Q}^H \end{bmatrix} (\mathbf{A}+\mathbf{E}) \begin{bmatrix} \mathbf{x} \ \mathbf{Q} \end{bmatrix} = \begin{bmatrix} \lambda + \mathbf{x}^H\mathbf{E}\mathbf{x} & \mathbf{x}^H\mathbf{E}\mathbf{Q} \\ \mathbf{Q}^H\mathbf{E}\mathbf{x} & \mathbf{Q}^H(\mathbf{A}+\mathbf{E})\mathbf{Q} \end{bmatrix} .$$

Since $(\mathbf{A}+\mathbf{E})\widetilde{\mathbf{x}} = \widetilde{\mathbf{x}}\widetilde{\lambda}$, we have

$$\begin{aligned} \begin{bmatrix} \lambda + \mathbf{x}^H\mathbf{E}\mathbf{x} & \mathbf{x}^H\mathbf{E}\mathbf{Q} \\ \mathbf{Q}^H\mathbf{E}\mathbf{x} & \mathbf{Q}^H(\mathbf{A}+\mathbf{E})\mathbf{Q} \end{bmatrix} \begin{bmatrix} 1 \\ \mathbf{p} \end{bmatrix} &= \mathbf{H}(\mathbf{A}+\mathbf{E})\mathbf{H} \begin{bmatrix} 1 \\ \mathbf{p} \end{bmatrix} = \mathbf{H}(\mathbf{A}+\mathbf{E})\widetilde{\mathbf{x}} \\ &= \mathbf{H}\widetilde{\mathbf{x}}\widetilde{\lambda} = \begin{bmatrix} 1 \\ \mathbf{p} \end{bmatrix} \widetilde{\lambda} . \end{aligned}$$

If $\mathbf{E}$ is small, we expect that $\mathbf{p}$ is small. These equations say that

$$\widetilde{\lambda} = \lambda + \mathbf{x}^H\mathbf{E}\mathbf{x} + \mathbf{x}^H\mathbf{E}\mathbf{Q}\mathbf{p} \approx \lambda + \mathbf{x}^H\mathbf{E}\mathbf{x} ,$$

and

$$\mathbf{p}\lambda = \mathbf{Q}^H\mathbf{E}\mathbf{x} + \mathbf{Q}^H\mathbf{A}\mathbf{Q}\mathbf{p} + \mathbf{Q}^H\mathbf{E}\mathbf{Q}\mathbf{p} \approx \mathbf{Q}^H\mathbf{E}\mathbf{x} + \mathbf{Q}^H\mathbf{A}\mathbf{Q}\mathbf{p} ,$$

where the approximations ignore presumed second-order terms. If λ is not an eigenvalue of the deflated matrix $\mathbf{Q}^H\mathbf{A}\mathbf{Q}$ (i.e., the multiplicity of λ as an eigenvalue of $\mathbf{A}$ is one), then

$$\mathbf{p} \approx (\mathbf{I}\lambda - \mathbf{Q}^H\mathbf{A}\mathbf{Q})^{-1}\mathbf{Q}^H\mathbf{E}\mathbf{x} .$$

Under this condition, we see that

$$\|\widetilde{\mathbf{x}} - \mathbf{x}\|_2 = \left\| \mathbf{H} \begin{bmatrix} 1 \\ \mathbf{0} \end{bmatrix} - \mathbf{H} \begin{bmatrix} 1 \\ \mathbf{p} \end{bmatrix} \right\|_2 = \|\mathbf{Q}\mathbf{p}\|_2 = \|\mathbf{p}\|_2 \leq \|(\mathbf{I}\lambda - \mathbf{Q}^H\mathbf{A}\mathbf{Q})^{-1}\|_2 \, \|\mathbf{Q}^H\mathbf{E}\mathbf{x}\|_2 .$$

We also have

$$|\widetilde{\lambda} - \lambda| \approx |\mathbf{x}^H\mathbf{E}\mathbf{x}| .$$

These rough approximations show that if another eigenvalue of $\mathbf{A}$ is close to λ, then $\|(\mathbf{I}\lambda - \mathbf{Q}^H\mathbf{A}\mathbf{Q})^{-1}\|_2$ will be large and the perturbed eigenvector will not necessarily be close to $\mathbf{x}$. This should be expected: for multiple eigenvalues, the eigenvectors can be chosen as arbitrary orthogonal vectors within the nullspace of $\mathbf{A} - \mathbf{I}\lambda$.

1.3.2.2 Eigenvalue Perturbations

In the remainder of this section, we would like to derive estimates that do not depend on approximation. We will begin by estimating eigenvalues.

Lemma 1.3.5 *Let* $\mathbf{A}$ *and* $\widetilde{\mathbf{A}}$ *be* $n \times n$ *Hermitian matrices with eigenvalues* λ_i *and* $\widetilde{\lambda}_i$*, respectively. Denote the eigenvalues of* $\widetilde{\mathbf{A}} - \mathbf{A}$ *by* ε_i*. Assume that all eigenvalues have been ordered by index from smallest to largest. Then for all* $1 \le i \le n$

$$\varepsilon_1 \le \widetilde{\lambda}_i - \lambda_i \le \varepsilon_n \text{ and} \tag{1.18a}$$

$$\left|\widetilde{\lambda}_i - \lambda_i\right| \le \|\widetilde{\mathbf{A}} - \mathbf{A}\|_2 \,. \tag{1.18b}$$

Proof The Spectral Theorem 1.3.1 shows that $\mathbf{A} = \mathbf{U}\boldsymbol{\Lambda}\mathbf{U}^H$ where $\mathbf{U}$ is unitary and $\boldsymbol{\Lambda}$ is diagonal. Given an index $i \in [1, n]$, partition

$$\mathbf{U} = \begin{bmatrix} \mathbf{U}_i & \mathbf{U}_{n-i} \end{bmatrix}$$

where $\mathbf{U}_i$ is $n \times i$. Then the Minimax Theorem 1.3.2 proves that

$$\begin{aligned}
\widetilde{\lambda}_i &\le \max_{\mathbf{y} \neq \mathbf{0}} \frac{(\mathbf{U}_i\mathbf{y})^H \widetilde{\mathbf{A}} (\mathbf{U}_i\mathbf{y})}{(\mathbf{U}_i\mathbf{y})^H (\mathbf{U}_i\mathbf{y})} \\
&\le \max_{\mathbf{y} \neq \mathbf{0}} \frac{(\mathbf{U}_i\mathbf{y})^H \mathbf{A} (\mathbf{U}_i\mathbf{y})}{(\mathbf{U}_i\mathbf{y})^H (\mathbf{U}_i\mathbf{y})} + \max_{\mathbf{y} \neq \mathbf{0}} \frac{(\mathbf{U}_i\mathbf{y})^H \left(\widetilde{\mathbf{A}} - \mathbf{A}\right) (\mathbf{U}_i\mathbf{y})}{(\mathbf{U}_i\mathbf{y})^H (\mathbf{U}_i\mathbf{y})} \\
&= \lambda_i + \max_{\mathbf{x} \neq \mathbf{0}} \frac{\mathbf{x}^H \left(\widetilde{\mathbf{A}} - \mathbf{A}\right) \mathbf{x}}{\mathbf{x} \cdot \mathbf{x}} \le \lambda_i + \varepsilon_n \,.
\end{aligned}$$

If $\mathbf{E} \equiv \widetilde{\mathbf{A}} - \mathbf{A}$, then $\mathbf{A} = \widetilde{\mathbf{A}} - \mathbf{E}$ and the largest eigenvalue of $-\mathbf{E}$ is $-\varepsilon_1$. By switching the roles of $\mathbf{A}$ and $\widetilde{\mathbf{A}}$, the preceding calculation also implies that

$$\lambda_i \le \widetilde{\lambda}_i - \varepsilon_1 \,.$$

This proves claim (1.18a).

Next, Lemma 1.3.3 implies that

$$-\varepsilon_1 \ge -\|\widetilde{\mathbf{A}} - \mathbf{A}\|_2 \text{ and } \varepsilon_n \le \|\widetilde{\mathbf{A}} - \mathbf{A}\|_2 \,.$$

Combining these inequalities with (1.18a) proves claim (1.18b).

The previous lemma provide fairly simple bounds on perturbations in eigenvalues due to perturbations in Hermitian matrices. However, the next example, due to [illegible] will show that perturbations in eigenvectors are more difficult to estimate.

Example 1.3.1 Let

$$\widetilde{\mathbf{U}} = \begin{bmatrix} 1 & 1 & -\varepsilon \\ 1 & -1 & \varepsilon \\ 0 & 2\varepsilon & 1 \end{bmatrix} \begin{bmatrix} \sqrt{2} & & \\ & \sqrt{2+4\varepsilon^2} & \\ & & \sqrt{1+2\varepsilon^2} \end{bmatrix}^{-1}$$

and

$$\widetilde{\widetilde{\mathbf{U}}} = \begin{bmatrix} 1/2 & 1 & -\varepsilon \\ 1 & -1/2 & 2\varepsilon \\ -5\varepsilon/2 & 0\varepsilon & 1 \end{bmatrix} \begin{bmatrix} \sqrt{5/4+25\varepsilon^2/4} & & \\ & \sqrt{5/4} & \\ & & \sqrt{1+5\varepsilon^2} \end{bmatrix}^{-1}$$

Note that both $\widetilde{\mathbf{U}}$ and $\widetilde{\widetilde{\mathbf{U}}}$ are unitary. Next, define

$$\widetilde{\boldsymbol{\Lambda}} = \begin{bmatrix} 1+\varepsilon & & \\ & 1-\varepsilon & \\ & & 2 \end{bmatrix}, \; \widetilde{\mathbf{A}} = \widetilde{\mathbf{U}}\widetilde{\boldsymbol{\Lambda}}\widetilde{\mathbf{U}}^H \text{ and } \widetilde{\widetilde{\mathbf{A}}} = \widetilde{\widetilde{\mathbf{U}}}\widetilde{\boldsymbol{\Lambda}}\widetilde{\widetilde{\mathbf{U}}}^H .$$

These three matrices can be considered to be small perturbations of

$$\mathbf{A} = \begin{bmatrix} 1 & & \\ & 1 & \\ & & 2 \end{bmatrix} = \boldsymbol{\Lambda} .$$

In fact,

$$\widetilde{\mathbf{A}} - \mathbf{A} = \begin{bmatrix} \varepsilon^2+\varepsilon^2 & \varepsilon-\varepsilon^2+\varepsilon^3 & -\varepsilon-\varepsilon^2 \\ \varepsilon-\varepsilon^2+\varepsilon^3 & \varepsilon^2+\varepsilon^3 & \varepsilon+\varepsilon^2 \\ -\varepsilon-\varepsilon^2 & \varepsilon+\varepsilon^2 & -2\varepsilon^2-2\varepsilon^3 \end{bmatrix} \frac{1}{1+2\varepsilon^2}$$

and

$$\widetilde{\widetilde{\mathbf{A}}} - \mathbf{A} = \begin{bmatrix} -3\varepsilon+5\varepsilon^2-20\varepsilon^3 & 4\varepsilon+10\varepsilon^2+10\varepsilon^3 & 5\varepsilon-5\varepsilon^2 \\ 4\varepsilon+10\varepsilon^2+10\varepsilon^3 & 3\varepsilon+20\varepsilon^2-5\varepsilon^3 & 10\varepsilon-10\varepsilon^2 \\ 5\varepsilon-5\varepsilon^2 & 10\varepsilon-10\varepsilon^2 & -25\varepsilon^2+25\varepsilon^3 \end{bmatrix} \frac{1}{5+25\varepsilon^2} .$$

The eigenvectors of $\widetilde{\mathbf{A}}$ and $\widetilde{\widetilde{\mathbf{A}}}$ are the columns of $\widetilde{\mathbf{U}}$ and $\widetilde{\widetilde{\mathbf{U}}}$, respectively, and correspond to identical eigenvalues. Inspection shows that the eigenvectors corresponding to either of the eigenvalues $1 \pm \varepsilon$ are not close. This is possible because the matrix **A** has a double eigenvalue, with corresponding invariant space given by the span of the first two axis vectors.

1.3.2.3 Eigenvector Perturbations

The next lemma will provide an estimate in the perturbation of an invariant space of an eigenvector for an isolated eigenvalue.

Lemma 1.3.6 *Suppose that* **A** *and* $\widetilde{\mathbf{A}}$ *are* $n \times n$ *Hermitian matrices. Assume that*

$$\mathbf{U} = \begin{bmatrix} \mathbf{U}_1 & \mathbf{U}_2 \end{bmatrix}$$

is a unitary matrix, and that the range of the $n \times k$ *matrix* $\mathbf{U}_1$ *is invariant under* **A**. *Let*

$$\mathbf{U}^H \mathbf{A} \mathbf{U} = \begin{bmatrix} \mathbf{B}_{11} & \\ & \mathbf{B}_{22} \end{bmatrix} \text{ and } \mathbf{U}^H \left(\widetilde{\mathbf{A}} - \mathbf{A} \right) \mathbf{U} = \begin{bmatrix} \mathbf{E}_{11} & {\mathbf{E}_{21}}^H \\ \mathbf{E}_{21} & \mathbf{E}_{22} \end{bmatrix} .$$

Assume that $\mathbf{B}_{11}$ *and* $\mathbf{B}_{22}$ *have no common eigenvalues, that*

$$\delta \equiv \min_{\substack{\lambda_{11} \text{ an eigenvalue of } \mathbf{B}_{11} \\ \lambda_{22} \text{ an eigenvalue of } \mathbf{B}_{22}}} |\lambda_{11} - \lambda_{22}| - \|\mathbf{E}_{11}\|_F - \|\mathbf{E}_{22}\|_F > 0 , \tag{1.19}$$

and that

$$2\|\mathbf{E}_{21}\|_F < \delta . \tag{1.20}$$

Then there is an $(n-k) \times k$ *matrix* **P** *satisfying*

$$\|\mathbf{P}\|_F \le 2 \frac{\|\mathbf{E}_{21}\|_F}{\delta} \tag{1.21}$$

so that the columns of $\mathbf{W}_1 = (\mathbf{U}_1 + \mathbf{U}_2 \mathbf{P}) \left(\mathbf{I} + \mathbf{P}^H \mathbf{P} \right)^{-1/2}$ *span a subspace invariant under* $\widetilde{\mathbf{A}}$. *If, in addition, we have*

$$\|\mathbf{E}_{11}\|_F + \|\mathbf{E}_{22}\|_F \le \frac{1}{2} \min_{\substack{\lambda_{11} \text{ an eigenvalue of } \mathbf{B}_{11} \\ \lambda_{22} \text{ an eigenvalue of } \mathbf{B}_{22}}} |\lambda_{11} - \lambda_{22}| ,$$

then

$$\mathit{dist}(\mathscr{R}(\mathbf{U}_1), \mathscr{R}(\mathbf{W}_1)) \le \frac{4\|\mathbf{E}_{21}\|_F}{\min_{\substack{\lambda_{11} \text{ an eigenvalue of } \mathbf{B}_{11} \\ \lambda_{22} \text{ an eigenvalue of } \mathbf{B}_{22}}} |\lambda_{11} - \lambda_{22}|} ,$$

Proof First, we will show that for Hermitian matrices $\mathbf{B}_{11}$ and $\mathbf{B}_{22}$,

$$\inf_{\|\mathbf{P}\|_F=1} \|\mathbf{P}\mathbf{B}_{11} - \mathbf{B}_{22}\mathbf{P}\|_F = \min_{\substack{\lambda_i \text{ an eigenvalue of } \mathbf{B}_{11} \\ \lambda_j \text{ an eigenvalue of } \mathbf{B}_{22}}} |\lambda_i - \lambda_j| \; . \tag{1.22}$$

Recall that Lemma 1.2.23 showed that

$$\inf_{\|\mathbf{P}\|_F=1} \|\mathbf{P}\mathbf{B}_{11} - \mathbf{B}_{22}\mathbf{P}\|_F \geq \min_{\substack{\lambda_i \text{ an eigenvalue of } \mathbf{B}_{11} \\ \lambda_j \text{ an eigenvalue of } \mathbf{B}_{22}}} |\lambda_i - \lambda_j| \; .$$

The Spectral Theorem 1.3.1 guarantees that there are unitary matrices $\mathbf{Q}_{11}$ and $\mathbf{Q}_{22}$ and diagonal matrices $\boldsymbol{\Lambda}_{11}$ and $\boldsymbol{\Lambda}_{22}$ so that $\mathbf{B}_{11} = \mathbf{Q}_{11}\boldsymbol{\Lambda}_{11}\mathbf{Q}_{11}{}^H$ and $\mathbf{B}_{22} = \mathbf{Q}_{22}\boldsymbol{\Lambda}_{22}\mathbf{Q}_{22}{}^H$. Suppose that the minimal separation of the eigenvalues of $\mathbf{B}_{11}$ and $\mathbf{B}_{22}$ is between the ith diagonal entry of $\boldsymbol{\Lambda}_{11}$ and the jth diagonal entry of $\boldsymbol{\Lambda}_{22}$. Let

$$\mathbf{P} = \mathbf{Q}_{22}\mathbf{e}_j\mathbf{e}_i{}^H\mathbf{Q}_{11}{}^H \; .$$

Then

$$\|\mathbf{P}\mathbf{B}_{11} - \mathbf{B}_{22}\mathbf{P}\|_F = \|\mathbf{Q}_{22}{}^H(\mathbf{P}\mathbf{B}_{11} - \mathbf{B}_{22}\mathbf{P})\mathbf{Q}_{11}\|_F = \|\mathbf{e}_j\mathbf{e}_i{}^H\boldsymbol{\Lambda}_{11} - \boldsymbol{\Lambda}_{22}\mathbf{e}_j\mathbf{e}_i{}^H\|_F$$
$$= |\mathbf{e}_i{}^H\boldsymbol{\Lambda}_{11}\mathbf{e}_i - \mathbf{e}_j{}^H\boldsymbol{\Lambda}_{22}\mathbf{e}_j|\|\mathbf{e}_j\mathbf{e}_i{}^H\|_F = |\mathbf{e}_i{}^H\boldsymbol{\Lambda}_{11}\mathbf{e}_i - \mathbf{e}_j{}^H\boldsymbol{\Lambda}_{22}\mathbf{e}_j| \; .$$

This proves the claim (1.22).

Note that Eq. (1.19) is equivalent to the first supposition of the Stewart Perturbation Theorem 1.2.3, and inequality (1.20) is equivalent to the second assumption (1.13). Since both hypotheses of the Stewart Perturbation Theorem are satisfied, we conclude that the matrix $\mathbf{P}$ exists as claimed, and that $\mathbf{P}$ satisfies inequality (1.21), namely

$$\|\mathbf{P}\|_F \leq 2\frac{\|\mathbf{E}_{21}\|_F}{\delta} \; .$$

Next, suppose that

$$\|\mathbf{E}_{11}\|_F + \|\mathbf{E}_{22}\|_F \leq \frac{1}{2} \min_{\substack{\lambda_i \text{ an eigenvalue of } \mathbf{B}_{11} \\ \lambda_j \text{ an eigenvalue of } \mathbf{B}_{22}}} |\lambda_i - \lambda_j| \; ,$$

Then

$$\min_{\substack{\lambda_i \text{ an eigenvalue of } \mathbf{B}_{11} \\ \lambda_j \text{ an eigenvalue of } \mathbf{B}_{22}}} |\lambda_i - \lambda_j|$$
$$\leq 2 \min_{\substack{\lambda_i \text{ an eigenvalue of } \mathbf{B}_{11} \\ \lambda_j \text{ an eigenvalue of } \mathbf{B}_{22}}} |\lambda_i - \lambda_j| - 2\|\mathbf{E}_{11}\|_F - 2\|\mathbf{E}_{22}\|_F = 2\delta \; ,$$

so

$$2\frac{\|\mathbf{E}_{21}\|_F}{\delta} \le 4\frac{\|\mathbf{E}_{21}\|_F}{\min_{\substack{\lambda_i \text{ an eigenvalue of } \mathbf{B}_{11} \\ \lambda_j \text{ an eigenvalue of } \mathbf{B}_{22}}} \left|\lambda_i - \lambda_j\right|} . \tag{1.23}$$

All that remains is to show that

$$\text{dist}(\mathscr{R}(\mathbf{U}_1), \mathscr{R}(\mathbf{W}_1)) \le \|\mathbf{P}\|_F .$$

Let $\mathbf{P} = \mathbf{Q}\boldsymbol{\Sigma}\mathbf{V}^H$ be the singular value decomposition of $\mathbf{P}$. Recall from the Definition 1.4 and Lemma 1.2.17 that

$$\begin{aligned}
&\text{dist}(\mathscr{R}(\mathbf{U}_1), \mathscr{R}(\mathbf{W}_1)) = \left\|\mathbf{U}_2{}^H\mathbf{W}_1\right\|_2 = \left\|\mathbf{P}(\mathbf{I} + \mathbf{P}^H\mathbf{P})^{-1/2}\right\|_2 \\
&= \left\|\mathbf{Q}\boldsymbol{\Sigma}\mathbf{V}^H\left(\mathbf{I} + \mathbf{V}\boldsymbol{\Sigma}^2\mathbf{V}^H\right)^{-1/2}\right\|_2 = \left\|\mathbf{Q}\boldsymbol{\Sigma}\mathbf{V}^H\mathbf{V}\left(\mathbf{I} + \boldsymbol{\Sigma}^2\right)^{-1/2}\mathbf{V}^H\right\|_2 \\
&\le \left\|\mathbf{Q}\boldsymbol{\Sigma}\mathbf{V}^H\right\|_2 = \|\mathbf{P}\|_2 \le \|\mathbf{P}\|_F .
\end{aligned}$$

Combining this inequality with (1.21) and (1.23) gives us the final claim in the lemma.

1.3.2.4 *A Posteriori* Errors

Our next lemma, which can be found in Parlett [137, p. 69], shows that the "residual" in an eigenvector equation can be used to determine an *a posteriori* estimate of the error in an approximate eigenvalue.

Lemma 1.3.7 *Suppose that* $\mathbf{A}$ *is an Hermitian matrix,* $\widetilde{\mathbf{x}}$ *is a nonzero vector and* $\widetilde{\lambda}$ *is a scalar. Then*

$$\min_{\lambda \text{ eigenvalue of } \mathbf{A}} \left|\widetilde{\lambda} - \lambda\right| \le \frac{\|\mathbf{A}\widetilde{\mathbf{x}} - \widetilde{\mathbf{x}}\widetilde{\lambda}\|_2}{\|\widetilde{\mathbf{x}}\|_2} .$$

Proof If $\widetilde{\lambda}$ is an eigenvalue of $\mathbf{A}$, the claim is obvious because both sides of the claimed inequality are zero. Otherwise, $\mathbf{A} - \mathbf{I}\widetilde{\lambda}$ is nonsingular, and $\widetilde{\mathbf{x}} = \left(\mathbf{A} - \mathbf{I}\widetilde{\lambda}\right)^{-1}\left(\mathbf{A} - \mathbf{I}\widetilde{\lambda}\right)\widetilde{\mathbf{x}}$. It follows that

$$\|\widetilde{\mathbf{x}}\|_2 \le \left\|\left(\mathbf{A} - \mathbf{I}\widetilde{\lambda}\right)^{-1}\right\|_2 \left\|\mathbf{A}\widetilde{\mathbf{x}} - \widetilde{\mathbf{x}}\widetilde{\lambda}\right\|_2$$

then Lemma 1.3.3 gives us

$$\le \max_{\lambda \text{ eigenvalue of } \mathbf{A}} \frac{1}{|\lambda - \widetilde{\lambda}|} \left\| \mathbf{A}\widetilde{\mathbf{x}} - \widetilde{\mathbf{x}}\widetilde{\lambda} \right\|_2$$

This inequality is equivalent to the claimed result.

Here is a similar result, which may also be found in Parlett [137, p. 222].

Lemma 1.3.8 *Suppose that* $\mathbf{A}$ *is an* $n \times n$ *Hermitian matrix. Let* $\widetilde{\mathbf{x}}$ *be a nonzero* n*-vector and* $\widetilde{\lambda}$ *be a scalar. Let* λ *be the eigenvalue of* $\mathbf{A}$ *closest to* $\widetilde{\lambda}$*, with unit eigenvector* $\mathbf{x}$*. Let*

$$\mathbf{U} = \begin{bmatrix} \mathbf{x} \; \mathbf{Q} \end{bmatrix}$$

be a unitary matrix and $\boldsymbol{\Lambda}$ *be a diagonal matrix so that*

$$\mathbf{A} \begin{bmatrix} \mathbf{x} \; \mathbf{Q} \end{bmatrix} = \begin{bmatrix} \mathbf{x} \; \mathbf{Q} \end{bmatrix} \begin{bmatrix} \lambda & \\ & \boldsymbol{\Lambda} \end{bmatrix} .$$

Then the orthogonal projection of $\widetilde{\mathbf{x}}$ *onto the space orthogonal to* $\mathbf{x}$ *has norm satisfying*

$$\left\| \mathbf{Q}\mathbf{Q}^H \widetilde{\mathbf{x}} \right\|_2 \le \frac{\left\| \mathbf{A}\widetilde{\mathbf{x}} - \widetilde{\mathbf{x}}\widetilde{\lambda} \right\|_2}{\min_{\lambda' \text{ an eigenvalue of } \boldsymbol{\Lambda}} |\lambda' - \widetilde{\lambda}|} . \tag{1.24a}$$

In particular, if we choose $\widetilde{\lambda}$ *to be the Rayleigh quotient*

$$\widetilde{\lambda} = \frac{\widetilde{\mathbf{x}}^H \mathbf{A} \widetilde{\mathbf{x}}}{\widetilde{\mathbf{x}} \cdot \widetilde{\mathbf{x}}} ,$$

then

$$\left| \widetilde{\lambda} - \lambda \right| \le \frac{1}{\min_{\lambda' \text{ an eigenvalue of } \boldsymbol{\Lambda}} \left| \lambda' - \widetilde{\lambda} \right|} \left(\frac{\left\| \mathbf{A}\widetilde{\mathbf{x}} - \widetilde{\mathbf{x}}\widetilde{\lambda} \right\|_2}{\|\widetilde{\mathbf{x}}\|_2} \right)^2 . \tag{1.24b}$$

Proof We can scale $\widetilde{\mathbf{x}}$ so that

$$\widetilde{\mathbf{x}} = \mathbf{x} + \mathbf{Q}\mathbf{y}$$

and note that $\mathbf{y} = \mathbf{Q}^H \widetilde{\mathbf{x}}$. Then

$$\mathbf{A}\widetilde{\mathbf{x}} - \widetilde{\mathbf{x}}\widetilde{\lambda} = (\mathbf{A} - \mathbf{I}\widetilde{\lambda})(\mathbf{x} + \mathbf{Q}\mathbf{y}) = \mathbf{x}(\lambda - \widetilde{\lambda}) + \mathbf{Q}(\boldsymbol{\Lambda} - \mathbf{I}\widetilde{\lambda})\mathbf{y} .$$

Next, the Pythagorean Theorem 3.2.2 in Chap. 3 of Volume I implies that

$$\left\| \mathbf{A}\widetilde{\mathbf{x}} - \widetilde{\mathbf{x}}\widetilde{\lambda} \right\|_2^2 = |\lambda - \widetilde{\lambda}|^2 + \mathbf{y}^H(\boldsymbol{\Lambda} - \mathbf{I}\widetilde{\lambda})^2\mathbf{y} \tag{1.25}$$

$$\geq \min_{\lambda' \text{ an eigenvalue of } \boldsymbol{\Lambda}} |\lambda' - \widetilde{\lambda}|^2 \, \|\mathbf{y}\|_2^2$$

This inequality implies the claim (1.24a).

Next, suppose that $\widetilde{\lambda}$ is the Rayleigh quotient for $\widetilde{\mathbf{x}}$. Then

$$(\lambda - \widetilde{\lambda}) + \mathbf{y}^H(\boldsymbol{\Lambda} - \mathbf{I}\widetilde{\lambda})\mathbf{y} = \begin{bmatrix} 1 & \mathbf{y}^H \end{bmatrix} \begin{bmatrix} \lambda & \\ & \boldsymbol{\Lambda} \end{bmatrix} \begin{bmatrix} 1 \\ \mathbf{y} \end{bmatrix} - \begin{bmatrix} 1 & \mathbf{y}^H \end{bmatrix} \begin{bmatrix} 1 \\ \mathbf{y} \end{bmatrix} \widetilde{\lambda}$$

$$= \begin{bmatrix} 1 & \mathbf{y}^H \end{bmatrix} \mathbf{U}^H\mathbf{A}\mathbf{U} \begin{bmatrix} \lambda & \\ & \boldsymbol{\Lambda} \end{bmatrix} \begin{bmatrix} 1 \\ \mathbf{y} \end{bmatrix} - \begin{bmatrix} 1 & \mathbf{y}^H \end{bmatrix} \mathbf{U}^H\mathbf{U} \begin{bmatrix} 1 \\ \mathbf{y} \end{bmatrix} \widetilde{\lambda}$$

$$= \widetilde{\mathbf{x}}^H\mathbf{A}\widetilde{\mathbf{x}} - \widetilde{\mathbf{x}}^H\widetilde{\mathbf{x}}\widetilde{\lambda} = 0 \,. \tag{1.26}$$

This implies that

$$(\widetilde{\lambda} - \lambda)\|\widetilde{\mathbf{x}}\|_2^2 = (\widetilde{\lambda} - \lambda)(1 + \mathbf{y} \cdot \mathbf{y}) = \mathbf{y}^H(\boldsymbol{\Lambda} - \mathbf{I}\widetilde{\lambda})\mathbf{y} + \mathbf{y}^H\mathbf{I}(\widetilde{\lambda} - \lambda)\mathbf{y}$$

$$= \mathbf{y}^H(\boldsymbol{\Lambda} - \mathbf{I}\lambda)\mathbf{y} \,. \tag{1.27}$$

So, Eq. (1.25) gives us

$$\frac{\|\mathbf{A}\widetilde{\mathbf{x}} - \widetilde{\mathbf{x}}\widetilde{\lambda}\|_2^2}{\|\widetilde{\mathbf{x}}\|_2^2} = \frac{|\widetilde{\lambda} - \lambda|^2 + \mathbf{y}^H(\boldsymbol{\Lambda} - \mathbf{I}\widetilde{\lambda})^2\mathbf{y}}{\|\widetilde{\mathbf{x}}\|_2^2}$$

then we can multiply the numerator and denominator by $\widetilde{\lambda} - \lambda$ and apply Eq. (1.26) to obtain

$$= \frac{|\widetilde{\lambda} - \lambda|^2\mathbf{y}^H(\boldsymbol{\Lambda} - \mathbf{I}\widetilde{\lambda})\mathbf{y} + \mathbf{y}^H(\boldsymbol{\Lambda} - \mathbf{I}\widetilde{\lambda})^2\mathbf{y}(\widetilde{\lambda} - \lambda)}{(\widetilde{\lambda} - \lambda)\|\widetilde{\mathbf{x}}\|_2^2}$$

$$= \frac{|\widetilde{\lambda} - \lambda| \left| \mathbf{y}^H \left\{ (\boldsymbol{\Lambda} - \mathbf{I}\widetilde{\lambda})(\widetilde{\lambda} - \lambda) + (\boldsymbol{\Lambda} - \mathbf{I}\widetilde{\lambda})^2 \right\} \mathbf{y} \right|}{(\widetilde{\lambda} - \lambda)\|\widetilde{\mathbf{x}}\|_2^2}$$

afterward, Eq. (1.27) produces

$$= \frac{|\widetilde{\lambda} - \lambda| \left| \mathbf{y}^H(\boldsymbol{\Lambda} - \mathbf{I}\lambda)(\boldsymbol{\Lambda} - \mathbf{I}\widetilde{\lambda})\mathbf{y} \right|}{\mathbf{y}^H(\boldsymbol{\Lambda} - \mathbf{I}\lambda)\mathbf{y}}$$

and finally, since there are no eigenvalues of $\boldsymbol{\Lambda}$ between λ and $\widetilde{\boldsymbol{\lambda}}$, we get

$$\geq |\widetilde{\lambda} - \lambda| \min_{\lambda' \text{ an eigenvalue of } \boldsymbol{\Lambda}} |\lambda' - \widetilde{\lambda}| \,.$$

This inequality is equivalent to the second claim (1.24b).

Inequality (1.24a) clarifies the sense in which small residuals imply that an approximate eigenvector nearly lies in the direction of the true eigenvector. Specifically, the residual needs to be small compared to the difference between the approximate eigenvalue and the next-to-closest true eigenvalue. Inequality (1.24b) implies that the error in a Rayleigh quotient as an approximation to an eigenvalue is proportional to the *square* of the residual for the corresponding approximate eigenvector. Thus Rayleigh quotients provide exceptionally good approximate values for eigenvalues.

Exercise 1.3.9 Let

$$\mathbf{A} = \begin{bmatrix} 1+2\varepsilon & 1 \\ 1 & 1-2\varepsilon \end{bmatrix} \text{ and } \widetilde{\mathbf{A}} = \begin{bmatrix} 1+2\varepsilon+3\varepsilon^2 & 1 \\ 1 & 1-\varepsilon+3\varepsilon^2 \end{bmatrix} .$$

1. Find the eigenvalues and unit eigenvectors for both matrices.
2. verify Lemma 1.3.5 for these matrices.
3. Find the distance between the invariant subspaces of $\mathbf{A}$ corresponding to its two eigenvalues, and verify Lemma 1.3.6.
4. Let λ be the smallest eigenvalue of $\mathbf{A}$, $\widetilde{\lambda}$ be the smallest eigenvalue of $\widetilde{\mathbf{A}}$, and $\widetilde{\mathbf{x}}$ be the corresponding eigenvector. Verify Lemma 1.3.7.
5. Verify Lemma 1.3.8 in the same way.

Exercise 1.3.10 Repeat the previous exercise for

$$\mathbf{A} = \begin{bmatrix} 2 & \varepsilon^2 \\ \varepsilon^2 & -2\varepsilon^2 \end{bmatrix} \text{ and } \widetilde{\mathbf{A}} = \begin{bmatrix} 2+3\varepsilon^2 & \varepsilon^2 \\ \varepsilon^2 & \varepsilon^2 \end{bmatrix} .$$

1.3.3 Symmetric Tridiagonal Matrices

When we studied systems of linear equations, we found that general problems could be reduced to triangular linear systems by Gaussian elimination. When we studied least squares problems, we found that general problems could be reduced to orthogonal or triangular linear systems by either successive reflection or successive orthogonal projection. For the symmetric eigenvalue problem, we will find a way to reduce the problem to finding eigenvalues and eigenvectors of **symmetric tridiagonal** matrices.

An $n \times n$ symmetric tridiagonal matrix has the form

$$\mathbf{T} = \begin{bmatrix} \alpha_1 & \beta_1 & & \\ \beta_1 & \alpha_2 & \ddots & \\ & \ddots & \ddots & \beta_{n-1} \\ & & \beta_{n-1} & \alpha_n \end{bmatrix} ,$$

with all real entries. Such a matrix $\mathbf{T}$ has at most $3n - 2$ nonzero entries, and only $2n - 1$ distinct nonzero entries.

If one of the sub-diagonal entries β_i is zero, then $\mathbf{T}$ has the form

$$\mathbf{T} = \begin{bmatrix} \mathbf{T}_i & \\ & \mathbf{T}_{n-i} \end{bmatrix}$$

where $\mathbf{T}_i$ and $\mathbf{T}_{n-i}$ are symmetric and tridiagonal. If λ is an eigenvalue of $\mathbf{T}_i$ with eigenvector $\mathbf{x}$, then

$$\mathbf{T} \begin{bmatrix} \mathbf{x} \\ 0 \end{bmatrix} = \begin{bmatrix} \mathbf{T}_i \mathbf{x} \\ 0 \end{bmatrix} = \begin{bmatrix} \mathbf{x}\lambda \\ 0 \end{bmatrix} = \begin{bmatrix} \mathbf{x} \\ 0 \end{bmatrix} \lambda .$$

Thus λ is also an eigenvalue of $\mathbf{T}$. In other words, we can reduce the problem of finding eigenvalues of $\mathbf{T}$ to the collection of problems of finding eigenvalues of symmetric tridiagonal matrices with nonzero off-diagonal entries.

Exercise 1.3.11 Let $\mathbf{T}$ be an Hermitian tridiagonal matrix.

1. Show that the diagonal entries of $\mathbf{T}$ are real.
2. If β_i is the subdiagonal entry of $\mathbf{T}$ in the ith column, let

$$\pi_i = \begin{cases} \overline{\beta_i}/|\beta_i|, & \beta_i \neq 0 \\ 1, & \beta_i = 0 \end{cases}$$

and define the matrix

$$\mathbf{P} = \begin{bmatrix} 1 & & & & \\ & \pi_1 & & & \\ & & \pi_1 \pi_2 & & \\ & & & \ddots & \\ & & & & \pi_1 \ldots \pi_{n-1} \end{bmatrix} .$$

Show that $\mathbf{P}$ is unitary.

3. Show that

$$\mathbf{PTP}^{-1} = \begin{bmatrix} \alpha_1 & |\beta_1| & & \\ |\beta_1| & \alpha_2 & \ddots & \\ & \ddots & \ddots & |\beta_{n-1}| \\ & & |\beta_{n-1}| & \alpha_n \end{bmatrix} .$$

This result shows that any Hermitian tridiagonal matrix is unitarily similar to a symmetric tridiagonal matrix with nonnegative sub-diagonal entries.

1.3.4 Sturm Sequences

For theoretical purposes, it is useful to perform a nonunitary change of basis before studying the eigenvalues of a real symmetric tridiagonal matrix $\mathbf{T}$. Let

$$\mathbf{T} = \begin{bmatrix} \alpha_1 & \beta_1 & & \\ \beta_1 & \alpha_2 & \ddots & \\ & \ddots & \ddots & \beta_{n-1} \\ & & \beta_{n-1} & \alpha_n \end{bmatrix} ,$$

and

$$\mathbf{D} = \begin{bmatrix} 1 & & & & \\ & \beta_1 & & & \\ & & \beta_1\beta_2 & & \\ & & & \ddots & \\ & & & & \beta_1\beta_2\ldots\beta_{n-1} \end{bmatrix} .$$

Then

$$\widetilde{\mathbf{T}} \equiv \mathbf{DTD}^{-1} = \begin{bmatrix} \alpha_1 & 1 & & \\ \beta_1^2 & \alpha_2 & \ddots & \\ & \ddots & \ddots & 1 \\ & & \beta_{n-1}^2 & \alpha_n \end{bmatrix} .$$

Since $\widetilde{\mathbf{T}}$ and $\mathbf{T}$ are similar, they have the same eigenvalues. More precisely, $\mathbf{Tp} = \mathbf{p}\lambda$ if and only if $\widetilde{\mathbf{T}}\mathbf{Dp} = \mathbf{Dp}\lambda$.

Suppose that $\widetilde{\mathbf{p}}$ is an eigenvector of $\widetilde{\mathbf{T}}$ with eigenvalue λ. Then $(\widetilde{\mathbf{T}}-\mathbf{I}\lambda)\widetilde{\mathbf{p}} = \mathbf{0}$, or

$$\mathbf{0} = \begin{bmatrix} \alpha_1-\lambda & 1 & & & \\ \beta_1^2 & \alpha_2-\lambda & 1 & & \\ & \beta_2^2 & \alpha_3-\lambda & \ddots & \\ & & \ddots & \ddots & 1 \\ & & & \beta_{n-1}^2 & \alpha_n-\lambda \end{bmatrix} \begin{bmatrix} \widetilde{\pi}_0 \\ \widetilde{\pi}_1 \\ \widetilde{\pi}_2 \\ \vdots \\ \widetilde{\pi}_{n-1} \end{bmatrix} .$$

The individual equations

$$\begin{aligned} 0 &= (\alpha_1-\lambda)\widetilde{\pi}_0 + \widetilde{\pi}_1 \\ 0 &= \beta_1^2\widetilde{\pi}_0 + (\alpha_2-\lambda)\widetilde{\pi}_1 + \widetilde{\pi}_2 \\ 0 &= \beta_2^2\widetilde{\pi}_1 + (\alpha_3-\lambda)\widetilde{\pi}_2 + \widetilde{\pi}_3 \\ &\vdots \\ 0 &= \beta_{n-1}^2\widetilde{\pi}_{n-2} + (\alpha_n-\lambda)\widetilde{\pi}_{n-1} . \end{aligned}$$

These can be written as a **three-term recurrence** for the entries of $\widetilde{\mathbf{p}}$:

$$\begin{aligned} &\text{given } \widetilde{\pi}_0 \neq 0 \text{ (say, } \widetilde{\pi}_0 = 1) \\ &\widetilde{\pi}_1(\lambda) = (\lambda - \alpha_{k+1})\widetilde{\pi}_0 \\ &\text{for } 1 \le k < n, \widetilde{\pi}_{k+1} = (\lambda - \alpha_{k+1})\widetilde{\pi}_k - \beta_k^2\widetilde{\pi}_{k-1} . \end{aligned}$$

Note that λ is an eigenvalue of $\widetilde{\mathbf{T}}$ if and only if $\widetilde{\pi}_n = 0$.

The three-term recurrence defines a sequence of polynomials $\{\widetilde{\pi}_k(\lambda), 0 \le k \le n\}$, called the **Sturm sequence** polynomials. For each k, $\widetilde{\pi}_k(\lambda)$ is a polynomial of degree k.

Example 1.3.2 Suppose that

$$\mathbf{T} = \begin{bmatrix} 2 & -1 & & \\ -1 & 2 & \ddots & \\ & \ddots & \ddots & -1 \\ & & -1 & 2 \end{bmatrix} .$$

(This symmetric tridiagonal matrix commonly arises in the discretization of a two-point boundary value problem for the ordinary differential equation $-u''(x) = f(x)$.)

Then the Sturm sequence polynomials for **T** are

$$\begin{aligned}
\widetilde{\pi}_0 &= 1 \\
\widetilde{\pi}_1(\lambda) &= (\lambda - 2)\widetilde{\pi}_0 = \lambda - 2 \,, \\
\widetilde{\pi}_2(\lambda) &= (\lambda - 2)\widetilde{\pi}_1(\lambda) - \widetilde{\pi}_0 = (\lambda - 2)^2 - 1 \,, \\
\widetilde{\pi}_3(\lambda) &= (\lambda - 2)\widetilde{\pi}_2(\lambda) - \widetilde{\pi}_1(\lambda) = (\lambda - 2)[(\lambda - 2)^2 - 1] - (\lambda - 2) \\
&= (\lambda - 2)[(\lambda - 2)^2 - 2] \,, \\
&\vdots
\end{aligned}$$

It turns out that we can find a simple formula for the zeros of the Sturm sequence polynomials. Note that the Gerschgorin circle theorem says that the eigenvalues of **T** lie in the circles $|\lambda - 2| \leq 2$ and $|\lambda - 2| \leq 1$. With considerable foresight, for $0 < \lambda < 4$ we can define θ by

$$\lambda - 2 = 2\cos\theta \,.$$

Since

$$\sin(k+2)\theta = 2\cos\theta \sin(k+1)\theta - \sin k\theta = (\lambda - 2)\sin(k+1)\theta - \sin k\theta$$

we see that

$$\widetilde{\pi}_k(\lambda) = \frac{\sin(k+1)\theta}{\sin\theta} \,, \; 0 \leq k \leq n$$

satisfies the Sturm sequence. The zeros of $\widetilde{\pi}_k(\lambda)$ are

$$\theta_{k,j} = \frac{j\widetilde{\pi}}{k+1} \,, \; 1 \leq j \leq k \,.$$

Thus the eigenvalues of the tridiagonal matrix in this example are

$$\lambda_{n,j} = 2 + 2\cos\frac{j\widetilde{\pi}}{n+1} \,, \; 1 \leq j \leq n \,.$$

Sturm sequence polynomials have the following important properties.

Lemma 1.3.9 *Given real scalars $\alpha_1, \ldots, \alpha_2$ and nonzero real scalars $\beta_1, \ldots, \beta_{n-1}$, define the polynomials $\pi_k(\lambda)$ for $0 \leq k \leq n$ by*

$$\pi_{-1} = 0 \,, \; \pi_0 = 1 \,,$$
$$\textit{for } 0 \leq k < n, \pi_{k+1} = (\lambda - \alpha_{k+1})\pi_k - \beta_k^2 \pi_{k-1} \,.$$

Then for $0 \leq k \leq n$ the highest order term in $\pi_k(\lambda)$ is λ^k. Furthermore, for $1 \leq k \leq n$, $\pi_k(\lambda)$ has k distinct zeros

$$\lambda_{k,1} > \lambda_{k,2} > \ldots > \lambda_{k,k} \,;$$

the zeros of π_k **interlace the zeros** *of* π_{k+1}*:*

$$\lambda_{k+1,1} > \lambda_{k,1} > \lambda_{k+1,2} > \lambda_{k,2} > \ldots > \lambda_{k,k} > \lambda_{k+1,k+1} \; ; \tag{1.28a}$$

and π_k *alternates sign at the zeros of* π_{k+1}*:*

$$\text{for } 1 \leq j \leq k+1, \; (-1)^j \pi_k(\lambda_{k+1,j}) < 0 \; . \tag{1.28b}$$

Finally, for $0 \leq k \leq n$*, the polynomials* π_k *and* π_{k+1} *agree in sign on the intervals* $\lambda_{k+1,j} < \lambda < \lambda_{k,j-1}$ *for* $2 \leq j \leq k+1$ *and on the unbounded interval* $\lambda > \lambda_{k+1,1}$*; and the polynomials* π_k *and* π_{k+1} *disagree in sign on the intervals* $\lambda_{k,j} < \lambda < \lambda_{k+1,j}$ *for* $1 \leq j \leq k$ *and on the unbounded interval* $\lambda_{k+1,k+1} > \lambda$*.*

Proof First, we will use induction to show that the highest order term in $\pi_k(\lambda)$ is λ^k. Note that $\pi_0(\lambda) = 1 = \lambda^0$ and $\pi_1(\lambda) = \lambda - \alpha_1$, so the claim is true for $k = 0$ and $k = 1$. Inductively, assume that $\pi_j(\lambda)$ has highest-order term λ^j for all $1 \leq j \leq k$. Since

$$\pi_{k+1}(\lambda) = (\lambda - \alpha_{k+1})\pi_k(\lambda) - \beta_k^2 \pi_{k-1}(\lambda) \; ,$$

we see that the highest-order term in $\pi_{k+1}(\lambda)$ is λ^{k+1}.

Next, we will use induction to prove that π_k has k distinct zeros that interlace the zeros of π_{k+1}, and that π_k alternates sign at the zeros of π_{k+1}.

First, we will establish the inductive hypotheses for $k = 1$. Since $\pi_1(\lambda) = \lambda - \alpha_1$, it follows that $\pi_1(\lambda)$ has one zero, namely $\lambda_{11} = \alpha_1$. If we evaluate $\pi(\lambda_{11})$, we get

$$\pi_2(\lambda_{11}) = (\lambda_{11} - \alpha_2)\pi_1(\lambda_{11}) - \beta_1^2 = -\beta_1^2 < 0 \; .$$

On the other hand, as $|\lambda|$ becomes very large, the highest-order term in $\pi_2(\lambda)$ dominates the lower-order terms. Thus, $\pi_2(\lambda) > 0$ for sufficiently large values of $|\lambda|$. Since $\pi_2(\lambda)$ is continuous, and since there are three values of λ, in descending order, at which $\pi_2(\lambda)$ is positive, then negative, then positive, it follows that $\pi_2(\lambda)$ must have two distinct zeros. There must be a zero λ_{21} greater than λ_{11}, and a zero λ_{22} less than λ_{11}. In other words, $\lambda_{21} > \lambda_{11} > \lambda_{22}$. Since $\pi_1(\lambda)$ is linear with highest-order term λ, $\pi_1(\lambda) > 0$ for $\lambda > \lambda_{11}$ and $\pi_1(\lambda) < 0$ for $\lambda < \lambda_{11}$. In particular, $\pi_1(\lambda_{21}) > 0$ and $\pi_1(\lambda_{22}) < 0$. Thus, $(-1)^j \pi_1(\lambda_{2j}) < 0$ for $j = 1, 2$.

Next, we will extend the induction. Suppose that all three parts of our inductive claim are true for Sturm sequence polynomials of degree 1 through k; then we will prove that the conclusion is true for $k+1$. If λ_{kj} is a zero of $\pi_k(\lambda)$, then

$$\pi_{k+1}(\lambda_{kj}) = (\lambda_{kj} - \alpha_{k+1})\pi_k(\lambda_{kj}) - \beta_k^2 \pi_{k-1}(\lambda_{kj}) = -\beta_k^2 \pi_{k-1}(\lambda_{kj}) \; .$$

The third part of the inductive hypothesis shows that for $1 \leq j \leq k$,

$$(-1)^j \pi_{k+1}(\lambda_{kj}) = -\beta_k^2 (-1)^j \pi_{k-1}(\lambda_{kj}) > 0 \; .$$

This shows that $\pi_{k+1}(\lambda)$ alternates sign at the k distinct zeros of $\pi_k(\lambda)$. Since $\pi_{k+1}(\lambda)$ is continuous, it must have $k-1$ zeros interlaced between the zeros of $\pi_k(\lambda)$. Note that $\pi_{k+1}(\lambda_{k1}) < 0$. Also, $\pi_{k+1}(\lambda)$ becomes large and positive as λ becomes large and positive, because the highest-order term λ^{k+1} dominates all the other terms in $\pi_{k+1}(\lambda)$. Thus $\pi_{k+1}(\lambda)$ must have a zero greater than λ_{k1}. Similarly, $(-1)^k \pi_{k+1}(\lambda_{kk}) > 0$ and $(-1)^{k+1}\pi_{k+1}(\lambda) > 0$ for λ sufficiently negative. This shows that $\pi_{k+1}(\lambda)$ has a zero less than λ_{kk}. It follows that $\pi_{k+1}(\lambda)$ has $k+1$ distinct zeros, and that the zeros of $\pi_k(\lambda)$ interlace the zeros of $\pi_{k+1}(\lambda)$. To show that $\pi_k(\lambda)$ alternates sign at the zeros of $\pi_{k+1}(\lambda)$, we first write $\pi_k(\lambda)$ in factored form:

$$\pi_k(\lambda) = \prod_{i=1}^{k} (\lambda - \lambda_{ki}) \; .$$

If we evaluate $\pi_k(\lambda_{k+1,j})$, then the terms $\lambda_{k+1,j} - \lambda_{ki}$ in the product will be negative for $1 \le i < j$ and positive for $j \le i \le k$. If we multiply each of the former terms by -1, the result will be positive:

$$(-1)^{j-1}\pi_k(\lambda_{k+1,j}) > 0 \; .$$

We can multiply by -1 to complete the inductive proof of the third inductive claim.

The proof of the claim that π_k and π_{k+1} agree in sign on certain intervals and disagree in sign on other intervals is delicate to write but easy to illustrate. Consider the case $k = 4$, shown in Fig. 1.3. As we have seen in the previous induction, both $\pi_k(\lambda)$ and $\pi_{k+1}(\lambda)$ alternate sign between their zeros. In addition, the zeros of $\pi_k(\lambda)$ lie between the zeros of $\pi_{k+1}(\lambda)$. The final claim of the lemma follows straightforwardly from these facts.

Theorem 1.3.3 (Sturm Sequence) *Given real scalars $\alpha_1, \ldots, \alpha_n$ and nonzero real scalars $\beta_1, \ldots, \beta_{n-1}$, define the polynomials $\pi_k(\lambda)$ for $0 \le k \le n$ by*

$$\pi_{-1} = 0 \; , \; \pi_0 = 1 \; ,$$

$$\text{for } 0 \le k < n, \pi_{k+1} = (\lambda - \alpha_{k+1})\pi_k - |\beta_k|^2 \pi_{k-1} \; .$$

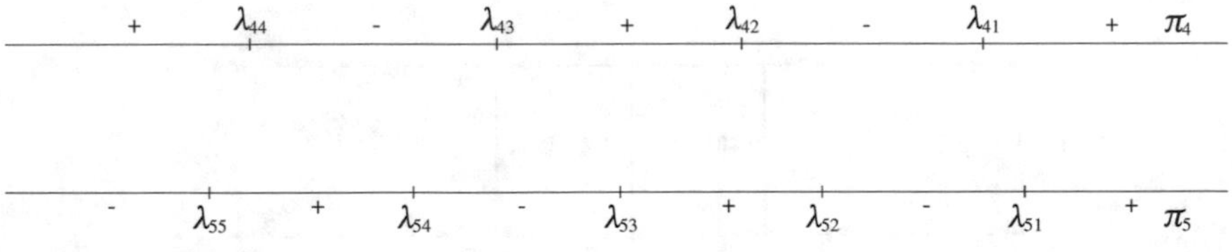

Fig. 1.3 Sign alternation in Sturm sequence polynomials

Then for $1 \leq k \leq n$ *the number of zeros of* π_k *that are greater than* λ *is equal to the number of disagreements in sign between consecutive members of the Sturm sequence* $\{\pi_i(\lambda) : 0 \leq k \leq n\}$*. Our convention is that if* $\pi_i(\lambda) = 0$*, then the sign of* $\pi_i(\lambda)$ *is taken to be the same as the sign of* $\pi_{i-1}(\lambda)$*.*

Proof We will prove this result by induction. Consider the case $k = 1$. If $\pi_0(\lambda) = 1$ and $\pi_1(\lambda)$ disagree in sign, then $0 > \pi_1(\lambda) = \lambda - \lambda_{11}$, so there is one zero of $\pi_1(\lambda)$ that is greater than λ. If $\pi_0(\lambda) = 1$ and $\pi_1(\lambda)$ agree in sign, then $0 \leq \pi_1(\lambda) = \lambda - \lambda_{11}$, so there are no zeros of $\pi_1(\lambda)$ that are greater than λ.

Inductively, we assume that the theorem is true for Sturm sequence polynomials of degree 1 through k, and we want to prove the result for $k + 1$. Suppose that we compute $\pi_0(\lambda), \ldots, \pi_k(\lambda)$ and find m disagreements in sign. The inductive hypothesis shows that there are exactly m zeros of $\pi_k(\lambda)$ that are greater than λ. There are three cases:

1. If $1 \leq m < k$, then

$$\lambda_{k,m+1} \leq \lambda < \lambda_{km} .$$

Next we evaluate $\pi_{k+1}(\lambda)$ and find its sign. If $\pi_{k+1}(\lambda)$ is nonzero and has the same sign as $\pi_k(\lambda)$, then the previous lemma shows us that

$$\lambda_{k+1,m+1} \leq \lambda < \lambda_{km} .$$

This situation is illustrated in Fig. 1.4. In this case, there are exactly m zeros of $\pi_{k+1}(\lambda)$ that are greater than λ, as was claimed. On the other hand, if $\pi_{k+1}(\lambda)$ is nonzero and has the opposite sign of $\pi_k(\lambda)$, then the Lemma 1.3.9 shows us that

$$\lambda_{k,m+1} \leq \lambda < \lambda_{k+1,m+1} .$$

(The $\leq$ sign is necessary to handle the possibility that $\pi_k(\lambda) = 0$.) In this case, $\pi_{k+1}(\lambda)$ has $m+1$ zeros greater than λ, and there are $m+1$ disagreements in sign in the Sturm sequence up to $\pi_{k+1}(\lambda)$. In the special case when $\pi_{k+1}(\lambda = 0)$, then $\lambda = \lambda_{k+1,m+1}$ and there are exactly m zeros of $\pi_{k+1}(\lambda)$ that are strictly greater than λ. Our sign convention gives $\pi_{k+1}(\lambda)$ the same sign as $\pi_k(\lambda)$, so there are

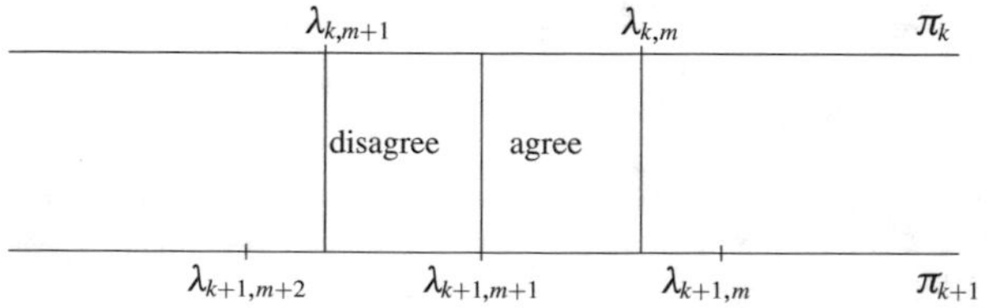

Fig. 1.4 Sturm polynomials have at least one sign disagreement

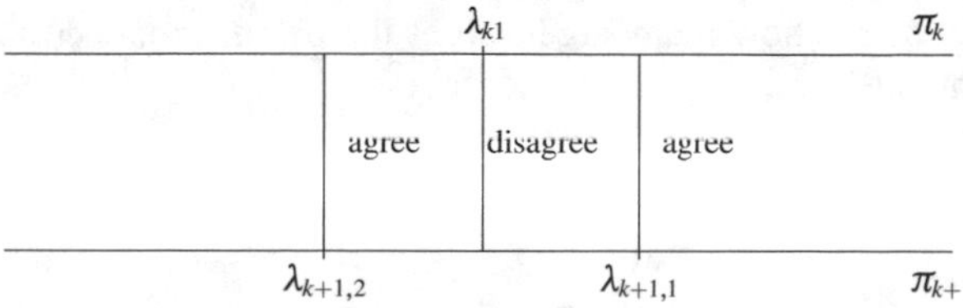

Fig. 1.5 Sturm polynomials have no disagreement

m disagreements in sign in the Sturm sequence. This verifies the inductive step in this case.

2. If $m = 0$, then we find no disagreements in sign in the Sturm sequence up to $\pi_k(\lambda)$. The previous lemma shows us that

$$\lambda_{k,1} \leq \lambda .$$

This situation is illustrated in Fig. 1.5: If $\pi_{k+1}(\lambda) < 0$, then

$$\lambda_{k1} \leq \lambda < \lambda_{k+1,1}$$

so there is one disagreement in sign and one zero of $\pi_{k+1}(\lambda)$ that is greater than λ. Otherwise,

$$\lambda_{k+1,1} \leq \lambda$$

and there are no disagreements in sign in the Sturm sequence.

3. In the remaining case we have $m = k$. Then the previous lemma shows us that

$$\lambda_{k,k} > \lambda .$$

If $\pi_{k+1}(\lambda)$ disagrees in sign with $\pi_k(\lambda)$, then

$$\lambda_{k+1,k+1} > \lambda .$$

In this case, there are $k + 1$ disagreements in sign and $k + 1$ zeros of $\pi_{k+1}(\lambda)$ greater than λ. If $\pi_{k+1}(\lambda)$ agrees in sign with $\pi_k(\lambda)$, then

$$\lambda_{k+1,k+1} \leq \lambda \leq \lambda_{kk} .$$

In this case, there are k disagreements in sign and k zeros of $\pi_{k+1}(\lambda)$ greater than λ.

Example 1.3.3 Let us apply Sturm sequences to the tridiagonal matrix in Example 1.3.2 The diagonal entries α_k are all 2, and the off-diagonal entries β_k are all -1. The Gerschgorin circle theorem says that all of the eigenvalues lie in the interval

$0 \leq \lambda \leq 4$. Let us see how many eigenvalues the Sturm sequence says are greater than 0. We compute

Sturm polynomial	sign
$\pi_0(0) = 1$	+
$\pi_1(0) = -2$	−
$\pi_2(0) = 3$	+
$\pi_3(0) = -4$	−
$\pi_4(0) = 5$	+

There are four sign changes in this sequence, so there are four eigenvalues of **T** that are greater than 0.

Next, let us see how many eigenvalues of **T** are greater than 4:

Sturm polynomial	sign
$\pi_0(4) = 1$	+
$\pi_1(4) = 2$	+
$\pi_2(4) = 3$	+
$\pi_3(4) = 4$	+
$\pi_4(4) = 5$	+

There are no sign changes in this sequence, so **T** has no eigenvalues greater than 4.

Let us see how many eigenvalues of **T** are greater than 2:

Sturm polynomial	sign
$\pi_0(2) = 1$	+
$\pi_1(2) = 0$	+ (use previous sign)
$\pi_2(2) = -1$	−
$\pi_3(2) = 0$	− (use previous sign)
$\pi_4(2) = 1$	+

Note that $\pi_1(2)$ is given the sign of $\pi_0(2)$, and $\pi_3(2)$ is given the sign of $\pi_2(2)$. There are two sign changes in this sequence, so there are two zeros greater than 2.

Example 1.3.4 **Legendre polynomials** are defined by the recurrence

$$p_{-1}(\lambda) = 0$$

$$p_0(\lambda) = 1$$

$$\text{for } k \geq 0\,,\; p_{k+1}(\lambda) = \frac{2k+1}{k+1}\lambda p_k(\lambda) - \frac{k}{k+1}p_{k-1}(\lambda)\,.$$

This recurrence suggests that we define the scalars

$$\alpha_k = 0 \text{ for } 0 \le k \le n \text{ and}$$

$$\beta_k = \frac{k}{2k+1} \text{ and } \gamma_k = \frac{k}{2k-1} \text{ for } 0 \le k < n \; ,$$

form the $(n+1)$-vector

$$\mathbf{p}(\lambda) = \begin{bmatrix} p_0(\lambda) \\ \vdots \\ p_n(\lambda) \end{bmatrix}$$

and the $(n+1) \times (n+1)$ tridiagonal matrix

$$\widetilde{\mathbf{T}} = \begin{bmatrix} \alpha_0 & \gamma_0 & & & \\ \beta_0 & \alpha_1 & \gamma_1 & & \\ & \beta_1 & \alpha_2 & \ddots & \\ & & \ddots & \ddots & \gamma_{n-1} \\ & & & \beta_{n-1} & \alpha_n \end{bmatrix} .$$

Then $(\widetilde{\mathbf{T}} - \mathbf{I}\lambda)\mathbf{p}(\lambda) = \mathbf{0}$ whenever λ is a zero of $p_{n+1}(\lambda)$.

Note that $\widetilde{\mathbf{T}}$ is not symmetric. However, since the products $\beta_k\gamma_k$ of its off-diagonal entries are positive, we can form the diagonal matrix

$$\mathbf{D} = \begin{bmatrix} \delta_0 & & & \\ & \delta_1 & & \\ & & \ddots & \\ & & & \delta_n \end{bmatrix}$$

where

$$\delta_0 = 1$$

$$\text{for } 0 \le k < n \; , \; \delta_{k+1} = \delta_k \sqrt{\beta_k\gamma_k} / |\beta_k| \; .$$

Then

$$\mathbf{T} = \mathbf{D}\widetilde{\mathbf{T}}\mathbf{D}^{-1} = \begin{bmatrix} \alpha_0 & \gamma_0\delta_0/\delta_1 & & & \\ \beta_0\delta_1/\delta_0 & \alpha_1 & \gamma_1\delta_2/\delta_1 & & \\ & \beta_1\delta_2/\delta_1 & \alpha_2 & \ddots & \\ & & \ddots & \ddots & \gamma_{n-1}\delta_{n-1}/\delta_n \\ & & & \beta_{n-1}\delta_n/\delta_{n-1} & \alpha_n \end{bmatrix}$$

$$= \begin{bmatrix} \alpha_0 & \sqrt{\beta_0\gamma_0} & & & \\ \sqrt{\beta_0\gamma_0} & \alpha_1 & \sqrt{\beta_1\gamma_1} & & \\ & \sqrt{\beta_1\gamma_1} & \alpha_2 & \ddots & \\ & & \ddots & \ddots & \sqrt{\beta_{n-1}\gamma_{n-1}} \\ & & & \sqrt{\beta_{n-1}\gamma_{n-1}} & \alpha_n \end{bmatrix}$$

is easily seen to be symmetric. Furthermore,

$$(\mathbf{T} - \mathbf{I}\lambda)\mathbf{D}\mathbf{p}(\lambda) = \mathbf{D}(\widetilde{\mathbf{T}} - \mathbf{I}\lambda)\mathbf{D}^{-1}\mathbf{D}\mathbf{p}(\lambda) = \mathbf{D}(\widetilde{\mathbf{T}} - \mathbf{I}\lambda)\mathbf{p}(\lambda) = \mathbf{D}\mathbf{0} = \mathbf{0}$$

whenever λ is a zero of $p_{n+1}(\lambda)$. Thus the Sturm Sequence Theorem 1.3.3 applies to Legendre polynomials.

Readers may experiment with the JavaScript program for **orthogonal polynomials** This program will allow the user to check that the claims in Lemma 1.3.9 and Theorem 1.3.3 apply to various kinds of orthogonal polynomials.

In MATLAB, readers might plot Legendre polynomials as follows:

```
x=zeros(100,1); for i=1:100, x(i)=4*i/100; end
p=zeros(100,5);
for i=1:100, p(i,1)=1; p(i,2)=(x(i)-2);
for j=3:5,   p(i,j)=(x(i)-2)*p(i,j-1)-p(i,j-2); end; end
plot(x,p(:,1),'y-',x,p(:,2),'m-',x,p(:,3),'c-',x,p(:,4),'r-',x,p(:,5),'g-')
```

Exercise 1.3.12 Find the eigenvalues and eigenvectors of the following matrices.

1.
$$\begin{bmatrix} 2 & 1 \\ 1 & 2 \end{bmatrix}$$

2.
$$\begin{bmatrix} 2 & -1 \\ -1 & 2 \end{bmatrix}$$

3.
$$\begin{bmatrix} 4 & 1 & 0 \\ 1 & 3 & 1 \\ 0 & 1 & 2 \end{bmatrix}$$

4.
$$\begin{bmatrix} \alpha & \beta & 0 \\ \beta & \alpha & \beta \\ 0 & \beta & \alpha \end{bmatrix}$$

Exercise 1.3.13 Use the Sturm Sequence Theorem 1.3.3 to prove the following facts about orthogonal polynomials.

1. The odd-order Legendre polynomials all have $\lambda = 0$ as a root.
2. The zeros of the Chebyshev polynomials ($\alpha_k = 0$ and $\beta_k = \frac{1}{2} = \gamma_k$) all lie between -1 and 1.
3. The Hermite polynomials ($\alpha_k = 0$ and $\beta_k = k$, $\gamma_k = \frac{1}{2}$) of degree greater than one each have a negative zero.
4. The Laguerre polynomials ($\alpha_k = 2k + 1$ and $\beta_k = -k = \gamma_k$) have no positive zeros.

Exercise 1.3.14 Use the Gerschgorin Circle Theorem 1.2.2 to find intervals that contain the zeros of the following polynomials of order $1 \leq n \leq 5$:

1. Legendre polynomials;
2. Chebyshev polynomials;
3. Hermite polynomials;
4. Laguerre polynomials.

Check those intervals using the Sturm Sequence Theorem 1.3.3, then plot the polynomials on those intervals:

1.3.5 Bisection

Our next goal is to develop a simple method to find an individual eigenvalue of a real symmetric tridiagonal matrix. Since a zero sub-diagonal entry would allow us to reduce the size of the problem, we assume that all of the sub-diagonal entries are nonzero.

Suppose that we want to find the j'th eigenvalue of an $n \times n$ symmetric tridiagonal matrix $\mathbf{T}$. We assume that we have found an interval $(\underline{\lambda}, \overline{\lambda})$ that contains this eigenvalue. For example, we may have evaluated the Sturm sequence at $\underline{\lambda}$ and found at least j sign changes; and we may have evaluated the Sturm sequence at $\overline{\lambda}$ and found at most $j - 1$ sign changes.

Next, we follow the ideas developed in Sect. 5.3 in Chap. 5 of Volume I, applying bisection to halve the interval containing the eigenvalue. We compute the interval midpoint $\lambda = \frac{1}{2}(\underline{\lambda}+\overline{\lambda})$, and evaluate the Sturm sequence at λ. If the Sturm sequence at λ has at least j sign changes, then we replace $\underline{\lambda}$ with λ and continue. Otherwise, we replace $\overline{\lambda}$ with λ and continue.

Example 1.3.5 Suppose that we want to find the next-to-largest eigenvalue of

$$\mathbf{T} = \begin{bmatrix} 6 & -1 & & & \\ -1 & 5 & -1 & & \\ & -1 & 4 & -1 & \\ & & -1 & 3 & -1 \\ & & & -1 & 2 \end{bmatrix}$$

The Gerschgorin Circle Theorem 1.2.2 can be used to show that all of the eigenvalues of **T** fall in the interval $1 \leq \lambda \leq 7$. Using bisection, we compute the Sturm sequence polynomials at a sequence of values for λ:

λ	1	7	4	5.5	4.75	5.125
π_0	1	1	1	1	1	1
π_1	−5	1	−2	−0.5	−1.25	−0.875
π_2	19	1	1	−1.25	−0.6875	−1.109
π_3	−52	2	2	−1.375	0.734	−0.373
π_4	85	7	1	−2.188	1.973	0.316
π_5	−33	33	0	−6.281	4.690	1.362
Number sign changes	5	0	2	1	2	2

We conclude from this table that the second eigenvalue lies in the interval $(5.125, 5.5)$. MATLAB computes the next-to-largest eigenvalue of this matrix to be 5.2077, which agrees with our conclusion from bisection.

In LAPACK, bisection is performed on symmetric tridiagonal matrices by routines `_laebz`. See, for example, dlaebz.f. This subroutine is normally called directly from routines `_stebz`, and indirectly from routines `_stevr` or `_stevx`. See, for example, dstebz.f, dstevr.f or dstevx.f.

The bisection algorithm can be used to prove the following lemma.

Lemma 1.3.10 *Let* **T** *be the real symmetric tridiagonal matrix*

$$\mathbf{T} = \begin{bmatrix} \alpha_1 & \beta_1 & & \\ \beta_1 & \alpha_2 & \ddots & \\ & \ddots & \ddots & \beta_{n-1} \\ & & \beta_{n-1} & \alpha_n \end{bmatrix},$$

where the scalars $\beta_1, \ldots, \beta_{n-1}$ *are all nonzero. Then* **T** *has n distinct eigenvalues and n corresponding eigenvectors.*

Proof Recall that the diagonal matrix

$$D = \begin{bmatrix} 1 & & & & \\ & \beta_1 & & & \\ & & \beta_1\beta_2 & & \\ & & & \ddots & \\ & & & & \beta_1\beta_2\ldots\beta_{n-1} \end{bmatrix}$$

is such that $\widetilde{\mathbf{T}} = D\mathbf{T}D^{-1}$ is tridiagonal with ones above the diagonal and positive entries below the diagonal. Using bisection and Sturm sequences applied to $\widetilde{\mathbf{T}}$, we can find n distinct eigenvalues $\lambda_1, \ldots, \lambda_n$ of $\widetilde{\mathbf{T}}$, and corresponding eigenvectors $\mathbf{p}_1, \ldots, \mathbf{p}_n$. Then the eigenvectors $\mathbf{p}_j$ have the form

$$\mathbf{p}_j = \begin{bmatrix} \pi_0(\lambda_j) \\ \vdots \\ \pi_{n-1}(\lambda_j) \end{bmatrix} ,$$

where $\pi_i(\lambda)$ is a Sturm sequence polynomial, and λ_j is a zero of $\pi_n(\lambda)$. Then $\mathbf{x}_j = D^{-1}\mathbf{p}_j$ satisfies

$$\mathbf{T}\mathbf{x}_j = (D^{-1}\widetilde{\mathbf{T}}D)(D^1\mathbf{p}_j) = D^{-1}\mathbf{p}_j\lambda_j = \mathbf{x}_j\lambda_j ,$$

so $\mathbf{x}_j$ is an eigenvector of $\mathbf{T}$ corresponding to λ_j.
The reader should also recall that Lemma 1.3.1 showed that eigenvectors of an Hermitian matrix corresponding to distinct eigenvalues are orthogonal.

Interested readers may experiment with the following JavaScript program for **bisection.** The user can enter the symmetric tridiagonal matrix diagonal and sub-diagonal. The program will plot the Sturm sequence polynomials, and display the **bracketing** intervals for bisection iteration. The list of available colors for the polynomial plot limits this program to matrices of size 6 or less.

Exercise 1.3.15 Use bisection to find the largest zero of the fourth-degree Hermite polynomial with an error of at most 0.05.

Exercise 1.3.16 Program the bisection method and use it to find the 20th eigenvalue (from smallest to largest) of the 50×50 tridiagonal matrix with diagonal entries 2 and off-diagonal entries -1.

1.3.6 Power Method

The bisection method has the advantage of being able to select a particular eigenvalue by its order. The disadvantage of bisection is slow convergence, since the accuracy is guaranteed to increase by only one bit per iteration. In this section, we will develop a different method that is capable of achieving high accuracy more rapidly. The speed of convergence will depend on the separation of the eigenvalues, but we will be able to perform shifts and work with inverses (indirectly) to improve the separation.

1.3.6.1 Forward Iteration

Suppose that $n > 1$ is an integer. Let $\mathbf{A}$ be an $n \times n$ Hermitian matrix, and let its eigenvalue of largest modulus have multiplicity one. These assumptions imply that this dominant eigenvalue is nonzero. Furthermore, the Spectral Theorem 1.3.1 guarantees that there is a unitary matrix **U** so that

$$\mathbf{A}\mathbf{U} = \mathbf{U}\boldsymbol{\Lambda} ,$$

where $\boldsymbol{\Lambda}$ is diagonal. Without loss of generality, we may assume that the first diagonal entry of $\boldsymbol{\Lambda}$ has maximum modulus.

Let the unit n-vector $\widetilde{\mathbf{x}}$ be some initial guess for the eigenvector associated with this dominant eigenvalue, and let

$$\mathbf{w} = \mathbf{U}^H\widetilde{\mathbf{x}}$$

have components ω_i. Then

$$\mathbf{A}\widetilde{\mathbf{x}} = \mathbf{A}\mathbf{U}\mathbf{w} = \mathbf{U}\boldsymbol{\Lambda}\mathbf{w} = \sum_{i=1}^{n} \mathbf{u}_i\lambda_i\omega_i = \left\{\mathbf{u}_1\omega_1 + \sum_{i=2}^{n} \mathbf{u}_i\omega_i\frac{\lambda_i}{\lambda_1}\right\}\lambda_1 .$$

It follows that for any integer $k \geq 1$ we have

$$\mathbf{A}^k\widetilde{\mathbf{x}} = \mathbf{U}\boldsymbol{\Lambda}^k\mathbf{w} = \left\{\mathbf{u}_1\omega_1 + \sum_{i=2}^{n} \mathbf{u}_i\omega_i\left(\frac{\lambda_i}{\lambda_1}\right)^k\right\}\lambda_1^k .$$

If $\omega_1 \neq 0$, then for large k we see that

$$\mathbf{A}^k\widetilde{\mathbf{x}} \approx \mathbf{u}_1\omega_1\lambda_1^k .$$

The error in this approximation depends on how rapidly the terms $(\lambda_i/\lambda_1)^k$ tend to zero for $i > 1$, relative to ω_1.

These computations lead to the **power method**, which can be described by the following

Algorithm 1.3.1 (Power Method)

given $\widetilde{\mathbf{x}}$ with $\|\widetilde{\mathbf{x}}\|_2 = 1$

until converged

$$\widetilde{\mathbf{y}} = \mathbf{A}\widetilde{\mathbf{x}}$$

$$\widetilde{\lambda} = \widetilde{\mathbf{x}}\cdot\widetilde{\mathbf{y}}$$

$$\widetilde{\mathbf{x}} = \frac{\widetilde{\mathbf{y}}}{\|\widetilde{\mathbf{y}}\|_2}$$

If $\mathbf{u}_1\cdot\widetilde{\mathbf{x}} \neq 0$, then $\widetilde{\mathbf{x}} \to \pm\mathbf{u}_1$ and $\widetilde{\lambda} \to \lambda_1$.

Example 1.3.6 Let us approximate the dominant eigenvalue of

$$\mathbf{A} = \begin{bmatrix} 2 & -1 & \\ -1 & 2 & -1 \\ & -1 & 2 \end{bmatrix},$$

using

$$\widetilde{\mathbf{x}}_0 = \begin{bmatrix} 1 \\ 1 \\ 1 \end{bmatrix} \frac{1}{\sqrt{3}} .$$

In the first iteration, we compute

$$\widetilde{\mathbf{y}}_1 = \mathbf{A}\widetilde{\mathbf{x}}_0 = \begin{bmatrix} 1 \\ 0 \\ 1 \end{bmatrix} \frac{1}{\sqrt{3}}, \widetilde{\mathbf{y}}_1 \cdot \widetilde{\mathbf{x}}_0 = \frac{2}{3} \approx 0.667, \widetilde{\mathbf{x}}_1 = \widetilde{\mathbf{y}}_1 \frac{1}{\|\widetilde{\mathbf{y}}_1\|} = \begin{bmatrix} 1 \\ 0 \\ 1 \end{bmatrix} \frac{1}{\sqrt{2}} \approx \begin{bmatrix} 0.7071 \\ 0 \\ 0.7071 \end{bmatrix} .$$

In the second iteration, we compute

$$\widetilde{\mathbf{y}}_2 = \begin{bmatrix} 2 \\ -2 \\ 2 \end{bmatrix} \frac{1}{\sqrt{2}} , \widetilde{\mathbf{y}}_2 \cdot \widetilde{\mathbf{x}}_1 = 2 , \widetilde{\mathbf{x}}_2 = \begin{bmatrix} 1 \\ -1 \\ 1 \end{bmatrix} \frac{1}{\sqrt{3}} \approx \begin{bmatrix} 0.5774 \\ -0.5774 \\ 0.5774 \end{bmatrix} .$$

In the third step, we compute

$$\widetilde{\mathbf{y}}_3 = \begin{bmatrix} 3 \\ -4 \\ 3 \end{bmatrix} \frac{1}{\sqrt{3}} , \widetilde{\mathbf{y}}_3 \cdot \widetilde{\mathbf{x}}_2 = \frac{10}{3} \approx 3.3333 , \widetilde{\mathbf{x}}_3 = \begin{bmatrix} 3 \\ -4 \\ 3 \end{bmatrix} \frac{1}{\sqrt{34}} \approx \begin{bmatrix} 0.5145 \\ -0.6860 \\ 0.5145 \end{bmatrix} .$$

In the fourth step, we compute

$$\widetilde{\mathbf{y}}_4 = \begin{bmatrix} 10 \\ -14 \\ 10 \end{bmatrix} \frac{1}{\sqrt{34}} , \widetilde{\mathbf{y}}_4 \cdot \widetilde{\mathbf{x}}_3 = \frac{58}{17} \approx 3.4118 , \widetilde{\mathbf{x}}_4 = \begin{bmatrix} 5 \\ -7 \\ 5 \end{bmatrix} \frac{1}{\sqrt{99}} \approx \begin{bmatrix} 0.5025 \\ -0.7035 \\ 0.5025 \end{bmatrix} .$$

In the fifth step, we compute

$$\widetilde{\mathbf{y}}_5 = \begin{bmatrix} 17 \\ -24 \\ 17 \end{bmatrix} \frac{1}{\sqrt{99}} , \widetilde{\mathbf{y}}_5 \cdot \widetilde{\mathbf{x}}_4 = \frac{338}{99} \approx 3.4141 , \widetilde{\mathbf{x}}_5 = \begin{bmatrix} 17 \\ -24 \\ 17 \end{bmatrix} \frac{1}{\sqrt{1154}} \approx \begin{bmatrix} 0.5004 \\ -0.7065 \\ 0.5004 \end{bmatrix} .$$

The exact eigenvalue is $\lambda = 2 + \sqrt{2} \approx 3.4142$, with eigenvector

$$\mathbf{x} = \begin{bmatrix} \frac{1}{2} \\ -\sqrt{\frac{1}{2}} \\ \frac{1}{2} \end{bmatrix} \approx \begin{bmatrix} 0.5 \\ -0.7071 \\ 0.5 \end{bmatrix} .$$

1.3.6.2 Inverse Iteration

Since the power method allowed us to find the largest eigenvalue of $\mathbf{A}$, it is reasonable to use powers of $\mathbf{A}^{-1}$ to find the eigenvalue of smallest modulus. Suppose that λ_n is the eigenvalue of smallest modulus, that it has multiplicity one, and is nonzero. Then

$$0 < |\lambda_n| < |\lambda_{n-1}| \leq \ldots \leq |\lambda_1| .$$

As before, given a unit vector $\widetilde{\mathbf{x}}$, let $\mathbf{w} = \mathbf{U}^H\widetilde{\mathbf{x}}$. Then

$$\mathbf{A}^{-k}\widetilde{\mathbf{x}} = \left\{ \sum_{i=1}^{n-1} \mathbf{u}_i\omega_i \left(\frac{\lambda_n}{\lambda_i}\right)^k + \mathbf{u}_n\omega_n \right\} \lambda_n^{-k} ,$$

and for large k,

$$\mathbf{A}^{-k}\widetilde{\mathbf{x}} \approx \mathbf{u}_n \, \omega_n \lambda_n^{-k} .$$

These computations lead to the following

Algorithm 1.3.2 (Inverse Iteration)

given $\widetilde{\mathbf{x}}$ with $\|\widetilde{\mathbf{x}}\|_2 = 1$

until converged

solve $\mathbf{A}\widetilde{\mathbf{y}} = \widetilde{\mathbf{x}}$

$$\widetilde{\lambda} = \frac{1}{\widetilde{\mathbf{x}} \cdot \widetilde{\mathbf{y}}}$$

$$\widetilde{\mathbf{x}} = \frac{\widetilde{\mathbf{y}}}{\|\widetilde{\mathbf{y}}\|_2}$$

If $\mathbf{u}_n \cdot \widetilde{\mathbf{x}} \neq 0$, then $\widetilde{\mathbf{x}} \to \pm\mathbf{u}_n$ and $\widetilde{\lambda} \to \lambda_n$, with error depending on the ratios λ_n/λ_i for $i < n$. Note that if $\mathbf{A}$ is Hermitian and tridiagonal, then linear systems involving $\mathbf{A}$ can be solved very rapidly using the techniques in Sect. 3.13.5 in Chap. 3 of Volume I.

1.3.6.3 Shifted Inverse Iteration

Our next version of the power method is far more useful than the previous two versions. If σ is close to an eigenvalue of $\mathbf{A}$, then $\mathbf{A} - \mathbf{I}\sigma$ has a small eigenvalue $\lambda_i - \sigma$. Suppose that

$$|\lambda_i - \sigma| < |\lambda_j - \sigma| \text{ for } j \neq i\ .$$

Then for large k we have

$$(\mathbf{A} - \mathbf{I}\sigma)^{-k}\widetilde{\mathbf{x}} = \left\{ \sum_{j \neq i} \mathbf{u}_j \omega_j \left(\frac{\lambda_i - \sigma}{\lambda_j - \sigma} \right)^k + \mathbf{u}_i \omega_i \right\} (\lambda_i - \sigma)^{-k} \approx \mathbf{u}_i \omega_i (\lambda_i - \sigma)^{-k}\ ,$$

with error depending on the ratios $(\lambda_i - \sigma)/(\lambda_j - \sigma)$ for $j \neq i$. These computations lead to the following

Algorithm 1.3.3 (Shifted Inverse Iteration)

given $\widetilde{\mathbf{x}}$ with $\|\widetilde{\mathbf{x}}\|_2 = 1$

until converged

solve $(\mathbf{A} - \mathbf{I}\sigma)\widetilde{\mathbf{y}} = \widetilde{\mathbf{x}}$

$$\widetilde{\lambda} = \frac{1}{\widetilde{\mathbf{x}} \cdot \widetilde{\mathbf{y}}} + \sigma$$

$$\widetilde{\mathbf{x}} = \frac{\widetilde{\mathbf{y}}}{\|\widetilde{\mathbf{y}}\|_2}$$

If $\mathbf{u}_i \cdot \widetilde{\mathbf{x}} \neq 0$, then $\widetilde{\mathbf{x}} \rightarrow \pm\mathbf{u}_i$ and $\widetilde{\lambda} \rightarrow \lambda_i - \sigma$.

Note that the closer that σ gets to λ_i, the more nearly singular $\mathbf{A} - \mathbf{I}\sigma$ becomes. Thus we can expect significant rounding errors in Algorithm 1.3.3 while solving for $\widetilde{\mathbf{y}}$. Let us study these errors. We compute $\widetilde{\mathbf{y}}$ by first factoring $\mathbf{A} = \mathbf{L}\mathbf{R}$, then we forward- and back-solve. This means that

$$\mathbf{L}\mathbf{R} = \mathbf{A} - \mathbf{I}\sigma + \mathbf{E} \text{ and } \mathbf{L}\mathbf{R}\widetilde{\mathbf{y}} = \widetilde{\mathbf{x}}\ .$$

Thus we are really applying the exact power method to the approximate matrix $(\mathbf{A} - \mathbf{I}\sigma + \mathbf{E})^{-1}$; at convergence we have

$$(\mathbf{A} - \mathbf{I}\sigma + \mathbf{E})\widetilde{\mathbf{y}} = \widetilde{\mathbf{y}}/\|\widetilde{\mathbf{y}}\|_2\ .$$

Based on our perturbation analysis in Lemma 1.3.5, we know that the error in the eigenvalue computed by inverse iteration will be proportional to $\|\mathbf{E}\|_F$. Furthermore, Lemma 1.3.6 shows that the distance between the invariant spaces for the eigenvalue closest to the shift will be proportional to $\|\mathbf{E}\|_F$ and inversely proportional to the separation of the shifted eigenvalue from the others. These results imply that although the numerical solution $\widetilde{\mathbf{y}}$ of $(\mathbf{A} - \mathbf{I}\sigma)\widetilde{\mathbf{y}} = \widetilde{\mathbf{x}}$ will be inaccurate, if $\mathbf{u}_i$ is a well-conditioned eigenvector of $\mathbf{A}$ then most of the error will lie in the magnitude of $\widetilde{\mathbf{y}}$, not in its direction.

In LAPACK, shifted inverse iteration is performed on symmetric tridiagonal matrices by routines `_stein`. See, for example, dstein.f. This LAPACK routine uses the Gram-Schmidt reorthogonalization process, described in Sect. 6.8.6 in Chap. 6 of Volume I, to determine mutually orthogonal eigenvectors for nearly equal eigenvalues.

1.3.6.4 Rayleigh Quotient Iteration

Note that if $\widetilde{\mathbf{x}}$ is close to being an eigenvector of $\mathbf{A}$, then inequality (1.24b) shows that the Rayleigh quotient $\widetilde{\mathbf{x}}^\top \mathbf{A}\widetilde{\mathbf{x}}/\widetilde{\mathbf{x}}\cdot\widetilde{\mathbf{x}}$ is a very good approximation to the corresponding eigenvalue. Here $\mathbf{a}^\top$ represents the **transpose** of the vector $\mathbf{a}$. Similarly, $\mathbf{A}^\top$ represents the **transpose** of the matrix $\mathbf{A}$. This suggests that during shifted inverse iteration we can adjust the shift σ by using the Rayleigh quotient

$$\sigma = \frac{\widetilde{\mathbf{x}}^\top \mathbf{A}\widetilde{\mathbf{x}}}{\widetilde{\mathbf{x}}\cdot\widetilde{\mathbf{x}}} .$$

As $\widetilde{\mathbf{x}} \to \mathbf{u}_i$ we find that $\sigma \to \lambda_i$ and the convergence of shifted inverse iteration is enhanced. These ideas lead to the following

Algorithm 1.3.4 (Rayleigh Quotient Iteration)

$$
\begin{aligned}
&\text{given } \widetilde{\mathbf{x}} \text{ with } \|\widetilde{\mathbf{x}}\|_2 = 1\\
&\text{until converged}\\
&\quad \widetilde{\lambda} = \widetilde{\mathbf{x}}\cdot(\mathbf{A}\widetilde{\mathbf{x}})\\
&\quad \text{solve } (\mathbf{A} - \mathbf{I}\widetilde{\lambda})\mathbf{y} = \widetilde{\mathbf{x}}\\
&\quad \widetilde{\mathbf{x}} = \mathbf{y}/\|\mathbf{y}\|_2
\end{aligned}
$$

The convergence of Rayleigh quotient iteration has been proved by Parlett [137, pp. 72–79] in the following very useful result.

Theorem 1.3.4 (Rayleigh Quotient Iteration) *Suppose that* $\mathbf{A}$ *is an Hermitian matrix. Let the sequences* $\{\widetilde{\lambda}_k\}_{k=1}^\infty$ *and* $\{\widetilde{\mathbf{x}}_k\}_{k=0}^\infty$ *be generated by the Rayleigh Quotient Iteration in Algorithm 1.3.4. Then* $\widetilde{\lambda}_k$ *converges to a scalar* λ*, and (except for certain initial conditions that are unstable with respect to numerical perturbations)* $\widetilde{\mathbf{x}}_k$ *converges to an eigenvector* $\mathbf{x}$ *of* $\mathbf{A}$ *corresponding to the eigenvalue* λ*. Furthermore, for all* $k \geq 0$ *we have*

$$\left\|(\mathbf{A} - \mathbf{I}\widetilde{\lambda}_{k+1})\widetilde{\mathbf{x}}_{k+1}\right\|_2 \leq \left\|(\mathbf{A} - \mathbf{I}\widetilde{\lambda}_k)\widetilde{\mathbf{x}}_k\right\|_2 ,$$

and the eigenvector approximations converge cubically*:*

$$\lim_{k\to\infty} \frac{\|(\mathbf{I}-\mathbf{x}\mathbf{x}^\top)\widetilde{\mathbf{x}}_k\|_2}{\|(\mathbf{I}-\mathbf{x}\mathbf{x}^\top)\widetilde{\mathbf{x}}_{k+1}\|_2^3} \leq 1$$

Interested readers may experiment with the following JavaScript program for the **power method.** The user can enter the diagonal and sub-diagonal of a symmetric tridiagonal matrix, and choose a shift. The program will perform either the shifted power method, or shifted inverse power method, with an option to use Rayleigh quotients in the shifted inverse power method after the initial user-specified shift. The Rayleigh quotient iterations demonstrate cubic convergence very effectively.

Exercise 1.3.17 Program the power method to find the largest eigenvalue and corresponding eigenvector for the 10×10 Hilbert matrix

$$\mathbf{A}_{ij} = \frac{1}{1+i+j}\,, \quad 0 \leq i,j < 10\,.$$

Choose the initial guess for the eigenvector to be $\mathbf{x} = \mathbf{e}$. Also describe how to use Lemma 1.3.7 to estimate the error in the eigenvalue approximation. You may find it useful to read the paper by Fettis and Caslin [65].

Exercise 1.3.18 Program the inverse power method to find the smallest eigenvalue and corresponding eigenvector for the 10×10 Hilbert matrix. Choose the initial guess for the eigenvector to be $\mathbf{x} = \mathbf{e}$. How accurate is your final result?

Exercise 1.3.19 Use Rayleigh quotient iteration to find the eigenvalue of the 10×10 Hilbert matrix that is closest to 0.34. Use Lemmas 1.3.7 and 1.3.8 to estimate the accuracy in your final result.

1.3.7 QL and QR Algorithms

In this section, we will develop the basic ideas behind a very useful technique for finding all of the eigenvalues of a tridiagonal matrix. Our discussion will follow the development in Parlett [137, Chap. 8].

1.3.7.1 QR Algorithm

The basic approach is described in the following lemma.

Lemma 1.3.11 *Suppose that* $\mathbf{T}_0$ *is an* $n \times n$ *real symmetric tridiagonal matrix and* σ *is a real scalar. Factor*

$$\mathbf{T}_0 - \mathbf{I}\sigma = \mathbf{Q}\mathbf{R}\,,$$

where

$$Q = \mathbf{G}_{12}{}^\top \cdot \ldots \cdot \mathbf{G}_{n-1,n}{}^\top$$

is a product of Givens rotations (see Sect. 6.9 in Chap. 6 of Volume I), and $\mathbf{R}$ *is right-triangular. Define the* **QR transform** *of* $\mathbf{T}_0$ *to be*

$$\mathbf{T}_1 \equiv \mathbf{R}Q + \mathbf{I}\sigma \ .$$

Then $\mathbf{T}_1$ *is tridiagonal and symmetric; furthermore*

$$\mathbf{T}_1 = Q^\top \mathbf{T}_0 Q \ ,$$

so $\mathbf{T}_1$ *is unitarily similar to* $\mathbf{T}_0$*. Furthermore,*

$$Q\mathbf{e}_1\mathbf{e}_1{}^\top\mathbf{R}\mathbf{e}_1 = (\mathbf{T}_0 - \mathbf{I}\sigma)\mathbf{e}_1 \text{ and}$$
$$\mathbf{e}_n\mathbf{e}_n{}^\top\mathbf{R}\mathbf{e}_n = (\mathbf{T}_0 - \mathbf{I}\sigma)Q\mathbf{e}_n \ .$$

In other words, the first column of Q *is determined by shifted iteration on* $\mathbf{T}_0$ *applied to the first axis vector, and the last column of* Q *is determined by shifted inverse iteration on* $\mathbf{T}_0$ *applied to the last axis vector.*

Proof Since $\mathbf{T}_0$ is tridiagonal, the Givens $Q\mathbf{R}$ factorization process in Sect. I.6.9 produces a right-triangular matrix $\mathbf{R}$ that has a diagonal and one super-diagonal. Because the rotation $\mathbf{G}_{j,j+1}$ has nonzero entries only in the rows and columns with those subscripts, we see that

$$\mathbf{R}Q = \mathbf{R}\mathbf{G}_{12}{}^\top \cdot \ldots \cdot \mathbf{G}_{n-1,n}{}^\top$$

has at most one sub-diagonal and two super-diagonals. Since Q is orthogonal,

$$\mathbf{T}_1 = \mathbf{R}Q + \mathbf{I}\sigma = Q^\top(\mathbf{T}_0 - \mathbf{I}\sigma)Q + \mathbf{I}\sigma = Q^\top\mathbf{T}_0 Q$$

is symmetric, similar to $\mathbf{T}_0$, and has at most one sub-diagonal and two super-diagonals. Because this matrix is symmetric, the second super-diagonal must vanish, and $\mathbf{T}_1$ must be tridiagonal.

Since $Q\mathbf{R} = \mathbf{T}_0 - \mathbf{I}\sigma$, it follows that

$$(\mathbf{T}_0 - \mathbf{I}\sigma)\mathbf{e}_1 = Q\mathbf{R}\mathbf{e}_1 = Q\mathbf{e}_1\mathbf{e}_1{}^\top\mathbf{R}\mathbf{e}_1 \ .$$

This implies that the first column of Q is determined by shifted iteration on $\mathbf{T}_0$ applied to the first axis vector. By taking the transpose to get $\mathbf{T}_0 - \mathbf{I}\sigma = \mathbf{R}^\top Q^\top$, we can multiply by Q on the right to get

$$(\mathbf{T}_0 - \mathbf{I}\sigma)Q = \mathbf{R}^\top \ .$$

This implies that

$$(\mathbf{T}_0 - \mathbf{I}\sigma)\mathbf{Q}\mathbf{e}_n = \mathbf{R}^\top \mathbf{e}_n = \mathbf{e}_n \mathbf{e}_n{}^\top \mathbf{R} \mathbf{e}_n \; .$$

This equation implies that the last column of $\mathbf{Q}$ is determined by shifted inverse iteration on $\mathbf{T}_0$ applied to the last axis vector.

1.3.7.2 QL Algorithm

Next, we will develop a useful modification of the $\mathbf{QR}$ transform. Let

$$\mathbf{J} = \begin{bmatrix} \mathbf{e}_n & \dots & \mathbf{e}_1 \end{bmatrix}$$

be the permutation matrix that reverses order. Next, define

$$\mathbf{U} = \mathbf{J}\mathbf{Q}\mathbf{J} \text{ and } \mathbf{L} = \mathbf{J}\mathbf{R}\mathbf{J} \; .$$

Then $\mathbf{U}$ is also orthogonal, and $\mathbf{L}$ is lower triangular with at most one sub-diagonal. It follows that

$$\mathbf{J}(\mathbf{T}_0 - \mathbf{I}\sigma)\mathbf{J} = \mathbf{U}\mathbf{L} \; .$$

In other words, the factorization of $\mathbf{T}_0 - \mathbf{I}\sigma$ could proceed either from top to bottom (the $\mathbf{QR}$ version) or from bottom to top (the $\mathbf{QL}$ version). In practice, the $\mathbf{QL}$ factorization corresponds to applying Givens rotations so that

$$\mathbf{G}_{12} \cdot \ldots \cdot \mathbf{G}_{n-1,n}(\mathbf{T}_0 - \mathbf{I}\sigma) = \mathbf{L} \; .$$

1.3.7.3 QR Transform

Suppose that we apply the $\mathbf{QR}$ transform iteratively, obtaining the following

Algorithm 1.3.5 (QR Transform)

given $\mathbf{T}_0$ real, symmetric and tridiagonal

for $0 \leq k$

  choose a real scalar σ_k

  determine $\mathbf{Q}_k$ so that $\mathbf{Q}_k{}^\top(\mathbf{T}_k - \mathbf{I}\sigma_k) = \mathbf{R}_k$

  form $\mathbf{T}_{k+1} = \mathbf{Q}_k{}^\top \mathbf{T}_k \mathbf{Q}_k$

Then we have the following results.

Lemma 1.3.12 *Given a real symmetric tridiagonal matrix* $\mathbf{T}_0$*, define the sequence of real symmetric tridiagonal matrices* $\{\mathbf{T}_k\}_{k=0}^{\infty}$ *by the* **QR** *Transform Algorithm 1.3.5. Define the orthogonal matrices*

$$\mathbf{P}_0 = \mathbf{I} \text{ and } \mathbf{P}_k = \mathbf{P}_{k-1}\mathbf{Q}_k \text{ for } k \geq 1 \ .$$

Then

$$\mathbf{T}_k = \mathbf{P}_{k-1}{}^{\top}\mathbf{T}\mathbf{P}_{k-1} \ , \tag{1.29}$$

$$\mathbf{P}_k\mathbf{e}_1\mathbf{e}_1{}^{\top}\mathbf{R}_k\mathbf{e}_1 = (\mathbf{T} - \mathbf{I}\sigma_k)\mathbf{P}_{k-1}\mathbf{e}_1 \text{ and} \tag{1.30}$$

$$\mathbf{P}_{k-1}\mathbf{e}_n\mathbf{e}_n{}^{\top}\mathbf{R}_k\mathbf{e}_n = (\mathbf{T} - \mathbf{I}\sigma_k)\mathbf{P}_k\mathbf{e}_n \ , \tag{1.31}$$

so the first and last columns of $\mathbf{P}_k$ *are determined by shifted iteration and shifted inverse iteration, respectively.*

Proof It is easy to see that $\mathbf{T}_k$, given by (1.29), satisfies the recurrence

$$\mathbf{T}_{k+1} = \mathbf{Q}_k{}^{\top}\mathbf{T}_k\mathbf{Q}_k \ .$$

As a result, $\mathbf{P}_{k-1}\mathbf{T}_k = \mathbf{T}\mathbf{P}_{k-1}$ and

$$\mathbf{P}_k\mathbf{R}_k = \mathbf{P}_{k-1}\mathbf{Q}_k\mathbf{R}_k = \mathbf{P}_{k-1}(\mathbf{T}_k - \mathbf{I}\sigma_k) = (\mathbf{T} - \mathbf{I}\sigma_k)\mathbf{P}_{k-1} \ .$$

Multiplying both sides times the first axis vector gives us the claim (1.30). Taking the transpose of this equation gives us

$$\mathbf{R}_k{}^{\top}\mathbf{P}_k{}^{\top} = \mathbf{P}_{k-1}{}^{\top}(\mathbf{T} - \mathbf{I}\sigma_k) \ .$$

We can multiply on the left by $\mathbf{P}_{k-1}$ and on the right by $\mathbf{P}_k$ to get

$$\mathbf{P}_{k-1}\mathbf{R}_k{}^{\top} = (\mathbf{T} - \mathbf{I}\sigma_k)\mathbf{P}_k \ .$$

Then the claim (1.31) follows by operating both sides of this equation on the last axis vector.

Note that a similar lemma can be developed for the **QL** version of the algorithm. In such a case, the first column of the orthogonal transformation corresponds to shifted inverse iteration, and the last column corresponds to shifted iteration.

LAPACK routines `_sterf` or `_steqr` perform either the **QL** or **QR** transform.

1.3.7.4 Wilkinson Shift

For the **QL** version of the algorithm, the shift σ_k is often chosen to be the eigenvalue of the leading 2×2 submatrix that is closest to the first diagonal entry. For the **QR** version, the shift is chosen to be the eigenvalue of the trailing 2×2 submatrix that is closest to the last diagonal entry. This is called the **Wilkinson shift**, and can be computed for the QL version by the following

Algorithm 1.3.6 (Wilkinson Shift)

$$\delta = (\alpha_2 - \alpha_1)/2$$

$$\sigma = \alpha_1 - \text{sign}(\delta)\frac{|\beta_1|^2}{|\delta| + \sqrt{\delta^2 + |\beta_1|^2}} \; .$$

Parlett [137, p. 152] shows that the QL algorithm applied to a tridiagonal matrix with the Wilkinson shift always converges, in the sense that the first subdiagonal entry of the transformed tridiagonal matrix tends to zero. This result implies convergence for the Wilkinson shift in the QR version of the algorithm, as well.

1.3.7.5 Chasing the Bulge

Next, let us discuss how to form $\mathbf{Q}^\top \mathbf{T}_k \mathbf{Q}$ without computing $\mathbf{R}$, and while maintaining symmetry and tridiagonal data storage. Suppose that $\mathbf{T}_k$ has diagonal entries α_i for $1 \le i \le n$, and sub-diagonal entries β_i for $1 \le i < n$. Given a shift σ_k, we can find a Givens rotation $\mathbf{G}_{12}$ so that

$$\mathbf{G}_{12} \begin{bmatrix} \alpha_1 - \sigma_k \\ \beta_1 \end{bmatrix} = \begin{bmatrix} \varrho_1 \\ 0 \end{bmatrix} .$$

Then the pattern of nonzero entries in computing $\mathbf{G}_{12}\mathbf{T}_k\mathbf{G}_{12}{}^\top$ is

$$\mathbf{G}_{12} \begin{bmatrix} \times & \times & & & \\ \times & \times & \times & & \\ & \times & \times & \times & \\ & & \times & \times & \times \\ & & & \times & \times \end{bmatrix} \mathbf{G}_{12}{}^\top = \begin{bmatrix} * & * & + & & \\ * & * & * & & \\ + & * & \times & \times & \\ & & \times & \times & \times \\ & & & \times & \times \end{bmatrix} .$$

Here the symbol $\times$ indicates an original entry of $\mathbf{T}_k$, while a symbol $*$ indicates a modified entry and $+$ indicates an additional entry. We will **chase the bulge** by choosing additional rotations to move the unwanted additional entries down the extra sub-diagonal.

After the first Givens rotation $\mathbf{G}_{12}$, let us store the value of the (3, 1) entry in a temporary location. Next, we choose $\mathbf{G}_{23}$ to zero out the new $+$ entry:

$$\mathbf{G}_{23} \begin{bmatrix} * & * & + & & \\ * & * & * & & \\ + & * & & \times & \\ & & \times & \times & \times \\ & & & \times & \times \end{bmatrix} \mathbf{G}_{23}{}^\top = \begin{bmatrix} * & * & 0 & & \\ * & * & * & + & \\ 0 & * & * & * & \\ & + & * & \times & \times \\ & & & \times & \times \end{bmatrix} .$$

The symbol 0 indicates that a zero entry has been achieved as needed, but a new + value has appeared outside the tridiagonal storage. We choose $\mathbf{G}_{34}$ to zero out that entry:

$$\mathbf{G}_{34} \begin{bmatrix} * & * & & & \\ * & * & * & + & \\ & * & * & * & \\ & + & * & \times & \times \\ & & & \times & \times \end{bmatrix} \mathbf{G}_{34}{}^{\top} = \begin{bmatrix} * & * & & & \\ * & * & * & 0 & \\ & * & * & * & + \\ & 0 & * & * & * \\ & & + & * & \times \end{bmatrix} .$$

With the final Givens rotation, we achieve tridiagonal storage again:

$$\mathbf{G}_{45} \begin{bmatrix} * & * & & & \\ * & * & * & & \\ & * & * & * & + \\ & & * & * & * \\ & & + & * & \times \end{bmatrix} \mathbf{G}_{45}{}^{\top} = \begin{bmatrix} * & * & & & \\ * & * & * & & \\ & * & * & * & 0 \\ & & * & * & * \\ & & 0 & * & * \end{bmatrix} .$$

1.3.7.6 Implicit QR

There are various ways to implement the symmetric application of the Givens rotations. These alternatives are discussed very thoroughly in a paper by Gates and Gragg [77]. This paper shows that there are ways to use the formulas for the entries of the Givens rotations and their symmetric application to $\mathbf{T}_k$ to avoid square roots. LAPACK routines `_sterf` and `_steqr` use the Pal-Walker-Kahan algorithm, which was first published by Parlett [137, p. 164ff]. This algorithm takes the following form.

Algorithm 1.3.7 (Pal-Walker-Kahan Implicit QR)

for $1 \leq j \leq n$
  $b_j = \beta_j^2$ /* store in β_j */
$c_0 = 1\ ,\ s_0 = 0\ ,\ \tau_1 = \alpha_1 - \sigma\ ,\ p_1 = \tau_1^2$
for $1 \leq j < n$
  $r_j = p_j + b_j$
  if $k > 1$ $b_{j-1} = s_{j-1} r_j$
  $c_j = p_j / r_j\ ,\ s_j = b_j / r_j$
  $\tau_{j+1} = c_j(\alpha_{j+1} - \sigma) - s_j \tau_j\ ,\ \alpha_j = \alpha_{j+1} + \tau_j - \tau_{j+1}$
  if $c_j > 0$ then $p_{j+1} = \tau_{j+1}^2 / c_j$ else $p_{j+1} = c_{j-1} b_j$
$b_{n-1} = s_{n-1} p_n\ ,\ \alpha_n = \tau_n + \sigma$

1.3.7.7 Implementation Details

We would like to discuss three more issues about the QL and QR algorithms for finding eigenvalues of symmetric tridiagonal matrices. First, a sub-diagonal entry of $\mathbf{T}$ is considered to be effectively zero when

$$\beta_j^2 \leq \varepsilon^2 |\alpha_j \alpha_{j+1}| ,$$

where ε is machine precision. In such a case, the original tridiagonal matrix $\mathbf{T}$ is partitioned as

$$\mathbf{T} = \begin{bmatrix} \mathbf{T}_j & \\ & \mathbf{T}_{n-j} \end{bmatrix} ,$$

where $\mathbf{T}_j$ is $j \times j$. Then the algorithm computes the eigenvalues of $\mathbf{T}_j$ and $\mathbf{T}_{n-j}$ separately, and merges the two sets of eigenvalues at the end.

Second, the algorithm makes a choice between performing a QL transform or a QR transform with each iteration. If $|\alpha_n| < |\alpha_1|$ then the algorithm uses the QR transform (which proceeds from first to last), otherwise it uses QL (which proceeds from last to first).

Finally, if eigenvectors of $\mathbf{T}$ are desired, then it is necessary to form the orthogonal matrices $\mathbf{P}_k$ in Algorithm 1.3.5. The matrices of eigenvectors formed in this way are numerically orthogonal; for a proof of this claim, see either Francis [73] or Kublanovskaya [115]. In LAPACK, these eigenvectors are accumulated in routines `_steqr`. See, for example, dsteqr.f. If eigenvectors are not desired but the user wants to find all eigenvalues of a symmetric tridiagonal matrix, then the user should call LAPACK routines `_sterf`. See, for example, dsterf.f.

Interested readers may experiment with the following JavaScript program for the **QL/QR algorithm.** Users can enter the symmetric tridiagonal matrix diagonal and sub-diagonal. At each stage of the iteration, if the first diagonal entry is smaller than the last, then the program will perform a QR iteration; otherwise it will perform a QL iteration. The rapid convergence of the iteration is evident from the convergence of the sub-diagonal entries to zero.

Exercise 1.3.20 Find all of the eigenvalues of the 100×100 symmetric tridiagonal matrix

$$\mathbf{A} = \begin{bmatrix} 2 & -1 & & & \\ -1 & 2 & \ddots & & \\ & \ddots & \ddots & \ddots & \\ & & \ddots & 2 & -1 \\ & & & -1 & 2 \end{bmatrix} .$$

This matrix arises in finite difference discretization of the two-point boundary value problem $-u''(x) = f(x)$ with boundary conditions $u(0) = 0 = u(1)$. How accurate is the smallest eigenvalue?

Exercise 1.3.21 Find all of the eigenvalues of the 100×100 symmetric tridiagonal matrix

$$\mathbf{A} = \begin{bmatrix} 1 & -1 & & & \\ -1 & 2 & \ddots & & \\ & \ddots & \ddots & \ddots & \\ & & \ddots & 2 & -1 \\ & & & -1 & 1 \end{bmatrix} .$$

This matrix arises in finite difference discretization of the two-point boundary value problem $-u''(x) = f(x)$ with boundary conditions $u'(0) = 0 = u'(1)$. How accurate is the smallest eigenvalue?

1.3.8 Divide and Conquer

In this section, we will discuss our second computational approach for finding the eigenvalues and eigenvectors of a real symmetric tridiagonal matrix. This new technique is based on the divide and conquer method, which is a standard algorithm design paradigm in computer science. The specific application of this approach to the symmetric tridiagonal eigenvalue problem is due to Cuppen [42]. The basic principle is to decompose a large matrix $\mathbf{T}$ in the form

$$\mathbf{T} = \begin{bmatrix} \mathbf{T}_1 & \\ & \mathbf{T}_2 \end{bmatrix} + \begin{bmatrix} \mathbf{e}_{j+1} \\ \mathbf{e}_1 \end{bmatrix} \beta_j \begin{bmatrix} {\mathbf{e}_{j+1}}^\top & {\mathbf{e}_1}^\top \end{bmatrix} ,$$

where β_j is the jth sub-diagonal entry of $\mathbf{T}$. Afterward, we find orthogonal matrices $\mathbf{Q}_1$ and $\mathbf{Q}_2$, and diagonal matrices $\boldsymbol{\Lambda}_1$ and $\boldsymbol{\Lambda}_2$ so that

$$\mathbf{T}_1\mathbf{Q}_1 = \mathbf{Q}_1\boldsymbol{\Lambda}_1 \text{ and } \mathbf{T}_2\mathbf{Q}_2 = \mathbf{Q}_2\boldsymbol{\Lambda}_2 .$$

Then we have

$$\begin{bmatrix} {\mathbf{Q}_1}^\top & \\ & {\mathbf{Q}_2}^\top \end{bmatrix} \left\{ \begin{bmatrix} \mathbf{T}_1 & \\ & \mathbf{T}_2 \end{bmatrix} + \begin{bmatrix} \mathbf{e}_{j+1} \\ \mathbf{e}_1 \end{bmatrix} \beta_j \begin{bmatrix} {\mathbf{e}_{j+1}}^\top & {\mathbf{e}_1}^\top \end{bmatrix} \right\} \begin{bmatrix} \mathbf{Q}_1 & \\ & \mathbf{Q}_2 \end{bmatrix}$$
$$= \begin{bmatrix} \boldsymbol{\Lambda}_1 & \\ & \boldsymbol{\Lambda}_2 \end{bmatrix} + \begin{bmatrix} {\mathbf{Q}_1}^\top \mathbf{e}_{j+1} \\ {\mathbf{Q}_2}^\top \mathbf{e}_1 \end{bmatrix} \beta_j \begin{bmatrix} {\mathbf{e}_{j+1}}^\top \mathbf{Q}_1 & {\mathbf{e}_1}^\top \mathbf{Q}_2 \end{bmatrix} .$$

This is a diagonal matrix plus a symmetric outer product.

The following lemma describes how we can compute the eigenvalues and eigenvectors of such a matrix.

Lemma 1.3.13 *Suppose that* $\mathbf{D}$ *is an* $n \times n$ *real diagonal matrix with non-increasing diagonal entries,* $\mathbf{b}$ *is a nonzero n-vector and* β *is a real nonzero scalar. Then there is an* $n \times n$ *orthogonal matrix* $\mathbf{Q}$*, diagonal matrices* $\mathbf{D}_1$ *and* $\mathbf{D}_2$ *so that the diagonal entries of* $\mathbf{D}_1$ *are strictly decreasing, and a vector* $\mathbf{z}_1$ *with no zero components so that*

$$\mathbf{Q}^\top \left\{\mathbf{D} + \mathbf{b}\beta\mathbf{b}^\top\right\} \mathbf{Q} = \begin{bmatrix} \mathbf{D}_1 & \\ & \mathbf{D}_2 \end{bmatrix} + \begin{bmatrix} \mathbf{z}_1 \\ \mathbf{0} \end{bmatrix} \beta \begin{bmatrix} \mathbf{z}_1{}^\top & \mathbf{0}^\top \end{bmatrix} . \tag{1.32}$$

Furthermore, if $\mathbf{x}_1$ *is an eigenvector of* $\mathbf{D}_1 + \mathbf{z}_1\beta\mathbf{z}_1{}^\top$ *with eigenvalue* λ_1*, then* $\mathbf{D}_1 - \mathbf{I}\lambda_1$ *is nonsingular,* $\mathbf{z}_1{}^\top\mathbf{x}_1 \neq 0$*,* $\mathbf{x}_1$ *is a nonzero scalar multiple of* $(\mathbf{D}_1 - \mathbf{I}\lambda_1)^{-1}\mathbf{z}_1$*, and* λ_1 *is a solution of the* **secular equation**

$$0 = \phi(\lambda) \equiv 1 + \beta\mathbf{z}_1{}^\top(\mathbf{D}_1 - \mathbf{I}\lambda)^{-1}\mathbf{z}_1 . \tag{1.33}$$

Finally, the eigenvalues of $\mathbf{D}_1 + \mathbf{z}_1\beta\mathbf{z}_1{}^\top$ *interlace the eigenvalues of* $\mathbf{D}_1$*, with one eigenvalue greater than the first eigenvalue of* $\mathbf{D}_1$ *when* $\beta > 0$*, or less than the last eigenvalue of* $\mathbf{D}_1$ *when* $\beta < 0$*.*

Proof Let the diagonal entries of $\mathbf{D}$ be δ_j for $1 \leq j \leq n$, and let the components of $\mathbf{b}$ be β_j. Suppose that i is the first row index such that $\delta_{i+1} = \delta_i$. Choose a Givens rotation $\mathbf{G}_{i,i+1}$ as in Sect. 6.9 in Chap. 6 of Volume I so that

$$\mathbf{G}_{i,i+1} \begin{bmatrix} \beta_i \\ \beta_{i+1} \end{bmatrix} = \begin{bmatrix} \varrho_i \\ 0 \end{bmatrix} .$$

Then appropriate partitioning shows that

$$\begin{aligned}
&\begin{bmatrix} \mathbf{I} & & \\ & \mathbf{G}_{i,i+1} & \\ & & \mathbf{I} \end{bmatrix} \left\{\mathbf{D} + \mathbf{b}\beta\mathbf{b}^\top\right\} \begin{bmatrix} \mathbf{I} & & \\ & \mathbf{G}_{i,i+1}{}^\top & \\ & & \mathbf{I} \end{bmatrix} \\
&= \begin{bmatrix} \mathbf{I} & & \\ & \mathbf{G}_{i,i+1} & \\ & & \mathbf{I} \end{bmatrix} \left\{ \begin{bmatrix} \mathbf{D}' & & \\ & \mathbf{I}\delta_i & \\ & & \mathbf{D}'' \end{bmatrix} \right. \\
&\left. + \begin{bmatrix} \mathbf{b}_1 \\ \mathbf{e}_1\beta_i + \mathbf{e}_2\beta_{i+1} \\ \mathbf{b}_2 \end{bmatrix} \beta \begin{bmatrix} \mathbf{b}_1{}^\top & \beta_i\mathbf{e}_1{}^\top + \beta_{i+1}\mathbf{e}_2{}^\top & \mathbf{b}_2{}^\top \end{bmatrix} \right\} \begin{bmatrix} \mathbf{I} & & \\ & \mathbf{G}_{i,i+1}{}^\top & \\ & & \mathbf{I} \end{bmatrix} \\
&= \begin{bmatrix} \mathbf{D}' & & \\ & \mathbf{I}\delta_i & \\ & & \mathbf{D}'' \end{bmatrix} + \begin{bmatrix} \mathbf{b}_1 \\ \mathbf{e}_1\varrho_i \\ \mathbf{b}_2 \end{bmatrix} \beta \begin{bmatrix} \mathbf{b}_1{}^\top & \varrho_i\mathbf{e}_1{}^\top & \mathbf{b}_2{}^\top \end{bmatrix} .
\end{aligned}$$

This is a diagonal matrix plus a symmetric outer product, such that the vector in the outer product is zero in the location corresponding to the first repeated eigenvalue. We may continue processing the outer product in this way, until we obtain

$$\mathbf{G}\left\{\mathbf{D}+\mathbf{b}\beta\mathbf{b}^\top\right\}\mathbf{G}^\top=\mathbf{D}+\mathbf{c}\beta\mathbf{c}^\top ,$$

where the ith component of $\mathbf{c}$ is zero whenever δ_i is a repeated diagonal entry of $\mathbf{D}$.

Next, let i the first index such that the ith component of $\mathbf{c}$ is zero. Let

$$\mathbf{P}=\left[\mathbf{e}_2 \ldots \mathbf{e}_{n-i+1}\ \mathbf{e}_1\right]$$

be the permutation of $n-i+1$ items that moves the first to last, and all others up one position. Again, appropriate partitioning shows that

$$\begin{aligned}
&\begin{bmatrix}\mathbf{I} & \\ & \mathbf{P}^\top\end{bmatrix}\left\{\mathbf{D}+\mathbf{c}\beta\mathbf{c}^\top\right\}\begin{bmatrix}\mathbf{I} & \\ & \mathbf{P}\end{bmatrix}\\
&=\begin{bmatrix}\mathbf{I} & \\ & \mathbf{P}^\top\end{bmatrix}\left\{\begin{bmatrix}\mathbf{D}' & \\ & \mathbf{D}''\end{bmatrix}+\begin{bmatrix}\mathbf{c}_1\\ \mathbf{c}_2\end{bmatrix}\beta\left[\mathbf{c}_1{}^\top\ \mathbf{c}_2{}^\top\right]\right\}\begin{bmatrix}\mathbf{I} & \\ & \mathbf{P}\end{bmatrix}\\
&=\begin{bmatrix}\mathbf{D}' & \\ & \mathbf{P}^\top\mathbf{D}''\mathbf{P}\end{bmatrix}+\begin{bmatrix}\mathbf{c}_1\\ \mathbf{P}^\top\mathbf{c}_2\end{bmatrix}\beta\left[\mathbf{c}_1{}^\top\ \mathbf{c}_2{}^\top\mathbf{P}\right]=\begin{bmatrix}\mathbf{D}''' & \\ & \delta_i\end{bmatrix}+\begin{bmatrix}\mathbf{z}\\ 0\end{bmatrix}\beta\left[\mathbf{z}^\top\ 0\right] .
\end{aligned}$$

We may continue this process, reordering the entries of the diagonal matrix and the outer product, until we obtain the claimed form (1.32).

Next, suppose that $\mathbf{x}_1$ is an eigenvector of $\mathbf{D}_1+\mathbf{z}_1\beta\mathbf{z}_1{}^\top$ with eigenvalue λ_1. We will prove by contradiction that $\mathbf{D}_1-\mathbf{I}\lambda_1$ is nonsingular. If $\mathbf{D}_1-\mathbf{I}\lambda_1$ were singular, then there would be an index i so that $\lambda_1=\delta_i$. Since

$$(\mathbf{D}_1-\mathbf{I}\lambda_1)\mathbf{x}_1+\mathbf{z}_1\beta\mathbf{z}_1\cdot\mathbf{x}_1=0 , \tag{1.34}$$

we have

$$0=\mathbf{e}_i{}^\top\left\{(\mathbf{D}_1-\mathbf{I}\lambda_1)\mathbf{x}_1+\mathbf{z}\beta\mathbf{z}_1\cdot\mathbf{x}_1\right\}=\mathbf{e}_i\cdot\mathbf{z}_1\beta\mathbf{z}_1\cdot\mathbf{x}_1 .$$

Since β is nonzero and the components of $\mathbf{z}_1$ are nonzero, we conclude that $\mathbf{z}_1\cdot\mathbf{x}_1=0$. Thus $\mathbf{D}_1\mathbf{x}_1=\mathbf{x}_1\lambda_1=\mathbf{x}_1\delta_i$. Since the eigenvalues of $\mathbf{D}_1$ are distinct, this equation implies that $\mathbf{x}_1=\mathbf{e}_i$. But then we must have $0=\mathbf{z}_1\cdot\mathbf{x}_1=\mathbf{z}_1\cdot\mathbf{e}_i$. Since the components of $\mathbf{z}_1$ are nonzero, we have a contradiction. We conclude that $\mathbf{D}_1-\mathbf{I}\lambda_1$ is nonsingular.

Next, we will prove by contradiction that $\mathbf{z}_1\cdot\mathbf{x}_1\neq 0$. Otherwise, we would have

$$(\mathbf{D}_1-\mathbf{I}\lambda_1)\mathbf{x}_1=\mathbf{z}_1\beta\mathbf{z}_1\cdot\mathbf{x}_1=0 .$$

Since $\mathbf{x}_1$ is nonzero and $\mathbf{D}_1 - \mathbf{I}\lambda_1$ is nonsingular, we have a contradiction. We conclude that $\mathbf{z}_1 \cdot \mathbf{x}_1 \neq 0$.

Since $(\mathbf{D}_1 - \mathbf{I}\lambda_1)\mathbf{x}_1 = \mathbf{z}_1\beta\mathbf{z}_1 \cdot \mathbf{x}_1$ where $\beta\mathbf{z}_1 \cdot \mathbf{x}_1 \neq 0$, we conclude that $\mathbf{x}_1$ is a nonzero scalar multiple of $(\mathbf{D}_1 - \mathbf{I}\lambda_1)^{-1}\mathbf{z}_1$. Then we can multiply the eigenvector equation for $\mathbf{x}_1$ on the left by $\mathbf{z}_1{}^\top(\mathbf{D}_1 - \mathbf{I}\lambda_1)^{-1}$ to get

$$0 = \mathbf{z}_1 \cdot \mathbf{x}_1 \left\{1 + \mathbf{z}_1{}^\top(\mathbf{D}_1 - \mathbf{I}\lambda_1)^{-1}\mathbf{z}_1\beta\right\} = \mathbf{z}_1 \cdot \mathbf{x}_1\phi(\lambda_1) ,$$

where the function ϕ was defined in the secular equation (1.33). We can rewrite ϕ in the form

$$\phi(\lambda) = 1 + \beta \sum_{i=1}^{k} \frac{\zeta_i^2}{\delta_i - \lambda} ,$$

where the components of $\mathbf{z}_1$ are ζ_i for $1 \leq i \leq k$, and note that

$$\phi'(\lambda) = \beta \sum_{i=1}^{k} \frac{\zeta_i^2}{(\delta_i - \lambda)^2} .$$

Thus ϕ is monotone between its poles, which are the diagonal entries of $\mathbf{D}_1$. The interlacing property follows immediately from this observation.

The rounding errors and computational complexity of the divide and conquer method were analyzed by Cuppen [42]. Numerical experiments by Cuppen indicate that the work involved in the divide and conquer method is typically proportional to $n \log n$, while the work with the *QLQR* method is roughly proportional to $6n^2$. The divide and conquer method appears to be more efficient for $n \geq 150$, for computations in which both eigenvalues and eigenvectors are desired.

The divide and conquer algorithm is available in LAPACK routines `_stedc`. See, for example, dstedc.f. The divide and conquer algorithm is actually performed in LAPACK routines `_laed0`, and the secular equation itself is solved in LAPACK routines `_laed4`. See, for example, dlaed0 and dlaed4. Routines `_laed0` form a tree of submatrices by recursive bisection until all submatrices have no more than 26 rows and columns. The eigenvalues and eigenvectors of these smallest submatrices are determined by the QL/QR algorithm, and then the divide and conquer approach described above is used to determine the eigenvalues and eigenvectors of parent matrices in the tree. For more details, examine the code for routines `_laed0`.

Interested readers may experiment with the following JavaScript program for the **divide and conquer algorithm.** Users can enter the symmetric tridiagonal matrix diagonal and sub-diagonal. The program is similar to LAPACK routine `dstedc`, but differs by sub-dividing the matrix into sub-matrices that are either 1×1 or 2×2, as described by the first "partition after indices" block. The eigenvalues (and eigenvectors) of these small matrices are easily determined. Afterward, the program solves the secular equation to find the eigenvalues of the larger matrices formed by coarser partition lines, until the eigenvalues of the original matrix are found.

Exercise 1.3.22 Consider the symmetric tridiagonal matrix

$$\mathbf{T} = \begin{bmatrix} 2 & -1 & & \\ -1 & 2 & -1 & \\ & -1 & 2 & -2 \\ & & -1 & 2 \end{bmatrix}$$

1. Use one step of divide and conquer to split the matrix into two 2×2 matrices plus a rank one outer product.
2. Next, find the eigenvalues of each of the two 2×2 matrices.
3. Solve the secular equation for the eigenvalues of the original symmetric tridiagonal matrix.

Exercise 1.3.23 Use divide and conquer to find all of the eigenvalues of the 100×100 symmetric tridiagonal matrix

$$\mathbf{A} = \begin{bmatrix} 2 & -1 & & & \\ -1 & 2 & \ddots & & \\ & \ddots & \ddots & \ddots & \\ & & \ddots & 2 & -1 \\ & & & -1 & 2 \end{bmatrix} .$$

How accurate is the smallest eigenvalue?

Exercise 1.3.24 Use divide and conquer to find all of the eigenvalues of the 100×100 symmetric tridiagonal matrix

$$\mathbf{A} = \begin{bmatrix} 1 & -1 & & & \\ -1 & 2 & \ddots & & \\ & \ddots & \ddots & \ddots & \\ & & \ddots & 2 & -1 \\ & & & -1 & 1 \end{bmatrix} .$$

How accurate is the smallest eigenvalue?

1.3.9 dqds

Dhillon and Parlett [54, p. 859] claim that "Divide and conquer is a fast method much of the time but can reduce to an $O(n^3)$ process for rather uniform eigenvalue distributions." In order to compute a numerically orthogonal set of eigenvectors for a full set of eigenvalues of a symmetric tridiagonal matrix, these authors developed a new idea called the **relatively robust representation**. The details of this development are complicated. We will summarize the approach, but suggest that readers examine the Dhillon Parlett paper for details.

Recall our discussion of shifted inverse iteration in Sect. 1.3.6. Given a very accurate approximation to an eigenvalue λ of a symmetric tridiagonal matrix $\mathbf{T}$, we can use shifted inverse iteration to compute the corresponding eigenvector $\mathbf{x}$ as a vector in the nullspace of $\mathbf{T} - \mathbf{I}\lambda$. In order to reveal this null vector, we might want to find a shift σ so that the factorization

$$\mathbf{T} - \mathbf{I}\sigma = \mathbf{L}\mathbf{D}\mathbf{L}^\top$$

does not involve much element growth. Such factorizations were discussed in Sect. 3.13.3 in Chap. 3 of Volume I.

Following this initial factorization, we will compute additional factorizations of the form

$$\mathbf{L}\mathbf{D}\mathbf{L}^\top - \mathbf{I}\widehat{\lambda} = \mathbf{L}_+\mathbf{D}_+\mathbf{L}_+{}^\top = \mathbf{U}_-\mathbf{D}_-\mathbf{U}_-{}^\top .$$

Here $\mathbf{L}_+$ is unit lower **bidiagonal,** $\mathbf{U}_-$ is unit upper bidiagonal and both $\mathbf{D}_-$ and $\mathbf{D}_+$ are diagonal. Let

$$\mathbf{L} = \begin{bmatrix} 1 & & & \\ \ell_1 & 1 & & \\ & \ddots & \ddots & \\ & & \ell_{n-1} & 1 \end{bmatrix} , \; \mathbf{D} = \begin{bmatrix} \delta_1 & & & \\ & \delta_2 & & \\ & & \ddots & \\ & & & \delta_n \end{bmatrix}$$

and

$$\mathbf{L}_+ = \begin{bmatrix} 1 & & & \\ \ell_{+,1} & 1 & & \\ & \ddots & \ddots & \\ & & \ell_{+,n-1} & 1 \end{bmatrix} , \; \mathbf{D}_+ = \begin{bmatrix} \delta_{+,1} & & & \\ & \delta_{+,2} & & \\ & & \ddots & \\ & & & \delta_{+,n} \end{bmatrix} .$$

Then we can compute the entries of $\mathbf{L}_+$ and $\mathbf{D}_+$ by the **differential form** of the **stationary qd transform**:

Algorithm 1.3.8 (dstqds)

$$
\begin{aligned}
&s_1 = -\sigma \\
&\text{for } 1 \leq k < n \\
&\quad \delta_{+,k} = s_k + \delta_k \\
&\quad \ell_{+,k} = (\delta_k \ell_k)/\delta_{+,k} \\
&\quad s_{k+1} = \ell_{+,k} \ell_k s_k - \sigma \\
&\delta_{+,n} = s_n + \delta_n
\end{aligned}
$$

Similarly, we let

$$
\mathbf{U}_- = \begin{bmatrix} 1 & \upsilon_{-,1} & & \\ & 1 & \ddots & \\ & & \ddots & \upsilon_{-,n-1} \\ & & & 1 \end{bmatrix}, \quad \mathbf{D}_- = \begin{bmatrix} \delta_{-,1} & & & \\ & \delta_{-,2} & & \\ & & \ddots & \\ & & & \delta_{-,n} \end{bmatrix},
$$

so that we can compute the entries of $\mathbf{U}_-$ and $\mathbf{D}_-$ by the differential form of the **progressive qd transform**:

Algorithm 1.3.9 (dqds)

$$
\begin{aligned}
&p_n = \delta_n - \sigma \\
&\text{for } n > k \geq 1 \\
&\quad \delta_{-,k+1} = \delta_k \ell_k^2 + p_{k+1} \\
&\quad t = \delta_k/\delta_{-,k+1} \\
&\quad \upsilon_{-,k} = \ell_k \, t \\
&\quad p_k = p_{k+1} t - \sigma \\
&\delta_{-,1} = p_1
\end{aligned}
$$

We might also compute a **twisted factorization** of the form

$$
\mathbf{L}\mathbf{D}\mathbf{L}^\top - \mathbf{I}\widehat{\lambda} = \mathbf{N}_k \mathbf{D}_k \mathbf{N}_k^\top \equiv \begin{bmatrix} \mathbf{L}_k & \mathbf{e}_k \upsilon_k \mathbf{e}_1^\top \\ \mathbf{0} & \mathbf{U}_k \end{bmatrix} \begin{bmatrix} \mathbf{D}_{k,1} & \\ & \mathbf{D}_{k,2} \end{bmatrix} \begin{bmatrix} \mathbf{L}_k^\top & \mathbf{0} \\ \mathbf{e}_1 \upsilon_k \mathbf{e}_k^\top & \mathbf{U}_k \end{bmatrix}, \tag{1.35}
$$

where $\mathbf{L}_k$ is $k \times k$ unit lower bidiagonal and $\mathbf{U}_k$ is unit upper bidiagonal.

Dhillon and Parlett [54, p. 884ff] show a small value of

$$
\gamma_k \equiv \mathbf{e}_k^\top \mathbf{D}_{k,1} \mathbf{e}_k
$$

indicates that $\mathbf{L}\mathbf{D}\mathbf{L}^\top - \mathbf{I}\widehat{\lambda}$ is nearly singular. If the dstqds Algorithm 1.3.8 and the dqds Algorithm 1.3.9 have been performed, then for a particular choice of k the twisted factorization is given by

$$\mathbf{L}_k = \mathbf{L}_{+,1} \;,\; \mathbf{U}_k = \mathbf{U}_{-,2} \;,\; \mathbf{D}_{k,2} = \mathbf{D}_{-,2} \text{ and } \mathbf{D}_{k,1} = \mathbf{D}_{+,1} - \mathbf{e}_k \delta_{-,k+1} \upsilon_k{}^2 \mathbf{e}_k{}^\top \;.$$

Thus the only new quantity is γ_k, which should be computed by the following algorithm

Algorithm 1.3.10 (dtwqds)

if $k < n$ then

$\quad \gamma_k = s_k + p_{k+1}(\delta_k / \delta_{-,k+1})$

else

$\quad \gamma_k = s_n + \delta_n$

The index k is chosen to make $|\gamma_k|$ as small as possible.

Given a twisted factorization (1.35), we can easily solve for $\mathbf{z}^{(k)}$ in the linear system

$$\mathbf{N}_k \mathbf{D}_k \mathbf{N}_k{}^\top \mathbf{z}^{(k)} = \mathbf{e}_k \gamma_k \;.$$

We will see that the solution process involves only multiplication. In partitioned form, this linear system is

$$\begin{bmatrix} \mathbf{L}_k & \mathbf{e}_k \upsilon_k \mathbf{e}_1{}^\top \\ & \mathbf{U}_k \end{bmatrix} \begin{bmatrix} \mathbf{D}_{k,1} & \\ & \mathbf{D}_{k,2} \end{bmatrix} \begin{bmatrix} \mathbf{L}_k{}^\top & \\ \mathbf{e}_1 \upsilon_k \mathbf{e}_k{}^\top & \mathbf{U}_k{}^\top \end{bmatrix} \begin{bmatrix} \mathbf{z}_1^{(k)} \\ \mathbf{z}_2^{(k)} \end{bmatrix} = \begin{bmatrix} \mathbf{e}_k \gamma_k \\ \mathbf{0} \end{bmatrix} .$$

We can "invert" the two left-hand matrices in this equation to get

$$\begin{bmatrix} \mathbf{L}_k{}^\top & \\ \mathbf{e}_1 \upsilon_k \mathbf{e}_k{}^\top & \mathbf{U}_k{}^\top \end{bmatrix} \begin{bmatrix} \mathbf{z}_1^{(k)} \\ \mathbf{z}_2^{(k)} \end{bmatrix} = \begin{bmatrix} \mathbf{e}_k \\ \mathbf{0} \end{bmatrix} ,$$

since the kth diagonal entry of $\mathbf{D}_{k,1}$ is γ_k. If $\mathbf{D}_{k,1}$ and $\mathbf{D}_{k,2}$ are invertible, then we will back-solve

$$\mathbf{L}_k{}^\top \mathbf{z}_1^{(k)} = \mathbf{e}_k$$

and forward-solve

$$\mathbf{U}_k{}^\top \mathbf{z}_2^{(k)} = -\mathbf{e}_1 \upsilon_k \mathbf{e}_k{}^\top \mathbf{z}_1^{(k)}$$

by the following **getvec** algorithm

Algorithm 1.3.11 (getvec)

$$
\begin{aligned}
&\zeta_k = 1 \\
&\text{for } k > j \geq 1 \\
&\quad \zeta_j = -\ell_j \zeta_{j+1} \\
&\text{for } k \leq j < n \\
&\quad \zeta_{j+1} = -\upsilon_j \zeta_j
\end{aligned}
$$

In this algorithm ζ_j is the jth component of $\mathbf{z}^{(k)}$. Dhillon and Parlett [54, p. 866] provide a modification of this algorithm that can be used even when $\mathbf{D}_{k,1}$ and/or $\mathbf{D}_{k,2}$ are singular.

The following result, due to Dhillon and Parlett [54, p. 886], shows that the residual $\left(\mathbf{L}\mathbf{D}\mathbf{L}^\top - \mathbf{I}\widehat{\lambda}\right)\mathbf{z}^{(k)}$ is small for at least one choice of k.

Lemma 1.3.14 *Suppose that $n > 1$ is an integer, that $\mathbf{L}\mathbf{D}\mathbf{L}^\top$ is a symmetric bidiagonal factorization of an $n \times n$ real symmetric tridiagonal matrix. Assume that we are given a real scalar $\widehat{\lambda}$ that is not equal to an eigenvalue of $\mathbf{L}\mathbf{D}\mathbf{L}^\top$. For any integer $k \in [1, n]$, let*

$$
\mathbf{L}\mathbf{D}\mathbf{L}^\top - \mathbf{I}\widehat{\lambda} = \mathbf{N}_k \mathbf{D}_k {\mathbf{N}_k}^\top
$$

be a twisted factorization, and let $\mathbf{z}^{(k)}$ solve

$$
\mathbf{N}_k \mathbf{D}_k {\mathbf{N}_k}^\top \mathbf{z}^{(k)} = \mathbf{e}_k \gamma_k \;,
$$

where γ_k is the kth diagonal entry of $\mathbf{D}_k$. Then for any eigenvalue index j there is an twisted factorization index k so that the eigenvector residual for $\mathbf{z}^{(k)}$ satisfies

$$
\frac{\left\| \left(\mathbf{L}\mathbf{D}\mathbf{L}^\top - \mathbf{I}\widehat{\lambda}\right) \mathbf{z}^{(k)} \right\|_2}{\left\| \mathbf{z}^{(k)} \right\|_2} \leq \sqrt{n} \left| \lambda_j - \widehat{\lambda} \right| \;. \tag{1.36}
$$

This lemma says that a good shift will produce a small value of γ_k in some twisted factorization, and that the eigenvector residual for the corresponding vector $\mathbf{z}^{(k)}$ will be small.

It is easy to see that the Rayleigh quotient for $\mathbf{z}^{(k)}$ is $\gamma_k / \|\mathbf{z}^{(k)}\|_2^2$. Dhillon and Parlett [54, p. 886] show that this Rayleigh quotient can be used to produce a shift that leads to rapid reduction in the eigenvector residual.

In a companion paper [53], Dhillon and Parlett describe how to use the relatively robust representations of a tridiagonal matrix to compute eigenvalues and numerically orthogonal eigenvectors. This problem can be complicated by the presence of eigenvalue clusters, which prevent the selection of a shift to produce a gap between other eigenvalues. These authors choose a shift that is close to the cluster so that for at least one eigenvalue in the cluster the gap relative to the shift exceeds some tolerance. Careful rounding error analysis shows that eigenvectors associated with eigenvalues with large gaps are numerically orthogonal, and that repetition of the algorithm to handle all eigenvalues in one or more clusters also produces numerically orthogonal eigenvectors. Both of the Dhillon and Parlett papers are excellent examples of careful rounding error analysis driving algorithm innovation.

The relatively robust representations are used to develop a **representation tree**. This is a data structure in which each item has a set or list or vector of child items, and a parent item (except for the root item). Each item in the representation tree is associated with a relatively robust representation and some set of eigenvalue indices. The conditional element growth of the relatively robust representation for each representation tree item is estimated, and the child items are selected by gaps in the eigenvalues.

In LAPACK, the eigenvalues and eigenvectors of a symmetric tridiagonal matrix are computed by relatively robust representations in routines `_stevr`. See, for example, dstevr.f. The computational work is actually performed in routine `_stemr`, which calls a number of routines, including routine `_larrv` to determine the representation tree.

1.3.10 Solver Comparison

An interesting comparison of the LAPACK symmetric tridiagonal eigenvalue problem solvers has been performed by Demmel et al. [49]. This technical report is also available online at netlib. These authors drew the following conclusions:

1. Divide and conquer (LAPACK routines `_stedc`) and relatively robust representations (LAPACK routines `_stemr`) are generally much faster than *QL*/*QR* (LAPACK routines `_steqr`) and bisection and inverse iteration (LAPACK routines `_stebz` and `_stein`) on large matrices.
2. The relatively robust representation routines `_stemr` almost always perform the fewest floating point operations, but at a lower floating point operation rate than all the other algorithms.
3. The exact performance of the relatively robust representation routines `_stemr` and the divide and conquer routines `_stedc` strongly depends on the matrix at hand.
4. The divide and conquer routines `_stedc` and *QL*/*QR* routines `_steqr` are the most accurate algorithms with observed accuracy $O(\sqrt{n}\varepsilon)$. The accuracy of bisection/inverse iteration routines `_stebz`/`_stein` and the relative robust representation routines `_stemr` is generally $O(n\varepsilon)$.
5. The relatively robust representation routines `_stemr` are preferable to the bisection/inverse iteration routines `_stebz`/`_stein` for finding a subset of the full set of eigenvalues.

We should also remark that if a given symmetric tridiagonal matrix $\mathbf{T}$ is positive, then the preferred LAPACK routine for finding its eigenvalues and eigenvectors is `_pteqr`. See, for example, dpteqr.f. This routine performs a Cholesky factorization of $\mathbf{T}$, calls `_bdsqr` to compute the singular values of the Cholesky factor, and then squares the singular values to obtain the eigenvalues of $\mathbf{T}$. We will discuss the singular value decomposition in Sect. 1.5.

Exercise 1.3.25 Form the 1000×1000 matrix

$$\mathbf{T} = \begin{bmatrix} 0 & 1/2 & & \\ 1/2 & 0 & \ddots & \\ & \ddots & \ddots & 1/2 \\ & & 1/2 & 0 \end{bmatrix}$$

corresponding to the Chebyshev polynomials.

1. Use LAPACK bisection routine `dstebz` combined with the inverse iteration routine `dstein` to compute all of the eigenvalues $\boldsymbol{\Lambda}$ and eigenvectors $\mathbf{X}$ of $\mathbf{T}$.
2. Use LAPACK QL/QR routine `dsteqr` to compute all of the eigenvalues $\boldsymbol{\Lambda}$ and eigenvectors $\mathbf{X}$ of $\mathbf{T}$. Measure the time required to compute the eigenvalues and eigenvectors, and compute $\|\mathbf{T}\mathbf{X} - \mathbf{X}\boldsymbol{\Lambda}\|_F$.
3. Use LAPACK divide and conquer routine `dstedc` to compute all of the eigenvalues $\boldsymbol{\Lambda}$ and eigenvectors $\mathbf{X}$ of $\mathbf{T}$. Measure the time required to compute the eigenvalues and eigenvectors, and compute $\|\mathbf{T}\mathbf{X} - \mathbf{X}\boldsymbol{\Lambda}\|_F$.
4. Use LAPACK relatively robust representation routine `dstevr` to compute all of the eigenvalues $\boldsymbol{\Lambda}$ and eigenvectors $\mathbf{X}$ of $\mathbf{T}$. Measure the time required to compute the eigenvalues and eigenvectors, and compute $\|\mathbf{T}\mathbf{X} - \mathbf{X}\boldsymbol{\Lambda}\|_F$.
5. Compare these LAPACK routines for accuracy and speed.

1.3.11 Householder Tridiagonalization

In Sects. 1.3.5–1.3.9, we developed a variety of algorithms for computing the eigenvalues and eigenvectors of a real symmetric tridiagonal matrix. Our goal in this section is to show that any Hermitian matrix can be transformed to symmetric tridiagonal form by a unitary similarity transformation. In fact, the unitary matrix can be chosen to be a product of Householder reflectors, computed by Algorithm 6.7.2 in Chap. 6 of Volume I. Let us describe this process in more detail.

We begin by partitioning

$$\mathbf{A} = \begin{bmatrix} \alpha & \mathbf{a}^H \\ \mathbf{a} & \mathbf{B} \end{bmatrix} .$$

We choose the $(n-1) \times (n-1)$ Householder reflector $\widetilde{\mathbf{H}}_1$ so that

$$\widetilde{\mathbf{H}}_1{}^H \mathbf{a} = \pm \mathbf{e}_{n-1} \|\mathbf{a}\|_2 .$$

Lemma 6.7.4 in Chap. 6 of Volume I shows that $\widetilde{\mathbf{H}}_1$ is unitary. We let

$$\mathbf{H}_1 \quad \begin{bmatrix} 1 & \\ & \mathbf{H}_1 \end{bmatrix}$$

Then

$$\mathbf{H}_1{}^H \mathbf{A} \mathbf{H}_1 = \begin{bmatrix} \alpha & \mathbf{a}^H \widetilde{\mathbf{H}}_1 \\ \widetilde{\mathbf{H}}_1^H \mathbf{a} & \widetilde{\mathbf{H}}_{n-1}{}^H \mathbf{B} \widetilde{\mathbf{H}}_1 \end{bmatrix} = \begin{bmatrix} \alpha & \pm\|\mathbf{a}\|_2 \mathbf{e}_1{}^H \\ \pm\mathbf{e}_1 \|\mathbf{a}\|_2 & \widetilde{\mathbf{H}}_1{}^H \mathbf{B} \widetilde{\mathbf{H}}_1 \end{bmatrix} .$$

This matrix is Hermitian, with a real entry in the first sub-diagonal position, and zeros below the first sub-diagonal in the first column.

We continue, in a similar fashion, to zero out entries below the first sub-diagonal in the other columns. Suppose that after k steps we have found Householder reflectors $\mathbf{H}_k, \ldots, \mathbf{H}_1$ so that

$$\mathbf{H}_k{}^H \cdot \ldots \cdot \mathbf{H}_1{}^H \mathbf{A} \mathbf{H}_1 \cdot \ldots \cdot \mathbf{H}_k = \begin{bmatrix} \mathbf{T}_k & \mathbf{e}_k \nu_k \mathbf{e}_1{}^H \\ \mathbf{e}_1 \nu_k \mathbf{e}_k{}^H & \mathbf{C} \end{bmatrix} ,$$

where $\mathbf{T}_k$ is a $k \times k$ real symmetric tridiagonal matrix, ν_k is real and $\mathbf{C}$ is Hermitian. We partition

$$\mathbf{C} = \begin{bmatrix} \alpha & \mathbf{a}^H \\ \mathbf{a} & \mathbf{B} \end{bmatrix}$$

and find a Householder reflector $\widetilde{\mathbf{H}}_{k+1}$ so that

$$\widetilde{\mathbf{H}}_{k+1}{}^H \mathbf{a} = \pm\mathbf{e}_1 \|\mathbf{a}\|_2 .$$

Then we let

$$\mathbf{H}_{k+1} = \begin{bmatrix} \mathbf{I} & \\ & \widetilde{\mathbf{H}}_{k+1} \end{bmatrix}$$

and note that

$$\begin{aligned} &\mathbf{H}_{k+1}{}^H \mathbf{H}_k{}^H \cdot \ldots \cdot \mathbf{H}_1{}^H \mathbf{A} \mathbf{H}_1 \cdot \ldots \cdot \mathbf{H}_k \mathbf{H}_{k+1} \\ &= \begin{bmatrix} \mathbf{I} & & \\ & 1 & \\ & & \widetilde{\mathbf{H}}_{k+1}{}^H \end{bmatrix} \begin{bmatrix} \mathbf{T}_k & \mathbf{e}_k \nu_k & \mathbf{0} \\ \nu_k \mathbf{e}_k{}^H & \alpha & \mathbf{a}^H \\ \mathbf{0} & \mathbf{a} & \mathbf{B} \end{bmatrix} \begin{bmatrix} \mathbf{I} & & \\ & 1 & \\ & & \widetilde{\mathbf{H}}_{k+1} \end{bmatrix} \\ &= \begin{bmatrix} \mathbf{T}_k & \mathbf{e}_k \nu_k & \mathbf{0} \\ \nu_k \mathbf{e}_k{}^H & \alpha & \pm\|\mathbf{a}\|_2 \mathbf{e}_1{}^H \\ \mathbf{0} & \pm\mathbf{e}_1 \|\mathbf{a}\|_2 & \widetilde{\mathbf{H}}_{k+1}{}^H \mathbf{B} \widetilde{\mathbf{H}}_{k+1} \end{bmatrix} . \end{aligned}$$

This matrix is Hermitian, with real sub-diagonal entries in the first $k+1$ entries, and zeros below the sub-diagonal in these columns.

Note that when $k = n-2$, there is no need to zero entries below the sub-diagonal. However, the Householder reflector $\widetilde{\mathbf{H}}_{n-1}$ may still be useful, because it produces a real sub-diagonal entry in this case.

At each step of the algorithm we need to compute a term of the form

$$\mathbf{H}^H\mathbf{B}\mathbf{H} = \left(\mathbf{I} - \mathbf{u}\tau\mathbf{u}^H\right)\mathbf{B}\left(\mathbf{I} - \mathbf{u}\tau\mathbf{u}^H\right) = \mathbf{B} - \mathbf{B}\mathbf{u}\tau\mathbf{u}^H - \mathbf{u}\tau\mathbf{u}^H\mathbf{B} + \mathbf{u}\tau^2\mathbf{u}^H\mathbf{B}\mathbf{u}\mathbf{u}^H \; .$$

We can compute this matrix as follows:

$$\mathbf{y} = \mathbf{B}\mathbf{u}\tau \;,\; \omega = \mathbf{u}\cdot\mathbf{y}\tau/2 \text{ and } \mathbf{H}^H\mathbf{B}\mathbf{H} = \mathbf{B} - (\mathbf{y} - \mathbf{u}\omega)\,\mathbf{u}^H - \mathbf{u}(\mathbf{y} - \mathbf{u}\omega)^H \; .$$

Thus the Householder tridiagonalization process can be performed by the following

Algorithm 1.3.12 (Householder Tridiagonalization)

for $1 \leq k \leq n-2$
  find $\mathbf{u}$ and τ so that $\left(\mathbf{I} - \mathbf{u}\tau\mathbf{u}^H\right)\mathbf{A}_{k+1:n,k} = \mathbf{e}_1\varrho_k$ /* Algorithm 6.7.2 in Chap. 6 of Volume I */
  $\mathbf{y} = \mathbf{A}_{k+1:n,k+1:n}\mathbf{u}\tau$ /* LAPACK BLAS routine _gemv */
  $\omega_k = \mathbf{u}\cdot\mathbf{y}\;\tau/2$ /* LAPACK BLAS routine _dot or _dotc */
  $\mathbf{A}_{k+1:n,k+1:n} = \mathbf{A}_{k+1:n,k+1:n} - \mathbf{y}\mathbf{u}^H - \mathbf{u}\mathbf{y}^H$ /* LAPACK BLAS routine _ syr2k or _her2k */

In LAPACK, reduction to symmetric tridiagonal form is accomplished by routines `_sytrd` (for real matrices) or `_hetrd` (for complex matrices). See, for example, dsytrd.f. This routine uses block organization to make use of Level 3 BLAS routines. The work for individual blocks is performed in routines `_sytd2` or `_hetd2`, which involve code similar to the discussion above.

1.3.12 Lanczos Process

In this section, we will present a second method for reducing a Hermitian matrix to tridiagonal form. This method is called the **Lanczos process** [117]. The idea is to find successive columns of an unitary matrix $\mathbf{U}$ so that

$$\mathbf{A}\mathbf{U} = \mathbf{U}\mathbf{T}_n$$

where

$$\mathbf{T}_n = \begin{bmatrix} \alpha_1 & \beta_1 & & \\ \beta_1 & \alpha_2 & \ddots & \\ & \ddots & \ddots & \beta_{n-1} \\ & & \beta_{n-1} & \alpha_n \end{bmatrix}$$

is real and tridiagonal.

We can choose any unit vector $\mathbf{u}_1$ for the first column of $\mathbf{U}$. Then the first diagonal entry of $\mathbf{T}_n$ must be

$$\alpha_1 = \mathbf{e}_1{}^H \mathbf{T}_n \mathbf{e}_1 = \mathbf{e}_1{}^H \mathbf{U}^H \mathbf{A} \mathbf{U} \mathbf{e}_1 = \mathbf{u}_1{}^H \mathbf{A} \mathbf{u}_1$$

It is easy to see that this equation implies that

$$\mathbf{v}_1 \equiv \mathbf{A}\mathbf{u}_1 - \mathbf{u}_1 \alpha_1 \perp \mathbf{u}_1 \; .$$

We take

$$\beta_1 = \|\mathbf{v}_1\|_2 \; .$$

If $\beta_1 \neq 0$, then we can define

$$\mathbf{u}_2 = \mathbf{v}_1 \frac{1}{\beta_1} \; ,$$

and find that

$$\mathbf{A}\mathbf{U}\mathbf{e}_1 = \mathbf{A}\mathbf{u}_1 = \mathbf{u}_1 \alpha_1 + \mathbf{u}_2 \beta_1 = \mathbf{U}\mathbf{T}_n \mathbf{e}_1 \; .$$

If $\beta_1 = 0$, then we can choose $\mathbf{u}_2$ to be any unit vector orthogonal to $\mathbf{u}_1$.

Inductively, suppose that we have found $k-1$ columns of $\mathbf{T}_n$ and k columns of $\mathbf{U}$. In other words, we have found two sets of real scalars $\{\alpha_1, \ldots, \alpha_{k-1}\}$ and $\{\beta_1, \ldots, \beta_{k-1}\}$, and orthonormal vectors $\mathbf{u}_1, \ldots, \mathbf{u}_k$ so that for $1 \leq i \leq k-1$

$$\mathbf{A}\mathbf{U}\mathbf{e}_i = \mathbf{U}\mathbf{T}_k \mathbf{e}_i \; .$$

Here $\mathbf{U} = [\mathbf{u}_1, \ldots, \mathbf{u}_k]$ and

$$\mathbf{T}_k = \begin{bmatrix} \alpha_1 & \beta_1 & & \\ \beta_1 & \alpha_2 & \ddots & \\ & \ddots & \ddots & \beta_{k-1} \\ & & \beta_{k-1} & \alpha_k \end{bmatrix} .$$

We define

$$\alpha_k = \mathbf{u}_k{}^H \mathbf{A} \mathbf{u}_k \; , \; \mathbf{v}_k = \mathbf{A}\mathbf{u}_k - \mathbf{u}_k \alpha_k - \mathbf{u}_{k-1} \beta_{k-1} \text{ and } \beta_k = \|\mathbf{v}_k\|_2 \; .$$

Then it is easy to see that for $1 \leq i \leq k$, we have $\mathbf{v}_k \perp \mathbf{u}_i$. In fact, for $1 \leq i \leq k-2$ we get

$$\mathbf{u}_i{}^H \mathbf{v}_k = \mathbf{u}_i{}^H \mathbf{A} \mathbf{u}_k = \mathbf{u}_k{}^H \mathbf{A} \mathbf{u}_i = \mathbf{u}_k{}^H \mathbf{U}\mathbf{T} \mathbf{e}_i = \mathbf{u}_k{}^H (\mathbf{u}_{i-1} \beta_{i-1} + \mathbf{u}_i \alpha_i + \mathbf{u}_{i+1} \beta_i) = 0 \; ,$$

while for $i = k - 1$, we find that

$$\mathbf{u}_{k-1}{}^H\mathbf{v}_k = \mathbf{u}_{k-1}{}^H\mathbf{A}\mathbf{u}_k - \beta_{k-1} = \mathbf{u}_k{}^H\mathbf{A}\mathbf{u}_{k-1} - \beta_{k-1} = \mathbf{u}_k{}^H\mathbf{U}\mathbf{T}\mathbf{e}_{k-1} - \beta_{k-1}$$
$$= \mathbf{u}_k{}^H(\mathbf{u}_{k-2}\beta_{k-2} + \mathbf{u}_{k-1}\alpha_{k-1} + \mathbf{u}_k\beta_{k-1}) - \beta_{k-1} = 0 \; ,$$

and finally for $i = k$, we see that

$$\mathbf{u}_k{}^H\mathbf{v}_k = \mathbf{u}_k{}^H\mathbf{A}\mathbf{u}_k - \alpha_k = 0 \; .$$

If $\beta_k \neq 0$, we let

$$\mathbf{u}_{k+1} = \mathbf{v}_k\frac{1}{\beta_k} \; ;$$

otherwise, we let $\mathbf{u}_{k+1}$ be any unit vector orthogonal to $\mathbf{u}_1, \ldots, \mathbf{u}_k$.

We can summarize the Lanczos process for an Hermitian matrix $\mathbf{A}$ with the following algorithm.

Algorithm 1.3.13 (Lanczos Process)

given $\mathbf{u}_1$ with $\|\mathbf{u}_1\|_2 = 1$
for $1 \leq k \leq n$
  $\mathbf{w}_k = \mathbf{A}\mathbf{u}_k$
  $\alpha_k = \mathbf{u}_k \cdot \mathbf{w}_k$
  if $k < n$
    $\mathbf{v}_k = \mathbf{w}_k - \mathbf{u}_k\alpha_k$
    $\beta_k = \|\mathbf{v}_k\|_2$
    if $\beta_k \neq 0$
      $\mathbf{u}_{k+1} = \mathbf{v}_k/\beta_k$
    else
      choose a unit vector $\mathbf{u}_{k+1} \perp \mathbf{u}_1, \ldots, \mathbf{u}_k$

The previous algorithm proves the following theorem.

Theorem 1.3.5 (Lanczos) *If* $\mathbf{A}$ *is an* $n \times n$ *Hermitian matrix and* $\mathbf{u}_1$ *is a unit* n*-vector, then there is an* $n \times n$ *unitary matrix* $\mathbf{U}$ *so that* $\mathbf{U}\mathbf{e}_1 = \mathbf{u}_1$ *and* $\mathbf{T} = \mathbf{U}^H\mathbf{A}\mathbf{U}$ *is real, symmetric and tridiagonal. If the subdiagonal entries of* $\mathbf{T}$ *are nonzero, then the matrix* $\mathbf{U}$ *is uniquely determined by the choice of the first column of* $\mathbf{U}$ *and the choice of the signs of the off-diagonal entries of* $\mathbf{T}$.

Example 1.3.7 Let us apply the Lanczos process to

$$\mathbf{A} = \begin{bmatrix} 4 & 3 & 2 & 1 \\ 3 & 4 & 3 & 2 \\ 2 & 3 & 4 & 3 \\ 1 & 2 & 3 & 4 \end{bmatrix} \text{ with } \mathbf{u}_1 = \begin{bmatrix} 1/2 \\ 1/2 \\ 1/2 \\ 1/2 \end{bmatrix} .$$

In the first step we compute

$$\mathbf{A}\mathbf{u}_1 = \begin{bmatrix} 5 \\ 6 \\ 6 \\ 5 \end{bmatrix} , \alpha_1 = \mathbf{u}_1{}^H\mathbf{A}\mathbf{u}_1 = 11 , \mathbf{u}_1 = \mathbf{A}\mathbf{u} - \mathbf{u}_1\alpha_1 = \begin{bmatrix} 1/2 \\ 1/2 \\ 1/2 \\ 1/2 \end{bmatrix} , \beta_1 = \|\mathbf{u}_1\| = 1.$$

At the next step, we compute

$$\mathbf{u}_2 = \mathbf{u}_1\frac{1}{\beta_1} = \begin{bmatrix} -1/2 \\ 1/2 \\ 1/2 \\ -1/2 \end{bmatrix} , \mathbf{A}\mathbf{u}_2 = \begin{bmatrix} 0 \\ 1 \\ 1 \\ 0 \end{bmatrix} , \alpha_2 = \mathbf{u}_2{}^H\mathbf{A}\mathbf{u}_2 = 1 ,$$

$$\mathbf{u}_2 = \mathbf{A}\mathbf{u}_2 - \mathbf{u}_2\alpha_2 - \mathbf{u}_1\beta_1 = \begin{bmatrix} 0 \\ 0 \\ 0 \\ 0 \end{bmatrix} , \beta_2 = \|\mathbf{u}_2\| = 0 .$$

Since $\beta_2 = 0$, we must choose $\mathbf{u}_3 \perp \mathbf{u}_1, \mathbf{u}_2$. We choose

$$\mathbf{u}_3 = \begin{bmatrix} -1/2 \\ 1/2 \\ -1/2 \\ 1/2 \end{bmatrix} , \mathbf{A}\mathbf{u}_3 = \begin{bmatrix} -1 \\ 0 \\ 0 \\ 1 \end{bmatrix} , \alpha_3 = \mathbf{u}_3{}^H\mathbf{A}\mathbf{u}_3 = 1 ,$$

$$\mathbf{u}_3 = \mathbf{A}\mathbf{u}_3 - \mathbf{u}_3\alpha_3 - \mathbf{u}_2\beta_2 = \begin{bmatrix} -1/2 \\ -1/2 \\ 1/2 \\ 1/2 \end{bmatrix} , \beta_3 = \|\mathbf{u}_3\| = 1 .$$

Finally,

$$\mathbf{u}_4 = \begin{bmatrix} -1/2 \\ -1/2 \\ 1/2 \\ 1/2 \end{bmatrix} , \mathbf{A}\mathbf{u}_4 = \begin{bmatrix} -2 \\ -1 \\ 1 \\ 2 \end{bmatrix} , \alpha_4 = \mathbf{u}_4{}^H\mathbf{A}\mathbf{u}_4 = 3 .$$

Then with $\mathbf{U} = [\mathbf{u}_1, \mathbf{u}_2, \mathbf{u}_3, \mathbf{u}_4]$ we have

$$\mathbf{U}^H\mathbf{A}\mathbf{U} = \mathbf{T} = \begin{bmatrix} \alpha_1 & \beta_1 & & \\ \beta_1 & \alpha_2 & \beta_2 & \\ & \beta_2 & \alpha_3 & \beta_2 \\ & & \beta_3 & \alpha_4 \end{bmatrix} = \begin{bmatrix} 1 & 1 & 1 & \\ & 1 & 1 & 0 \\ & & 0 & 1 & 1 \\ & & & 1 & 3 \end{bmatrix} .$$

For implementations of the Lanczos algorithm, see Netlib for Lanczos routine lanczs, Laso routine diwla, Linalg routine laqmr and underwood, and Napack routine lancz. See Lanczos algorithm for a MATLAB implementation.

Exercise 1.3.26 Use the Lanczos process to tridiagonalize the following matrices. In each case, choose the first vector to be $\mathbf{u}_1 = \mathbf{e}\frac{1}{\sqrt{n}}$.

1.

$$\begin{bmatrix} 1 & 0 & 1 \\ 0 & 1 & 0 \\ 1 & 0 & 1 \end{bmatrix}$$

2.

$$\begin{bmatrix} & & 1 & 2 \\ & & 2 & 1 \\ 1 & 2 & & \\ 2 & 1 & & \end{bmatrix}$$

3.

$$\begin{bmatrix} 2 & -1 & 0 & -1 \\ -1 & 2 & -1 & 0 \\ 0 & -1 & 2 & -1 \\ -1 & 0 & -1 & 2 \end{bmatrix}$$

Exercise 1.3.27 Use the tridiagonalization of the first matrix in the previous exercise to find the eigenvector corresponding to the eigenvalue $\lambda = 1$.

Exercise 1.3.28 First, get a copy of the matrix BCSSTRUC4. Next, reduce this matrix to symmetric tridiagonal form, using an appropriate LAPACK routine. Afterward, use the Lanczos process to reduce BCSSTRUC4 to tridiagonal form. Which tridiagonalization process is faster? Which tridiagonalization process could take advantage of the sparsity of the nonzero entries of BCSSTRUC4?

1.3.13 Symmetric Eigenvalue Problems

LAPACK basically provides four algorithms for finding the eigenvalues or eigenvectors of an Hermitian matrix.

- Routines `_syev` or `_heev` call `_steqr` to perform the Pal-Walker-Kahan version of the *QL* and *QR* algorithms; see Sect. 1.3.7.
- Routines `_syevd` or `_heevd` call `_stedc` to perform the divide and conquer algorithm; see Sect. 1.3.8.
- Routines `_syevr` or `_heevr` use the relatively robust representation in routines `_stemr`; if necessary, they also call `_stebz` to perform bisection and `_stein` to perform inverse iteration; see Sects. 1.3.9, 1.3.5 and 1.3.6.
- Routines `_syevx` or `_heevx` call `_steqr` to perform the Pal-Walker-Kahan version of the *QL* and *QR* algorithms; if necessary, they also call `_stebz` to perform bisection and `_stein` to perform inverse iteration.

All of these routines begin by scaling the matrix, and then calling `_sytrd` or `_hetrd` to reduce the matrix to real symmetric tridiagonal form:

$$\mathbf{U}^H\mathbf{A}\mathbf{U} = \mathbf{T} .$$

Once the symmetric tridiagonal eigenvalue problem has been solved, we have an orthogonal matrix $\mathbf{Q}$ and a real diagonal matrix $\boldsymbol{\Lambda}$ so that

$$\mathbf{T}\mathbf{Q} = \mathbf{Q}\boldsymbol{\Lambda} .$$

These two equations imply that

$$\mathbf{A}(\mathbf{U}\mathbf{Q}) = (\mathbf{U}\mathbf{Q})\boldsymbol{\Lambda} .$$

In other words, the eigenvectors of $\mathbf{A}$ are the columns of the matrix $\mathbf{U}\mathbf{Q}$, which can be computed by successively multiplying $\mathbf{Q}$ by the Householder reflectors that were used to determine $\mathbf{U}$. In LAPACK, this matrix is typically assembled in routines `_ormtr` or `_unmtr`.

The MATLAB function eig will compute the eigenvalues and eigenvectors of a symmetric matrix. For sparse symmetric matrices, use eigs. Although the MATLAB command `eig` is the same command used to find the eigenvalues of a general matrix, MATLAB apparently performs a symmetry check on the matrix before choosing the algorithm to use.

The GSL routines gsl_eigen_symm and gsl_eigen_symmv will compute the eigenvalues and the eigenvectors of a real symmetric matrix. For a complex Hermitian matrix, use gsl_eigen_herm and gsl_eigen_hermv.

Exercise 1.3.29 First, get a copy of the matrix BCSSTRUC4. Next, use an appropriate LAPACK routine to find the smallest eigenvalue and corresponding eigenvector of this matrix. How can you measure the accuracy in the eigenvalue and eigenvector? Afterward, use an appropriate LAPACK routine to find all of the eigenvalues of BCSSTRUC4. How does the smallest eigenvalue obtained by this routine compare to the previously determined smallest eigenvalue?

1.4 General Matrices

So far, we have examined some basic theory for eigenvalue problems in Sect. 1.2, and developed several sophisticated algorithms for finding eigenvalues and eigenvectors of Hermitian matrices in Sect. 1.3. We have postponed the study of general eigenvalue problems until now, because this subject is significantly more difficult. New theory in Sect. 1.4.1.1 will show that it is possible for some square matrices to have at most one eigenvector for each distinct eigenvalue; in other words, a repeated eigenvalue of a non-Hermitian matrix may have an eigenvector deficiency. While all Hermitian matrices have a unitary similarity transformation to diagonal form, in Sect. 1.4.1.3 we will see that for any square matrix there is a unitary similarity transformation to right triangular form; the eigenvalues of the original matrix will be evident on the diagonal of the triangular matrix. Also recall that for Hermitian matrices, our eigenvalue algorithms began by performing a unitary similarity transformation to real symmetric tridiagonal form. For general matrices, the eigenvalue algorithms will begin by performing a unitary similarity transformation to upper Hessenberg form, which is described in Sect. 1.4.8.

For general matrices, the most effective scheme for computing eigenvalues is the QR algorithm, described in Sect. 1.4.8.4. It is based on the power method, and requires some clever manipulation to discover complex eigenvalues while performing real arithmetic on real matrices. Before developing this algorithm, we will develop some important theory for eigenvalues of general square matrices, including a perturbation analysis in Sect. 1.4.2.

1.4.1 Theory

In this section, we will examine some of the difficulties that can occur in searching for eigenvalues and eigenvectors of general matrices. Knowledge of these difficulties will guide us in trying to find simpler forms for the eigenvalue problems, much as we were able to reduce the symmetric eigenvalue problem to tridiagonal form.

1.4.1.1 Eigenvector Deficiency

Recall that the Fundamental Theorem of Algebra 1.2.1 showed that the characteristic polynomial $\det(\mathbf{A}-\mathbf{I}\lambda)$ for an $n \times n$ matrix $\mathbf{A}$ with complex entries has n complex roots, counting multiplicity. Also recall that Corollary 1.2.1 showed that every $n \times n$

matrix **A** has n complex eigenvalues, counting multiplicity. As the next example will show, this does not necessarily mean that **A** has n linearly independent eigenvectors.

Example 1.4.1 Consider the matrix

$$\mathbf{J} = \begin{bmatrix} 1 & 1 \\ 0 & 1 \end{bmatrix} .$$

The characteristic polynomial of **J** is $\det(\mathbf{J} - \mathbf{I}\lambda) = (1 - \lambda)^2$. This quadratic has a double root $\lambda = 1$. In order to find a corresponding eigenvector, we want to find $\mathbf{z} \in \mathscr{N}(\mathbf{J} - \mathbf{I}\lambda)$ where

$$\mathbf{J} - \mathbf{I}\lambda = \begin{bmatrix} 1-\lambda & 1 \\ 0 & 1-\lambda \end{bmatrix} = \begin{bmatrix} 0 & 1 \\ 0 & 0 \end{bmatrix} .$$

Since $\mathscr{N}(\mathbf{J} - \mathbf{I}\lambda)$ is the span of the first axis vector $\mathbf{e}_1$, there is only one linearly independent eigenvector corresponding to the double eigenvalue $\lambda = 1$.

1.4.1.2 Eigenvalues

Example 1.4.1 suggests the following definition.

Definition 1.4.1 An $n \times n$ matrix **A** is **defective** if and only if it has an eigenvalue of multiplicity k with fewer than k linearly independent eigenvectors. Otherwise, **A** is said to be **non-defective**.

The following lemma is somewhat analogous to Lemma 1.3.1 for Hermitian matrices.

Lemma 1.4.1 *Suppose that the $n \times n$ matrix **A** has k eigenvectors $\mathbf{z}_1, \ldots, \mathbf{z}_k$ corresponding to k distinct eigenvalues $\lambda_1, \ldots, \lambda_k$. Then the set $\{\mathbf{z}_1, \ldots, \mathbf{z}_k\}$ is linearly independent.*

Proof We will prove the claim by induction. The claim is trivial for $k = 1$, since a set consisting of a single nonzero vector is linearly independent. Inductively, suppose that the claim is true for up to $k-1$ eigenvectors corresponding to distinct eigenvalues. Let $\mathbf{z}_1, \ldots, \mathbf{z}_k$ be eigenvectors corresponding to distinct eigenvalues $\lambda_1, \ldots, \lambda_k$ and suppose that some linear combination of these vectors is zero:

$$\mathbf{0} = \sum_{j=1}^{k-1} \mathbf{z}_j \alpha_j + \mathbf{z}_k \alpha_k .$$

Then multiplication by **A** gives us

$$\mathbf{0} = \sum_{j=1}^{k-1} \mathbf{z}_j \lambda_j \alpha_j + \mathbf{z}_k \lambda_k \alpha_k .$$

If we multiply the former equation by λ_k and subtract the latter equation, we obtain

$$\mathbf{0} = (\sum_{j=1}^{k-1} \mathbf{z}_j\alpha_j)\lambda_k - \sum_{j=1}^{k-1} \mathbf{z}_j\lambda_j\alpha_j = \sum_{j=1}^{k-1} \mathbf{z}_j\alpha_j(\lambda_k - \lambda_j) \ .$$

Since $\{\mathbf{z}_1, \ldots, \mathbf{z}_{k-1}\}$ is linearly independent and the λ_j are distinct, the inductive hypothesis says that we must have $\alpha_j = 0$ for $1 \leq j \leq k-1$. We are left with $\mathbf{z}_k\alpha_k = \mathbf{0}$; since $\mathbf{z}_k \neq \mathbf{0}$ this implies that $\alpha_k = 0$. In other words, the set $\{\mathbf{z}_1, \ldots, \mathbf{z}_k\}$ is linearly independent. This completes the inductive proof of the claim.

Lemma 1.4.1 has the following useful corollary.

Corollary 1.4.1 *If all of the eigenvalues of a square matrix* $\mathbf{A}$ *are distinct, then there is a nonsingular matrix* $\mathbf{Z}$ *and a diagonal matrix* $\boldsymbol{\Lambda}$ *so that* $\mathbf{AZ} = \mathbf{Z}\boldsymbol{\Lambda}$.

Proof If the $n \times n$ matrix $\mathbf{A}$ has n distinct eigenvalues $\lambda_1, \ldots, \lambda_n$, then it has n eigenvectors $\mathbf{z}_1, \ldots, \mathbf{z}_n$. Lemma 1.4.1 shows that these eigenvectors are linearly independent, and Lemma 3.2.9 in Chap. 3 of Volume I shows that the matrix $\mathbf{Z} = \begin{bmatrix} \mathbf{z}_1 & \ldots & \mathbf{z}_n \end{bmatrix}$ is nonsingular. If we define

$$\boldsymbol{\Lambda} = \begin{bmatrix} \lambda_1 & & \\ & \ddots & \\ & & \lambda_n \end{bmatrix}$$

then $\mathbf{AZ} = \mathbf{Z}\boldsymbol{\Lambda}$.

However, we can obtain a stronger result that is somewhat analogous to the Spectral Theorem 1.3.1.

Theorem 1.4.1 *Suppose that* $\mathbf{A}$ *is a square matrix. Then* $\mathbf{A}$ *is non-defective if and only if there is a nonsingular matrix* $\mathbf{Z}$ *such that* $\mathbf{Z}^{-1}\mathbf{AZ}$ *is diagonal.*

Proof Let us prove that a matrix is similar to a diagonal matrix if and only if it is non-defective. Suppose that $\mathbf{Z}^{-1}\mathbf{AZ} = \boldsymbol{\Lambda}$ where $\boldsymbol{\Lambda}$ is diagonal and $\mathbf{Z}$ is nonsingular. Then $\mathbf{AZ} = \mathbf{Z}\boldsymbol{\Lambda}$, so each column of $\mathbf{Z}$ is an eigenvector of $\mathbf{A}$. Thus $\mathbf{A}$ is non-defective.

Next, we will show that a non-defective matrix is similar to a diagonal matrix. Suppose that $\mathbf{A}$ is non-defective. Let $\lambda_1, \ldots, \lambda_r$ denote the distinct eigenvalues of $\mathbf{A}$, and let $m_1, \ldots, m_r$ denote their multiplicities. Thus $n = \sum_{i=1}^r m_r$. Because $\mathbf{A}$ is non-defective, λ_i has m_i linearly independent eigenvectors given by the columns of some $n \times m_i$ matrix $\mathbf{Z}_i$. Thus for all $1 \leq i \leq r$ we see that $\mathscr{N}(\mathbf{Z}_i) = \{\mathbf{0}\}$, and $\mathbf{AZ}_i = \mathbf{Z}_i\lambda_i$. Define

$$\mathbf{Z} = [\mathbf{Z}_1, \ldots, \mathbf{Z}_r] \text{ and } \boldsymbol{\Lambda} = \begin{bmatrix} \mathbf{I}\lambda_1 & & \\ & \ddots & \\ & & \mathbf{I}\lambda_r \end{bmatrix} .$$

Then $\mathbf{Z}$ and $\boldsymbol{\Lambda}$ are $n \times n$ matrices, and $\mathbf{AZ} = \mathbf{Z}\boldsymbol{\Lambda}$.

All that remains is to show that the square matrix $\mathbf{Z}$ is nonsingular. To do so, we will show that $\mathscr{N}(\mathbf{Z}) = \{\mathbf{0}\}$. Suppose that

$$\mathbf{0} = \mathbf{Z}\mathbf{c} = [\mathbf{Z}_1 \; \ldots \; \mathbf{Z}_r] \begin{bmatrix} \mathbf{c}_1 \\ \vdots \\ \mathbf{c}_r \end{bmatrix} = \sum_{i=1}^{r} \mathbf{Z}_i \mathbf{c}_i \; .$$

We will prove by contradiction that all of the vectors $\mathbf{c}_i$ are zero. If one or more of the vectors $\mathbf{c}_i$ is nonzero, we may reorder the indices so that $\mathbf{c}_i = \mathbf{0}$ for $k < i \le r$. Then each vector $\mathbf{Z}_i\mathbf{c}_i$ for $1 \le i \le k$ is an eigenvector of $\mathbf{A}$ with eigenvalue λ_i, and $\sum_{i=1}^{k} \mathbf{Z}_i\mathbf{c}_i$ is a linear combination of linearly independent eigenvectors of $\mathbf{A}$ with distinct eigenvalues. However, Lemma 1.4.1 proved that eigenvectors corresponding to distinct eigenvalues are linearly independent, so the assumption that any $\mathbf{c}_i$ is nonzero leads to a contradiction. Since each $\mathbf{Z}_i$ has linearly independent columns and since $\mathbf{Z}_i\mathbf{c}_i = \mathbf{0}$ for all i, we conclude that $\mathbf{c}_i = \mathbf{0}$ for all i. This shows that $\mathscr{N}(\mathbf{Z}) = \{\mathbf{0}\}$.

Theorem 1.4.1 suggests the following definition.

Definition 1.4.2 A matrix square $\mathbf{A}$ is **diagonalizable** if and only if there is a nonsingular matrix $\mathbf{Z}$ so that $\mathbf{Z}^{-1}\mathbf{A}\mathbf{Z}$ is diagonal.
With this definition, Theorem 1.4.1 says that $\mathbf{A}$ is non-defective if and only if it is diagonalizable.

1.4.1.3 Schur Decomposition

Corollary 1.2.1 showed that any $n \times n$ matrix $\mathbf{A}$ has at least one eigenvalue λ, and Lemma 1.2.1 showed that $\mathbf{A}$ has at least one eigenvector $\mathbf{z}$ corresponding to any eigenvalue λ. The proof of the next theorem shows that we can use this information to simplify the task of finding the other eigenvalues of $\mathbf{A}$.

Theorem 1.4.2 (Schur Decomposition) *For every $n \times n$ matrix $\mathbf{A}$ there is an $n \times n$ unitary matrix $\mathbf{U}$ and an $n \times n$ right-triangular matrix $\mathbf{R}$ so that $\mathbf{U}^H\mathbf{A}\mathbf{U} = \mathbf{R}$.*

Proof We will prove this theorem by induction on the matrix order n. The theorem is trivial for $n = 1$, because in that case we can take $\mathbf{U} = 1$ and $\mathbf{R} = \mathbf{A}$. Inductively, assume that the theorem is true for all matrices of order at most $n - 1$.

Suppose that $\mathbf{z}$ is an eigenvector of an $n \times n$ matrix $\mathbf{A}$ with eigenvalue λ. Since $\mathbf{z} \neq \mathbf{0}$, we can scale $\mathbf{z}$ so that $\|\mathbf{z}\|_2 = 1$. The results in Lemma 6.7.4 in Chap. 6 of Volume I show that we can find a unitary matrix $\mathbf{U}$ with first column equal to $\mathbf{z}$. This means that we can partition

$$\mathbf{U} = \begin{bmatrix} \mathbf{z} & \mathbf{V} \end{bmatrix} .$$

Then

$$\mathbf{A}\mathbf{U} = \mathbf{A}\begin{bmatrix} \mathbf{z} & \mathbf{V} \end{bmatrix} = \begin{bmatrix} \mathbf{z}\lambda & \mathbf{A}\mathbf{V} \end{bmatrix} ,$$

and

$$\begin{bmatrix} \mathbf{z}^H \\ \mathbf{V}^H \end{bmatrix} \mathbf{A} \begin{bmatrix} \mathbf{z} & \mathbf{V} \end{bmatrix} = \begin{bmatrix} \mathbf{z}^H \\ \mathbf{V}^H \end{bmatrix} \begin{bmatrix} \mathbf{z}\lambda & \mathbf{AV} \end{bmatrix} = \begin{bmatrix} \lambda & \mathbf{z}^H\mathbf{AV} \\ \mathbf{0} & \mathbf{V}^H\mathbf{AV} \end{bmatrix} .$$

In this way, we have constructed a unitary change of basis that reduces **A** to a block-triangular matrix. Lemma 1.2.3 shows that the eigenvalues of the block-triangular matrix are either λ or an eigenvector of the $(n-1) \times (n-1)$ matrix $\mathbf{V}^H\mathbf{AV}$. This process of reducing the size of the eigenvalue problem is called **deflation**.

By the inductive hypothesis, there is a unitary matrix W and a right-triangular matrix **R** so that

$$\mathrm{W}^H \mathbf{V}^H \mathbf{AV} \mathrm{W} = \mathbf{R} .$$

Then

$$\mathrm{U} \begin{bmatrix} 1 & \\ & \mathrm{W} \end{bmatrix} = \begin{bmatrix} \mathbf{z} & \mathbf{V} \end{bmatrix} \begin{bmatrix} 1 & \\ & \mathrm{W} \end{bmatrix}$$

is a product of unitary matrices and therefore unitary. Furthermore,

$$\begin{bmatrix} 1 & \\ & \mathrm{W}^H \end{bmatrix} \begin{bmatrix} \mathbf{z}^H \\ \mathbf{V}^H \end{bmatrix} \mathbf{A} \begin{bmatrix} \mathbf{z} & \mathbf{V} \end{bmatrix} \begin{bmatrix} 1 & \\ & \mathrm{W} \end{bmatrix} = \begin{bmatrix} 1 & \\ & \mathrm{W}^H \end{bmatrix} \begin{bmatrix} \lambda & \mathbf{z}^H\mathbf{AV} \\ \mathbf{0} & \mathbf{V}^H\mathbf{AV} \end{bmatrix} \begin{bmatrix} 1 & \\ & \mathrm{W} \end{bmatrix}$$
$$= \begin{bmatrix} \lambda & \mathbf{z}^H\mathbf{AV}\mathrm{W} \\ \mathbf{0} & \mathbf{R} \end{bmatrix}$$

is right-triangular. This completes the induction.

The advantage of the Schur decomposition is that it can be accomplished by unitary matrices, which are very numerically stable for machine computation. The difficulty with the Schur decomposition is that it requires complex arithmetic for real matrices.

Recall from Lemma 1.2.5 that whenever a complex scalar λ is an eigenvalue of a real matrix **A**, then its complex conjugate $\overline{\lambda}$ is also an eigenvalue. Eventually, we will learn how to transform a real matrix to a block right-triangular form, in which the diagonal blocks are 1×1 or 2×2, and the 2×2 blocks have complex conjugate eigenvalues. Let us begin by defining such block right-triangular matrices.

Definition 1.4.3 The square matrix **R** is **right quasi-triangular** if and only if it has the partitioned form

$$\mathbf{R} = \begin{bmatrix} \mathbf{R}_{11} & \mathbf{R}_{12} & \ldots & \mathbf{R}_{1k} \\ & \mathbf{R}_{22} & \ldots & \mathbf{R}_{2k} \\ & & \ddots & \vdots \\ & & & \mathbf{R}_{kk} \end{bmatrix}$$

where each diagonal block $\mathbf{R}_{ii}$ is either 1×1 or 2×2.

The following variation of the Schur Decomposition Theorem 1.4.2 removes the need for complex arithmetic with real matrices.

Theorem 1.4.3 (Real Schur Decomposition) *For every real $n \times n$ matrix* $\mathbf{A}$ *there is an real $n \times n$ orthogonal matrix* $\mathbf{Q}$ *and a real $n \times n$ right* **quasi-triangular matrix** $\mathbf{R}$ *so that* $\mathbf{Q}^\top \mathbf{A} \mathbf{Q} = \mathbf{R}$.

Proof We will prove this theorem by induction on the matrix order n. The theorem is trivial if $n = 1$, because we can take $\mathbf{Q} = 1$ and $\mathbf{R} = \mathbf{A}$. Inductively, we assume that the theorem is true for all matrices of order at most $n - 1$.

Suppose that $\mathbf{z}$ is an eigenvector of an $n \times n$ matrix $\mathbf{A}$ with eigenvalue λ. There are two cases: either λ is real, or its imaginary part is nonzero.

If λ is real, then we have $\mathbf{A}\mathbf{z} = \mathbf{z}\lambda$ and the complex conjugate of this equation gives us $\mathbf{A}\overline{\mathbf{z}} = \overline{\mathbf{z}}\lambda$. We can write $\mathbf{z} = \mathbf{x} + \mathbf{y}i$, where either the real part $\mathbf{x}$ or the imaginary part $\mathbf{y}$ must be nonzero. If

$$\mathbf{0} \neq \mathbf{x} = (\mathbf{z} + \overline{\mathbf{z}})/2 \,,$$

then we have

$$\mathbf{A}\mathbf{x} = \mathbf{A}\left(\mathbf{z} + \overline{\mathbf{z}}\right)/2 = \left(\mathbf{z}\lambda + \overline{\mathbf{z}}\lambda\right)/2 = \mathbf{x}\lambda \,,$$

so $\mathbf{x}$ is a real eigenvector of $\mathbf{A}$. Otherwise,

$$\mathbf{0} \neq \mathbf{y} = (\mathbf{z} - \overline{\mathbf{z}})/2$$

is a real eigenvector of $\mathbf{A}$, because

$$\mathbf{A}\mathbf{y} = \mathbf{A}\left(\mathbf{z} - \overline{\mathbf{z}}\right)/2 = \left(\mathbf{z}\lambda - \overline{\mathbf{z}}\lambda\right)/2 = \mathbf{y}\lambda \,,$$

Thus, without loss of generality we may assume that $\mathbf{z}$ is real.

We can scale the real nonzero vector $\mathbf{z}$ so that $\|\mathbf{z}\|_2 = 1$. The results in Lemma 6.7.3 in Chap. 6 of Volume I show that we can find a orthogonal matrix $\mathbf{U}$ with first column equal to $\mathbf{z}$. Thus, we can partition

$$\mathbf{U} = \begin{bmatrix} \mathbf{z} & \mathbf{V} \end{bmatrix} .$$

Then

$$\mathbf{A}\mathbf{U} = \mathbf{A}\begin{bmatrix} \mathbf{z}\ \mathbf{V} \end{bmatrix} = \begin{bmatrix} \mathbf{z}\lambda\ \mathbf{A}\mathbf{V} \end{bmatrix} ,$$

and

$$\begin{bmatrix} \mathbf{z}^\top \\ \mathbf{V}^\top \end{bmatrix} \mathbf{A}\left[\mathbf{z}\ \mathbf{V}\right] = \begin{bmatrix} \mathbf{z}^\top \\ \mathbf{V}^\top \end{bmatrix} \left[\mathbf{z}\lambda\ \mathbf{A}\mathbf{V}\right] = \begin{bmatrix} \lambda & \mathbf{z}^\top\mathbf{A}\mathbf{V} \\ \mathbf{0} & \mathbf{V}^\top\mathbf{A}\mathbf{V} \end{bmatrix} .$$

In this way, we have constructed a unitary change of basis that reduces $\mathbf{A}$ to a block-triangular matrix. Lemma 1.2.3 shows that the eigenvalues of the block-triangular matrix are either λ or an eigenvector of the $(n-1) \times (n-1)$ matrix $\mathbf{V}^\top \mathbf{A} \mathbf{V}$.

If λ is not real and has eigenvector $\mathbf{z}$, then $\mathbf{A}\mathbf{z} = \mathbf{z}\lambda$ and we can take complex conjugates to obtain $\mathbf{A}\overline{\mathbf{z}} = \overline{\mathbf{z}\lambda}$. Let us write $\mathbf{z} = \mathbf{x} + \mathbf{y}\mathrm{i}$ where $\mathbf{x}$ and $\mathbf{y}$ are real n-vectors. Note that $\mathbf{y}$ must be nonzero, otherwise $\mathbf{A}\mathbf{x} = \mathbf{x}\lambda$ would imply that λ is real. It is also easy to see that $\mathbf{x}$ must be nonzero. If we write $\lambda = \alpha + \mathrm{i}\beta$ where α and β are real, then

$$\mathbf{A}(\mathbf{x} + \mathbf{y}\mathrm{i}) = \mathbf{A}\mathbf{z} = \mathbf{z}\lambda = (\mathbf{x} + \mathbf{y}\mathrm{i})(\alpha + \mathrm{i}\beta) = (\mathbf{x}\alpha - \mathbf{y}\beta) + (\mathbf{y}\alpha + \mathbf{x}\beta)\mathrm{i}\,.$$

The real and imaginary parts of this equation give us

$$\mathbf{A}\begin{bmatrix}\mathbf{x} & \mathbf{y}\end{bmatrix} = \begin{bmatrix}\mathbf{x} & \mathbf{y}\end{bmatrix}\begin{bmatrix}\alpha & \beta \\ -\beta & \alpha\end{bmatrix}\,. \tag{1.37}$$

It is easy to see that the eigenvalues of the 2×2 matrix

$$\mathbf{P} = \begin{bmatrix}\alpha & \beta \\ -\beta & \alpha\end{bmatrix}$$

are $\alpha \pm \mathrm{i}\beta$, which are either λ or $\overline{\lambda}$.

Following the ideas in Sect. 6.7 in Chap. 6 of Volume I, we can perform a Householder QR factorization to find an $n \times n$ real orthogonal matrix $\mathbf{H}$ and a nonsingular 2×2 right triangular matrix $\mathbf{U}$ so that

$$\begin{bmatrix}\mathbf{x} & \mathbf{y}\end{bmatrix} = \mathbf{H}\begin{bmatrix}\mathbf{U} \\ \mathbf{0}\end{bmatrix}\,.$$

Let us partition

$$\mathbf{H}^\top \mathbf{A} \mathbf{H} = \begin{bmatrix}\mathbf{B} & \mathbf{C} \\ \mathbf{D} & \mathbf{E}\end{bmatrix}$$

where $\mathbf{B}$ is 2×2. Then we can multiply Eq. (1.37) on the left by $\mathbf{H}^\top$ to get

$$\begin{bmatrix}\mathbf{B} & \mathbf{C} \\ \mathbf{D} & \mathbf{E}\end{bmatrix}\begin{bmatrix}\mathbf{U} \\ \mathbf{0}\end{bmatrix} = \begin{bmatrix}\mathbf{U} \\ \mathbf{0}\end{bmatrix}\mathbf{P}\,.$$

Since $\mathbf{U}$ is nonsingular, this equation implies that $\mathbf{D} = \mathbf{0}$. Thus we have found an orthogonal matrix $\mathbf{H}$ so that

$$\mathbf{H}^\top \mathbf{A} \mathbf{H} = \begin{bmatrix}\mathbf{B} & \mathbf{C} \\ & \mathbf{E}\end{bmatrix}$$

is block right-triangular. Furthermore,

$$\mathbf{B} = \mathbf{U}\mathbf{P}\mathbf{U}^{-1}$$

is similar to **P**, so its eigenvalues are either λ or $\overline{\lambda}$.

At this point in the proof, if λ is real or not, we have found an orthogonal matrix H so that

$$\mathrm{H}^\top \mathbf{A} \mathrm{H} = \begin{bmatrix} \mathbf{B} & \mathbf{C} \\ & \mathbf{E} \end{bmatrix}$$

where **B** is either 1×1 or 2×2. By the inductive hypothesis, there is a real orthogonal matrix Q and a right quasi-triangular matrix **R** so that

$$\mathrm{Q}^\top \mathbf{E} \mathrm{Q} = \mathbf{R} \,.$$

Then

$$\begin{bmatrix} \mathbf{I} & \\ & \mathrm{Q}^\top \end{bmatrix} \mathrm{H}^\top \mathbf{A} \mathrm{H} \begin{bmatrix} \mathbf{I} & \\ & \mathrm{Q} \end{bmatrix} = \begin{bmatrix} \mathbf{B} & \mathbf{C}\mathrm{Q} \\ & \mathrm{Q}^\top \mathbf{E} \mathrm{Q} \end{bmatrix} = \begin{bmatrix} \mathbf{B} & \mathbf{C}\mathrm{Q} \\ & \mathbf{R} \end{bmatrix}$$

is right quasi-triangular. This completes the induction.

The following classical result is an interesting consequence of the Schur decomposition.

Theorem 1.4.4 (Cayley-Hamilton) *If* **A** *is a square matrix with characteristic polynomial*

$$p(\lambda) = \det(\mathbf{A} - \mathbf{I}\lambda) \,,$$

then

$$p(\mathbf{A}) = \mathbf{0} \,. \tag{1.38}$$

Proof Suppose that **R** is a right-triangular matrix. Example 1.2.7 shows that the eigenvalues of **R** are its diagonal entries, and that its characteristic polynomial is

$$p(\lambda) \equiv \det(\mathbf{R} - \mathbf{I}\lambda) = \prod_{i=1}^{n} (\varrho_{ii} - \lambda) \,.$$

Denote the entries of **R** by ϱ_{ij}. We claim that

$$\mathbf{e}_i{}^\top \prod_{k=i}^{n} (\mathbf{I}\varrho_{kk} - \mathbf{R}) = 0 \text{ for } 1 \le i \le n \,.$$

This claim is obviously true for $i = n$, since the last row of $\mathbf{I}\varrho_{nn} - \mathbf{R}$ is zero. Assume inductively that the claim is true for row indices greater than i. Then

$$\mathbf{e}_i{}^\top \prod_{k=i}^{n} (\mathbf{I}\varrho_{kk} - \mathbf{R}) = \mathbf{e}_i{}^\top (\mathbf{I}\varrho_{ii} - \mathbf{R}) \prod_{k=i+1}^{n} (\mathbf{I}\varrho_{kk} - \mathbf{R})$$

$$= -\sum_{j=i+1}^{n} \varrho_{ij}\mathbf{e}_j{}^\top \prod_{k=i+1}^{n} (\mathbf{I}\varrho_{kk} - \mathbf{R})$$

$$= -\left[\sum_{j=i+1}^{n} \varrho_{ij}\mathbf{e}_j{}^\top \prod_{k=j}^{n} (\mathbf{I}\varrho_{kk} - \mathbf{R})\right] \prod_{k=i+1}^{j-1} (\mathbf{I}\varrho_{kk} - \mathbf{R}) = \mathbf{0} \ .$$

This completes the induction. It follows that

$$\mathbf{e}_i{}^\top p(\mathbf{R}) = \mathbf{e}_i{}^\top \prod_{k=1}^{n} (\mathbf{I}\varrho_{kk} - \mathbf{R}) = \left[\mathbf{e}_i{}^\top \prod_{k=i}^{n} (\mathbf{I}\varrho_{kk} - \mathbf{R})\right] \prod_{k=1}^{i-1} (\mathbf{I}\varrho_{kk} - \mathbf{R})$$

$$= \mathbf{0}^\top \prod_{k=1}^{i-1} (\mathbf{I}\varrho_{kk} - \mathbf{R}) = \mathbf{0}^\top$$

for all $1 \leq i \leq n$. Since the rows of $p(\mathbf{R})$ are all zero, we conclude that $p(\mathbf{R}) = \mathbf{0}$.

Given a general square matrix $\mathbf{A}$, we can use the Schur Decomposition Theorem 1.4.2 to factor $\mathbf{U}^H\mathbf{A}\mathbf{U} = \mathbf{R}$ where $\mathbf{U}$ is unitary and $\mathbf{R}$ is right-triangular. Lemma and Corollary 3.2.3 in Chap. 3 of Volume I now imply that the characteristic polynomial of $\mathbf{A}$ is

$$p(\lambda) = \det(\mathbf{A} - \mathbf{I}\lambda) = \det\left(\mathbf{U}\mathbf{R}\mathbf{U}^H - \mathbf{I}\lambda\right) = \det\left(\mathbf{U}\left[\mathbf{R} - \mathbf{I}\lambda\right]\mathbf{U}^H\right)$$

then Lemma 3.2.17 in Chap. 3 of Volume I yields

$$= \det\left(\mathbf{U}\right)\det\left(\mathbf{R} - \mathbf{I}\lambda\right)\det\left(\mathbf{U}^H\right)$$

and finally Corollary 3.2.3 in Chap. 3 of Volume I gives us

$$= \det\left(\mathbf{R} - \mathbf{I}\lambda\right) \ .$$

In other words, the characteristic polynomial of $\mathbf{A}$ is the characteristic polynomial of $\mathbf{R}$. Consequently,

$$p(\mathbf{A}) = p\left(\mathbf{U}\mathbf{R}\mathbf{U}^H\right) = \prod_{i=1}^{n}\left(\mathbf{I}\varrho_{ii} - \mathbf{U}\mathbf{R}\mathbf{U}^H\right) = \mathbf{U}\prod_{i=1}^{n}\left(\mathbf{I}\varrho_{ii} - \mathbf{R}\right)\mathbf{U}^H = \mathbf{U}p\left(\mathbf{R}\right)\mathbf{U}^H$$

$$= \mathbf{U}\mathbf{0}\mathbf{U}^H = \mathbf{0} \ .$$

The Schur Decomposition Theorem 1.4.2 has the following interesting consequence which can also be found in Stewart [164, p. 284].

Theorem 1.4.5 *Suppose that* $\mathbf{A}$ *is an* $n \times n$ *matrix and* $\varepsilon > 0$*. Then there is a nonsingular matrix* $\mathbf{X}$ *and a diagonal matrix* $\boldsymbol{\Lambda}$ *so that*

$$\left\|\mathbf{X}^{-1}\mathbf{A}\mathbf{X} - \boldsymbol{\Lambda}\right\|_\infty \leq \varepsilon \,.$$

Proof The claim is obvious for $n = 1$, so we will assume that $n > 1$. Using the Schur Decomposition Theorem 1.4.2, we can find a unitary $n \times n$ matrix $\mathbf{U}$ and an $n \times n$ right-triangular matrix $\mathbf{R}$ with entries ϱ_{ij} so that

$$\mathbf{U}^H\mathbf{A}\mathbf{U} = \mathbf{R} \,.$$

Next, choose a scalar $\delta \leq 1$ so that for all $1 \leq i < j \leq n$

$$\delta|\varrho_{ij}|(n-1) \leq \varepsilon \,.$$

Finally, define the diagonal matrix $\mathbf{D} = \operatorname{diag}\left(1, \delta, \ldots, \delta^{n-1}\right)$ and the matrix $\mathbf{X} = \mathbf{U}\mathbf{D}$. Then

$$\mathbf{X}^{-1}\mathbf{A}\mathbf{X} = \mathbf{D}^{-1}\mathbf{R}\mathbf{D} = \begin{bmatrix} \lambda_1 & \varrho_{12}\delta & \varrho_{13}\delta^2 & \ldots & \varrho_{1n}\delta^{n-1} \\ & \lambda_2 & \varrho_{23}\delta & \ldots & \varrho_{2n}\delta^{n-2} \\ & & \lambda_3 & \ldots & \varrho_{3n}\delta^{n-3} \\ & & & \ddots & \vdots \\ & & & & \lambda_n \end{bmatrix} .$$

Further,

$$\left\|\mathbf{X}^{-1}\mathbf{A}\mathbf{X} - \boldsymbol{\Lambda}\right\|_\infty = \max_{1\leq i\leq n} \sum_{j=i+1}^{n} \left|\varrho_{ij}\delta^{j-i}\right| \leq \max_{1\leq i\leq n} \sum_{j=i+1}^{n} \frac{\varepsilon}{n-1}\delta^{j-i-1}$$
$$\leq \frac{\varepsilon}{n-1} \max_{1\leq i\leq n} (n-i) = \varepsilon \,.$$

This theorem implies that for any square matrix $\mathbf{A}$ there is a nonsingular matrix $\mathbf{X}$ and a diagonal matrix $\boldsymbol{\Lambda}$ so that the residual $\mathbf{A}\mathbf{X} - \mathbf{X}\boldsymbol{\Lambda}$ is small:

$$\|\mathbf{A}\mathbf{X} - \mathbf{X}\boldsymbol{\Lambda}\|_\infty = \left\|\mathbf{X}\left(\mathbf{X}^{-1}\mathbf{A}\mathbf{X} - \boldsymbol{\Lambda}\right)\right\|_\infty \leq \|\mathbf{X}\|_\infty \left\|\mathbf{X}^{-1}\mathbf{A}\mathbf{X} - \boldsymbol{\Lambda}\right\|_\infty \leq \|\mathbf{X}\|_\infty\varepsilon$$
$$= \varepsilon\|\mathbf{U}\mathbf{D}\|_\infty \leq \varepsilon\|\mathbf{U}\|_\infty\|\mathbf{D}\|_\infty \leq \varepsilon\sqrt{n} \,.$$

Unfortunately, the condition number of $\mathbf{X}$ could be very large:

$$\begin{aligned}\|\mathbf{X}\|_\infty &= \|\mathbf{U}\mathbf{D}\|_\infty \leq \|\mathbf{U}\|_\infty \|\mathbf{D}\|_\infty \leq \sqrt{n} \text{ and} \\ \left\|\mathbf{X}^{-1}\right\|_\infty &= \left\|\mathbf{D}^{-1}\mathbf{U}^H\right\|_\infty \leq \left\|\mathbf{D}^{-1}\right\|_\infty \left\|\mathbf{U}^H\right\|_\infty \leq \delta^{1-n}\sqrt{n} \\ &\leq \left\{\frac{(n-1)\max_{1\leq i<j\leq n}|\varrho_{ij}|}{\varepsilon}\right\}^{n-1} \sqrt{n}\,.\end{aligned}$$

The Schur decomposition can also be used to bound powers of a matrix, as is shown by the following result from Golub and van Loan [84, p. 370].

Lemma 1.4.2 *Let* $\mathbf{A}$ *be an* $n \times n$ *matrix, and let* $\lambda_{\max}$ *and* $\lambda_{\min}$ *be the largest and smallest absolute values of eigenvalues of* $\mathbf{A}$. *Let* $\mathbf{U}^H\mathbf{A}\mathbf{U} = \mathbf{R}$ *be the Schur decomposition of* $\mathbf{A}$, *described in Theorem 1.4.2, and let*

$$\mathbf{R} = \boldsymbol{\Lambda} + \mathbf{S}$$

where $\boldsymbol{\Lambda}$ *is diagonal and* $\mathbf{S}$ *is strictly upper triangular. Then for all* $\mu \geq 0$ *and for all* $k \geq 0$ *we have*

$$\left\|\mathbf{A}^k\right\|_2 \leq (1+\mu)^{n-1}\left\{|\lambda_{\max}| + \frac{\|\mathbf{S}\|_F}{1+\mu}\right\}^k . \tag{1.39}$$

Furthermore, if $\lambda_{\min} > 0$ *and*

$$\mu > \frac{\|\mathbf{S}\|_F}{\lambda_{\min}} - 1\,, \tag{1.40}$$

then for all $k \geq 0$ *we have*

$$\left\|\mathbf{A}^{-k}\right\|_2 \leq (1+\mu)^{n-1}\left\{|\lambda_{\min}| - \frac{\|\mathbf{S}\|_F}{1+\mu}\right\}^{-k} . \tag{1.41}$$

Proof Given $\mu \geq 0$, define the $n \times n$ diagonal matrix

$$\mathbf{M} = \operatorname{diag}(1, 1+\mu, \ldots, (1+\mu)^{n-1})\,.$$

Then

$$\mathbf{e}_i{}^H\mathbf{M}\mathbf{S}\mathbf{M}^{-1}\mathbf{e}_j = (1+\mu)^{i-1}\mathbf{e}_i{}^H\mathbf{S}\mathbf{e}_j(1+\mu)^{1-j} = (1+\mu)^{i-j}\mathbf{e}_i{}^H\mathbf{S}\mathbf{e}_j\,,$$

and since $\mathbf{S}$ is strictly upper triangular, we have

$$\left\|\mathbf{M}\mathbf{S}\mathbf{M}^{-1}\right\|_F^2 = \sum_{j=2}^{n}\sum_{i=1}^{j-1}\frac{\mathbf{S}_{ij}^2}{(1+\mu)^{2(j-i)}} \leq \frac{1}{(1+\mu)^2}\sum_{j=2}^{n}\sum_{i=1}^{j-1}\mathbf{S}_{ij}^2 = \|\mathbf{S}\|_F^2/(1+\mu)^2\,.$$

It follows that

$$\left\|\mathbf{A}^k\right\|_2 = \left\|\mathbf{R}^k\right\|_2 = \left\|(\Lambda+\mathbf{S})^k\right\|_2 = \left\|\mathbf{M}^{-1}\left[\mathbf{M}(\Lambda+\mathbf{S})\mathbf{M}^{-1}\right]^k\mathbf{M}\right\|_2$$
$$= \left\|\mathbf{M}^{-1}\left[\Lambda+\mathbf{MSM}^{-1}\right]^k\mathbf{M}\right\|_2 \leq \|\mathbf{M}\|_2\left\|\mathbf{M}^{-1}\right\|_2\left\{\|\Lambda\|_2+\left\|\mathbf{MSM}^{-1}\right\|_2\right\}^k$$
$$\leq (1+\mu)^{n-1}\left\{\lambda_{\max}+\frac{\|\mathbf{S}\|_F}{1+\mu}\right\}^k .$$

This proves (1.39).

Next, suppose that $\lambda_{\min} > 0$. Then

$$\left\|\Lambda^{-1}\mathbf{MSM}^{-1}\right\|_2 \leq \frac{\left\|\mathbf{MSM}^{-1}\right\|_2}{\lambda_{\min}} \leq \frac{\left\|\mathbf{MSM}^{-1}\right\|_F}{\lambda_{\min}} \leq \frac{\|\mathbf{S}\|_F}{\lambda_{\min}(1+\mu)} .$$

Choose $\mu > 0$ so that inequality (1.40) is satisfied. Then

$$\left\|\Lambda^{-1}\mathbf{MSM}^{-1}\right\|_2 < 1 ,$$

and

$$\left\|\mathbf{A}^{-k}\right\|_2 = \left\|\mathbf{R}^{-k}\right\|_2 = \left\|\mathbf{M}^{-1}\left[\Lambda+\mathbf{MSM}^{-1}\right]^{-k}\mathbf{M}\right\|_2$$
$$= \left\|\mathbf{M}^{-1}\left[\left(\Lambda+\mathbf{MSM}^{-1}\right)^{-1}\right]^k\mathbf{M}\right\|_2 = \left\|\mathbf{M}^{-1}\left[\left(\mathbf{I}+\Lambda^{-1}\mathbf{MSM}^{-1}\right)^{-1}\Lambda^{-1}\right]^k\mathbf{M}\right\|_2$$

so inequality (3.23) in Chap. 3 of Volume I gives us

$$\leq \|\mathbf{M}\|_2\left\|\mathbf{M}^{-1}\right\|_2\left[\frac{\left\|\Lambda^{-1}\right\|_2}{1-\left\|\Lambda^{-1}\mathbf{MSM}^{-1}\right\|_2}\right]^k \leq (1+\mu)^{n-1}\left\{\frac{1/\lambda_{\min}}{1-\|\mathbf{S}\|_F/[\lambda_{\min}(1+\mu)]}\right\}^k .$$

This inequality is equivalent to the claim (1.41).

In LAPACK, the Schur decomposition of a matrix is computed by routines `_gees` or `_geesx`. See, for example, dgees.f. In MATLAB, the command schur will compute the Schur decomposition of a square matrix.

1.4.1.4 Nonunitary Reductions

The Schur decomposition in Theorem 1.4.2 reveals the eigenvalues, but not the eigenvectors. The latter are harder to find, but can be determined from invariant subspaces associated with distinct eigenvalues. These invariant subspaces can be determined by nonunitary similarity transformation, as shown in the next lemma taken from Golub and van Loan [84, p. 352].

Lemma 1.4.3 *Suppose that the* $n \times n$ *matrix* $\mathbf{T}$ *can be partitioned as*

$$\mathbf{T} = \begin{bmatrix} \mathbf{T}_{11} & \mathbf{T}_{12} \\ & \mathbf{T}_{22} \end{bmatrix} ,$$

where $\mathbf{T}_{11}$ *is* $p \times p$. *For any* $p \times (n-p)$ *matrix* $\mathbf{X}$, *define the linear transformation*

$$\phi(\mathbf{X}) = \mathbf{T}_{11}\mathbf{X} - \mathbf{X}\mathbf{T}_{22} .$$

Then ϕ *is nonsingular if and only if* $\mathbf{T}_{11}$ *and* $\mathbf{T}_{22}$ *have no eigenvalues in common. If* ϕ *is nonsingular, let* $\mathbf{Z}$ *solve*

$$\phi(\mathbf{Z}) = -\mathbf{T}_{12} .$$

Then

$$\begin{bmatrix} \mathbf{I} & -\mathbf{Z} \\ & \mathbf{I} \end{bmatrix} \begin{bmatrix} \mathbf{T}_{11} & \mathbf{T}_{12} \\ & \mathbf{T}_{22} \end{bmatrix} \begin{bmatrix} \mathbf{I} & \mathbf{Z} \\ & \mathbf{I} \end{bmatrix} = \begin{bmatrix} \mathbf{T}_{11} & \\ & \mathbf{T}_{22} \end{bmatrix} .$$

Proof First, we will show that if ϕ is singular, then $\mathbf{T}_{11}$ and $\mathbf{T}_{22}$ have a shared eigenvalue. Suppose that $\phi(\mathbf{X}) = \mathbf{0}$ for some $\mathbf{X} \neq \mathbf{0}$. Let $\mathbf{X}$ have singular value decomposition

$$\mathbf{X} = \mathbf{U} \begin{bmatrix} \boldsymbol{\Sigma} & \mathbf{0} \\ \mathbf{0} & \mathbf{0} \end{bmatrix} \mathbf{V}^H$$

where $\boldsymbol{\Sigma}$ is an $r \times r$ diagonal matrix with positive diagonal entries and both $\mathbf{U}$ and $\mathbf{V}$ are unitary. Since $\mathbf{X} \neq \mathbf{0}$, we must have $r \geq 1$.

Let

$$\mathbf{A} = \mathbf{U}^H\mathbf{T}_{11}\mathbf{U} = \begin{bmatrix} \mathbf{A}_{11} & \mathbf{A}_{12} \\ \mathbf{A}_{21} & \mathbf{A}_{22} \end{bmatrix} \text{ and } \mathbf{B} = \mathbf{V}^H\mathbf{T}_{22}\mathbf{V} = \begin{bmatrix} \mathbf{B}_{11} & \mathbf{B}_{12} \\ \mathbf{B}_{21} & \mathbf{B}_{22} \end{bmatrix}$$

where $\mathbf{A}_{11}$ and $\mathbf{B}_{11}$ are $r \times r$. Then the equation $\mathbf{0} = \phi(\mathbf{X}) = \mathbf{T}_{11}\mathbf{X} - \mathbf{X}\mathbf{T}_{22}$ implies that

$$\begin{bmatrix} \mathbf{A}_{11}\boldsymbol{\Sigma} & \mathbf{0} \\ \mathbf{A}_{21}\boldsymbol{\Sigma} & \mathbf{0} \end{bmatrix} = \begin{bmatrix} \mathbf{A}_{11} & \mathbf{A}_{12} \\ \mathbf{A}_{21} & \mathbf{A}_{22} \end{bmatrix} \begin{bmatrix} \boldsymbol{\Sigma} & \mathbf{0} \\ \mathbf{0} & \mathbf{0} \end{bmatrix} = \begin{bmatrix} \boldsymbol{\Sigma} & \mathbf{0} \\ \mathbf{0} & \mathbf{0} \end{bmatrix} \begin{bmatrix} \mathbf{B}_{11} & \mathbf{B}_{12} \\ \mathbf{B}_{21} & \mathbf{B}_{22} \end{bmatrix} = \begin{bmatrix} \boldsymbol{\Sigma}\mathbf{B}_{11} & \boldsymbol{\Sigma}\mathbf{B}_{12} \\ \mathbf{0} & \mathbf{0} \end{bmatrix} .$$

This equation implies that $\mathbf{A}_{21} = \mathbf{0}$, $\mathbf{B}_{12} = \mathbf{0}$ and $\mathbf{A}_{11} = \boldsymbol{\Sigma}\mathbf{B}_{11}\boldsymbol{\Sigma}^{-1}$ is similar to $\mathbf{B}_{11}$. Since $r \geq 1$, the matrices $\mathbf{A}_{11}$ and $\mathbf{B}_{11}$ must have an eigenvalue in common. Since $\mathbf{A}$ has been shown to be block upper triangular, and $\mathbf{B}$ has been shown to be block lower triangular, the eigenvalues of $\mathbf{A}_{11}$ and $\mathbf{B}_{11}$ are eigenvalues of $\mathbf{A}$ and $\mathbf{B}$, respectively. Since $\mathbf{A}$ is unitarily similar to $\mathbf{T}_{11}$ and $\mathbf{B}$ is unitarily similar to $\mathbf{T}_{22}$, the two diagonal blocks of $\mathbf{T}$ must share an eigenvalue.

Next, we will show that if $\mathbf{T}_{11}$ and $\mathbf{T}_{22}$ have a shared eigenvalue λ, then ϕ is singular. Suppose that $\mathbf{T}_{11}\mathbf{x} = \mathbf{x}\lambda$ and $\mathbf{y}^H\mathbf{T}_{22} = \lambda\mathbf{y}^H$, where $\mathbf{x}$ and $\mathbf{y}$ are nonzero. Then

$$\phi\left(\mathbf{x}\mathbf{y}^H\right) = \mathbf{T}_{11}\mathbf{x}\mathbf{y}^H - \mathbf{x}\mathbf{y}^H\mathbf{T}_{22} = \mathbf{x}\lambda\mathbf{y}^H - \mathbf{x}\lambda\mathbf{y}^H = \mathbf{0} .$$

Finally, if ϕ is nonsingular, then we can solve $\phi(\mathbf{Z}) = -\mathbf{T}_{12}$ for $\mathbf{Z}$ and see that

$$\begin{bmatrix} \mathbf{I} & -\mathbf{Z} \\ & \mathbf{I} \end{bmatrix} \begin{bmatrix} \mathbf{T}_{11} & \mathbf{T}_{12} \\ & \mathbf{T}_{22} \end{bmatrix} \begin{bmatrix} \mathbf{I} & \mathbf{Z} \\ & \mathbf{I} \end{bmatrix} = \begin{bmatrix} \mathbf{I} & -\mathbf{Z} \\ & \mathbf{I} \end{bmatrix} \begin{bmatrix} \mathbf{T}_{11} & \mathbf{T}_{11}\mathbf{Z} + \mathbf{T}_{12} \\ & \mathbf{T}_{22} \end{bmatrix}$$
$$= \begin{bmatrix} \mathbf{T}_{11} & \mathbf{T}_{11}\mathbf{Z} + \mathbf{T}_{12} - \mathbf{Z}\mathbf{T}_{22} \\ & \mathbf{T}_{22} \end{bmatrix} = \begin{bmatrix} \mathbf{T}_{11} & \\ & \mathbf{T}_{22} \end{bmatrix} .$$

This lemma can easily be generalized to reduce any given matrix to **block diagonal** form, with the diagonal blocks corresponding to distinct eigenvalues.

Lemma 1.4.3 also suggests the following definition.

Definition 1.4.4 Suppose that $\mathbf{A}$ is an $m \times m$ matrix and $\mathbf{B}$ is an $n \times n$ matrix. Then the **separation** of $\mathbf{A}$ and $\mathbf{B}$ is

$$\text{sep}(\mathbf{A}, \mathbf{B}) = \min_{\substack{\mathbf{X}_{m\times n} \\ \mathbf{X}\neq\mathbf{0}}} \frac{\|\mathbf{A}\mathbf{X} - \mathbf{X}\mathbf{B}\|_F}{\|\mathbf{X}\|_F} . \tag{1.42}$$

Given an $m \times p$ matrix $\mathbf{A}$, $q \times n$ matrix $\mathbf{B}$ and $m \times n$ matrix $\mathbf{C}$, an equation for the $p \times q$ matrix $\mathbf{X}$ of the form

$$\mathbf{A}\mathbf{X} - \mathbf{X}\mathbf{B} = \mathbf{C}$$

is called a **Sylvester equation**. In the special case where $\mathbf{A}$ and $\mathbf{B}$ are square matrices in quasi-triangular form (see Definition 1.4.3), the Sylvester equation can be solved by LAPACK routines `_trsyl`. See, for example, dtrsyl.f. It is also possible to solve the Sylvester equation in MATLAB.

1.4.1.5 Jordan Form

For Hermitian matrices, we found that there are unitary similarity transformations to tridiagonal and diagonal forms. In Sect. 1.3.11 we saw that the former is easy to compute via Householder transformations, while the Spectral Theorem 1.3.1 indicates that the latter requires that we find all of the eigenvectors and eigenvalues of the matrix. For non-defective (not necessarily Hermitian) matrices, Theorem 1.4.1 showed that there are nonsingular similarity transformations to diagonal form. For general matrices, Theorem 1.4.2 shows that there is a unitary similarity transformation to right-triangular form, although this similarity transformation is not easy to compute because it involves finding all of the eigenvalues. In this section, we will discuss a "simpler" canonical form to which we can reduce any matrix by a similarity transformation. This canonical form is very popular in mathematics, but is unstable in floating point arithmetic.

We require the following definition before we present the theorem and some examples.

Definition 1.4.5 A **Jordan block J** has the form $\mathbf{J} = \mathbf{I}\lambda + \mathbf{N}$ where $\mathbf{N}$ is the **nilpotent** matrix

$$\mathbf{N} = \begin{bmatrix} 0 & 1 & & \\ & 0 & \ddots & \\ & & \ddots & 1 \\ & & & 0 \end{bmatrix} .$$

Lemma 1.4.4 *The nullspace of the nilpotent matrix* $\mathbf{N}$ *is the span of the first axis vector* $\mathbf{e}_1$*. Consequently, an* $n \times n$ *Jordan block* $\mathbf{J} = \mathbf{I}\lambda + \mathbf{N}$ *has only the single eigenvector* $\mathbf{e}_1$ *corresponding to its eigenvalue* λ*, which has multiplicity* n*.*

Proof It is easy to see that $\mathbf{e}_1 \in \mathscr{N}(\mathbf{N})$. Since for any vector $\mathbf{z}$ we have

$$\mathbf{N}\mathbf{z} = \begin{bmatrix} \zeta_2 \\ \vdots \\ \zeta_n \\ 0 \end{bmatrix} ,$$

it is also easy to see that $\mathbf{N}\mathbf{z} = \mathbf{0}$ implies that the second through last entries of $\mathbf{z}$ are zero.

The characteristic polynomial $\det(\mathbf{I}\lambda + \mathbf{N}) = 0$ shows that λ is the only eigenvalue of $\mathbf{J}$, and that it has multiplicity n. Since the nullspace of $\mathbf{J} - \mathbf{I}\lambda$ is the span of $\mathbf{e}_1$, there is only one eigenvector of $\mathbf{J}$ corresponding to the eigenvalue λ.

A proof of the following theorem may be found in Halmos [92, p. 113] or Herstein [94, p. 258].

Theorem 1.4.6 (Jordan Decomposition) *Suppose that* $\mathbf{A}$ *is an* $n \times n$ *matrix. Assume that its characteristic polynomial has been factored in the form*

$$\det(\mathbf{I}\lambda - \mathbf{A}) = \prod_{i=1}^{r} (\lambda - \lambda_i)^{n_i} ,$$

where λ_i *are distinct eigenvalues of* $\mathbf{A}$*. Also assume that its* **minimal polynomial** *(i.e., the polynomial* $\mu(\lambda)$ *of minimum degree so that* $\mu(\mathbf{A}) = \mathbf{0}$*) has been factored in the form*

$$\mu(\mathbf{A}) = \prod_{i=1}^{r} (\lambda - \lambda_i)^{m_i} .$$

Then there is a nonsingular $n \times n$ matrix $\mathbf{X}$ and an $n \times n$ **block diagonal** *matrix $\mathbf{J}$ so that*

$$\mathbf{A}\mathbf{X} = \mathbf{X}\mathbf{J} ,$$

where we can partition

$$\mathbf{J} = \begin{bmatrix} \mathbf{J}_1 & & \\ & \ddots & \\ & & \mathbf{J}_r \end{bmatrix}$$

and the only eigenvalue of $\mathbf{J}_i$ is λ_i. Furthermore, each diagonal block $\mathbf{J}_i$ can be partitioned as

$$\mathbf{J}_i = \begin{bmatrix} \mathbf{J}_{i,1} & & \\ & \ddots & \\ & & \mathbf{J}_{i,k_i} \end{bmatrix}$$

where each $\mathbf{J}_{i,j}$ is a Jordan block corresponding to a distinct linearly independent eigenvector of $\mathbf{A}$. For each eigenvalue λ_i of $\mathbf{A}$, there is at least one Jordan block $\mathbf{J}_{i,j}$ of order m_i, all the other Jordan blocks for this eigenvalue have order at most m_i, and the sum over j of the orders of all of the Jordan blocks $\mathbf{J}_{i,j}$ is n_i.

Here is a rough outline of the proof of the Jordan Decomposition Theorem. Note that the Jordan Decomposition Theorem is trivial if the matrix is non-defective, because in that case the matrix is diagonalizable. Further, note that defective matrices must have at least one eigenvalue of multiplicity greater than 1, since matrices with distinct eigenvalues are diagonalizable. Thus, it is sufficient to consider defective matrices with only one eigenvalue. We can subtract a multiple of the identity from such matrices and consider matrices with only zero for an eigenvalue. The remainder of the proof involves a technical discussion about the dimensions of the ranges of powers of the matrix. Instead of going through the details, we will present several examples.

Example 1.4.2 The matrix

$$\mathbf{J} = \begin{bmatrix} 2 & 1 & & & & & \\ & 2 & & & & & \\ & & 2 & 1 & & & \\ & & & 2 & 1 & & \\ & & & & 2 & & \\ & & & & & 4 & 1 \\ & & & & & & 4 \end{bmatrix}$$

is block-diagonal, and each block is a Jordan block. Note that the same eigenvalue appears in multiple Jordan blocks.

Example 1.4.3 Consider

$$\mathbf{A} = \begin{bmatrix} 2 & 0 & 1 \\ & 2 & 0 \\ & & 2 \end{bmatrix} .$$

Since $\mathbf{A}$ is right-triangular, we know that 2 is the only eigenvalue of $\mathbf{A}$, and thus it has multiplicity 3. Let us find the eigenvectors of $\mathbf{A}$:

$$\mathbf{0} = (\mathbf{A} - \mathbf{I}2)\mathbf{x} = \begin{bmatrix} 0 & 0 & 1 \\ & 0 & 0 \\ & & 0 \end{bmatrix} \begin{bmatrix} \xi_1 \\ \xi_2 \\ \xi_3 \end{bmatrix} \Rightarrow \mathbf{x} = \mathbf{e}_1 \text{ or } \mathbf{e}_2 .$$

Since $\mathbf{A}$ has only two linearly independent eigenvectors, $\mathbf{A}$ is defective. Thus, $\mathbf{A}$ is not similar to a diagonal matrix; it must have a non-trivial Jordan canonical form.

Since any Jordan block has only one eigenvector, the Jordan canonical form for $\mathbf{A}$ must involve a 1×1 block and a 2×2 block. We can choose the Jordan canonical form to be

$$\mathbf{J} = \begin{bmatrix} 2 & & \\ & 2 & 1 \\ & & 2 \end{bmatrix} .$$

At this point, we seek vectors $\mathbf{x}_1, \mathbf{x}_2$ and $\mathbf{x}_3$ so that

$$\mathbf{A} \begin{bmatrix} \mathbf{x}_1 & \mathbf{x}_2 & \mathbf{x}_3 \end{bmatrix} = \begin{bmatrix} \mathbf{x}_1 & \mathbf{x}_2 & \mathbf{x}_3 \end{bmatrix} \mathbf{J} .$$

This equation says that

$$\mathbf{A}\mathbf{x}_3 = \mathbf{x}_2 + \mathbf{x}_3 2 \Longleftrightarrow (\mathbf{A} - \mathbf{I}2)\mathbf{x}_3 = \mathbf{x}_2 .$$

Since the range of $\mathbf{A} - \mathbf{I}2$ is the span of $\mathbf{e}_1$, we must have $\mathbf{x}_2 = \mathbf{e}_1$. Then one solution of

$$\mathbf{e}_1 = (\mathbf{A} - \mathbf{I}2)\mathbf{x}_3 = \begin{bmatrix} 0 & 0 & 1 \\ 0 & 0 & 0 \\ 0 & 0 & 0 \end{bmatrix} \begin{bmatrix} \xi_1 \\ \xi_2 \\ \xi_3 \end{bmatrix} = \begin{bmatrix} \xi_3 \\ 0 \\ 0 \end{bmatrix}$$

is $\mathbf{x}_3 = \mathbf{e}_3$. Thus $\mathbf{A} \begin{bmatrix} \mathbf{e}_2 & \mathbf{e}_1 & \mathbf{e}_3 \end{bmatrix} = \begin{bmatrix} \mathbf{e}_2 & \mathbf{e}_1 & \mathbf{e}_3 \end{bmatrix} \mathbf{J}$.

The Jordan Decomposition Theorem has the following useful consequence.

Theorem 1.4.7 *Suppose that* $\mathbf{A}$ *is an* $n \times n$ *matrix and* $\varepsilon > 0$. *Then there is a nonsingular matrix* $\mathbf{X}$ *and a diagonal matrix* $\boldsymbol{\Lambda}$ *so that*

$$\mathbf{U} = \mathbf{X}^{-1}\mathbf{A}\mathbf{X} - \boldsymbol{\Lambda}$$

is upper bi-diagonal, with entries $\upsilon_{i,j}$ *satisfying*

$$\max_{1 \le i < n} |\upsilon_{i,i+1}| \le \varepsilon \; .$$

Proof The claim is obvious for $n = 1$, so we will assume that $n > 1$. Using the Jordan Decomposition Theorem 1.4.6, we can find a nonsingular $n \times n$ matrix $\mathbf{Y}$ and an $n \times n$ Jordan canonical form $\mathbf{J}$ so that

$$\mathbf{Y}^{-1}\mathbf{A}\mathbf{Y} = \mathbf{J} \; .$$

Next, define the diagonal matrix $\mathbf{D} = \operatorname{diag}\left(1, \varepsilon, \ldots, \varepsilon^{n-1}\right)$ and the matrix $\mathbf{X} = \mathbf{Y}\mathbf{D}$. Then

$$\mathbf{X}^{-1}\mathbf{A}\mathbf{X} = \mathbf{D}^{-1}\mathbf{J}\mathbf{D} = \begin{bmatrix} \lambda_1 & \eta_{12}\varepsilon & & & \\ & \lambda_2 & \eta_{23}\varepsilon & & \\ & & \lambda_3 & \ddots & \\ & & & \ddots & \eta_{n-1,n}\varepsilon \\ & & & & \lambda_n \end{bmatrix} .$$

Here $\eta_{i,i+1}$ is a super-diagonal entry of $\mathbf{J}$, and is either zero or one. It follows that the first upper diagonal entries of $\mathbf{U} = \mathbf{X}^{-1}\mathbf{A}\mathbf{X} - \boldsymbol{\Lambda}$ are either 0 or ε.

It is important to note that the Jordan canonical form of a matrix is not stable under arbitrary numerical perturbations. This is because small perturbations to a defective matrix can cause all of the eigenvalues to become distinct, which implies that the perturbed matrix is non-defective. Thus, it is impossible to compute the Jordan form using floating point arithmetic. For more information, see Golub and Wilkinson [85].

However, if a matrix has integer or rational entries, it is possible to compute its Jordan canonical form symbolically. MATLAB is able to find the Jordan canonical form in such a case. Here is a MATLAB session to compute a **Jordan Canonical Form**:

```
>> A = sym('[ 0 -3 1 2; -2 1 -1 2; -2 1 -1 2; -2 -3 1 4]')
>> [X,J]=jordan(A)
```

Unfortunately, some versions of MATLAB do not include the symbolic manipulation package. Instead, readers might be able to run `maple` or `mathematica`.

Here is a Maple session to compute a **Jordan Canonical Form**:

```
> with(linalg):
> A = matrix(4,4,  0 -3 1 2  -2 1 -1 2  -2 1 -1 2  -2 -3 1 4);
> J=jordan(A,'X');
> print(X);
```

1.4.1.6 Companion Matrices

Definition 1.4.6 An $n \times n$ matrix of the form

$$\mathbf{C} = \begin{bmatrix} -\gamma_{n-1} & \dots & -\gamma_1 & -\gamma_0 \\ 1 & \dots & 0 & 0 \\ \vdots & \ddots & \vdots & \vdots \\ 0 & \dots & 1 & 0 \end{bmatrix} \tag{1.43}$$

is called a **companion matrix**.

The interesting feature of companion matrices is that they are easily related to polynomials.

Lemma 1.4.5 *If* $\mathbf{C}$ *is an* $n \times n$ *companion matrix, then*

$$\det(\mathbf{I}\lambda - \mathbf{C}) = \lambda^n + \sum_{j=0}^{n-1} \gamma_j \lambda^j .$$

Proof This is an easy consequence of the Laplace Expansion Theorem 3.2.4 in Chap. 3 of Volume I.

Lemma 1.4.6 *A companion matrix* $\mathbf{C}$ *has at most one eigenvector for any eigenvalue.*

Proof If $\lambda = 0$ is an eigenvalue of a companion matrix $\mathbf{C}$ with eigenvector $\mathbf{x}$, then the form of the companion matrix in Eq. (1.43) implies that all but the last entry of $\mathbf{x}$ must be zero. However, Lemma 1.4.5 implies that $\gamma_0 = 0$, so $\mathbf{x}$ must be the last axis vector. This is the only possible eigenvector for a zero eigenvalue, and any Jordan block corresponding to $\lambda = 0$ must have order equal to the multiplicity of zero as a root of the characteristic polynomial.

If $\lambda \neq 0$ is an eigenvalue of $\mathbf{C}$ with eigenvector $\mathbf{x}$, then at least one component of $\mathbf{x}$ must be nonzero. Let the components of $\mathbf{x}$ be ξ_j for $0 \leq j < n$, and assume that $\mathbf{x}$ has been scaled so that $\xi_j = 1$. Then Eq. (1.43) for the form of a companion matrix, and the eigenvector equation $\mathbf{C}\mathbf{x} = \mathbf{x}\lambda$ imply that

$$\xi_i = \begin{cases} \lambda^{j-i}, \, 0 \leq i \leq j \\ \lambda^{i-j}, \, j \leq i < n \end{cases} .$$

In other words, any eigenvector of $\mathbf{C}$ corresponding to the nonzero eigenvalue λ is a scalar multiple of $\mathbf{z}_\lambda$ where $\mathbf{z}_\lambda{}^\top = \left[\lambda^{n-1} \ldots 1\right]$. Thus there is only one linearly independent eigenvector of $\mathbf{C}$ for any nonzero eigenvalue.

Example 1.4.4 Consider the companion matrix

$$\mathbf{C} = \begin{bmatrix} 7 & -16 & 12 \\ 1 & 0 & 0 \\ 0 & 1 & 0 \end{bmatrix} .$$

Then

$$\det(\mathbf{C} - \mathbf{I}\lambda) = -(\lambda^3 - 7\lambda_2 + 16\lambda - 12) = (2 - \lambda)^2(3 - \lambda) .$$

Thus we can take the Jordan canonical form to be

$$\mathbf{J} = \begin{bmatrix} 2 & 1 & \\ & 2 & \\ & & 3 \end{bmatrix} .$$

The eigenvector $\mathbf{x}_3$ for $\lambda = 3$ solves

$$\mathbf{0} = (\mathbf{C} - \mathbf{I}3)\mathbf{x}_3 = \begin{bmatrix} 4 & -16 & 12 \\ 1 & -3 & 0 \\ 0 & 1 & -3 \end{bmatrix} \begin{bmatrix} \xi_{1,3} \\ \xi_{2,3} \\ \xi_{3,3} \end{bmatrix} .$$

The only linearly independent solution is

$$\mathbf{x}_3 = \begin{bmatrix} 9 \\ 3 \\ 1 \end{bmatrix} .$$

The eigenvector $\mathbf{x}_1$ for $\lambda = 2$ solves

$$\mathbf{0} = (\mathbf{C} - \mathbf{I}2)\mathbf{x}_1 = \begin{bmatrix} 5 & -16 & 12 \\ 1 & -2 & 0 \\ 0 & 1 & -2 \end{bmatrix} \begin{bmatrix} \xi_{1,1} \\ \xi_{2,1} \\ \xi_{3,1} \end{bmatrix} .$$

The only linearly independent solution is

$$\mathbf{x}_1 = \begin{bmatrix} 4 \\ 2 \\ 1 \end{bmatrix}$$

We want to find $\mathbf{X} = \begin{bmatrix} \mathbf{x}_1 & \mathbf{x}_2 & \mathbf{x}_3 \end{bmatrix}$ so that $\mathbf{CX} = \mathbf{XJ}$. All that remains is to determine $\mathbf{x}_2$. The Jordan form implies that $\mathbf{Cx}_2 = \mathbf{x}_1 + \mathbf{x}_2 2$; equivalently, $(\mathbf{C} - \mathbf{I}2)\mathbf{x}_2 = \mathbf{x}_1$. Thus we solve

$$\begin{bmatrix} 5 & -16 & 12 \\ 1 & -2 & 0 \\ 0 & 1 & -2 \end{bmatrix} \begin{bmatrix} \xi_{1,2} \\ \xi_{2,2} \\ \xi_{3,2} \end{bmatrix} = \begin{bmatrix} 4 \\ 2 \\ 1 \end{bmatrix} .$$

One solution is

$$\mathbf{x}_2 = \begin{bmatrix} 8 \\ 3 \\ 1 \end{bmatrix}$$

In summary,

$$\mathbf{CX} = \begin{bmatrix} 7 & -16 & 12 \\ 1 & 0 & 0 \\ 0 & 1 & 0 \end{bmatrix} \begin{bmatrix} 4 & 8 & 9 \\ 2 & 3 & 3 \\ 1 & 1 & 1 \end{bmatrix} = \begin{bmatrix} 4 & 8 & 9 \\ 2 & 3 & 3 \\ 1 & 1 & 1 \end{bmatrix} \begin{bmatrix} 2 & 1 & \\ & 2 & \\ & & 3 \end{bmatrix} = \mathbf{XJ} .$$

Companion matrices can be useful in solving **recurrence relations**. These are equations of the form

$$\eta_k + \sum_{j=0}^{k-1} \gamma_j \eta_j = 0 . \tag{1.44}$$

In order to find general solutions of this equation, we will form the matrix

$$\widehat{\mathbf{C}} = \begin{bmatrix} 0 & 1 & \dots & 0 \\ \vdots & \vdots & \ddots & \vdots \\ 0 & 0 & \dots & 1 \\ -\gamma_0 & -\gamma_1 & \dots & -\gamma_{n-1} \end{bmatrix} .$$

Note that if $\mathbf{P}$ is the permutation matrix such that $\mathbf{Pe}_j = \mathbf{e}_{n-1-j}$ for $0 \le j < n$, then $\mathbf{C} = \mathbf{P}^\top \widehat{\mathbf{C}} \mathbf{P}$ is a companion matrix.

Let us define the polynomial

$$p(\lambda) = \lambda^n + \sum_{j=0}^{n-1} \gamma_j \lambda^j = \det(\mathbf{I}\lambda - \mathbf{C}) = \det(\mathbf{P}^\top [\mathbf{I}\lambda - \mathbf{C}]\mathbf{P}) = \det(\mathbf{I}\lambda - \widehat{\mathbf{C}}) .$$

It is easy to see that if $p(\lambda) = 0$, then $\widehat{\mathbf{C}}$ has eigenvalue λ and the vector $\mathbf{x}$ with components

$$\xi_j = \lambda^j \text{ for } 0 \leq j < n$$

is a corresponding eigenvector. This statement is valid even if $\lambda = 0$, in which case $\mathbf{x} = \mathbf{e}_0$ is the first axis vector.

If $\lambda = 0$ is a zero of $p(\lambda)$ with multiplicity m, then $0 = \gamma_0 = \ldots = \gamma_{m-1}$. Consequently, for $0 \leq k < m$ the sequence $\left\{\eta_j^{(k)}\right\}_{j=0}^{n}$ with

$$\eta_j^{(k)} = \delta_{jk} = \begin{cases} 1, & j = k \\ 0, & \text{otherwise} \end{cases}$$

is a solution of the recurrence relation (1.44). Correspondingly, we can write this fact in the matrix-vector form

$$\widehat{\mathbf{C}}\mathbf{e}_0 = \mathbf{0} \text{ and } \widehat{\mathbf{C}}\mathbf{e}_k = \mathbf{e}_{k-1} \text{ for } 1 \leq k < m \, ,$$

or collectively as

$$\widehat{\mathbf{C}}\left[\mathbf{e}_0, \ldots, \mathbf{e}_{m-1}\right] = \left[\mathbf{e}_0, \ldots, \mathbf{e}_{m-1}\right] \begin{bmatrix} 0 & 1 & \ldots & 0 \\ & \ddots & \ddots & \vdots \\ & & \ddots & 1 \\ & & & 0 \end{bmatrix} .$$

This is of the form $\widehat{\mathbf{C}}\mathbf{X} = \mathbf{X}\mathbf{J}$ where $\mathbf{J}$ is a Jordan block.

If $\lambda \neq 0$ is a zero of $p(\lambda)$ with multiplicity m, then for $0 \leq k < m$, the sequence $\left\{\eta_j^{(k)}\right\}_{j=0}^{n}$ with

$$\eta_j^{(k)} = \binom{j}{k} \lambda^{j-k}$$

is a solution of the recurrence relation (1.44), because

$$\binom{n}{k} \lambda^{n-k} + \sum_{j=0}^{n-1} \gamma_j \binom{j}{k} \lambda^{j-k} = \frac{1}{k!} D^k p(\lambda) = 0$$

for $0 \leq k < m$. Let the vector $\mathbf{x}_k$ have components

$$\xi_j^{(k)} = \binom{j}{k} \lambda^{j-k} \, .$$

Since

$$\binom{j+1}{k} = \binom{j}{k-1} + \binom{j}{k} ,$$

we see that

$$\widehat{\mathbf{C}}\mathbf{x}_0 = \mathbf{x}_0 \lambda \text{ and } \widehat{\mathbf{C}}\mathbf{x}_k = \mathbf{x}_{k-1} + \mathbf{x}_k \lambda \text{ for } 1 \leq k < m .$$

This can be written collectively as

$$\widehat{\mathbf{C}}\left[\mathbf{x}_0, \ldots, \mathbf{x}_{m-1}\right] = \left[\mathbf{x}_0, \ldots, \mathbf{x}_{m-1}\right] \begin{bmatrix} \lambda & 1 & \ldots & 0 \\ & \ddots & \ddots & \vdots \\ & & \ddots & 1 \\ & & & \lambda \end{bmatrix} .$$

This is also of the form $\widehat{\mathbf{C}}\mathbf{X} = \mathbf{X}\mathbf{J}$ where $\mathbf{J}$ is a Jordan block.

We can factor

$$p(\lambda) = \prod_{i=1}^{r} (\lambda - \lambda_i)^{m_i} ,$$

and note that

$$\sum_{i=1}^{r} m_i = n .$$

For each eigenvalue λ_i with multiplicity m_i, we have found m_i linearly independent solutions of the recurrence relation (1.44). The general solution of this recurrence relation is an arbitrary linear combination of the various solutions for distinct eigenvalues. Let us choose scalars $\alpha_{i,\ell}$ for $1 \leq i \leq r$ and $0 \leq \ell < m_i$. If all the roots of $p(\lambda)$ are nonzero, we can write the general solution of the recurrence relation in the form

$$\eta_j = \sum_{i=1}^{r} \sum_{\ell=0}^{m_i-1} \binom{j}{\ell} \lambda^{j-\ell} \alpha_{i,\ell} .$$

Otherwise, without loss of generality we assume that $\lambda_1 = 0$ and write the general solution of the recurrence relation in the form

$$\eta_j = \sum_{\ell=0}^{m_1-1} \delta_{j\ell} \alpha_{1,\ell} + \sum_{i=2}^{r} \sum_{\ell=0}^{m_i-1} \binom{j}{\ell} \lambda^{j-\ell} \alpha_{i,\ell} .$$

Exercise 1.4.1 Find a similarity transformation that diagonalizes the matrix

$$\begin{bmatrix} 1 & 1 \\ 0 & 1+\varepsilon \end{bmatrix} .$$

Discuss the dependence of this transformation on ε as $\varepsilon \to 0$.

Exercise 1.4.2 Determine whether

$$\begin{bmatrix} 0 & 1 & 0 & 0 \\ 0 & 0 & 1 & 0 \\ 0 & 0 & 0 & 1 \\ -1 & -4 & -6 & -4 \end{bmatrix}$$

is diagonalizable.

Exercise 1.4.3 A matrix $\mathbf{A}$ is **normal** if and only if $\mathbf{A}^H\mathbf{A} = \mathbf{A}\mathbf{A}^H$. Prove that a normal matrix that is also triangular is diagonal.

Exercise 1.4.4 Prove that a matrix is normal if and only if there is a unitary matrix $\mathbf{U}$ such that $\mathbf{U}^H\mathbf{A}\mathbf{U}$ is diagonal. (You may want to use the Spectral Theorem 1.3.1.)

Exercise 1.4.5 Prove that permutation matrices are normal.

Exercise 1.4.6 Let $\mathbf{x}$ and $\mathbf{y}$ be n-vectors such that $\mathbf{y} \cdot \mathbf{x} = 1$. Show that there are $n \times n$ matrices $\mathbf{X}$ and $\mathbf{Y}$ so that $\mathbf{X}\mathbf{e}_1 = \mathbf{x}$, $\mathbf{Y}\mathbf{e}_1 = \mathbf{y}$ and $\mathbf{Y}^H = \mathbf{X}^{-1}$.

Exercise 1.4.7 Let λ be an eigenvalue of $\mathbf{A}$ with multiplicity 1. Construct a matrix $\mathbf{X}$ so that

$$\mathbf{X}^{-1}\mathbf{A}\mathbf{X} = \begin{bmatrix} \lambda & \\ & \mathbf{C} \end{bmatrix}$$

is block diagonal. (Hint: Let $\mathbf{A}\mathbf{x} = \mathbf{x}\lambda$ and $\mathbf{A}^H\mathbf{y} = \mathbf{y}\overline{\lambda}$; then use the previous exercise.)

Exercise 1.4.8 Prove that if $\mathbf{U}$ is unitary and triangular, then $\mathbf{U}$ is diagonal.

Exercise 1.4.9 Use the previous exercise and the Schur decomposition to prove that every unitary matrix is diagonalizable.

Exercise 1.4.10 Given $\mathbf{z}$ so that $\mathbf{z}^H\mathbf{A} = \lambda\mathbf{z}^H$, show how to construct unitary matrix $\mathbf{U}$ so that

$$\mathbf{U}^H\mathbf{A}\mathbf{u} = \begin{bmatrix} \lambda & \\ \mathbf{a} & \mathbf{B} \end{bmatrix} .$$

Exercise 1.4.11 Suppose that $\lambda_1, \ldots, \lambda_n$ are the eigenvalues of $\mathbf{A}$, and $\mathbf{x}_1$ is an eigenvector of $\mathbf{A}$ corresponding to λ_1. If $\lambda_1 \neq 0$ and $\mathbf{u}_1 \cdot \mathbf{x}_1 = \lambda_1$, show that the eigenvalues of $\mathbf{A} - \mathbf{u}\mathbf{u}^\top$ are $0, \lambda_2, \ldots, \lambda_n$. (Hint: if $\mathbf{x}_i$ is an eigenvector of $\mathbf{A}$ with eigenvalue λ_i, show that there is a scalar α_i so that $\mathbf{x}_i - \mathbf{x}_1\alpha_i$ is an eigenvector of $\mathbf{A} - \mathbf{u}\mathbf{u}^\top$.)

Exercise 1.4.12 Find the eigenvalues and eigenvectors of the following matrices.

1.
$$\begin{bmatrix} 1 & 1 \\ -1 & 3 \end{bmatrix}$$

2.
$$\begin{bmatrix} 2 & 1 & 0 \\ 0 & 2 & 0 \\ 0 & 0 & -3 \end{bmatrix}$$

3.
$$\begin{bmatrix} 1 & 1 & 0 \\ 0 & 2 & 0 \\ 0 & 0 & -3 \end{bmatrix}$$

4.
$$\begin{bmatrix} 7 & 1 & 2 \\ -1 & 7 & 0 \\ 1 & -1 & 6 \end{bmatrix}$$

5.
$$\begin{bmatrix} 5 & 4 & 3 \\ -1 & 0 & -3 \\ 1 & -2 & 1 \end{bmatrix}$$

6.
$$\begin{bmatrix} 11 & -4 & -5 \\ 21 & -8 & -11 \\ 3 & -1 & 0 \end{bmatrix}$$

Exercise 1.4.13 Find a Jordan canonical form $\mathbf{J}$ for each of the following matrices $\mathbf{A}$, and find a nonsingular matrix $\mathbf{X}$ so that $\mathbf{A}\mathbf{X} = \mathbf{X}\mathbf{J}$.

1.
$$\begin{bmatrix} -3 & 3 & -2 \\ -7 & 6 & -3 \\ 1 & -1 & 2 \end{bmatrix}$$

2.

$$\begin{bmatrix} 0 & 1 & -1 \\ -4 & 4 & -2 \\ -2 & 1 & 1 \end{bmatrix}$$

3.

$$\begin{bmatrix} 0 & -1 & -1 \\ -3 & -1 & -2 \\ 7 & 5 & 6 \end{bmatrix}$$

4.

$$\begin{bmatrix} 0 & -3 & 1 & 2 \\ -2 & 1 & -1 & 2 \\ -2 & 1 & -1 & 2 \\ -2 & -3 & 1 & 4 \end{bmatrix}$$

Exercise 1.4.14 Prove Lemma 1.4.5. (Hint: use expansion by minors.)

Exercise 1.4.15 Let **A** be an $n \times n$ matrix.

1. Prove that $\mathscr{N}\left(\mathbf{A}^k\right) \subset \mathscr{N}\left(\mathbf{A}^{k+1}\right)$ for $1 \leq k$.
2. Prove that if $\text{rank}\left(\mathbf{A}^m\right) = \text{rank}\left(\mathbf{A}^{m+1}\right)$ for some integer $m > 0$, then $\text{rank}\left(\mathbf{A}^k\right) = \text{rank}\left(\mathbf{A}^m\right)$ for all $k \geq m$.
3. Prove that if $\text{rank}\left(\mathbf{A}^m\right) = \text{rank}\left(\mathbf{A}^{m+1}\right)$ for some integer $m > 0$, then $\mathscr{N}\left(\mathbf{A}^k\right) = \mathscr{N}\left(\mathbf{A}^m\right)$ for all $k \geq m$.
4. If λ is an eigenvalue of **A** and $\text{rank}\left((\mathbf{A} - \mathbf{I}\lambda)^m\right) = \text{rank}\left((\mathbf{A} - \mathbf{I}\lambda)^{m+1}\right)$ for some positive integer m, show that

$$\begin{aligned} \mathscr{K}_\lambda &\equiv \{n-\text{vectors } \mathbf{z} : \text{ there exists a positive integer } p \text{ so that } (\mathbf{A} - \mathbf{I}\lambda)^p\mathbf{z} = \mathbf{0}\} \\ &= \mathscr{N}\left((\mathbf{A} - \mathbf{I}\lambda)^m\right) . \end{aligned}$$

1.4.2 Perturbation Analysis

In Sect. 1.3.2, we examined perturbations in eigenvalues and invariant subspaces due to perturbations in an Hermitian matrix. Our goal in this section is to study such perturbations for general square matrices. Again, we will begin by approximating eigenvalue and eigenvector perturbations due to a matrix perturbation. Afterward, we will prove the Bauer-Fike theorem, which bounds perturbations in eigenvalues due to perturbations in non-defective square matrices. We will also use the Schur decomposition to bound eigenvalue perturbations. At the end of this section, we will compute a bound on the distance between invariant subspaces due to matrix perturbations. For additional discussion of perturbations of eigenvalue problems, we recommend Bhatia [17], Golub and van Loan [84, p. 357ff] and Wilkinson [186, Chap. 2]

Let us begin our approximation of eigenvalue and eigenvector perturbations. Suppose that $\mathbf{A}$ is an $n \times n$ matrix. Let the $\mathbf{x}$ be an eigenvector of $\mathbf{A}$ with eigenvalue λ, and assume that $\|\mathbf{x}\|_2 = 1$. Also let $\mathbf{y}$ be an eigenvector of $\mathbf{A}^H$ with eigenvalue $\overline{\lambda}$, again with $\|\mathbf{y}\|_2 = 1$. Suppose that $\mathbf{E}$ is an $n \times n$ matrix with small entries. Assume that there is an n-vector $\widetilde{\mathbf{x}}$ and a scalar $\widetilde{\lambda}$ so that $(\mathbf{A}+\mathbf{E})\widetilde{\mathbf{x}} = \widetilde{\mathbf{x}}\widetilde{\lambda}$. We will use $\mathbf{E}$ to estimate $|\widetilde{\lambda} - \lambda|$ and $\|\widetilde{\mathbf{x}} - \mathbf{x}\|_2$.

As in the proof by deflation of the Schur Decomposition Theorem 1.4.2, we can find an Hermitian unitary matrix $\mathbf{H}$ whose first column is $\mathbf{x}$, and partition

$$\mathbf{H} = \begin{bmatrix} \mathbf{x} & \mathbf{U} \end{bmatrix} .$$

Assume that $\widetilde{\mathbf{x}}$ has been scaled so that $\|\widetilde{\mathbf{x}}\|_2 = 1$. Then

$$\mathbf{H}\widetilde{\mathbf{x}} = \mathbf{H}^H\widetilde{\mathbf{x}} = \begin{bmatrix} \mathbf{x}^H \\ \mathbf{U}^H \end{bmatrix} \widetilde{\mathbf{x}} = \begin{bmatrix} \mathbf{x} \cdot \widetilde{\mathbf{x}} \\ \mathbf{U}^H\widetilde{\mathbf{x}} \end{bmatrix} \equiv \begin{bmatrix} \omega \\ \mathbf{q} \end{bmatrix} ,$$

where

$$\mathbf{q} \equiv \mathbf{U}^H\widetilde{\mathbf{x}} = \mathbf{U}^H(\widetilde{\mathbf{x}} - \mathbf{x})$$

is presumed to be small, and $\omega = \mathbf{x} \cdot \widetilde{\mathbf{x}}$ is presumed to be close to one. Since $\mathbf{H}$ is unitary, we also have

$$\widetilde{\mathbf{x}} = \mathbf{H} \begin{bmatrix} \omega \\ \mathbf{q} \end{bmatrix} = \mathbf{x}\omega + \mathbf{U}\mathbf{q} .$$

If $\omega \neq 0$, we can solve this equation for $\mathbf{x}$ to get

$$\mathbf{x} = (\widetilde{\mathbf{x}} - \mathbf{U}\mathbf{q}) / \omega .$$

If $\mathbf{E}$ and $\mathbf{q}$ are small, ω is close to one and $\widetilde{\lambda}$ is close to λ, then

$$\begin{aligned} \mathbf{x}(\widetilde{\lambda} - \lambda) &= (\widetilde{\mathbf{x}} - \mathbf{U}\mathbf{q}) \frac{\widetilde{\lambda}}{\omega} - \mathbf{x}\lambda = (\mathbf{A}+\mathbf{E})\widetilde{\mathbf{x}}\frac{1}{\omega} - \mathbf{U}\mathbf{q}\frac{\widetilde{\lambda}}{\omega} - \mathbf{A}\mathbf{x} \\ &= (\mathbf{A}+\mathbf{E})(\mathbf{x}+\mathbf{U}\mathbf{q})\frac{1}{\omega} - \mathbf{U}\mathbf{q}\frac{\widetilde{\lambda}}{\omega} - \mathbf{A}\mathbf{x} \approx \mathbf{E}\mathbf{x} + \mathbf{A}\mathbf{U}\mathbf{q} - \mathbf{U}\mathbf{q}\lambda . \end{aligned} \tag{1.45}$$

We can multiply on the left by $\mathbf{y}^H$ and divide by $\mathbf{y} \cdot \mathbf{x}$ to get

$$\widetilde{\lambda} - \lambda \approx \left[\mathbf{y}^H\mathbf{E}\mathbf{x} + (\mathbf{y}^H\mathbf{A} - \lambda\mathbf{y}^H)\mathbf{U}\mathbf{q}\right] \frac{1}{\mathbf{y} \cdot \mathbf{x}} = \frac{\mathbf{y}^H\mathbf{E}\mathbf{x}}{\mathbf{y} \cdot \mathbf{x}} \le \frac{\|\mathbf{E}\|_2}{\mathbf{y} \cdot \mathbf{x}} . \tag{1.46}$$

This is the dominant term in the approximation to the error in the eigenvalue due to the matrix perturbation $\mathbf{E}$. This approximation suggests that we define

$$\kappa(\lambda) \equiv \frac{1}{\mathbf{y} \cdot \mathbf{x}}$$

to be the **condition number of the eigenvalue** λ.

We can also multiply Eq. (1.45) on the left by $\mathbf{U}^H$ to get

$$\mathbf{0} \approx \mathbf{U}^H \mathbf{E} \mathbf{x} + (\mathbf{U}^H \mathbf{A} \mathbf{U} - \mathbf{I}\lambda)\mathbf{q} \; .$$

If λ is not an eigenvalue of the deflated matrix $\mathbf{U}^H \mathbf{A} \mathbf{U}$, then

$$\widetilde{\mathbf{x}} - \mathbf{x} = \mathbf{U}\mathbf{q} \approx -\mathbf{U}(\mathbf{U}^H \mathbf{A} \mathbf{U} - \mathbf{I}\lambda)^{-1} \mathbf{U}^H \mathbf{E} \mathbf{x} \; . \qquad (1.47)$$

Thus the perturbation in the eigenvector may not be small whenever λ has multiplicity greater than one, or is close to another eigenvalue of $\mathbf{A}$.

Example 1.4.5 Let $\mathbf{J} = \mathbf{I} + \mathbf{N}$ where $\mathbf{N}$ is the $n \times n$ nilpotent matrix in Definition 1.4.5. For any $\varepsilon > 0$ let

$$\widetilde{\mathbf{J}} = \mathbf{J} - \mathbf{e}_n \varepsilon \mathbf{e}_1{}^\top \; .$$

Then the characteristic polynomial of $\widetilde{\mathbf{J}}$ is

$$\det\left(\mathbf{I}\lambda - \widetilde{\mathbf{J}}\right) = (\lambda - 1)^n - \varepsilon \; .$$

Thus the eigenvalues of $\widetilde{\mathbf{J}}$ all satisfy

$$\left|\widetilde{\lambda} - 1\right| = \sqrt[n]{\varepsilon} \; .$$

In this example, the only eigenvector of $\mathbf{J}$ is $\mathbf{x} = \mathbf{e}_1$, and the only eigenvector of $\mathbf{J}^H$ is $\mathbf{y} = \mathbf{e}_n$. This means that $\mathbf{y} \cdot \mathbf{x} = 0$, and the perturbation discussion above does not apply to this example.

The eigenvalue $\widetilde{\lambda} = 1 + \sqrt[n]{\varepsilon}$ of $\widetilde{\mathbf{J}}$ has eigenvector

$$\widetilde{\mathbf{x}} = \begin{bmatrix} 1 \\ \varepsilon^{1/n} \\ \vdots \\ \varepsilon^{(n-1)/n} \end{bmatrix} ,$$

where we have scaled $\widetilde{\mathbf{x}}$ so that $\mathbf{x} \cdot \widetilde{\mathbf{x}} = 1$. The particular small perturbation in this example does not produce necessarily small perturbations in either the eigenvalue or the eigenvector.

However, the next theorem shows that better results are possible for diagonalizable matrices.

Theorem 1.4.8 (Bauer-Fike) *[13] Given a square matrix* $\mathbf{A}$*, suppose that there is a nonsingular matrix* $\mathbf{Z}$ *and a diagonal matrix* $\boldsymbol{\Lambda}$ *so that*

$$\mathbf{A}\mathbf{Z} = \mathbf{Z}\boldsymbol{\Lambda} .$$

If $\widetilde{\mathbf{A}}$ *is the same size as* $\mathbf{A}$*, then for any eigenvalue* $\widetilde{\lambda}$ *of* $\widetilde{\mathbf{A}}$ *and for any vector norm index* $p \in [1, \infty]$ *we have*

$$\min_{\lambda \text{ an eigenvalue of } \mathbf{A}} \left|\widetilde{\lambda} - \lambda\right| \leq \|\mathbf{Z}\|_p \left\|\mathbf{Z}^{-1}\right\|_p \left\|\widetilde{\mathbf{A}} - \mathbf{A}\right\|_p .$$

Proof The claim is obviously true if $\widetilde{\lambda}$ is an eigenvalue of $\mathbf{A}$, so we will assume that it is not. If $\widetilde{\lambda}$ is an eigenvalue of $\widetilde{\mathbf{A}}$ but not an eigenvalue of $\mathbf{A}$, then

$$\left(\boldsymbol{\Lambda} - \mathbf{I}\widetilde{\lambda}\right)^{-1} \mathbf{Z}^{-1} \left(\widetilde{\mathbf{A}} - \mathbf{I}\widetilde{\lambda}\right) \mathbf{Z} = \mathbf{I} + \left(\boldsymbol{\Lambda} - \mathbf{I}\widetilde{\lambda}\right)^{-1} \mathbf{Z}^{-1} \left(\widetilde{\mathbf{A}} - \mathbf{A}\right) \mathbf{Z}$$

is singular. The contrapositive of Lemma 3.6.1 in Chap. 3 of Volume I implies that

$$1 \leq \left\|\left(\boldsymbol{\Lambda} - \mathbf{I}\widetilde{\lambda}\right)^{-1} \mathbf{Z}^{-1} \left(\widetilde{\mathbf{A}} - \mathbf{A}\right) \mathbf{Z}\right\|_p \leq \left\|\left(\boldsymbol{\Lambda} - \mathbf{I}\widetilde{\lambda}\right)^{-1}\right\|_p \left\|\mathbf{Z}^{-1}\right\|_p \left\|\widetilde{\mathbf{A}} - \mathbf{A}\right\|_p \|\mathbf{Z}\|_p$$

$$= \max_{1 \leq i \leq n} \frac{1}{\left|\lambda_i - \widetilde{\lambda}\right|} \|\mathbf{Z}\|_p \left\|\mathbf{Z}^{-1}\right\|_p \left\|\widetilde{\mathbf{A}} - \mathbf{A}\right\|_p .$$

Another bound on eigenvalue perturbation may be obtained from the Schur decomposition.

Theorem 1.4.9 (Golub-vanLoan) *[84, p. 358] Suppose that the square matrix* $\mathbf{A}$ *has the Schur decomposition*

$$\mathbf{U}^H \mathbf{A} \mathbf{U} = \mathbf{R}$$

where $\mathbf{U}$ *is unitary and* $\mathbf{R}$ *is right-triangular. Let*

$$\mathbf{R} = \mathbf{D} + \mathbf{N}$$

where $\mathbf{D}$ *is diagonal and* $\mathbf{N}$ *is strictly right-triangular. Let* $|\mathbf{N}|$ *be the matrix of absolute values of components of* $\mathbf{N}$*, and assume that* p *is the smallest integer so that* $|\mathbf{N}|^p = \mathbf{0}$*. If* $\widetilde{\mathbf{A}}$ *is another matrix of the same size as* $\mathbf{A}$*, define*

$$\theta = \|\widetilde{\mathbf{A}} - \mathbf{A}\|_2 \sum_{k=0}^{p-1} \|\mathbf{N}\|_2^k .$$

Then for any eigenvalue $\widetilde{\lambda}$ *of* $\widetilde{\mathbf{A}}$ *we have*

$$\min_{\lambda \text{ an eigenvalue of } \mathbf{A}} |\widetilde{\lambda} - \lambda| \leq \max\{\theta, \theta^{1/p}\} .$$

Proof The claim is obviously true if $\widetilde{\lambda}$ is an eigenvalue of $\mathbf{A}$, so we will consider the case in which it is not. If $\widetilde{\lambda}$ is an eigenvalue of $\widetilde{\mathbf{A}}$ but not an eigenvalue of $\mathbf{A}$, then

$$\left(\mathbf{I}\widetilde{\lambda} - \mathbf{A}\right)^{-1}\left(\widetilde{\mathbf{A}} - \mathbf{I}\widetilde{\lambda}\right) = \mathbf{I} - \left(\mathbf{I}\widetilde{\lambda} - \mathbf{A}\right)^{-1}\left(\widetilde{\mathbf{A}} - \mathbf{A}\right)$$

is singular. The contrapositive of Lemma 3.6.1 in Chap. 3 of Volume I implies that

$$\begin{aligned}
1 &\leq \left\|(\mathbf{I}\widetilde{\lambda} - \mathbf{A})^{-1}(\widetilde{\mathbf{A}} - \mathbf{A})\right\|_2 \leq \left\|\left(\mathbf{I}\widetilde{\lambda} - \mathbf{A}\right)^{-1}\right\|_2 \|\widetilde{\mathbf{A}} - \mathbf{A}\|_2 \\
&= \left\|\mathbf{U}^H\left(\mathbf{I}\widetilde{\lambda} - \mathbf{A}\right)^{-1}\mathbf{U}\right\|_2 \|\widetilde{\mathbf{A}} - \mathbf{A}\|_2 = \left\|\left(\mathbf{I}\widetilde{\lambda} - \mathbf{R}\right)^{-1}\right\|_2 \|\widetilde{\mathbf{A}} - \mathbf{A}\|_2 \\
&= \left\|\left(\left[\mathbf{I}\widetilde{\lambda} - \mathbf{D}\right] - \mathbf{N}\right)^{-1}\right\|_2 \|\widetilde{\mathbf{A}} - \mathbf{A}\|_2 . \qquad (1.48)
\end{aligned}$$

Since $\mathbf{I}\widetilde{\lambda} - \mathbf{D}$ is diagonal and $|\mathbf{N}|_p = \mathbf{0}$, it is easy to check that

$$\left(\left[\mathbf{I}\widetilde{\lambda} - \mathbf{D}\right]^{-1}\mathbf{N}\right)^p = \mathbf{0} .$$

Then the following Neumann series is finite:

$$\begin{aligned}
\left(\left[\mathbf{I}\widetilde{\lambda} - \mathbf{D}\right] - \mathbf{N}\right)^{-1} &= \left(\left[\mathbf{I}\widetilde{\lambda} - \mathbf{D}\right]\left\{\mathbf{I} - \left[\mathbf{I}\widetilde{\lambda} - \mathbf{D}\right]^{-1}\mathbf{N}\right\}\right)^{-1} \\
&= \left\{\mathbf{I} - \left[\mathbf{I}\widetilde{\lambda} - \mathbf{D}\right]^{-1}\mathbf{N}\right\}^{-1}\left[\mathbf{I}\widetilde{\lambda} - \mathbf{D}\right]^{-1} \\
&= \left\{\sum_{k=0}^{p-1}\left(\left[\mathbf{I}\widetilde{\lambda} - \mathbf{D}\right]^{-1}\mathbf{N}\right)^k\right\}\left[\mathbf{I}\widetilde{\lambda} - \mathbf{D}\right]^{-1} .
\end{aligned}$$

If we define

$$\delta = \frac{1}{\left\|\left[\mathbf{I}\widetilde{\lambda} - \mathbf{D}\right]^{-1}\right\|_2} = \min_{\lambda \text{ an eigenvalue of } \mathbf{A}} \left|\widetilde{\lambda} - \lambda\right| ,$$

then

$$\left\| \left(\left[\mathbf{I}\widetilde{\lambda} - \mathbf{D} \right] - \mathbf{N} \right)^{-1} \right\|_2 \leq \left\| \left[\mathbf{I}\widetilde{\lambda} - \mathbf{D} \right]^{-1} \right\|_2 \sum_{k=0}^{p-1} \left\| \left[\mathbf{I}\widetilde{\lambda} - \mathbf{D} \right]^{-1} \mathbf{N} \right\|_2^k$$

$$\leq \frac{1}{\delta} \sum_{k=0}^{p-1} \left(\frac{\|\mathbf{N}\|_2}{\delta} \right)^k . \tag{1.49}$$

We now have two cases. If $\delta > 1$, then the previous inequality implies that

$$\left\| \left(\left[\mathbf{I}\widetilde{\lambda} - \mathbf{D} \right] - \mathbf{N} \right)^{-1} \right\|_2 \leq \frac{1}{\delta} \sum_{k=0}^{p-1} \|\mathbf{N}\|_2^k .$$

substituting this inequality into (1.48) produces

$$1 \leq \left\| \left(\left[\mathbf{I}\widetilde{\lambda} - \mathbf{D} \right] - \mathbf{N} \right)^{-1} \right\|_2 \left\| \widetilde{\mathbf{A}} - \mathbf{A} \right\|_2 \leq \frac{1}{\delta} \left\| \widetilde{\mathbf{A}} - \mathbf{A} \right\|_2 \sum_{k=0}^{p-1} \|\mathbf{N}\|_2^k = \frac{\theta}{\delta} ,$$

from which we conclude that $\delta \leq \theta$.

On the other hand, if $\delta \leq 1$, then inequality (1.49) implies that

$$\left\| \left(\left[\mathbf{I}\widetilde{\lambda} - \mathbf{D} \right] - \mathbf{N} \right)^{-1} \right\|_2 \leq \frac{1}{\delta^p} \sum_{k=0}^{p-1} \|\mathbf{N}\|_2^k .$$

substituting this inequality into (1.48) produces

$$1 \leq \left\| \left(\left[\mathbf{I}\widetilde{\lambda} - \mathbf{D} \right] - \mathbf{N} \right)^{-1} \right\|_2 \left\| \widetilde{\mathbf{A}} - \mathbf{A} \right\|_2 \leq \frac{1}{\delta^p} \left\| \widetilde{\mathbf{A}} - \mathbf{A} \right\|_2 \sum_{k=0}^{p-1} \|\mathbf{N}\|_2^k = \frac{\theta}{\delta^p} ,$$

from which we conclude that $\delta \leq \theta^{1/p}$. These two cases prove the claimed result.

The previous results in this section have bounded errors in eigenvalues. Our next theorem could be used to bound the errors in the invariant subspace associated with one or more eigenvalues. The error in the invariant subspace will be measured by its distance from the true invariant subspace.

Theorem 1.4.10 *Suppose that* $\mathbf{A}$ *and* $\widetilde{\mathbf{A}}$ *are* $n \times n$ *matrices. Assume that*

$$\mathbf{U} = \left[\mathbf{U}_1 \ \mathbf{U}_2 \right]$$

is an $n \times n$ *unitary matrix, and that the range of the* $n \times k$ *matrix* $\mathbf{U}_1$ *is an invariant subspace under* $\mathbf{A}$. *Define*

$$\begin{bmatrix} \mathbf{B}_{11} & \mathbf{B}_{12} \\ \mathbf{0} & \mathbf{B}_{22} \end{bmatrix} = \begin{bmatrix} {\mathbf{U}_1}^H \\ {\mathbf{U}_2}^H \end{bmatrix} \mathbf{A} \left[\mathbf{U}_1 \ \mathbf{U}_2 \right] \text{ and } \begin{bmatrix} \widetilde{\mathbf{B}}_{11} & \widetilde{\mathbf{B}}_{12} \\ \widetilde{\mathbf{B}}_{21} & \widetilde{\mathbf{B}}_{22} \end{bmatrix} = \begin{bmatrix} {\mathbf{U}_1}^H \\ {\mathbf{U}_2}^H \end{bmatrix} \widetilde{\mathbf{A}} \left[\mathbf{U}_1 \ \mathbf{U}_2 \right] .$$

Suppose that

$$\delta = \inf_{\|\mathbf{P}\|_2=1} \|\mathbf{P}\mathbf{B}_{11} - \mathbf{B}_{22}\mathbf{P}\|_2 - \|\widetilde{\mathbf{B}}_{11} - \mathbf{B}_{11}\|_2 - \|\widetilde{\mathbf{B}}_{22} - \mathbf{B}_{22}\|_2 > 0 \ ,$$

and assume that

$$4\|\widetilde{\mathbf{B}}_{21}\|_2 \left(\|\mathbf{B}_{12}\|_2 + \|\widetilde{\mathbf{B}}_{12} - \mathbf{B}_{12}\|_2\right) \leq \delta^2 \ . \tag{1.50}$$

Then there is an $(n-k) \times k$ *matrix* $\mathbf{P}$ *satisfying*

$$\|\mathbf{P}\|_2 \leq 2\frac{\|\widetilde{\mathbf{B}}_{21}\|_2}{\delta} \ ,$$

and such that the columns of $\mathbf{W}_1 = (\mathbf{U}_1 + \mathbf{U}_2\mathbf{P})\left(\mathbf{I} + \mathbf{P}^H\mathbf{P}\right)^{-1/2}$ *are an orthonormal basis for an invariant subspace of* $\widetilde{\mathbf{A}}$*. Furthermore, the distance between the range of* $\mathbf{U}_1$ *and the range of* $\mathbf{W}_1$ *satisfies*

$$dist\left(\mathscr{R}\left(\mathbf{U}_1\right), \mathscr{R}\left(\mathbf{W}_1\right)\right) \leq 2\frac{\left\|\widetilde{\mathbf{B}}_{21} - \mathbf{B}_{21}\right\|_2}{\delta} \ .$$

Proof We can use the Stewart Perturbation Theorem 1.2.3 to show that there is an $(n-k) \times k$ matrix $\mathbf{P}$ satisfying

$$\|\mathbf{P}\|_2 \leq 2\frac{\|\widetilde{\mathbf{B}}_{21}\|_2}{\delta} \ ,$$

and such that the columns of $\mathbf{W}_1 = (\mathbf{U}_1 + \mathbf{U}_2\mathbf{P})\left(\mathbf{I} + \mathbf{P}^H\mathbf{P}\right)^{-1/2}$ form an orthonormal basis for a subspace invariant under $\widetilde{\mathbf{A}}$. Next, recall that Lemma 1.2.17 showed that the distance between the ranges of $\mathbf{U}_1$ and $\mathbf{W}_1$ is

$$\text{dist}\left(\mathscr{R}\left(\mathbf{U}_1\right), \mathscr{R}\left(\mathbf{W}_1\right)\right) = \left\|{\mathbf{U}_2}^H\mathbf{W}_1\right\|_2 = \left\|\mathbf{P}\left(\mathbf{I} + \mathbf{P}^H\mathbf{P}\right)^{-1/2}\right\|_2 \ .$$

If the singular value decomposition of $\mathbf{P}$ is

$$\mathbf{P} = \mathbf{Q}\boldsymbol{\Sigma}\mathbf{V}^H \ ,$$

then it is easy to see that

$$\left\|\mathbf{P}\left(\mathbf{I}+\mathbf{P}^H\mathbf{P}\right)^{-1/2}\right\|_2 = \left\|\mathbf{Q}\boldsymbol{\Sigma}\mathbf{V}^H\left(\mathbf{I} + \mathbf{V}\boldsymbol{\Sigma}^2\mathbf{V}^H\right)^{-1/2}\right\|_2 = \left\|\mathbf{Q}\boldsymbol{\Sigma}\left(\mathbf{I} + \boldsymbol{\Sigma}^2\right)^{-1/2}\mathbf{V}^H\right\|_2$$
$$\leq \left\|\mathbf{Q}\boldsymbol{\Sigma}\mathbf{V}^H\right\|_2 = \|\mathbf{P}\|_2 \ .$$

The claimed inequality bounding the distance between $\mathscr{R}(\mathbf{U}_1)$ and $\mathscr{R}(\mathbf{W}_1)$ now follows easily.

Recall from Lemma 1.2.23 that $\inf_{\|\mathbf{P}\|_2=1} \|\mathbf{P}\mathbf{B}_{11} - \mathbf{B}_{22}\mathbf{P}\|_2$ can be bounded by the minimum difference between eigenvalues of the diagonal blocks $\mathbf{B}_{11}$ and $\mathbf{B}_{22}$. Thus the bound in Theorem 1.4.10 on the distance between the invariant subspaces is proportional to $\|\widetilde{\mathbf{B}}_{21}\|_2$, which measures the amount by which the original invariant subspace fails to be invariant under the perturbed matrix. This bound is also inversely proportional to δ, which is bounded below by the minimum difference of eigenvalues of the original matrix associated with the proposed invariant subspace and any other eigenvalue of $\mathbf{A}$, minus the norm of the perturbation in the diagonal blocks $\mathbf{B}_{11}$ and $\mathbf{B}_{22}$.

1.4.3 Condition Numbers

Next, we would like to discuss condition number estimates for eigenvalues and eigenvectors that are computed by LAPACK. Suppose that λ is an eigenvalue of a quasi-triangular matrix $\mathbf{T}$, with right eigenvector $\mathbf{x}$ and left eigenvector $\mathbf{y}$:

$$\mathbf{T}\mathbf{x} = \mathbf{x}\lambda \text{ and } \mathbf{y}^H\mathbf{T} = \lambda\mathbf{y}^H .$$

Let $\mathbf{E}$ is a perturbation to $\mathbf{T}$ with singular value decomposition

$$\mathbf{E} = \mathbf{U}\boldsymbol{\Sigma}\mathbf{V}^H$$

and maximum singular value $\sigma_{\max}(\mathbf{E})$, then approximation (1.46) implies that the corresponding eigenvalue $\widetilde{\lambda}$ of $\mathbf{T}+\mathbf{E}$ satisfies

$$\frac{|\widetilde{\lambda}-\lambda|}{\sigma_{\max}(\mathbf{T})} \approx \frac{|\mathbf{y}^H\mathbf{E}\mathbf{x}|}{|\mathbf{y}\cdot\mathbf{x}|\sigma_{\max}(\mathbf{T})} = \frac{|\mathbf{y}^H\mathbf{U}\boldsymbol{\Sigma}\mathbf{V}^H\mathbf{x}|}{|\mathbf{y}\cdot\mathbf{x}|\sigma_{\max}(\mathbf{T})}$$
$$\le \frac{\left\|\mathbf{U}^H\mathbf{y}\right\|_2 \sigma_{\max}(\mathbf{E}) \left\|\mathbf{V}^H\mathbf{x}\right\|_2}{|\mathbf{y}\cdot\mathbf{x}|\sigma_{\max}(\mathbf{T})} = \frac{\|\mathbf{y}\|_2\|\mathbf{x}\|_2}{|\mathbf{y}\cdot\mathbf{x}|}\frac{\sigma_{\max}(\mathbf{E})}{\sigma_{\max}(\mathbf{T})} .$$

Consequently, in LAPACK the reciprocal condition number of λ is estimated to be

$$\frac{1}{s(\lambda)} = \frac{\mathbf{y}\cdot\mathbf{x}}{\|\mathbf{y}\|_2\|\mathbf{x}\|_2} .$$

If the matrix $\begin{bmatrix}\mathbf{x} & \mathbf{W}\end{bmatrix}$ is unitary, then approximation (1.47) implies that the corresponding eigenvector $\widetilde{\mathbf{x}}$ of $\mathbf{T}+\mathbf{E}$ satisfies

$$\frac{\|\widetilde{\mathbf{x}}-\mathbf{x}\|_2}{\|\mathbf{x}\|_2} \approx \frac{\left\|\left(\mathbf{W}\mathbf{W}^H\mathbf{T}\mathbf{W}-\mathbf{I}\lambda\right)^{-1}\mathbf{W}^H\mathbf{E}\mathbf{x}\right\|_2}{\|\mathbf{x}\|_2}$$
$$\le \left\|\left(\mathbf{W}^H\mathbf{T}\mathbf{W}-\mathbf{I}\lambda\right)^{-1}\right\|_2 \left\|\mathbf{W}^H\mathbf{E}\right\|_2 \le \frac{\|\mathbf{E}\|_2}{\sigma_{\min}(\mathbf{W}^H\mathbf{T}\mathbf{W}-\mathbf{I}\lambda)} .$$

Thus the desired reciprocal condition number is the smallest singular value of $\mathbf{W}^H\mathbf{T}\mathbf{W} - \mathbf{I}\lambda$, which is the same as $\text{sep}(\lambda, \mathbf{T}_{22})$. LAPACK approximates this by the one-norm of $(\mathbf{T}_{22} - \mathbf{I}\lambda)^{-1}$.

For quasi-triangular matrices arising from a computed Schur decomposition, these reciprocal condition numbers can be computed by routines `_trsna`. See, for example, dtrsna.f.

Exercise 1.4.16 Let

$$\mathbf{R} = \begin{bmatrix} \varrho_{11} & \varrho_{12} \\ & \varrho_{22} \end{bmatrix},$$

with $\varrho_{11} \neq \varrho_{22}$. Since ϱ_{11} and ϱ_{22} are eigenvalues of $\mathbf{R}$, find their condition numbers as a function of the components of $\mathbf{R}$.

Exercise 1.4.17 Suppose that

$$\mathbf{A} = \begin{bmatrix} \alpha & \mathbf{a}^H \\ & \mathbf{B} \end{bmatrix}.$$

1. If α is not an eigenvalue of $\mathbf{B}$, show that the condition number of α as a eigenvalue of $\mathbf{A}$ is $\sqrt{1 + \|(\mathbf{B} - \mathbf{I}\alpha)^{-1}\mathbf{a}\|_2^2}$.
2. Show that the condition number is at least as large as $\sqrt{1 + \|\mathbf{a}\|_2^2/\sigma^2}$, where $\sigma = \text{sep}(\alpha, \mathbf{B})$ is the separation, defined in Eq. (1.42).

Exercise 1.4.18 Let

$$\mathbf{A} = \begin{bmatrix} 0 & -1 & -1 & -1 & -1 & -1 \\ & 1 & -1 & -1 & -1 & -1 \\ & & 1 & -1 & -1 & -1 \\ & & & 1 & -1 & -1 \\ & & & & 1 & -1 \\ & & & & & -1 \end{bmatrix}.$$

Compute the left and right eigenvectors corresponding to the zero eigenvalue, and determine the condition number of this eigenvalue.

Exercise 1.4.19 Let

$$\mathbf{A} = \begin{bmatrix} 20 & 20 & & & & \\ & 19 & 20 & & & \\ & & 18 & 20 & & \\ & & & \ddots & \ddots & \\ & & & & 2 & 20 \\ & & & & & 1 \end{bmatrix}.$$

Find the condition numbers of the eigenvalues of this matrix.

1.4.4 A Posteriori Estimates

As we saw in Lemma 1.3.7, it is common to use the eigenvector residual to measure the error in a computed eigenvector-eigenvalue pair. However, for non-Hermitian matrices, we need to be a bit more careful when we describe what the residual norm measures.

Lemma 1.4.7 *Suppose that* $\mathbf{A}$ *is a square matrix,* $\widetilde{\mathbf{x}}$ *is a nonzero vector and* $\widetilde{\lambda}$ *is a scalar. Then*

$$\frac{1}{\left\|\left(\mathbf{A}-\mathbf{I}\widetilde{\lambda}\right)^{-1}\right\|_2} \leq \frac{\|\mathbf{A}\widetilde{\mathbf{x}}-\widetilde{\mathbf{x}}\widetilde{\lambda}\|_2}{\|\widetilde{\mathbf{x}}\|_2} .$$

Proof The proof of this lemma is contained in the proof of Lemma 1.3.7;

Later, in Lemma 1.5.5 we will see that the 2-norm of a matrix is equal to its largest singular value. As a result, Lemma 1.4.7 shows that the residual norm provides an upper bound for the smallest singular value of $\mathbf{A}-\mathbf{I}\widetilde{\lambda}$.

1.4.5 Rayleigh Quotients

Given an approximate eigenvector $\widetilde{\mathbf{x}}$ for a square matrix $\mathbf{A}$, the following lemma, which can also be found in Stewart [164, p. 280] provides an optimal value for the corresponding approximate eigenvalue.

Lemma 1.4.8 *Suppose that* $\mathbf{A}$ *is an* $n \times n$ *matrix and* $\widetilde{\mathbf{x}}$ *is an* n*-vector. Then the* **Rayleigh quotient**

$$\widetilde{\lambda} = \frac{\widetilde{\mathbf{x}}^H \mathbf{A}\widetilde{\mathbf{x}}}{\widetilde{\mathbf{x}}^H\widetilde{\mathbf{x}}}$$

minimizes the residual norm $\|\mathbf{A}\widetilde{\mathbf{x}}-\widetilde{\mathbf{x}}\mu\|_2$ *over all choices of the scalar* μ.

Proof Corollary 6.3.1 in Chap. 6 of Volume I shows that the solution of the normal equations

$$\widetilde{\mathbf{x}}^H\widetilde{\mathbf{x}}\mu = \widetilde{\mathbf{x}}^H\mathbf{A}\widetilde{\mathbf{x}} .$$

solves the given least squares problem for μ. The solution of the normal equations is the Rayleigh quotient.

1.4.6 Power Method

At this point in our discussion of eigenvalues for general matrices, we have examined some theory, and proved some bounds on perturbations in eigenvalues and associated invariant subspaces. We are now ready to develop numerical methods for finding eigenvalues and eigenvectors.

In Sect. 1.3.6, we discussed the power method, inverse iteration and shifted inverse iteration for finding eigenvectors of Hermitian matrices. The same ideas can be used to find eigenvectors of general matrices. Our discussion of the power method, which is drawn from Wilkinson [186, p. 570ff], will necessarily depend on whether there exists a basis of eigenvectors for the given matrix.

1.4.6.1 Forward Iteration

The power method was previously presented in Sect. 1.3.6.1 and applied to Hermitian matrices. However, this algorithm is applicable to general square matrices. For convenience, we recall the

Algorithm 1.4.1 (Power Method)

$$
\begin{aligned}
&\text{given } \widetilde{\mathbf{x}} \text{ with } \|\widetilde{\mathbf{x}}\|_2 = 1 \\
&\text{until converged} \\
&\quad \widetilde{\mathbf{y}} = \mathbf{A}\widetilde{\mathbf{x}} \\
&\quad \widetilde{\lambda} = \widetilde{\mathbf{x}} \cdot \widetilde{\mathbf{y}} \\
&\quad \widetilde{\mathbf{x}} = \widetilde{\mathbf{y}} / \|\widetilde{\mathbf{y}}\|_2
\end{aligned}
$$

We will begin by examining the power method for a non-defective matrix with a unique dominant eigenvalue. Let $\mathbf{A}$ be an $n \times n$ matrix with $n > 1$, and suppose that there is a nonsingular matrix $\mathbf{X}$ and a diagonal matrix $\boldsymbol{\Lambda}$ so that

$$\mathbf{A}\mathbf{X} = \mathbf{X}\boldsymbol{\Lambda} \; .$$

Furthermore, suppose that the eigenvalues λ_j of $\mathbf{A}$, which are the diagonal entries of $\boldsymbol{\Lambda}$, satisfy

$$\lambda_1 = \ldots = \lambda_r \text{ and } |\lambda_r| > |\lambda_{r+1}| \geq \ldots \geq |\lambda_n| \; .$$

Given a unit n-vector $\widetilde{\mathbf{x}}$, consider the following Let $\mathbf{w}$ solve

$$\mathbf{X}\mathbf{w} = \widetilde{\mathbf{x}} \; .$$

We will assume that the components of $\mathbf{w}$ are ω_i, and that $\omega_i \neq 0$ for some component $1 \leq i \leq r$. For any integer $k \geq 1$ we have

$$\mathbf{A}^k\widetilde{\mathbf{x}} = \mathbf{A}^k\mathbf{X}\mathbf{w} = \mathbf{X}\mathbf{\Lambda}^k\mathbf{w} = \sum_{i=1}^{n} \mathbf{x}_i\lambda_i^k\omega_i = \lambda_1^k\left[\sum_{i=1}^{r} \mathbf{x}_i\omega_i + \sum_{i=r+1}^{n} \mathbf{x}_i\left(\frac{\lambda_i}{\lambda_1}\right)^k\omega_i\right] .$$

As k becomes large, the right-hand side tends to a scalar multiple of $\sum_{i=1}^{r} \mathbf{x}_i\omega_i$, which is an eigenvector of $\mathbf{A}$.

On the other hand, the next example will show that if the dominant eigenvalue is not unique, then the power method may not converge.

Example 1.4.6 Consider the matrix

$$\mathbf{A} = \begin{bmatrix} 1 & 0 \\ 0 & -1 \end{bmatrix} .$$

It is easy to see that

$$\mathbf{A}^k\widetilde{\mathbf{x}} = \begin{bmatrix} 1 & 0 \\ 0 & -1 \end{bmatrix}^k \begin{bmatrix} \xi_1 \\ \xi_2 \end{bmatrix} = \begin{bmatrix} \xi_1 \\ (-1)^k\xi_2 \end{bmatrix} .$$

In this case, the power method converges if and only if $\xi_2 = 0$.

A more common difficulty for the power method is that general real matrices may have a dominant eigenvalue that appears as a complex conjugate pairs. The next example will show that the power method will fail to converge in such a case.

Example 1.4.7 Consider the Givens rotation

$$\mathbf{G} = \begin{bmatrix} \cos\theta & \sin\theta \\ -\sin\theta & \cos\theta \end{bmatrix} .$$

Let the initial vector for the power method be

$$\widetilde{\mathbf{x}} = \begin{bmatrix} \sin\phi \\ \cos\phi \end{bmatrix} .$$

Then

$$\mathbf{G}\widetilde{\mathbf{x}} = \begin{bmatrix} \cos\theta & \sin\theta \\ -\sin\theta & \cos\theta \end{bmatrix} \begin{bmatrix} \sin\phi \\ \cos\phi \end{bmatrix} = \begin{bmatrix} \sin\theta\cos\phi + \cos\theta\sin\phi \\ \cos\theta\cos\phi - \sin\theta\sin\phi \end{bmatrix} = \begin{bmatrix} \sin(\theta+\phi) \\ \cos(\theta+\phi) \end{bmatrix} .$$

Similar calculations show that

$$\mathbf{G}^k\widetilde{\mathbf{x}} = \begin{bmatrix} \sin(k\theta+\phi) \\ \cos(k\theta+\phi) \end{bmatrix} .$$

This expression shows that the power method will not converge. Non-convergence is due to the fact that the eigenvalues of $\mathbf{G}$ are $\lambda = \mathrm{e}^{\pm \mathrm{i}\theta}$, which both have modulus one; in other words, there is no dominant eigenvalue.

Our next example will show that an eigenvector deficiency may lead to slow convergence of the power method.

Example 1.4.8 Consider the $n \times n$ Jordan block

$$\mathbf{J} = \mathbf{I}\lambda + \mathbf{N} ,$$

where $\lambda \neq 0$ and $\mathbf{N}$ is the nilpotent matrix in Definition 1.4.5. Then for any axis vector $\mathbf{e}_j$ with $1 \leq j \leq n$ and any power $k \geq n$ we have

$$\begin{aligned}
\mathbf{J}^k \mathbf{e}_j &= (\mathbf{I}\lambda + \mathbf{N})^k \mathbf{e}_j = \sum_{i=0}^{n-1} \mathbf{N}^i \mathbf{e}_j \binom{k}{i} \lambda^{k-i} = \sum_{i=j-n}^{j-1} \mathbf{e}_{j-i} \binom{k}{i} \lambda^{k-i} = \sum_{\ell=1}^{n} \mathbf{e}_\ell \binom{k}{j-\ell} \lambda^{k+\ell-j} \\
&= \binom{k}{j-1} \lambda^{k+1-j} \left\{ \mathbf{e}_1 + \sum_{\ell=2}^{n} \mathbf{e}_\ell \lambda^{\ell-1} \prod_{m=1}^{\ell-1} \frac{j-m}{k+\ell+1-j-m} \right\} \\
&\approx \binom{k}{j-1} \lambda^{k+1-j} \left\{ \mathbf{e}_1 + \mathbf{e}_2 \frac{j-1}{k+2-j} \lambda \right\} .
\end{aligned}$$

In this example, the power method converges, but the convergence is not geometric.

1.4.6.2 Shifted Inverse Iteration

In order to find the eigenvector corresponding to the smallest eigenvalue of $\mathbf{A}$, we can use **inverse iteration**, which was described by Algorithm 1.3.3. Also, we can find the eigenvector corresponding to the eigenvalue closest to some scalar σ by performing the shifted inverse iteration Algorithm 1.3.3. These two variations of the power method are subject to the same difficulties with multiple dominant eigenvalues and eigenvector deficiencies as the power method. This is because shifted inverse iteration is actually the power method applied to the shifted matrix $(\mathbf{A} - \mathbf{I}\sigma)^{-1}$.

1.4.6.3 Rayleigh Quotients

The convergence of shifted inverse iteration can be enhanced through the use of **Rayleigh quotients**. Following Stewart [164, p. 346f], we will develop some approximations to illustrate this enhanced convergence. Suppose that $\widetilde{\mathbf{x}}$ is an approximate eigenvector for $\mathbf{A}$, with

$$\|\widetilde{\mathbf{x}}\|_2 = 1 .$$

Define the Rayleigh quotient for $\widetilde{\mathbf{x}}$ to be

$$\widetilde{\lambda} = \frac{\widetilde{\mathbf{x}}^H \mathbf{A} \widetilde{\mathbf{x}}}{\widetilde{\mathbf{x}}^H \widetilde{\mathbf{x}}} = \widetilde{\mathbf{x}}^H \mathbf{A} \widetilde{\mathbf{x}} \,.$$

Shifted inverse iteration with the Rayleigh quotient as the eigenvalue approximation determines a new vector $\widetilde{\mathbf{y}}$ by the equation

$$(\mathbf{A} - \mathbf{I}\widetilde{\lambda})\widetilde{\mathbf{y}} = \widetilde{\mathbf{x}} \,.$$

Let $\mathbf{H}$ be a unitary matrix with first column $\widetilde{\mathbf{x}}$:

$$\mathbf{H} = \begin{bmatrix} \widetilde{\mathbf{x}} \; \mathbf{U} \end{bmatrix} \,.$$

We can compute and partition

$$\mathbf{H}^H \mathbf{A} \mathbf{H} = \begin{bmatrix} \widetilde{\lambda} & \mathbf{x}^H \mathbf{A} \mathbf{U} \\ \mathbf{U}^H \mathbf{A} \widetilde{\mathbf{x}} & \mathbf{U}^H \mathbf{A} \mathbf{U} \end{bmatrix} \text{ and } \mathbf{v} \equiv \mathbf{H}^H \widetilde{\mathbf{y}} = \begin{bmatrix} 1 \\ \mathbf{p} \end{bmatrix} \nu \text{ where } \nu = \widetilde{\mathbf{x}} \cdot \widetilde{\mathbf{y}} \,. \tag{1.51}$$

If $\widetilde{\mathbf{x}}$ is close to being an eigenvector of $\mathbf{A}$, then we expect that $\widetilde{\mathbf{y}}$ should have roughly the same direction as $\widetilde{\mathbf{x}}$, so $\nu = \widetilde{\mathbf{x}} \cdot \widetilde{\mathbf{y}}$ should be nonzero.

The eigenvector residual norm from Lemma 1.4.7 can be evaluated as

$$\begin{aligned} \varrho \equiv \frac{\left\| \mathbf{A}\widetilde{\mathbf{x}} - \widetilde{\mathbf{x}}\widetilde{\lambda} \right\|_2}{\|\widetilde{\mathbf{x}}\|_2} &= \left\| \left[\mathbf{H}^H \left(\mathbf{A} - \mathbf{I}\widetilde{\lambda} \right) \mathbf{H} \right] \left[\mathbf{H}^H \widetilde{\mathbf{x}} \right] \right\|_2 \\ &= \left\| \begin{bmatrix} 0 & \widetilde{\mathbf{x}}^H \mathbf{A} \mathbf{U} \\ \mathbf{U}^H \mathbf{A} \widetilde{\mathbf{x}} & \mathbf{U}^H \left(\mathbf{A} - \mathbf{I}\widetilde{\lambda} \right) \mathbf{U} \end{bmatrix} \begin{bmatrix} 1 \\ \mathbf{0} \end{bmatrix} \right\|_2 = \left\| \mathbf{U}^H \mathbf{A} \widetilde{\mathbf{x}} \right\|_2 \,. \end{aligned}$$

If $\widetilde{\mathbf{x}}$ is nearly an eigenvector of $\mathbf{A}$, then $\mathbf{A}\widetilde{\mathbf{x}}$ should have nearly the same direction as $\widetilde{\mathbf{x}}$. In such a case, $\mathbf{A}\widetilde{\mathbf{x}}$ should be nearly orthogonal to the columns of $\mathbf{U}$, because those columns were determined to be orthogonal to $\widetilde{\mathbf{x}}$. Thus the right-hand side in the equation for ϱ should be small.

We can also compute

$$\begin{aligned} \begin{bmatrix} 1 \\ \mathbf{0} \end{bmatrix} &= \mathbf{H}^H \widetilde{\mathbf{x}} = \mathbf{H}^H \left(\mathbf{A} - \mathbf{I}\widetilde{\lambda} \right) \widetilde{\mathbf{y}} = \left[\mathbf{H}^H \left(\mathbf{A} - \mathbf{I}\widetilde{\lambda} \right) \mathbf{H} \right] \mathbf{H}^H \widetilde{\mathbf{y}} \\ &= \begin{bmatrix} 0 & \widetilde{\mathbf{x}}^H \mathbf{A} \mathbf{U} \\ \mathbf{U}^H \mathbf{A} \widetilde{\mathbf{x}} & \mathbf{U}^H \left(\mathbf{A} - \mathbf{I}\widetilde{\lambda} \right) \mathbf{U} \end{bmatrix} \begin{bmatrix} 1 \\ \mathbf{p} \end{bmatrix} \nu = \begin{bmatrix} \widetilde{\mathbf{x}}^H \mathbf{A} \mathbf{U} \mathbf{p} \\ \mathbf{U}^H \mathbf{A} \widetilde{\mathbf{x}} + \mathbf{U}^H \left(\mathbf{A} - \mathbf{I}\widetilde{\lambda} \right) \mathbf{U} \mathbf{p} \end{bmatrix} \nu \end{aligned}$$

The bottom partition of this equation implies that

$$\mathbf{U}^H \mathbf{A} \widetilde{\mathbf{x}} = \mathbf{U}^H \left(\mathbf{I}\widetilde{\lambda} - \mathbf{A} \right) \mathbf{U} \mathbf{p} \,. \tag{1.52}$$

If no other eigenvalue of $\mathbf{A}$ is near the Rayleigh quotient $\widetilde{\lambda}$, then Eq. (1.52) shows that $\mathbf{p}$ should be small.

Recalling the Definition (1.51) of $\mathbf{v}$, we can determine a unitary matrix $\mathbf{Q}$ for which the first column is $\mathbf{v}/\|\mathbf{v}\|_2$:

$$\mathbf{Q} = \begin{bmatrix}\mathbf{v}/\|\mathbf{v}\|_2 & \mathbf{W}\end{bmatrix} .$$

Since $\mathbf{p}$ is small, we can approximate $\mathbf{Q}$ by

$$\widetilde{\mathbf{Q}} = \begin{bmatrix}1 & -\mathbf{p}^H \\ \mathbf{p} & \mathbf{I}\end{bmatrix} .$$

This is because

$$\widetilde{\mathbf{Q}}^H\widetilde{\mathbf{Q}} = \begin{bmatrix}1 + \|\mathbf{p}\|_2^2 & \mathbf{0}^H \\ \mathbf{0} & \mathbf{I} + \mathbf{p}\mathbf{p}^H\end{bmatrix}$$

agrees with the identity matrix to second order in $\mathbf{p}$, and $\widetilde{\mathbf{Q}}$ has the same first column as $\mathbf{Q}$. It follows that the remaining columns of $\mathbf{Q}$ can be approximated by

$$\mathbf{W} \approx \begin{bmatrix}-\mathbf{p}^H \\ \mathbf{I}\end{bmatrix} .$$

Let

$$\widetilde{\mu} \equiv \frac{\widetilde{\mathbf{y}}^H\mathbf{A}\widetilde{\mathbf{y}}}{\widetilde{\mathbf{y}}^H\widetilde{\mathbf{y}}}$$

be the Rayleigh quotient for the new vector $\widetilde{\mathbf{y}}$ determined by shifted inverse iteration from the original vector $\widetilde{\mathbf{x}}$. Then the product of our two unitary matrices is

$$\mathbf{H}\mathbf{Q} = \begin{bmatrix}\mathbf{H}\mathbf{v}/\|\mathbf{H}\mathbf{v}\|_2 & \mathbf{H}\mathbf{W}\end{bmatrix} = \begin{bmatrix}\widetilde{\mathbf{y}}/\|\widetilde{\mathbf{y}}\|_2 & \mathbf{H}\mathbf{W}\end{bmatrix} .$$

It follows that

$$(\mathbf{H}\mathbf{Q})^H\mathbf{A}(\mathbf{H}\mathbf{Q}) = \begin{bmatrix}\widetilde{\mathbf{y}}^H/\|\widetilde{\mathbf{y}}\|_2 \\ (\mathbf{H}\mathbf{W})^H\end{bmatrix}\mathbf{A}\begin{bmatrix}\widetilde{\mathbf{y}}/\|\widetilde{\mathbf{y}}\|_2 & \mathbf{H}\mathbf{W}\end{bmatrix}$$

$$= \begin{bmatrix}\widetilde{\mu} & \widetilde{\mathbf{y}}^H\mathbf{A}\mathbf{H}\mathbf{W}/\|\widetilde{\mathbf{y}}\|_2 \\ \mathbf{W}^H\mathbf{H}^H\mathbf{A}\widetilde{\mathbf{y}}/\|\widetilde{\mathbf{y}}\|_2 & \mathbf{W}^H\mathbf{H}^H\mathbf{A}\mathbf{H}\mathbf{W}\end{bmatrix} .$$

This equation allows us to compute the eigenvector residual norm for the new vector $\widetilde{\mathbf{y}}$ in shifted inverse iteration:

$$\frac{\|\mathbf{A}\widetilde{\mathbf{y}}-\widetilde{\mathbf{y}}\widetilde{\mu}\|_2}{\|\widetilde{\mathbf{y}}\|_2} = \left\|(\mathbf{HQ})^H\left(\mathbf{A}-\mathbf{I}\widetilde{\mu}\right)(\mathbf{HQ})(\mathbf{HQ})^H\widetilde{\mathbf{y}}/\,\|\widetilde{\mathbf{y}}\|_2\right\|_2$$

$$= \left\|\begin{bmatrix} 0 & \mathbf{y}^H\mathbf{AHW}/\|\widetilde{\mathbf{y}}\|_2 \\ \mathbf{W}^H\mathbf{H}^H\mathbf{A}\widetilde{\mathbf{y}}/\|\widetilde{\mathbf{y}}\|_2 & \mathbf{W}^H\mathbf{H}^H(\mathbf{A}-\mathbf{I}\widetilde{\mu})\mathbf{HW}\end{bmatrix}\mathbf{Q}^H\mathbf{v}/\,\|\mathbf{v}\|_2\right\|_2$$

$$= \left\|\begin{bmatrix} 0 & \mathbf{y}^H\mathbf{AHW}/\|\widetilde{\mathbf{y}}\|_2 \\ \mathbf{W}^H\mathbf{H}^H\mathbf{A}\widetilde{\mathbf{y}}/\|\widetilde{\mathbf{y}}\|_2 & \mathbf{W}^H\mathbf{H}^H(\mathbf{A}-\mathbf{I}\widetilde{\mu})\mathbf{HW}\end{bmatrix}\begin{bmatrix}1\\ \mathbf{0}\end{bmatrix}\right\|_2$$

$$= \frac{\left\|\mathbf{W}^H\mathbf{H}^H\mathbf{A}\widetilde{\mathbf{y}}\right\|_2}{\|\widetilde{\mathbf{y}}\|_2} = \frac{\left\|\mathbf{W}^H\mathbf{H}^H\mathbf{AHv}\right\|_2}{\|\mathbf{v}\|_2}\,.$$

$$\approx \frac{\left\|\begin{bmatrix}-\mathbf{p} & \mathbf{I}\end{bmatrix}\begin{bmatrix}\widetilde{\lambda} & \mathbf{x}^H\mathbf{AU} \\ \mathbf{U}^H\mathbf{A}\widetilde{\mathbf{x}} & \mathbf{U}^H\mathbf{AU}\end{bmatrix}\begin{bmatrix}1\\ \mathbf{p}\end{bmatrix}\right\|_2}{\left\|\begin{bmatrix}1\\ \mathbf{p}\end{bmatrix}\right\|_2} \approx \left\|\mathbf{U}^H\mathbf{A}\widetilde{\mathbf{x}}+\mathbf{U}^H\mathbf{AUp}-\mathbf{p}\widetilde{\lambda}-\mathbf{p}\mathbf{x}^H\mathbf{AUp}\right\|_2$$

then Eq. (1.52) gives us

$$= \|\mathbf{p}\|_2\left|\mathbf{x}^H\mathbf{AUp}\right|$$

afterward, the Cauchy inequality (3.15) in Chap. 3 of Volume Iimplies that

$$\le \|\mathbf{p}\|_2^2\left\|\mathbf{U}^H\mathbf{A}^H\widetilde{\mathbf{x}}\right\|_2$$

and finally we can use Eq. (1.52) again to get

$$\le \left\|\left(\mathbf{I}\widetilde{\lambda}-\mathbf{U}^H\mathbf{AU}\right)^{-1}\mathbf{U}^H\mathbf{A}\widetilde{\mathbf{x}}\right\|_2^2\left\|\mathbf{U}^H\mathbf{A}^H\widetilde{\mathbf{x}}\right\|_2$$

$$\le \left\|\left(\mathbf{I}\widetilde{\lambda}-\mathbf{U}^H\mathbf{AU}\right)^{-1}\right\|_2^2\left\|\mathbf{U}^H\mathbf{A}^H\widetilde{\mathbf{x}}\right\|_2\left\|\mathbf{U}^H\mathbf{A}\widetilde{\mathbf{x}}\right\|_2^2 = \left\|\left(\mathbf{I}\widetilde{\lambda}-\mathbf{U}^H\mathbf{AU}\right)^{-1}\right\|_2^2\left\|\mathbf{U}^H\mathbf{A}^H\widetilde{\mathbf{x}}\right\|_2\varrho^2.$$

This inequality shows that the eigenvector residual norm for the shifted inverse iteration update $\widetilde{\mathbf{y}}$ is approximately proportional to the square of the eigenvector residual norm for $\widetilde{\mathbf{x}}$. Thus we should expect quadratic convergence of shifted inverse iteration when the Rayleigh quotient is used.

If $\mathbf{A}$ is Hermitian, then $\mathbf{U}^H\mathbf{A}^H\widetilde{\mathbf{x}} = \mathbf{U}^H\mathbf{A}\widetilde{\mathbf{x}}$, and the inequality above shows that the new eigenvector residual norm is proportional to the cube of the old eigenvector residual norm. For Hermitian matrices, shifted inverse iteration using the Rayleigh quotient converges cubically.

Interested readers may experiment with the following JavaScript program for the **power method.** Enter the matrix entries by rows. Entries within a row are separated by a comma, and rows are separated by a semi-colon. The program allows the reader to select a shift, and to choose between the power method, inverse power method, or Rayleigh quotient iteration.

Exercise 1.4.20 Apply the power method (Algorithm 1.4.1) to find the largest eigenvalue and corresponding eigenvector of

$$\mathbf{A} = \begin{bmatrix} 6 & 5 & -5 \\ 2 & 6 & -2 \\ 2 & 5 & -1 \end{bmatrix} .$$

Use $\mathbf{x} = \begin{bmatrix} -1 & 1 & 1 \end{bmatrix}$ to begin.

Exercise 1.4.21 Apply the inverse power method to find the smallest eigenvalue and corresponding eigenvector of the matrix **A** in the previous exercise.

Exercise 1.4.22 Apply the shifted inverse power method to find the eigenvalue closest to 3 for the matrix **A** in the first exercise.

Exercise 1.4.23 Let

$$\mathbf{B} = \begin{bmatrix} 0 & 2 \\ -2 & 0 \end{bmatrix} .$$

What are the eigenvalues of **B**? Will the power method converge for this matrix? If we use the shifted inverse power method, can we use real shifts and expect convergence?

1.4.7 Orthogonal Iteration

Next, we will generalize the power method to compute invariant subspaces of general matrices. This corresponds to applying the power method to a collection of linearly independent vectors simultaneously. We will use the ideas of this section in Sect. 1.4.8.5.

Suppose that **A** is an $n \times n$ matrix, and consider the following.

Algorithm 1.4.2 (Orthogonal Iteration)

given an $n \times r$ matrix $\widetilde{\mathbf{U}}$ with $\widetilde{\mathbf{U}}^H \widetilde{\mathbf{U}} = \mathbf{I}$

until converged

compute $\mathbf{Z} = \mathbf{A}\widetilde{\mathbf{U}}$

factor $\mathbf{Z} = \widetilde{\mathbf{U}}\widetilde{\mathbf{R}}$ where $\widetilde{\mathbf{R}}$ is right triangular and $\widetilde{\mathbf{U}}$ has orthonormal columns

This algorithm generalizes the power method 1.3.1, because with each iteration the first column of $\widetilde{\mathbf{U}}$ is identical to the vector produced by the power method.

The following lemma describes the convergence of orthogonal iteration. This lemma is taken from Golub and van Loan [84, p. 368], and is similar to a discussion in Parlett and Poole [138].

Lemma 1.4.9 *Suppose that* $\mathbf{A}$ *is an* $n \times n$ *matrix with* $n \geq 2$, *and assume that the Schur decomposition of* $\mathbf{A}$ *is*

$$\mathbf{U}^H \mathbf{A} \mathbf{U} = \mathbf{R}$$

where $\mathbf{U}$ *is unitary and*

$$\mathbf{R} = \boldsymbol{\Lambda} + \mathbf{S}$$

is right-triangular. Here $\boldsymbol{\Lambda}$ *is diagonal and* $\mathbf{S}$ *is strictly upper triangular. Assume that the eigenvalues of* $\mathbf{A}$ *(which are the diagonal entries of* $\boldsymbol{\Lambda}$*) satisfy*

$$|\lambda_1| \geq \ldots \geq |\lambda_r| > |\lambda_{r+1}| \geq \ldots \geq |\lambda_n| \; .$$

Partition

$$\mathbf{U} = \begin{bmatrix} \mathbf{U}_1 & \mathbf{U}_2 \end{bmatrix} \text{ and } \mathbf{R} = \begin{bmatrix} \mathbf{R}_{11} & \mathbf{R}_{12} \\ & \mathbf{R}_{22} \end{bmatrix} \, ,$$

where $\mathbf{U}_1$ *is* $n \times r$ *and* $\mathbf{R}_{11}$ *is* $r \times r$. *Choose the real scalar* μ *to satisfy*

$$\mu > \frac{\|\mathbf{S}\|_F}{|\lambda_r|} - 1 \; . \tag{1.53}$$

Next, let the $n \times r$ *matrix* $\widetilde{\mathbf{U}}_0$ *have orthonormal columns:*

$$\widetilde{\mathbf{U}}_0{}^H \widetilde{\mathbf{U}}_0 = \mathbf{I} \, ,$$

and assume that

$$\delta_0 \equiv dist\left(\mathscr{R}\left(\mathbf{U}_1\right) \, , \, \mathscr{R}\left(\widetilde{\mathbf{U}}_0\right)\right) < 1 \, ,$$

where the distance between two subspaces was defined in Definition 1.2.7. Let $\left\{\widetilde{\mathbf{U}}_k\right\}_{k=0}^{\infty}$ *be the sequence of* $n \times r$ *matrices with orthonormal columns generated by the orthogonal iteration Algorithm 1.4.2. Then*

$$\mathrm{dist}\left(\mathscr{R}(\mathbf{U}_1)\,,\,\mathscr{R}\left(\widetilde{\mathbf{U}}_k\right)\right) \le \frac{\delta_0}{\sqrt{1-\delta_0^2}}(1+\mu)^{n-2}\left[\frac{|\lambda_{r+1}|+\|\mathbf{S}\|_F/(1+\mu)}{|\lambda_r|-\|\mathbf{S}\|_F/(1+\mu)}\right]^k$$
$$\times\left\{1+\frac{\|\mathbf{R}_{12}\|_F}{sep(\mathbf{R}_{11},\mathbf{R}_{22})}\right\}\;. \tag{1.54}$$

Recall that the separation of two matrices was presented in Definition 1.4.4.

Proof Let $\{\mathbf{R}_k\}_{k=1}^{\infty}$ be the sequence of right-triangular matrices generated by the orthogonal iteration Algorithm 1.4.2. Initially, we have $\mathbf{A}\widetilde{\mathbf{U}}_0 = \widetilde{\mathbf{U}}_1\mathbf{R}_1$. Assume inductively that

$$\mathbf{A}^{k-1}\widetilde{\mathbf{U}}_0 = \widetilde{\mathbf{U}}_{k-1}\mathbf{R}_{k-1}\cdot\ldots\cdot\mathbf{R}_1\;,$$

for some $k \ge 2$. Then

$$\mathbf{A}^k\widetilde{\mathbf{U}}_0 = \mathbf{A}\mathbf{A}^{k-1}\widetilde{\mathbf{U}}_0 = \mathbf{A}\widetilde{\mathbf{U}}_{k-1}\mathbf{R}_{k-1}\cdot\ldots\cdot\mathbf{R}_1 = \widetilde{\mathbf{U}}_k\mathbf{R}_k\mathbf{R}_{k-1}\cdot\ldots\cdot\mathbf{R}_1$$

is a QR factorization of $\mathbf{A}^k\widetilde{\mathbf{U}}_0$. The Schur decomposition $\mathbf{U}^H\mathbf{A}\mathbf{U} = \mathbf{R}$ of $\mathbf{A}$ implies that

$$\mathbf{R}^k\mathbf{U}^H\widetilde{\mathbf{U}}_0 = \mathbf{U}^H\mathbf{A}^k\widetilde{\mathbf{U}}_0 = \mathbf{U}^H\widetilde{\mathbf{U}}_k\mathbf{R}_k\cdot\ldots\cdot\mathbf{R}_1\;.$$

Since $\mathbf{R}_{11}$ and $\mathbf{R}_{22}$ have no common eigenvalues, Lemma 1.4.3 shows that the equation

$$\mathbf{R}_{11}\mathbf{X} - \mathbf{X}\mathbf{R}_{22} = -\mathbf{R}_{12}$$

has a unique solution. Furthermore,

$$\begin{bmatrix}\mathbf{I} & -\mathbf{X}\\ & \mathbf{I}\end{bmatrix}\begin{bmatrix}\mathbf{R}_{11} & \mathbf{R}_{12}\\ & \mathbf{R}_{22}\end{bmatrix}\begin{bmatrix}\mathbf{I} & \mathbf{X}\\ & \mathbf{I}\end{bmatrix} = \begin{bmatrix}\mathbf{R}_{11} & \\ & \mathbf{R}_{22}\end{bmatrix}$$

is block diagonal. Then Eq. (1.4.7) implies that

$$\begin{bmatrix}\mathbf{R}_{11}^k & \\ & \mathbf{R}_{22}^k\end{bmatrix}\begin{bmatrix}\mathbf{I} & -\mathbf{X}\\ & \mathbf{I}\end{bmatrix}\mathbf{U}^H\widetilde{\mathbf{U}}_0 = \begin{bmatrix}\mathbf{I} & -\mathbf{X}\\ & \mathbf{I}\end{bmatrix}\mathbf{U}^H\widetilde{\mathbf{U}}_k\mathbf{R}_k\cdot\ldots\cdot\mathbf{R}_1\;.$$

Thus

$$\mathbf{R}_{11}^k\left(\mathbf{U}_1{}^H\widetilde{\mathbf{U}}_0 - \mathbf{X}\mathbf{U}_2{}^H\widetilde{\mathbf{U}}_0\right) = \left(\mathbf{U}_1{}^H\widetilde{\mathbf{U}}_k - \mathbf{X}\mathbf{U}_2{}^H\widetilde{\mathbf{U}}_k\right)\mathbf{R}_k\cdot\ldots\cdot\mathbf{R}_1 \tag{1.55}$$

and

$$\mathbf{R}_{22}^k \mathrm{U}_2{}^H \widetilde{\mathbf{U}}_0 = \mathrm{U}_2{}^H \widetilde{\mathbf{U}}_k \mathbf{R}_k \cdot \ldots \cdot \mathbf{R}_1 \,. \tag{1.56}$$

Since $\mathbf{I} + \mathbf{X}\mathbf{X}^H$ is Hermitian and positive, it has a Cholesky factorization (see Sect. 3.13.3 in Chap. 3 of Volume I)

$$\mathbf{I} + \mathbf{X}\mathbf{X}^H = \mathrm{L}\mathrm{L}^H$$

where $\mathbf{L}$ is left-triangular. Furthermore, the smallest singular value of L is necessarily greater than or equal to one. Let $\mathbf{Z}$ solve

$$\mathbf{Z}\mathrm{L}^H = \mathrm{U}\begin{bmatrix} \mathbf{I} \\ -\mathbf{X}^H \end{bmatrix} = \mathrm{U}_1 - \mathrm{U}_2\mathbf{X}^H \,.$$

Then $\mathscr{R}(\mathbf{Z}) = \mathscr{R}\left(\mathrm{U}_1 - \mathrm{U}_2\mathbf{X}^H\right)$, and

$$\mathbf{Z}^H\mathbf{Z} = \left\{\mathrm{U}\begin{bmatrix} \mathbf{I} \\ -\mathbf{X}^H \end{bmatrix}\mathrm{L}^{-H}\right\}^H \left\{\mathrm{U}\begin{bmatrix} \mathbf{I} \\ -\mathbf{X}^H \end{bmatrix}\mathrm{L}^{-H}\right\}$$
$$= \mathrm{L}^{-1}\left[\mathbf{I}\; -\mathbf{X}\right]\mathrm{U}^H\mathrm{U}\begin{bmatrix} \mathbf{I} \\ -\mathbf{X}^H \end{bmatrix}\mathrm{L}^{-H} = \mathrm{L}^{-1}\left(\mathbf{I} + \mathbf{X}\mathbf{X}^H\right)\mathrm{L}^{-H} = \mathbf{I} \,.$$

We conclude that the columns of $\mathbf{Z}$ form an orthonormal basis for the range of $\mathrm{U}_1 - \mathrm{U}_2\mathbf{X}^H$.

We also have

$$\mathbf{A}^H\left[\mathrm{U}_1 - \mathrm{U}_2\mathbf{X}^H\; \mathrm{U}_2\right] = \mathbf{A}^H\left[\mathrm{U}_1\; \mathrm{U}_2\right]\begin{bmatrix} \mathbf{I} & \\ -\mathbf{X}^H & \mathbf{I} \end{bmatrix}$$
$$= \left[\mathrm{U}_1\; \mathrm{U}_2\right]\begin{bmatrix} \mathbf{R}_{11}{}^H & \\ \mathbf{R}_{12}{}^H & \mathbf{R}_{22}{}^H \end{bmatrix}\begin{bmatrix} \mathbf{I} & \\ -\mathbf{X}^H & \mathbf{I} \end{bmatrix} = \left[\mathrm{U}_1\; \mathrm{U}_2\right]\begin{bmatrix} \mathbf{I} & \\ -\mathbf{X}^H & \mathbf{I} \end{bmatrix}\begin{bmatrix} \mathbf{R}_{11}{}^H & \\ & \mathbf{R}_{22}{}^H \end{bmatrix}$$
$$= \left[\mathrm{U}_1 - \mathrm{U}_2\mathbf{X}^H\; \mathrm{U}_2\right]\begin{bmatrix} \mathbf{R}_{11}{}^H & \\ & \mathbf{R}_{22}{}^H \end{bmatrix}$$

Thus the columns of $\mathrm{U}_1 - \mathrm{U}_2\mathbf{X}^H$ form a basis for the invariant space of $\mathbf{A}^H$ corresponding to the dominant eigenvalues. In the previous paragraph, we showed that the columns of $\mathbf{Z}$ form an orthonormal basis for this same invariant subspace.

Since $\left[\mathbf{Z}\; \mathrm{U}_2\right]$ is unitary,

$$\begin{bmatrix} \mathbf{W}_1 \\ \mathbf{W}_2 \end{bmatrix} = \begin{bmatrix} \mathbf{Z}^H \\ \mathrm{U}_2{}^H \end{bmatrix}\widetilde{\mathbf{U}}_0$$

is a matrix with orthonormal columns. Furthermore, $\mathbf{W}_1$ is $r \times r$. The CS Decomposition Theorem 6.11.1 in Chap. 6 of Volume I implies that

$$\sigma_{\min}(\mathbf{W}_1)^2 = 1 - \sigma_{\max}(\mathbf{W}_2)^2 = 1 - \left\| \mathbf{U}_2{}^H \widetilde{\mathbf{U}}_0 \right\|_2^2 \geq 1 - \delta_0^2 > 0$$

Thus

$$\mathbf{U}_1{}^H \widetilde{\mathbf{U}}_0 - \mathbf{X} \mathbf{U}_2{}^H \widetilde{\mathbf{U}}_0 = \left[\mathbf{I} \; -\mathbf{X} \right] \begin{bmatrix} \mathbf{U}_1{}^H \\ \mathbf{U}_2{}^H \end{bmatrix} \widetilde{\mathbf{U}}_0 = \mathbf{L} \mathbf{Z}^H \widetilde{\mathbf{U}}_0$$

is nonsingular, and

$$\left\| \left(\mathbf{U}_1{}^H \widetilde{\mathbf{U}}_0 - \mathbf{X} \mathbf{U}_2{}^H \widetilde{\mathbf{U}}_0 \right)^{-1} \right\|_2 = \left\| \left(\mathbf{Z}^H \widetilde{\mathbf{U}}_0 \right)^{-1} \mathbf{L}^{-1} \right\|_2$$

$$\leq \left\| \left(\mathbf{Z}^H \widetilde{\mathbf{U}}_0 \right)^{-1} \right\|_2 = \frac{1}{\sigma_{\min}(\mathbf{Z}^H \widetilde{\mathbf{U}}_0)} \leq \frac{1}{\sqrt{1 - \delta_0^2}} . \tag{1.57}$$

Now Eq. (1.56) implies that

$$\mathbf{U}_2{}^H \widetilde{\mathbf{U}}_k = \mathbf{R}_{22}{}^k \mathbf{U}_2{}^H \widetilde{\mathbf{U}}_0 \left(\mathbf{R}_k \cdot \ldots \cdot \mathbf{R}_1 \right)^{-1}$$

and Eq. (1.55) gives us

$$= \mathbf{R}_{22}{}^k \mathbf{U}_2{}^H \widetilde{\mathbf{U}}_0 \left(\mathbf{U}_1{}^H \widetilde{\mathbf{U}}_0 - \mathbf{X} \mathbf{U}_2{}^H \widetilde{\mathbf{U}}_0 \right)^{-1} \mathbf{R}_{11}{}^{-k} \left(\mathbf{U}_1{}^H \widetilde{\mathbf{U}}_0 - \mathbf{X} \mathbf{U}_2{}^H \widetilde{\mathbf{U}}_0 \right) .$$

Then Lemma 1.2.17 and the previous equation imply that

$$\operatorname{dist}\left(\mathscr{R}(\mathbf{U}_1), \mathscr{R}(\widetilde{\mathbf{U}}_k) \right) = \left\| \mathbf{U}_2{}^H \widetilde{\mathbf{U}}_k \right\|_2$$

$$= \left\| \mathbf{R}_{22}^k \mathbf{U}_2{}^H \widetilde{\mathbf{U}}_0 \left(\mathbf{U}_1{}^H \widetilde{\mathbf{U}}_0 - \mathbf{X} \mathbf{U}_2{}^H \widetilde{\mathbf{U}}_0 \right)^{-1} \mathbf{R}_{11}^{-k} \left(\mathbf{U}_1{}^H \widetilde{\mathbf{U}}_0 - \mathbf{X} \mathbf{U}_2{}^H \widetilde{\mathbf{U}}_0 \right) \right\|_2$$

$$\leq \left\| \mathbf{R}_{22}^k \right\|_2 \left\| \mathbf{U}_2{}^H \widetilde{\mathbf{U}}_0 \right\|_2 \left\| \left(\mathbf{U}_1{}^H \widetilde{\mathbf{U}}_0 - \mathbf{X} \mathbf{U}_2{}^H \widetilde{\mathbf{U}}_0 \right)^{-1} \right\|_2 \left\| \mathbf{R}_{11}^{-k} \right\|_2 \left\| \left(\mathbf{U}_1{}^H \widetilde{\mathbf{U}}_0 - \mathbf{X} \mathbf{U}_2{}^H \widetilde{\mathbf{U}}_0 \right) \right\|_2$$

then Lemma 1.2.17 and inequality (1.57) give us

$$\leq \frac{\delta_0}{\sqrt{1 - \delta_0^2}} \left\| \mathbf{R}_{22}^k \right\|_2 \left\| \mathbf{R}_{11}^{-k} \right\|_2 \left\| \mathbf{U}_1{}^H \widetilde{\mathbf{U}}_0 - \mathbf{X} \mathbf{U}_2{}^H \widetilde{\mathbf{U}}_0 \right\|_2 .$$

Inequality (1.39) implies that

$$\left\| \mathbf{R}_{22}^k \right\|_2 \leq (1 + \mu)^{n-r-1} \left\{ |\lambda_{r+1}| + \frac{\|\mathbf{S}\|_F}{1 + \mu} \right\}^k ,$$

and inequality (1.41) implies that

$$\left\|\mathbf{R}_{11}^{-k}\right\|_2 \leq (1+\mu)^{r-1}\left\{|\lambda_r| - \frac{\|\mathbf{S}\|_f}{1+\mu}\right\}^{-k} .$$

We also see that

$$\left\|\mathbf{U}_1{}^H\widetilde{\mathbf{U}}_0 - \mathbf{X}\mathbf{U}_2{}^H\widetilde{\mathbf{U}}_0\right\|_2 \leq \left\|\mathbf{U}_1{}^H\widetilde{\mathbf{U}}_0\right\|_2 + \|\mathbf{X}\|_2\left\|\mathbf{U}_2{}^H\widetilde{\mathbf{U}}_0\right\|_2 \leq 1 + \|\mathbf{X}\|_2$$
$$= 1 + \frac{\|\mathbf{R}_{12}\|_F}{\|\mathbf{R}_{11}\mathbf{X} - \mathbf{X}\mathbf{R}_{22}\|_F / \|\mathbf{X}\|_F} \leq 1 + \frac{\|\mathbf{R}_{12}\|_F}{\min_{\mathbf{Y}\neq\mathbf{0}}\|\mathbf{R}_{11}\mathbf{Y} - \mathbf{Y}\mathbf{R}_{22}\|_F / \|\mathbf{Y}\|_F} .$$

We can combine the last four inequalities to get the claimed result.

1.4.8 Upper Hessenberg Matrices

Many computational techniques in linear algebra involve reductions to canonical forms that greatly simplify the work. For example, when we solved systems of linear equations in Sect. 3.7 in Chap. 3 of Volume I, we used Gaussian factorization to reduce a given linear system to a triangular or trapezoidal system. For least squares problems, we used successive reflection in Sect. 6.7 in Chap. 6 of Volume I to reduce the problem to a triangular or trapezoidal system. Recently, when we wanted to find the eigenvalues of an Hermitian matrix, we use successive reflection in Sect. 1.3.3 to reduce the problem to a real symmetric tridiagonal form. However, for general eigenvalue problems we will need a new special form, which is described in the following definition.

Definition 1.4.7 An $n \times n$ square matrix $\mathbf{H}$ is said to be **upper Hessenberg** if and only if $\mathbf{H}_{ij} = 0$ whenever $1 \leq j < i - 1$.
In other words, upper Hessenberg matrices have at most one nonzero sub-diagonal.

1.4.8.1 Householder Reduction

General square matrices can be transformed to upper Hessenberg form by similarity transformations involving Householder reflectors. To begin this similarity transformation, suppose that we partition a given square matrix $\mathbf{A}$ in the form

$$\mathbf{A} = \begin{bmatrix} \alpha_{11} & \mathbf{a}_{12}{}^H \\ \mathbf{a}_{21} & \mathbf{A}_{22} \end{bmatrix} .$$

Using the ideas in Lemma 6.7.3 in Chap. 6 of Volume I, we can find a Householder reflector $\mathbf{U}_1$ so that

$$\mathbf{U}_1{}^H\mathbf{a}_{21} = \mathbf{e}_1\eta_1 .$$

Then

$$\begin{bmatrix} 1 & \\ & \mathbf{U}_1{}^H \end{bmatrix} \mathbf{A} \begin{bmatrix} 1 & \\ & \mathbf{U}_1 \end{bmatrix} = \begin{bmatrix} \alpha_{11} & \mathbf{a}_{12}{}^H \mathbf{U}_1 \\ \mathbf{U}_1{}^H \mathbf{a}_{21} & \mathbf{U}_1{}^H \mathbf{A}_{22} \mathbf{U}_1 \end{bmatrix} = \begin{bmatrix} \alpha_{11} & \mathbf{a}_{12}{}^H \mathbf{U}_1 \\ \mathbf{e}_1 \eta_1 & \mathbf{U}_1 \mathbf{A}_{22} \mathbf{U}_1{}^H \end{bmatrix}$$

has zeros in the first column below the first sub-diagonal, and is unitarily similar to $\mathbf{A}$.

Suppose that after k steps of this process, we have found Hermitian unitary matrices $\widetilde{\mathbf{U}}_1, \ldots, \widetilde{\mathbf{U}}_k$, a $k \times k$ upper Hessenberg matrix $\mathbf{H}_{11}$ and a scalar η_k so that

$$\widetilde{\mathbf{U}}_k{}^H \ldots \widetilde{\mathbf{U}}_1{}^H \mathbf{A} \widetilde{\mathbf{U}}_1 \ldots \widetilde{\mathbf{U}}_k = \begin{bmatrix} \mathbf{H}_k & \mathbf{a}_{12} & \mathbf{A}_{13} \\ \eta_k \mathbf{e}_k{}^H & \alpha_{22} & \mathbf{a}_{23}{}^H \\ & \mathbf{a}_{32} & \mathbf{A}_{33} \end{bmatrix} .$$

This matrix has zeros below the first diagonal in the first k columns. We can find a Householder reflector $\mathbf{U}_{k+1}$ so that

$$\mathbf{U}_{k+1}{}^H \mathbf{a}_{32} = \mathbf{e}_1 \eta_{k+1} .$$

Then

$$\begin{bmatrix} \mathbf{I} & & \\ & 1 & \\ & & \mathbf{U}_{k+1}{}^H \end{bmatrix} \widetilde{\mathbf{U}}_k{}^H \ldots \widetilde{\mathbf{U}}_1{}^H \mathbf{A} \widetilde{\mathbf{U}}_1 \ldots \widetilde{\mathbf{U}}_k \begin{bmatrix} \mathbf{I} & & \\ & 1 & \\ & & \mathbf{U}_{k+1} \end{bmatrix}$$
$$= \begin{bmatrix} \mathbf{H}_k & \mathbf{a}_{12} & \mathbf{A}_{13} \mathbf{U}_{k+1} \\ \eta_k \mathbf{e}_k{}^H & \alpha_{22} & \mathbf{a}_{23}{}^H \mathbf{U}_{k+1} \\ & \mathbf{U}_{k+1}{}^H \mathbf{a}_{32} & \mathbf{U}_{k+1}{}^H \mathbf{A}_{33} \mathbf{U}_{k+1} \end{bmatrix}$$
$$= \begin{bmatrix} \mathbf{H}_k & \mathbf{a}_{12} & \mathbf{A}_{13} \mathbf{U}_{k+1} \\ \eta_k \mathbf{e}_k{}^H & \alpha_{22} & \mathbf{a}_{23}{}^H \mathbf{U}_{k+1} \\ & \mathbf{e}_1 \eta_{k+1} & \mathbf{U}_{k+1}{}^H \mathbf{A}_{33} \mathbf{U}_{k+1} \end{bmatrix} .$$

This matrix has zeros below the first diagonal in the first $k+1$ columns.

The effect of rounding errors due to the use of Householder reflectors in computing a Hessenberg or tridiagonal form is discussed by Wilkinson [186, p. 351]. Wilkinson recommends scaling the matrix before performing the Hessenberg reduction, in order to control the growth of rounding errors. Such ideas are discussed by Parlett and Reinsch [139, p. 315ff], and are based on ideas by Osborne [134]. In LAPACK, scaling of a matrix is performed by routines `_lascl` and `_gebal`. The former routine is designed to avoid overflow or underflow due to very large or very small entries, while the latter routine implements the Parlett and Reinsch algorithm for balancing a matrix, which was discussed in the previous paragraph.

In LAPACK, reduction to upper Hessenberg form is accomplished by routines `_gehrd`. See, for example, dgehrd.f. This routine uses block organization to make use of Level 3 BLAS routines. The work for individual blocks is performed in routine `_gehd2`, which develops a representation for a product of Householder reflectors similar to the discussion in Sect. 6.7.1.4 in Chap. 6 of Volume I. For a detailed discussion of this block upper Hessenberg reduction algorithm, see Quintana-Orti and van de Geijn [146].

1.4.8.2 Determinant

The following lemma contains an algorithm (see Wilkinson [186, p. 426]) for computing the determinant of an upper Hessenberg matrix.

Lemma 1.4.10 *Suppose that* $\mathbf{H}$ *is an* $n \times n$ *upper Hessenberg matrix with nonzero sub-diagonal entries* $\eta_{j+1,j}$ *for* $1 \leq j < n$*, and let* λ *be a scalar. Then there is an* $(n-1)$*-vector* $\mathbf{x}$ *and a scalar* κ *so that*

$$(\mathbf{H} - \mathbf{I}\lambda)\begin{bmatrix}\mathbf{x} \\ 1\end{bmatrix} = \mathbf{e}_1\kappa \; ,$$

and

$$\det(\mathbf{H} - \mathbf{I}\lambda) = (-1)^n \kappa \prod_{j=1}^{n-1} \eta_{j+1,j} \; . \tag{1.58}$$

Proof Let partition

$$\mathbf{H} - \mathbf{I}\lambda = \begin{bmatrix}\mathbf{p}_{11}{}^H & \pi_{12} \\ \mathbf{P}_{21} & \mathbf{p}_{22}\end{bmatrix} ,$$

and note that $\mathbf{P}_{21}$ is right-triangular with nonzero diagonal entries. Thus there is a unique vector $\mathbf{x}$ so that

$$\mathbf{P}_{21}\mathbf{x} = -\mathbf{p}_{22} \; ,$$

from which we may compute

$$\kappa = \mathbf{p}_{11} \cdot \mathbf{x} + \pi_{12} \; .$$

These equations imply that

$$(\mathbf{H} - \mathbf{I}\lambda)\begin{bmatrix}\mathbf{x} \\ 1\end{bmatrix} = \begin{bmatrix}\mathbf{p}_{11}{}^H & \pi_{12} \\ \mathbf{P}_{21} & \mathbf{p}_{22}\end{bmatrix}\begin{bmatrix}\mathbf{x} \\ 1\end{bmatrix} = \begin{bmatrix}\kappa \\ \mathbf{0}\end{bmatrix} .$$

Since $\det(\mathbf{H} - \mathbf{I}\lambda)$ is an alternating multilinear functional, Definition 3.2.23 in Chap. 3 of Volume I implies that

$$\det\begin{bmatrix}\mathbf{p}_{11}{}^H & \kappa \\ \mathbf{P}_{21} & \mathbf{0}\end{bmatrix} = \det\begin{bmatrix}\mathbf{p}_{11}{}^H & \mathbf{p}_{11} \cdot \mathbf{x} + \pi_{12} \\ \mathbf{P}_{21} & \mathbf{P}_{21}\mathbf{x} + \mathbf{p}_{22}\end{bmatrix} = \det\begin{bmatrix}\mathbf{p}_{11}{}^H & \pi_{12} \\ \mathbf{P}_{21} & \mathbf{p}_{22}\end{bmatrix} = \det(\mathbf{H} - \mathbf{I}\lambda) \; .$$

Then Lemma 3.2.15 in Chap. 3 of Volume I and the Laplace expansion Theorem 3.2.4 in Chap. 3 of Volume I imply that

$$\det(\mathbf{H} - \mathbf{I}\lambda) = (-1)^n \det \begin{bmatrix} \kappa & \mathbf{p}_{11}^H \\ \mathbf{0} & \mathbf{P}_{21} \end{bmatrix} = (-1)^n \kappa \det \mathbf{P}_{21} - (-1)^n \kappa \prod_{j=1}^{n-1} \eta_{j+1,j} .$$

This lemma shows that we may compute the determinant of $\mathbf{H} - \mathbf{I}\lambda$ by solving $(\mathbf{H} - \mathbf{I}\lambda)\mathbf{z} = \mathbf{e}_1$ for $\mathbf{z}$, defining $\kappa = 1/\zeta_n$ to be the reciprocal of the last entry of $\mathbf{z}$, and then computing the determinant as in (1.58).

1.4.8.3 QR Factorization

In Sect. 1.3.7 we used a QR factorization of an Hermitian tridiagonal matrix to develop a very effective algorithm for finding all of its eigenvalues. We will generalize this approach to upper Hessenberg matrices in this section.

First, we will discuss the use of Givens transformations to reduce an upper Hessenberg matrix $\mathbf{H}$ to right-triangular form. To begin this process, we partition

$$\mathbf{H} = \begin{bmatrix} \mathbf{h}_{11} & \mathbf{H}_{12} \\ & \mathbf{H}_{22} \end{bmatrix} ,$$

where $\mathbf{h}_{11}$ is a 2-vector. Following the ideas in Sect. 6.9 in Chap. 6 of Volume I, we can find a Givens rotation $\mathbf{G}_{12}$ and a scalar ϱ_1 so that

$$\mathbf{G}_{12}\mathbf{h}_{11} = \mathbf{e}_1\varrho_1 .$$

Then

$$\begin{bmatrix} \mathbf{G}_{12} & \\ & \mathbf{I} \end{bmatrix} \mathbf{H} = \begin{bmatrix} \mathbf{e}_1\varrho_1 & \mathbf{G}_{12}\mathbf{H}_{12} \\ & \mathbf{H}_{22} \end{bmatrix} .$$

Inductively, assume that after $k-1$ steps of this process, we have found unitary matrices $\widetilde{\mathbf{G}}_1, \ldots, \widetilde{\mathbf{G}}_{k-1}$ and a $(k-1) \times (k-1)$ right-triangular matrix $\mathbf{R}_{11}$ so that

$$\widetilde{\mathbf{G}}_{k-1} \ldots \widetilde{\mathbf{G}}_1 \mathbf{H} = \begin{bmatrix} \mathbf{R}_{11} & \mathbf{r}_{12} & \mathbf{R}_{13} \\ & \mathbf{h}_{22} & \mathbf{H}_{23} \\ & & \mathbf{H}_{33} \end{bmatrix} .$$

Here $\mathbf{h}_{22}$ is a 2-vector. We choose a Given transformation $\mathbf{G}_{k,k+1}$ so that

$$\mathbf{G}_{k,k+1}\mathbf{h}_{22} = \mathbf{e}_1\varrho_k .$$

Then

$$\begin{bmatrix} \mathbf{I} & & \\ & G_{k,k+1} & \\ & & \mathbf{I} \end{bmatrix} \widetilde{G}_{k-1} \ldots \widetilde{G}_1 \mathbf{H} = \begin{bmatrix} \mathbf{R}_{11} & \mathbf{r}_{12} & \mathbf{R}_{13} \\ & \mathbf{e}_1 \varrho_k & G_{k,k+1} \mathbf{H}_{23} \\ & & \mathbf{H}_{33} \end{bmatrix} .$$

This validates the inductive hypothesis. The process continues in this way until we find that $\widetilde{G}_{n-1} \ldots \widetilde{G}_1 \mathbf{H}$ is right-triangular.

Next, we will prove the curious fact that the QR factorization of an upper Hessenberg matrix is almost uniquely determined by the first column of $\mathbf{Q}$. Variants of this theorem can be found in Stewart [164, p. 368] and Golub and van Loan [84, p. 381]. We will use this theorem in Sect. 1.4.8.5.

Theorem 1.4.11 (Implicit Q) *Let* $\mathbf{A}$ *be an* $n \times n$ *matrix. Suppose that the unitary matrices* $\mathbf{U}$ *and* $\mathbf{W}$ *are such that both* $\mathbf{U}^H\mathbf{A}\mathbf{U}$ *and* $\mathbf{W}^H\mathbf{A}\mathbf{W}$ *are upper Hessenberg. Furthermore, assume that both* $\mathbf{U}$ *and* $\mathbf{W}$ *have the same first column. Let* k *be the first column index such that sub-diagonal entry* $(k+1,k)$ *of* $\mathbf{U}^H\mathbf{A}\mathbf{U}$ *is zero; if no sub-diagonal entries of this matrix are zero then we take* $k = n$*. Partition*

$$\mathbf{U} = \begin{bmatrix} \mathbf{U}_1 & \mathbf{U}_2 \end{bmatrix} \text{ and } \mathbf{W} = \begin{bmatrix} \mathbf{W}_1 & \mathbf{W}_2 \end{bmatrix}$$

where $\mathbf{U}_1$ *and* $\mathbf{W}_1$ *are both* $n \times k$*. Then there is a unitary diagonal matrix* $\mathbf{D}$ *so that* $\mathbf{W}_1 = \mathbf{U}_1\mathbf{D}$*, and the* $(k+1,k)$ *entry of* $\mathbf{W}^H\mathbf{A}\mathbf{W}$ *is zero.*

Proof Define the unitary matrix

$$\mathbf{V} = \mathbf{W}^H\mathbf{U} ,$$

and note that the first column of this matrix is

$$\mathbf{V}\mathbf{e}_1 = \mathbf{W}^H\mathbf{U}\mathbf{e}_1 = \mathbf{W}^H\mathbf{W}\mathbf{e}_1 = \mathbf{e}_1 .$$

If we define

$$\mathbf{H} = \mathbf{U}^H\mathbf{A}\mathbf{U} \text{ and } \widetilde{\mathbf{H}} = \mathbf{W}^H\mathbf{A}\mathbf{W} ,$$

then we have

$$\widetilde{\mathbf{H}}\mathbf{V} = \mathbf{W}^H\mathbf{A}\mathbf{W}\mathbf{W}^H\mathbf{U} = \mathbf{W}^H\mathbf{A}\mathbf{U} = \mathbf{W}^H\mathbf{U}\mathbf{H} = \mathbf{V}\mathbf{H} .$$

We will prove inductively that for $1 \leq j < k$ we have $\mathbf{V}\mathbf{e}_j = \mathbf{e}_j\delta_j$ for some scalar δ_j of modulus one. This inductive claim is obviously true for $j = 1$. Assume inductively that the claim is true for all column indices less than j, where $1 < j < k$. Then $\mathbf{V}\mathbf{e}_j$ is orthogonal to $\mathbf{V}\mathbf{e}_i = \mathbf{e}_i\delta_i$ for all $1 \leq i < j$, so the first $j-1$ components of $\mathbf{V}\mathbf{e}_j$ are zero. The equation $\widetilde{\mathbf{H}}\mathbf{V}\mathbf{e}_{j-1} = \mathbf{V}\mathbf{H}\mathbf{e}_{j-1}$ can be expanded to produce

$$\left(\sum_{i=1}^{j-1} \mathbf{e}_i \widetilde{\mathbf{H}}_{i,j-1}\right) \delta_{j-1} = \sum_{i=1}^{j} \mathbf{V}\mathbf{e}_i \mathbf{H}_{ij} = \sum_{i=1}^{j-1} \mathbf{e}_i \delta_i \mathbf{H}_{ij} + \mathbf{V}\mathbf{e}_j \mathbf{H}_{j,j-1}$$

The first $j-1$ components of this equation give us

$$\widetilde{\mathbf{H}}_{i,j-1}\delta_{j-1} = \delta_i \mathbf{H}_{i,j-1} \; .$$

All that remains is the equation

$$\mathbf{V}\mathbf{e}_j \mathbf{H}_{j,j-1} = \mathbf{e}_j \widetilde{\mathbf{H}}_{j,j-1}\delta_{j-1} \; .$$

Since $j < k$, we must have $\mathbf{H}_{j,j-1} \neq 0$, so we can solve for $\mathbf{V}\mathbf{e}_j$ to get

$$\mathbf{V}\mathbf{e}_j = \mathbf{e}_j \frac{\widetilde{\mathbf{H}}_{j,j-1}\delta_{j-1}}{\mathbf{H}_{j,j-1}} \equiv \mathbf{e}_j \delta_j \; .$$

Since $\mathbf{V}\mathbf{e}_j$ is a unit vector, we must have $|\delta_j| = 1$. This completes the induction.

Note that when $j = k$ we have

$$0 = \mathbf{V}\mathbf{e}_k \mathbf{H}_{k,k-1} = \mathbf{e}_k \widetilde{\mathbf{H}}_{k,k-1}\delta_{k-1} \; .$$

In this case, we conclude that $\widetilde{\mathbf{H}}_{k,k-1} = 0$.

We have shown that

$$\mathbf{V} = \begin{bmatrix} \mathbf{D} & \mathbf{0} \\ \mathbf{0} & \mathbf{E} \end{bmatrix}$$

where $\mathbf{D} = \text{diag}(\delta_1, \ldots, \delta_k)$. Thus

$$\begin{bmatrix} \mathbf{U}_1 & \mathbf{U}_2 \end{bmatrix} = \mathbf{U} = \mathbf{W}\mathbf{V} = \begin{bmatrix} \mathbf{W}_1 & \mathbf{W}_2 \end{bmatrix} \begin{bmatrix} \mathbf{D} & \mathbf{0} \\ \mathbf{0} & \mathbf{E} \end{bmatrix} = \begin{bmatrix} \mathbf{W}_1\mathbf{E} & \mathbf{W}_2\mathbf{E} \end{bmatrix} \; .$$

This completes the proof of the theorem.

1.4.8.4 QR Iteration

At this point, we are ready to develop the QR algorithm for computing the eigenvalues and eigenvectors of an upper Hessenberg matrix. This algorithm begins by shifting and factoring

$$\mathbf{H}_0 - \mathbf{I}\sigma = \mathbf{Q}\mathbf{R} \; ,$$

followed by forming

$$\mathbf{H}_1 = \mathbf{R}\mathbf{Q} + \mathbf{I}\sigma \ .$$

Note that

$$\mathbf{H}_1 = \left\{\mathbf{Q}^H (\mathbf{H}_0 - \mathbf{I}\sigma)\right\} \mathbf{Q} + \mathbf{I}\sigma = \mathbf{Q}^H \mathbf{H}_0 \mathbf{Q}$$

is similar to $\mathbf{H}_0$.

The matrix $\mathbf{H}_1 = \mathbf{Q}^H \mathbf{H}_0 \mathbf{Q}$ can be computed by applying a product of Householder transformations. Given an upper Hessenberg matrix $\mathbf{H}_0$ and a shift σ, we can partition

$$\mathbf{H}_0 - \mathbf{I}\sigma = \begin{bmatrix} \mathbf{h}_{11} & \mathbf{H}_{12} \\ & \mathbf{H}_{22} \end{bmatrix} ,$$

where $\mathbf{h}_{11}$ is a 2-vector. Next, we choose a Householder reflector $\mathbf{U}_1$ and a scalar ϱ_1 so that

$$\mathbf{U}_1{}^H \mathbf{h}_{11} = \mathbf{e}_1 \varrho_1 \ .$$

Let

$$\widetilde{\mathbf{U}}_1 = \begin{bmatrix} \mathbf{U}_1 & \\ & \mathbf{I} \end{bmatrix}; .$$

The pattern of nonzero entries in $\widetilde{\mathbf{U}}_1{}^H \mathbf{H}_0 \widetilde{\mathbf{U}}_1$ is

$$\widetilde{\mathbf{U}}_1{}^H \begin{bmatrix} \times & \times & \times & \times & \times \\ \times & \times & \times & \times & \times \\ & \times & \times & \times & \times \\ & & \times & \times & \times \\ & & & \times & \times \end{bmatrix} \widetilde{\mathbf{U}}_1 = \begin{bmatrix} * & * & * & * & * \\ * & * & * & * & * \\ + & * & \times & \times & \times \\ & & \times & \times & \times \\ & & & \times & \times \end{bmatrix}$$

Here the symbol $\times$ indicates an original entry of $\mathbf{H}_0$, while a symbol $*$ indicates a modified entry and $+$ indicates an additional entry. Note that the Householder reflector $\mathbf{U}_1$ affected only the first two rows and columns.

Next, we choose a Householder reflector $\mathbf{U}_2$ to zero out the new $+$ entry in the first subdiagonal position:

$$\widetilde{\mathbf{U}}_2{}^H \begin{bmatrix} * & * & * & * & * \\ * & * & * & * & * \\ + & * & \times & \times & \times \\ & & \times & \times & \times \\ & & & \times & \times \end{bmatrix} \widetilde{\mathbf{U}}_2 = \begin{bmatrix} * & * & * & * & * \\ * & * & * & * & * \\ 0 & * & * & * & * \\ & + & * & \times & \times \\ & & & \times & \times \end{bmatrix}$$

Note that the Householder reflector $\mathbf{U}_2$ affected entries in only the second and third rows and columns. We can continue to **chase the bulge** in this way until an upper Hessenberg form has been obtained:

$$\widetilde{\mathbf{U}_3}^H \begin{bmatrix} * & * & * & * & * \\ * & * & * & * & * \\ & * & * & * & * \\ & + & * & \times & \times \\ & & & \times & \times \end{bmatrix} \widetilde{\mathbf{U}_3} = \begin{bmatrix} * & * & * & * & * \\ * & * & * & * & * \\ & * & * & * & * \\ & 0 & * & * & * \\ & & + & * & \times \end{bmatrix}$$

and

$$\widetilde{\mathbf{U}_4}^H \begin{bmatrix} * & * & * & * & * \\ * & * & * & * & * \\ & * & * & * & * \\ & & * & * & * \\ & & + & * & \times \end{bmatrix} \widetilde{\mathbf{U}_4} = \begin{bmatrix} * & * & * & * & * \\ * & * & * & * & * \\ & * & * & * & * \\ & & * & * & * \\ & & 0 & * & * \end{bmatrix} .$$

In LAPACK, the QR algorithm is performed in routines `_lahqr`. See, for example, dlahqr.f.

The basic ideas behind the QR iteration are described in the following lemma.

Lemma 1.4.11 *Suppose that* $\mathbf{H}_0$ *is an* $n \times n$ *upper Hessenberg matrix and* σ *is a scalar. Let*

$$\mathbf{H}_0 - \mathbf{I}\sigma = \mathbf{Q}\mathbf{R} ,$$

where

$$\mathbf{Q} = \mathbf{G}_{12}{}^H \cdot \ldots \cdot \mathbf{G}_{n-1,n}{}^H$$

is a product of Givens rotations (see Sect. 6.9 in Chap. 6 of Volume I), and $\mathbf{R}$ *is right-triangular. Define the* **QR transform** *of* $\mathbf{H}_0$ *to be*

$$\mathbf{H}_1 \equiv \mathbf{R}\mathbf{Q} + \mathbf{I}\sigma = \mathbf{Q}^H \mathbf{H}_0 \mathbf{Q} .$$

Then $\mathbf{H}_1$ *is upper Hessenberg and unitarily similar to* $\mathbf{H}_0$*. Furthermore,*

$$\mathbf{Q}\mathbf{e}_1 \mathbf{e}_1{}^H \mathbf{R}\mathbf{e}_1 = (\mathbf{H}_0 - \mathbf{I}\sigma)\mathbf{e}_1 \text{ and}$$
$$\mathbf{e}_n \mathbf{e}_n{}^H \mathbf{R}\mathbf{e}_n = (\mathbf{H}_0 - \mathbf{I}\sigma)\mathbf{Q}\mathbf{e}_n .$$

In other words, the first column of $\mathbf{Q}$ *is determined by shifted iteration on* $\mathbf{H}_0$ *applied to the first axis vector, and the last column of* $\mathbf{Q}$ *is determined by shifted inverse iteration on* $\mathbf{H}_0$ *applied to the last axis vector.*

Proof Since $\mathbf{H}_0$ is upper Hessenberg, the Givens QR factorization process in Sect. 6.9 in Chap. 6 of Volume I produces a right-triangular matrix $\mathbf{R}$. Because the rotation $\mathbf{G}_{j,j+1}$ has nonzero entries only in the rows and columns with those subscripts, we see that

$$\mathbf{R}\mathbf{Q} = \mathbf{R}\mathbf{G}_{12}{}^H \cdot \ldots \cdot \mathbf{G}_{n-1,n}{}^H$$

has at most one sub-diagonal. Since $\mathbf{Q}$ is orthogonal,

$$\mathbf{H}_1 = \mathbf{R}\mathbf{Q} + \mathbf{I}\sigma = \mathbf{Q}^H(\mathbf{H}_0 - \mathbf{I}\sigma)\mathbf{Q} + \mathbf{I}\sigma = \mathbf{Q}^H\mathbf{H}_0\mathbf{Q}$$

is similar to $\mathbf{H}_0$, and has at most one sub-diagonal. Thus $\mathbf{H}_1$ is upper Hessenberg.

Since $\mathbf{Q}\mathbf{R} = \mathbf{H}_0 - \mathbf{I}\sigma$, it follows that

$$(\mathbf{H}_0 - \mathbf{I}\sigma)\mathbf{e}_1 = \mathbf{Q}\mathbf{R}\mathbf{e}_1 = \mathbf{Q}\mathbf{e}_1\mathbf{e}_1{}^H\mathbf{R}\mathbf{e}_1 \ .$$

This implies that the first column of $\mathbf{Q}$ is determined by shifted iteration on $\mathbf{H}_0$ applied to the first axis vector. By taking the hermitian to get $\mathbf{H}_0{}^H - \mathbf{I}\overline{\sigma} = \mathbf{R}^H\mathbf{Q}^H$, we can multiply by $\mathbf{Q}$ on the right to get

$$(\mathbf{H}_0{}^H - \mathbf{I}\overline{\sigma})\mathbf{Q} = \mathbf{R}^H \ .$$

This implies that

$$(\mathbf{H}_0{}^H - \mathbf{I}\overline{\sigma})\mathbf{Q}\mathbf{e}_n = \mathbf{R}^H\mathbf{e}_n = \mathbf{e}_n\overline{\mathbf{e}_n{}^H\mathbf{R}\mathbf{e}_n} \ .$$

This equation implies that the last column of $\mathbf{Q}$ is determined by shifted inverse iteration on $\mathbf{H}_0$ applied to the last axis vector.

1.4.8.5 Eigenvalues

In practice, we can find all of the eigenvalues of an upper Hessenberg matrix by applying the QR algorithm with shifts. The basic ideas and an overview of the convergence theory of this method can be found in survey papers by Watkins [181–184]. We will provide some of the important ideas, and leave further study to the reader.

First, we will connect the QR algorithm to both similarity transformations and the power method by the following lemma.

Lemma 1.4.12 *Suppose that* $\mathbf{H}_0$ *is an* $n \times n$ *matrix. Let the sequences of unitary matrices* $\{\mathbf{Q}_k\}_{k=1}^\infty$*, right-triangular matrices* $\{\mathbf{R}_k\}_{k=1}^\infty$ *and square matrices* $\{\mathbf{H}_k\}_{k=0}^\infty$ *be computed by the QR algorithm*

$$\textit{for } 1 \le k$$
$$\mathbf{H}_{k-1} = \mathbf{Q}_k\mathbf{R}_k$$
$$\mathbf{H}_k = \mathbf{R}_k\mathbf{Q}_k \ .$$

Let

$$\mathbf{U}_k = \mathbf{Q}_1 \cdot \ldots \cdot \mathbf{Q}_k \text{ and } \mathbf{T}_k = \mathbf{R}_k \cdot \ldots \cdot \mathbf{R}_1 .$$

Then for all $k \geq 1$ we have

$$\mathbf{H}_0^k = \mathbf{U}_k \mathbf{T}_k \text{ and } \mathbf{H}_k = {\mathbf{U}_k}^H \mathbf{H}_0 \mathbf{U}_k .$$

Proof We will prove the claims by induction. For $k = 1$, we have

$$\mathbf{H}_0^1 = \mathbf{Q}_1 \mathbf{R}_1$$

and

$$\mathbf{H}_1 = \mathbf{R}_1 \mathbf{Q}_1 = {\mathbf{Q}_1}^H \mathbf{H}_0 \mathbf{Q}_1 .$$

Thus we have verified the claims for $k = 1$. Assume inductively that the claims are true for $k - 1 \geq 1$. Then

$$\begin{aligned}
\mathbf{H}_0^k &= \mathbf{H}_0 \mathbf{H}_0^{k-1} = \mathbf{Q}_1 \mathbf{R}_1 \mathbf{U}_{k-1} \mathbf{T}_{k-1} = \mathbf{Q}_1 \mathbf{R}_1 \mathbf{Q}_1 \cdot \ldots \cdot \mathbf{Q}_{k-1} \mathbf{R}_{k-1} \cdot \ldots \cdot \mathbf{R}_1 \\
&= \mathbf{Q}_1 \mathbf{Q}_2 \mathbf{R}_2 \mathbf{Q}_2 \cdot \ldots \cdot \mathbf{Q}_{k-1} \mathbf{R}_{k-1} \cdot \ldots \cdot \mathbf{R}_1 = \ldots \\
&= \mathbf{Q}_1 \cdot \ldots \cdot \mathbf{Q}_{k-1} \mathbf{R}_{k-1} \mathbf{Q}_{k-1} \mathbf{R}_{k-1} \cdot \ldots \cdot \mathbf{R}_1 \\
&= \mathbf{Q}_1 \cdot \ldots \cdot \mathbf{Q}_{k-1} \mathbf{Q}_k \mathbf{R}_k \mathbf{R}_{k-1} \cdot \ldots \cdot \mathbf{R}_1 = \mathbf{U}_k \mathbf{T}_k .
\end{aligned}$$

Also,

$$\mathbf{H}_k = \mathbf{R}_k \mathbf{Q}_k = {\mathbf{Q}_k}^H \mathbf{H}_{k-1} \mathbf{Q}_k = {\mathbf{Q}_k}^H {\mathbf{U}_{k-1}}^H \mathbf{H}_0 \mathbf{H}_{k-1} \mathbf{Q}_k = {\mathbf{U}_k}^H \mathbf{H}_0 \mathbf{U}_k .$$

Consider the case in which the eigenvalues of an $n \times n$ upper Hessenberg matrix $\mathbf{H}$ have distinct moduli:

$$|\lambda_1| > |\lambda_2| > \ldots > |\lambda_n| .$$

Then the powers $\mathbf{H}^k$ are the results of applying orthogonal iteration to the identity matrix. This is like the power method, but with the additional feature that after multiplication by the matrix $\mathbf{H}$ the vectors are orthonormalized. If the eigenvalues have distinct moduli, Lemma 1.4.9 implies that these orthonormal vectors (i.e., the columns of $\mathbf{U}_k$) will converge to an orthonormal basis for the invariant space for $\mathbf{H}$. Under such circumstances, it follows that $\mathbf{H}_k = {\mathbf{U}_k}^H \mathbf{H} \mathbf{U}_k$ will converge to an upper triangular matrix.

If some eigenvalues have the same modulus, then $\mathbf{H}_k$ will converge to block triangular form, with blocks of sizes corresponding to the numbers of eigenvalues with equal modulus. In practice, rounding errors will make the eigenvalues of the numerical matrix distinct, except for the case of complex conjugate pairs of eigenvalues for a real matrix.

1.4.8.6 Shifts

The rate of convergence of the QR algorithm will depend on the ratios of moduli of successive eigenvalues of $\mathbf{H}$. We can enhance the contrast of these eigenvalue moduli by employing a shift. Since the eigenvalues of $\mathbf{H} - \mathbf{I}\sigma$ are $\lambda_i - \sigma$, if the shift σ is close to the nth eigenvalue of $\mathbf{H}$ then $|\lambda_n - \sigma|/|\lambda_{n-1} - \sigma|$ will be small and the $(n, n-1)$ entry of $\mathbf{H}$ will rapidly approach zero in the QR algorithm. Once a zero subdiagonal entry has been produced, the upper Hessenberg matrix becomes block right-triangular, and the remaining eigenvalues are eigenvalues of the diagonal blocks. Additional shifts can be used to speed convergence of sub-diagonal entries of the diagonal blocks. Lemma 1.4.11 describes how the QR algorithm can be modified to perform these shifts while maintaining similarity in the QR algorithm. Section 1.4.8.4 describes how these shifts can be performed implicitly, without explicitly subtracting the shift from the diagonal entries of $\mathbf{H}$ and without actually performing its QR factorization.

For proof of the convergence of the QR algorithm, see Wilkinson [186, p. 517ff], Parlett [136] or Watkins [185].

For complex matrices, LAPACK routines `_lahqr` use the **Wilkinson shift** [186, p. 512]. This is the eigenvalue of the final 2×2 diagonal block of $\mathbf{H}$ that is closest to the last diagonal entry. This shift may be computed by the following algorithm.

Algorithm 1.4.3 (Wilkinson Shift)

$$\sigma = \mathbf{H}_{n,n}$$

$$\omega = \sqrt{\mathbf{H}_{n-1,n}\mathbf{H}_{n,n-1}}$$

if $|\omega| > 0$

$$\xi = (\mathbf{H}_{n-1,n-1} - \sigma)/2$$

$$\alpha = \max\{|\omega|, |\xi|\}$$

$$\eta = \alpha\sqrt{(\xi/\alpha)^2 + (\eta/\alpha)^2}$$

if $|\xi| > 0$

$$\zeta = \xi/|\xi|$$

if $\Re e(\zeta)\Re e(\eta) + \Im m(\zeta)\Im m(\eta) < 0$

$$\eta = -\eta$$

$$\sigma = \mathbf{H}_{n,n} - \omega(\omega/(\xi + \eta))$$

We should also remark that Wilkinson [186, p. 505ff] and Golub and van Loan [84, p. 386f] show that choosing the shift to be the last diagonal element of **H** leads to quadratic convergence of the last subdiagonal entry of **H** to zero. In practice, the Wilkinson shift performs even better.

1.4.8.7 Double Shifts

For real matrices, we would like to avoid using complex arithmetic. However, this would seem to restrict our shifts in the QR algorithm to be real numbers, and make rapid convergence of the QR algorithm difficult when the given matrix has complex conjugate eigenvalues. Fortunately, we can use the **Francis double shift** and the Implicit Q Theorem 1.4.11 to perform shifts by a pair of complex conjugate numbers using real arithmetic.

Let us begin by determining when we should use real or complex shifts. Let

$$\tau = \frac{1}{2} \operatorname{tr} \begin{bmatrix} \mathbf{H}_{n-1,n-1} & \mathbf{H}_{n-1,n} \\ \mathbf{H}_{n,n-1} & \mathbf{H}_{nn} \end{bmatrix} = (\mathbf{H}_{n-1,n-1} + \mathbf{H}_{n,n})/2$$

and

$$\delta = \det\left(\begin{bmatrix} \mathbf{H}_{n-1,n-1} & \mathbf{H}_{n-1,n} \\ \mathbf{H}_{n,n-1} & \mathbf{H}_{nn} \end{bmatrix} - \mathbf{I}\tau \right) = (\mathbf{H}_{n-1,n-1} - \tau)(\mathbf{H}_{n,n} - \tau) - \mathbf{H}_{n,n-1}\mathbf{H}_{n-1,n} \; .$$

Then the eigenvalues of the final 2×2 diagonal block of **H** are

$$\lambda = \tau \pm \sqrt{-\delta} \; .$$

If $\delta < 0$ then these eigenvalues are real, then LAPACK routines `dlahqr` or `slahqr` will shift twice by whichever of these two eigenvalues is closest to $\mathbf{H}_{nn}$; this is the Wilkinson shift.

Otherwise, the eigenvalues of the last diagonal block are complex conjugates λ and $\overline{\lambda}$. Suppose that we were willing to do complex arithmetic at this point, in order to perform the QR algorithm. We would perform the following steps:

$$\begin{aligned}
&\text{factor } \mathbf{H}_0 - \mathbf{I}\lambda = \mathbf{Q}_1\mathbf{R}_1 \\
&\text{compute } \mathbf{H}_1 = \mathbf{R}_1\mathbf{Q}_1 + \mathbf{I}\lambda \\
&\text{factor } \mathbf{H}_1 - \mathbf{I}\overline{\lambda} = \mathbf{Q}_2\mathbf{R}_2 \\
&\text{compute } \mathbf{H}_2 = \mathbf{R}_2\mathbf{Q}_2 + \mathbf{I}\overline{\lambda}
\end{aligned}$$

It is easy to see that

$$\mathbf{H}_1 = \mathbf{R}_1 \mathbf{Q}_1 + \mathbf{I}\lambda = \mathbf{Q}_1{}^H (\mathbf{H}_0 - \mathbf{I}\lambda) \mathbf{Q}_1 + \mathbf{I}\lambda = \mathbf{Q}_1{}^H \mathbf{H}_0 \mathbf{Q}_1$$

and that

$$\mathbf{H}_2 = \mathbf{R}_2 \mathbf{Q}_2 + \mathbf{I}\overline{\lambda} = \mathbf{Q}_2{}^H \mathbf{Q}_1{}^H (\mathbf{H}_0 - \mathbf{I}\lambda) \mathbf{Q}_1 \mathbf{Q}_2 + \mathbf{I}\overline{\lambda} = \mathbf{Q}_2{}^H \mathbf{Q}_1{}^H \mathbf{H}_0 \mathbf{Q}_1 \mathbf{Q}_2 \,. \tag{1.59}$$

We also have

$$\begin{aligned}
\mathbf{Q}_1 \mathbf{Q}_2 \mathbf{R}_2 \mathbf{R}_1 &= \mathbf{Q}_1 (\mathbf{H}_1 - \mathbf{I}\overline{\lambda}) \mathbf{R}_1 = \mathbf{Q}_1 (\mathbf{H}_1 - \mathbf{I}\overline{\lambda}) \mathbf{Q}_1{}^H (\mathbf{H}_0 - \mathbf{I}\lambda) \\
&= \left(\mathbf{Q}_1{}^H \mathbf{H}_1 \mathbf{Q}_1 - \mathbf{I}\overline{\lambda} \right) (\mathbf{H}_0 - \mathbf{I}\lambda) \\
&= (\mathbf{H}_0 - \mathbf{I}\overline{\lambda})(\mathbf{H}_0 - \mathbf{I}\lambda) = \mathbf{H}_0^2 - \mathbf{H}_0(\lambda + \overline{\lambda}) + \mathbf{I}|\lambda|^2 \,.
\end{aligned}$$

This matrix is real, even though $\mathbf{Q}_1$, $\mathbf{Q}_2$, $\mathbf{R}_1$ and $\mathbf{R}_2$ may be complex. Since $\mathbf{R}_2\mathbf{R}_1$ is right-triangular, the first column of $\mathbf{Q}_1\mathbf{Q}_2$ is a unit vector in the direction of the first column of $\mathbf{H}_0^2 - \mathbf{H}_0(\lambda + \overline{\lambda}) + \mathbf{I}|\lambda|^2$.

If $\mathbf{H}$ is an upper Hessenberg matrix with entries $\mathbf{H}_{ij}$, a straightforward calculation shows that its first column is

$$\begin{aligned}
\mathbf{v} &\equiv \left\{ \mathbf{H}^2 - \mathbf{H}(\lambda + \overline{\lambda}) + \mathbf{I}|\lambda|^2 \right\} \mathbf{e}_1 \\
&= \mathbf{e}_1 \left\{ \mathbf{H}_{11}^2 + \mathbf{H}_{12}\mathbf{H}_{21} - \mathbf{H}_{11}(\lambda + \overline{\lambda}) + |\lambda|^2 \right\} + \mathbf{e}_2 \left\{ \mathbf{H}_{21}\mathbf{H}_{11} + \mathbf{H}_{22}\mathbf{H}_{21} - \mathbf{H}_{21}(\lambda + \overline{\lambda}) \right\} \\
&\quad + \mathbf{e}_3 \mathbf{H}_{32}\mathbf{H}_{21} \\
&= \mathbf{e}_1 \left\{ \mathbf{H}_{11}^2 + \mathbf{H}_{12}\mathbf{H}_{21} - \mathbf{H}_{11} 2\tau + \tau^2 + \delta \right\} + \mathbf{e}_2 \left\{ \mathbf{H}_{21}\mathbf{H}_{11} + \mathbf{H}_{22}\mathbf{H}_{21} - \mathbf{H}_{21} 2\tau \right\} \\
&\quad + \mathbf{e}_3 \mathbf{H}_{32}\mathbf{H}_{21} \,.
\end{aligned}$$

Once we compute this vector, we want to find a Householder reflector $\mathbf{U}_0$ so that

$$\mathbf{U}_0{}^H \mathbf{v} / \|\mathbf{v}\|_2 = \mathbf{e}_1 \,,$$

as described in Lemma 6.7.4 in Chap. 6 of Volume I. Since $\mathbf{U}_0$ is unitary, we also have $\mathbf{U}_0\mathbf{e}_1 = \mathbf{v}/\|\mathbf{v}\|_2$. Thus the first column of $\mathbf{U}_0$ is the same as the first column of $\mathbf{Q}_1\mathbf{Q}_2$.

In order to illustrate the computational process, suppose that $\mathbf{H}_0$ is 6×6:

$$\mathbf{H}_0 = \begin{bmatrix} \times & \times & \times & \times & \times & \times \\ \times & \times & \times & \times & \times & \times \\ & \times & \times & \times & \times & \times \\ & & \times & \times & \times & \times \\ & & & \times & \times & \times \\ & & & & \times & \times \end{bmatrix} .$$

Here the symbol $\times$ indicates an original entry of $\mathbf{H}_0$. Since $\mathbf{U}_0$ differs from the identity only in the first three rows and columns, we have

$$\mathbf{U}_0{}^H \mathbf{H}_0 \mathbf{U}_0 = \begin{bmatrix} * & * & * & * & * & * \\ * & * & * & * & * & * \\ + & * & * & * & * & * \\ + & + & * & \times & \times & \times \\ & & & \times & \times & \times \\ & & & & \times & \times \end{bmatrix} .$$

Here the symbol $*$ indicates a modified entry and $+$ indicates an additional entry. We can choose a Householder reflector $\mathbf{U}_1$ to zero out the $+$ entries in the first column below the sub-diagonal:

$$\mathbf{U}_1{}^H \mathbf{U}_0{}^H \mathbf{H}_0 \mathbf{U}_0 \mathbf{U}_1 = \begin{bmatrix} * & * & * & * & * & * \\ * & * & * & * & * & * \\ 0 & * & * & * & * & * \\ 0 & + & * & * & * & * \\ & + & + & * & \times & \times \\ & & & & \times & \times \end{bmatrix} .$$

Afterward, we can choose another Householder reflector $\mathbf{U}_2$ to zero out the $+$ entries in the second column below the sub-diagonal:

$$\mathbf{U}_2{}^H \mathbf{U}_1{}^H \mathbf{U}_0{}^H \mathbf{H}_0 \mathbf{U}_1 \mathbf{U}_2 = \begin{bmatrix} * & * & * & * & * & * \\ * & * & * & * & * & * \\ & * & * & * & * & * \\ & & 0 & * & * & * \\ & & 0 & + & * & * \\ & & + & + & * & \times \end{bmatrix} .$$

Next, we choose $\mathbf{U}_3$ to zero out the $+$ entries in the third column:

$$
{\mathbf{U}_3}^H {\mathbf{U}_2}^H {\mathbf{U}_1}^H {\mathbf{U}_0}^H \mathbf{H}_0 \mathbf{U}_0 \mathbf{U}_1 \mathbf{U}_2 \mathbf{U}_3 = \begin{bmatrix} * & * & * & * & * & * \\ * & * & * & * & * & * \\ & * & * & * & * & * \\ & & * & * & * & * \\ & & 0 & * & * & * \\ & & 0 & + & * & * \end{bmatrix} .
$$

Finally, we choose $\mathbf{U}_4$ to zero out the $+$ entry in the fourth column:

$$
{\mathbf{U}_4}^H {\mathbf{U}_3}^H {\mathbf{U}_2}^H {\mathbf{U}_1}^H {\mathbf{U}_0}^H \mathbf{H}_0 \mathbf{U}_0 \mathbf{U}_1 \mathbf{U}_2 \mathbf{U}_3 \mathbf{U}_4 = \begin{bmatrix} * & * & * & * & * & * \\ * & * & * & * & * & * \\ & * & * & * & * & * \\ & & * & * & * & * \\ & & & * & * & * \\ & & & 0 & * & * \end{bmatrix} .
$$

This matrix is upper Hessenberg, and is unitarily similar to $\mathbf{H}_0$ via a unitary matrix that is a product of Householder reflectors, but has the same first column as $Q_1 Q_2$. The Implicit Q Theorem 1.4.11 shows that this matrix is essentially the same as the upper Hessenberg matrix $\mathbf{H}_2$, which would have been obtained by explicitly performing the Francis double shift.

Additional details of the implementation of the QR algorithm for general matrices can be found in various papers. The stopping criterion used in LAPACK is due to Ahues and Tisseur, and can be found in the LAPACK Working Notes LAPACK Working Notes 122. The shift strategy and organization to employ LAPACK Level 3 BLAS routines is discussed by Braman et al. [22]. The deflation strategy is discussed by the same three authors in [23]. These ideas are implemented in the LAPACK routines `_hseqr`. Given an upper Hessenberg matrix $\mathbf{H}$, these routines determine either the Schur form (complex arithmetic) or the real Schur form (real arithmetic) for $\mathbf{H}$, and the unitary transformation used to achieve this form. For example, see dhseqr.f.

1.4.8.8 Eigenvectors

Once the eigenvalues of an upper Hessenberg matrix have been determined, the corresponding eigenvectors can be computed by shifted inverse iteration. In LAPACK, this computation is performed by routines `_hsein`. For example, see dhsein.f.

Interested readers may experiment with the following JavaScript program for the **real QR iteration.** The reader may enter a real square matrix entries by rows; entries within a row are separated by commas, and rows are separated by semicolons. Clicking on "Iterate" will cause the program to execute one step of the QR iteration, and clicking on "Start Over" will begin the iteration for new matrices. The user may find it interesting to experiment with the following matrices:

$$\begin{bmatrix} 14 & 2 & -2 & 6 \\ -2 & 14 & -6 & -2 \\ -2 & 6 & 14 & 2 \\ -6 & -2 & -2 & 14 \end{bmatrix} = \begin{bmatrix} 1 & 1 & 1 & 1 \\ 1 & -1 & 1 & -1 \\ 1 & 1 & -1 & -1 \\ 1 & -1 & -1 & 1 \end{bmatrix} \begin{bmatrix} 3 & -2 & & \\ 2 & 3 & & \\ & & 4 & 1 \\ & & -1 & 4 \end{bmatrix} \begin{bmatrix} 1 & 1 & 1 & 1 \\ 1 & -1 & 1 & -1 \\ 1 & 1 & -1 & -1 \\ 1 & -1 & -1 & 1 \end{bmatrix}^{-1} \times 4$$

$$\begin{bmatrix} 11 & -1 & 7 & -1 \\ 7 & 11 & -1 & -1 \\ -1 & -1 & 11 & 7 \\ -1 & 7 & -1 & 11 \end{bmatrix} = \begin{bmatrix} 1 & 1 & 1 & 1 \\ 1 & -1 & 1 & -1 \\ 1 & 1 & -1 & -1 \\ 1 & -1 & -1 & 1 \end{bmatrix} \begin{bmatrix} 4 & & & \\ & 3 & -2 & \\ & 2 & 3 & \\ & & & 1 \end{bmatrix} \begin{bmatrix} 1 & 1 & 1 & 1 \\ 1 & -1 & 1 & -1 \\ 1 & 1 & -1 & -1 \\ 1 & -1 & -1 & 1 \end{bmatrix}^{-1} \times 4$$

and

$$\begin{bmatrix} 0 & 1 & 0 & 0 \\ -4 & 4 & 0 & 0 \\ -5 & 2 & 1 & 1 \\ -6 & 3 & -4 & 5 \end{bmatrix} = \begin{bmatrix} 1 & & & \\ 2 & 1 & & \\ 3 & 2 & 1 & \\ 4 & 3 & 2 & 1 \end{bmatrix} \begin{bmatrix} 2 & 1 & & \\ & 2 & & \\ & & 3 & 1 \\ & & & 3 \end{bmatrix} \begin{bmatrix} 1 & & & \\ 2 & 1 & & \\ 3 & 2 & 1 & \\ 4 & 3 & 2 & 1 \end{bmatrix}^{-1}.$$

1.4.9 General Eigenproblems

Users can compute the eigenvalues and eigenvectors via the LAPACK routines `_geev` or `_geevx`. If only eigenvalues are desired, then users should use routines `_gees` or `_geesx`. MATLAB users can use the command eig. This MATLAB command is used for both Hermitian matrices and general matrices.

The steps in the LAPACK algorithm consist of the following.

1. First, the matrix is scaled and balanced, using LAPACK routines `_lascl` and `_gebal`. The former reduces the risk of underflow or overflow, while the latter improves the numerical conditioning of the eigenvalue computations. Given $\mathbf{A}$, this step finds a diagonal matrix $\boldsymbol{\Sigma}$ and computes $\mathbf{B} = \boldsymbol{\Sigma}\mathbf{A}\boldsymbol{\Sigma}^{-1}$. In MATLAB, balancing can be performed by specifying an appropriate option to the command `eig`.
2. Next, unitary similarity transformations are performed to reduce the matrix to upper Hessenberg form. Given $\mathbf{A}$, this step finds $\mathbf{U}$ and $\mathbf{H}$ so that $\mathbf{BU} = \mathbf{UH}$. This step is accomplished by calling routine `_gehrd`.
3. Afterward, the Schur form (complex arithmetic) or real Schur form (real arithmetic) of the upper Hessenberg matrix is computed by LAPACK routine `_hseqr`. The eigenvalues of the original complex matrix are the diagonal entries of the Schur triangular form; for real matrices the eigenvalues are the eigenvalues

of the 1×1 or 2×2 diagonal blocks of the real Schur form. This step finds a unitary matrix $\mathbf{W}$ and a right triangular or right quasi-triangular matrix $\mathbf{R}$ so that $\mathbf{HW} = \mathbf{WR}$.

4. The eigenvectors of the Schur decomposition can be computed by inverse iteration. This is accomplished by LAPACK routines `_trevc` and `_ormhr`. This step finds a nonsingular matrix $\mathbf{X}$ and a diagonal matrix $\boldsymbol{\Lambda}$ so that $\mathbf{RX} = \mathbf{X}\boldsymbol{\Lambda}$.
5. The eigenvectors of the original matrix are computed from the various transformations in the previous steps. Here we note that $\mathbf{B(UWX)} = \mathbf{UHWX} = \mathbf{UWRX} = \mathbf{(UWX)}\boldsymbol{\Lambda}$.
6. The eigenvectors are scaled to correspond to the original matrix: $\mathbf{A}\boldsymbol{\Sigma}^{-1}\mathbf{(UWX)} = \boldsymbol{\Sigma}^{-1}\mathbf{(UWX)}\boldsymbol{\Lambda}$.

The LAPACK routines `_geevx` and `_geesx` are also able to compute reciprocal condition numbers for the eigenvalues and eigenvectors.

For those readers who were interested in Sect. 3.12 in Chap. 3 of Volume I regarding object-oriented programming and linear algebra, we have provided several procedures named `eigenvalues` to compute eigenvalues and eigenvectors within the following classes of matrices: `SymmetricTridiagonal Matrix`, `UpperHessenbergMatrix`, `SymmetricBandMatrix`, `SquareMatrix` and `SymmetricMatrix`. In each of these matrix classes, the `eigenvalues` routine is essentially a wrapper around the corresponding LAPACK routines. Readers may examine LaDouble.C and LaDoubleComplex.C for implementation details.

Exercise 1.4.24 Get a copy of MBEAUSE. This is a 496×496 nonsymmetric matrix. Find the eigenvalues of this matrix, and find the condition number of the largest real eigenvalue.

1.5 Singular Value Decomposition

The singular value decomposition has already been very useful to us for a variety of topics. For example, in Sect. 3.5.2 in Chap. 3 of Volume I mentioned that singular values can be used to evaluate the 2-norm of a matrix, and showed how singular values can be used to prove that $\|\mathbf{A}\|_F \leq \sqrt{\operatorname{rank}(\mathbf{A})}\|\mathbf{A}\|_2$. We also demonstrated the existence of the pseudo-inverse of a matrix in Sect. 6.4 in Chap. 6 of Volume I by using the singular value decomposition. In Sect. 6.11 in Chap. 6 of Volume I, we showed how the singular value decomposition can be used to solve least squares problems, and to determine the effective rank of a matrix. In Sect. 6.11.2 in Chap. 6 of Volume I, we used the singular value decomposition to perform regularization of ill-posed least-squares problems. The CS decomposition was developed in Sect. 6.11.3 in Chap. 6 of Volume I by using the singular value decomposition.

In this chapter, we have used the singular value decomposition a number of times. One use was in the proof of Lemma 1.2.17, to evaluate the distance between subspaces. Another was in the proof of Lemma 1.2.19 to compute the principal angles between subspaces. In Lemma 1.3.6 we used singular values to estimate the perturbation in an invariant space, and in Lemma 1.4.3 we used singular values to solve the Sylvester equation. In Theorem 1.4.10 singular values were used to estimate the distance between invariant subspaces of a matrix and its perturbation and at the end of Sect. 1.4.2 singular values were employed to approximate condition numbers for eigenvalues and eigenvectors. Thus, the reader should have already developed some intuitive feel for the singular value decomposition, without fully appreciating its development.

Our goal in this section is to prove the existence of the singular value decomposition, and to develop an algorithm for its computation. We will examine the theoretical implications of the singular value decomposition, and discuss its practical application to least squares problems.

1.5.1 Existence

We will begin our discussion of the singular value decomposition with the following existence theorem. This theorem was used previously in Sect. 6.11 in Chap. 6 of Volume I.

Theorem 1.5.1 (Singular Value Decomposition) *Suppose that* $\mathbf{A}$ *is a nonzero* $m \times n$ *matrix. Then there is an integer* $r \in [1, \min\{m, n\}]$, *a unitary matrix*

$$\mathbf{U} = \begin{bmatrix} \mathbf{U}_1 & \mathbf{U}_2 \end{bmatrix}$$

where $\mathbf{U}_1$ *is* $m \times r$, *a unitary matrix*

$$\mathbf{V} = \begin{bmatrix} \mathbf{V}_1 & \mathbf{V}_2 \end{bmatrix}$$

where $\mathbf{V}_1$ *is* $n \times r$, *and an* $r \times r$ *real positive diagonal matrix* $\boldsymbol{\Sigma}$ *such that*

$$\mathbf{A} = \begin{bmatrix} \mathbf{U}_1 & \mathbf{U}_2 \end{bmatrix} \begin{bmatrix} \boldsymbol{\Sigma} & \mathbf{0} \\ \mathbf{0} & \mathbf{0} \end{bmatrix} \begin{bmatrix} {\mathbf{V}_1}^H \\ {\mathbf{V}_2}^H \end{bmatrix} = \mathbf{U}_1 \boldsymbol{\Sigma} {\mathbf{V}_1}^H . \quad (1.60)$$

Proof Since $\mathbf{A}^H\mathbf{A}$ is Hermitian and nonnegative, the Spectral Theorem 1.3.1 shows that its eigenvalues are real and nonnegative, and its eigenvectors can be chosen to be orthonormal. In particular, we can order the eigenvectors and eigenvalues so that

$$\mathbf{A}^H\mathbf{A}\mathbf{V} = \mathbf{V} \begin{bmatrix} \boldsymbol{\Sigma}^2 & 0 \\ 0 & 0 \end{bmatrix} ,$$

where $\boldsymbol{\Sigma}$ is a diagonal matrix with nonzero diagonal entries in non-increasing order, and $\mathbf{V}$ is unitary. We can partition

$$\mathbf{V} = [\mathbf{V}_1, \mathbf{V}_2]$$

where $\mathbf{V}_1$ is $n \times r$. Note that $\mathbf{A}\mathbf{V}_2 = \mathbf{0}$, since the eigenvector equation $\mathbf{A}\mathbf{V} = \mathbf{V}\begin{bmatrix} \boldsymbol{\Sigma} & \mathbf{0} \\ \mathbf{0} & \mathbf{0} \end{bmatrix}$ can be pre-multiplied by $\mathbf{V}^H$ to show that the columns of $\mathbf{A}\mathbf{V}_2$ have zero norm. Also note that

$$\boldsymbol{\Sigma}^{-1}{\mathbf{V}_1}^H\mathbf{A}^H\mathbf{A}\mathbf{V}_1\boldsymbol{\Sigma}^{-1} = \mathbf{I} ,$$

so the columns of

$$\mathbf{U}_1 \equiv \mathbf{A}\mathbf{V}_1\boldsymbol{\Sigma}^{-1}$$

are orthonormal. We can extend the columns of $\mathbf{U}_1$ to an orthonormal basis for all m-vectors, producing an unitary matrix $\mathbf{U} = [\mathbf{U}_1, \mathbf{U}_2]$. Then

$$\mathbf{U}^H\mathbf{A}\mathbf{V} = \begin{bmatrix} {\mathbf{U}_1}^H\mathbf{A}\mathbf{V}_1 & {\mathbf{U}_1}^H\mathbf{A}\mathbf{V}_2 \\ {\mathbf{U}_2}^H\mathbf{A}\mathbf{V}_1 & {\mathbf{U}_2}^H\mathbf{A}\mathbf{V}_2 \end{bmatrix}$$
$$= \begin{bmatrix} \boldsymbol{\Sigma}^{-1}{\mathbf{V}_1}^H\mathbf{A}^H\mathbf{A}\mathbf{V}_1 & {\mathbf{U}_1}^H\mathbf{0} \\ {\mathbf{U}_2}^H\mathbf{U}_1\boldsymbol{\Sigma} & {\mathbf{U}_2}^H\mathbf{0} \end{bmatrix} = \begin{bmatrix} \boldsymbol{\Sigma} & \mathbf{0} \\ \mathbf{0} & \mathbf{0} \end{bmatrix} .$$

The singular value decomposition suggests the following definition.

Definition 1.5.1 If the $m \times n$ matrix $\mathbf{A}$ has the singular value decomposition (1.60), then the $\min\{m, n\}$ diagonal entries of

$$\mathbf{U}^H\mathbf{A}\mathbf{V} = \begin{bmatrix} \boldsymbol{\Sigma} & \mathbf{0} \\ \mathbf{0} & \mathbf{0} \end{bmatrix}$$

are called the **singular values** of $\mathbf{A}$, the columns of $\mathbf{V}$ are called the **right singular vectors**, and the columns of $\mathbf{U}$ are called the **left singular vectors**.

Here is an interesting consequence of the proof of existence of singular values.

Corollary 1.5.1 *If $\{\lambda_i\}_{i=1}^n$ is the set of eigenvalues of the $n \times n$ Hermitian matrix* $\mathbf{A}$, *then the set of its singular values is $\{|\lambda_i|\}_{i=1}^n$.*

Proof Since the singular values of $\mathbf{A}$ are the square roots of the eigenvalues of $\mathbf{A}^H\mathbf{A} = \mathbf{A}^2$, we have $\sigma_i^2 = \lambda_i^2$. Since singular values are nonnegative, this implies that $\sigma_i = |\lambda_i|$.

In other words, an Hermitian matrix may have negative eigenvalues, but its singular values must be nonnegative.

1.5.2 Normal Matrices

Next, we will investigate a class of diagonalizable matrices for which the singular values are easily related to their eigenvalues.

Definition 1.5.2 A square matrix **A** is **normal** if and only if $\mathbf{A}^H\mathbf{A} = \mathbf{A}\mathbf{A}^H$. Mathematicians say that a normal matrix **commutes** with its Hermitian.

The next lemma is an interesting consequence of the definition of a normal matrix.

Lemma 1.5.1 *If* **A** *is normal and triangular, then* **A** *is diagonal.*

Proof If **A** is right-triangular, we may partition

$$\mathbf{A} = \begin{bmatrix} \alpha_{11} & {\mathbf{a}_{12}}^H \\ & \mathbf{A}_{22} \end{bmatrix} .$$

Then

$$\mathbf{A}^H\mathbf{A} = \begin{bmatrix} \alpha_{11} & \\ \mathbf{a}_{12} & {\mathbf{A}_{22}}^H \end{bmatrix} \begin{bmatrix} \alpha_{11} & {\mathbf{a}_{12}}^H \\ & \mathbf{A}_{22} \end{bmatrix} = \begin{bmatrix} \alpha_{11}^2 & \alpha_{11}{\mathbf{a}_{12}}^H \\ \mathbf{a}_{12}\alpha_{11} & \mathbf{a}_{12}{\mathbf{a}_{12}}^H + {\mathbf{A}_{22}}^H\mathbf{A}_{22} \end{bmatrix}$$

and

$$\mathbf{A}\mathbf{A}^H = \begin{bmatrix} \alpha_{11} & {\mathbf{a}_{12}}^H \\ & \mathbf{A}_{22} \end{bmatrix} \begin{bmatrix} \alpha_{11} & \\ \mathbf{a}_{12} & {\mathbf{A}_{22}}^H \end{bmatrix} = \begin{bmatrix} \alpha_{11}^2 + {\mathbf{a}_{12}}^H\mathbf{a}_{12} & {\mathbf{a}_{12}}^H{\mathbf{A}_{22}}^H \\ \mathbf{A}_{22}\mathbf{a}_{12} & \mathbf{A}_{22}{\mathbf{A}_{22}}^H \end{bmatrix} .$$

If **A** is normal, these two matrices are equal. Then the first diagonal entries of these two matrices imply that $\mathbf{a}_{12} = \mathbf{0}$. We may continue this diagnostic process to show that the entries above the diagonal in the remaining rows are zero as well.

A similar proof shows that a normal left-triangular matrix is diagonal.

The previous lemma has the following corollary, which characterizes normal matrices.

Corollary 1.5.2 *A square matrix* **A** *is normal if and only there is a unitary matrix* **U** *so that* $\mathbf{U}^H\mathbf{A}\mathbf{U}$ *is diagonal.*

Proof Let **A** have Schur decomposition $\mathbf{U}^H\mathbf{A}\mathbf{U} = \mathbf{R}$, where $\mathbf{U}$ is unitary and **R** is right-triangular. If **A** is normal, then

$$\begin{aligned} \mathbf{A}^H\mathbf{A} &= \mathbf{U}\mathbf{R}^H\mathbf{U}^H\mathbf{U}\mathbf{R}\mathbf{U}^H = \mathbf{U}\mathbf{R}^H\mathbf{R}\mathbf{U}^H \\ &= \mathbf{A}\mathbf{A}^H = \mathbf{U}\mathbf{R}\mathbf{U}^H\mathbf{U}\mathbf{R}^H\mathbf{U} = \mathbf{U}\mathbf{R}\mathbf{R}^H\mathbf{U} . \end{aligned}$$

This equation implies that **R** is normal, and then Lemma 1.5.1 implies that **R** is diagonal.

On the other hand, if there is a unitary matrix **U** so that $\mathbf{U}^H\mathbf{A}\mathbf{U} = \mathbf{D}$ is diagonal, then

$$\mathbf{A}^H\mathbf{A} = \mathbf{U}\overline{\mathbf{D}}\mathbf{U}^H\mathbf{U}\mathbf{D}\mathbf{U}^H = \mathbf{U}\overline{\mathbf{D}}\mathbf{D}\mathbf{U}^H$$
$$= \mathbf{U}\mathbf{D}\overline{\mathbf{D}}\mathbf{U}^H = \mathbf{U}\mathbf{D}\mathbf{U}^H\mathbf{U}\overline{\mathbf{D}}\mathbf{U}^H = \mathbf{A}\mathbf{A}^H ,$$

so **A** is normal.

The previous corollary allows us to characterize the singular values of normal matrices.

Corollary 1.5.3 *If* **A** *is a normal matrix, then its singular values are the moduli of its eigenvalues.*

Proof Since **A** is normal, Corollary 1.5.2 shows that there is a unitary matrix **U** and a diagonal matrix **D** so that $\mathbf{U}^H\mathbf{A}\mathbf{U} = \mathbf{D} \equiv \text{diag}(\delta_1, \ldots, \delta_n)$. It follows that

$$\mathbf{H}^H\mathbf{A} = \mathbf{U}\overline{\mathbf{D}}\mathbf{D}\mathbf{U}^H ,$$

so the eigenvalues of this matrix are the scalars $\overline{\delta_i}\delta_i = |\delta_i|^2$. The singular values of **A** are the square roots of these scalars.

Example 1.5.1 Let us determine the exact singular value decomposition of the Läuchli matrix from Example 6.5.4 in Chap. 6 of Volume I:

$$\mathbf{A} = \begin{bmatrix} 1 & 1 \\ \delta & 0 \\ 0 & \delta \end{bmatrix} .$$

Recall that δ is assumed to be small but positive. In order to determine the singular values of **A**, we will first find the eigenvalues of

$$\mathbf{A}^\top\mathbf{A} = \begin{bmatrix} 1+\delta^2 & 1 \\ 1 & 1+\delta^2 \end{bmatrix} .$$

We can express the eigenvalues and eigenvectors of this matrix by

$$\mathbf{A}^\top\mathbf{A}\mathbf{V} \equiv \begin{bmatrix} 1+\delta^2 & 1 \\ 1 & 1+\delta^2 \end{bmatrix} \begin{bmatrix} 1 & -1 \\ 1 & 1 \end{bmatrix} \frac{1}{\sqrt{2}} = \begin{bmatrix} 1 & -1 \\ 1 & 1 \end{bmatrix} \frac{1}{\sqrt{2}} \begin{bmatrix} 2+\delta^2 & 0 \\ 0 & \delta^2 \end{bmatrix} \equiv \mathbf{V}\boldsymbol{\Lambda}^2 .$$

Thus the singular values of **A** are $\sqrt{2+\delta^2}$ and δ, and the right singular vectors of **A** are the columns of **V**.

Next, we would like to find the left singular vectors of **A**. Let us define

$$\beta = \frac{1}{\sqrt{2+\delta^2}} .$$

Then it is easy to check that

$$\mathbf{A}\mathbf{A}^\top \mathbf{U} \equiv \begin{bmatrix} 2 & \delta & \delta \\ \delta & \delta^2 & 0 \\ \delta & 0 & \delta^2 \end{bmatrix} \begin{bmatrix} \beta\sqrt{2} & 0 & -\delta\beta \\ \delta\beta/\sqrt{2} & -1/\sqrt{2} & \beta \\ \delta\beta/\sqrt{2} & 1/\sqrt{2} & \beta \end{bmatrix}$$

$$= \begin{bmatrix} \beta\sqrt{2} & 0 & -\delta\beta \\ \delta\beta/\sqrt{2} & -1/\sqrt{2} & \beta \\ \delta\beta/\sqrt{2} & 1/\sqrt{2} & \beta \end{bmatrix} \begin{bmatrix} 2+\delta^2 & & \\ & \delta^2 & \\ & & 0 \end{bmatrix} \equiv \mathbf{U} \begin{bmatrix} \boldsymbol{\Lambda} & \\ & 0 \end{bmatrix} .$$

To check our singular value decomposition, we note that $\mathbf{V}^\top\mathbf{V} = \mathbf{I}$, that

$$\mathbf{U}^\top\mathbf{U} = \begin{bmatrix} \beta\sqrt{2} & \delta\beta/\sqrt{2} & \delta\beta/\sqrt{2} \\ 0 & -1/\sqrt{2} & 1/\sqrt{2} \\ -\delta\beta & \beta & \beta \end{bmatrix} \begin{bmatrix} \beta\sqrt{2} & 0 & -\delta\beta \\ \delta\beta/\sqrt{2} & -1/\sqrt{2} & \beta \\ \delta\beta/\sqrt{2} & 1/\sqrt{2} & \beta \end{bmatrix}$$

$$= \begin{bmatrix} 2\beta^2+\delta^2\beta^2 & 0 & 0 \\ 0 & 1 & 0 \\ 0 & 0 & 2\beta^2+\delta^2\beta^2 \end{bmatrix} = \mathbf{I} ,$$

and that

$$\mathbf{U}\boldsymbol{\Sigma}\mathbf{V}^\top \equiv \begin{bmatrix} \beta\sqrt{2} & \delta\beta/\sqrt{2} & \delta\beta/\sqrt{2} \\ 0 & -1/\sqrt{2} & 1/\sqrt{2} \\ -\delta\beta & \beta & \beta \end{bmatrix} \begin{bmatrix} 1/\beta & 0 \\ 0 & \delta \\ 0 & 0 \end{bmatrix} \begin{bmatrix} 1/\sqrt{2} & -1/\sqrt{2} \\ 1/\sqrt{2} & 1/\sqrt{2} \end{bmatrix} = \begin{bmatrix} 1 & 1 \\ \delta & 0 \\ 0 & \delta \end{bmatrix} \equiv \mathbf{A} .$$

Exercise 1.5.1 Use the singular value decomposition to show that for every square matrix $\mathbf{A}$ there is an Hermitian matrix $\mathbf{S}$ and a unitary matrix $\mathbf{Q}$ so that $\mathbf{A} = \mathbf{Q}\mathbf{S}$. This is called the **polar decomposition** of $\mathbf{A}$.

Exercise 1.5.2 Use MATLAB to find the singular values of the square Hilbert matrices of size 2 through 5.

1.5.3 Pseudo-Inverse

Lemma 1.5.2 *If the $m \times n$ matrix $\mathbf{A}$ has singular value decomposition $\mathbf{A} = \mathbf{U}_1\boldsymbol{\Sigma}\mathbf{V}_1{}^H$, then the pseudo-inverse of $\mathbf{A}$ is*

$$\mathbf{A}^\dagger = \mathbf{V}_1\boldsymbol{\Sigma}^{-1}\mathbf{U}_1{}^H .$$

Proof Let us verify the Penrose pseudo-inverseNote that

$$\mathbf{A}\mathbf{A}^\dagger = \left(\mathbf{U}_1\boldsymbol{\Sigma}\mathbf{V}_1{}^H\right)\left(\mathbf{V}_1\boldsymbol{\Sigma}^{-1}\mathbf{U}_1{}^H\right) = \mathbf{U}_1\mathbf{U}_1{}^H ,$$

and

$$\mathbf{A}^\dagger\mathbf{A} = \left(\mathbf{V}_1\boldsymbol{\Sigma}^{-1}\mathbf{U}_1{}^H\right)\left(\mathbf{U}_1\boldsymbol{\Sigma}\mathbf{V}_1{}^H\right) = \mathbf{V}_1\mathbf{V}_1{}^H \; .$$

These verify Penrose conditions (6.5c) and (6.5d) in Chap. 6 of Volume I. Next, we compute

$$\mathbf{A}^\dagger\mathbf{A}\mathbf{A}^\dagger = \left(\mathbf{V}_1\mathbf{V}_1{}^H\right)\mathbf{A}^\dagger = \mathbf{V}_1\mathbf{V}_1{}^H\left(\mathbf{V}_1\boldsymbol{\Sigma}\mathbf{U}_1{}^H\right) = \mathbf{V}_1\boldsymbol{\Sigma}\mathbf{U}_1{}^H = \mathbf{A}^\dagger \; ,$$

and

$$\mathbf{A}\mathbf{A}^\dagger\mathbf{A} = \mathbf{A}\left(\mathbf{V}_1\mathbf{V}_1{}^H\right) = \left(\mathbf{U}_1\boldsymbol{\Sigma}\mathbf{V}_1{}^H\right)\mathbf{V}_1\mathbf{V}_1{}^H = \mathbf{U}_1\boldsymbol{\Sigma}\mathbf{V}_1{}^H = \mathbf{A} \; .$$

These verify the first two Penrose pseudo-inverse conditions.

The following lemma shows that the Newton iteration for approximate division in Example 2.3.1.4 in Chap. 2 of Volume I can be generalized (see Ben-Israel and Cohen [15]) to compute a pseudo-inverse of a matrix. This iteration can be useful in certain large sparse computations, as in Beylkin et al. [16]. For most situations, the algorithm described in Sect. 1.5.10 is preferable to the algorithm in the following lemma.

Lemma 1.5.3 *Let* $\mathbf{A}$ *be a nonzero* $m \times n$ *matrix, and choose* $\alpha \in (0, 2/\sigma_{\max}(\mathbf{A})^2)$. *Let* $\mathbf{X}_0 = \mathbf{A}^H\alpha$, *and define the sequence* $\{\mathbf{X}_k\}_{k=0}^\infty$ *of* $n \times n$ *matrices by*

$$\mathbf{X}_{k+1} = \mathbf{X}_k 2 - \mathbf{X}_k\mathbf{A}\mathbf{X}_k \; .$$

Then $\mathbf{X}_k \to \mathbf{A}^\dagger$, *and*

$$\left(\mathbf{X}_{k+1} - \mathbf{A}^\dagger\right)\mathbf{A} = -\left\{\left(\mathbf{X}_k - \mathbf{A}^\dagger\right)\mathbf{A}\right\}^2 \; .$$

Proof If $\mathbf{A}$ has singular value decomposition

$$\mathbf{A} = \mathbf{U}_1\boldsymbol{\Sigma}\mathbf{V}_1{}^H$$

where $\boldsymbol{\Sigma}$ is $r \times r$ diagonal and nonsingular, then

$$\mathbf{X}_0 = \mathbf{V}_1\boldsymbol{\Lambda}_0\mathbf{U}_1{}^H$$

where

$$\boldsymbol{\Lambda}_0 = \boldsymbol{\Sigma}\alpha \; .$$

We claim that for all $k \geq 0$ there is a positive diagonal matrix $\boldsymbol{\Lambda}_k$ so that

$$\mathbf{X}_k = \mathbf{V}_1\boldsymbol{\Lambda}_k\mathbf{U}_1{}^H \; . \tag{1.61}$$

This has been shown to be true for $k = 0$. Assume inductively that this claim is true for $k \geq 0$. Then

$$\begin{aligned}\mathbf{X}_{k+1} &= \mathbf{X}_k 2 - \mathbf{X}_k \mathbf{A} \mathbf{X}_k = \mathbf{V}_1 \boldsymbol{\Lambda}_k 2 \mathbf{U}_1{}^H - \left(\mathbf{V}_1 \boldsymbol{\Lambda}_k \mathbf{U}_1{}^H\right)\left(\mathbf{U}_1 \boldsymbol{\Sigma} \mathbf{V}_1{}^H\right)\left(\mathbf{V}_1 \boldsymbol{\Lambda}_k \mathbf{U}_1{}^H\right) \\ &= \mathbf{V}_1 \left\{2\boldsymbol{\Lambda}_k - \boldsymbol{\Lambda}_k^2 \boldsymbol{\Sigma}\right) \mathbf{U}_1{}^H ,\end{aligned}$$

so we can take

$$\boldsymbol{\Lambda}_{k+1} = 2\boldsymbol{\Lambda}_k - \boldsymbol{\Lambda}_k^2 \boldsymbol{\Sigma} .$$

This completes the induction and proves the claim (1.61).

Next, we note that for all $k \geq 0$ we have

$$\mathbf{A}^\dagger \mathbf{A} \mathbf{X}_k = \left(\mathbf{V}_1 \boldsymbol{\Sigma}^{-1} \mathbf{U}_1{}^H\right)\left(\mathbf{U}_1 \boldsymbol{\Sigma} \mathbf{V}_1{}^H\right) \mathbf{X}_k = \mathbf{V}_1 \mathbf{V}_1{}^H \mathbf{X}_k = \mathbf{X}_k$$

and

$$\mathbf{X}_k \mathbf{A} \mathbf{A}^\dagger = \mathbf{X}_k \left(\mathbf{U}_1 \boldsymbol{\Sigma} \mathbf{V}_1{}^H\right)\left(\mathbf{V}_1 \boldsymbol{\Sigma}^{-1} \mathbf{U}_1{}^H\right) = \mathbf{X}_k \mathbf{U}_1 \mathbf{U}_1{}^H = \mathbf{X}_k .$$

It follows that

$$\begin{aligned}\mathbf{X}_{k+1} - \mathbf{A}^\dagger &= \mathbf{X}_k 2 - \mathbf{X}_k \mathbf{A} \mathbf{X}_k - \mathbf{A}^\dagger = -\left(\mathbf{X}_k \mathbf{A} - \mathbf{I}\right) \mathbf{A}^\dagger \left(\mathbf{A} \mathbf{X}_k - \mathbf{I}\right) \\ &= -\left(\mathbf{X}_k \mathbf{A} - \mathbf{I}\right) \mathbf{A}^\dagger \mathbf{A} \mathbf{A}^\dagger \left(\mathbf{A} \mathbf{X}_k - \mathbf{I}\right) \\ &= -\left(\mathbf{X}_k \mathbf{A} \mathbf{A}^\dagger - \mathbf{A}^\dagger\right) \mathbf{A} \left(\mathbf{A}^\dagger \mathbf{A} \mathbf{X}_k - \mathbf{A}^\dagger\right) \\ &= -\left(\mathbf{X}_k - \mathbf{A}^\dagger\right) \mathbf{A} \left(\mathbf{X}_k - \mathbf{A}^\dagger\right) ,\end{aligned}$$

which in turn implies that

$$\left(\mathbf{X}_{k+1} - \mathbf{A}^\dagger\right) \mathbf{A} = -\left\{\left(\mathbf{X}_k - \mathbf{A}^\dagger\right) \mathbf{A}\right\}^2 .$$

This equation and (1.61) imply that

$$\boldsymbol{\Lambda}_{k+1} \boldsymbol{\Sigma} - \mathbf{I} = -\left(\boldsymbol{\Lambda}_k \boldsymbol{\Sigma} - \mathbf{I}\right)^2 .$$

Thus $\boldsymbol{\Lambda}_k$ converges quadratically to $\boldsymbol{\Sigma}^{-1}$, provided that the diagonal entries $\lambda_i^{(0)} = \sigma_i \alpha$ of $\boldsymbol{\Lambda}_0$ of $\boldsymbol{\Lambda}_0$ satisfy

$$1 > |\lambda_i^{(0)} \sigma_i - 1| = |1 - \sigma_i^2 \alpha| .$$

This inequality is satisfied if and only if $\alpha < 2/\sigma_{\max}(\mathbf{A})^2$.

Interested reader may experiment with the following JavaScript program for the **Newton pseudo-inverse iteration.** The reader can enter the matrix by rows, with entries of an individual row separate by commas, and rows separated by semicolons. It is necessary to choose an appropriate value for the initial scale factor α. The program will compute the residuals in the four Penrose pseudo-inverse conditions (6.5) in Chap. 6 of Volume I,and the Frobenius norm of the change in the approximate pseudo-inverse. For the Läuchli matrix of Example 6.5.4 in Chap. 6 of Volume I, the iteration requires roughly 60 iterations to converge. For well-conditioned matrices such as

$$\mathbf{A} = \begin{bmatrix} 1 & 3 \\ 2 & 2 \\ 3 & 1 \end{bmatrix} ,$$

the iteration converges in about 10 iterations.

Exercise 1.5.3 Get a copy of the 207 × 277 matrix WM1. Use the iteration in Lemma 1.5.3 to compute the pseudoinverse.

1.5.4 Range and Nullspace

The LR Theorem 3.7.1 in Chap. 3 of Volume I can be used to generate a trapezoidal basis for the columns of a nonzero matrix; if $\mathbf{Q}^\top \mathbf{A}\mathbf{P} = \mathbf{L}\mathbf{R}$ where $\mathbf{Q}$ and $\mathbf{P}$ are permutation matrices, $\mathbf{L}$ is left trapezoidal and $\mathbf{R}$ is right trapezoidal, then a basis for $\mathscr{R}(\mathbf{A})$ is given by the columns of $\mathbf{QL}$. Similarly, $\mathscr{N}(\mathbf{A}) = \mathscr{N}(\mathbf{R}\mathbf{P}^\top)$. Alternatively, a Householder QR factorization could be used to generate an orthonormal basis for $\mathscr{R}(\mathbf{A})$, as discussed in Sect. 6.7.3 in Chap. 6 of Volume I. In the next lemma, we will show how to use a singular value decomposition to generate orthonormal bases for both $\mathscr{R}(\mathbf{A})$ and The results of this lemma were used to prove Lemma 1.2.19. $\mathscr{N}(\mathbf{A})$.

Lemma 1.5.4 *Suppose that* $\mathbf{A}$ *has singular value decomposition*

$$\mathbf{A} = \begin{bmatrix} \mathbf{U}_1 & \mathbf{U}_2 \end{bmatrix} \begin{bmatrix} \boldsymbol{\Sigma} & \mathbf{0} \\ \mathbf{0} & \mathbf{0} \end{bmatrix} \begin{bmatrix} {\mathbf{V}_1}^H \\ {\mathbf{V}_2}^H \end{bmatrix} ,$$

where $\begin{bmatrix} \mathbf{U}_1 & \mathbf{U}_2 \end{bmatrix}$ *and* $\begin{bmatrix} \mathbf{V}_1 & \mathbf{V}_2 \end{bmatrix}$ *are unitary, and* $\boldsymbol{\Sigma}$ *is* $r \times r$ *diagonal with positive diagonal entries. Then*

$$\mathscr{R}(\mathbf{A}) = \mathscr{R}(\mathbf{U}_1) \text{ and } \mathscr{N}(\mathbf{A}) = \mathscr{R}(\mathbf{V}_2) ,$$

and

$$\mathscr{R}(\mathbf{A}^H) = \mathscr{R}(\mathbf{V}_1) \text{ and } \mathscr{N}(\mathbf{A}^H) = \mathscr{R}(\mathbf{U}_2) .$$

Proof Let us show that $\mathscr{R}(\mathbf{A}) \subset \mathscr{R}(\mathbf{U}_1)$. If $\mathbf{Ax} \in \mathscr{R}(\mathbf{A})$, then

$$\mathbf{Ax} = \begin{bmatrix} \mathbf{U}_1 & \mathbf{U}_2 \end{bmatrix} \begin{bmatrix} \boldsymbol{\Sigma} & \mathbf{0} \\ \mathbf{0} & \mathbf{0} \end{bmatrix} \begin{bmatrix} \mathbf{V}_1{}^H \\ \mathbf{V}_2{}^H \end{bmatrix} \mathbf{x} = \mathbf{U}_1 \boldsymbol{\Sigma} \mathbf{V}_1{}^H \mathbf{x} ,$$

so $\mathbf{Ax} \in \mathscr{R}(\mathbf{U}_1)$. Next, let us show that $\mathscr{R}(\mathbf{U}_1) \subset \mathscr{R}(\mathbf{A})$. If $\mathbf{U}_1\mathbf{y} \in \mathscr{R}(\mathbf{U}_1)$, we can define $\mathbf{x} = \mathbf{V}_1 \boldsymbol{\Sigma}^{-1}\mathbf{y}$ and see that

$$\mathbf{U}_1\mathbf{y} = \mathbf{U}_1 \boldsymbol{\Sigma} \mathbf{V}_1{}^H \mathbf{x} = \mathbf{Ax} \in \mathscr{R}(\mathbf{A}) .$$

Thus $\mathscr{R}(\mathbf{A}) = \mathscr{R}(\mathbf{U}_1)$, and $\dim \mathscr{R}(\mathbf{A}) = \dim \mathscr{R}(\mathbf{U}_1) = r$.

Next, note that

$$\mathbf{AV}_2 = \begin{bmatrix} \mathbf{U}_1 & \mathbf{U}_2 \end{bmatrix} \begin{bmatrix} \boldsymbol{\Sigma} & \mathbf{0} \\ \mathbf{0} & \mathbf{0} \end{bmatrix} \begin{bmatrix} \mathbf{V}_1{}^H \\ \mathbf{V}_2{}^H \end{bmatrix} \mathbf{V}_2 = \begin{bmatrix} \mathbf{U}_1 & \mathbf{U}_2 \end{bmatrix} \begin{bmatrix} \boldsymbol{\Sigma} & \mathbf{0} \\ \mathbf{0} & \mathbf{0} \end{bmatrix} \begin{bmatrix} \mathbf{0} \\ \mathbf{I} \end{bmatrix} = \mathbf{0} .$$

Thus the $n - r$ columns of $\mathbf{V}_2$ all lie in $\mathscr{N}(\mathbf{A})$. The Fundamental Theorem of Linear Algebra 3.2.3 in Chap. 3 of Volume I implies that these columns form a basis for $\mathscr{N}(\mathbf{A})$. The remaining claims follow from the transpose of the singular value decomposition.

1.5.5 Norms

Our next lemma provides a way to evaluate the matrix 2-norm and Frobenius norm in terms of the singular values. This result was mentioned in Sect. 3.5.2 in Chap. 3 of Volume I and in Sect. 3.5.2 in Chap. 3 of Volume I.

Lemma 1.5.5 *For any matrix* $\mathbf{A}$ *with nonzero singular values* $\sigma_1 \geq \ldots \geq \sigma_r$ *we have* $\|\mathbf{A}\|_2 = \sigma_1$ *and* $\|\mathbf{A}\|_F = \sqrt{\sum_{i=1}^r \sigma_i^2}$.

Proof Let $\mathbf{A} = \mathbf{U}\boldsymbol{\Sigma}\mathbf{V}^H$ be the singular value decomposition of $\mathbf{A}$. Then the Definition 3.5.3 in Chap. 3 of Volume I of $\|\mathbf{A}\|_2$ gives us

$$\|\mathbf{A}\|_2 = \max_{\mathbf{x} \neq \mathbf{0}} \frac{\|\mathbf{Ax}\|_2}{\|\mathbf{x}\|_2} = \max_{\mathbf{x} \neq \mathbf{0}} \frac{\|\mathbf{U}\boldsymbol{\Sigma}\mathbf{V}^H\mathbf{x}\|_2}{\|\mathbf{V}^H\mathbf{x}\|_2} = \max_{\mathbf{x} \neq \mathbf{0}} \frac{\|\boldsymbol{\Sigma}\mathbf{V}^H\mathbf{x}\|_2}{\|\mathbf{V}^H\mathbf{x}\|_2} \leq \sigma_1 \max_{\mathbf{x} \neq \mathbf{0}} .$$

This upper bound is achieved by $\mathbf{x} = \mathbf{Ve}_1$.

Lemma 3.5.9 in Chap. 3 of Volume I implies the second equation in the following chain:

$$\|\mathbf{A}\|_F^2 = \|\mathbf{U}\boldsymbol{\Sigma}\mathbf{V}^H\|_F^2 = \|\boldsymbol{\Sigma}\|_F^2 = \sum_{i=1}^r \sigma_i^2 .$$

1.5.6 Minimax Theorem

Our next theorem is similar to the Minimax Theorem 1.3.2 for eigenvalues. This theorem will be used in Sect. 1.5.7 to develop a perturbations theory for singular values.

Theorem 1.5.2 (Minimax) *If* $\mathbf{A}$ *is an* $m \times n$ *matrix,* $p = \min\{m, n\}$ *and* $\sigma_1 \geq \ldots \geq \sigma_p$ *are the singular values of* $\mathbf{A}$*, then*

$$\sigma_i = \min_{\dim \mathscr{S}=n-i+1} \max_{\substack{\mathbf{x}\in\mathscr{S} \\ \mathbf{x}\neq\mathbf{0}}} \frac{\|\mathbf{A}\mathbf{x}\|_2}{\|\mathbf{x}\|_2} = \max_{\dim \mathscr{S}=i} \min_{\substack{\mathbf{x}\in\mathscr{S} \\ \mathbf{x}\neq\mathbf{0}}} \frac{\|\mathbf{A}\mathbf{x}\|_2}{\|\mathbf{x}\|_2}$$

Proof Since $\|\mathbf{A}\mathbf{x}\|_2^2 = \mathbf{x}^H\mathbf{A}^H\mathbf{A}\mathbf{x}$ and the eigenvalues of $\mathbf{A}^H\mathbf{A}$ are the squares of the singular values of $\mathbf{A}$, the claim follows from the Minimax Theorem 1.3.2. The dimensions of the subspaces in this theorem differ from those in Theorem 1.3.2, because this theorem orders the singular values from largest to smallest, while the minimax theorem ordered the eigenvalues from smallest to largest.

Corollary 1.5.4 *If* $\mathbf{A}$ *is an* $m \times n$ *matrix,* $p = \min\{m, n\}$ *and* $\sigma_1 \geq \ldots \geq \sigma_p$ *are the singular values of* $\mathbf{A}$*, then*

$$\sigma_i = \min_{\dim \mathscr{S}=n-i+1} \max_{\substack{\mathbf{x}\in\mathscr{S},\,\mathbf{x}\neq\mathbf{0} \\ \mathbf{y}\neq\mathbf{0}}} \frac{|\mathbf{y}\cdot\mathbf{A}\mathbf{x}|}{\|\mathbf{x}\|_2\|\mathbf{y}\|_2} .$$

Proof From Theorem 1.5.2 we have

$$\sigma_i = \min_{\dim \mathscr{S}=n-i+1} \max_{\substack{\mathbf{x}\in\mathscr{S} \\ \mathbf{x}\neq\mathbf{0}}} \frac{\|\mathbf{A}\mathbf{x}\|_2}{\|\mathbf{x}\|_2}$$

then Corollary 3.5.1 in Chap. 3 of Volume I gives us

$$= \min_{\dim \mathscr{S}=n-i+1} \max_{\substack{\mathbf{x}\in\mathscr{S} \\ \mathbf{x}\neq\mathbf{0}}} \frac{1}{\|\mathbf{x}\|_2} \max_{\mathbf{y}\neq\mathbf{0}} \frac{|\mathbf{y}\cdot\mathbf{A}\mathbf{x}|}{\|\mathbf{y}\|_2} .$$

The following theorem shows conclusively that the smallest singular value of a matrix measures the 2-norm distance to the nearest matrix of lower rank. This theorem was originally proved by Eckhart and Young [59].

Theorem 1.5.3 (Eckhart-Young) *Suppose that* $\mathbf{A}$ *is an* $m \times n$ *matrix,* $p = \min\{m, n\}$ *and* $\sigma_1 \geq \ldots \sigma_p$ *are the singular values of* $\mathbf{A}$*. Let* $\mathbf{v}_1, \ldots, \mathbf{v}_p$ *be the corresponding right singular vectors, and* $\mathbf{u}_1, \ldots, \mathbf{u}_p$ *be the corresponding left* *singular vectors. Then for all* $k \in [0, r)$ *we have*

$$\left\| \mathbf{A} - \sum_{i=1}^{k} \mathbf{v}_i \sigma_i \mathbf{u}_i{}^H \right\|_2 = \sigma_{k+1} = \min_{rank(\mathbf{B})=k} \|\mathbf{A} - \mathbf{B}\|_2 \ .$$

Proof Let the singular value decomposition of $\mathbf{A}$ be $\mathbf{A} = \mathbf{U}\boldsymbol{\Sigma}\mathbf{V}^H$. We can partition

$$\mathbf{U} = \begin{bmatrix} \mathbf{U}_1 & \mathbf{U}_2 \end{bmatrix} \ , \ \mathbf{V} = \begin{bmatrix} \mathbf{V}_1 & \mathbf{V}_2 \end{bmatrix} \text{ and } \boldsymbol{\Sigma} = \begin{bmatrix} \boldsymbol{\Sigma}_1 & \mathbf{0} \\ \mathbf{0} & \boldsymbol{\Sigma}_2 \end{bmatrix} ,$$

where $\mathbf{U}_1$ is $m \times k$, $\mathbf{V}_1$ is $n \times k$ and $\boldsymbol{\Sigma}_1$ is $k \times k$. Then

$$\left\| \mathbf{A} - \sum_{i=1}^{k} \mathbf{v}_i \sigma_i \mathbf{u}_i{}^H \right\|_2 = \left\| \begin{bmatrix} \mathbf{U}_1 & \mathbf{U}_2 \end{bmatrix} \begin{bmatrix} \boldsymbol{\Sigma}_1 & \mathbf{0} \\ \mathbf{0} & \boldsymbol{\Sigma}_2 \end{bmatrix} \begin{bmatrix} \mathbf{V}_1{}^H \\ \mathbf{V}_2{}^H \end{bmatrix} - \mathbf{U}_1 \boldsymbol{\Sigma}_1 \mathbf{V}_1{}^H \right\|_2$$

$$= \left\| \mathbf{U}_2 \boldsymbol{\Sigma}_2 \mathbf{V}_2{}^H \right\|_2 = \|\boldsymbol{\Sigma}_2\|_2 = \sigma_{k+1} \ .$$

This proves the first claim.

Since $\mathbf{U}_1 \boldsymbol{\Sigma}_1 \mathbf{V}_1{}^H$ has rank k and $\|\mathbf{A} - \mathbf{U}_1 \boldsymbol{\Sigma}_1 \mathbf{V}_1{}^H\|_2 = \sigma_{k+1}$, it follows that

$$\min_{\text{rank}(\mathbf{B})=k} \|\mathbf{A} - \mathbf{B}\|_2 \leq \sigma_{k+1} \ .$$

Suppose that the $m \times n$ matrix $\mathbf{B}$ has rank k. The Fundamental Theorem of Linear Algebra 3.2.3 in Chap. 3 of Volume I implies that $\dim \mathscr{N}(\mathbf{B}) = n - k$, and Lemmas 3.2.4 and 3.2.7 in Chap. 3 of Volume I imply that this subspace has a basis consisting of $n - k$ n-vectors. Let the columns of the $n \times (n - k)$ matrix $\mathbf{Z}$ be the members of this basis. Then Lemma 3.2.5 in Chap. 3 of Volume I shows that the $n \times (n + 1)$ matrix

$$\mathbf{C} \equiv \begin{bmatrix} \mathbf{V}_1 & \mathbf{V}_2 \mathbf{e}_1 & \mathbf{Z} \end{bmatrix}$$

has a non-trivial nullspace. If $\mathbf{x}$, ξ and $\mathbf{y}$ are not all zero and

$$\mathbf{V}_1 \mathbf{x} + \mathbf{V}_2 \mathbf{e}_1 \xi - \mathbf{Z} \mathbf{y} = \mathbf{0} \ ,$$

then

$$\mathbf{z} \equiv \mathbf{Z} \mathbf{y} = \mathbf{V}_1 \mathbf{x} + \mathbf{V}_2 \mathbf{e}_1 \xi$$

is such that $\mathbf{B}\mathbf{z} = \mathbf{0}$, $\mathbf{z} \neq \mathbf{0}$ and

$$\mathbf{A}\mathbf{z} = \mathbf{U}_1 \boldsymbol{\Sigma}_1 \mathbf{V}_1{}^H \mathbf{z} + \mathbf{U}_2 \boldsymbol{\Sigma}_2 \mathbf{V}_2{}^H \mathbf{z} \ .$$

We may scale $\mathbf{z}$ so that $1 = \|\mathbf{z}\|_2^2 = \|\mathbf{x}\|_2 + |\xi|^2$. It follows that

$$\|\mathbf{A}-\mathbf{B}\|_2^2 \geq \|(\mathbf{A}-\mathbf{B})\mathbf{z}\|_2^2 = \|\mathbf{A}\mathbf{z}\|_2^2$$
$$= \left\| \begin{bmatrix} \mathbf{U}_1 & \mathbf{U}_2 \end{bmatrix} \begin{bmatrix} \boldsymbol{\Sigma}_1 & \mathbf{0} \\ \mathbf{0} & \boldsymbol{\Sigma}_2 \end{bmatrix} \begin{bmatrix} {\mathbf{V}_1}^H \\ {\mathbf{V}_2}^H \end{bmatrix} \begin{bmatrix} \mathbf{V}_1 & \mathbf{V}_2 \end{bmatrix} \begin{bmatrix} \mathbf{x} \\ \mathbf{e}_1\xi \end{bmatrix} \right\|_2 = \left\| \begin{bmatrix} \boldsymbol{\Sigma}_1 & \mathbf{0} \\ \mathbf{0} & \boldsymbol{\Sigma}_2 \end{bmatrix} \begin{bmatrix} \mathbf{x} \\ \mathbf{e}_1\xi \end{bmatrix} \right\|_2^2$$
$$\geq \sigma_{k+1}^2 \left\| \begin{bmatrix} \mathbf{x} \\ \mathbf{e}_1\xi \end{bmatrix} \right\|_2^2 = \sigma_{k+1}^2 \ .$$

This completes the proof of the second claim.

1.5.7 Perturbation Theory

Once again, we will perform the steps in scientific computing applied to the determination of singular values. We have proved their existence in Theorem 1.5.1, and will examine their sensitivity to perturbations in this section. Algorithms for computing singular values will be presented in Sects. 1.5.9 and 1.5.10.

We will begin by bounding the perturbations in singular values. This result is similar to the bound on perturbations of eigenvalues of Hermitian matrices in Lemma 1.3.5.

Lemma 1.5.6 *Suppose that* $\mathbf{A}$ *and* $\widetilde{\mathbf{A}}$ *are two* $m \times n$ *matrices, and let* $p = \min\{m, n\}$. *Assume that the singular values of* $\mathbf{A}$ *are* $\sigma_1 \geq \ldots \geq \sigma_p$, *and the singular values of* $\widetilde{\mathbf{A}}$ *are* $\widetilde{\sigma}_1 \geq \ldots \geq \widetilde{\sigma}_p$. *Then for all* $1 \leq i \leq p$ *we have*

$$|\widetilde{\sigma}_i - \sigma_i| \leq \left\|\widetilde{\mathbf{A}} - \mathbf{A}\right\|_2 \ .$$

Proof Let $\mathscr{S}$ be a subspace of n-vectors with $\dim \mathscr{S} = n - i + 1$. The Minimax Theorem 1.5.2 implies that

$$\sigma_i \leq \max_{\substack{\mathbf{x}\in\mathscr{S} \\ \mathbf{x}\neq\mathbf{0}}} \frac{\|\mathbf{A}\mathbf{x}\|_2}{\|\mathbf{x}\|_2} \leq \max_{\substack{\mathbf{x}\in\mathscr{S} \\ \mathbf{x}\neq\mathbf{0}}} \frac{\left\|\widetilde{\mathbf{A}}\mathbf{x}\right\|_2}{\|\mathbf{x}\|_2} + \max_{\substack{\mathbf{x}\in\mathscr{S} \\ \mathbf{x}\neq\mathbf{0}}} \frac{\left\|(\mathbf{A}-\widetilde{\mathbf{A}})\mathbf{x}\right\|_2}{\|\mathbf{x}\|_2} \leq \max_{\substack{\mathbf{x}\in\mathscr{S} \\ \mathbf{x}\neq\mathbf{0}}} \frac{\left\|\widetilde{\mathbf{A}}\mathbf{x}\right\|_2}{\|\mathbf{x}\|_2} + \|\mathbf{A}-\widetilde{\mathbf{A}}\|_2 \ .$$

We can take the minimum over all $\mathscr{S}$ to get

$$\sigma_i \leq \widetilde{\sigma}_i + \|\mathbf{A}-\widetilde{\mathbf{A}}\|_2; .$$

Then we can interchange $\mathbf{A}$ and $\widetilde{\mathbf{A}}$ to get

$$\widetilde{\sigma}_i \leq \sigma_i + \|\widetilde{\mathbf{A}}-\mathbf{A}\|_2; .$$

Combining the last two inequalities proves the claim.

Next, we will bound the perturbation in singular vectors associated with distinct singular values.

Theorem 1.5.4 (Stewart Singular Subspace Perturbation) *Let* $\mathbf{A}$ *and* $\widetilde{\mathbf{A}}$ *be* $m \times n$ *matrices. Suppose that there are unitary matrices*

$$\mathbf{U} = \begin{bmatrix} \mathbf{U}_1 & \mathbf{U}_2 \end{bmatrix} \text{ and } \mathbf{V} = \begin{bmatrix} \mathbf{V}_1 & \mathbf{V}_2 \end{bmatrix} ,$$

with $\mathbf{U}_1$ $m \times r$ *and* $\mathbf{V}_1$ $n \times r$, *so that*

$$\begin{bmatrix} \mathbf{B}_{11} & \mathbf{0} \\ \mathbf{0} & \mathbf{B}_{22} \end{bmatrix} = \begin{bmatrix} \mathbf{U}_1{}^H \\ \mathbf{U}_2{}^H \end{bmatrix} \mathbf{A} \begin{bmatrix} \mathbf{V}_1 & \mathbf{V}_2 \end{bmatrix} .$$

Suppose that the singular values $\sigma_j^{(1)}$ *of* $\mathbf{B}_{11}$ *and* $\sigma_i^{(2)}$ *of* $\mathbf{B}_{22}$ *are such that*

$$\delta = \min_{i,j} \left| \sigma_j^{(1)} - \sigma_i^{(2)} \right| - \left\| \mathbf{U}_1{}^H (\widetilde{\mathbf{A}} - \mathbf{A}) \mathbf{V}_1 \right\|_F - \left\| \mathbf{U}_2{}^H (\widetilde{\mathbf{A}} - \mathbf{A}) \mathbf{V}_2 \right\|_F > 0 .$$

Also, assume that

$$\left\| \mathbf{U}_1{}^H (\widetilde{\mathbf{A}} - \mathbf{A}) \mathbf{V}_2 \right\|_F^2 + \left\| \mathbf{U}_2{}^H (\widetilde{\mathbf{A}} - \mathbf{A}) \mathbf{V}_1 \right\|_F^2 < \frac{\sqrt{2}}{4} \delta^2 .$$

There there exists an $(m-r) \times r$ *matrix* $\mathbf{Q}$ *and an* $(n-r) \times r$ *matrix* $\mathbf{P}$ *so that*

$$\begin{aligned} &\begin{bmatrix} \mathbf{I} & \mathbf{Q}^H \\ -\mathbf{Q} & \mathbf{I} \end{bmatrix} \begin{bmatrix} \mathbf{U}_1{}^H \\ \mathbf{U}_2{}^H \end{bmatrix} \widetilde{\mathbf{A}} \begin{bmatrix} \mathbf{V}_1 & \mathbf{V}_2 \end{bmatrix} \begin{bmatrix} \mathbf{I} & -\mathbf{P}^H \\ \mathbf{P} & \mathbf{I} \end{bmatrix} \\ &= \begin{bmatrix} (\mathbf{U}_1 + \mathbf{U}_2 \mathbf{Q})^H \widetilde{\mathbf{A}} (\mathbf{V}_1 + \mathbf{V}_2 \mathbf{P}) & \mathbf{0} \\ \mathbf{0} & (\mathbf{U}_2 - \mathbf{U}_1 \mathbf{Q}^H)^H \widetilde{\mathbf{A}} (\mathbf{V}_2 - \mathbf{V}_1 \mathbf{P}^H) \end{bmatrix} . \end{aligned}$$

Furthermore,

$$\|\mathbf{Q}\|_F^2 + \|\mathbf{P}\|_F^2 \le 4 \frac{\left\| \mathbf{U}_1{}^H \left(\widetilde{\mathbf{A}} - \mathbf{A}\right) \mathbf{V}_2 \right\|_F^2 + \left\| \mathbf{U}_2{}^H \left(\widetilde{\mathbf{A}} - \mathbf{A}\right) \mathbf{V}_1 \right\|_F^2}{\delta^2} .$$

The final estimate in this theorem shows that the perturbation in the singular vectors is bounded by the size of the perturbation in the matrix *divided by* the quantity δ, which bounds the separation between the singular values.

Proof Without loss of generality, we may assume that $m \ge n$; otherwise, we can work with the Hermitians of $\mathbf{A}$ and $\widetilde{\mathbf{A}}$. Define

$$\begin{bmatrix} \widetilde{\mathbf{B}}_{11} & \widetilde{\mathbf{B}}_{12} \\ \widetilde{\mathbf{B}}_{21} & \widetilde{\mathbf{B}}_{22} \end{bmatrix} = \begin{bmatrix} \mathbf{U}_1{}^H \\ \mathbf{U}_2{}^H \end{bmatrix} \widetilde{\mathbf{A}} \begin{bmatrix} \mathbf{V}_1 & \mathbf{V}_2 \end{bmatrix} .$$

Note that

$$\widetilde{\mathbf{B}}_{12} = \mathbf{U}_1{}^H \widetilde{\mathbf{A}} \mathbf{V}_2 = \mathbf{U}_1{}^H \left(\widetilde{\mathbf{A}} - \mathbf{A}\right) \mathbf{V}_2$$

and

$$\widetilde{\mathbf{B}}_{21} = \mathbf{U}_2{}^H \widetilde{\mathbf{A}} \mathbf{V}_1 = \mathbf{U}_2{}^H \left(\widetilde{\mathbf{A}} - \mathbf{A}\right) \mathbf{V}_1 \; .$$

Given an arbitrary $(m - r) \times r$ matrix $\mathbf{Q}$ and an arbitrary $(n - r) \times r$ matrix $\mathbf{P}$, we have

$$\begin{aligned}
&\begin{bmatrix} \mathbf{I} & \mathbf{Q}^H \\ -\mathbf{Q} & \mathbf{I} \end{bmatrix} \begin{bmatrix} \mathbf{U}_1{}^H \\ \mathbf{U}_2{}^H \end{bmatrix} \widetilde{\mathbf{A}} \begin{bmatrix} \mathbf{V}_1 & \mathbf{V}_2 \end{bmatrix} \begin{bmatrix} \mathbf{I} & -\mathbf{P}^H \\ \mathbf{P} & \mathbf{I} \end{bmatrix} \\
&= \begin{bmatrix} \mathbf{I} & \mathbf{Q}^H \\ -\mathbf{Q} & \mathbf{I} \end{bmatrix} \begin{bmatrix} \widetilde{\mathbf{B}}_{11} & \widetilde{\mathbf{B}}_{12} \\ \widetilde{\mathbf{B}}_{21} & \widetilde{\mathbf{B}}_{22} \end{bmatrix} \begin{bmatrix} \mathbf{I} & -\mathbf{P}^H \\ \mathbf{P} & \mathbf{I} \end{bmatrix} \\
&= \begin{bmatrix} \widetilde{\mathbf{B}}_{11} + \widetilde{\mathbf{B}}_{12}\mathbf{P} + \mathbf{Q}^H\widetilde{\mathbf{B}}_{21} + \mathbf{Q}^H\widetilde{\mathbf{B}}_{22}\mathbf{P} & \widetilde{\mathbf{B}}_{12} - \widetilde{\mathbf{B}}_{11}\mathbf{P}^H + \mathbf{Q}^H\widetilde{\mathbf{B}}_{22} - \mathbf{Q}^H\widetilde{\mathbf{B}}_{21}\mathbf{P}^H \\ \widetilde{\mathbf{B}}_{21} + \widetilde{\mathbf{B}}_{22}\mathbf{P} - \mathbf{Q}\widetilde{\mathbf{B}}_{11} - \mathbf{Q}\widetilde{\mathbf{B}}_{12}\mathbf{P} & \widetilde{\mathbf{B}}_{22} - \widetilde{\mathbf{B}}_{21}\mathbf{P}^H - \mathbf{Q}\widetilde{\mathbf{B}}_{12} + \mathbf{Q}\widetilde{\mathbf{B}}_{11}\mathbf{P}^H \end{bmatrix} \; .
\end{aligned}$$

In order for the off-diagonal blocks to be zero, we want to choose $\mathbf{Q}$ and $\mathbf{P}$ so that

$$\mathbf{Q}\widetilde{\mathbf{B}}_{11} - \widetilde{\mathbf{B}}_{22}\mathbf{P} = \widetilde{\mathbf{B}}_{21} - \mathbf{Q}\widetilde{\mathbf{B}}_{21}\mathbf{P}^H \text{ and} \mathbf{P}\widetilde{\mathbf{B}}_{11}{}^H - \widetilde{\mathbf{B}}_{22}{}^H\mathbf{Q} = \widetilde{\mathbf{B}}_{12}{}^H - \mathbf{P}\widetilde{\mathbf{B}}_{21}{}^H\mathbf{Q} \; .$$

This suggests that we define the linear operator $\mathscr{T}$ by

$$\mathscr{T}\left(\begin{bmatrix} \mathbf{Q} \\ \mathbf{P} \end{bmatrix}\right) \equiv \begin{bmatrix} \mathbf{Q}\widetilde{\mathbf{B}}_{11} - \widetilde{\mathbf{B}}_{22}\mathbf{P} \\ \mathbf{P}\widetilde{\mathbf{B}}_{11}{}^H - \widetilde{\mathbf{B}}_{22}{}^H\mathbf{Q} \end{bmatrix} ,$$

and the nonlinear operator ϕ by

$$\phi\left(\begin{bmatrix} \mathbf{Q} \\ \mathbf{P} \end{bmatrix}\right) = \begin{bmatrix} \mathbf{Q}\widetilde{\mathbf{B}}_{12}\mathbf{P} \\ \mathbf{P}\widetilde{\mathbf{B}}_{21}{}^H\mathbf{Q} \end{bmatrix} .$$

We want to find $\mathbf{Q}$ and $\mathbf{P}$ so that

$$\mathscr{T}\left(\begin{bmatrix} \mathbf{Q} \\ \mathbf{P} \end{bmatrix}\right) = \begin{bmatrix} \widetilde{\mathbf{B}}_{21} \\ \widetilde{\mathbf{B}}_{12}{}^H \end{bmatrix} - \phi\left(\begin{bmatrix} \mathbf{Q} \\ \mathbf{P} \end{bmatrix}\right) .$$

We will use Lemma 1.2.21 to solve this equation. We will define the inner product

$$\left(\begin{bmatrix} \mathbf{Q}_1 \\ \mathbf{P}_1 \end{bmatrix} , \begin{bmatrix} \mathbf{Q}_2 \\ \mathbf{P}_2 \end{bmatrix}\right) = \operatorname{tr}\left(\mathbf{Q}_1{}^H\mathbf{Q}_2\right) + \operatorname{tr}\left(\mathbf{P}_1{}^H\mathbf{P}_2\right) \, ,$$

where tr(**M**) is the trace of the square matrix **M**. The norm associated with this inner product is square root of

$$\left\| \begin{bmatrix} \mathbf{Q} \\ \mathbf{P} \end{bmatrix} \right\|^2 = \|\mathbf{Q}\|_F^2 + \|\mathbf{P}\|_F^2 .$$

Let the singular value decompositions of $\widetilde{\mathbf{B}}_{11}$ and $\widetilde{\mathbf{B}}_{22}$ be

$$\widetilde{\mathbf{B}}_{11} = \widetilde{\mathbf{U}}_1 \widetilde{\boldsymbol{\Sigma}}_1 \widetilde{\mathbf{V}}_1{}^H \text{ and } \widetilde{\mathbf{B}}_{22} = \widetilde{\mathbf{U}}_2 \widetilde{\boldsymbol{\Sigma}}_2 \widetilde{\mathbf{V}}_2{}^H ,$$

where $\widetilde{\mathbf{U}}_1$, $\widetilde{\mathbf{V}}_1$, $\widetilde{\mathbf{U}}_2$ and $\widetilde{\mathbf{V}}_2$ are unitary. Given **Q** and **P**, define

$$\mathbf{Q}' = \widetilde{\mathbf{U}}_2{}^H \mathbf{Q} \widetilde{\mathbf{U}}_1 \text{ and } \mathbf{P}' = \widetilde{\mathbf{V}}_2{}^H \mathbf{P} \widetilde{\mathbf{V}}_1 .$$

Then

$$\left\| \mathscr{T}\left(\begin{bmatrix} \mathbf{Q} \\ \mathbf{P} \end{bmatrix} \right) \right\|^2 = \|\mathbf{Q}\widetilde{\mathbf{B}}_{11} - \widetilde{\mathbf{B}}_{22}\mathbf{P}\|_F^2 + \left\| \mathbf{P}\widetilde{\mathbf{B}}_{11}{}^H - \mathbf{B}_{22}{}^H \mathbf{Q} \right\|_F^2$$

$$= \left\| \mathbf{Q}\widetilde{\mathbf{U}}_1 \widetilde{\boldsymbol{\Sigma}}_1 \widetilde{\mathbf{V}}_1{}^H - \widetilde{\mathbf{U}}_2 \widetilde{\boldsymbol{\Sigma}}_2 \widetilde{\mathbf{V}}_2{}^H \mathbf{P} \right\|_F^2 + \left\| \mathbf{P}\widetilde{\mathbf{V}}_1 \widetilde{\boldsymbol{\Sigma}}_1{}^\top \widetilde{\mathbf{U}}_1{}^H - \widetilde{\mathbf{V}}_2 \widetilde{\boldsymbol{\Sigma}}_2{}^\top \widetilde{\mathbf{U}}_2{}^H \mathbf{Q} \right\|_F^2$$

$$= \|\mathbf{Q}'\widetilde{\boldsymbol{\Sigma}}_1 - \widetilde{\boldsymbol{\Sigma}}_2 \mathbf{P}'\|_F^2 + \left\| \mathbf{P}'\widetilde{\boldsymbol{\Sigma}}_1{}^\top - \widetilde{\boldsymbol{\Sigma}}_2{}^\top \mathbf{Q}' \right\|_F^2$$

$$= \sum_{i=1}^{n-r} \sum_{j=1}^{r} \left\{ \left(\mathbf{Q}'_{ij} \widetilde{\sigma}_j^{(1)} - \widetilde{\sigma}_i^{(2)} \mathbf{P}'_{ij} \right)^2 + \left(\mathbf{P}'_{ij} \widetilde{\sigma}_j^{(1)} - \widetilde{\sigma}_i^{(2)} \mathbf{Q}'_{ij} \right)^2 \right\}$$

$$+ \sum_{i=n+1-r}^{m-r} \sum_{j=1}^{r} \left(\mathbf{Q}'_{ij} \widetilde{\sigma}_j^{(1)} \right)^2 .$$

Note that

$$(\alpha\sigma_1 - \sigma_2\beta)^2 + (\beta\sigma_1 - \sigma_2\alpha)^2 = \begin{bmatrix} \alpha & \beta \end{bmatrix} \begin{bmatrix} \sigma_1^2 + \sigma_2^2 & -2\sigma_1\sigma_2 \\ -2\sigma_1\sigma_2 & \sigma_1^2 + \sigma_2^2 \end{bmatrix} \begin{bmatrix} \alpha \\ \beta \end{bmatrix} ,$$

and that the eigenvalues of the matrix in this quadratic form are $\lambda = (\sigma_1 \pm \sigma_2)^2$. If $m > n$ then we define $\widetilde{\sigma}_{n+1-r}^{(2)} = 0$. It follows that

$$\left\| \mathscr{T}\left(\begin{bmatrix} \mathbf{Q} \\ \mathbf{P} \end{bmatrix} \right) \right\|^2$$

$$\geq \sum_{i=1}^{n-r} \sum_{j=1}^{r} \left\{ \left(\widetilde{\sigma}_j^{(1)} - \widetilde{\sigma}_i^{(2)} \right)^2 \left[(\mathbf{Q}'_{ij})^2 + (\mathbf{P}'_{ij})^2 \right] \right\} + \sum_{i=n+1-r}^{m-r} \sum_{j=1}^{r} \left(\widetilde{\sigma}_j^{(1)} \right)^2 (\mathbf{Q}'_{ij})^2$$

$$\geq \min_{\substack{1\leq i\leq n+1-r\\1\leq j\leq r}} \left(\widetilde{\sigma}_j^{(1)} - \sigma\widetilde{\sigma}2)\right)^2 \left\{\sum_{i=1}^{m-r}\sum_{j=1}^{r}(\mathbf{Q}'_{ij})^2 + \sum_{i=1}^{n-r}\sum_{j=1}^{r}(\mathbf{P}'_{ij})^2\right\}$$

$$= \min_{\substack{1\leq i\leq n+1-r\\1\leq j\leq r}} \left(\widetilde{\sigma}_j^{(1)} - \widetilde{\sigma}_i^{(2)}\right)^2 \{\|\mathbf{Q}'\|_F^2 + \|\mathbf{P}'\|_F^2\}$$

$$= \min_{\substack{1\leq i\leq n+1-r\\1\leq j\leq r}} \left(\widetilde{\sigma}_j^{(1)} - \widetilde{\sigma}_i^{(2)}\right)^2 \{\|\mathbf{Q}\|_F^2 + \|\mathbf{P}\|_F^2\} .$$

Thus

$$\|\mathscr{T}^{-1}\|^2 = \max_{\begin{bmatrix}\mathbf{Q}\\\mathbf{P}\end{bmatrix}\neq\mathbf{0}} \frac{\left\|\begin{bmatrix}\mathbf{Q}\\\mathbf{P}\end{bmatrix}\right\|^2}{\left\|\mathscr{T}\left(\begin{bmatrix}\mathbf{Q}\\\mathbf{P}\end{bmatrix}\right)\right\|^2} \geq \frac{1}{\min_{\substack{1\leq i\leq n+1-r\\1\leq j\leq r}} \left(\widetilde{\sigma}_j^{(1)} - \widetilde{\sigma}_i^{(2)}\right)^2} .$$

Next, we note that

$$\left\|\phi\left(\begin{bmatrix}\mathbf{Q}\\\mathbf{P}\end{bmatrix}\right)\right\|^2 = \|\mathbf{Q}\widetilde{\mathbf{B}}_{12}\mathbf{P}\|_F^2 + \left\|\mathbf{P}\widetilde{\mathbf{B}}_{21}{}^H\mathbf{Q}\right\|_F^2 \leq \left(\|\widetilde{\mathbf{B}}_{12}\|_F^2 + \|\widetilde{\mathbf{B}}_{21}\|_F^2\right)\|\mathbf{Q}\|_F^2\|\mathbf{P}\|_F^2$$

$$\leq \frac{1}{4}\left(\|\widetilde{\mathbf{B}}_{12}\|_F^2 + \|\widetilde{\mathbf{B}}_{21}\|_F^2\right)\left(\|\mathbf{Q}\|_F^2 + \|\mathbf{P}\|_F^2\right)^2 ,$$

so

$$\left\|\phi\left(\begin{bmatrix}\mathbf{Q}\\\mathbf{P}\end{bmatrix}\right)\right\| \leq \frac{1}{2}\left\|\begin{bmatrix}\widetilde{\mathbf{B}}_{12}\\\widetilde{\mathbf{B}}_{21}{}^H\end{bmatrix}\right\|\left\|\begin{bmatrix}\mathbf{Q}\\\mathbf{P}\end{bmatrix}\right\|^2 .$$

Furthermore,

$$\left\|\phi\left(\begin{bmatrix}\mathbf{Q}\\\mathbf{P}\end{bmatrix}\right) - \phi\left(\begin{bmatrix}\widetilde{\mathbf{Q}}\\\widetilde{\mathbf{P}}\end{bmatrix}\right)\right\|^2 = \left\|\begin{bmatrix}\mathbf{Q}\widetilde{\mathbf{B}}_{12}\mathbf{P} - \widetilde{\mathbf{Q}}\widetilde{\mathbf{B}}_{12}\widetilde{\mathbf{P}}\\\mathbf{P}\widetilde{\mathbf{B}}_{21}{}^H\mathbf{Q} - \widetilde{\mathbf{P}}\widetilde{\mathbf{B}}_{21}{}^H\widetilde{\mathbf{Q}}\end{bmatrix}\right\|^2$$

$$= \|\mathbf{Q}\widetilde{\mathbf{B}}_{12}\mathbf{P} - \widetilde{\mathbf{Q}}\widetilde{\mathbf{B}}_{12}\widetilde{\mathbf{P}}\|_F^2 + \left\|\mathbf{P}\widetilde{\mathbf{B}}_{21}{}^H\mathbf{Q} - \widetilde{\mathbf{P}}\widetilde{\mathbf{B}}_{21}{}^H\widetilde{\mathbf{Q}}\right\|_F^2$$

$$= \|\mathbf{Q}\widetilde{\mathbf{B}}_{12}(\mathbf{P} - \widetilde{\mathbf{P}}) + (\mathbf{Q} - \widetilde{\mathbf{Q}})\widetilde{\mathbf{B}}_{12}\widetilde{\mathbf{P}}\|_F^2 + \left\|\mathbf{P}\widetilde{\mathbf{B}}_{21}{}^H(\mathbf{Q} - \widetilde{\mathbf{Q}}) + (\mathbf{P} - \widetilde{\mathbf{P}})\widetilde{\mathbf{B}}_{21}{}^H\widetilde{\mathbf{Q}}\right\|_F^2$$

$$\leq \left\{\|\mathbf{Q}\|_F\|\widetilde{\mathbf{B}}_{12}\|_F\|\mathbf{P} - \widetilde{\mathbf{P}}\|_F + \|\mathbf{Q} - \widetilde{\mathbf{Q}}\|_F\|\widetilde{\mathbf{B}}_{12}\|_F\|\widetilde{\mathbf{P}}\|_F\right\}^2$$

$$+ \left\{\|\mathbf{P}\|_F\|\widetilde{\mathbf{B}}_{21}\|_F\|\mathbf{Q} - \widetilde{\mathbf{Q}}\|_F + \|\mathbf{P} - \widetilde{\mathbf{P}}\|_F\|\widetilde{\mathbf{B}}_{21}\|_F\|\widetilde{\mathbf{Q}}\|_F\right\}^2$$

$$\leq 2\|\widetilde{\mathbf{B}}_{12}\|_F^2\left\{\|\mathbf{Q} - \widetilde{\mathbf{Q}}\|_F^2 + \|\mathbf{P} - \widetilde{\mathbf{P}}\|_F^2\right\}\max\left\{\|\mathbf{Q}\|_F^2 , \|\widetilde{\mathbf{P}}\|_F^2\right\}$$

$$
\begin{aligned}
&+ 2\|\widetilde{\mathbf{B}}_{21}\|_F^2 \left\{\|\mathbf{Q}-\widetilde{\mathbf{Q}}\|_F^2 + \|\mathbf{P}-\widetilde{\mathbf{P}}\|_F^2\right\} \max\left\{\|\mathbf{P}\|_F^2 + \|\widetilde{\mathbf{Q}}\|_F^2\right\} \\
&\leq 2\left\{\|\widetilde{\mathbf{B}}_{12}\|_F^2 + \|\widetilde{\mathbf{B}}_{21}\|_F^2\right\} \left\{\|\mathbf{Q}-\widetilde{\mathbf{Q}}\|_F^2 + \|\mathbf{P}-\widetilde{\mathbf{P}}\|_F^2\right\} \max\left\{\|\mathbf{Q}\|_F^2, \|\widetilde{\mathbf{P}}\|_F^2, \|\widetilde{\mathbf{Q}}\|_F^2, \|\widetilde{\mathbf{P}}\|_F^2\right\} \\
&\leq 2\left\|\begin{bmatrix}\widetilde{\mathbf{B}}_{12} \\ \widetilde{\mathbf{B}}_{21}{}^H\end{bmatrix}\right\|^2 \left\|\begin{bmatrix}\mathbf{Q}-\widetilde{\mathbf{Q}} \\ \mathbf{P}-\widetilde{\mathbf{P}}\end{bmatrix}\right\|^2 \max\left\{\left\|\begin{bmatrix}\mathbf{Q} \\ \mathbf{P}\end{bmatrix}\right\|^2, \left\|\begin{bmatrix}\widetilde{\mathbf{Q}} \\ \widetilde{\mathbf{P}}\end{bmatrix}\right\|^2\right\} ,
\end{aligned}
$$

so

$$
\left\|\phi\left(\begin{bmatrix}\mathbf{Q} \\ \mathbf{P}\end{bmatrix}\right) - \phi\left(\begin{bmatrix}\widetilde{\mathbf{Q}} \\ \widetilde{\mathbf{P}}\end{bmatrix}\right)\right\| \leq 2\frac{1}{\sqrt{2}} \left\|\begin{bmatrix}\widetilde{\mathbf{B}}_{12} \\ \widetilde{\mathbf{B}}_{21}{}^H\end{bmatrix}\right\| \left\|\begin{bmatrix}\mathbf{Q}-\widetilde{\mathbf{Q}} \\ \mathbf{P}-\widetilde{\mathbf{P}}\end{bmatrix}\right\| \max\left\{\left\|\begin{bmatrix}\mathbf{Q} \\ \mathbf{P}\end{bmatrix}\right\|, \left\|\begin{bmatrix}\widetilde{\mathbf{Q}} \\ \widetilde{\mathbf{P}}\end{bmatrix}\right\|\right\} .
$$

Thus the assumptions of Lemma 1.2.21 have been verified with

$$
\eta = \frac{1}{\sqrt{2}} \left\|\begin{bmatrix}\widetilde{\mathbf{B}}_{12} \\ \widetilde{\mathbf{B}}_{21}{}^H\end{bmatrix}\right\| .
$$

If

$$
1 > 4\eta \left\|\begin{bmatrix}\widetilde{\mathbf{B}}_{12} \\ \widetilde{\mathbf{B}}_{21}{}^H\end{bmatrix}\right\| \|\mathscr{T}^{-1}\|^2 = \frac{4}{\sqrt{2}} \left\|\begin{bmatrix}\widetilde{\mathbf{B}}_{12} \\ \widetilde{\mathbf{B}}_{21}{}^H\end{bmatrix}\right\|^2 \frac{1}{\min_{\substack{1\leq i\leq n+1-r \\ 1\leq j\leq r}} \left(\widetilde{\sigma}_j^{(1)} - \widetilde{\sigma}_i^{(2)}\right)^2} ,
$$

then this lemma guarantees that the desired **Q** and **P** exist with

$$
\left\|\begin{bmatrix}\mathbf{Q} \\ \mathbf{P}\end{bmatrix}\right\| < 2\left\|\mathscr{T}^{-1}\begin{bmatrix}\widetilde{\mathbf{B}}_{12} \\ \widetilde{\mathbf{B}}_{21}{}^H\end{bmatrix}\right\| \leq 2\frac{\left\|\begin{bmatrix}\widetilde{\mathbf{B}}_{12} \\ \widetilde{\mathbf{B}}_{21}{}^H\end{bmatrix}\right\|}{\min_{\substack{1\leq i\leq n+1-r \\ 1\leq j\leq r}} \left|\widetilde{\sigma}_j^{(1)} - \widetilde{\sigma}_i^{(2)}\right|} .
$$

The conclusion of the theorem can be obtained by using Lemma 1.5.6 to bound the singular values of $\widetilde{\mathbf{A}}$ in terms of the singular values of $\mathbf{A}$.

Exercise 1.5.4 Let

$$
\mathbf{A} = \begin{bmatrix}1 & 1 \\ 0 & 0 \\ 0 & 0\end{bmatrix} \text{ and } \widetilde{\mathbf{A}} = \begin{bmatrix}1 & 1 \\ \delta & 0 \\ 0 & \delta\end{bmatrix} .
$$

The singular value decompositions of these two matrices can be found in Example 1.5.1.

1. Compute the differences of the singular values for these two matrices, and verify the conclusion of Lemma 1.5.6.

2. If $\mathbf{U}$ and $\mathbf{V}$ are the matrices of singular vectors for $\mathbf{A}$, and $\widetilde{\mathbf{U}}$ and $\widetilde{\mathbf{V}}$ are the matrices of singular vectors for $\widetilde{\mathbf{A}}$, find $\mathbf{Q}$ and $\mathbf{P}$ so that

$$\widetilde{\mathbf{U}} = \mathbf{U} \begin{bmatrix} \mathbf{I} & -\mathbf{Q}^H \\ \mathbf{Q} & \mathbf{I} \end{bmatrix} \text{ and } \widetilde{\mathbf{V}} = \mathbf{V} \begin{bmatrix} \mathbf{I} & -\mathbf{P}^H \\ \mathbf{P} & \mathbf{I} \end{bmatrix} .$$

3. Use the result of the previous step to verify the conclusion of the Stewart Singular Subspace Perturbation Theorem 1.5.4.

1.5.8 Householder Bidiagonalization

So far, we have proved the existence of singular values, and studied their sensitivity to perturbations. Now we are ready to develop an algorithm for computing singular values and vectors. We will see in Sect. 1.5.9 that it is easy to compute singular values of **bidiagonal matrices**, much as it was easy to compute eigenvalues of symmetric tridiagonal matrices. Our goal in this section is to show how to transform a given rectangular matrix to **bidiagonal form**.

Suppose that we are given an $m \times n$ matrix $\mathbf{A}$ with $m \geq n$. In the first step of Householder bidiagonalization, we partition

$$\mathbf{A} = \begin{bmatrix} \mathbf{a} & \mathbf{B} \end{bmatrix}$$

and use Algorithm 6.7.2 in Chap. 6 of Volume I to find a Householder transformation $\mathbf{H}_1$ so that

$$\mathbf{H}_1{}^H \mathbf{a} = \mathbf{e}_1 \alpha_1 ,$$

where α_1 is real. Afterward, we partition

$$\mathbf{H}_1{}^H \mathbf{A} = \begin{bmatrix} \mathbf{e}_1 \alpha_1 & \mathbf{H}_1{}^H \mathbf{B} \end{bmatrix} = \begin{bmatrix} \alpha_1 & \mathbf{c}^H \\ \mathbf{0} & \mathbf{C} \end{bmatrix}$$

and choose a second Householder transformation $\widetilde{\mathbf{H}}_1$ so that

$$\widetilde{\mathbf{H}}_1{}^H \mathbf{c} = \mathbf{e}_1 \beta_1 ,$$

where β_1 is real. Then

$$\mathbf{H}_1{}^H \mathbf{A} \widetilde{\mathbf{H}}_1 = \begin{bmatrix} \alpha_1 & \beta_1 \mathbf{e}_1{}^\top \\ \mathbf{0} & \mathbf{C}\widetilde{\mathbf{H}}_1 \end{bmatrix} .$$

The process continues by applying the same approach to the smaller matrix $\mathbf{C}\widetilde{\mathbf{H}}_1$. Eventually we obtain

$$\mathbf{H}_n{}^H \dots \mathbf{H}_1{}^H \mathbf{A} \widetilde{\mathbf{H}}_1 \dots \widetilde{\mathbf{H}}_{n-1} = \begin{bmatrix} \alpha_1 & \beta_1 & \dots & 0 \\ \vdots & \ddots & \ddots & \vdots \\ 0 & \dots & \ddots & \beta_{n-1} \\ 0 & \dots & \dots & \alpha_n \\ \mathbf{0} & \dots & \dots & \mathbf{0} \end{bmatrix} = \begin{bmatrix} \mathbf{D} \\ 0 \end{bmatrix}$$

where $\mathbf{D}$ is upper bidiagonal and real.

If $m < n$, then we begin by partitioning

$$\mathbf{A} = \begin{bmatrix} \mathbf{a}^H \\ \mathbf{B} \end{bmatrix} ,$$

and find a Householder transformation $\widetilde{\mathbf{H}}_1$ so that

$$\widetilde{\mathbf{H}}_1{}^H \mathbf{a} = \mathbf{e}_1 \alpha_1 ,$$

where α_1 is real. Afterward, we partition

$$\mathbf{A}\widetilde{\mathbf{H}}_1 = \begin{bmatrix} \alpha_1 \mathbf{e}_1{}^\top \\ \mathbf{B}\widetilde{\mathbf{H}}_1 \end{bmatrix} = \begin{bmatrix} \alpha_1 & \mathbf{0}^\top \\ \mathbf{c} & \mathbf{C} \end{bmatrix}$$

and choose a second Householder transformation $\mathbf{H}_1$ so that

$$\mathbf{H}_1{}^H \mathbf{c} = \mathbf{e}_1 \beta_1 ,$$

where β_1 is real. Then

$$\mathbf{H}_1{}^H \mathbf{A} \widetilde{\mathbf{H}}_1 = \begin{bmatrix} \alpha_1 & \mathbf{0}^\top \\ \mathbf{e}_1 \beta_1 & \mathbf{H}_1{}^H \mathbf{C} \end{bmatrix} .$$

The process continues by applying the same approach to the smaller matrix $\mathbf{H}_1 \mathbf{C}$. Eventually we obtain

$$\mathbf{H}_{m-1}{}^H \dots \mathbf{H}_1{}^H \mathbf{A} \widetilde{\mathbf{H}}_1 \dots \widetilde{\mathbf{H}}_m = \begin{bmatrix} \alpha_1 & \dots & 0 & 0 & \mathbf{0} \\ \beta_1 & \ddots & \vdots & \vdots & \vdots \\ \vdots & \ddots & \ddots & \vdots & \vdots \\ 0 & \dots & \beta_{m-1} & \alpha_m & \mathbf{0} \end{bmatrix} = \begin{bmatrix} \mathbf{D} & \mathbf{0} \end{bmatrix} ,$$

where $\mathbf{D}$ is lower bidiagonal and real.

LAPACK performs bidiagonalization in routines `_gebrd`. For example, see dgebrd.f.

Exercise 1.5.5 Transform

$$\mathbf{A} = \begin{bmatrix} 1 & 1 \\ \delta & 0 \\ 0 & \delta \end{bmatrix}$$

to bidiagonal form using exact arithmetic. Afterward, assuming that δ is so small that $\mathrm{fl}(1+\delta^2) = 1$, determine the bidiagonal form in floating point arithmetic.

1.5.9 Implicit QR Algorithm

After transforming a given matrix to bidiagonal form, the next step is to find the singular value decomposition of the $n \times n$ real bidiagonal matrix $\mathbf{D}$. Before describing this step, we remark that if $\mathbf{D}$ is $n \times n$ lower bidiagonal then we can choose a sequence of Givens rotations to zero out the subdiagonal entries:

$$\widetilde{\mathbf{D}} = \mathbf{G}_{n-1,n} \cdot \ldots \cdot \mathbf{G}_{1,2}\mathbf{D} \,.$$

It is easy to see that $\widetilde{\mathbf{D}}$ is upper bidiagonal. Without loss of generality, we may assume that the original matrix $\mathbf{D}$ is upper bidiagonal.

Recall that the singular values of $\mathbf{D}$ are the square roots of the eigenvalues of $\mathbf{D}^\top\mathbf{D}$. In the QR algorithm for $\mathbf{D}^\top\mathbf{D}$, we would choose a shift σ and formally factor $\mathbf{D}^\top\mathbf{D} - \mathbf{I}\sigma = \mathbf{V}\mathbf{R}$ where $\mathbf{V}$ is unitary and $\mathbf{R}$ is upper bidiagonal. Then we would form the tridiagonal matrix $\mathbf{R}\mathbf{V} + \mathbf{I}\sigma$. These computations would be performed by the implicit QR algorithm as described in Sect. 1.3.7. In the remainder of this section, we will show that we can avoid forming $\mathbf{D}^\top\mathbf{D}$, and implicitly achieve the same result by working directly on $\mathbf{D}$.

If we were to follow the symmetric QR algorithm of Sect. 1.3.7, the shift would be chosen to be that eigenvalue of the last 2×2 diagonal block of $\mathbf{D}^\top\mathbf{D}$ that is closest to the last diagonal entry. Then we could choose a Givens rotation $\widetilde{\mathbf{G}}_{12}$ to zero out the first super-diagonal entry of

$$\mathbf{D}^\top\mathbf{D} - \mathbf{I}\sigma = \begin{bmatrix} \alpha_1 & & & \\ \beta_1 & \alpha_2 & & \\ & \ddots & \ddots & \\ & & \beta_{n-1} & \alpha_n \end{bmatrix}^\top \begin{bmatrix} \alpha_1 & & & \\ \beta_1 & \alpha_2 & & \\ & \ddots & \ddots & \\ & & \beta_{n-1} & \alpha_n \end{bmatrix} - \mathbf{I}\sigma$$

$$= \begin{bmatrix} \alpha_1^2 - \sigma & \alpha_1\beta_1 & & \\ \beta_1\alpha_1 & |\beta_1|^2 + |\alpha_2|^2 - \sigma & \ddots & \\ & \ddots & \ddots & \alpha_{n-1}\beta_{n-1} \\ & & \beta_{n-1}\alpha_{n-1} & |\beta_{n-1}|^2 + |\alpha_n|^2 - \sigma \end{bmatrix} .$$

In other words, we would choose $\widetilde{\mathbf{G}}_{12}$ so that

$$\widetilde{\mathbf{G}}_{12}{}^\top \begin{bmatrix} |\alpha_1|^2 - \sigma \\ \alpha_1\beta_1 \end{bmatrix} = \begin{bmatrix} \varrho_1 \\ 0 \end{bmatrix} .$$

Afterward, we would choose additional Givens rotations to return to tridiagonal form.

However, LAPACK uses a slightly different strategy. The shift is chosen to be the square of the singular value σ of the last 2×2 diagonal block of $\mathbf{D}$ that is closest to the last diagonal entry of $\mathbf{D}$. The first Givens rotation $\widetilde{\mathbf{G}}_{12}$ is chosen to zero out the first super-diagonal entry of $\mathbf{D}^\top\mathbf{D} - \mathbf{I}\sigma^2$. In other words, we choose $\widetilde{\mathbf{G}}_{12}$ so that

$$\widetilde{\mathbf{G}}_{12}{}^\top \begin{bmatrix} (\alpha_1^2 - \sigma^2)/\alpha_1 \\ \beta_1 \end{bmatrix} = \begin{bmatrix} \varrho_1 \\ 0 \end{bmatrix} .$$

Of course, if $\alpha_1 = 0$ then the first super-diagonal entry of $\mathbf{D}^\top\mathbf{D}$ is zero, and we can choose $\widetilde{\mathbf{G}}_{12} = \mathbf{I}$. In the remainder of the implicit QR algorithm for the singular values, we merely **chase the bulge**. Symbolically, we begin with

$$\mathbf{D} = \begin{bmatrix} \times & \times & & & \\ & \times & \times & & \\ & & \times & \times & \\ & & & \times & \times \\ & & & & \times \end{bmatrix}$$

and find that the initial Givens transformation produces

$$\mathbf{D}\widetilde{\mathbf{G}}_{12} = \begin{bmatrix} * & * & & & \\ + & * & \times & & \\ & & \times & \times & \\ & & & \times & \times \\ & & & & \times \end{bmatrix}$$

Next, we choose a Givens transformation $\mathbf{G}_{12}$ to zero the new subdiagonal entry:

$$\mathbf{G}_{12}{}^\top\mathbf{D}\widetilde{\mathbf{G}}_{12} = \begin{bmatrix} * & * & + & & \\ 0 & * & * & & \\ & & \times & \times & \\ & & & \times & \times \\ & & & & \times \end{bmatrix}$$

We can find a Givens transformation $\widetilde{\mathbf{G}}_{23}$ to zero out the new nonzero entry in the second super-diagonal: subdiagonal entry:

$$\mathbf{G}_{12}{}^{\top}\mathbf{D}\widetilde{\mathbf{G}}_{12}\widetilde{\mathbf{G}}_{23} = \begin{bmatrix} * & * & 0 & & \\ & * & * & & \\ & + & * & \times & \\ & & & \times & \times \\ & & & & \times \end{bmatrix}$$

At this point, the submatrix consisting of all but the first row and column has the same form as $\mathbf{D}\widetilde{\mathbf{G}}_{12}$. Thus we can continue to find additional Givens rotations so that

$$\mathbf{G}_{n-1,n}{}^{\top} \cdot \ldots \cdot \mathbf{G}_{12}{}^{\top}\mathbf{D}\widetilde{\mathbf{G}}_{12} \cdot \ldots \cdot \widetilde{\mathbf{G}}_{n-1,n} = \begin{bmatrix} * & * & 0 & & \\ & * & * & & \\ & & * & * & \\ & & & * & * \\ & & & & * \end{bmatrix} .$$

Thus we have found unitary matrices $\mathbf{U}$ and $\mathbf{V}$ so that

$$\widetilde{\mathbf{D}} = \mathbf{U}^{\top}\mathbf{D}\mathbf{V}$$

is bidiagonal. Furthermore,

$$\widetilde{\mathbf{D}}^{\top}\widetilde{\mathbf{D}} = \mathbf{V}^{\top}\mathbf{D}^{\top}\mathbf{D}\mathbf{V}$$

is unitarily similar to $\mathbf{D}^{\top}$, and the first column of $\mathbf{V}$ is the same as the first column of the unitary matrix that would be used in the implicit QR algorithm for $\mathbf{D}^{\top}\mathbf{D}$. As a result, at some point this process should produce an very nearly diagonal matrix $\widetilde{\mathbf{D}}$.

The implicit QR algorithm to find the singular values of a bidiagonal matrix is performed in LAPACK routines `_bdsqr`. Consider, for example, dbdsqr.f. The shift in this algorithm is computed in LAPACK routines `_las2`. See, for example, dlas2.f.

1.5.10 General Singular Values

We can find the singular values and vectors of a given $m \times n$ matrix $\mathbf{A}$ as follows. First, we use the ideas in Sect. 1.5.8 to find unitary matrices $\mathbf{U}_1$ and $\mathbf{V}_1$ so that

$$\mathbf{U}_1{}^H\mathbf{A}\mathbf{V}_1 = \mathbf{B}$$

is bidiagonal. Afterward, we use the implicit QR algorithm as described in Sect. 1.5.9 to find unitary matrices $\mathbf{U}_2$ and $\mathbf{V}_2$ so that

$$\mathbf{U}_2{}^H\mathbf{B}\mathbf{V}_2 = \mathbf{D}$$

is diagonal. Then we find that

$$\mathbf{A} = \mathbf{U}_1\mathbf{U}_2\mathbf{D}\mathbf{V}_2{}^H\mathbf{V}_1{}^H$$

is the singular value decomposition of $\mathbf{A}$.

LAPACK computes the singular value decomposition via the implicit QR algorithm in routines `_gesvd`. For example, examine dgesvd.f. LAPACK also provides routines `_gesdd` to compute the singular value decomposition via the SVD divide and conquer algorithm, and routines `_gejsv` to compute the singular value decomposition via a Jacobi iteration (see Drmac and Veselic [57]). MATLAB users can compute the singular value decomposition of a matrix by using the command svd.

For those readers who were interested in Sect. 3.12 in Chap. 3 of Volume I regarding object-oriented programming and linear algebra, we have provided a `SingularValueDecomposition` class. Readers may examine SingularValueDecomposition.H and SingularValueDecomposition.C for implementation details. This class is currently designed to solve least squares problems and perform regularization, as described in Sect. 6.11 in Chap. 6 of Volume I.

Exercise 1.5.6 Find the singular value decomposition

$$\mathbf{A} = \mathbf{U}\boldsymbol{\Sigma}\mathbf{V}^H$$

of the 20×20 Hilbert matrix

$$\mathbf{A}_{ij} = \frac{1}{1+i+j}\,, \quad 0 \le i,j < 20\,.$$

Then multiply the factors together to compute

$$\widetilde{\mathbf{A}} \equiv \mathbf{U}\boldsymbol{\Sigma}\mathbf{V}^H\,,$$

and use Lemma 1.5.6 to estimate the errors in the computed singular values. What is the minimal number of significant digits in any singular value?

Get a copy of the 3140×3140 matrix PSIMGR_1. Find its singular value decomposition, and use Lemma 1.5.6 to find the errors in the singular values.

1.5.11 Least Squares Problems

Recall that in Sect. 6.11.1 in Chap. 6 of Volume I we discussed the use of singular value decompositions in solving least squares problems. At that time, we were not aware of the Eckhart-Young Theorem 1.5.3 connecting the singular values to matrix rank. In this section, we would like to discuss how to use a desired condition number for the given matrix to select its effective rank, and solve an associated least squares problem.

Given an $m \times n$ matrix $\mathbf{A}$, suppose that we have determined its singular value decomposition

$$\mathbf{A} = \mathbf{U}\boldsymbol{\Sigma}\mathbf{V}^H ,$$

where $\boldsymbol{\Sigma}$ is an $r \times r$ nonsingular diagonal matrix. In order to find the smallest vector $\mathbf{x}$ that minimizes

$$\|\mathbf{A}\mathbf{x} - \mathbf{b}\|_2^2 = \|\mathbf{U}\boldsymbol{\Sigma}\mathbf{V}^H\mathbf{x} - \mathbf{b}\|_2^2 = \|\boldsymbol{\Sigma}\mathbf{V}^H\mathbf{x} - \mathbf{U}^H\mathbf{b}\|_2^2 + \|\left(\mathbf{I} - \mathbf{U}\mathbf{U}^H\right)\mathbf{b}\|_2^2 ,$$

we compute

$$\mathbf{x} = \mathbf{V}\boldsymbol{\Sigma}^{-1}\mathbf{U}^H\mathbf{b} .$$

In general, numerical computation of the singular value decomposition will return $\min\{m, n\}$ nonzero singular values for an arbitrary $m \times n$ matrix. In order to select only those singular values that are meaningful, it is common to select a condition number for $\mathbf{A}$ and use only those singular values that correspond to the chosen condition number. If the singular values are ordered from largest to smallest

$$\sigma_1 \geq \ldots \geq \sigma_{\min\{m,n\}} ,$$

then the effective rank is r where r is the largest index so that

$$\frac{\sigma_r}{\sigma_1} \leq \frac{1}{\kappa(\mathbf{A})} .$$

LAPACK uses the singular value decomposition to solve least squares problems in routines `_gelss`. See, for example, dgelss.f. This approach is the basis for the `solve` and `regularize` functions in the `SingularValueDecomposition` class. See SingularValueDecomposition.H, SingularValueDecomposition.C and LaDouble.C.

1.6 Linear Recurrences

In this section, we will use eigenvalues and eigenvectors to solve linear recurrences, which are described as follows.

Definition 1.6.1 Given an $n \times n$ matrix $\mathbf{A}$, an n-vector $\mathbf{t}_0$ and n-vectors $\mathbf{b}_k$ for $k \geq 0$, a constant coefficient **linear recurrence** has the form

$$\mathbf{t}_{k+1} = \mathbf{A}\mathbf{t}_k + \mathbf{b}_k \ .$$

These arise quite naturally in the numerical solution of differential equations [93, Chap. 3] [116, p. 9ff], Markov processes [64, v. 1, p. 419ff] and combinatorics [150, p. 194ff] [175, p. 279ff].

1.6.1 Solution of Linear Recurrences

Our first lemma in this section will show that the solution of constant coefficient linear recurrences involves powers of matrices. In Sect. 1.6.2, we will discuss how these matrix powers can be computed easily from certain canonical forms.

Lemma 1.6.1 *Given an $n \times n$ matrix $\mathbf{A}$, an n-vector $\mathbf{t}_0$ and a sequence of n-vectors $\{\mathbf{b}_k\}_{k=0}^{\infty}$, the solution of the linear recurrence*

$$\mathbf{t}_{k+1} = \mathbf{A}\mathbf{t}_k + \mathbf{b}_k \textit{ for } k \geq 0$$

is

$$\mathbf{t}_k = \mathbf{A}^k\mathbf{t}_0 + \sum_{j=0}^{k-1} \mathbf{A}^{k-j-1}\mathbf{b}_k j \textit{ for } k \geq 0 \ .$$

Proof We will prove the result by induction. The result is obvious for $k = 0$ and $k = 1$. Inductively, assume that the result is true for $k \geq 1$. We will prove that it is true for $k + 1$:

$$\mathbf{t}_{k+1} = \mathbf{A}\mathbf{t}_k + \mathbf{b}_k k = \mathbf{A}(\mathbf{A}^k\mathbf{t}_0 + \sum_{j=0}^{k-1} \mathbf{A}^{k-j-1}\mathbf{b}_k j) + \mathbf{b}_k k$$

$$= \mathbf{A}^{k+1}\mathbf{t}_0 + \sum_{j=0}^{k-1} \mathbf{A}^{k-j}\mathbf{b}_k j + \mathbf{b}_k k = \mathbf{A}^{k+1}\mathbf{t}_0 + \sum_{j=0}^{k} \mathbf{A}^{k-j}\mathbf{b}_k j \ .$$

Of course, the conclusion of this lemma requires us to compute matrix powers. It is conceivable that the computation of a matrix power is at least as difficult as the direct computation of the linear recurrence. However, analysis of the stability of the linear recurrence in Sect. 1.6.4, or examination of the asymptotic behavior of the linear recurrence for large recurrence indices k will follow easily from a careful study of analytical expressions for matrix powers.

1.6.2 Powers of Matrices

In computing the matrix powers, we will make use of the following important theorem.

Theorem 1.6.1 (Binomial Expansion) *If λ is a scalar and $\mathbf{A}$ is a square matrix, then for all nonnegative integers k we have*

$$(\mathbf{I}\lambda + \mathbf{A})^k = \sum_{j=0}^{k} \mathbf{A}^j \lambda^{k-j} \binom{k}{j} .$$

Proof It is easy to verify the **Pascal triangle** identity:

$$\binom{k-1}{j} + \binom{k-1}{j-1} = \frac{(k-1)!}{j!(k-j)!} [(k-j)+j] = \binom{k}{j} .$$

Next, we note that the binomial expansion is trivial for $k = 0$, and easy for $k = 1$:

$$\sum_{j=0}^{1} \mathbf{A}^j \lambda^{1-j} \binom{1}{j} = \mathbf{A}^0 \lambda \binom{1}{0} + \mathbf{A}^1 \lambda^0 \binom{1}{1} = \mathbf{I}\lambda + \mathbf{A} .$$

Assume inductively that the binomial expansion is true for $k - 1 \geq 0$. Then

$$\begin{aligned}
(\mathbf{I}\lambda + \mathbf{A})^k &= (\mathbf{I}\lambda + \mathbf{A}) \sum_{j=0}^{k-1} \mathbf{A}_j \lambda^{k-1-j} \binom{k-1}{j} \\
&= \sum_{j=0}^{k-1} \mathbf{A}^j \lambda^{k-j} \binom{k-1}{j} + \sum_{j=0}^{k-1} \mathbf{A}_{j+1} \lambda^{k-1-j} \binom{k-1}{j} \\
&= \mathbf{I}\lambda_k + \sum_{j=1}^{k-1} \mathbf{A}_j \lambda^{k-j} \binom{k-1}{j} + \sum_{\ell=1}^{k} \mathbf{A}^\ell \lambda^{k-\ell} \binom{k-1}{\ell-1} \\
&= \mathbf{I}\lambda_k + \sum_{j=1}^{k-1} \mathbf{A}^j \lambda^{k-j} \left\{ \binom{k-1}{j} + \binom{k-1}{j-1} \right\} + \mathbf{A}^k = \sum_{j=0}^{k} \mathbf{A}^j \lambda^{k-j} \binom{k}{j} .
\end{aligned}$$

In the remainder of this section, we will compute powers of matrices. Canonical forms will help us to simplify these computations.

1.6.2.1 Non-defective Matrices

First, suppose that $\mathbf{A}$ is non-defective (i.e., diagonalizable). Then there is a nonsingular matrix $\mathbf{X}$ and a diagonal matrix $\boldsymbol{\Lambda}$ so that

$$\mathbf{A}\mathbf{X} = \mathbf{X}\boldsymbol{\Lambda} .$$

It follows that for $k \geq 0$,

$$\mathbf{A}^k = (\mathbf{X}\boldsymbol{\Lambda}\mathbf{X}^{-1}) \cdot \ldots \cdot (\mathbf{X}\boldsymbol{\Lambda}\mathbf{X}^{-1}) = \mathbf{X}\boldsymbol{\Lambda}^k\mathbf{X}^{-1} .$$

Since $\boldsymbol{\Lambda}^k$ is the diagonal matrix with entries given by the k'th power of the diagonal entries of $\boldsymbol{\Lambda}$, it is easy to compute $\boldsymbol{\Lambda}^k$. As a result, it is easy to compute $\mathbf{A}^k$. Furthermore, Lemma 1.6.1 shows that the solution of the constant coefficient linear recurrence

$$\mathbf{t}_{k+1} = \mathbf{A}\mathbf{t}_k + \mathbf{b}_k$$

can be computed by

$$\mathbf{t}_k = \mathbf{X}\left\{\boldsymbol{\Lambda}^k\mathbf{u}_0 + \sum_{j=0}^{k-1}\boldsymbol{\Lambda}^{k-j-1}\mathbf{c}_j\right\} ,$$

where

$$\mathbf{X}\mathbf{u}_0 = \mathbf{t}_0 \text{ and } \mathbf{X}\mathbf{c}_j = \mathbf{b}_j \text{ for } 0 \leq j < k .$$

1.6.2.2 Defective Matrices

If $\mathbf{A}$ is defective, then we can use the Jordan canonical form to compute powers of $\mathbf{A}$. Suppose that $\mathbf{A}\mathbf{X} = \mathbf{X}\mathbf{J}$ where $\mathbf{X}$ is nonsingular and $\mathbf{J}$ is a Jordan canonical form. Then $\mathbf{A}^k = \mathbf{X}\mathbf{J}^k\mathbf{X}^{-1}$. This shows that if we can compute the powers of Jordan blocks, then we can compute powers of a defective matrix.

Suppose that $\mathbf{J} = \mathbf{I}\lambda + \mathbf{N}$ is a Jordan block. Then the binomial expansion Theorem 1.6.1 shows us that

$$\mathbf{J}^k = (\mathbf{I}\lambda + \mathbf{N})^k = \sum_{\ell=1}^{k}\mathbf{N}^\ell\binom{k}{\ell}\lambda^{k-\ell} .$$

By definition of $\mathbf{N}$, we have

$$\mathbf{N}\mathbf{e}_j = \begin{cases} \mathbf{e}_{j+1}, & 1 \le j \le n-1 \\ \mathbf{0}, & j = n \end{cases} .$$

A simple induction then shows that for $1 \le k < n$ we have

$$\mathbf{N}^k\mathbf{e}_j = \begin{cases} \mathbf{e}_{j+k}, & 1 \le j \le n-k \\ \mathbf{0}, & n-k < j \le n \end{cases} ,$$

and that $\mathbf{N}^k = \mathbf{0}$ for $k \ge n$. This gives us the following lemma.

Lemma 1.6.2 *If $\mathbf{J} = \mathbf{I}\lambda + \mathbf{N}$ is an $n \times n$ Jordan block, then the only nonzero entries of $\mathbf{J}^k$ occur for $1 \le i \le j \le \min\{ i+k, n\}$, in which case the i, j entry of $\mathbf{J}^k$ is*

$$\mathbf{e}_i{}^\top \mathbf{J}^k \mathbf{e}_j = \binom{k}{j-i} \lambda^{k-j+i} .$$

Note that if the eigenvalue λ of a Jordan block satisfies $|\lambda| < 1$, then $\mathbf{J}^k \to \mathbf{0}$ as $k \to \infty$.

1.6.3 Examples

The classical example of a linear recurrence is the following mathematical curiosity.

Example 1.6.1 (Fibonacci Numbers) Fibonacci numbers ϕ_k are defined by $\phi_{-1} = 0$, $\phi_0 = 1$ and the recurrence

$$\phi_{k+1} = \phi_k + \phi_{k-1} \text{ for } k \ge 0 .$$

It is easy to compute some new terms in the series:

$$\phi_1 = 1 \, , \; \phi_2 = 2 \, , \; \phi_3 = 3 \, , \; \phi_4 = 5 \, , \; \ldots .$$

We would like to understand how ϕ_k behaves for large k. To do so, we will write the Fibonacci recurrence in matrix-vector form, and use matrix powers to solve the resulting constant coefficient linear recurrence.

We begin by defining the vector $\mathbf{f}_k$ and matrix $\mathbf{A}$ by

$$\mathbf{f}_k \equiv \begin{bmatrix} \phi_k \\ \phi_{k-1} \end{bmatrix} , \; \mathbf{A} = \begin{bmatrix} 1 & 1 \\ 1 & 0 \end{bmatrix} .$$

Then we can write the Fibonacci recurrence in the form

$$\mathbf{f}_0 = \begin{bmatrix}1\\0\end{bmatrix} \text{ and } \mathbf{f}_{k+1} = \mathbf{A}\mathbf{f}_k \text{ for } k > 0 \,.$$

It is easy to see that the solution of the recurrence is

$$\mathbf{f}_k = \mathbf{A}^k \mathbf{f}_0 \,.$$

However, we would like to find a simple way to compute large powers of the matrix **A**.

Since **A** is symmetric, it is diagonalizable. The eigenvalues of **A** satisfy the characteristic equation

$$0 = \det(\mathbf{A} - \mathbf{I}\lambda) = \lambda^2 - \lambda - 1 \,.$$

Thus the larger eigenvalue is

$$\lambda_1 = \frac{1}{2}(1 + \sqrt{5}) \,;$$

the other eigenvalue is $\lambda_2 = -1/\lambda_1$. It is easy to see that

$$\mathbf{A}\mathbf{X} = \begin{bmatrix}1 & 1\\1 & 0\end{bmatrix}\begin{bmatrix}\lambda_1 & -1\\1 & \lambda_1\end{bmatrix} = \begin{bmatrix}\lambda_1 & -1\\1 & \lambda_1\end{bmatrix}\begin{bmatrix}\lambda_1 & 0\\0 & -1/\lambda_1\end{bmatrix} = \mathbf{X}\boldsymbol{\Lambda}$$

gives the diagonalization of **A**. Then

$$\mathbf{f}_k = \mathbf{A}^k\mathbf{f}_0 = \mathbf{X}\boldsymbol{\Lambda}^k\mathbf{X}^{-1}\mathbf{f}_0 = \begin{bmatrix}\lambda_1 & -1\\1 & \lambda_1\end{bmatrix}\begin{bmatrix}\lambda_1^k & 0\\0 & (-1/\lambda_1)^k\end{bmatrix}\begin{bmatrix}\lambda_1 & -1\\1 & \lambda_1\end{bmatrix}^{-1}\begin{bmatrix}1\\0\end{bmatrix}$$
$$= \begin{bmatrix}\lambda_1^k & -(-1/\lambda_1)^k\\\lambda_1^k & -(1/\lambda_1)^{k-1}\end{bmatrix}\begin{bmatrix}\lambda_1\\-1\end{bmatrix}\frac{1}{\lambda_1^2+1} \,.$$

This gives us a formula for ϕ_k:

$$\phi_k = \frac{\lambda_1^{k+2} + (-1/\lambda_1)^k}{\lambda_1^2 + 1} = \frac{\lambda_1^{k+1} - (-1/\lambda_1)^{k+1}}{\lambda_1 - (-1/\lambda_1)} \,.$$

Note that for large values of k,

$$\frac{\phi_{k+1}}{\phi_k} \approx \lambda_1 = \frac{1+\sqrt{5}}{2} \,.$$

Example 1.6.2 (Four Step Linear Recurrence) Suppose that we want to solve

$$\sigma_k = 7\sigma_{k-1} - 16\sigma_{k-2} + 12\sigma_{k-3} \; .$$

We can write this as a linear recurrence by defining the vector

$$\mathbf{s}_k = \begin{bmatrix} \sigma_k \\ \sigma_{k-1} \\ \sigma_{k-2} \end{bmatrix}$$

and the companion matrix

$$\mathbf{C} = \begin{bmatrix} 7 & -16 & 12 \\ 1 & 0 & 0 \\ 0 & 1 & 0 \end{bmatrix} \; .$$

Then we have

$$\mathbf{s}_k = \begin{bmatrix} \sigma_k \\ \sigma_{k-1} \\ \sigma_{k-2} \end{bmatrix} = \begin{bmatrix} 7 & -16 & 12 \\ 1 & 0 & 0 \\ 0 & 1 & 0 \end{bmatrix} \begin{bmatrix} \sigma_{k-1} \\ \sigma_{k-2} \\ \sigma_{k-3} \end{bmatrix} = \mathbf{C}\mathbf{s}_{k-1} \; .$$

In order to solve this recurrence, we want to find the eigenvalues of $\mathbf{C}$. Its characteristic polynomial is

$$\det(\mathbf{C} - \mathbf{I}\lambda) = -(\lambda^3 - 7\lambda^2 + 16\lambda + 12) = (2 - \lambda)^2(3 - \lambda) \; .$$

Since this companion matrix has a multiple eigenvalue, Lemma 1.4.6 shows that its Jordan canonical form has a nontrivial Jordan block. We can compute the similarity transformation to canonical form and get

$$\mathbf{C}\mathbf{X} = \begin{bmatrix} 7 & -16 & 12 \\ 1 & 0 & 0 \\ 0 & 1 & 0 \end{bmatrix} \begin{bmatrix} 4 & 8 & 9 \\ 2 & 3 & 3 \\ 1 & 1 & 1 \end{bmatrix} = \begin{bmatrix} 4 & 8 & 9 \\ 2 & 3 & 3 \\ 1 & 1 & 1 \end{bmatrix} \begin{bmatrix} 2 & 1 & 0 \\ 0 & 2 & 0 \\ 0 & 0 & 3 \end{bmatrix} = \mathbf{X}\mathbf{J} \; .$$

Then the solution to the linear recurrence is

$$\mathbf{s}_k = \mathbf{X}\mathbf{J}^k\mathbf{X}^{-1}\mathbf{s}_0$$

where for $k \geq 0$ Lemma 1.6.2 shows that

$$\mathbf{J}^k = \begin{bmatrix} 2^k & k2^{k-1} & 0 \\ 0 & 2^k & 0 \\ 0 & 0 & 3^k \end{bmatrix} \; .$$

1.6.4 Stability

Next, we would like to study the **stability** of solutions to linear recurrences. Suppose that we have the linear recurrence $\mathbf{t}_{k+1} = \mathbf{A}\mathbf{t}_k$ for $k \geq 0$ with some initial value $\mathbf{t}_0$. If we perturb the initial value to $\widetilde{\mathbf{t}_0}$, we would like to know if the linear recurrence amplifies the initial error $\|\widetilde{\mathbf{t}_0} - \mathbf{t}_0\|$. This suggests that we compute

$$\left\|\widetilde{\mathbf{t}_k} - \mathbf{t}_k\right\| = \left\|\mathbf{A}^k(\widetilde{\mathbf{t}_0} - \mathbf{t}_0)\right\| \leq \|\mathbf{A}\|^k \left\|\widetilde{\mathbf{t}_0} - \mathbf{t}_0\right\| .$$

If $\|\mathbf{A}\| < 1$, then the linear recurrence reduces the norm of the initial perturbation as k increases. However, we can be a bit more careful in our discussion.

Suppose that every eigenvalue of $\mathbf{A}$ satisfies $|\lambda| < 1$. Let $\mathbf{A} = \mathbf{X}\mathbf{J}\mathbf{X}^{-1}$ be a Jordan canonical form for $\mathbf{A}$, and let $\mathbf{d}_k = \mathbf{X}^{-1}(\widetilde{\mathbf{t}_k} - \mathbf{t}_k)$. Then

$$\|\mathbf{d}_k\| = \left\|\mathbf{J}^k\mathbf{d}_0\right\| \to 0 \text{ as } k \to \infty .$$

It follows that $\|\widetilde{\mathbf{t}_k} - \mathbf{t}_k\| \to 0$ as well.

Note that if $\mathbf{A}$ has at least one eigenvalue satisfying $|\lambda| > 1$, then some perturbation of the initial value $\mathbf{t}_0$ will be magnified. In particular, suppose that $\mathbf{x}$ is an eigenvector of $\mathbf{A}$ corresponding to such an eigenvalue λ, and that ε is small but nonzero. We can choose

$$\widetilde{\mathbf{t}_0} = \mathbf{t}_0 + \mathbf{x}\varepsilon .$$

Then

$$\widetilde{\mathbf{t}_k} - \mathbf{t}_k = \mathbf{A}^k(\widetilde{\mathbf{t}_0} - \mathbf{t}_0) = \mathbf{x}\lambda^k\varepsilon .$$

As $k \to \infty$, the perturbation in the solution of the linear recurrence becomes arbitrarily large.

The case when the largest eigenvalue of $\mathbf{A}$ has modulus exactly equal to one is more difficult. If none of the eigenvalues of $\mathbf{A}$ with modulus 1 have nontrivial Jordan blocks, then $\mathbf{d}_k = \mathbf{J}^k\mathbf{d}_0$ shows that the components of the perturbation cannot grow as k increases. In such a case, some components of $\mathbf{d}_k$ may have constant modulus as k increases. The more interesting case occurs when all of the eigenvalues of $\mathbf{A}$ lie on or within the unit circle, but $\mathbf{A}$ has an eigenvalue λ with modulus one and a nontrivial Jordan block. In this case $\mathbf{d}_k = \mathbf{J}^k\mathbf{d}_0$ together with Lemma 1.6.2 shows that some components of $\mathbf{d}_k$ will grow at least linearly with k.

These observations suggest the following definition.

Definition 1.6.2 The linear recurrence $\mathbf{t}_{k+1} = \mathbf{A}\mathbf{t}_k + \mathbf{d}_k$ is

- **stable** if and only if for every eigenvalue λ of $\mathbf{A}$ we have $|\lambda| < 1$,
- **unstable** if and only if there is at least one eigenvalue λ of $\mathbf{A}$ satisfying $|\lambda| > 1$,

- **neutrally stable** if and only for all eigenvalues λ of $\mathbf{A}$ we have $|\lambda| \leq 1$ and whenever $|\lambda| = 1$ the corresponding Jordan block for λ is 1×1, and
- **weakly unstable** if and only if for every eigenvalue λ of $\mathbf{A}$ we have $|\lambda| \leq 1$, but there is at least one eigenvalue λ with $|\lambda| = 1$ and the corresponding Jordan block is larger than 1×1.

Exercise 1.6.1

Consider the difference equation

$$\frac{y_{k+1} - y_k}{\Delta t} = -10 y_k$$

with $y_0 = 1$. This difference equation approximates the ordinary differential equation $y'(t) = -10y(t)$. Write this as a linear recurrence, and determine conditions on the timestep Δt so that the recurrence is stable, unstable or neutrally stable.

Exercise 1.6.2

Consider the systems of difference equations

$$\frac{u_i^{(k+1)} - u_i^{(k)}}{\Delta t} = \frac{\frac{u_{i+1}^{(k)} - u_i^{(k)}}{\Delta x} - \frac{u_i^{(k)} - u_{i-1}^{(k)}}{\Delta x}}{\Delta x}$$

for $1 \leq i < n = 1/\Delta x$, with $u_i^{(0)}$ given for $0 \leq i \leq n$, and $u_0^{(k)} = 0 = u_n^{(k)}$ for $0 \leq k$. This difference equation approximates the partial differential equation

$$\frac{\partial u}{\partial t} = \frac{\partial^2 u}{\partial x^2}$$

with boundary conditions $u(0,t) = 0 = u(1,t)$ for $t > 0$. Write the system of difference equations as a linear recurrence, and determine an upper bound on Δt in terms of Δx so that the linear recurrence is stable. (Hint: either use the results in Example 1.3.2 to get an exact expression for the eigenvalues of the matrix in the linear recurrence, or use the Gerschgorin Circle Theorem 1.2.2.)

Exercise 1.6.3

Suppose that for $k \geq 2$ we have

$$u_{k+1} = (u_k + u_{k-1})/2 .$$

Write this as a linear recurrence, and find the limit of u_k as $k \to \infty$ in terms of u_0 and u_1.

Exercise 1.6.4

Find the analytical solution to the recurrence

$$a_{k+1} = 6a_k - 9a_{k-1}$$

with $a_0 = 1$ and $a_1 = 2$.

Exercise 1.6.5

Let a_k be the number of cars that are at most 1 year old, b_k the number of cars that are at most 2 years old, and c_k the number of cars that are at most 3 years old. Suppose that 10% of cars must be replaced when they are 2 years old, and all cars must be replaced when they are 3 years old.

1. Show that

$$\begin{bmatrix} a_{k+1} \\ b_{k+1} \\ c_{k+1} \end{bmatrix} = \begin{bmatrix} 0 & 0.1 & 1 \\ 1 & 0 & 0 \\ 0 & 0.9 & 0 \end{bmatrix} \begin{bmatrix} a_k \\ b_k \\ c_k \end{bmatrix} .$$

2. If all cars are new initially, what will be the limiting age distribution of cars at large time?

1.7 Functions of Matrices

In Sect. 1.6 we learned how to compute powers of matrices in order to solve linear recurrences. In this section, we will generalize these ideas to compute analytic functions of matrices, such as exponentials or square roots. Matrix exponentials are useful in studying systems of ordinary differential equations [37, p. 75ff], [109, p. 40ff]. The matrix square root arises in connection with the matrix polar decomposition, which is useful in mechanics [121, p. 178], [124, p. 51].

1.7.1 Convergent Matrix Series

In the previous section, we used canonical form of matrices to compute powers of matrices. We can generalize this notion to functions defined by power series.

Definition 1.7.1 Suppose that the function ϕ has the Taylor series

$$\phi(\lambda) = \sum_{k=0}^{\infty} \lambda^k \phi_k$$

that converges for $|\lambda| < \varrho$. Then for any $n \times n$ matrix $\mathbf{A}$ for which any eigenvalue λ of $\mathbf{A}$ satisfies $|\lambda| < \varrho$ we define

$$\phi(\mathbf{A}) = \sum_{k=0}^{\infty} \mathbf{A}^k \phi_k .$$

The following lemma provides a way to compute an analytical function of a matrix.

Lemma 1.7.1 *If* **A** *is a square matrix and* $\mathbf{A} = \mathbf{X}\mathbf{J}\mathbf{X}^{-1}$ *where* **J** *is the Jordan canonical form of* **A**, *then*

$$\phi(\mathbf{A}) = \mathbf{X}\left(\sum_{k=0}^{\infty} \mathbf{J}^k \phi_k\right)\mathbf{X}^{-1} = \mathbf{X}\phi(\mathbf{J})X^{-1} .$$

If the original power series converges for $|\lambda| < \varrho$, *then* $\phi(\mathbf{A})$ *converges provided that all of the eigenvalues of* **A** *have modulus less than* ϱ.

Proof First, let $\mathbf{J} = \mathbf{I}\lambda + \mathbf{N}$ be a Jordan block, as described in Definition 1.4.5. The Taylor series for ϕ gives us

$$\phi(\mathbf{J}) = \sum_{k=0}^{\infty} \mathbf{J}^k \phi_k = \sum_{k=0}^{\infty} (\mathbf{I}\lambda + \mathbf{N})^k \phi_k$$

then the binomial expansion Theorem 1.6.1 leads to

$$= \sum_{k=0}^{\infty} \sum_{j=0}^{k} \mathbf{N}^j \binom{k}{j} \lambda^{k-j} \phi_k = \sum_{j=0}^{\infty} \sum_{k=j}^{\infty} \mathbf{N}^j \binom{k}{j} \lambda^{k-j} \phi_k = \sum_{j=0}^{\infty} \mathbf{N}^j \frac{\phi^{(j)}(\lambda)}{j!} ,$$

where $\phi^{(j)}(\lambda)$ is the *j*th derivative of $\phi(\lambda)$. Since **N** is nilpotent, this final series sum is finite, and therefore converges.

For a general square matrix **A**, Theorem 1.4.6 implies that there is a nonsingular matrix **X** and a Jordan canonical form **J** so that $\mathbf{A} = \mathbf{X}\mathbf{J}\mathbf{X}^{-1}$. If all of the eigenvalues of **A** have modulus less than ϱ, then the beginning of this proof shows that $\phi(\mathbf{J})$ converges. It follows that

$$\phi(\mathbf{A}) = \phi\left(\mathbf{X}\mathbf{J}\mathbf{X}^{-1}\right) = \sum_{k=0}^{\infty} \left(\mathbf{X}\mathbf{J}\mathbf{X}^{-1}\right)^k \phi_k = \mathbf{X}\left(\sum_{k=0}^{\infty} \mathbf{J}^k \phi_k\right)\mathbf{X}^{-1} = \mathbf{X}\phi(\mathbf{J})\mathbf{X}^{-1}$$

also converges.

1.7.2 Matrix Exponentials

One of the most useful matrix functions is the **matrix exponential**

$$\exp(\mathbf{A}) = \sum_{k=0}^{\infty} \mathbf{A}^k \frac{1}{k!} .$$

The following lemma will be useful in discussing the importance of the matrix exponential to ordinary differential equations in Sect. 3.3.1 in Chap. 3 of Volume III.

Theorem 1.7.1 (Matrix Exponential) *Suppose that* $\mathbf{A}$ *is an* $n \times n$ *matrix. Then the infinite series* $\exp(\mathbf{A})$ *converges. Also, if* τ *is a scalar then*

$$\frac{\mathrm{d}}{\mathrm{d}\tau}\exp(\mathbf{A}\tau) = \mathbf{A}\exp(\mathbf{A}\tau) = \exp(\mathbf{A}\tau)A \; .$$

If $\mathbf{J} = \mathbf{I}\lambda + \mathbf{N}$ *is an* $n \times n$ *Jordan block, then* $\exp(\mathbf{J}\tau)$ *is right triangular; in particular, if* $1 \leq i \leq j \leq n$ *then the* i, j *entry of* $\exp(\mathbf{J}\tau)$ *is*

$$\mathbf{e}_i \cdot \exp(\mathbf{J}\tau)\mathbf{e}_j = \frac{\tau^{j-i}}{(j-i)!}\exp(\lambda\tau) \; .$$

Furthermore, $\exp(0) = \mathbf{I}$*, and* $\exp(-\mathbf{A}) = \exp(\mathbf{A})^{-1}$*. Finally, if* $\mathbf{A}$ *and* $\mathbf{B}$ *are two square matrices of the same size such that* $\mathbf{AB} = \mathbf{BA}$*, then*

$$\exp\left(\mathbf{A} + \mathbf{B}\right) = \exp(\mathbf{A})\exp(\mathbf{B}) \; .$$

Proof First, we note that

$$\|\exp(\mathbf{A}\tau)\| \leq \sum_{k=0}^{\infty}\left\|\mathbf{A}^k\frac{\tau^k}{k!}\right\| \leq \sum_{k=0}^{\infty}\|\mathbf{A}\|^k\frac{|\tau|^k}{k!} = \mathbf{e}^{\|\mathbf{A}\||\tau|} \; .$$

This shows that the matrix exponential is a convergent series. This should be expected, since the exponential function has a Taylor series with an finite radius of convergence.

Next, let us calculate

$$\frac{\mathrm{d}}{\mathrm{d}\tau}\exp(\mathbf{A}\tau) = \frac{\mathrm{d}}{\mathrm{d}\tau}\left(\sum_{k=0}^{\infty}\mathbf{A}^k\frac{\tau^k}{k!}\right) = \sum_{k=1}^{\infty}\mathbf{A}^k\frac{\tau^{k-1}}{(k-1)!}$$

$$= \mathbf{A}\sum_{j=0}^{\infty}\mathbf{A}^j\frac{\tau^j}{j!} = \mathbf{A}\exp(\mathbf{A}\tau) \; .$$

We can also see that this is equal to $\exp(\mathbf{A}\tau)\mathbf{A}$ by pulling $\mathbf{A}$ out on the right at the beginning of the previous equation line.

For a Jordan block $\mathbf{J}$, we compute

$$\exp(\mathbf{J}\tau) = \sum_{k=1}^{\infty}(\mathbf{I}\lambda + \mathbf{N})^k\frac{\tau^k}{k!} = \sum_{k=0}^{\infty}\left(\sum_{j=0}^{k}\mathbf{N}^j\lambda^{k-j}\binom{k}{j}\right)\frac{\tau^k}{k!}$$

$$= \sum_{j=0}^{\infty}\mathbf{N}^j\sum_{k=j}^{\infty}\frac{\tau^k}{k!}\frac{k!}{j!(k-j)!}\lambda^{k-j} = \sum_{j=0}^{\infty}\mathbf{N}^j\frac{\tau^j}{j!}\sum_{i=0}^{\infty}\frac{\lambda^i\tau^i}{i!} = \sum_{j=0}^{\infty}\mathbf{N}^j\frac{\tau^j}{j!}\mathbf{e}^{\lambda\tau} \; .$$

Substituting $\tau = 0$ in the previous result shows that $\exp(\mathbf{0}) = \mathbf{I}$.

We can also multiply

$$\exp(\mathbf{A})\exp(-\mathbf{A}) = \sum_{k=0}^{\infty}\sum_{j=0}^{\infty}\mathbf{A}^{k+j}\frac{(-1)^j}{k!j!} = \sum_{k=0}^{\infty}\sum_{i=k}^{\infty}\mathbf{A}^i\frac{(-1)^{i-k}}{k!(i-k)!}$$

$$= \sum_{i=0}^{\infty}\mathbf{A}^i\frac{1}{i!}\sum_{k=0}^{i}(-1)^{i-k}\binom{i}{k} = \sum_{i=0}^{\infty}\mathbf{A}^i\frac{1}{i!}(1-1)^i = \mathbf{A}^0 = \mathbf{I}\,.$$

This equation shows that $\exp(-\mathbf{A}) = \exp(\mathbf{A})^{-1}$.

To prove the final claim, we note that

$$\exp(\mathbf{A})\exp(\mathbf{B}) = \left(\sum_{i=0}^{\infty}\frac{1}{i!}\mathbf{A}^i\right)\left(\sum_{j=0}^{\infty}\frac{1}{j!}\mathbf{B}^j\right) = \sum_{i=0}^{\infty}\sum_{j=0}^{\infty}\frac{1}{i!j!}\mathbf{A}^i\mathbf{B}^j$$

$$= \sum_{i=0}^{\infty}\sum_{k=i}^{\infty}\frac{1}{i!(k-i)!}\mathbf{A}^i\mathbf{B}^{k-i} = \sum_{k=0}^{\infty}\sum_{i=0}^{k}\frac{1}{i!(k-i)!}\mathbf{A}^i\mathbf{B}^{k-i} = \sum_{k=0}^{\infty}\frac{1}{k!}\sum_{i=0}^{k}\binom{k}{i}\mathbf{A}^i\mathbf{B}^{k-i}$$

and since $\mathbf{AB} = \mathbf{BA}$), we get

$$= \sum_{k=0}^{\infty}\frac{1}{k!}(\mathbf{A}+\mathbf{B})^k = \exp(\mathbf{A}+\mathbf{B})\ .$$

The next theorem will be useful to our study of initial value problems in Chap. 3 in Volume III.

Theorem 1.7.2 (Matrix Logarithm) *Suppose that* $\mathbf{A}$ *is a nonsingular* $n\times n$ *matrix. Then there is an* $n \times n$ *matrix* $\mathbf{B}$ *so that*

$$\mathbf{A} = \exp(\mathbf{B})\ .$$

Proof If $\mathbf{J}$ is a nonsingular $m \times m$ Jordan block, then we can write

$$\mathbf{J} = \mathbf{I}\lambda + \mathbf{N}$$

where $\lambda \neq 0$ and $\mathbf{N}$ is the $m \times m$ nilpotent matrix. Recall that

$$\log(1+x) = -\sum_{k=1}^{\infty}\frac{(-x)^k}{k} \text{ for } |x| < 1\ .$$

This suggests that we define

$$\mathbf{L} = \log\left(\mathbf{I} + \mathbf{N}\frac{1}{\lambda}\right) \equiv -\sum_{k=1}^{\infty}\frac{(-\lambda)^k}{k}\mathbf{N}^k\ . = -\sum_{k=1}^{m-1}\frac{(-\lambda)^k}{k}\mathbf{N}^k$$

where the last equation is true because $\mathbf{N}^m = \mathbf{0}$. Since

$$1 + x = \exp(\log(1 + x)) = \sum_{j=0}^{\infty} \frac{1}{j!} [\log(1 + x)]^j ,$$

it follows that if the right-hand side of this equation is expanded in powers of x, all terms must vanish except for the two lowest-order terms. We conclude that

$$\mathbf{I} + \mathbf{N}\frac{1}{\lambda} = \exp(\mathbf{L}) .$$

It follows that

$$\mathbf{J} = (\mathbf{I}\lambda) \left(\mathbf{I} + \mathbf{N}\frac{1}{\lambda} \right) = \exp\left(\mathbf{I}\log\lambda\right) \exp\left(\mathbf{L}\right)$$

and since $\mathbf{I}\log\lambda$ commutes with $\mathbf{L}$, Theorem 1.7.1 gives us

$$= \exp\left(\mathbf{I}\log\lambda + \mathbf{L}\right) .$$

Thus any nonsingular Jordan block is the exponential of a matrix.

For a general nonsingular matrix $\mathbf{A}$, there is a nonsingular matrix $\mathbf{X}$ so that

$$\mathbf{A}\mathbf{X} = \mathbf{X}\mathbf{J}$$

where $\mathbf{J}$ is block diagonal and the diagonal matrices are nonsingular Jordan blocks. From our work in the previous paragraph, we can find a block diagonal matrix $\mathbf{M}$ so that

$$\mathbf{J} = \exp(\mathbf{M}) .$$

Then

$$\mathbf{A} = \mathbf{X}\mathbf{J}\mathbf{X}^{-1} = \mathbf{X}\exp(\mathbf{M})\mathbf{X}^{-1} = \exp\left(\mathbf{X}\mathbf{M}\mathbf{X}^{-1}\right) .$$

In a comprehensive review paper [126], Moler and Van Loan have discussed the deficiencies of various techniques for computing matrix exponentials. Because of these difficulties, LAPACK does not contain a routine to compute a matrix exponential, nor does the GNU Scientific Library. On the other hand, MATLAB provides the command expm.

Exercise 1.7.1 Compute the exponential of

$$\mathbf{A} = \begin{bmatrix} 1 & 0 \\ 0 & -1 \end{bmatrix} .$$

Exercise 1.7.2 If γ and σ are scalars such that $|\gamma|^2 + |\sigma|^2 = 1$, let

$$\mathbf{G} = \begin{bmatrix} \gamma & \sigma \\ -\overline{\sigma} & \overline{\gamma} \end{bmatrix}$$

be the corresponding plane rotation. Compute $\exp(\mathbf{G})$. (Hint: see Example 1.4.7 for a calculation of $\mathbf{G}^k$ when γ and σ are real.)

Exercise 1.7.3 Suppose that $\mathbf{H}$ is an Hermitian elementary reflector. Compute $\exp(\mathbf{H})$. (Hint: you may find the Taylor series for $\cosh(1)$ and $\sinh(1)$ to be useful.)

Exercise 1.7.4 Compute $\exp(\mathbf{A})$ where

$$\mathbf{A} = \begin{bmatrix} 5 & -6 & -6 \\ -1 & 4 & 2 \\ 3 & -6 & -4 \end{bmatrix} .$$

Exercise 1.7.5 Compute $\exp(\mathbf{A}t)$ where

$$\mathbf{A} = \begin{bmatrix} 3 & 1 & -1 \\ 2 & 2 & -1 \\ 2 & 2 & 0 \end{bmatrix} .$$

1.7.3 Matrix Square Roots

If $\mathbf{A}$ is Hermitian and nonnegative, then the Spectral Theorem 1.3.1 shows that there is a unitary matrix $\mathbf{Q}$ and a diagonal matrix $\boldsymbol{\Lambda}$ with nonnegative diagonal entries, such that

$$\mathbf{A} = \mathbf{Q}\boldsymbol{\Lambda}\mathbf{Q}^H .$$

If we define $\sqrt{\boldsymbol{\Lambda}}$ to be the diagonal matrix with diagonal entries $\sqrt{\boldsymbol{\Lambda}}_{ii} = \sqrt{\boldsymbol{\Lambda}_{ii}}$, and let

$$\mathbf{S} = \mathbf{Q}\sqrt{\boldsymbol{\Lambda}}\mathbf{Q}^H ,$$

then

$$\mathbf{S}^2 = \mathbf{Q}\sqrt{\boldsymbol{\Lambda}}^2\mathbf{Q}^H = \mathbf{Q}\boldsymbol{\Lambda}\mathbf{Q}^H = \mathbf{A} .$$

This suggests that we write $\mathbf{S} = \sqrt{\mathbf{A}}$. In general, if $\mathbf{A}$ is a nonnegative Hermitian matrix then we will require $\sqrt{\mathbf{A}}$ to have nonnegative eigenvalues.

It is interesting to note that the nilpotent matrix

$$\mathbf{N} = \begin{bmatrix} 0 & 1 \\ 0 & 0 \end{bmatrix}$$

does not have a square root. To see this, note that

$$\begin{bmatrix} \alpha & \gamma \\ \beta & \delta \end{bmatrix}^2 = \begin{bmatrix} \alpha^2 + \beta\gamma & \gamma(\alpha+\delta) \\ \beta(\alpha+\delta) & \beta\gamma + \delta^2 \end{bmatrix} .$$

If this matrix squared were equal to $\mathbf{N}$, then we would have $\beta(\alpha+\delta) = 0$, so either $\beta = 0$ or $\alpha + \delta = 0$. In the latter case, we would have $\gamma(\alpha+\delta) = 0$, so the square of this matrix could not be $\mathbf{N}$. In the former case, the diagonal entries of the matrix squared would imply that $\alpha = 0 = \delta$, so again the entry above the diagonal could not be zero.

Lemma 1.7.2 *If* $\mathbf{A}$ *is nonsingular, then there is a nonsingular matrix* $\mathbf{S}$ *so that* $\mathbf{S}^2 = \mathbf{A}$ *and all of the eigenvalues of* $\mathbf{S}$ *have nonnegative real part.*

Proof Let

$$\phi\left(\varrho e^{i\theta}\right) = \sqrt{\varrho} e^{i\theta/2} \text{ where } -\pi \le \theta < \pi .$$

In other words, ϕ is the square root function with branch cut taken along the negative real axis. If $\mathbf{J} = \mathbf{I}\varrho e^{i\theta} + \mathbf{N}$ is a Jordan block with $\varrho \neq 0$, let

$$\mathbf{S} = \sum_{j=0}^{n-1} \mathbf{N}^j \frac{\phi^{(j)}(\lambda)}{j!} .$$

The proof of Lemma 1.7.1 shows that $\mathbf{S} = \phi(\mathbf{J})$, so the diagonal entries of $\mathbf{S}$ have nonnegative real part. However, we would like to show that $\mathbf{S}^2 = \mathbf{J}$. To do so, we recall **Leibniz's rule** for high-order differentiation of a product of functions:

$$\frac{d^\ell(uv)}{dx^\ell} = \sum_{k=0}^{\ell} \binom{\ell}{k} \frac{d^k u}{dx^k} \frac{d^{\ell-k} v}{dx^{\ell-k}} .$$

Then we compute

$$\mathbf{S}^2 = \left(\sum_{j=0}^{\infty} \mathbf{N}^j \frac{\phi^{(j)}(\lambda)}{j!} \right) \left(\sum_{k=0}^{\infty} \mathbf{N}^k \frac{\phi^{(k)}(\lambda)}{k!} \right) = \sum_{k=0}^{\infty} \sum_{\ell=k}^{\infty} \mathbf{N}^\ell \frac{\phi^{(k)}(\lambda)\phi^{(\ell-k)}(\lambda)}{k!(\ell-k)!}$$

$$= \sum_{\ell=0}^{\infty} \mathbf{N}^\ell \frac{1}{\ell!} \sum_{k=0}^{\ell} \binom{\ell}{k} \phi^{(k)}(\lambda)\phi^{(\ell-k)}(\lambda) = \sum_{\ell=0}^{\infty} \mathbf{N}^\ell \frac{1}{\ell!} \frac{d^\ell \phi(\lambda)^2}{d\lambda^\ell} = \mathbf{N}^0 \frac{1}{0!}\lambda + \mathbf{N}^1 \frac{1}{1!}$$

$$= \mathbf{I}\lambda + \mathbf{N} = \mathbf{J} .$$

More generally, if $\mathbf{A}$ is nonsingular, then the Jordan Decomposition Theorem 1.4.6 implies that there is a nonsingular matrix $\mathbf{X}$ and a Jordan canonical form $\mathbf{J}$ in which all Jordan blocks have nonzero eigenvalues, so that

$$\mathbf{A} = \mathbf{X}\mathbf{J}\mathbf{X}^{-1} \, .$$

Then

$$\mathbf{S} = \mathbf{X}\phi(\mathbf{J})\mathbf{X}^{-1}$$

is such that

$$\mathbf{S}^2 = \mathbf{X}\phi(\mathbf{J})^2\mathbf{X}^{-1} = \mathbf{X}\mathbf{J}\mathbf{X}^{-1} = \mathbf{A} \, ,$$

and all of the eigenvalues of $\mathbf{S}$ have nonnegative real part.
Of course, the Jordan decomposition of a matrix is numerically unstable, and can not be determined in floating point arithmetic.

Björck and Hammarling [19] and Higham [98] have shown how to use the Schur decomposition of a matrix to compute its **square root**. Higham [99] discussed several iterations for computing square roots of nonsingular matrices. Similar to the Newton iteration in Example 2.3.1.5 in Chap. 2 of Volume I, the iteration

$$\begin{aligned} &\mathbf{S}_0 = \mathbf{A} \\ &\text{for } 0 \leq k \; \mathbf{S}_{k+1} = (\mathbf{S}_k + \mathbf{S}_k^{-1}\mathbf{A})/2 \end{aligned} \tag{1.62}$$

converges quadratically to a square root of $\mathbf{A}$, provided that the eigenvalues of $\mathbf{A}$ are all real and positive. Unfortunately, this iteration is not necessarily numerically stable, and rounding errors are amplified by the iteration unless the eigenvalues satisfy the condition

$$\lambda_j \leq 9\lambda_i$$

for all eigenvalue indices i and j [97]. A better iteration, due to Denman and Beavers [50], takes the form

$$\begin{aligned} &\mathbf{S}_0 = \mathbf{A} \, , \; \mathbf{R}_0 = \mathbf{I} \\ &\text{for } 0 \leq k \\ &\quad \mathbf{S}_{k+1} = (\mathbf{S}_k + \mathbf{R}k^{-1})/2 \\ &\quad \mathbf{R}_{k+1} = (\mathbf{R}_k + \mathbf{S}_k^{-1})/2 \, . \end{aligned}$$

Then $\mathbf{S}_k \rightarrow \sqrt{\mathbf{A}}$ and $\mathbf{R}_k \rightarrow \sqrt{\mathbf{A}^{-1}}$.

MATLAB users may use the command sqrtm to compute the square root of a nonsingular matrix. LAPACK and the GNU Scientific Library do not have functions to compute matrix square roots.

The reader may experiment with the following JavaScript program for **matrix square roots.** A square matrix may be entered by rows, with row entries separated by commas, and rows separated by semi-colons. The reader may choose between the Newton iteration or the Denman and Beavers iteration. The program displays the approximate square root (and its inverse for the Denman and Beavers iteration), as well as the errors in the approximations.

Exercise 1.7.6 Compute $\sqrt{\mathbf{A}}$ where

$$\mathbf{A} = \begin{bmatrix} 4 & 1 & & \\ & 4 & & \\ & & 9 & 1 \\ & & & 9 \end{bmatrix} .$$

Exercise 1.7.7 The 2×2 Hadamard matrix is

$$\mathbf{H}_2 = \begin{bmatrix} 1 & 1 \\ 1 & -1 \end{bmatrix} .$$

Show that $\sqrt{\mathbf{H}_2}$ cannot have all real entries.

Exercise 1.7.8 Compute $\sqrt{\mathbf{H}_4}$ where

$$\mathbf{H}_4 = \begin{bmatrix} 1 & 1 & 1 & 1 \\ 1 & -1 & 1 & -1 \\ 1 & 1 & -1 & -1 \\ 1 & -1 & -1 & 1 \end{bmatrix} = \begin{bmatrix} \mathbf{H}_2 & \mathbf{H}_2 \\ \mathbf{H}_2 & -\mathbf{H}_2 \end{bmatrix} .$$

Exercise 1.7.9 Apply the Newton iteration (1.62) to

$$\mathbf{A} = \begin{bmatrix} 85 & -51 & -75 & 45 \\ -51 & 85 & 45 & -75 \\ -75 & 45 & 85 & -51 \\ 45 & -75 & -51 & 85 \end{bmatrix} = \begin{bmatrix} 1 & 1 & 1 & 1 \\ 1 & -1 & 1 & -1 \\ 1 & 1 & -1 & -1 \\ 1 & -1 & -1 & 1 \end{bmatrix} \begin{bmatrix} 1 & & & \\ & 4 & & \\ & & 16 & \\ & & & 64 \end{bmatrix} \begin{bmatrix} 1 & 1 & 1 & 1 \\ 1 & -1 & 1 & -1 \\ 1 & 1 & -1 & -1 \\ 1 & -1 & -1 & 1 \end{bmatrix} .$$

Plot $\|\mathbf{S}_k^2 - \mathbf{A}\|_F$ and $\|\mathbf{S}_k - \sqrt{\mathbf{A}}\|_F$ as a function of k.

Chapter 2
Iterative Linear Algebra

One thing that being a scientist has taught me is that you can never be certain about anything. You never know the truth. You can only approach it and hope to get a bit nearer to it each time. You iterate towards the truth. You don't know it.

James Lovelock, quoted by Lorrie Goldstein in
Green 'drivel' exposed, Toronto Sun, June 22, 2012

Abstract This chapter describes iterative methods for approximately solving systems of linear equations. This discussion begins by presenting the concept of a sparse matrix, where it arises and how it might be represented in a computer. Next, simple methods based on iterative improvement are presented, along with termination criteria for the iteration. Afterwards, more elaborate gradient methods are examined, such as conjugate gradients and biconjugate gradients. The chapter proceeds tominimum resiual methods, and ends with a fairly detailed discussion of multigrid methods.

2.1 Overview

In Chaps. 3 and 6 in Volume I, we studied matrix factorizations for solving systems of linear equations. These factorizations were used in Chap. 1 to develop iterative methods for finding eigenvalues and eigenvectors. In this chapter, we will use our knowledge of eigenvalues and eigenvectors to develop iterative methods for solving systems of linear equations. These new methods will be particularly useful for solving large sparse systems of linear equations, which often arise in the numerical solution of partial differential equations.

Our goals in this chapter are to develop several basic iterative methods for solving linear systems. The most rudimentary schemes will be based on iterative improvement. Gradient and minimum residual methods are more recently developed

---

Additional Material: The details of the computer programs referred in the text are available in the Springer website (http://extras.springer.com/2018/978-3-319-69107-7) for authorized users.

J.A. Trangenstein, *Scientific Computing*, Texts in Computational Science and Engineering 19, https://doi.org/10.1007/978-3-319-69107-7_2

iterative methods, and are in common use by contemporary software packages. We will end the chapter with a discussion of multigrid methods. These methods are very efficient, often finding accurate solutions by using work proportional to the number of unknowns. The curious feature of multigrid methods is that they make essential use of the old iterative improvement methods, in combination with multigrid operators called prolongations and restrictions. In this way, our discussion of iterative methods will come full circle.

A careful comparison of iterative methods with direct methods (i.e., Gaussian factorization) often depends on the problem selection. For a general discussion of iterative methods versus direct methods in the numerical solution of elliptic and parabolic partial differential equation, see Trangenstein [173, Sect. 3.1]. For steady-state problems, under reasonable conditions the cited text proves that Gaussian elimination employing band storage is more efficient than a simple iterative method only in one spatial dimension. This is consistent with empirical observations (see, for example, Aziz and Settari [8, p. 207]) that direct methods are preferable only for 1D or small 2D problems, and iterative methods are more efficient for large 2D problems or essentially all 3D problems.

For more detailed discussion of iterative methods for linear algebra, we suggest books by Axelsson [7], Barrett et al. [10], Greenbaum [88], Hackbusch [90], Hageman and Young [91], Kelley [107], Saad [154], Varga [177], Wachspress [178], Young [189], and Chap. 3 of Trangenstein [173]. The book by Barrett et al. is also available online. For iterative linear algebra software, we recommend the Barrett et al. templates, PETSc (Portable, Extensible Toolkit for Scientific Computation) Sparse Linear Solvers and MATLAB sparse matrix operations.

2.2 Sparse Matrices

An $m \times n$ matrix is **sparse** if it has far fewer than $m \times n$ nonzero entries. Sparse matrices commonly arise in discretization of partial differential equations, but may also arise in other problem areas, such as networks of electrical circuits or people.

Because sparse matrices have few nonzero entries, it would be disadvantageous to represent them as square arrays, thereby storing all the zero entries in the array. There are several techniques for storing sparse matrices. We will use the **Yale sparse matrix format**. The nonzero entries of the sparse matrix $\mathbf{A}$ are stored by rows in a single array $\mathbf{a}$ of scalars; the length of this array is the number of nonzero entries in $\mathbf{A}$. A second array $\mathbf{c}$ of integers stores the corresponding column indices for each value in the first array; the length of this array is the number of nonzero entries of $\mathbf{A}$. A third array $\mathbf{s}$ of integers stores for each row the index in the second array of the first column number for that row; this array has length equal to one plus the number of rows of $\mathbf{A}$. In order to multiply the sparse matrix $\mathbf{A}$ times a vector $\mathbf{x}$, we would perform the following algorithm.

Algorithm 2.2.1 (Sparse Matrix Times Vector)

for $0 \leq i < m$

$Ax[i] = 0$

for $\mathbf{s}[i] \leq k < \mathbf{s}[i+1]$

$Ax[i] += \mathbf{a}[k] * \mathbf{x}[\mathbf{c}[k]]$

Example 2.2.1 The matrix

$$\mathbf{A} = \begin{bmatrix} 2 & -1 & & & -1 \\ -1 & 2 & -1 & & \\ & -1 & 2 & -1 & \\ & & -1 & 2 & -1 \\ -1 & & & -1 & 2 \end{bmatrix}$$

could be represented by the sparse matrix storage arrays

$$\mathbf{a} = [2\,,\, -1\,,\, -1\,,\, -1\,,\, 2\,,\, -1\,,\, -1\,,\, 2\,,\, -1\,,\, -1\,,\, 2\,,\, -1\,,\, -1\,,\, -1\,,\, 2]$$
$$\mathbf{c} = [0\,,\, 1\,,\, 4\,,\, 0\,,\, 1\,,\, 2\,,\, 1\,,\, 2\,,\, 3\,,\, 2\,,\, 3\,,\, 4\,,\, 0\,,\, 3\,,\, 4]$$
$$\mathbf{s} = [0\,,\, 3\,,\, 6\,,\, 9\,,\, 12\,,\, 15]$$

A C^{++} class to implement this sparse matrix storage scheme has been implemented in the files SparsityPattern.H and SparsityPattern.C. Another C^{++} class to implement sparse matrices can be found in the files SparseMatrix.H and SparseMatrix.C. These C^{++} classes were copied from the files that were published by Trangenstein [173] and were in turn adopted from the deal.II SparseMatrix class.

Exercise 2.2.1 Suppose that matrix entries require twice as many bits of computer memory as row and column indices. Determine the amount of computer memory required by the Yale sparse matrix format to store an $m \times n$ matrix with k nonzero entries, and determine an upper bound on the number of nonzero entries for the Yale sparse matrix format to require less memory than storing the full matrix.

2.3 Neumann Series

Suppose that $\mathbf{A}$ is an $m \times m$ matrix. Many basic iterative methods for solving the linear system $\mathbf{A}\mathbf{x} = \mathbf{b}$ are designed as follows. First, we choose a convenient matrix $\mathbf{C}$ and an initial guess $\widetilde{\mathbf{x}}_0$, and then we compute the iterates

$$\mathbf{r}_n = b - \mathbf{A}\widetilde{\mathbf{x}}_n \;, \quad \widetilde{\mathbf{x}}_{n+1} = \widetilde{\mathbf{x}}_{(n)} + \mathbf{C}\mathbf{r}_n \;.$$

Note that the errors in the solution satisfy

$$\widetilde{\mathbf{x}}_{n+1} - \mathbf{x} = \widetilde{\mathbf{x}}_n - x + \mathbf{CA}\,(\mathbf{x} - \widetilde{\mathbf{x}}_n) = (\mathbf{I} - \mathbf{CA})(\widetilde{\mathbf{x}}_n - \mathbf{x}) \,.$$

This implies that

$$\widetilde{\mathbf{x}}_n - \mathbf{x} = (\mathbf{I} - \mathbf{CA})^n(\widetilde{\mathbf{x}}_0 - \mathbf{x}) \,.$$

Next, note that the residuals satisfy

$$\mathbf{r}_{n+1} = \mathbf{b} - \mathbf{A}\widetilde{\mathbf{x}}_{n+1} = \mathbf{b} - \mathbf{A}\,(\widetilde{\mathbf{x}}_n - \mathbf{C}\mathbf{r}_n) = \mathbf{r}_n - \mathbf{AC}\mathbf{r}_n = (\mathbf{I} - \mathbf{AC})\mathbf{r}_n \,,$$

so

$$\mathbf{r}_{n+\ell} = (\mathbf{I} - \mathbf{AC})^\ell \mathbf{r}_n \,.$$

It follows that the changes in the solution satisfy

$$\widetilde{\mathbf{x}}_{n+\ell+1} - \widetilde{\mathbf{x}}_{n+\ell} = \mathbf{C}\mathbf{r}_{n+\ell} = \mathbf{C}(\mathbf{I} - \mathbf{AC})^\ell \mathbf{r}_n \,.$$

If this iterative improvement algorithm converges, we must have

$$\mathbf{x} - \widetilde{\mathbf{x}}_n = \sum_{\ell=0}^{\infty}(\widetilde{\mathbf{x}}_{n+\ell+1} - \widetilde{\mathbf{x}}_{n+\ell}) = \mathbf{C}\left[\sum_{\ell=0}^{\infty}(\mathbf{I} - \mathbf{AC})^\ell\right]\mathbf{r}_n = \left[\sum_{\ell=0}^{\infty}(\mathbf{I} - \mathbf{CA})^\ell\right]\mathbf{C}\mathbf{r}_n \,.$$

Thus the convergence of these methods will be related to the convergence of the **Neumann series** $\sum_{n=0}^{\infty}(\mathbf{I} - \mathbf{AC})^n$ or $\sum_{n=0}^{\infty}(\mathbf{I} - \mathbf{CA})^n$.

Since the matrices we will consider are not necessarily diagonalizable, we will use the Schur decomposition of Theorem 1.4.2 to bound norms of matrix powers. Numerically, this approach is more robust than the use of Jordan canonical forms, because the latter are unstable in floating-point computations.

The next lemma will be used in Lemma 2.3.1 to show that the Neumann series converges, and in Lemma 2.4.1 to show that iterative improvement convergences.

Lemma 2.3.1 *If* $\mathbf{A}$ *is a square matrix with spectral radius* $\varrho(\mathbf{A}) < 1$, *then* $\|\mathbf{A}^n\|_2 \to 0$ *as* $n \to \infty$.

Proof Choose $\theta \geq 0$ so that

$$1 + \theta > \frac{2\|\mathbf{U}\|_F}{1 - \varrho(\mathbf{A})} \,.$$

Then

$$\varrho(\mathbf{A}) + \frac{\|\mathbf{U}\|_F}{1 + \theta} < \frac{1 + \varrho(\mathbf{A})}{2} < 1 \,.$$

It follows from Lemma 1.4.2 that

$$\|\mathbf{A}^n\|_2 \leq (1+\theta)^{m-1}\left(\frac{1+\varrho(\mathbf{A})}{2}\right)^n .$$

Since the right-hand side in this inequality tends to zero as n tends to infinity, the corollary is proved.

The next corollary will be used in Corollary 2.3.2 and in Lemma 2.4.8 to understand stopping criteria for iterative improvement.

Corollary 2.3.1 *Suppose that* $\mathbf{A}$ *is a square matrix.*

1. *If* $\varrho(\mathbf{A}) < 1$, *then* $\mathbf{I}-\mathbf{A}$ *is nonsingular, and* $\sum_{n=0}^{\infty}\mathbf{A}^n = (\mathbf{I}-\mathbf{A})^{-1}$.
2. *If* $\sum_{n=0}^{\infty}\mathbf{A}^n$ *converges, then* $\varrho(\mathbf{A}) < 1$.

Proof Let us prove the first claim. If we choose θ as in the proof of Corollary 2.3.1, then

$$\left\|\sum_{n=0}^{N}\mathbf{A}^n\right\|_2 \leq \sum_{n=0}^{N}\|\mathbf{A}^n\|_2 \leq (1+\theta)^{n-1}\sum_{n=0}^{N}\left[\frac{1+\varrho(\mathbf{A})}{2}\right]^n \leq \frac{(1+\theta)^{n-1}}{1-[1+\varrho(\mathbf{A})]/2} .$$

Thus the Neumann series $\sum_{n=0}^{\infty}\mathbf{A}^n$ converges absolutely. Since Corollary 2.3.1 shows that

$$(\mathbf{I}-\mathbf{A})\sum_{n=0}^{N}\mathbf{A}^n - \mathbf{I} = -\mathbf{A}^{N+1} \to 0 \text{ as } N \to \infty ,$$

we see that

$$(\mathbf{I}-\mathbf{A})\sum_{n=0}^{\infty}\mathbf{A}^n = \mathbf{I} .$$

Thus $\mathbf{I}-\mathbf{A}$ is nonsingular, and the first result is proved.

Next, we will prove the second claim. Suppose that $\mathbf{A}\mathbf{x} = \mathbf{x}\lambda$ with $\mathbf{x} \neq 0$. Then

$$\left(\sum_{n=0}^{\infty}\mathbf{A}^n\right)\mathbf{x} = \left(\sum_{n=0}^{\infty}\lambda^n\right)\mathbf{x} .$$

Since the geometric series converges if and only if $|\lambda| < 1$, the claim is proved.

The next corollary tells us about Neumann series for nonnegative matrices that are not too large.

Corollary 2.3.2 ([7, p. 213]) *If* $\mathbf{A}$ *is a square matrix with real nonnegative entries and* $\varrho(\mathbf{A}) < 1$, *then* $\mathbf{I}-\mathbf{A}$ *is nonsingular and* $(\mathbf{I}-\mathbf{A})^{-1}$ *has all nonnegative entries.*

Proof Since $\varrho(\mathbf{A}) < 1$, Corollary 2.3.1 shows that the Neumann series $\sum_{n=0}^{\infty} \mathbf{A}^n$ converges to $(\mathbf{I} - \mathbf{A})^{-1}$. Since $\mathbf{A} \geq 0$ implies that all of the terms in the series have nonnegative entries, the matrix $(\mathbf{I} - \mathbf{A})^{-1}$ is nonnegative.

Exercise 2.3.1 Let $\triangle x = 1/m$ and define the $m \times m$ matrix

$$\mathbf{A} = \frac{1}{\triangle x^2} \begin{bmatrix} 2 & -1 & & \\ -1 & \ddots & \ddots & \\ & \ddots & \ddots & -1 \\ & & -1 & 2 \end{bmatrix} .$$

Show that $\mathbf{I} - \mathbf{A}/2 \geq 0$ and $\varrho(\mathbf{I} - \mathbf{A}/2) < 1$. What can you conclude from Corollary 2.3.2?

2.4 Iterative Improvement

Suppose that we want to solve a linear system $\mathbf{A}\mathbf{x} = \mathbf{b}$, and we have an initial guess $\widetilde{\mathbf{x}}_0$ for the solution $\mathbf{x}$. Given a **preconditioner** $\mathbf{C}$ that is convenient for computation, we can compute a sequence of approximations $\widetilde{\mathbf{x}}_n$ to the solution by means of the following

Algorithm 2.4.1 (Iterative Improvement)

$$\widetilde{\mathbf{x}}_{n+1} = \widetilde{\mathbf{x}}_n + \mathbf{C}[\mathbf{b} - \mathbf{A}\widetilde{\mathbf{x}}_n] . \tag{2.1}$$

We have already used this idea in Sect. 3.9.2 in Chap. 3 of Volume I, Sects. 6.7.6 and 6.8.7 in Chap. 6 of Volume I to reduce rounding errors in solving linear systems via matrix factorization. In those sections, the matrix $\mathbf{C}$ corresponded to some product of matrix factors computed in floating point arithmetic. Since many iterative methods for solving systems of linear equations will also take the form of iterative improvement, it is important for us to understand the general properties of such iterations.

2.4.1 Theory

We begin with a general result that characterizes the conditions under which iterative improvement will converge.

Lemma 2.4.1 ([189, p. 77]) *Suppose that* $\mathbf{A}$ *and* $\mathbf{C}$ *are* $m \times m$ *matrices, and* $\mathbf{b}$ *is an* m*-vector. Then the iterative improvement algorithm* (2.1) *converges for any initial guess if and only if the spectral radius of* $\mathbf{I} - \mathbf{C}\mathbf{A}$ *satisfies* $\varrho(\mathbf{I} - \mathbf{C}\mathbf{A}) < 1$.

Proof Suppose that $\varrho(\mathbf{I}-\mathbf{CA}) < 1$. Let us denote the error in the approximation by $\mathbf{d}_n = \widetilde{\mathbf{x}}_n - \mathbf{x}$. Then $\mathbf{r}_n = \mathbf{A}\widetilde{\mathbf{x}}_n - \mathbf{A}\mathbf{x} = \mathbf{A}\mathbf{d}_n$, and the error after one more step of the iteration is

$$\mathbf{d}_{n+1} = \widetilde{\mathbf{x}}_{n+1} - \mathbf{x} = [\widetilde{\mathbf{x}}_n - \mathbf{C}\mathbf{r}_n] - \mathbf{x} = \mathbf{d}_n - \mathbf{CA}\mathbf{d}_n = [\mathbf{I}-\mathbf{CA}]\mathbf{d}_n .$$

The solution of this linear recurrence is $\mathbf{d}_n = (\mathbf{I}-\mathbf{CA})^n\mathbf{e}_0$, and Corollary 2.3.1 shows that $\mathbf{d}_n \to 0$.

To prove the converse, suppose that $\varrho(\mathbf{I}-\mathbf{CA}) \geq 1$. Then there exists an eigenvector $\mathbf{z}$ of $\mathbf{I}-\mathbf{CA}$ with eigenvalue λ such that $|\lambda| \geq 1$. Let $\widetilde{\mathbf{x}}_0 = \mathbf{x}+\mathbf{z}$. Then $\mathbf{d}_0 = \widetilde{\mathbf{x}}_0 - \mathbf{x} = \mathbf{z}$ so $\mathbf{d}_n = (\mathbf{I}-\mathbf{CA})^n\mathbf{d}_0 = \mathbf{z}\lambda^n$. It follows that $\|\mathbf{d}_n\| = \|\mathbf{z}\|\ |\lambda|^n \geq \|\mathbf{z}\|$, so iterative improvement with this initial guess does not converge.

In many cases, it is difficult to determine if the preconditioner satisfies $\varrho(\mathbf{I}-\mathbf{CA}) < 1$. The next lemma provides alternative convergence conditions in cases where the preconditioner $\mathbf{C}$ and the original matrix $\mathbf{A}$ satisfy certain nonnegativity conditions.

Lemma 2.4.2 ([7, Corollary 6.17]) *Suppose that* $\mathbf{A}$ *and* $\mathbf{C}$ *are real* $m\times m$ *matrices,* $\mathbf{I}-\mathbf{CA} \geq \mathbf{0}$*,* $\mathbf{C}$ *is nonsingular and* $\mathbf{C} \geq \mathbf{0}$*. Then the iterative improvement algorithm is convergent for any initial guess if and only if* $\mathbf{A}$ *is nonsingular and* $\mathbf{A}^{-1} \geq \mathbf{0}$*.*

Proof Suppose that iterative improvement is convergent for any initial guess. Then Lemma 2.4.1 shows that $\varrho(\mathbf{I}-\mathbf{CA}) < 1$. Consequently, Corollary 2.3.2 implies that $\mathbf{CA} = \mathbf{I}-(\mathbf{I}-\mathbf{CA})$ is nonsingular. Since $\mathbf{I}-\mathbf{CA} \geq \mathbf{0}$, Corollary 2.3.1 implies that

$$(\mathbf{CA})^{-1} = \sum_{n=0}^{\infty}(\mathbf{I}-\mathbf{CA})^n \geq \mathbf{0} .$$

Because $\mathbf{C}$ and $\mathbf{CA}$ are nonsingular, $\mathbf{A}$ is nonsingular. Finally, since $\mathbf{C} \geq \mathbf{0}$ and $(\mathbf{CA})^{-1} \geq \mathbf{0}$, we see that

$$\mathbf{A}^{-1} = (\mathbf{CA})^{-1}\mathbf{C} \geq \mathbf{0} .$$

This proves the first direction of the lemma.

To prove the converse, we are again given that $\mathbf{I}-\mathbf{CA} \geq \mathbf{0}$, $\mathbf{C}$ is nonsingular and $\mathbf{C} \geq \mathbf{0}$. Suppose in addition that $\mathbf{A}$ is nonsingular, and $\mathbf{A}^{-1} \geq \mathbf{0}$. For any matrix $\mathbf{B}$, the finite geometric series satisfies

$$\left(\sum_{j=0}^{n}\mathbf{B}^j\right)(\mathbf{I}-\mathbf{B}) = \mathbf{I}-\mathbf{B}^{n+1} .$$

If $\mathbf{B} = \mathbf{I}-\mathbf{CA}$, it follows that $\mathbf{C} = (\mathbf{I}-\mathbf{B})\mathbf{A}^{-1}$ and

$$\left(\sum_{j=0}^{n}(\mathbf{I}-\mathbf{CA})^j\right)\mathbf{C} = \mathbf{A}^{-1} - (\mathbf{I}-\mathbf{CA})^{n+1}\mathbf{A}^{-1} .$$

Since both $\mathbf{I} - \mathbf{CA} \geq \mathbf{0}$ and $\mathbf{A}^{-1} \geq \mathbf{0}$, it follows that, for any $\mathbf{v} > \mathbf{0}$,

$$\left(\sum_{j=0}^{n} (\mathbf{I} - \mathbf{CA})^j \right) \mathbf{Cv} = \mathbf{A}^{-1}\mathbf{v} - (\mathbf{I} - \mathbf{CA})^{n+1}\mathbf{A}^{-1}\mathbf{v} \leq \mathbf{A}^{-1}\mathbf{v} \, .$$

Because $\mathbf{Cv} > \mathbf{0}$, this geometric series is nonnegative and bounded above; therefore it converges. This implies that $(\mathbf{I} - \mathbf{CA})^n \to \mathbf{0}$ as $n \to \infty$.

The following lemma will be used to discuss the relative efficiency of iterative improvement algorithms for solving linear systems.

Lemma 2.4.3 ([189, p. 87]) *Suppose that* $\| \cdot \|$ *is a norm on m-vectors, and that the* $m \times m$ *matrices* $\mathbf{A}$ *and* $\mathbf{C}$ *are such that* $\|\mathbf{I} - \mathbf{CA}\| < 1$. *Then the iterative improvement algorithm using these two matrices converges. Further, given any* $\varepsilon < 1$, *the number of iterations of iterative improvement required to reduce the error by a factor of* ε *is at most* $\log(\varepsilon) / \log(\|\mathbf{I} - \mathbf{CA}\|)$.

Proof If $\mathbf{x}_0 - \mathbf{x}$ is the error in the initial guess, the error after n iterations of iterative improvement satisfies

$$\|(\mathbf{I} - \mathbf{CA})^n (\mathbf{x}_0 - \mathbf{x}) \| \leq \|\mathbf{I} - \mathbf{CA}\|^n \|\mathbf{x}_0 - \mathbf{x}\| \, .$$

It follows that if $\|\mathbf{I} - \mathbf{CA}\| < 1$, then the iteration converges. If $\|\mathbf{I} - \mathbf{CA}\|^n \leq \varepsilon$, then the error after n iterations will be reduced by a factor of ε or less. Taking logarithms of this inequality, we get $n \log(\|\mathbf{I} - \mathbf{CA}\|) \leq \log \varepsilon$, which is easily solved for n to prove the lemma.

2.4.2 Richardson's Iteration

Our first iterative improvement method is highly inefficient on its own. The principal reason for studying this method is because it is commonly used as a smoother in the multigrid method, described in Sect. 2.7.

Richardson's iteration is iterative improvement with preconditioner $\mathbf{C} = \mathbf{I}/\mu$, where μ is an appropriately chosen scalar. This algorithm can be written in the form

Algorithm 2.4.2 (Richardson's Iteration)

$$\mathbf{r} = \mathbf{b} - \mathbf{A}\widetilde{\mathbf{x}}$$
$$\widetilde{\mathbf{x}} = \widetilde{\mathbf{x}} + \mathbf{r}/\mu$$

A C^{++} program to implement Richardson's iteration is available as ir.h.

The following example shows that Richardson's iteration is equivalent to explicit timestepping to steady state of a related ordinary differential equation.

Example 2.4.1 Suppose that we approximate the system of ordinary differential equations

$$\mathbf{u}'(t) = -\mathbf{A}\mathbf{u}(t) + \mathbf{b} ,$$

by the finite difference approximation

$$\frac{u_{n+1} - u_n}{\Delta t} = -\mathbf{A}u_n + \mathbf{b} .$$

This discretization can be rewritten in the form of a Richardson iteration

$$u_{n+1} = u_n + (\mathbf{b} - \mathbf{A}u_n)\Delta t$$

in which $\mu = 1/\Delta t$. The steady state of the ordinary differential equation satisfies

$$0 = \mathbf{u}'(\infty) = -\mathbf{A}\mathbf{u}(\infty) + \mathbf{b} ,$$

which implies that $\mathbf{A}\mathbf{u}(\infty) = \mathbf{b}$.

The next lemma provides conditions that will guarantee the convergence of Richardson's iteration.

Lemma 2.4.4 *Suppose that the eigenvalues of the square matrix* $\mathbf{A}$ *are positive, and let* $\mu > \varrho(\mathbf{A})/2$. *Then Richardson's iteration converges from any initial guess.*

Proof The assumptions imply that $-\mu < \lambda - \mu < \mu$ for all eigenvalues λ of $\mathbf{A}$. These inequalities can also be written as $|1 - \lambda/\mu| < 1$. If $\mathbf{x}$ is an eigenvector of $\mathbf{A}$ with eigenvalue λ, then $\mathbf{x}$ is an eigenvector of $\mathbf{I}-\mathbf{A}/\mu$ with eigenvalue $1-\lambda/\mu$. Thus the spectral radius of $\mathbf{I} - \mathbf{A}/\mu$ is strictly less than one, and Lemma 2.4.1 guarantees that the iteration will converge from any initial guess.

It is not hard to see that the smallest value for $\varrho(\mathbf{I} - \mathbf{A}/\mu)$ is obtained by taking $\mu = \frac{1}{2}(\lambda_m + \lambda_1)$, where $\lambda_1 \leq \lambda \leq \lambda_m$ are the eigenvalues of $\mathbf{A}$. With this choice, if λ is any eigenvalue of $\mathbf{A}$ then

$$\left|1 - \frac{\lambda}{\mu}\right| \leq \frac{\lambda_m - \lambda_1}{\lambda_m + \lambda_1} .$$

For this choice of μ, components of the error in iterative improvement corresponding to the smallest and largest eigenvalues will be reduced the least, and the components of the error due to the average eigenvalue will be reduced the most. If, on the other hand, we choose $\mu = \lambda_m$, then for all eigenvalues λ of $\mathbf{A}$ we have

$$\left|1 - \frac{\lambda}{\mu}\right| \leq 1 - \frac{\lambda_1}{\lambda_m} .$$

With this choice of μ, we find that Richardson's iteration will make the greatest reduction in the component of the error associated with the largest eigenvalue, and

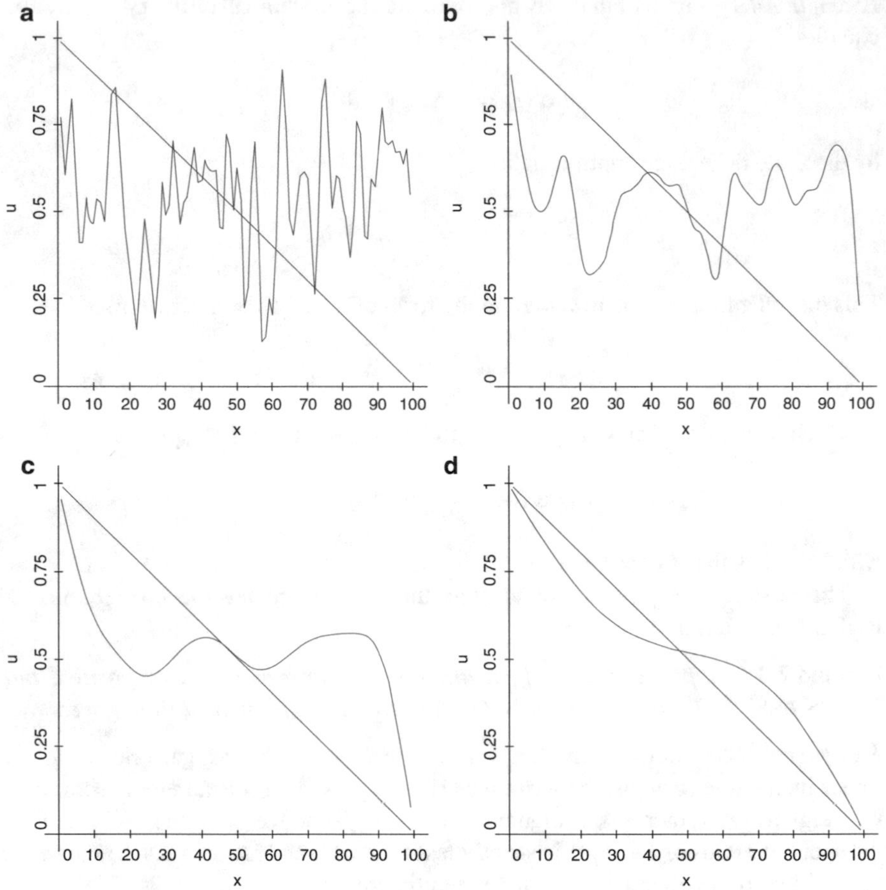

Fig. 2.1 Computed solution with Richardson's iteration, 100 grid cells, $\mu = \varrho(\mathbf{A})$, and random initial guess; the straight line is the exact solution. (**a**) 1 iteration. (**b**) 10 iterations. (**c**) 100 iterations. (**d**) 1000 iterations

the smallest reduction in the component of the error associated with the smallest eigenvalue. It is this latter feature of Richardson's iteration that makes it potentially useful in multigrid methods.

Figures 2.1 and 2.2 show results for Richardson's iteration in solving

$$\begin{bmatrix} 2 & -1 & & \\ -1 & 2 & \ddots & \\ & \ddots & \ddots & -1 \\ & & -1 & 2 \end{bmatrix} \begin{bmatrix} \xi_1 \\ \xi_2 \\ \vdots \\ \xi_m \end{bmatrix} = \begin{bmatrix} 1 \\ 0 \\ \vdots \\ 0 \end{bmatrix}$$

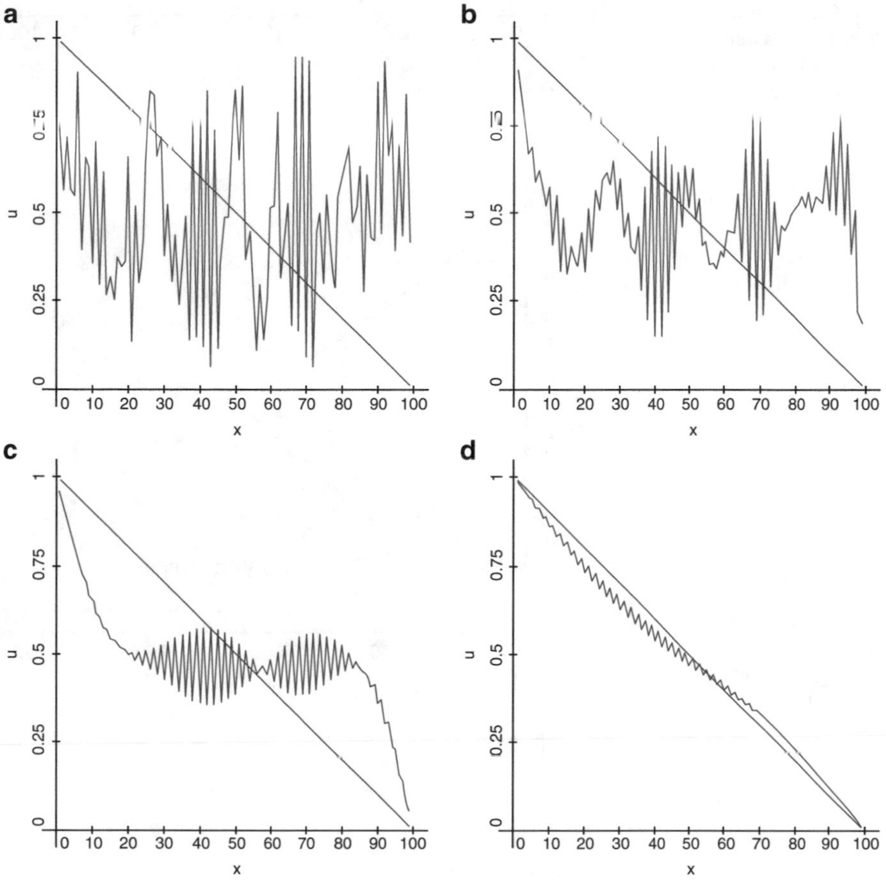

Fig. 2.2 Computed solution with Richardson's iteration, 100 grid cells, $\mu = \frac{1}{2}\varrho(\mathbf{A})$, and random initial guess; the straight line is the exact solution. (**a**) 1 iteration. (**b**) 10 iterations. (**c**) 100 iterations. (**d**) 1000 iterations

with $\mu = 4 \approx \lambda_m$ and $\mu = 2 \approx \frac{1}{2}(\lambda_m + \lambda_1)$. In both cases, the initial guess is chosen to be random and uniformly distributed in 100 grid cells of equal length contained in [0, 1]. Figures 2.3 and 2.4 show the reduction of the errors in these iterations. Note that the larger value of μ produces a smoother but less accurate solution than the smaller value of μ.

A program to perform Richardson's iteration (and other iterative solvers) for Laplace's equation in one dimension can be found in IterativeImprovementMain.C. This program calls Fortran subroutine `richardson` in iterative_improvement.f. The two-dimensional Laplacian is solved in IterativeError2D.C. This program calls Fortran subroutine `richardson` in file iterative_improvement_2d.f. By setting the number of grid cells in one coordinate direction to zero, the user can perform a mesh refinement study, and compare the performance of several iterative methods.

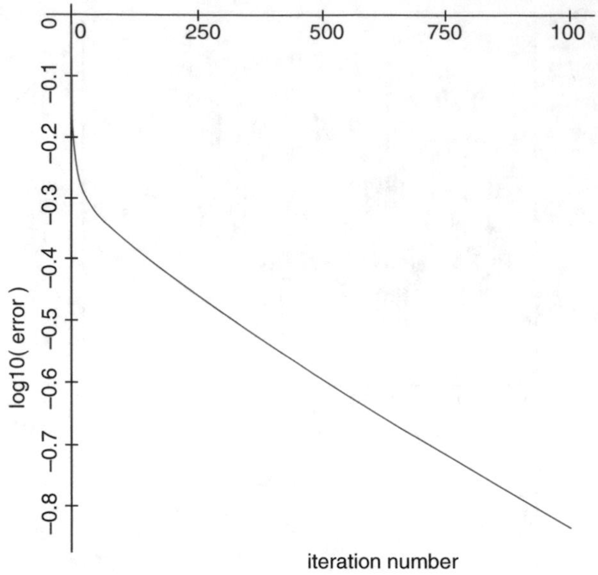

Fig. 2.3 Error in Richardson's iteration, $\mu = \varrho(\mathbf{A})$: log error versus iteration number

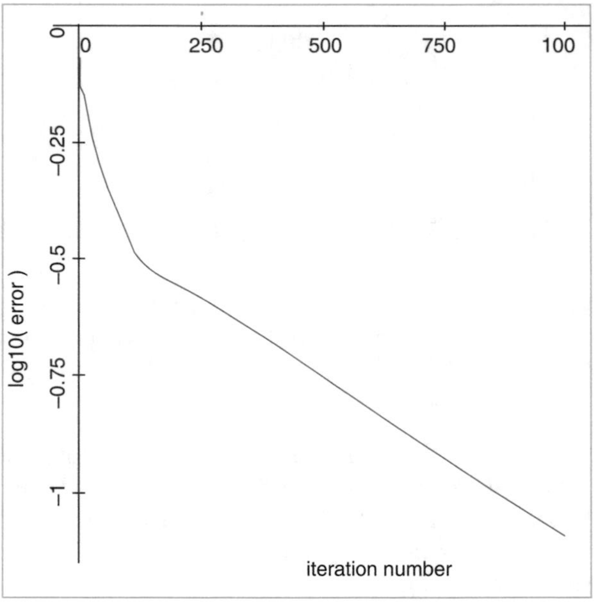

Fig. 2.4 Error in Richardson's iteration, $\mu = \frac{1}{2}\varrho(\mathbf{A})$: log error versus iteration number

Readers may also experiment with the JavaScript 1D Richardson iteration program **richardson.html** This program allows the user to select the size of the matrix and the parameter μ. The program displays the current iteration number, together with the maximum residual $\|\mathbf{r}\|_\infty$, the maximum error $\|\mathbf{x} - \widetilde{\mathbf{x}}_k\|_\infty$ and the improvement ratio $\|\mathbf{x} - \widetilde{\mathbf{x}}_k\|_\infty / \|\mathbf{x} - \widetilde{\mathbf{x}}_{k-1}\|_\infty$.

Exercise 2.4.1 Suppose that we discretize the heat equation using backward Euler time integration and centered differences in space. In 1D, this gives us the linear system

$$\frac{\mathbf{u}_i^{n+1} - \mathbf{u}_i^n}{\Delta t} = D \frac{(\mathbf{u}_{i+1}^{n+1} - \mathbf{u}_i^{n+1})/\Delta x - (\mathbf{u}_i^{n+1} - \mathbf{u}_{i-1}^{n+1})/\Delta x}{\Delta x} \tag{2.2}$$

for the new solution $\mathbf{u}^{n+1}$, given the old solution $\mathbf{u}^n$. In 2D, the corresponding linear system is

$$\begin{aligned}\frac{\mathbf{u}_{ij}^{n+1} - \mathbf{u}_{ij}^n}{\Delta t} &= D \frac{(\mathbf{u}_{i+1,j}^{n+1} - \mathbf{u}_{ij}^{n+1})/\Delta x - (\mathbf{u}_{ij}^{n+1} - \mathbf{u}_{i-1,j}^{n+1})/\Delta x}{\Delta x} \\ &+ D \frac{(\mathbf{u}_{i,j+1}^{n+1} - \mathbf{u}_{ij}^{n+1})/dy - (\mathbf{u}_{ij}^{n+1} - \mathbf{u}_{i,j-1}^{n+1})/dy}{\Delta y} .\end{aligned} \tag{2.3}$$

The corresponding linear system in 3D is

$$\begin{aligned}\frac{\mathbf{u}_{ijk}^{n+1} - \mathbf{u}_{ijk}^n}{\Delta t} &= D \frac{(\mathbf{u}_{i+1,jk}^{n+1} - \mathbf{u}_{ijk}^{n+1})/\Delta x - (\mathbf{u}_{ijk}^{n+1} - \mathbf{u}_{i-1,jk}^{n+1})/\Delta x}{\Delta x} \\ &+ D \frac{(\mathbf{u}_{i,j+1,k}^{n+1} - \mathbf{u}_{ijk}^{n+1})/dy - (\mathbf{u}_{ijk}^{n+1} - \mathbf{u}_{i,j-1,k}^{n+1})/dy}{\Delta y} \\ &+ D \frac{(\mathbf{u}_{ij,k+1}^{n+1} - \mathbf{u}_{ijk}^{n+1})/dz - (\mathbf{u}_{ijk}^{n+1} - \mathbf{u}_{ij,k-1}^{n+1})/dz}{\Delta z} .\end{aligned} \tag{2.4}$$

If we use want to use Richardson's iteration to solve these linear systems, how should we choose μ in 1D, 2D and 3D?

Exercise 2.4.2 The Crank-Nicolson scheme for the heat equation uses second-order centered differences in time. In 1D, this gives us the linear system

$$\begin{aligned}\frac{\mathbf{u}_i^{n+1} - \mathbf{u}_i^n}{\Delta t} &= \frac{D}{2} \frac{(\mathbf{u}_{i+1}^{n+1} - \mathbf{u}_i^{n+1})/\Delta x - (\mathbf{u}_i^{n+1} - \mathbf{u}_{i-1}^{n+1})/\Delta x}{\Delta x} \\ &+ \frac{D}{2} \frac{(\mathbf{u}_{i+1}^{n} - \mathbf{u}_i^{n})/\Delta x - (\mathbf{u}_i^{n} - \mathbf{u}_{i-1}^{n})/\Delta x}{\Delta x}\end{aligned} \tag{2.5}$$

for the new solution $\mathbf{u}^{n+1}$, given the old solution $\mathbf{u}^n$. In 2D, the corresponding linear system is

$$\begin{aligned}\frac{\mathbf{u}_{ij}^{n+1} - \mathbf{u}_{ij}^n}{\Delta t} &= \frac{D}{2} \frac{(\mathbf{u}_{i+1,j}^{n+1} - \mathbf{u}_{ij}^{n+1})/\Delta x - (\mathbf{u}_{ij}^{n+1} - \mathbf{u}_{i-1,j}^{n+1})/\Delta x}{\Delta x} \\ &+ \frac{D}{2} \frac{(\mathbf{u}_{i,j+1}^{n+1} - \mathbf{u}_{ij}^{n+1})/\Delta y - (\mathbf{u}_{ij}^{n+1} - \mathbf{u}_{i,j-1}^{n+1})/\Delta y}{\Delta y}\end{aligned}$$

$$+ \frac{D}{2} \frac{(\mathbf{u}_{i+1,j}^n - \mathbf{u}_{ij}^n)/\triangle x - (\mathbf{u}_{ij}^n - \mathbf{u}_{i-1,j}^n)/\triangle x}{\triangle x}$$
$$+ \frac{D}{2} \frac{(\mathbf{u}_{i,j+1}^n - \mathbf{u}_{ij}^n)/\triangle y - (\mathbf{u}_{ij}^n - \mathbf{u}_{i,j-1}^n)/\triangle y}{\triangle y} . \quad (2.6)$$

The corresponding linear system in 3D is

$$\frac{\mathbf{u}_{ijk}^{n+1} - \mathbf{u}_{ijk}^n}{\triangle t} = \frac{D}{2} \frac{(\mathbf{u}_{i+1,jk}^{n+1} - \mathbf{u}_{ijk}^{n+1})/\triangle x - (\mathbf{u}_{ijk}^{n+1} - \mathbf{u}_{i-1,jk}^{n+1})/\triangle x}{\triangle x}$$
$$+ \frac{D}{2} \frac{(\mathbf{u}_{i,j+1,k}^{n+1} - \mathbf{u}_{ijk}^{n+1})/\triangle y - (\mathbf{u}_{ijk}^{n+1} - \mathbf{u}_{i,j-1,k}^{n+1})/\triangle y}{\triangle y}$$
$$+ \frac{D}{2} \frac{(\mathbf{u}_{ij,k+1}^{n+1} - \mathbf{u}_{ijk}^{n+1})/\triangle z - (\mathbf{u}_{ijk}^{n+1} - \mathbf{u}_{ij,k-1}^{n+1})/\triangle z}{\triangle z}$$
$$+ \frac{D}{2} \frac{(\mathbf{u}_{i+1,jk}^{n} - \mathbf{u}_{ijk}^{n})/\triangle x - (\mathbf{u}_{ijk}^{n} - \mathbf{u}_{i-1,jk}^{n})/\triangle x}{\triangle x}$$
$$+ \frac{D}{2} \frac{(\mathbf{u}_{i,j+1,k}^{n} - \mathbf{u}_{ijk}^{n})/\triangle y - (\mathbf{u}_{ijk}^{n} - \mathbf{u}_{i,j-1,k}^{n})/\triangle y}{\triangle y}$$
$$+ \frac{D}{2} \frac{(\mathbf{u}_{ij,k+1}^{n} - \mathbf{u}_{ijk}^{n})/\triangle z - (\mathbf{u}_{ijk}^{n} - \mathbf{u}_{ij,k-1}^{n})/\triangle z}{\triangle z} . \quad (2.7)$$

If we use want to use Richardson's iteration to solve these linear systems, how should we choose μ in 1D, 2D and 3D?

Exercise 2.4.3 Get a copy of matrix **E20R0100** from MatrixMarket. This is a sparse nonsymmetric matrix, and should be represented in computer memory without storing the zero entries. If this matrix is denoted by $\mathbf{A}$, let $\mathbf{x}$ be the vector of ones, and $\mathbf{b} = \mathbf{A}\mathbf{x}$. Choose the initial guess $\widetilde{\mathbf{x}}$ for Richardson's iteration to have random entries chosen from a uniform probability distribution on $(0, 1)$. Program Richardson's iteration to solve $\mathbf{A}\mathbf{x} = \mathbf{b}$, and plot the logarithm base 10 of the error $\|\widetilde{\mathbf{x}} - \mathbf{x}\|_\infty$ versus iteration number. At what order of error does the iteration stop making progress? Relate this error size to the condition number of this matrix, which is estimated to be on the order of 10^{10}.

2.4.3 Jacobi Iteration

Richardson's iteration uses a particularly crude preconditioner, namely $\mathbf{C} = \mathbf{I}/\mu$. Our next attempt at iterative improvement will use a preconditioner that is somewhat more closely related to the given matrix.

Let us write $\mathbf{A} = \mathbf{D} - \mathbf{L} - \mathbf{U}$, where $\mathbf{D}$ is diagonal, $\mathbf{L}$ is strictly lower triangular and $\mathbf{U}$ is strictly upper triangular. If $\mathbf{D}$ has nonzero diagonal entries, then **Jacobi iteration** is the iterative improvement algorithm with preconditioner $\mathbf{C} = \mathbf{D}^{-1}$. In other words, the Jacobi algorithm takes the form

$$\widetilde{\mathbf{x}}_{n+1} = \widetilde{\mathbf{x}}_n + \mathbf{D}^{-1}\left[\mathbf{b} - \mathbf{A}\widetilde{\mathbf{x}}_n\right] .$$

This algorithm can be written in the form

Algorithm 2.4.3 (Jacobi Iteration)

$$\mathbf{r} = \mathbf{b} - \mathbf{A}\widetilde{\mathbf{x}}$$
$$\widetilde{\mathbf{x}} = \widetilde{\mathbf{x}} + \mathbf{D}^{-1}\mathbf{r} .$$

Note that all components of the residual are computed before any components of $\widetilde{\mathbf{x}}$ are updated.

There are several practical circumstances under which the Jacobi iteration will converge.

Lemma 2.4.5 ([177, p. 79]) *If the square matrix* $\mathbf{A}$ *is strictly diagonally dominant, then the Jacobi iteration Algorithm 2.4.3 converges for any initial guess.*

Proof Note that the spectral radius of $\mathbf{I} - \mathbf{D}^{-1}\mathbf{A}$ satisfies

$$\varrho(\mathbf{I} - \mathbf{D}^{-1}\mathbf{A}) \leq \|\mathbf{D}^{-1}(\mathbf{D} - \mathbf{A})\|_\infty = \max_i \sum_{j \neq i} \left|\mathbf{A}_{ij}/\mathbf{A}_{ii}\right| < 1 .$$

Then Lemma 2.4.1 proves that the iteration converges.

In order to apply the Jacobi preconditioner to a general positive matrix, we will need to modify the iteration with a weighting factor. Let $\nu(\mathbf{A})$ be the maximum number of nonzero entries in any row of $\mathbf{A}$. Choose γ so that $0 < \gamma < 2/\nu(\mathbf{A})$. Then the weighted Jacobi iteration is

Algorithm 2.4.4 (Weighted Jacobi Iteration)

$$\mathbf{r} = \mathbf{b} - \mathbf{A}\widetilde{\mathbf{x}}$$
$$\widetilde{\mathbf{x}} = \widetilde{\mathbf{x}} + \mathbf{D}^{-1}\mathbf{r}\gamma .$$

Lemma 5.1 by Bramble [24, p. 74] shows that $\mathbf{C} = \mathbf{D}^{-1}\gamma$ is a preconditioner for $\mathbf{A}$ that is useful as a smoother in the multigrid algorithm. For more information, see Sect. 2.7.1. Some additional results regarding the convergence of the weighted Jacobi iteration for Hermitian positive matrices can be found in Hackbusch [90, p. 87ff] and Young [189, p. 109ff].

In Fig. 2.5 we show the computed solution for the Jacobi iteration for the same problem as in Figs. 2.1 and 2.2. Figure 2.6 shows the error in the Jacobi iteration as a

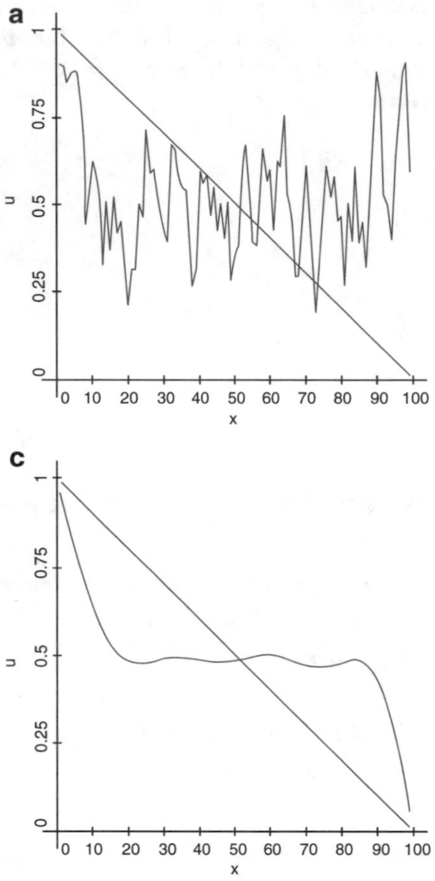

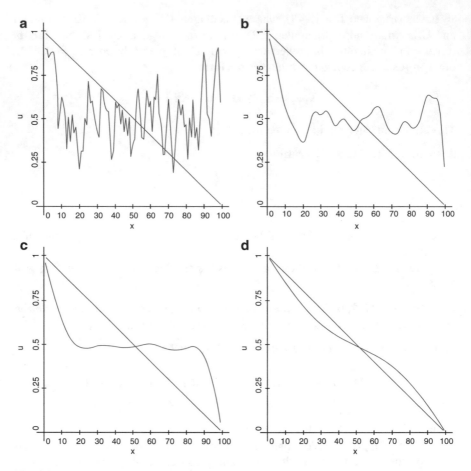

Fig. 2.5 Computed solution with Jacobi iteration, $\gamma = 2/3$, random initial guess; the straight line is the exact solution. (**a**) 1 iteration. (**b**) 10 iterations. (**c**) 100 iterations. (**d**) 1000 iterations

function of the number of iterations. For this particular problem, the Jacobi iteration is the same as Richardson's iteration with $\mu = 2$.

A Fortran 77 implementation of the Jacobi iteration can be found in files Jacobi.f and JacobiREVCOM.f. at `netlib.org`. A C^{++} program to implement the Jacobi iteration as a preconditioner appears as routine `preconditionJacobi` of class `SparseMatrix` in files SparseMatrix.H and SparseMatrix.C.

A program to perform Jacobi's iteration as an iterative solver can be found in IterativeImprovementMain.C. This program calls a Fortran subroutine `jacobi` in iterative_improvement.f. A related algorithm, typically used for smoothing in multigrid, is given by `jacobi_omega`, also in `iterative_improvement.f`. The Laplacian in two dimensions is solved in IterativeError2D.C. This program calls Fortran subroutine `jacobi_omega` in iterative_improvement_2d.f.

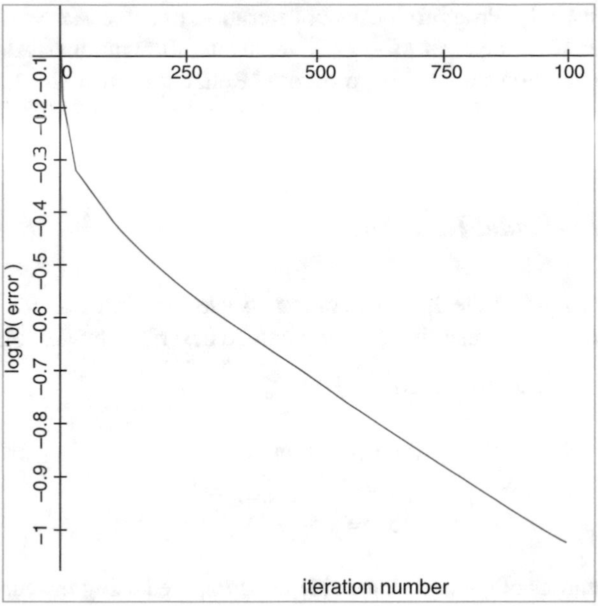

Fig. 2.6 Error in Jacobi iteration with $\gamma = 2/3$; log error versus iteration number

Readers may also experiment with the JavaScript 1D Jacobi iteration program **jacobi.html** This program allows the user to select the size of the matrix and the parameter γ. The program displays the current iteration number, together with the maximum residual $\|\mathbf{r}\|_\infty$, the maximum error $\|\mathbf{x}-\widetilde{\mathbf{x}}_k\|_\infty$ and the improvement ratio $\|\mathbf{x}-\widetilde{\mathbf{x}}_k\|_\infty / \|\mathbf{x}-\widetilde{\mathbf{x}}_{k-1}\|_\infty$.

Exercise 2.4.4 Repeat the previous exercise for the Crank-Nicolson scheme, described by Eqs. (2.5)–(2.7).

Exercise 2.4.5 Get a copy of matrix **BCSSTM19** from MatrixMarket. This is a sparse, diagonally dominant symmetric positive matrix, and should be represented in computer memory without storing the zero entries. If this matrix is denoted by $\mathbf{A}$, let $\mathbf{x}$ be the vector of ones, and $\mathbf{b} = \mathbf{A}\mathbf{x}$. Choose the initial guess $\widetilde{\mathbf{x}}$ for the Jacobi iteration to have random entries chosen from a uniform probability distribution on $(0, 1)$. Program the Jacobi iteration to solve $\mathbf{A}\mathbf{x} = \mathbf{b}$, and plot the logarithm base 10 of the error $\|\widetilde{\mathbf{x}} - \mathbf{x}\|_\infty$ versus iteration number. At what magnitude of error does the iteration stop making progress? Relate this error size to the condition number of this matrix, which is estimated to be on the order of 10^5.

Exercise 2.4.6 Get a copy of matrix **BCSSTK14** from MatrixMarket. This is a sparse, symmetric positive matrix that is not diagonally-dominant, and should be represented in computer memory without storing the zero entries. If this matrix is denoted by $\mathbf{A}$, let $\mathbf{x}$ be the vector of ones, and $\mathbf{b} = \mathbf{A}\mathbf{x}$. Choose the initial guess $\widetilde{\mathbf{x}}$ for the Jacobi iteration to have random entries chosen from a uniform probability distribution on $(0,1)$. Program the Jacobi iteration to solve $\mathbf{A}\mathbf{x} = \mathbf{b}$, and plot the logarithm base 10 of the error $\|\widetilde{\mathbf{x}} - \mathbf{x}\|_\infty$ versus iteration number. At what order of error does the iteration stop making progress? Relate this error size to the condition number of this matrix, which is estimated to be on the order of 10^{10}.

2.4.4 *Gauss-Seidel Iteration*

It is natural to modify the Jacobi iteration to use new information as soon as it becomes available. The resulting iteration can be described by the following

Algorithm 2.4.5 (Gauss-Seidel)

$$\begin{aligned}
&\text{for } 1 \le i \le m \\
&\quad r = \mathbf{b}_i - \textstyle\sum_{j=1}^m \mathbf{A}_{ij}\widetilde{\mathbf{x}}_j \\
&\quad \widetilde{\mathbf{x}}_i = \widetilde{\mathbf{x}}_i + r/\mathbf{A}_{ii}
\end{aligned} \tag{2.8}$$

Note that the entries of the right-hand side are computed using the current entries of $\widetilde{\mathbf{x}}$, and that the entries r of the residual do not have to be stored as a vector. Thus the Gauss-Seidel iteration is simpler to program, and requires less computer storage, than the Jacobi iteration. We will also see that the Gauss-Seidel iteration tends to converge more rapidly than the Jacobi iteration.

In order to describe the Gauss-Seidel iteration mathematically, we will split the matrix $\mathbf{A}$ into its diagonal, strictly lower and strictly upper triangular parts:

$$\mathbf{A} = \mathbf{D} - \mathbf{L} - \mathbf{U} \,.$$

Let $\mathbf{x}$ be the true solution to $\mathbf{A}\mathbf{x} = \mathbf{b}$. Since $\mathbf{b} = \mathbf{A}\mathbf{x} = (\mathbf{D} - \mathbf{L} - \mathbf{U})\mathbf{x}$, the Gauss-Seidel algorithm can be rewritten in the form

$$(\mathbf{D} - \mathbf{L})\,(\mathbf{x} - \widetilde{\mathbf{x}}_{n+1}) - \mathbf{U}\,(\mathbf{x} - \widetilde{\mathbf{x}}_n) = \mathbf{0} \,.$$

This can be rewritten as

$$\mathbf{x} - \widetilde{\mathbf{x}}_{n+1} = -(\mathbf{D} - \mathbf{L})^{-1}\mathbf{U}\,(\mathbf{x} - \widetilde{\mathbf{x}}_n) = \left[\mathbf{I} - (\mathbf{D} - \mathbf{L})^{-1}\mathbf{A}\right](\mathbf{x} - \widetilde{\mathbf{x}}_n) \,.$$

In terms of our discussion of iterative improvement in Sect. 2.4, this implies that the Gauss-Seidel preconditioner is $\mathbf{C} = (\mathbf{D} - \mathbf{L})^{-1}$. Lemma 2.4.1 implies that the Gauss-Seidel iteration will converge if and only if $\varrho(\mathbf{I} - \mathbf{C}\mathbf{A}) = \varrho\left((\mathbf{D} - \mathbf{L})^{-1}\mathbf{U}\right) < 1$. In practice, this condition is difficult to verify *a priori*. Fortunately, we have other ways to determine that Gauss-Seidel iteration will converge.

Lemma 2.4.6 ([90, p. 90], [177, p. 84]) *If* $\mathbf{A}$ *is Hermitian and positive, then the Gauss-Seidel iteration converges for any initial guess.*

Proof Since $\mathbf{A}$ is symmetric, we have $\mathbf{U} = \mathbf{L}^H$ in the splitting $\mathbf{A} = \mathbf{D} - \mathbf{L} - \mathbf{U}$. Because $\mathbf{A}$ is positive, $\mathbf{D}$ is positive. Let $\mathbf{M} = \mathbf{D}^{-1/2}\mathbf{L}\mathbf{D}^{-1/2}$, and note that

$$\mathbf{D}^{1/2}(\mathbf{D}-\mathbf{L})^{-1}\mathbf{L}^H\mathbf{D}^{-1/2} = \left(\mathbf{I} - \mathbf{D}^{-1/2}\mathbf{L}\mathbf{D}^{-1/2}\right)^{-1}\mathbf{D}^{-1/2}\mathbf{L}^H\mathbf{D}^{-1/2} = (\mathbf{I} - \mathbf{M})^{-1}\mathbf{M}^H \; .$$

Suppose that λ is an eigenvalue of $(\mathbf{I} - \mathbf{M})^{-1}\mathbf{M}^H$ with eigenvector $\mathbf{x}$ satisfying $\|\mathbf{x}\|^2 = 1$. Then $\mathbf{M}^H\mathbf{x} = (\mathbf{I} - \mathbf{M})\mathbf{x}\lambda$, so

$$\mathbf{x} \cdot \mathbf{M}^H\mathbf{x} = (1 - \mathbf{x} \cdot \mathbf{M}\mathbf{x})\,\lambda \; .$$

Let $\mathbf{x}\cdot\mathbf{M}\mathbf{x} = a+ib$ where a and b are real scalars. Since $\mathbf{D}^{-1/2}\mathbf{A}\mathbf{D}^{-1/2} = \mathbf{I}-\mathbf{M}-\mathbf{M}^H$ is positive,

$$0 < \mathbf{x} \cdot \mathbf{D}^{-1/2}\mathbf{A}\mathbf{D}^{-1/2}\mathbf{x} = \mathbf{x} \cdot (\mathbf{I} - \mathbf{M} - \mathbf{M}^H)\mathbf{x} = 1 - 2a \; .$$

We can solve for the eigenvalue λ to get

$$|\lambda|^2 = \left|\frac{\mathbf{x} \cdot \mathbf{M}^H\mathbf{x}}{1 - \mathbf{x} \cdot \mathbf{M}\mathbf{x}}\right|^2 = \left|\frac{a - bi}{1 - a - bi}\right|^2 = \frac{a^2 + b^2}{1 - 2a + a^2 + b^2} < 1 \; .$$

Thus $\varrho\left([\mathbf{I} - \mathbf{M}]^{-1}\mathbf{M}^H\right) < 1$. Since $[\mathbf{I}-\mathbf{M}]^{-1}\mathbf{M}^H$ is similar to $\mathbf{I}-\mathbf{C}\mathbf{A} = (\mathbf{D}-\mathbf{L})^{-1}\mathbf{L}^H$, Lemma 2.4.1 completes the proof.

Recall from Lemma 2.4.1 that iterative improvement algorithms of the form $\widetilde{\mathbf{x}}_{n+1} = \widetilde{\mathbf{x}}_n - \mathbf{C}[\mathbf{A}\widetilde{\mathbf{x}}_n - \mathbf{b}]$ converge if and only if $\varrho(\mathbf{I}-\mathbf{C}\mathbf{A}) < 1$. The Stein-Rosenberg theorem [177, p. 76] shows that when $\mathbf{A} = \mathbf{D} - \mathbf{L} - \mathbf{U}$ is split into diagonal, strictly lower triangular and strictly upper triangular parts, and when $\mathbf{I} - \mathbf{D}^{-1}\mathbf{A} \geq \mathbf{0}$, then

$$\varrho(\mathbf{I} - \mathbf{D}^{-1}\mathbf{A}) > \varrho(\mathbf{I} - [\mathbf{D} - \mathbf{L}]^{-1}\mathbf{A}) \; .$$

In other words, under these assumptions Gauss-Seidel iteration converges more rapidly than Jacobi iteration. Axelsson [7, p. 238] claims that for matrices such as those that arise from discretization of the heat equation, $-\log \varrho(\mathbf{I} - \mathbf{C}\mathbf{A})$ is about twice as large for Gauss-Seidel iteration as it is for Jacobi iteration. Although this means that Gauss-Seidel iteration would take about half as many iterations as Jacobi to converge, the number of iterations is still too large to make implicit integration of the heat equation competitive with explicit time integration. However, Gauss-Seidel iteration is useful as a smoother in multigrid iterations.

In Fig. 2.7 we show the computed solution for the Gauss-Seidel iteration for the same problem as in Figs. 2.1, 2.2 and 2.5. Note that the error smooths quickly, as in the Richardson iteration with $\mu = \varrho(\mathbf{A})$. However, Fig. 2.8 shows that the error also reduces more quickly than either Richardson's iteration or the Jacobi iteration.

There are several variants of the Gauss-Seidel iteration. In the **Gauss-Seidel to-fro** iteration, the order of the transversal of the unknowns is reversed from one iteration to the next:

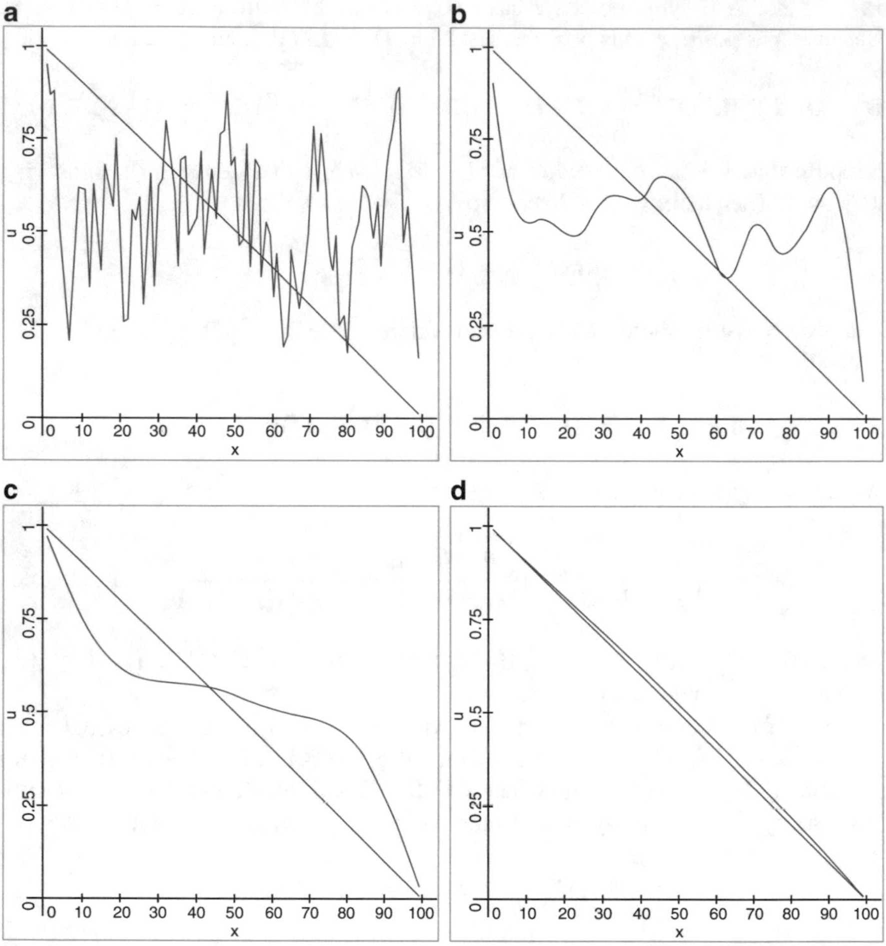

Fig. 2.7 Computed solution with Gauss-Seidel iteration for Laplace equation with random initial guess; the straight line is the exact solution. (**a**) 1 iteration. (**b**) 10 iterations. (**c**) 100 iterations. (**d**) 1000 iterations

Algorithm 2.4.6 (Gauss-Seidel To-Fro)

$$
\begin{aligned}
&\text{for } 1 \leq i \leq m \\
&\quad r = \mathbf{e}_i \cdot (\mathbf{b} - \mathbf{A}\widetilde{\mathbf{x}}) \\
&\quad \mathbf{e}_i \cdot \widetilde{\mathbf{x}} = \mathbf{e}_i \cdot \widetilde{\mathbf{x}} + r/\mathbf{A}_{ii} \\
&\text{for } m \geq i \geq 1 \\
&\quad r = \mathbf{e}_i \cdot (\mathbf{b} - \mathbf{A}\widetilde{\mathbf{x}}) \\
&\quad \mathbf{e}_i \cdot \widetilde{\mathbf{x}} = \mathbf{e}_i \cdot \widetilde{\mathbf{x}} + \mathbf{r}/\mathbf{A}_{ii} \,.
\end{aligned}
\tag{2.9}
$$

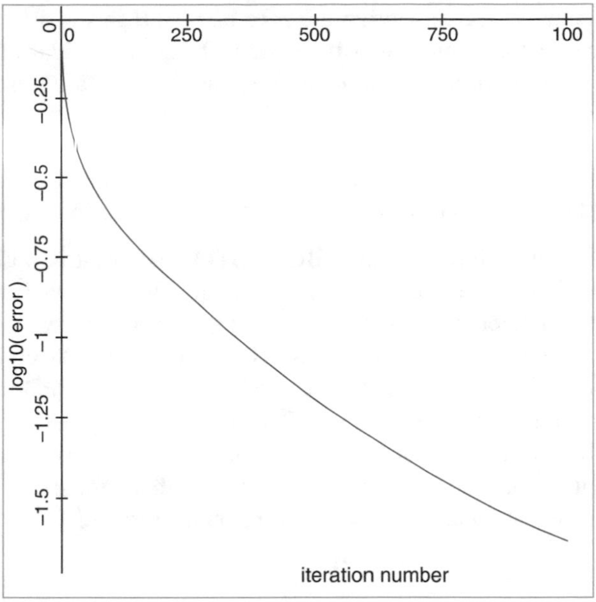

Fig. 2.8 Error in Gauss-Seidel iteration: log error versus iteration number

This helps to remove the bias of the Gauss-Seidel iteration toward one end of the problem domain. Another variant is to cycle through the unknowns in "red-black" ordering, for regular grids that can be related to a checkerboard layout. The red unknowns could be processed first, followed by the black unknowns.

A C^{++} program to implement the Gauss-Seidel iteration as a preconditioner appears is available as routine `preconditionSOR` of class `SparseMatrix` in files SparseMatrix.H and SparseMatrix.C. The Gauss-Seidel iteration is not programmed explicitly; rather, it can be achieved by choosing the relaxation factor to be one in `preconditionSOR`.

A program to perform Gauss-Seidel iteration in one dimension can be found in IterativeImprovementMain.C. This program calls a Fortran subroutine `gauss_seidel_to_fro` in iterative_improvement.f. A related algorithm is given by `gauss_seidel_red_black` in the same file. The Laplacian in two dimensions is solved in IterativeError2D.C. This program calls Fortran subroutine `gauss_seidel_to_fro` in iterative_improvement_2d.f.

Readers may also experiment with the JavaScript 1D Gauss-Seidel iteration program **gauss_seidel.html.** This program allows the user to select the size of the matrix and the parameter γ. The program displays the current iteration number, together with the maximum residual $\|\mathbf{r}\|_\infty$, the maximum error $\|\mathbf{x} - \widetilde{\mathbf{x}}_k\|_\infty$ and the improvement ratio $\|\mathbf{x} - \widetilde{\mathbf{x}}_k\|_\infty / \|\mathbf{x} - \widetilde{\mathbf{x}}_{k-1}\|_\infty$.

Exercise 2.4.7 Get a copy of matrix **BCSSTM19** from MatrixMarket. This is a sparse, diagonally dominant symmetric positive matrix, and should be represented in computer memory without storing the zero entries. If this matrix is denoted by $\mathbf{A}$, let $\mathbf{x}$ be the vector of ones, and $\mathbf{b} = \mathbf{A}\mathbf{x}$. Choose the initial guess $\widetilde{\mathbf{x}}$ for the Gauss-Seidel iteration to have random entries chosen from a uniform probability distribution on $(0, 1)$. Program the Gauss-Seidel iteration to solve $\mathbf{A}\mathbf{x} = \mathbf{b}$, and plot the logarithm base 10 of the error $\|\widetilde{\mathbf{x}} - \mathbf{x}\|_\infty$ versus iteration number. At what order of error does the iteration stop making progress? Relate this error size to the condition number of this matrix, which is estimated to be on the order of 10^5.

Exercise 2.4.8 Get a copy of matrix **BCSSTK14** from MatrixMarket. This is a sparse, symmetric positive matrix that is not diagonally-dominant, and should be represented in computer memory without storing the zero entries. If this matrix is denoted by $\mathbf{A}$, let $\mathbf{x}$ be the vector of ones, and $\mathbf{b} = \mathbf{A}\mathbf{x}$. Choose the initial guess $\widetilde{\mathbf{x}}$ for the Gauss-Seidel iteration to have random entries chosen from a uniform probability distribution on $(0, 1)$. Program the Gauss-Seidel iteration to solve $\mathbf{A}\mathbf{x} = \mathbf{b}$, and plot the logarithm base 10 of the error $\|\widetilde{\mathbf{x}} - \mathbf{x}\|_\infty$ versus iteration number. At what order of error does the iteration stop making progress? Relate this error size to the condition number of this matrix, which is estimated to be on the order of 10^{10}.

2.4.5 *Successive Over-Relaxation*

It is common to modify the Gauss-Seidel iteration by including a relaxation parameter. Again, we split the matrix $\mathbf{A}$ into its diagonal, strictly lower and strictly upper triangular parts:

$$\mathbf{A} = \mathbf{D} - \mathbf{L} - \mathbf{U} .$$

If the residual in the midst of the Gauss-Seidel iteration is

$$\mathbf{r} = \mathbf{b} - (-\mathbf{L}\widetilde{\mathbf{x}}_{n+1} + \mathbf{D}\widetilde{\mathbf{x}}_n - \mathbf{U}\widetilde{\mathbf{x}}_n) ,$$

then relaxation of the Gauss-Seidel iteration would pick some scalar ω and take

$$\widetilde{\mathbf{x}}_{n+1} = \widetilde{\mathbf{x}}_n + \mathbf{D}^{-1}\mathbf{r}\omega .$$

The resulting algorithm takes the form

Algorithm 2.4.7 (SOR Iteration)

$$\begin{aligned}
&\text{for } 1 \le i \le m \\
&\quad r = \mathbf{e}_i \cdot (\mathbf{b} - \mathbf{A}\widetilde{\mathbf{x}}) \\
&\quad \mathbf{e}_i \cdot \widetilde{\mathbf{x}} = \mathbf{e}_i \cdot \widetilde{\mathbf{x}} + \omega r / \mathbf{A}_{ii}
\end{aligned} .$$

Again, we use the new values of $\widetilde{\mathbf{x}}$ as soon as they are computed. The term *over-relaxation* comes from the fact that the optimal value of the relaxation parameter ω will turn out to be greater than one.

In matrix-vector form, the **SOR iteration** can be written

$$\mathbf{D}\widetilde{\mathbf{x}}_{n+1} = \mathbf{D}\widetilde{\mathbf{x}}_n + [\mathbf{b} - \mathbf{D}\widetilde{\mathbf{x}}_n + \mathbf{L}\widetilde{\mathbf{x}}_{n+1} + \mathbf{U}\widetilde{\mathbf{x}}_n]\,\omega\ ,$$

from which it follows that

$$(\mathbf{D} - \mathbf{L}\omega)\widetilde{\mathbf{x}}_{n+1} = (\mathbf{D} - \mathbf{L}\omega)\widetilde{\mathbf{x}}_n + (\mathbf{b} - \mathbf{D}\widetilde{\mathbf{x}}_n + \mathbf{L}\widetilde{\mathbf{x}}_n + \mathbf{U}\widetilde{\mathbf{x}}_n)\,\omega\ ,$$

which is equivalent to

$$\widetilde{\mathbf{x}}_{n+1} = \widetilde{\mathbf{x}}_n + (\mathbf{D} - \mathbf{L}\omega)^{-1}\,(\mathbf{b} - \mathbf{A}\widetilde{\mathbf{x}}_n)\,\omega\ .$$

Similarly, the error $\mathbf{d}_n = \widetilde{\mathbf{x}}_n - \mathbf{x}$ satisfies

$$\mathbf{D}\mathbf{d}_{n+1} = \mathbf{D}\mathbf{d}_n - (\mathbf{D}\mathbf{d}_n - \mathbf{L}\mathbf{d}_{n+1} - \mathbf{U}\mathbf{d}_n)\omega$$

or

$$\mathbf{d}_{n+1} = (\mathbf{D} - \mathbf{L}\omega)^{-1}[\mathbf{D}(1-\omega) + \mathbf{U}\omega]\mathbf{d}_n\ .$$

This suggests that we should study the eigenvalues of

$$\mathbf{G}_\omega \equiv (\mathbf{D} - \mathbf{L}\omega)^{-1}[\mathbf{D}(1-\omega) + \mathbf{U}\omega]\ . \tag{2.10}$$

The next lemma, due to Kahan [177, p. 107], provides a restriction on the useful relaxation parameters.

Lemma 2.4.7 *Suppose that* $\mathbf{G}_\omega$ *is defined by* (2.10)*, where* $\mathbf{D}$ *is diagonal and nonsingular,* $\mathbf{L}$ *is strictly lower triangular and* $\mathbf{U}$ *is strictly upper triangular. Then*

$$\varrho(\mathbf{G}_\omega) \geq |\omega - 1|\ ,$$

so the SOR iteration diverges if either $\omega < 0$ *or* $\omega > 2$.

Proof Since the determinant of $\mathbf{G}_\omega$ is the product of its eigenvalues λ_i,

$$\begin{aligned}\prod_i \lambda_i^m &= \det\{(\mathbf{I} - \mathbf{D}^{-1}\mathbf{L}\omega)^{-1}[\mathbf{I}(1-\omega) + \mathbf{D}^{-1}\mathbf{U}\omega]\} \\ &= \det(\mathbf{I} - \mathbf{D}^{-1}\mathbf{L}\omega)^{-1}\det[\mathbf{I}(1-\omega) + \mathbf{D}^{-1}\mathbf{U}\omega] \\ &= \det[\mathbf{I}(1-\omega) + \mathbf{D}^{-1}\mathbf{U}\omega] = (1-\omega)^m\ .\end{aligned}$$

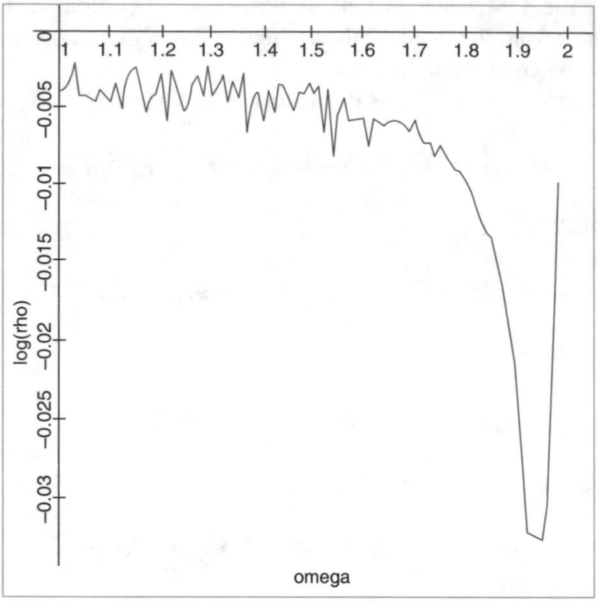

Fig. 2.9 Spectral radius in SOR iteration: log radius versus over-relaxation factor ω

It follows that at least one eigenvalue must satisfy $|\lambda_i| \geq |1 - \omega|$.
A theorem due to Ostrowski [135] shows that SOR converges for any Hermitian positive matrix and arbitrary initial guesses whenever $0 < \omega < 2$.

Wachspress [178] discusses ways to find the optimal relaxation parameter ω. These approaches are seldom used nowadays, due to the development of other iterative linear solvers.

In Fig. 2.9 we show the spectral radius for the SOR iteration as a function of the over-relaxation factor, for the same problem as in Figs. 2.3 and 2.4. This figure shows that the spectral radius is very sensitive to the choice of ω near the optimal value. In Fig. 2.10 we show the computed solution for $\omega = 1.94$, which is close to the optimal value. Note that Fig. 2.11 shows that the solution converges rapidly, but Fig. 2.10 shows that it is not smoothed rapidly.

A Fortran 77 implementation of the SOR iteration can be found in files SOR.f and SORREVCOM.f at `netlib.org`. A C^{++} program to implement the successive over-relaxation (SOR) iteration as a preconditioner appears as routine `preconditionSOR` of class `SparseMatrix` in files SparseMatrix.H and SparseMatrix.C.

A C^{++} program to perform SOR iteration in one dimension can be found in SORMain.C. This program calls a Fortran subroutine `sor` in iterative_improvement.f. The `SORMain.C` main program can plot the spectral radius of the iteration versus the over-relaxation factor.

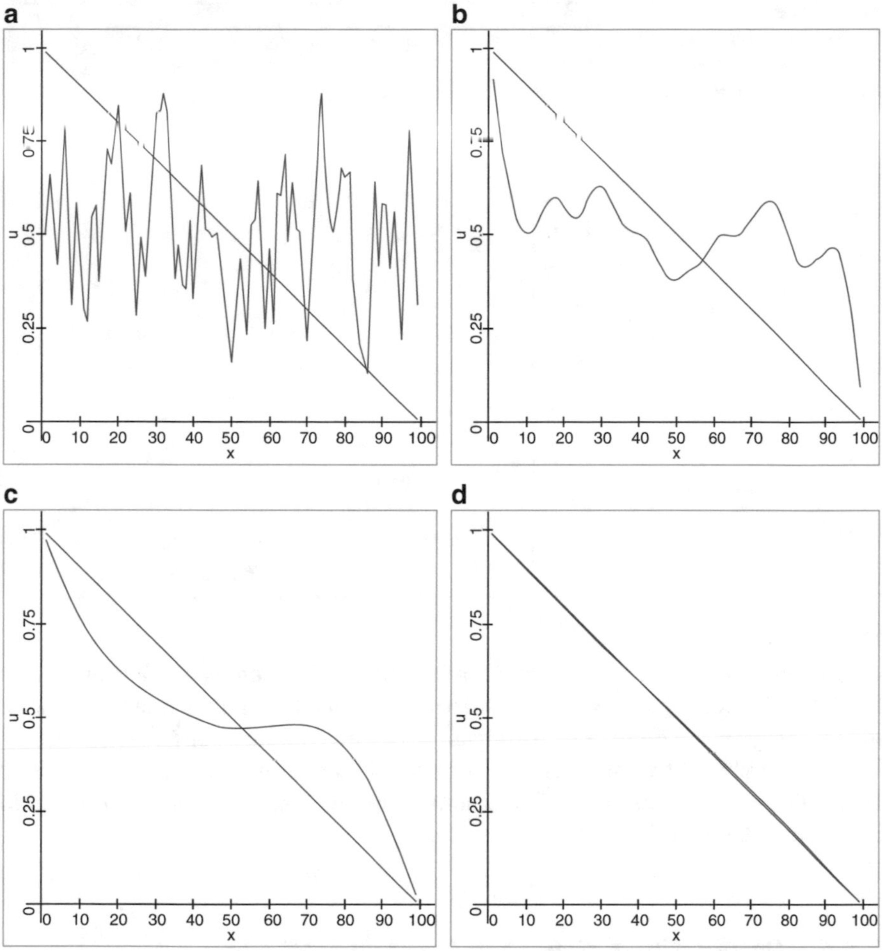

Fig. 2.10 Computed solution with Gauss-Seidel iteration for Laplace equation with random initial guess, 100 grid cells, $\omega = 1.94$; the straight line is the exact solution. (**a**) 1 iteration. (**b**) 10 iterations. (**c**) 100 iterations. (**d**) 1000 iterations

Readers may also experiment with the JavaScript 1D SOR iteration program sor.html. This program allows the user to select the size of the matrix and the parameter γ. The program displays the current iteration number, together with the maximum residual $\|\mathbf{r}\|_\infty$, the maximum error $\|\mathbf{x}-\widetilde{\mathbf{x}}_k\|_\infty$ and the improvement ratio $\|\mathbf{x}-\widetilde{\mathbf{x}}_k\|_\infty / \|\mathbf{x}-\widetilde{\mathbf{x}}_{k-1}\|_\infty$.

Exercise 2.4.9 Get a copy of matrix **E20R0100** from MatrixMarket. This is a sparse nonsymmetric matrix, and should be represented in computer memory without storing the zero entries. If this matrix is denoted by **A**, let **x** be the vector of ones, and $\mathbf{b} = \mathbf{A}\mathbf{x}$. Choose the initial guess $\widetilde{\mathbf{x}}$ for Richardson's iteration to have

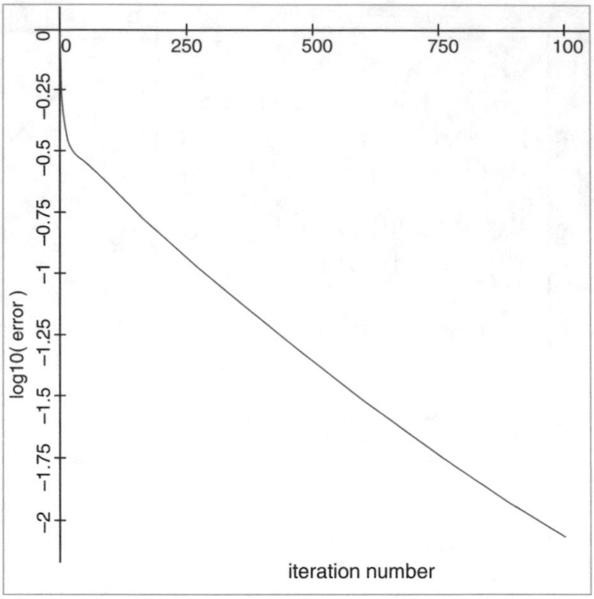

Fig. 2.11 Error in Gauss-Seidel iteration, 100 grid cells, $\omega = 1.94$: log error versus iteration number

random entries chosen from a uniform probability distribution on $(0, 1)$. Program the SOR iteration to solve $\mathbf{A}\mathbf{x} = \mathbf{b}$, and experiment to find the optimal relaxation factor ω to two significant digits. Plot the logarithm base 10 of the error $\|\widetilde{\mathbf{x}} - \mathbf{x}\|_\infty$ versus iteration number. At what order of error does the iteration stop making progress? Relate this error size to the condition number of this matrix, which is estimated to be on the order of 10^{10}.

2.4.6 Termination Criteria for Iterative Methods

Typically, when we solve $\mathbf{A}\mathbf{x} = \mathbf{b}$ numerically, we iterate until we believe that the numerical solution is close to the true solution. Theorem 3.6.1 in Chap. 3 of Volume I suggests that for well-conditioned linear systems, we can terminate an iterative method when the residual becomes small relative to the right-hand side of the linear system. However, if the linear system is not well conditioned then relatively small residuals may not imply small relative errors in the numerical solution.

Stopping based on the change in the solution is more tricky, as the next lemma shows.

Lemma 2.4.8 *Suppose that* $\mathbf{A}$ *and* $\mathbf{C}$ *are* $m \times m$ *matrices, that* $\mathbf{b}$ *and* $\widetilde{\mathbf{x}}_0$ *are m-vectors, and that* $\mathbf{x}$ *satisfies* $\mathbf{A}\mathbf{x} = \mathbf{b}$*. Let* $\|\cdot\|$ *be a norm on m-vectors, and use this same notation to denote the subordinate matrix norm on* $m \times m$ *matrices. Assume that there exists* $\varrho < 1$ *so that* $\|\mathbf{I} - \mathbf{C}\mathbf{A}\| \le \varrho$*. If the iterates* $\widetilde{\mathbf{x}}_n$ *are defined by the* *iterative improvement iteration* $\widetilde{\mathbf{x}}_{n+1} = \widetilde{\mathbf{x}}_n + \mathbf{C}(\mathbf{b} - \mathbf{A}\widetilde{\mathbf{x}}_n)$, *then*

$$\|\widetilde{\mathbf{x}}_{n+1} - \mathbf{x}\| \leq \frac{\varrho}{1-\varrho} \|\widetilde{\mathbf{x}}_{n+1} - \widetilde{\mathbf{x}}_n\| .$$

Proof As we have seen previously in Lemma 2.4.1,

$$\widetilde{\mathbf{x}}_{n+1} - \mathbf{x} = (\mathbf{I} - \mathbf{C}\mathbf{A})(\widetilde{\mathbf{x}}_n - \mathbf{x}) .$$

It follows that

$$\begin{aligned}(\mathbf{I} - \mathbf{C}\mathbf{A})(\widetilde{\mathbf{x}}_n - \widetilde{\mathbf{x}}_{n+1}) &= (\mathbf{I} - \mathbf{C}\mathbf{A})(\widetilde{\mathbf{x}}_n - \mathbf{x}) - (\mathbf{I} - \mathbf{C}\mathbf{A})(\widetilde{\mathbf{x}}_{n+1} - \mathbf{x}) \\ &= \widetilde{\mathbf{x}}_{n+1} - \mathbf{x} - (\mathbf{I} - \mathbf{C}\mathbf{A})(\widetilde{\mathbf{x}}_{n+1} - \mathbf{x}) = \mathbf{C}\mathbf{A}(\widetilde{\mathbf{x}}_{n+1} - \mathbf{x}) .\end{aligned}$$

Next, we note that Corollary 2.3.1 implies that

$$(\mathbf{C}\mathbf{A})^{-1} = [\mathbf{I} - (\mathbf{I} - \mathbf{C}\mathbf{A})]^{-1} = \sum_{j=0}^{\infty} (\mathbf{I} - \mathbf{C}\mathbf{A})^j .$$

As a result,

$$\left\|(\mathbf{C}\mathbf{A})^{-1}\right\| \leq \sum_{j=0}^{\infty} \|(\mathbf{I} - \mathbf{C}\mathbf{A})^j\| \leq \sum_{j=0}^{\infty} \varrho^j = \frac{1}{1-\varrho} .$$

Putting our results together, we obtain

$$\begin{aligned}\|\widetilde{\mathbf{x}}_{n+1} - \mathbf{x}\| &= \|(\mathbf{C}\mathbf{A})^{-1}[\mathbf{C}\mathbf{A}(\widetilde{\mathbf{x}}_{n+1} - \mathbf{x})]\| = \|(\mathbf{C}\mathbf{A})^{-1}(\mathbf{I} - \mathbf{C}\mathbf{A})(\widetilde{\mathbf{x}}_n - \widetilde{\mathbf{x}}_{n+1})\| \\ &\leq \|(\mathbf{C}\mathbf{A})^{-1}\| \|\mathbf{I} - \mathbf{C}\mathbf{A}\| \|\widetilde{\mathbf{x}}_n - \widetilde{\mathbf{x}}_{n+1}\| \leq \frac{\varrho}{1-\varrho} \|\widetilde{\mathbf{x}}_n - \widetilde{\mathbf{x}}_{n+1}\| .\end{aligned}$$

If we could estimate $\|\mathbf{I} - \mathbf{C}\mathbf{A}\|$, then we could use this lemma to safely terminate an iterative improvement algorithm based on changes in the solution. For example, if $\|\mathbf{I} - \mathbf{C}\mathbf{A}\| = 1 - 10^{-2}$ and we want an absolute error in the solution of at most ε, then we could stop when the change in the solution is at most $10^{-2}\varepsilon$. The problem is that we often have no reliable *a priori* estimate for the spectral radius of $\mathbf{I} - \mathbf{C}\mathbf{A}$.

However, the iterative improvement iteration is related to the power method, discussed in Sect. 1.4.6. If

$$\mathbf{I} - \mathbf{C}\mathbf{A} = \mathbf{X}\boldsymbol{\Lambda}\mathbf{X}^{-1}$$

is diagonalizable, then

$$\widetilde{\mathbf{x}}_n - \mathbf{x} = (\mathbf{I} - \mathbf{C}\mathbf{A})^n \left(\widetilde{\mathbf{x}}_0 - \mathbf{x}\right) = \mathbf{X}\boldsymbol{\Lambda}^n\mathbf{X}^{-1} \left(\widetilde{\mathbf{x}}_0 - \mathbf{x}\right) .$$

We expect that for large iteration numbers n that $\widetilde{\mathbf{x}}^{(n)} - \mathbf{x}$ lies in the direction of the dominant eigenvector of $\mathbf{I} - \mathbf{CA}$, corresponding to the eigenvalue of maximum modulus. If the eigenvalues and eigenvectors of $\mathbf{I} - \mathbf{CA}$ are arranged largest first, then

$$\begin{aligned}\widetilde{\mathbf{x}}_{n+1} - \widetilde{\mathbf{x}}_n &= [\widetilde{\mathbf{x}}_{n+1} - \mathbf{x}] - [\widetilde{\mathbf{x}}_n - \mathbf{x}] \\ &\approx \mathbf{X}\mathbf{e}_1 \lambda_1^{n+1} \mathbf{e}_1 \cdot \mathbf{X}^{-1} (\widetilde{\mathbf{x}}_0 - \mathbf{x}) - \mathbf{X}\mathbf{e}_1 \lambda_1^n \mathbf{e}_1 \cdot \mathbf{X}^{-1} (\widetilde{\mathbf{x}}_0 - \mathbf{x}) \\ &= \left[\mathbf{X}\mathbf{e}_1 (\lambda_1 - 1) \mathbf{e}_1 \cdot \mathbf{X}^{-1} (\widetilde{\mathbf{x}}_0 - \mathbf{x})\right] \lambda_1^n .\end{aligned}$$

If $1 > |\lambda_1| = \varrho(\mathbf{I} - \mathbf{CA})$ and $\mathbf{e}_1 \cdot \mathbf{X}^{-1} (\widetilde{\mathbf{x}}_0 - \mathbf{x}) \neq 0$, then it follows that

$$\frac{\|\widetilde{\mathbf{x}}_{n+1} - \widetilde{\mathbf{x}}_n\|}{\|\widetilde{\mathbf{x}}_n - \widetilde{\mathbf{x}}_{n-1}\|} \approx \frac{\|\mathbf{X}\mathbf{e}_1\| \, |\lambda_1 - 1| \, \left|\mathbf{e}_1 \cdot \mathbf{X}^{-1} (\widetilde{\mathbf{x}}_0 - \mathbf{x})\right| \, |\lambda_1^n|}{\|\mathbf{X}\mathbf{e}_1\| \, |\lambda_1 - 1| \, \left|\mathbf{e}_1 \cdot \mathbf{X}^{-1} (\widetilde{\mathbf{x}}_0 - \mathbf{x})\right| \, |\lambda_1^{n-1}|} = |\lambda_1| = \varrho(\mathbf{I} - \mathbf{CA}) .$$

This ratio can be used to estimate ϱ in Lemma 2.4.8.

Three C^{++} classes to implement termination criteria for solver iterations can be found in classes `SolverControl`, `ReductionControl` and `RelativeErrorControl` in files Solver.H and Solver.C.

Exercise 2.4.10 Consider the one-dimensional heat equation

$$\frac{\partial u}{\partial t} = \frac{\partial^2 u}{\partial x^2} , \quad x \in (0,1), \quad t > 0$$

$$u(0,t) = 1 , \quad u(1,t) = 0 , \quad t > 0$$

$$u(x,0) = \begin{cases} 1, \, x < \frac{1}{2} , \\ 0, \, x > \frac{1}{2} . \end{cases}$$

1. Write a program to compute the analytical solution to this problem.
2. Program centered differences using forward Euler for this problem. Choose the timestep $\triangle t$ so that the scheme is positive according to its Fourier analysis.
3. Program centered differences and backward Euler for this problem. Use Gauss-Seidel iteration to solve the linear equations, and the solution at the previous timestep as the initial guess.
4. Discuss strategies for choosing $\triangle t$ for the implicit scheme.
5. Discuss strategies for terminating the Gauss-Seidel iteration in the implicit scheme.
6. For $\triangle x = 10^{-1}, 10^{-2}, \ldots, 10^{-6}$, plot the logarithm of the maximum error in these two numerical methods at $t = 1$ versus the logarithm of the computer time.

Exercise 2.4.11 Consider the two-dimensional Laplace equation

$$\frac{\partial^2 u}{\partial x^2} + \frac{\partial^2 u}{\partial y^2} = 0 \,, \quad 0 < x, y < 1 \,,$$

$$u(0, y) = 0 = u(1, y) \,, \quad 0 < y < 1$$

$$u(x, 0) = 0 \,, \quad u(x, 1) = \sin(\pi x) \sinh \pi \,, \quad 0 < x < 1 \,.$$

1. Use separation of variables to compute the analytical solution to this problem.
2. Program centered differences for this problem, in order to compute the residual in Gauss-Seidel.
3. Use Gauss-Seidel iteration to solve the linear equations, using zero for the initial guess.
4. For $\Delta x = \Delta y = 2^{-1}, 2^{-2}, \ldots, 2^{-7}$, plot minus the logarithm of the maximum error in the numerical solution on the vertical axis, versus logarithm of the computer time required to perform $1/(\Delta x \Delta y)$ iterations on the horizontal axis.
5. Assume that the smallest eigenvalue of the discrete Laplacian accurately approximates the smallest eigenvalue of the true Laplacian. Use your results from separation of variables to estimate the smallest eigenvalue of the true Laplacian, and use the Gerschgorin circle theorem to estimate the largest eigenvalue of the discrete Laplacian. Combine these to get an estimate of the condition number of the discrete Laplacian.
6. Normally, as we refine the mesh, we expect the discretization of the Laplacian to become more accurate. In order to achieve a more accurate solution in solving the linear system, we would need to use smaller tolerances in our error estimates as we refine the mesh. Suppose that the error in the exact solution of the linear system for the discrete Laplacian is $\varepsilon(\Delta x) = 4\Delta x^2$. Use your estimate of the condition number of the discrete Laplacian as a function of Δx to determine what relative error tolerance we should place on the residual in order that the error in the Gauss-Seidel iteration is no larger than $\varepsilon(\Delta x)$.

Exercise 2.4.12 Read about how to choose the optimal relaxation parameter for SOR in Wachspress [178], and describe how you would apply this approach to implicit centered differences for the heat equation.

Exercise 2.4.13 Read about Chebyshev acceleration of iterative methods in Golub and van Loan [84, p. 621] and Wachspress [178]. Describe the basic algorithm, and the modifications that must be made to SOR so that we can apply Chebyshev acceleration.

Exercise 2.4.14 Get a copy of matrix **E20R0100** from MatrixMarket. This is a sparse nonsymmetric matrix, and should be represented in computer memory without storing the zero entries. If this matrix is denoted by $\mathbf{A}$, let $\mathbf{x}$ be the vector of ones, and $\mathbf{b} = \mathbf{A}\mathbf{x}$. Choose the initial guess $\widetilde{\mathbf{x}}$ for Richardson's iteration to have random entries chosen from a uniform probability distribution on (0, 1). Program

Richardson's method to solve $\mathbf{A}\mathbf{x} = \mathbf{b}$, and plot

1. the logarithm base 10 of the error $\|\widetilde{\mathbf{x}} - \mathbf{x}\|_\infty$ versus iteration number,
2. the logarithm base 10 of the residual norm $\|\mathbf{A}\widetilde{\mathbf{x}} - \mathbf{b}\|_\infty$ versus iteration number, and
3. the logarithm base 10 of the solution increment $\|\widetilde{\mathbf{x}}^{(n+1)} - \widetilde{\mathbf{x}}^{(n)}\|_\infty$ versus iteration number.

At what order of error does the iteration stop making progress? How can you estimate the norm of ϱ in Lemma 2.4.8 from the norms of the solution increment? Given ϱ, a residual norm $\|\mathbf{A}\widetilde{\mathbf{x}}^{(n)} - \mathbf{b}\|_\infty$ and a solution increment norm $\|\widetilde{\mathbf{x}}^{(n+1)} - \widetilde{\mathbf{x}}^{(n)}\|_\infty$, could you use the ideas in Theorem 3.6.2 in Chap. 3 of Volume I to estimate the condition number of $\mathbf{A}$? Develop criteria to terminate Richardson's iteration when the error in the approximate solution is below a given tolerance.

Exercise 2.4.15 Get a copy of matrix **BCSSTM19** from MatrixMarket. This is a sparse, diagonally dominant symmetric positive matrix, and should be represented in computer memory without storing the zero entries. Repeat the previous exercise for this matrix, using the Jacobi iteration instead of Richardson's iteration.

Exercise 2.4.16 Get a copy of matrix **BCSSTK14** from MatrixMarket. This is a sparse, symmetric positive matrix that is not diagonally-dominant, and should be represented in computer memory without storing the zero entries. Repeat the previous exercise for this matrix.

Exercise 2.4.17 Get a copy of matrix **BCSSTM19** from MatrixMarket. Repeat the previous exercise for this matrix, using the Gauss-Seidel iteration instead of the Jacobi iteration.

2.5 Gradient Methods

So far, all of our previous iterative methods for solving systems of linear equations have been based on the iterative improvement iteration. In this section, we will discuss a completely different approach, provided that the matrix $\mathbf{A}$ in the linear system is Hermitian and positive. Instead of viewing the problem as a linear equation $\mathbf{A}\mathbf{x} = \mathbf{b}$, we will view it as an optimization problem to minimize $\phi(\mathbf{x})$ where

$$\phi(\mathbf{x}) \equiv \frac{1}{2}\mathbf{x}^H\mathbf{A}\mathbf{x} - \frac{1}{2}\mathbf{b}^H\mathbf{x} - \frac{1}{2}\mathbf{x}^H\mathbf{b} = \frac{1}{2}(\mathbf{b} - \mathbf{A}\mathbf{x})^H\mathbf{A}^{-1}(\mathbf{b} - \mathbf{A}\mathbf{x}) - \frac{1}{2}\mathbf{b}^H\mathbf{A}^{-1}\mathbf{b} . \tag{2.11}$$

It is easy to see that ϕ is real-valued, and $\phi(\mathbf{x}) \geq -\frac{1}{2}\mathbf{b}^H\mathbf{A}^{-1}\mathbf{b}$ for all vectors $\mathbf{x}$. Furthermore, the minimum of ϕ occurs at $\mathbf{x}$ where

$$\mathbf{0} = \mathbf{b} - \mathbf{A}\mathbf{x} \equiv \mathbf{r} .$$

Thus for Hermitian positive matrices, the minimization problem is equivalent to the original linear system.

It is also interesting to note that $\phi(\mathbf{x})$ has bounded level sets. To see this fact, we will use the Spectral Theorem 1.3.1 to write

$$\mathbf{A} = \mathbf{Q}\boldsymbol{\Lambda}\mathbf{Q}^H ,$$

where $\mathbf{Q}$ is a unitary matrix and $\boldsymbol{\Lambda}$ is diagonal with positive diagonal entries $\lambda_1 \leq \ldots \leq \lambda_m$. We also define

$$\mathbf{y} = \mathbf{Q}^H\mathbf{x} \quad \text{and} \quad \mathbf{c} = \boldsymbol{\Lambda}^{-1}\mathbf{Q}^H\mathbf{b} .$$

Then

$$\begin{aligned}\phi(\mathbf{x}) &= \frac{1}{2}\left(\mathbf{y}^H\mathbf{Q}^H\right)\left(\mathbf{Q}\boldsymbol{\Lambda}\mathbf{Q}^H\right)(\mathbf{Q}\mathbf{y}) - \left(\mathbf{c}^H\boldsymbol{\Lambda}\mathbf{Q}^H\right)(\mathbf{Q}\mathbf{y}) \\ &= \frac{1}{2}\mathbf{y}^H\boldsymbol{\Lambda}\mathbf{y} - \mathbf{c}^H\boldsymbol{\Lambda}\mathbf{y} = \frac{1}{2}(\mathbf{y}-\mathbf{c})^H\boldsymbol{\Lambda}(\mathbf{y}-\mathbf{c}) - \frac{1}{2}\mathbf{c}^H\mathbf{c} .\end{aligned}$$

If $\phi(\mathbf{x}) \leq \gamma$, it follows that

$$\begin{aligned}\lambda_1\|\mathbf{x}-\mathbf{Q}\mathbf{c}\|_2^2 = \lambda_1\|\mathbf{y}-\mathbf{c}\|_2^2 \leq \sum_{i=1}^m \lambda_i(\mathbf{y}_i-\mathbf{c}_i)^2 = (\mathbf{y}-\mathbf{c})^H\boldsymbol{\Lambda}(\mathbf{y}-\mathbf{c}) = 2\phi(\mathbf{x}) + \mathbf{c}^H\mathbf{c} \\ \leq 2\gamma + \|\mathbf{c}\|_2^2 .\end{aligned}$$

Since the diagonal entries of $\boldsymbol{\Lambda}$ are positive, this inequality provides a bound on points inside a level set of ϕ. A variation of this argument also shows that $\phi(\mathbf{x}) \to \infty$ as $\|\mathbf{x}\| \to \infty$.

We will develop several algorithms for minimizing $\phi(\mathbf{x})$. The first will minimize ϕ along the direction of steepest descent at each approximate solution $\mathbf{x}_n$, while the second algorithm will minimize $\phi(\mathbf{x})$ over subspaces of m-vectors of strictly increasing dimension. The first algorithm is guaranteed to reduce the value of ϕ at each step, and is therefore globally convergent. Certain expanding subspaces in the second algorithm will imply that the algorithm will terminate in at most m steps. For banded matrices $\mathbf{A}$ with $O(1)$ nonzero entries in each row, each step will involve $O(m)$ operations; in this case the total work will be at most $O(m^2)$ operations for banded Hermitian positive systems. In many cases, the second algorithm will produce good results in far fewer iterations than the first algorithm. The remaining algorithms will be developed to deal with non-Hermitian matrices.

2.5.1 Matrix Inner Product

We will begin by showing that an Hermitian positive matrix can generate an inner product and norm.

Lemma 2.5.1 *Suppose that the $m \times m$ matrix $\mathbf{A}$ is Hermitian and positive. Define the operator $[\cdot,\cdot]_{\mathbf{A}}$ mapping pairs of m-vectors to scalars by*

$$[\mathbf{z},\mathbf{y}]_{\mathbf{A}} = \mathbf{z}^H \mathbf{A} \mathbf{y} .$$

Then $[\cdot,\cdot]_{\mathbf{A}}$ is an ***inner product*** *on m-vectors. Conversely, if $[\cdot,\cdot]$ is an inner product on m-vectors, then there is an Hermitian positive matrix $\mathbf{A}$ so that for all m-vectors $\mathbf{y}$ and $\mathbf{z}$ we have $[\mathbf{z},\mathbf{y}] = \mathbf{z}^H\mathbf{A}\mathbf{y}$.*

Proof Suppose that the $m \times m$ matrix $\mathbf{A}$ is Hermitian and positive. Then for all m-vectors $\mathbf{y}$, $\mathbf{y}_1$, $\mathbf{y}_2$ and $\mathbf{z}$, and all scalars γ we have

$$\begin{aligned}
&[\mathbf{z},\mathbf{y}]_{\mathbf{A}} = \mathbf{z}^H\mathbf{A}\mathbf{y} = \overline{\mathbf{y}^H\mathbf{A}\mathbf{z}} = \overline{[\mathbf{y},\mathbf{z}]_{\mathbf{A}}} \,, \\
&[\mathbf{z},\mathbf{y}_1+\mathbf{y}_2]_{\mathbf{A}} = \mathbf{z}^H\mathbf{A}(\mathbf{y}_1+\mathbf{y}_2) = \mathbf{z}^H\mathbf{A}\mathbf{y}_1 + \mathbf{z}^H\mathbf{A}\mathbf{y}_2 = [\mathbf{z},\mathbf{y}_1]_{\mathbf{A}} + [\mathbf{z},\mathbf{y}_2]_{\mathbf{A}} \,, \\
&[\mathbf{z},\gamma\mathbf{y}]_{\mathbf{A}} = \mathbf{z}^H\mathbf{A}(\gamma\mathbf{y}) = \mathbf{z}^H\mathbf{A}\mathbf{y}\gamma = [\mathbf{z},\mathbf{y}]_{\mathbf{A}}\gamma \,, \\
&[\mathbf{y},\mathbf{y}]_{\mathbf{A}} = \mathbf{y}^H\mathbf{A}\mathbf{y} \ge 0 \quad \text{and} \\
&0 = [\mathbf{y},\mathbf{y}]_{\mathbf{A}} = \mathbf{y}^H\mathbf{A}\mathbf{y} \quad \Longrightarrow \quad \mathbf{y} = \mathbf{0} \,.
\end{aligned}$$

These conditions prove that $[\cdot,\cdot]_{\mathbf{A}}$ is an inner product, as defined in Halmos [92, p. 121].

Next, we prove the converse. Suppose that $[\cdot,\cdot]$ is an inner product on n-vectors. Define the components of an $n \times n$ matrix $\mathbf{A}$ by

$$\mathbf{A}_{ij} = [\mathbf{e}_i,\mathbf{e}_j] \,,$$

where $\mathbf{e}_i$ is the ith axis vector. Then for all real m-vectors $\mathbf{y}$ and $\mathbf{z}$ we have

$$[\mathbf{z},\mathbf{y}] = \left[\sum_{i=1}^m z_i\mathbf{e}_i, \sum_{j=1}^m y_j\mathbf{e}_j\right] = \sum_{i=1}^m\sum_{j=1}^m \overline{z_i} y_j [\mathbf{e}_i,\mathbf{e}_j] = \sum_{i=1}^m\sum_{j=1}^m \overline{z_i} y_j A_{ij} = \mathbf{z}^H\mathbf{A}\mathbf{y} \,.$$

Since

$$\mathbf{A}_{ji} = [\mathbf{e}_j,\mathbf{e}_i] = \overline{[\mathbf{e}_i,\mathbf{e}_j]} = \overline{A_{ij}} \,,$$

we see that **A** is Hermitian. Also, if **y** is an arbitrary m-vector, then

$$\mathbf{y}^H\mathbf{A}\mathbf{y} = [\mathbf{y},\mathbf{y}] \geq 0 \quad \text{and}$$
$$0 = \mathbf{y}^H\mathbf{A}\mathbf{y} = [\mathbf{y},\mathbf{y}] \quad \Longrightarrow \quad \mathbf{y} = \mathbf{0} .$$

These conditions imply that **A** is positive.
Lemma 2.5.1 suggests the following definition.

Definition 2.5.1 If the $m \times m$ matrix **A** is Hermitian and positive, then for all m-vectors **y** the **A-norm** of **y** is $\|\!|\mathbf{y}\|\!|_{\mathbf{A}} = \sqrt{\mathbf{y}^H\mathbf{A}\mathbf{y}}$.

2.5.2 Steepest Descent

Given a current iterate $\widetilde{\mathbf{x}}_n$, the steepest descent algorithm chooses the step to lie in the direction of the gradient of the objective function:

$$\widetilde{\mathbf{x}}_{n+1} = \widetilde{\mathbf{x}}_n + (\mathbf{b} - \mathbf{A}\widetilde{\mathbf{x}}_n)\,\alpha_n .$$

Here the real scalar α_n is chosen to minimize ϕ along the direction of steepest descent. In other words, if the objective along the direction of steepest descent is

$$\psi_n(\alpha) = \phi\left(\widetilde{\mathbf{x}}_n + [\mathbf{b} - \mathbf{A}\widetilde{\mathbf{x}}_n]\,\alpha\right) ,$$

then we choose $\alpha_n = \operatorname{argmin} \psi_n(\alpha)$. Here the argmin function implies that

$$\psi_n(\alpha_n) = \min_{\alpha \geq 0} \psi_n(\alpha) .$$

Let us find a formula for argmin $\psi(\alpha)$. If the current residual is $\mathbf{r} = \mathbf{b} - \mathbf{A}\widetilde{\mathbf{x}}$, then

$$\psi(\alpha) = \phi\left(\widetilde{\mathbf{x}} + \mathbf{r}\alpha\right) = \frac{1}{2}[\widetilde{\mathbf{x}} + \mathbf{r}\alpha]^H\mathbf{A}\,[\widetilde{\mathbf{x}} + \mathbf{r}\alpha] - \frac{1}{2}\mathbf{b}^H\,[\widetilde{\mathbf{x}} + \mathbf{r}\alpha] - \frac{1}{2}[\widetilde{\mathbf{x}} + \mathbf{r}\alpha]^H\mathbf{b}$$
$$= \phi\left(\widetilde{\mathbf{x}}\right) - \alpha\,\|\mathbf{r}\|_2^2 + \frac{\alpha^2}{2}\,\|\!|\mathbf{r}\|\!|_{\mathbf{A}}^2 .$$

At a minimum of ψ we have

$$0 = \psi_n'(\alpha) = -\,\|\mathbf{r}\|_2^2 + \alpha\,\|\!|\mathbf{r}\|\!|_{\mathbf{A}}^2 ,$$

so

$$\alpha_n = \frac{\|\mathbf{r}_n\|_2^2}{\|\!|\mathbf{r}_n\|\!|_{\mathbf{A}}^2} .$$

The steepest descent algorithm can now be written in the form

Algorithm 2.5.1 (Steepest Descent)

$$\mathbf{r}_n = \mathbf{b} - \mathbf{A}\widetilde{\mathbf{x}}_n \tag{2.12}$$

$$\mathbf{s}_n = \mathbf{A}\mathbf{r}_n \tag{2.13}$$

$$\alpha_n = \frac{\mathbf{r}_n \cdot \mathbf{r}_n}{\mathbf{r}_n \cdot \mathbf{s}_n} \tag{2.14}$$

$$\widetilde{\mathbf{x}}_{n+1} = \widetilde{\mathbf{x}}_n + \mathbf{r}_n\alpha_n \; . \tag{2.15}$$

This algorithm requires two matrix-vector multiplications per step, and the storage of three m-vectors.

In order to analyze the steepest descent algorithm, we will make the following assumptions in the remainder of this section.

Assumption 2.5.1 Suppose that the $m \times m$ matrix $\mathbf{A}$ is Hermitian and positive, with eigenvalues

$$0 < \lambda_1 \leq \lambda_2 \leq \ldots \leq \lambda_m \; .$$

Assume that $\mathbf{b}$ is an m-vector, and that $\mathbf{x}$ is the solution of $\mathbf{A}\mathbf{x} = \mathbf{b}$. Finally, given any m-vector $\widetilde{\mathbf{x}}^{(n)}$, assume that the residual $\mathbf{r}^{(n)}$, step length α_n and next steepest descent iterate $\widetilde{\mathbf{x}}^{(n+1)}$ are defined by the steepest descent algorithm (2.5.1).

Lemma 2.5.2 ([120, p. 151]) *Under Assumptions 2.5.1 we have*

$$|||\widetilde{\mathbf{x}}_{n+1} - \mathbf{x}|||_{\mathbf{A}}^2 = |||\widetilde{\mathbf{x}}_n - \mathbf{x}|||_{\mathbf{A}}^2 \left\{ 1 - \frac{\|\mathbf{r}_n\|_2^4}{|||\mathbf{r}_n|||_{\mathbf{A}}^2 \; |||\mathbf{r}_n|||_{\mathbf{A}^{-1}}^2} \right\} \; . \tag{2.16}$$

Proof Let us write $\mathbf{d}_n = \mathbf{x} - \widetilde{\mathbf{x}}_n$ and note that $\mathbf{A}\mathbf{d}_n = \mathbf{r}_n$. Then

$$\begin{aligned}
&|||\mathbf{d}_n|||_{\mathbf{A}}^2 - |||\mathbf{d}_{n+1}|||_{\mathbf{A}}^2 \\
&= [\mathbf{d}_n]^H \mathbf{A}\mathbf{d}_n - [\mathbf{d}_n - \mathbf{r}_n\alpha_n]^H \mathbf{A} [\mathbf{d}_n - \mathbf{r}_n\alpha_n] \\
&= \alpha_n [\mathbf{r}_n]^H \mathbf{A}\mathbf{d}_n + \alpha_n [\mathbf{d}_n]^H \mathbf{A}\mathbf{r}_n - \alpha_n^2 [\mathbf{r}_n] \mathbf{A}\mathbf{r}_n \\
&= 2\frac{\|\mathbf{r}_n\|_2^2}{|||\mathbf{r}_n|||_{\mathbf{A}}^2} \|\mathbf{r}_n\|_2^2 - \left[\frac{\|\mathbf{r}_n\|_2^2}{|||\mathbf{r}_n|||_{\mathbf{A}}^2} \right]^2 |||\mathbf{r}_n|||_{\mathbf{A}}^2 = \frac{\|\mathbf{r}_n\|_2^4}{|||\mathbf{r}_n|||_{\mathbf{A}}^2} \; .
\end{aligned}$$

Since

$$|||\mathbf{d}_n|||_{\mathbf{A}}^2 = (\mathbf{r}_n)^H \mathbf{A}^{-1} \mathbf{r}_n = |||\mathbf{r}_n|||_{\mathbf{A}^{-1}}^2 \; ,$$

we see that

$$\|\widetilde{\mathbf{x}}_{n+1} - \mathbf{x}\|_{\mathbf{A}}^2 = \|\widetilde{\mathbf{x}}_n - \mathbf{x}\|_{\mathbf{A}}^2 - \frac{\|\mathbf{r}_n\|_2^2}{\|\mathbf{r}_n\|_{\mathbf{A}}^2} = \|\widetilde{\mathbf{x}}_n - \mathbf{x}\|_{\mathbf{A}}^2 \left\{ 1 - \frac{\|\mathbf{r}_n\|_2^4}{\|\mathbf{r}_n\|_{\mathbf{A}}^2 \, \|\mathbf{r}_n\|_{\mathbf{A}^{-1}}^2} \right\} .$$

The next lemma will be used to establish a bound on the convergence rate for the steepest descent algorithm.

Lemma 2.5.3 (Kantorovich Inequality) *[102] Under Assumptions 2.5.1 we have that for all real nonzero m-vectors* $\mathbf{r}$

$$\frac{(\mathbf{r}^H\mathbf{A}\mathbf{r})\,(\mathbf{r}^H\mathbf{A}^{-1}\mathbf{r})}{(\mathbf{r}^H\mathbf{r})^2} \le \frac{(\lambda_1 + \lambda_m)^2}{4\lambda_1\lambda_m} = \left[\frac{\sqrt{\lambda_m/\lambda_1} + \sqrt{\lambda_1/\lambda_m}}{2} \right]^2 . \tag{2.17}$$

Proof First, note that

$$\sqrt{\frac{\lambda_1}{\lambda_m}} = \frac{\lambda_1}{\sqrt{\lambda_1\lambda_m}} \le \frac{\lambda_2}{\sqrt{\lambda_1\lambda_m}} \le \ldots \le \frac{\lambda_m}{\sqrt{\lambda_1\lambda_m}} = \sqrt{\frac{\lambda_m}{\lambda_1}} .$$

Next, consider the function $f(\varrho) = \varrho + (1/\varrho)$. Note that f is convex for $\varrho > 0$, with a minimum at $\varrho = 1$. Further,

$$f\left(\frac{\lambda_1}{\sqrt{\lambda_1\lambda_m}}\right) = \sqrt{\frac{\lambda_1}{\lambda_m}} + \sqrt{\frac{\lambda_m}{\lambda_1}} = f\left(\frac{\lambda_m}{\sqrt{\lambda_1\lambda_m}}\right) .$$

It follows from this equation and the previous inequality that, for all $1 \le i \le m$,

$$\frac{\lambda_i}{\sqrt{\lambda_1\lambda_m}} + \frac{\sqrt{\lambda_1\lambda_m}}{\lambda_i} = f\left(\frac{\lambda_i}{\sqrt{\lambda_1\lambda_m}}\right) \le f\left(\frac{\lambda_1}{\sqrt{\lambda_1\lambda_m}}\right) = f\left(\frac{\lambda_m}{\sqrt{\lambda_1\lambda_m}}\right) = \sqrt{\frac{\lambda_1}{\lambda_m}} + \sqrt{\frac{\lambda_m}{\lambda_1}} .$$

The spectral theorem shows that $\mathbf{A} = \mathbf{Q}\boldsymbol{\Lambda}\mathbf{Q}^H$, where the $m \times m$ matrix $\mathbf{Q}$ is unitary and $\boldsymbol{\Lambda}$ is diagonal with diagonal entries $\lambda_1, \ldots, \lambda_m$. The same orthogonal matrix diagonalizes $\mathbf{A}/\sqrt{\lambda_1\lambda_m} + \mathbf{A}^{-1}\sqrt{\lambda_1\lambda_m}$, which has eigenvalues $\lambda_i/\sqrt{\lambda_1\lambda_m} + \sqrt{\lambda_1\lambda_m}/\lambda_i$. All of these eigenvalues are bounded above by $\sqrt{\frac{\lambda_1}{\lambda_m}} + \sqrt{\frac{\lambda_m}{\lambda_1}}$. It follows that for all real nonzero m-vectors $\mathbf{r}$, if $\mathbf{s} = \mathbf{Q}^H\mathbf{r}$ then

$$\frac{\mathbf{r}^H\left[\mathbf{A}/\sqrt{\lambda_1\lambda_m} + \mathbf{A}^{-1}\sqrt{\lambda_1\lambda_m}\right]\mathbf{r}}{\mathbf{r}^H\mathbf{r}} = \frac{\sum_{i=1}^m \left(\lambda_i/\sqrt{\lambda_1\lambda_m} + \sqrt{\lambda_1\lambda_m}/\lambda_i\right)\mathbf{s}_i^2}{\sum_{i=1}^m \mathbf{s}_i^2}$$

$$\le \sqrt{\frac{\lambda_1}{\lambda_m}} + \sqrt{\frac{\lambda_m}{\lambda_1}} .$$

Using the inequality $4ab \leq (|a| + |b|)^2$ leads to

$$(\mathbf{r}^H \mathbf{A}\mathbf{r})\,(\mathbf{r}^H \mathbf{A}^{-1}\mathbf{r}) \leq \frac{1}{4}\left[\frac{\mathbf{r}^H\mathbf{A}\mathbf{r}}{\sqrt{\lambda_1\lambda_m}} + \sqrt{\lambda_1\lambda_m}\mathbf{r}^H\mathbf{A}^{-1}\mathbf{r}\right]^2 \leq \frac{1}{4}\left[\sqrt{\frac{\lambda_m}{\lambda_1}} + \sqrt{\frac{\lambda_1}{\lambda_m}}\right]^2 (\mathbf{r}^H\mathbf{r})^2 \; .$$

Inequality (2.17) follows immediately from this result.

The next theorem establishes an upper bound on the convergence rate for steepest descent.

Theorem 2.5.1 (Steepest Descent) *Under Assumptions 2.5.1, the steepest descent method converges to* $\mathbf{x}$ *for any initial guess* $\widetilde{\mathbf{x}}_0$*, and the error in the iterates satisfies*

$$\|\widetilde{\mathbf{x}}_{n+1} - \mathbf{x}\|_{\mathbf{A}} \leq \frac{\lambda_m - \lambda_1}{\lambda_m + \lambda_1}\|\widetilde{\mathbf{x}}_n - \mathbf{x}\|_{\mathbf{A}} \; .$$

Proof Equation (2.16) and the Kantorovich inequality (2.17) imply that

$$\begin{aligned}\|\widetilde{\mathbf{x}}_{n+1} - \mathbf{x}\|_{\mathbf{A}}^2 &= \|\widetilde{\mathbf{x}}_n - \mathbf{x}\|_{\mathbf{A}}^2\left\{1 - \frac{\|\mathbf{r}_n\|_2^2}{\|\mathbf{r}_n\|_{\mathbf{A}}^2\,\|\mathbf{r}_n\|_{\mathbf{A}^{-1}}^2}\right\} \\ &\leq \|\widetilde{\mathbf{x}}_n - \mathbf{x}\|_{\mathbf{A}}^2\left\{1 - \frac{4\lambda_1\lambda_m}{(\lambda_1+\lambda_m)^2}\right\} = \|\widetilde{\mathbf{x}}_n - \mathbf{x}\|_{\mathbf{A}}^2\left(\frac{\lambda_m - \lambda_1}{\lambda_m + \lambda_1}\right)^2 \; .\end{aligned}$$

Since the eigenvalues of the Hermitian positive matrix $\mathbf{A}$ are $0 < \lambda_1 \leq \lambda_m$, steepest descent reduces the $\mathbf{A}$-norm of the error by at least a factor of $(\lambda_m - \lambda_1)/(\lambda_m + \lambda_1) < 1$ with each iteration. Akaike [2] has shown that this convergence rate is exact, except when the iteration is begun at certain degenerate starting points. Since finite difference matrices typically have large ratios λ_m/λ_1 in order to approximate the infinite spectrum of the underlying differential operators, we expect the steepest descent method to converge very slowly for most of the problems discussed in this book. In practice, the problem with steepest descent is that it tends to search repetitively in the directions associated with the smallest and largest eigenvalues. We need an approach that will prevent the repetition of search in a prior direction.

2.5.3 Conjugate Gradients

The basic idea behind conjugate gradients, originally developed by Hestenes and Stiefel [96], (and closely related to an approach by Lanczos [118]) is to minimize $\phi(\mathbf{v})$, defined to be

$$\phi(\mathbf{x}) \equiv \frac{1}{2}\mathbf{x}^H\mathbf{A}\mathbf{x} - \frac{1}{2}\mathbf{b}^H\mathbf{x} - \frac{1}{2}\mathbf{x}^H\mathbf{b} = \frac{1}{2}(\mathbf{b} - \mathbf{A}\mathbf{x})^H\mathbf{A}^{-1}(\mathbf{b} - \mathbf{A}\mathbf{x}) - \frac{1}{2}\mathbf{b}^H\mathbf{A}^{-1}\mathbf{b}\; ,$$

by searching along an appropriate set of linearly independent directions. The following definition will describe what "appropriate" means in this case.

Definition 2.5.2 Suppose that the $m \times m$ matrix $\mathbf{A}$ is Hermitian positive, and that the m-vectors $\mathbf{p}_0, \ldots, \mathbf{p}_n$ are nonzero. Then $\mathbf{p}_0, \ldots, \mathbf{p}_n$ are $\mathbf{A}$-conjugate if and only if $i \neq j$ implies that

$$\left[\mathbf{p}_i, \mathbf{p}_j\right]_{\mathbf{A}} \equiv (\mathbf{p}_i)^H \mathbf{A} \mathbf{p}_j = 0 \,.$$

This definition allows us to state and prove the following lemma.

Lemma 2.5.4 *Suppose that the $m \times m$ matrix $\mathbf{A}$ is Hermitian and positive, and that the m-vectors $\mathbf{p}_0, \ldots, \mathbf{p}_n$ are $\mathbf{A}$-conjugate. Then $\mathbf{p}_0, \ldots, \mathbf{p}_n$ are linearly independent, and $n < m$.*

Proof Suppose that $\sum_{i=0}^{m} \mathbf{p}_i \alpha_i = 0$ is a linear combination of the $\mathbf{A}$-conjugate vectors. Since the vectors $\mathbf{p}_i$ are $\mathbf{A}$-conjugate, for all $0 \leq j \leq m$ we have

$$0 = \left[\mathbf{p}_j, \sum_{i=0}^{m} \mathbf{p}_i \alpha_i\right]_{\mathbf{A}} = \left[\mathbf{p}_j, \mathbf{p}_j\right]_{\mathbf{A}} \alpha_j \,.$$

Note that since $\mathbf{p}_j \neq \mathbf{0}$ and $\mathbf{A}$ is positive, $\left[\mathbf{p}_j, \mathbf{p}_j\right]_{\mathbf{A}} > 0$. We conclude that $\alpha_j = 0$ for all j. Since any set of more than n m-vectors is linearly dependent, we must have $n < m$.

The next theorem shows us how to use $\mathbf{A}$-conjugate vectors to minimize the quadratic form ϕ in (2.11).

Theorem 2.5.2 (Conjugate Direction) *[120, p. 170] Suppose that the $m \times m$ matrix $\mathbf{A}$ is Hermitian and positive, $\mathbf{b}$ is an m-vector, and $\mathbf{A}\mathbf{x} = \mathbf{b}$. Assume that the m-vectors $\mathbf{p}_0, \ldots, \mathbf{p}_{m-1}$ are $\mathbf{A}$-conjugate. Given any m-vector $\widetilde{\mathbf{x}}_0$, compute the scalars α_n and vectors $\widetilde{\mathbf{x}}_n$ by the*

Algorithm 2.5.2 (Conjugate Direction)

$$\begin{aligned}
&\textit{for } 0 \leq n < m \\
&\quad \mathbf{r}_n = \mathbf{b} - \mathbf{A}\widetilde{\mathbf{x}}_n \\
&\quad \alpha_n = \frac{\mathbf{p}_n \cdot \mathbf{r}_n}{\mathbf{p}_n \cdot \mathbf{A}\mathbf{p}_n} \\
&\quad \widetilde{\mathbf{x}}_{n+1} = \widetilde{\mathbf{x}}_n + \mathbf{p}_n \alpha_n \,.
\end{aligned}$$

Then $\mathbf{A}\widetilde{\mathbf{x}}_m = \mathbf{b}$ and $\widetilde{\mathbf{x}}_m = \mathbf{x}$.

Proof Since the vectors $\mathbf{p}_0, \ldots, \mathbf{p}_{m-1}$ form a basis for all m-vectors, there are scalars β_i so that

$$\mathbf{x} - \widetilde{\mathbf{x}}_0 = \sum_{i=0}^{m-1} \mathbf{p}_i \beta_i \,.$$

If we multiply this equation by $(\mathbf{p}_n)^H \mathbf{A}$, then we can solve for β_n to get

$$\beta_n = \frac{[\mathbf{p}_n, \mathbf{x} - \widetilde{\mathbf{x}}_0]_{\mathbf{A}}}{[\mathbf{p}_n, \mathbf{p}_n]_{\mathbf{A}}} = \frac{\mathbf{p}_n \cdot (\mathbf{b} - \mathbf{A}\widetilde{\mathbf{x}}_0)}{[\mathbf{p}_n, \mathbf{p}_n]_{\mathbf{A}}} \,.$$

This gives us a relation between the solution $\mathbf{x}$, the initial guess $\widetilde{\mathbf{x}}_0$ and the $\mathbf{A}$-conjugate vectors $\mathbf{p}_0, \ldots, \mathbf{p}_{m-1}$. Next, we will show inductively that $\widetilde{\mathbf{x}}_n$, defined by (2.5.2), satisfies

$$\widetilde{\mathbf{x}}_n = \widetilde{\mathbf{x}}_0 + \sum_{i=0}^{n-1} \mathbf{p}_i \beta_i \,.$$

This statement is obviously true for $n = 0$. Assuming that the statement is true for n, we will prove that it is true for $n + 1$. Note that the inductive hypothesis and the A-conjugacy of the $\mathbf{p}_j$ imply that

$$\mathbf{p}_n \cdot \mathbf{A}\left(\widetilde{\mathbf{x}}_n - \widetilde{\mathbf{x}}_0\right) = \sum_{i=0}^{n-1} [\mathbf{p}_n, \mathbf{p}_i]_{\mathbf{A}} \, \beta_i = 0 \,.$$

Thus

$$\begin{aligned}\alpha_n &= \frac{\mathbf{p}_n \cdot (\mathbf{b} - \mathbf{A}\widetilde{\mathbf{x}}_n)}{[\mathbf{p}_n, \mathbf{p}_n]_{\mathbf{A}}} = \frac{\mathbf{p}_n \cdot \mathbf{A}\left(\widetilde{\mathbf{x}}_0 - \widetilde{\mathbf{x}}_n\right) + \mathbf{p}_n \cdot (\mathbf{b} - \mathbf{A}\widetilde{\mathbf{x}}_0)}{[\mathbf{p}_n, \mathbf{p}_n]_{\mathbf{A}}} \\ &= \frac{\mathbf{p}_n \cdot (\mathbf{b} - \mathbf{A}\widetilde{\mathbf{x}}_0)}{[\mathbf{p}_n, \mathbf{p}_n]_{\mathbf{A}}} = \beta_n \,.\end{aligned}$$

The next theorem will show that algorithm (2.5.2) actually performs a minimization over expanding subspaces, with residuals orthogonal to the previous search vectors.

Theorem 2.5.3 (Expanding Subspace) *[120, p. 171] Suppose that the $m \times m$ matrix $\mathbf{A}$ is Hermitian positive, $\mathbf{b}$ is an m-vector, and the m-vectors $\mathbf{p}_0, \ldots, \mathbf{p}_{m-1}$ are $\mathbf{A}$-conjugate. Also suppose that given an m-vector $\widetilde{\mathbf{x}}_0$ we perform the algorithm* (2.5.2) *to compute the scalars α_n for $0 \leq n < m$ and vectors $\widetilde{\mathbf{x}}_n$ for $1 \leq n \leq m$. Then for $1 \leq n \leq m$, $\widetilde{\mathbf{x}}_n$ minimizes*

$$\phi(\mathbf{v}) \equiv \frac{1}{2}\mathbf{v}^H \mathbf{A}\mathbf{v} - \frac{1}{2}\mathbf{b}^H \mathbf{v} - \frac{1}{2}\mathbf{v}^H \mathbf{b}$$

over the sets

$$\mathscr{L}_n \equiv \{\widetilde{\mathbf{x}}_{n-1} + \mathbf{p}_{n-1}\beta : \beta \textit{ a real scalar }\} \textit{ and}$$

$$\mathscr{M}_n \equiv \left\{ \widetilde{\mathbf{x}}_0 + \sum_{i=0}^{n-1} \mathbf{p}_i\beta_i : \beta_i \textit{ a real scalar} \right\} .$$

Furthermore, for all $0 \le i < n$,

$$\mathbf{p}_i \cdot (\mathbf{b} - \mathbf{A}\widetilde{\mathbf{x}}_n) = 0 .$$

Proof Since $\mathscr{L}_n \subset \mathscr{M}_n$, it suffices to show that $\widetilde{\mathbf{x}}_n$ minimizes $\phi(\mathbf{v})$ over $\mathscr{M}_n$. At a minimum we want the matrix of second derivatives to be nonnegative; for $0 \le i, j < n$ the i, j entry of this matrix is

$$\frac{\partial^2 \phi(\widetilde{\mathbf{x}}_n)}{\partial\beta_i \partial\beta_j} = [\mathbf{p}_i, \mathbf{p}_j]_{\mathbf{A}} .$$

If $\mathbf{P} = [\mathbf{p}_0, \ldots, \mathbf{p}_{m-1}]$ is the matrix of $\mathbf{A}$-conjugate vectors, then the matrix of second derivatives of ϕ with respect to the β values is $\mathbf{P}^H\mathbf{A}\mathbf{P}$. Since $\mathbf{A}$ is positive and $\mathbf{P}$ is nonsingular, this matrix of second derivatives is positive. Thus, we need only show that the first-order conditions for a minimum are satisfied.

At a minimum of $\phi\left(\widetilde{\mathbf{x}}_0 + \sum_{i=0}^{n-1} \mathbf{p}_i\beta_i\right)$ over all possible choices of the coefficients β_i, we want the first derivatives with respect to the β_i to be zero. In other words, for all $0 \le i < n$ we require

$$0 = \frac{\partial\phi(\widetilde{\mathbf{x}}_n)}{\partial\beta_i} = \mathbf{p}_i \cdot \mathbf{A}\widetilde{\mathbf{x}}_n - \mathbf{b}^H\mathbf{p}_i = \mathbf{p}_i \cdot (\mathbf{A}\widetilde{\mathbf{x}}_n - \mathbf{b}) .$$

All that remains is to show that for all $\mathbf{v} \in \mathscr{M}_n$ and for all $0 \le i < n$

$$0 = \mathbf{p}_i \cdot (\mathbf{A}\widetilde{\mathbf{x}}_n - \mathbf{b}) . \tag{2.18}$$

We will prove this statement by induction on n.

For $n = 0$, $\mathscr{M}_n$ consists of a single vector and the set of first-order conditions for a minimum is empty. Assume inductively that the first-order conditions (2.18) are satisfied for all iterates from 0 to n. We will prove that Eqs. (2.18) are true for $n + 1$. Note that

$$\mathbf{A}\widetilde{\mathbf{x}}_{n+1} - \mathbf{b} = (\mathbf{A}\widetilde{\mathbf{x}}_n - \mathbf{b}) + \mathbf{A}\mathbf{p}_n\alpha_n .$$

Thus the definition of α_n shows that

$$\mathbf{p}_n \cdot (\mathbf{A}\widetilde{\mathbf{x}}_n - \mathbf{b}) + \mathbf{p}_n \cdot \mathbf{A}\mathbf{p}_n\alpha_n = 0 .$$

Furthermore, the inductive hypothesis and the **A**-conjugacy of the $\mathbf{p}_i$ show that for all $0 \leq i < n$

$$\mathbf{p}_i \cdot (\mathbf{A}\widetilde{\mathbf{x}}_n - \mathbf{b}) + \mathbf{p}_i \cdot \mathbf{A}\mathbf{p}_n \alpha_n = 0 \; .$$

This proves the first-order condition (2.18) for the minimum over $\mathscr{M}_n$.
The proof of the expanding subspace theorem suggests that we seek **A**-conjugate directions $\mathbf{p}_n$ that are related to the residuals $\mathbf{b} - \mathbf{A}\widetilde{\mathbf{x}}_n$, since the latter are orthogonal to all previous search directions and point in the directions of steepest descent.

The next theorem shows how to compute the **A**-conjugate vectors $\mathbf{p}_0, \ldots, \mathbf{p}_{m-1}$.

Theorem 2.5.4 (Conjugate Gradient) *[120, p. 174] Assume that the $m \times m$ matrix $\mathbf{A}$ is Hermitian positive, and that $\mathbf{b}$ and $\widetilde{\mathbf{x}}_0$ are m-vectors. Suppose that we perform the following*

Algorithm 2.5.3 (Conjugate Gradient)

$$\begin{aligned}
&\mathbf{p}_0 = \mathbf{r}_0 = \mathbf{b} - \mathbf{A}\widetilde{\mathbf{x}}_0 \; , \\
&\textit{while } 0 \leq n < m \textit{ and } \mathbf{r}_n \neq \mathbf{0} \\
&\quad \alpha_n = [\mathbf{p}_n \cdot \mathbf{r}_n] \, / \, [\mathbf{p}_n \cdot \mathbf{A}\mathbf{p}_n] \\
&\quad \widetilde{\mathbf{x}}_{n+1} = \widetilde{\mathbf{x}}_n + \mathbf{p}_n \alpha_n \\
&\quad \mathbf{r}_{n+1} = \mathbf{b} - \mathbf{A}\widetilde{\mathbf{x}}_{n+1} \\
&\quad \beta_n = -\, [\mathbf{p}_n \cdot \mathbf{A}\mathbf{r}_{n+1}] \, / \, [\mathbf{p}_n \cdot \mathbf{A}\mathbf{p}_n] \\
&\quad \mathbf{p}_{n+1} = \mathbf{r}_{n+1} + \mathbf{p}_n \beta_n \; .
\end{aligned} \tag{2.19}$$

Then if $\mathbf{r}_n \neq \mathbf{0}$:

1. *the span of the gradients is a* ***Krylov subspace***

$$\langle \mathbf{r}_0, \ldots, \mathbf{r}_n \rangle \equiv \left\{ \sum_{i=0}^{n} \mathbf{r}_i \gamma_i : \gamma_i \textit{ a scalar} \right\} = \mathscr{K}_{n+1} \equiv \left\langle \mathbf{A}^0 \mathbf{r}_0, \ldots, \mathbf{A}^n \mathbf{r}_0 \right\rangle \; ;$$

2. $\langle \mathbf{p}_0, \ldots, \mathbf{p}_n \rangle = \mathscr{K}_{n+1}$*; and*
3. *the vectors $\mathbf{p}_0, \ldots, \mathbf{p}_n$ are* **A***-conjugate.*

Proof We will prove all three results simultaneously by induction. All three are obvious for $n = 0$. Inductively, we will assume that they are true for all iterations up to and including n, and show that they are true for $n + 1$.

Inductively, we assume that there exist scalars $\varrho_0, \ldots, \varrho_n$ so that

$$\mathbf{r}_n = \sum_{i=0}^{n} \mathbf{A}^i \mathbf{r}_0 \varrho_i \; ,$$

and that there are scalars $\pi_0, \ldots, \pi_n$ so that

$$\mathbf{p}_n = \sum_{i=0}^{n} \mathbf{A}^i \mathbf{r}_0 \pi_i .$$

Next, note that the definition of the residual $\mathbf{r}_n$ and the iterative expression for $\widetilde{\mathbf{x}}_{n+1}$ imply that

$$\begin{aligned}
\mathbf{r}_{n+1} &\equiv \mathbf{b} - \mathbf{A}\widetilde{\mathbf{x}}_{n+1} = \mathbf{b} - \mathbf{A}(\widetilde{\mathbf{x}}_n + \mathbf{p}_n\alpha_n) = (\mathbf{b} - \mathbf{A}\widetilde{\mathbf{x}}_n) - \mathbf{A}\mathbf{p}_n\alpha_n \\
&= \mathbf{r}_n - \mathbf{A}\mathbf{p}_n\alpha_n = \sum_{i=0}^{n} \mathbf{A}^i \mathbf{r}_0 \varrho_i - \mathbf{A}\sum_{i=0}^{n} \mathbf{A}^i \mathbf{r}_0 \pi_i \\
&= \mathbf{A}^0 \mathbf{r}_0 \varrho_0 + \sum_{i=1}^{n} \mathbf{A}^i \mathbf{r}_0 \left[\varrho_i + \pi_{i-1}\right] + \mathbf{A}^{n+1} \mathbf{r}_0 \pi_n \in \mathscr{K}_{n+2} .
\end{aligned}$$

Since the inductive hypotheses guarantee that the hypotheses of the Expanding Subspace Theorem 2.5.3 are valid up to $\widetilde{\mathbf{x}}_{n+1}$, the expanding subspace theorem shows that $\mathbf{r}_{n+1}$ is orthogonal to $\mathbf{p}_0, \ldots, \mathbf{p}_n$. It follows that $\mathbf{r}_{n+1}$ is orthogonal to the Krylov subspace $\mathscr{K}_{n+1}$. Since the induction assumes that $\mathbf{r}_{n+1} \neq 0$, it follows that $\langle \mathbf{r}_0, \ldots, \mathbf{r}_{n+1} \rangle = \mathscr{K}_{n+2}$. This proves the first claim.

Next, we note that the algorithm (2.19) chooses $\mathbf{p}_{n+1} = \mathbf{r}_{n+1} + \mathbf{p}_n\beta_n$. Since the inductive hypothesis implies that $\mathbf{p}_n \in \mathscr{K}_{n+1}$, and since we have just proved that $\mathbf{r}_{n+1} \in \mathscr{K}_{n+2}$, it follows that $\mathbf{p}_{n+1} \in \mathscr{K}_{n+2}$. This proves the second claim.

By the evaluation of $\mathbf{p}_{n+1}$ in the algorithm (2.19) we find that for all $0 \leq i \leq n$

$$\mathbf{p}_i \cdot \mathbf{A}\mathbf{p}_{n+1} = \mathbf{p}_i \cdot \mathbf{A}\left(\mathbf{r}_{n+1} + \mathbf{p}_n\beta_n\right) .$$

When $i = n$, the definition of β_n shows that $\mathbf{p}_n \cdot \mathbf{A}\mathbf{p}_{n+1} = 0$. When $i < n$, the inductive hypothesis shows that $\mathbf{p}_i \cdot \mathbf{A}\mathbf{p}_n = 0$, and the Expanding Subspace Theorem 2.5.3 shows that $\mathbf{p}_i \cdot \mathbf{A}\mathbf{r}_{n+1} = 0$. This proves the final claim.

The following corollary gives us some alternative forms for computing the terms in the conjugate gradient algorithm (2.19).

Corollary 2.5.1 *Suppose that the real $m \times m$ matrix $\mathbf{A}$ is symmetric positive and that $\widetilde{\mathbf{x}}_0$ is a real m-vector. Suppose that we perform the algorithm* (2.19). *Then if* $\mathbf{r}_n \neq 0$,

$$\begin{aligned}
\mathbf{r}_{n+1} &= \mathbf{r}_n - \mathbf{A}\mathbf{p}_n\alpha_n , \\
\alpha_n &= \frac{\mathbf{r}_n \cdot \mathbf{r}_n}{\mathbf{p}_n \cdot \mathbf{A}\mathbf{p}_n} \quad \textit{and} \\
\beta_n &= \frac{\mathbf{r}_{n+1} \cdot \mathbf{r}_{n+1}}{\mathbf{r}_n \cdot \mathbf{r}_n} .
\end{aligned}$$

Proof The proof of the first claim was contained in the proof of the conjugate gradient theorem (2.19). The second claim is obvious for $n = 0$. To prove the second claim inductively for $n > 0$, note that the definition of $\mathbf{p}_n$ in algorithm (2.19) implies that

$$\mathbf{r}_n \cdot \mathbf{p}_n = \mathbf{r}_n \cdot (\mathbf{r}_n + \mathbf{p}_{n-1}\beta_{n-1}) \ .$$

Since the Expanding Subspace Theorem 2.5.3 shows that

$$\mathbf{r}_n \cdot \mathbf{p}_{n-1} = 0 \ ,$$

we see that

$$\alpha_n \equiv \frac{\mathbf{p}_n \cdot \mathbf{r}_n}{\mathbf{p}_n \cdot \mathbf{A}\mathbf{p}_n} = \frac{\mathbf{r}_n \cdot \mathbf{r}_n}{\mathbf{p}_n \cdot \mathbf{A}\mathbf{p}_n} \ .$$

This proves the second claim.

To prove the third claim, note that the Conjugate Gradient Theorem 2.5.4 implies that $\mathbf{r}_n \in \mathscr{K}_{n+1}$, and that the span of the search directions $\langle \mathbf{p}_0, \ldots, \mathbf{p}_n \rangle$ is $\mathscr{K}_{n+1}$. Since the Expanding Subspace Theorem 2.5.3 shows that $\mathbf{r}_{n+1}$ is orthogonal to $\langle \mathbf{p}_0, \ldots, \mathbf{p}_n \rangle$, it follows that

$$\mathbf{r}_{n+1} \cdot \mathbf{r}_n = 0 \ .$$

Since the first claim implies that

$$\mathbf{A}\mathbf{p}_n = (\mathbf{r}_n - \mathbf{r}_{n+1}) / \alpha_n \ ,$$

it follows that

$$\beta_n \equiv -\frac{\mathbf{p}_n \cdot \mathbf{A}\mathbf{r}_{n+1}}{\mathbf{p}_n \cdot \mathbf{A}\mathbf{p}_n} = -\frac{(\mathbf{r}_{n+1} - \mathbf{r}_n) \cdot \mathbf{r}_{n+1}}{\mathbf{p}_n \cdot \mathbf{A}\mathbf{p}_n} \frac{1}{\alpha_n} = \frac{\mathbf{r}_{n+1} \cdot \mathbf{r}_{n+1}}{\mathbf{p}_n \cdot \mathbf{A}\mathbf{p}_n} \frac{\mathbf{p}_n \cdot \mathbf{A}\mathbf{p}_n}{\mathbf{r}_n \cdot \mathbf{r}_n}$$
$$= \frac{\mathbf{r}_{n+1} \cdot \mathbf{r}_{n+1}}{\mathbf{r}_n \cdot \mathbf{r}_n} \ .$$

Several variations of the conjugate gradient algorithm have appeared in the literature. The most efficient and accurate form appears to be the following reported by Reid [148]:

Algorithm 2.5.4 (Reid's Conjugate Gradient)

$$
\begin{aligned}
&\mathbf{r} = \mathbf{b} - \mathbf{A}\widetilde{\mathbf{x}} \\
&\mathbf{p} = \mathbf{r} \\
&\gamma = \mathbf{p} \cdot \mathbf{r} \\
&\text{until convergence do} \\
&\quad \mathbf{z} = \mathbf{A}\mathbf{p} \\
&\quad \alpha = \gamma / \mathbf{p} \cdot \mathbf{z} \\
&\quad \widetilde{\mathbf{x}} = \widetilde{\mathbf{x}} + \mathbf{p}\alpha \\
&\quad \mathbf{r} = \mathbf{r} - \mathbf{z}\alpha \\
&\quad \delta = \mathbf{r} \cdot \mathbf{r} \\
&\quad \mathbf{p} = \mathbf{r} + \mathbf{p}\delta/\gamma \\
&\quad \gamma = \delta
\end{aligned}
$$

This form requires only one matrix-vector multiplication per iteration. The algorithm requires storage for four m-vectors, namely $\widetilde{\mathbf{x}}$, $\mathbf{p}$, $\mathbf{r}$ and $\mathbf{z}$.

The Conjugate Gradient Algorithm 2.5.4 should be terminated whenever one of the following four conditions is satisfied.

1. If $\alpha \leq 0$, then either the residual is zero or $\mathbf{A}$ is not positive.
2. If $\|\mathbf{p}\|_\infty \alpha$ is small compared to $\|\mathbf{x}\|_\infty$, then the conjugate gradient algorithm will make little change in its approximation to the solution.
3. If $\|\mathbf{r}\|_\infty$ is small compared to $\|\mathbf{b}\|_\infty$ (provided that $\mathbf{A}$ is reasonably well scaled), then the standard *a posteriori* error estimate (3.30) in Chap. 3 of Volume I, namely

$$
\frac{\|\widetilde{\mathbf{x}} - \mathbf{x}\|}{\|\mathbf{x}\|} \leq \kappa\left(\mathbf{A}\right) \frac{\|\mathbf{A}\widetilde{\mathbf{x}} - \mathbf{b}\|}{\|\mathbf{b}\|} ,
$$

   indicates that the relative error in the solution is as small as the conditioning the system will allow.
4. If more than m iterations have been performed, because the conjugate gradient algorithm should have converged to the exact solution in at most m iterations with exact arithmetic.

Next, let us examine some estimates for the convergence of the conjugate gradient iteration.

Theorem 2.5.5 *Suppose that the $m \times m$ matrix $\mathbf{A}$ is Hermitian and positive, $\mathbf{b}$ is an m-vector, and given the m-vector $\widetilde{\mathbf{x}}_0$ the vectors $\widetilde{\mathbf{x}}_n$ for $1 \leq n \leq m$ are computed by the conjugate gradient iteration* (2.19). *Then*

$$
\|\widetilde{\mathbf{x}}_{n+1} - \mathbf{x}\|_{\mathbf{A}}^2 = \min_{\substack{\text{polynomials } q \text{ of} \\ \text{degree at most } n}} \|[\mathbf{I} + \mathbf{A}q(\mathbf{A})]\left(\widetilde{\mathbf{x}}_0 - \mathbf{x}\right)\|_{\mathbf{A}}^2 .
$$

Proof The conjugate gradient algorithm (2.19) shows that $\widetilde{\mathbf{x}}_{n+1} - \widetilde{\mathbf{x}}_0$ is in the span of the vectors $\mathbf{p}_0, \ldots, \mathbf{p}_n$, and the Conjugate Gradient Theorem 2.5.4 shows that the span of the vectors $\mathbf{p}_0, \ldots, \mathbf{p}_n$ is $\mathscr{K}_{n+1}$. Thus $\widetilde{\mathbf{x}}_{n+1} = \widetilde{\mathbf{x}}_0 + q_n(\mathbf{A})\mathbf{r}_0$ for some polynomial q_n of degree at most n. This in turn implies that

$$\widetilde{\mathbf{x}}_{n+1} - \mathbf{x} = \widetilde{\mathbf{x}}_0 - \mathbf{x} + q_n(\mathbf{A})\mathbf{A}(\widetilde{\mathbf{x}}_0 - \mathbf{x}) = [\mathbf{I} + q_n(\mathbf{A})\mathbf{A}]\,(\widetilde{\mathbf{x}}_0 - \mathbf{x}) \ .$$

This result implies that

$$\|\widetilde{\mathbf{x}}_{n+1} - \mathbf{x}\|_{\mathbf{A}}^2 = \|[\mathbf{I} + q_n(\mathbf{A})\mathbf{A}]\,(\widetilde{\mathbf{x}}_0 - \mathbf{x})\|_{\mathbf{A}}^2 \ .$$

Next, note that the Expanding Subspace Theorem 2.5.3 implies that $\widetilde{\mathbf{x}}_{n+1}$ minimizes $\|\widetilde{\mathbf{x}} - \mathbf{x}\|_{\mathbf{A}}^2$ over all $\widetilde{\mathbf{x}} - \mathbf{x}_0 \in \langle \mathbf{p}_0, \ldots, \mathbf{p}_n \rangle$. The conjugate gradient theorem implies that $\widetilde{\mathbf{x}}_{n+1}$ minimizes $\|\widetilde{\mathbf{x}} - \mathbf{x}\|_{\mathbf{A}}^2$ over all $\widetilde{\mathbf{x}} \in \widetilde{\mathbf{x}}_0 + \mathscr{K}_{n+1}$, where $\mathscr{K}_{n+1}$ is the Krylov subspace $\mathscr{K}_n = \langle \mathbf{A}^0\mathbf{r}_0, \ldots, \mathbf{A}^n\mathbf{r}_0 \rangle$. Equivalently, this says that $\widetilde{\mathbf{x}}_{n+1}$ minimizes $\|\widetilde{\mathbf{x}} - \mathbf{x}\|_{\mathbf{A}}^2$ over all $\widetilde{\mathbf{x}} - \mathbf{x} = [\mathbf{I} + q(\mathbf{A})\mathbf{A}]\,(\widetilde{\mathbf{x}}_0 - \mathbf{x})$ for some polynomial q of degree at most n.

Corollary 2.5.2 *Suppose that the $m \times m$ matrix $\mathbf{A}$ is Hermitian and positive, and that given a m-vector $\widetilde{\mathbf{x}}_0$ the m-vectors $\widetilde{\mathbf{x}}_n$ for $1 \leq n \leq m$ are computed by the conjugate gradient iteration* (2.19). *Then*

$$\|\widetilde{\mathbf{x}}_n - \mathbf{x}\|_{\mathbf{A}} \leq \|\widetilde{\mathbf{x}}_0 - \mathbf{x}\|_{\mathbf{A}} \min_{\substack{\text{polynomials } q \text{ of} \\ \text{degree } \leq n \text{ with } q(0)=1}} \ \max_{\substack{\lambda \text{ an eigen-} \\ \text{value of } \mathbf{A}}} |q(\lambda)| \ .$$

Proof Since $\mathbf{A}$ is Hermitian and positive, the spectral theorem implies that there is an orthogonal matrix $\mathbf{Q}$ and a positive diagonal matrix $\boldsymbol{\Lambda}$ such that $\mathbf{AQ} = \mathbf{Q}\boldsymbol{\Lambda}$. Let us define $\mathbf{y}_0 = \mathbf{Q}^H(\widetilde{\mathbf{x}}_0 - \mathbf{x})$ and $q(\lambda) = 1 + q_{n-1}(\lambda)$. Then

$$\begin{aligned}\|\widetilde{\mathbf{x}}_0 - \mathbf{x}\|_{\mathbf{A}}^2 &= (\widetilde{\mathbf{x}}_0 - \mathbf{x})^H\mathbf{A}\,(\widetilde{\mathbf{x}}_0 - \mathbf{x}) = (\widetilde{\mathbf{x}}_0 - \mathbf{x})^H\mathbf{Q}\boldsymbol{\Lambda}\mathbf{Q}^H\,(\widetilde{\mathbf{x}}_0 - \mathbf{x}) \\ &= (\mathbf{y}_0)^H\boldsymbol{\Lambda}\mathbf{y}_0\end{aligned}$$

and

$$\begin{aligned}\|\widetilde{\mathbf{x}}_n - \mathbf{x}\|_{\mathbf{A}}^2 &= (\widetilde{\mathbf{x}}_0 - \mathbf{x})^H\mathbf{A}\,[\mathbf{I} + q_{n-1}(\mathbf{A})\mathbf{A}]^2\,(\widetilde{\mathbf{x}}_0 - \mathbf{x}) \\ &= (\mathbf{y}_0)^H\boldsymbol{\Lambda}\,[\mathbf{I} + q_{n-1}(\boldsymbol{\Lambda})\boldsymbol{\Lambda}]^2\,\mathbf{y}_0 = \sum_{i=1}^m q(\lambda_i)^2\lambda_i\eta_i^2 \leq \max_i q(\lambda_i)^2 \sum_{i=1}^m \lambda_i\eta_i^2 \\ &= \left[\max_i |q(\lambda_i)|\right]^2 \|\widetilde{\mathbf{x}}_0 - \mathbf{x}\|_{\mathbf{A}}^2 \ .\end{aligned}$$

Note that $q(\lambda) = 1 + q_{n-1}(\lambda)\lambda$ is a polynomial of degree at most n, and that $q(0) = 1$. The previous Theorem 2.5.5 verifies the use of the minimum in the conclusion of this corollary.

This corollary implies that if the eigenvalues of **A** occur in some number $n < m$ of tight clusters, then the iteration will converge in at most n iterations. This is true because we could let q be the polynomial with those cluster values as its zeros, and use the previous corollary to see that the error in conjugate gradients is nearly zero after n iterations.

Axelsson [7, Chapter 13] proves several additional estimates for the convergence rate of conjugate gradients. One way to estimate the bound in Corollary 2.5.2 is to assume that the eigenvalues λ of **A** satisfy $0 < a \leq \lambda \leq b$. Then

$$\min_{\substack{\text{polynomials } q \text{ of} \\ \text{degree } \leq n \text{ with } q(0)=1}} \left\{ \max_{\lambda \text{ an eigenvalue of } \mathbf{A}} |q(\lambda)| \right\} \leq \min_{\substack{\text{polynomials } q \text{ of} \\ \text{degree } \leq n \text{ with } q(0)=1}} \left\{ \max_{a \leq \lambda \leq b} |q(\lambda)| \right\}$$

$$= \frac{1}{T_n\left(\frac{b-a}{b+a}\right)} = \frac{2\sigma^n}{1+\sigma^{2n}} . \tag{2.20}$$

Here T_n is the **Chebyshev polynomial** of degree n and

$$\sigma \equiv \frac{\sqrt{b/a}-1}{\sqrt{b/a}+1} .$$

Recall that Chebyshev polynomials were discussed in the exercises of Sect. 1.3.4; these polynomials will also be discussed in Sect. 1.2.6 and 1.7.3 in Chap. 1 of Volume III. However, the estimate (2.20) is pessimistic in practice.

Axelsson [7, p. 591] summarizes his discussion of the convergence of conjugate gradients by claiming that there are three phases in the convergence. There is an initial sublinearly convergent phase, in which

$$\|\mathbf{x}_n - \mathbf{x}\|_{\mathbf{A}} = \|\mathbf{r}_n\|_{\mathbf{A}^{-1}} = O\left(\left[\frac{1}{n+1}\right]^2\right) .$$

Next, there is an intermediate linearly convergent phase, in which

$$\|\mathbf{x}_n - \mathbf{x}\|_{\mathbf{A}} = \|\mathbf{r}_n\|_{\mathbf{A}^{-1}} = O(\sigma^n) .$$

Finally, there is typically a superlinearly convergent phase, which may only be seen if the convergence tolerances are very small and the condition number of the matrix is large.

A program to perform conjugate gradients can be found in IterativeError2D.C. This program calls Fortran subroutine `matrix_multiply` in iterative_improvement_2d.f. Note that this Fortran routine looks much like a finite difference computation on a grid, rather than like a matrix-vector multiplication. The remainder of the conjugate gradient algorithm in procedure `runOnce` of file `IterativeError2D.C` uses LAPACK BLAS routines, or loops that ignore the grid structure. In this case, readers should select conjugate gradients and de-select

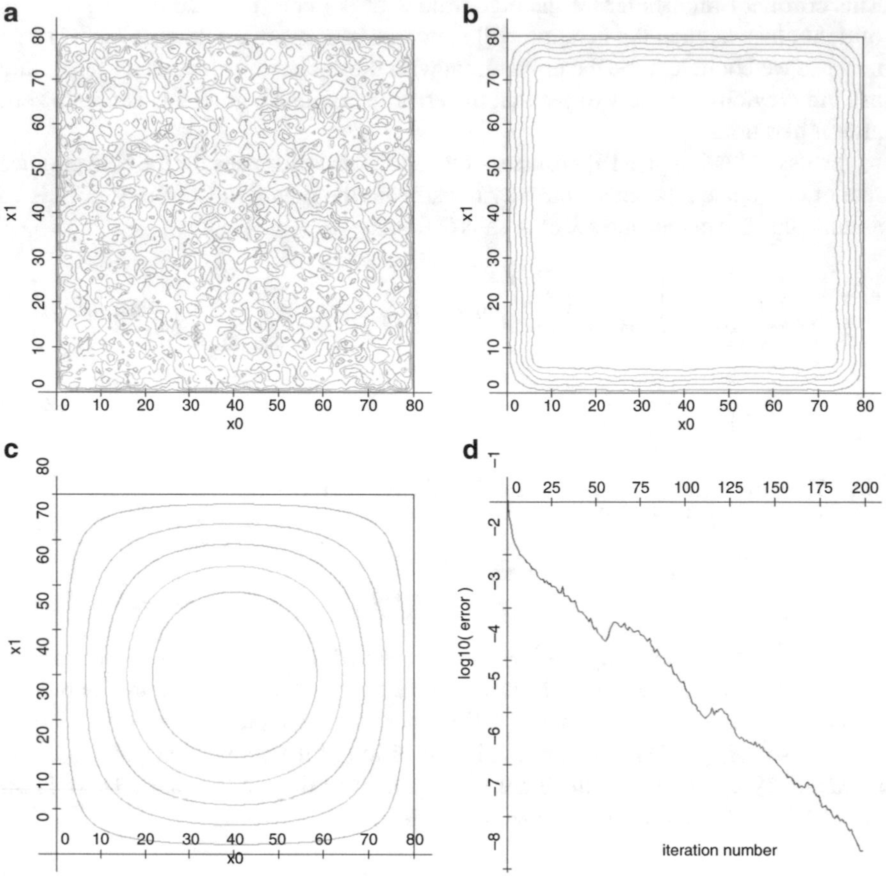

Fig. 2.12 Computed solution with conjugate gradient iteration, 80 × 80 grid cells, random initial guess. (a) 1 iteration. (b) 10 iterations. (c) 100 iterations. (d) error

the other iterative methods; readers should also select the "identity" preconditioner for conjugate gradients. The executable selects random values between zero and one for the initial solution values on the grid, and contours the numerical solution for each iteration. By setting the number of grid cells in one coordinate direction to zero, the user can perform a mesh refinement study, and compare the performance of several iterative methods. Some computational results for solving the Laplace equation with conjugate gradients are shown in Fig. 2.12.

Exercise 2.5.1 For the matrix and vector

$$\mathbf{A} = \begin{bmatrix} 2 & -1 & 0 & 0 \\ -1 & 2 & -1 & 0 \\ 0 & -1 & 2 & -1 \\ 0 & 0 & -1 & 2 \end{bmatrix} \qquad \mathbf{b} = \begin{bmatrix} 1 \\ 0 \\ 0 \\ 0 \end{bmatrix}$$

compute the conjugate gradient iterates beginning with the initial guess $\mathbf{x}_0 = \mathbf{0}$.

Exercise 2.5.2 Repeat the previous exercise with

$$\mathbf{A} = \begin{bmatrix} 1 & -1 & 0 & 0 \\ -1 & 2 & -1 & 0 \\ 0 & -1 & 2 & -1 \\ 0 & 0 & -1 & 1 \end{bmatrix} .$$

Exercise 2.5.3 Get a copy of matrix **BCSSTK14** from MatrixMarket. This is a sparse, symmetric positive matrix that is not diagonally-dominant, and should be represented in computer memory without storing the zero entries. If this matrix is denoted by $\mathbf{A}$, let $\mathbf{x}$ be the vector of ones, and $\mathbf{b} = \mathbf{A}\mathbf{x}$. Choose the initial guess $\widetilde{\mathbf{x}}$ for the Gauss-Seidel iteration to have random entries chosen from a uniform probability distribution on $(0, 1)$. Program the conjugate gradient method to solve $\mathbf{A}\mathbf{x} = \mathbf{b}$, and plot the logarithm base 10 of the error $\|\widetilde{\mathbf{x}} - \mathbf{x}\|_\infty$ versus iteration number. At what order of error does the iteration stop making progress? Relate this error size to the condition number of this matrix, which is estimated to be on the order of 10^{10}. Describe how you would choose convergence criteria for this iteration.

2.5.4 Preconditioned Conjugate Gradients

Theorem 2.5.5 indicates that the convergence of conjugate gradients should be improved if the eigenvalues of the matrix $\mathbf{A}$ are clustered in some way. In practice, this approximate clustering can be accomplished via the following approach. We can transform the system of equations $\mathbf{A}\mathbf{x} = \mathbf{b}$ to an equivalent system

$$\widehat{\mathbf{A}}\widehat{\mathbf{x}} \equiv \left(\mathbf{L}^{-1}\mathbf{A}\mathbf{L}^{-H}\right)\left(\mathbf{L}^{H}\mathbf{x}\right) = \mathbf{L}^{-1}\mathbf{b} \equiv \widehat{\mathbf{b}} ,$$

where $\mathbf{L}$ is some nonsingular matrix. In general, it is too expensive to find matrices $\mathbf{L}$ so that $\mathbf{L}^{-1}\mathbf{A}\mathbf{L}^{-H} \approx \mathbf{I}$; instead, we will look for matrices $\mathbf{L}$ so that systems involving $\mathbf{L}$ are easy to solve, and the eigenvalues of $\widehat{\mathbf{A}} = \mathbf{L}^{-1}\mathbf{A}\mathbf{L}^{-H}$ are more tightly clustered than the eigenvalues of $\mathbf{A}$. Also, it would help if the condition numbers satisfied $\kappa\left(\widehat{\mathbf{A}}\right) \ll \kappa\left(\mathbf{A}\right)$.

The matrix $\mathbf{Q} = \mathbf{L}\mathbf{L}^H$ will be called a **preconditioner** for $\mathbf{A}$ if it is used to improve the convergence of some basic iterative algorithm, such as conjugate gradients. Let us see how we could use a preconditioner in conjugate gradients. If we apply the preferred form (2.5.4) of the conjugate gradient algorithm to $\widehat{\mathbf{A}}\widehat{\mathbf{x}} = \widehat{\mathbf{b}}$, we obtain the following

Algorithm 2.5.5 (Preconditioned Conjugate Gradients)

CG for $\widehat{\mathbf{A}}$	correspondence	Preconditioned CG
$\widehat{\mathbf{r}} = \widehat{\mathbf{b}} - \widehat{\mathbf{A}}\widehat{\mathbf{x}}$	$= \mathbf{L}^{-1}\mathbf{r}$	$\mathbf{r} = \mathbf{b} - \mathbf{A}\mathbf{x}$
$\widehat{\mathbf{p}} = \widehat{\mathbf{r}}$	$= \mathbf{L}^{-1}\mathbf{r} = \mathbf{L}^H\mathbf{Q}^{-1}\mathbf{r} = \mathbf{L}^H\widetilde{\mathbf{p}}$	solve $\mathbf{Q}\widetilde{\mathbf{p}} = \mathbf{r}$
$\widehat{\gamma} = \widehat{\mathbf{p}}\cdot\widehat{\mathbf{r}}$	$= \widehat{\mathbf{p}}\cdot\widehat{\mathbf{p}} = \widetilde{\mathbf{p}}\cdot\mathbf{r} =$	$\widetilde{\gamma} = \widetilde{\mathbf{p}}\cdot\mathbf{r}$
until convergence do		until convergence do
$\widehat{\mathbf{z}} = \widehat{\mathbf{A}}\widehat{\mathbf{p}}$	$= \mathbf{L}^{-1}\mathbf{A}\widetilde{\mathbf{p}} = \mathbf{L}^{-1}\widetilde{\mathbf{z}}$	$\widetilde{\mathbf{z}} = \mathbf{A}\widetilde{\mathbf{p}}$
$\widehat{\alpha} = \widehat{\gamma}/\widehat{\mathbf{p}}\cdot\widehat{\mathbf{z}}$	$= \widehat{\mathbf{p}}\cdot\widehat{\mathbf{p}}/\widehat{\mathbf{p}}\cdot\widehat{\mathbf{A}}\widehat{\mathbf{p}} =$	$\widetilde{\alpha} = \widetilde{\gamma}/\widetilde{\mathbf{p}}\cdot\widetilde{\mathbf{z}}$
$\widehat{\mathbf{x}} = \widehat{\mathbf{x}} + \widehat{\mathbf{p}\alpha}$	$= \mathbf{L}^H\left(\widetilde{\mathbf{x}} + \widetilde{\mathbf{p}\alpha}\right)$	$\widetilde{\mathbf{x}} = \widetilde{\mathbf{x}} + \widetilde{\mathbf{p}\alpha}$
$\widehat{\mathbf{r}} = \widehat{\mathbf{r}} - \widehat{\mathbf{z}\alpha}$	$= \mathbf{L}^{-1}\left(\mathbf{r} - \widetilde{\mathbf{z}\alpha}\right)$	$\mathbf{r} = \mathbf{r} - \widetilde{\mathbf{z}\alpha}$
	$\mathbf{L}^{-H}\widetilde{\mathbf{r}} = \widetilde{\mathbf{y}}$	solve $\mathbf{Q}\widetilde{\mathbf{y}} = \mathbf{r}$
$\widehat{\delta} = \widehat{\mathbf{r}}\cdot\widehat{\mathbf{r}}$	$= \left(\mathbf{L}^{-1}\mathbf{r}\right)\cdot\left(\mathbf{L}^{-1}\mathbf{r}\right) = \mathbf{r}\cdot\mathbf{Q}^{-1}\mathbf{r} =$	$\widetilde{\delta} = \widehat{\mathbf{y}}\cdot\mathbf{r}$
$\widehat{\mathbf{p}} = \widehat{\mathbf{y}} + \widehat{\mathbf{p}\delta/\gamma}$	$= \mathbf{L}^H\left(\widetilde{\mathbf{y}} + \widetilde{\mathbf{p}\delta/\gamma}\right)$	$\widetilde{\mathbf{p}} = \widetilde{\mathbf{y}} + \widetilde{\mathbf{p}\delta/\gamma}$
$\widehat{\gamma} = \widehat{\delta}$	$=$	$\widetilde{\gamma} = \widetilde{\delta}$

Of course, the first column of this algorithm computes a solution $\widetilde{\mathbf{x}}$ to $\widetilde{\mathbf{A}\mathbf{x}} = \widetilde{\mathbf{b}}$, not the solution to $\mathbf{A}\mathbf{x} = \mathbf{b}$. We can compute many of the same quantities in the preconditioned algorithm, and the solution to the original linear system, by the algorithm in the third column.

Note that each iteration of the preconditioned algorithm involves one matrix-vector multiplication by the original matrix $\mathbf{A}$, and solution of one linear system with the preconditioner $\mathbf{Q}$. This algorithm also requires storage of five m-vectors, namely $\widetilde{\mathbf{p}}$, $\widetilde{\mathbf{r}}$, $\widetilde{\mathbf{x}}$, $\widetilde{\mathbf{y}}$ and $\widetilde{\mathbf{z}}$. In this preconditioned form of the algorithm, it does not matter if the preconditioner $\mathbf{Q}$ is factored as $\mathbf{Q} = \mathbf{L}\mathbf{L}^H$. It is important, however, that $\mathbf{Q}$ be symmetric and positive, and allow fast solutions of its linear systems.

The preconditioned conjugate gradient algorithm minimizes

$$\widetilde{\mathbf{r}}^H\widetilde{\mathbf{A}}^{-1}\widetilde{\mathbf{r}} = \mathbf{r}^H\mathbf{L}^{-H}\left(\mathbf{L}^{-1}\mathbf{A}\mathbf{L}^{-H}\right)^{-1}\mathbf{L}^{-1}\mathbf{r} = \mathbf{r}^H\mathbf{A}^{-1}\mathbf{r}\,,$$

which is the same objective function minimized by the (unpreconditioned) conjugate gradient algorithm. Here $\mathbf{r} = \mathbf{b} - \mathbf{A}\mathbf{x}$ and $\widetilde{\mathbf{r}} = \widetilde{\mathbf{b}} - \widetilde{\mathbf{A}\mathbf{x}} = \mathbf{L}^{-1}\mathbf{r}$. However, the step directions now lie in the preconditioned Krylov subspace $\left\langle\left(\mathbf{A}\mathbf{Q}^{-1}\right)^0\mathbf{r}_0,\ldots,\left(\mathbf{A}\mathbf{Q}^{-1}\right)^n\mathbf{r}_0\right\rangle$. Thus the preconditioned conjugate gradient iteration will find that the error at the n iteration is bounded by $2\widetilde{\sigma}^n/(1+\widetilde{\sigma}^{2k})$ where

$$\widetilde{\sigma} = \frac{\sqrt{\kappa\left(\mathbf{AQ}^{-1}\right)}-1}{\sqrt{\kappa\left(\mathbf{AQ}^{-1}\right)}+1} .$$

If the preconditioner $\mathbf{Q}$ is such that $\kappa\left(\mathbf{AQ}^{-1}\right) \ll \kappa\left(\mathbf{A}\right)$, then preconditioning will substantially reduce the number of iterations.

Golub and van Loan [84, p. 653] and Saad [154, p. 265ff] suggest several candidates for preconditioners. A very simple preconditioner is $\mathbf{Q} = \mathrm{diag}(\mathbf{A})$, the diagonal of $\mathbf{A}$; this corresponds to the **Jacobi iteration.** If $\mathbf{A}$ has block tridiagonal form

$$\mathbf{A} = \begin{bmatrix} \mathbf{A}_1 & \mathbf{B}_1 & \dots & \mathbf{0} \\ \mathbf{B}_1{}^H & \mathbf{A}_2 & \ddots & \vdots \\ \vdots & \ddots & \ddots & \mathbf{B}_{m_2-1} \\ \mathbf{0} & \dots & \mathbf{B}_{m_2-1}{}^H & \mathbf{A}_{m_2} \end{bmatrix} ,$$

where the diagonal blocks $\mathbf{A}_1, \dots, \mathbf{A}_{m_2}$ are banded, then the **block Jacobi preconditioner** takes

$$\mathbf{Q} = \begin{bmatrix} \mathbf{A}_1 & \mathbf{0} & \dots & \mathbf{0} \\ \mathbf{0} & \mathbf{A}_2 & \ddots & \vdots \\ \vdots & \ddots & \ddots & \mathbf{0} \\ \mathbf{0} & \dots & \mathbf{0} & \mathbf{A}_{m_2} \end{bmatrix} .$$

Also, if we write $\mathbf{A} = \mathbf{D} + \mathbf{U} + \mathbf{U}^H$ where $\mathbf{D}$ is diagonal and $\mathbf{U}$ is strictly upper triangular, we can let $\mathbf{Q}$ be given by the **symmetric successive over-relaxation (SSOR)** iteration

$$\mathbf{Q} = \left(\mathbf{D} + \mathbf{U}^H\omega\right)\mathbf{D}^{-1}\left(\mathbf{D} + \mathbf{U}\omega\right) .$$

Often, a multigrid iteration will give a useful preconditioner; see the Multigrid Convergence Theorem 2.7.1.

A C^{++} implementation of preconditioned conjugate gradients can be found in cg.h. The corresponding Fortran 77 implementation of preconditioned conjugate gradients can be found in files CG.f and CGREVCOM.f at `netlib.org`.

A program to perform preconditioned conjugate gradients for the solution of the 2D Laplace equation can be found in IterativeError2D.C. This program calls Fortran subroutine `matrix_multiply` in iterative_improvement_2d.f. Note that this Fortran routine looks much like a finite difference computation on a grid, rather than like a matrix-vector multiplication. The remainder of the conjugate gradient algorithm in procedure `runOnce` of file `IterativeError2D.C` uses

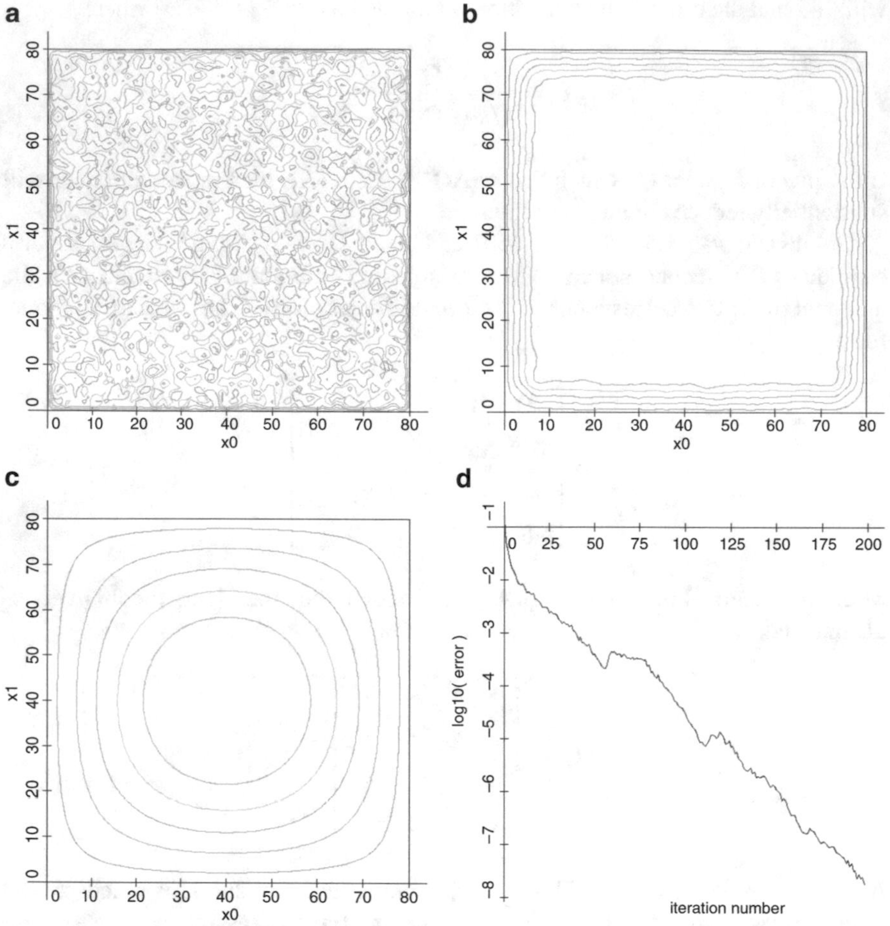

Fig. 2.13 Computed solution with conjugate gradient iteration preconditioned by block Jacobi, 80×80 grid cells, random initial guess. (a) 1 iteration. (b) 10 iterations. (c) 100 iterations. (d) error

LAPACK BLAS routines, or loops that ignore the grid structure. Readers can select several preconditioners for conjugate gradients, including no preconditioner (i.e., the identity matrix for the preconditioner). The executable selects random values between zero and one for the initial solution values on the grid, and contours the numerical solution for each iteration. By setting the number of grid cells in one coordinate direction to zero, the user can perform a mesh refinement study, and compare the performance of several iterative methods.

Some computational results with conjugate gradients preconditioned by block Jacobi iteration are shown in Fig. 2.13.

Exercise 2.5.4 Read about incomplete Cholesky factorization in [6, 7, 58]. A particularly sophisticated incomplete factorization is due to Jones and Plassman

[101], and is available as TOMS 740 in software at Netlib. Then describe how to use incomplete factorization as a preconditioner. For more details you may also see Golub and van Loan [84, p. 657].

2.5.5 *Biconjugate Gradients*

The Conjugate Gradient Algorithm 2.19 or 2.5.4 is suitable only for Hermitian positive matrices. For general nonsingular linear systems $\mathbf{A}\mathbf{x} = \mathbf{b}$, we could apply conjugate gradients to the Hermitian positive system

$$\mathbf{A}^H\mathbf{A}\mathbf{x} = \mathbf{A}^H\mathbf{b}$$

or the Hermitian positive system

$$\mathbf{A}\mathbf{A}^H\mathbf{y} = \mathbf{b} \text{ with } \mathbf{x} = \mathbf{A}^H\mathbf{y} ,$$

but the condition number of either linear system would be the square of the condition number of the original linear system. It would be useful to develop gradient methods that apply to general nonsingular systems without squaring the condition number.

2.5.5.1 Biorthogonalization

It is possible to generalize the Lanczos process of Sect. 1.3.12 to non-Hermitian matrices with the following

Algorithm 2.5.6 (Lanczos Biorthogonalization)

given $\mathbf{v}_1$ with $\|\mathbf{v}_1\|_2 = 1$ and $\mathbf{u}_1$ with $\mathbf{u}_1 \cdot \mathbf{v}_1 = 1$

for $1 \le k \le m$ {

$\quad \mathbf{z}_k = \mathbf{A}\mathbf{v}_k$

$\quad \delta_k = \mathbf{u}_k \cdot \mathbf{z}_k \qquad (2.21a)$

$\quad$ if $k = 1$ {

$\qquad \widehat{\mathbf{v}}_2 = \mathbf{z}_1 - \mathbf{v}_1\delta_1 \qquad (2.21b)$

$\qquad \widehat{\mathbf{u}}_2 = \mathbf{A}^H\mathbf{u}_1 - \mathbf{u}_1\overline{\delta_1} \qquad (2.21c)$

$\quad$ } else {

$\qquad \widehat{\mathbf{v}}_{k+1} = \mathbf{z}_k - \mathbf{v}_k\delta_k - \mathbf{v}_{k-1}\varrho_{k-1} \qquad (2.21d)$

$\qquad \widehat{\mathbf{u}}_{k+1} = \mathbf{A}^H\mathbf{u}_k - \mathbf{u}_k\overline{\delta_k} - \mathbf{u}_{k-1}\lambda_{k-1} \qquad (2.21e)$

$\quad$ }

$$\lambda_k = \|\widehat{\mathbf{v}}_{k+1}\|_2 \tag{2.21f}$$

if $\lambda_k > 0$ {

$$\mathbf{v}_{k+1} = \widehat{\mathbf{v}}_{k+1}/\lambda_k \tag{2.21g}$$

$$\varrho_k = \widehat{\mathbf{u}}_{k+1} \cdot \mathbf{v}_{k+1} \tag{2.21h}$$

if $\varrho_k = 0$ stop

$$\mathbf{u}_{k+1} = \widehat{\mathbf{u}}_{k+1}/\overline{\varrho k} \tag{2.21i}$$

} else {

$$\nu_{k+1} = \|\widehat{\mathbf{u}}_{k+1}\|_2$$

if $\nu_{k+1} > 0$ {

$$\text{choose } \mathbf{v}_{k+1} \perp \{\mathbf{u}_1, \ldots, \mathbf{u}_k\} \text{ with } \|\mathbf{v}_{k+1}\|_2 = 1 \text{ and } \widehat{\mathbf{u}}_{k+1} \cdot \mathbf{v}_{k+1} > 0 \tag{2.21j}$$

$$\varrho_k = \widehat{\mathbf{u}}_{k+1} \cdot \mathbf{v}_{k+1}$$

$$\mathbf{u}_{k+1} = \widehat{\mathbf{u}}_{k+1}/\varrho k \tag{2.21k}$$

} else {

$$\text{choose } \mathbf{v}_{k+1} \perp \{\mathbf{u}_1, \ldots, \mathbf{u}_k\} \text{ with } \|\mathbf{v}_{k+1}\|_2 = 1 \text{ and} \tag{2.21l}$$

$$\text{choose } \mathbf{u}_{k+1} \perp \{\mathbf{v}_1, \ldots, \mathbf{v}_k\} \text{ with } \mathbf{u}_{k+1} \cdot \mathbf{v}_{k+1} = 1 \tag{2.21m}$$

$$\varrho_k = 0$$

}

}

}

Each iteration of this algorithm requires two matrix-vector multiplications, one by and one by $\mathbf{A}^H$. The latter can be particularly expensive (if the matrix $\mathbf{A}$ is stored on distributed processors in such a way that the computation of $\mathbf{A}^H\mathbf{x}$ requires extra communication between processors) or very inconvenient (if $\mathbf{A}$ is the Jacobian matrix of a function $\mathbf{f}(\mathbf{x})$ and $\mathbf{A}\mathbf{x}$ is being approximated by a difference quotient as in Sect. 3.8.3). The algorithm requires storage of the upper Hessenberg matrix $\mathbf{H}$ and at least six m-vectors (if we are confident that no special steps will be needed), namely $\mathbf{v}_{k-1}$, $\mathbf{v}_k$, $\mathbf{v}_{k+1}$, $\mathbf{u}_{k-1}$, $\mathbf{u}_k$ and $\mathbf{u}_{k+1}$. The vector $\mathbf{z}_k = \mathbf{A}\mathbf{v}_k$ can share storage with $\mathbf{v}_{k+1}$ and $\mathbf{A}^H\mathbf{u}_k$ can share storage with $\mathbf{u}_{k+1}$.

In order to prove some important results about this algorithm, we will make the following useful assumptions, which will form the inductive hypotheses in the proof of Theorem 2.5.6.

Assumption 2.5.2 Assume that we have found sequences $\{\mathbf{v}_j\}_{j=1}^k$, $\{\mathbf{u}_j\}_{j=1}^k$, $\{\delta_j\}_{j=1}^{k-1}$, $\{\lambda_j\}_{j=1}^{k-1}$ and $\{\varrho_j\}_{j=1}^{k-1}$ so that

$$\|\mathbf{v}_j\|_2 = 1 \text{ for all } 1 \le j \le k \tag{2.22a}$$

$$\mathbf{u}_i \cdot \mathbf{v}_j = \begin{cases} 0, \ i \neq j \\ 1, \ i = j \end{cases} \text{ for all } 1 \le i, j \le k \tag{2.22b}$$

$$\mathbf{A}\mathbf{v}_j = \begin{cases} \mathbf{v}_1\delta_1 + \mathbf{v}_2\lambda_1, & 1 = j < k \\ \mathbf{v}_{j-1}\varrho_{j-1} + \mathbf{v}_j\delta_j + \mathbf{v}_{j+1}\lambda_j, & 1 < j < k \end{cases} \text{ and} \tag{2.22c}$$

$$\mathbf{A}^H\mathbf{u}_j = \begin{cases} \mathbf{u}_1\overline{\delta_1} + \mathbf{u}_2\overline{\varrho_1}, & 1 = j < k \\ \mathbf{u}_{j-1}\lambda_{j-1} + \mathbf{u}_j\overline{\delta_j} + \mathbf{u}_{j+1}\overline{\varrho_j}, & 1 < j < k \end{cases} . \tag{2.22d}$$

We also assume that $\lambda_j \ge 0$ for $1 \le j < k$.
Here is our first result.

Lemma 2.5.5 *Suppose that Assumptions 2.5.2 are satisfied. If* $\widehat{\mathbf{v}}_{k+1}$ *is defined by* (2.21b) *for* $k = 1$, *or by* (2.21d) *for* $k > 1$, *then we have*

$$\mathbf{u}_j \cdot \widehat{\mathbf{v}}_{k+1} = 0 \textit{ for } 1 \le j \le k \ . \tag{2.23}$$

Also, if $\widehat{\mathbf{u}}_{k+1}$ *is defined by* (2.21c) *for* $k = 1$, *or by* (2.21e) *for* $k > 1$, *then we have*

$$\widehat{\mathbf{u}}_{k+1} \cdot \mathbf{v}_j = 0 \textit{ for } 1 \le j \le k \ . \tag{2.24}$$

Proof Note that for $i \le j \le k$, Eq. (2.21b) or (2.21d) implies that

$$\mathbf{u}_j \cdot \widehat{\mathbf{v}}_{k+1} = \begin{cases} \mathbf{u}_j \cdot (\mathbf{A}\mathbf{v}_1 - \mathbf{v}_1\delta_1), & k = 1 \\ \mathbf{u}_j \cdot (\mathbf{A}\mathbf{v}_k - \mathbf{v}_k\delta_k - \mathbf{v}_{k-1}\varrho_{k-1}), & k > 1 \end{cases}$$

then the inductive assumption (2.22b) on $\mathbf{u}_i \cdot \mathbf{v}_j$ produces

$$= \begin{cases} \mathbf{u}_j \cdot \mathbf{A}\mathbf{v}_k, & 1 \le j < k-1 \\ \mathbf{u}_{k-1} \cdot \mathbf{A}\mathbf{v}_k - \varrho_{k-1}, & 1 \le j = k-1 \\ \mathbf{u}_k \cdot \mathbf{A}\mathbf{v}_k - \delta_k, & 1 \le j = k \end{cases}$$

afterward, the inductive assumption (2.22d) on $\mathbf{A}^H\mathbf{u}_j$ yields

$$= \begin{cases} (\mathbf{u}_1\overline{\delta_1} + \mathbf{u}_2\overline{\varrho_1}) \cdot \mathbf{v}_k, & 1 = j < k-1 \\ \left(\mathbf{u}_{j-1}\lambda_{j-1} + \mathbf{u}_j\overline{\delta_j} + \mathbf{u}_{j+1}\overline{\varrho_j}\right) \cdot \mathbf{v}_k, & 1 < j < k-1 \\ \left(\mathbf{u}_1\overline{\delta_1} + \mathbf{u}_2\overline{\varrho_1}\right) \cdot \mathbf{v}_2 - \varrho_1, & 1 = j = k-1 \\ \left(\mathbf{u}_{k-2}\lambda_{k-2} + \mathbf{u}_{k-1}\overline{\delta_{k-1}} + \mathbf{u}_k\overline{\varrho_{k-1}}\right) \cdot \mathbf{v}_k - \varrho_{k-1}, & 1 < j = k-1 \\ \delta_k - \delta_k, & 1 \le j = k \end{cases}$$

and finally the inductive assumption (2.22b) on $\mathbf{u}_i \cdot \mathbf{v}_j$ produces

$$= 0 .$$

Similarly, for $i \le j \le k$ Eq. (2.21c) or (2.21e) gives us

$$\widehat{\mathbf{u}}_{k+1} \cdot \mathbf{v}_j = \begin{cases} \left(\mathbf{A}^H \mathbf{u}_1 - \mathbf{u}_1 \overline{\delta_1}\right) \cdot \mathbf{v}_j, & k = 1 \\ \left(\mathbf{A}^H \mathbf{u}_k - \mathbf{u}_k \overline{\delta_k} - \mathbf{u}_{k-1} \lambda_{k-1}\right) \cdot \mathbf{v}_j, & k > 1 \end{cases}$$

then the inductive assumption (2.22b) on $\mathbf{u}_i \cdot \mathbf{v}_j$ produces

$$= \begin{cases} \mathbf{u}_k \cdot \mathbf{A} \mathbf{v}_j, & 1 \le j < k-1 \\ \mathbf{u}_k \cdot \mathbf{A} \mathbf{v}_{k-1} - \lambda_{k-1}, & 1 \le j = k-1 \\ \mathbf{u}_k \cdot \mathbf{A} \mathbf{v}_k - \delta_k, & 1 \le j = k \end{cases}$$

in addition, the inductive assumption (2.22c) on $\mathbf{A}\mathbf{v}_j$ yields

$$= \begin{cases} \mathbf{u}_k \cdot (\mathbf{v}_1 \delta_1 + \mathbf{v}_2 \lambda_1), & 1 = j < k-1 \\ \mathbf{u}_k \cdot \left(\mathbf{v}_{j-1} \varrho_{j-1} + \mathbf{v}_j \delta_j + \mathbf{v}_{j+1} \lambda_j\right), & 1 < j < k-1 \\ \mathbf{u}_2 \cdot (\mathbf{v}_1 \delta_1 + \mathbf{v}_2 \lambda_1) - \lambda_1, & 1 = j = k-1 \\ \mathbf{u}_k \cdot (\mathbf{v}_{k-2} \varrho_{k-2} + \mathbf{v}_{k-1} \delta_{k-1} + \mathbf{v}_k \lambda_{k-1}) - \lambda_{k-1}, & 1 < j = k-1 \\ \delta_k - \delta_k, & 1 \le j = k \end{cases}$$

and finally the inductive assumption (2.22b) on $\mathbf{u}_i \cdot \mathbf{v}_j$ produces

$$= 0 .$$

These results imply that $\widehat{\mathbf{v}}_{k+1} \perp \{\mathbf{u}_j\}_{j=1}^k$ and $\widehat{\mathbf{u}}_{k+1} \perp \{\mathbf{v}_j\}_{j=1}^k$. In particular, if $k = m$ we will see that $\widehat{\mathbf{v}}_{m+1} = \mathbf{0}$ and $\widehat{\mathbf{u}}_{m+1} = \mathbf{0}$, because both of these vectors will be orthogonal to a basis for m-vectors.

Our next lemma deals with the special steps in the Lanczos biorthogonalization algorithm.

Lemma 2.5.6 *Suppose that Assumptions 2.5.2 are satisfied. Then the special choices* (2.21j), (2.21l) *and* (2.21m) *in the Lanczos Biorthogonalization Algorithm 2.5.6 are possible.*

Proof Define and factor

$$\mathbf{V}_1 = \left[\mathbf{v}_1 \ldots \mathbf{v}_k\right] = \mathbf{Q}_1 \mathbf{R}_1$$

where

$$\mathbf{Q} = \left[\mathbf{Q}_1 \; \mathbf{Q}_2\right]$$

is unitary and $\mathbf{R}_1$ is $k \times k$ right-triangular. Also define

$$\mathbf{U}_1 = \left[\mathbf{u}_1 \ldots \mathbf{u}_k\right] = \left[\mathbf{Q}_1 \; \mathbf{Q}_2\right] \begin{bmatrix} \mathbf{X}_1 \\ \mathbf{X}_2 \end{bmatrix} ,$$

where $\mathbf{X}_1$ is $k \times k$. Then inductive assumption (2.22b) implies that

$$\mathbf{I} = \mathbf{U}_1{}^H\mathbf{V}_1 = \left[\mathbf{X}_1{}^H\ \mathbf{X}_2{}^H\right]\begin{bmatrix}\mathbf{Q}_1{}^H \\ \mathbf{Q}_2{}^H\end{bmatrix}\left[\mathbf{Q}_1\ \mathbf{Q}_2\right]\begin{bmatrix}\mathbf{R}_1 \\ \mathbf{0}\end{bmatrix} = \mathbf{X}_1{}^H\mathbf{R}_1\ ,$$

so $\mathbf{R}_1$ is nonsingular and $\mathbf{X}_1 = \mathbf{R}_1{}^{-H}$. We can write

$$\mathbf{v}_{k+1} = \mathbf{Q}\mathbf{b} = \left[\mathbf{Q}_1\ \mathbf{Q}_2\right]\begin{bmatrix}\mathbf{b}_1 \\ \mathbf{b}_2\end{bmatrix} \text{ and } \mathbf{u}_{k+1} = \mathbf{Q}\mathbf{a} = \left[\mathbf{Q}_1\ \mathbf{Q}_2\right]\begin{bmatrix}\mathbf{a}_1 \\ \mathbf{a}_2\end{bmatrix}\ . \tag{2.25}$$

The inductive assumption (2.22b) implies that

$$\mathbf{0} = \mathbf{V}_1{}^H\mathbf{u}_{k+1} = \mathbf{R}_1{}^H\mathbf{Q}_1{}^H\left[\mathbf{Q}_1\ \mathbf{Q}_2\right]\begin{bmatrix}\mathbf{a}_1 \\ \mathbf{a}_2\end{bmatrix} = \mathbf{R}_1{}^H\mathbf{a}_1\ , \tag{2.26}$$

so we must take $\mathbf{a}_1 = \mathbf{0}$. The inductive assumption (2.22b) also requires

$$\mathbf{0} = \mathbf{U}_1{}^H\mathbf{v}_{k+1} = \left[\mathbf{X}_1{}^H\ \mathbf{X}_2{}^H\right]\begin{bmatrix}\mathbf{Q}_1{}^H \\ \mathbf{Q}_2{}^H\end{bmatrix}\left[\mathbf{Q}_1\ \mathbf{Q}_2\right]\begin{bmatrix}\mathbf{b}_1 \\ \mathbf{b}_2\end{bmatrix} = \mathbf{R}_1^{-1}\mathbf{b}_1 + \mathbf{X}_2{}^H\mathbf{b}_2\ , \tag{2.27}$$

which implies that we must take

$$\mathbf{b}_1 = -\mathbf{R}\mathbf{X}_2{}^H\mathbf{b}_2\ .$$

Finally, the inductive assumption (2.22a) implies that we want

$$1 = \|\mathbf{v}_{k+1}\|_2^2 = \|\mathbf{b}_1\|_2^2 + \|\mathbf{b}_2\|_2^2 = \|\mathbf{R}\mathbf{X}_2{}^H\mathbf{b}_2\|_2^2 + \|\mathbf{b}_2\|_2^2 = \mathbf{b}_2{}^H\left(\mathbf{X}_2\mathbf{R}^H\mathbf{R}\mathbf{X}_2{}^H + \mathbf{I}\right)\mathbf{b}_2\ . \tag{2.28}$$

We can perform a Cholesky factorization

$$\mathbf{X}_2\mathbf{R}^H\mathbf{R}\mathbf{X}_2{}^H + \mathbf{I} = \mathbf{L}\mathbf{L}^H \tag{2.29}$$

and solve

$$\mathbf{L}^H\mathbf{b}_2 = \mathbf{c}$$

for any unit vector $\mathbf{c}$. Finally, the inductive hypothesis (2.22b) requires that

$$1 = \mathbf{u}_{k+1}\cdot\mathbf{v}_{k+1} = \mathbf{a}_2{}^H\mathbf{Q}_2{}^H\left[\mathbf{Q}_1\ \mathbf{Q}_2\right]\begin{bmatrix}\mathbf{b}_1 \\ \mathbf{b}_2\end{bmatrix} = \mathbf{a}_2{}^H\mathbf{b}_2\ . \tag{2.30}$$

This can be satisfied by choosing $\mathbf{a}_2 = \mathbf{b}_2/\|\mathbf{b}_2\|_2^2$.

Let us verify that the choice (2.21j) is possible for $\mathbf{v}_{k+1}$ and α_{k+1} when $\lambda_k = 0$ but $\|\widehat{\mathbf{u}}_{k+1}\|_2 > 0$. Since $\widehat{\mathbf{u}}_{k+1}$ is determined, we can compute $\widehat{\mathbf{a}}_1$ and $\widehat{\mathbf{a}}_2$ using the second equation in (2.25). Equations (2.24) and (2.26) imply that $\widehat{\mathbf{a}}_1 = \mathbf{0}$. Since

$$0 < \|\widehat{\mathbf{u}}_{k+1}\|_2^2 = \|\mathbf{a}_1\|_2^2 + \|\mathbf{a}_2\|_2^2 = \|\mathbf{a}_2\|_2^2 ,$$

we conclude that $\mathbf{a}_2 \neq \mathbf{0}$. Using the Cholesky factorization (2.29), we can compute

$$\mathbf{b}_2 = \mathbf{L}^{-H}\mathbf{L}^{-1}\widehat{\mathbf{a}}_2 / \left\|\mathbf{L}^{-1}\widehat{\mathbf{a}}_2\right\|_2 \text{ and } \mathbf{b}_1 = -\mathbf{R}_1\mathbf{X}_2{}^H\mathbf{b}_2$$

and use the first equation in (2.25), to compute $\mathbf{v}_{k+1}$. Equation (2.27) shows that $\mathbf{v}_{k+1} \perp \{\mathbf{u}_1, \ldots, \mathbf{u}_k\}$, and Eq. (2.28) proves that $\|\mathbf{v}_{k+1}\| = 1$. We also note that

$$\varrho_k \equiv \widehat{\mathbf{u}}_{k+1} \cdot \mathbf{v}_{k+1} = \widehat{\mathbf{a}}_2 \cdot \mathbf{b}_2 = \left\|\mathbf{L}^{-1}\widehat{\mathbf{a}}_2\right\|_2 > 0 .$$

Let us verify that the choices (2.21l) and (2.21m) are possible when $\lambda_k = 0$ and $\|\widehat{\mathbf{u}}_{k+1}\|_2 = 0$. Using the Cholesky factorization (2.29), we can compute

$$\mathbf{b}_2 = \mathbf{L}^{-H}\mathbf{e}_1 \text{ and } \mathbf{b}_1 = -\mathbf{R}_1\mathbf{X}_2{}^H\mathbf{b}_2$$

then compute $\mathbf{v}_{k+1}$ by the first equation in (2.25). We can also compute

$$\mathbf{a}_1 = \mathbf{0} \text{ and } \mathbf{a}_2 = \mathbf{b}_2/\|\mathbf{b}_2\|_2^2$$

and compute $\mathbf{u}_{k+1}$ by the second equation in (2.25). Equation (2.26) shows that $\mathbf{u}_{k+1} \perp \{\mathbf{v}_1, \ldots, \mathbf{v}_k\}$, and Eq. (2.27) shows that $\mathbf{v}_{k+1} \perp \{\mathbf{u}_1, \ldots, \mathbf{u}_k\}$. Equations (2.28) and (2.29) show that $\|\mathbf{v}_{k+1}\|_2 = 1$, and Eq. (2.30) verifies that $\mathbf{u}_{k+1} \cdot \mathbf{v}_{k+1} = 1$.

Our next six lemmas will verify the inductive hypotheses for a subsequent step of the Lanczos biorthogonalization algorithm.

Lemma 2.5.7 *The vector $\mathbf{v}_{k+1}$ computed by the Lanczos Biorthogonalization Algorithm 2.5.6 satisfies $\|\mathbf{v}_{k+1}\|_2 = 1$.*

Proof There are two cases. If $\lambda_k > 0$ then Eqs. (2.21g) and (2.21f) imply that

$$\|\mathbf{v}_{k+1}\|_2 = \|\widehat{\mathbf{v}}_{k+1}\|_2 / \lambda_k = \|\widehat{\mathbf{v}}_{k+1}\|_2 / \|\widehat{\mathbf{v}}_{k+1}\|_2 = 1 .$$

Otherwise, the condition $\|\mathbf{v}_{k+1}\|_2 = 1$ is enforced by the special choices (2.21j) or (2.21l).

Lemma 2.5.8 *Suppose that Assumptions 2.5.2 are satisfied. Then the vector $\mathbf{v}_{k+1}$ computed by the Lanczos Biorthogonalization Algorithm 2.5.6 satisfies*

$$\mathbf{u}_j \cdot \mathbf{v}_{k+1} = 0 \text{ for } 1 \leq j \leq k . \tag{2.31}$$

Proof Whenever $\lambda_k = 0$, the special choices (2.21j) or (2.21l) guarantee that $\mathbf{v}_{k+1} \perp \{\mathbf{u}_1, \ldots, \mathbf{u}_k\}$. Otherwise $\lambda_k > 0$, and for $1 \leq j \leq k$ Eq. (2.21g) gives us

$$\mathbf{u}_j \cdot \mathbf{v}_{k+1} = \mathbf{u}_j \cdot \widehat{\mathbf{v}}_{k+1}/\lambda_k$$

then Eq. (2.21b) or (2.21d) produces

$$= \begin{cases} \mathbf{u}_j \cdot (\mathbf{A}\mathbf{v}_1 - \mathbf{v}_1\delta_1)/\lambda_k, & k = 1 \\ \mathbf{u}_j \cdot (\mathbf{A}\mathbf{v}_k - \mathbf{v}_k\delta_k - \mathbf{v}_{k-1}\varrho_{k-1})/\lambda_k, & k > 1 \end{cases}$$

then the inductive hypothesis (2.22b) on $\mathbf{u}_i \cdot \mathbf{v}_j$ implies that

$$= \begin{cases} \mathbf{u}_j \cdot \mathbf{A}\mathbf{v}_k/\lambda_k, & 1 \leq j \leq k-2 \\ (\mathbf{u}_{k-1} \cdot \mathbf{A}\mathbf{v}_k - \varrho_{k-1})/\lambda_k, & j = k-1 \\ (\mathbf{u}_k \cdot \mathbf{A}\mathbf{v}_k - \delta_k)/\lambda_k, & j = k \end{cases}$$

then the inductive hypothesis (2.22d) gives us

$$= \begin{cases} \left(\mathbf{u}_1\overline{\delta_1} + \mathbf{u}_2\overline{\varrho_2}\right) \cdot \mathbf{v}_k/\lambda_k, & 1 = j \leq k-2 \\ \left(\mathbf{u}_{j-1}\lambda_{j-1} + \mathbf{u}_j\overline{\delta_j} + \mathbf{u}_{j+1}\overline{\varrho_j}\right) \cdot \mathbf{v}_k/\lambda_k, & 1 < j \leq k-2 \\ \left(\left[\mathbf{u}_1\overline{\delta_1} + \mathbf{u}_2\overline{\varrho_2}\right] \cdot \mathbf{v}_2 - \varrho_1\right)/\lambda_2, & 1 = j = k-1 \\ \left(\left[\mathbf{u}_{k-2}\lambda_{k-1} + \mathbf{u}_{k-1}\overline{\delta_{k-1}} + \mathbf{u}_k\overline{\varrho_{k-1}}\right] \cdot \mathbf{v}_k - \varrho_{k-1}\right)/\lambda_k, & 1 < j = k-1 \\ (\mathbf{u}_k \cdot \mathbf{A}\mathbf{v}_k - \delta_k)/\lambda_k, & j = k \end{cases}$$

then the inductive hypothesis (2.22b) on $\mathbf{u}_i \cdot \mathbf{v}_j$ implies that

$$= \begin{cases} 0, & 1 \leq j \leq k-2 \\ (\varrho_{k-1} - \varrho_{k-1})/\lambda_k, & 1 \leq j = k-1 \\ (\mathbf{u}_k \cdot \mathbf{A}\mathbf{v}_k - \delta_k)/\lambda_k, & 1 \leq j = k \end{cases}$$

and then we use Eq. (2.21a) to get

$$= \begin{cases} 0, & 1 \leq j \leq k-1 \\ (\delta_k - \delta_k)/\lambda_k, & 1 \leq j = k \end{cases} = 0\,.$$

This proves (2.31) when $\lambda_k > 0$.

Lemma 2.5.9 *Suppose that Assumptions 2.5.2 are satisfied. If the vector* $\mathbf{u}_{k+1}$ *is computed by the Lanczos Biorthogonalization Algorithm 2.5.6, then it satisfies*

$$\mathbf{u}_{k+1} \cdot \mathbf{v}_j = 0 \text{ for } 1 \leq j \leq k\,. \tag{2.32}$$

Proof There are three ways to compute $\mathbf{u}_{k+1}$ in the Lanczos biorthogonalization algorithm. If $\lambda_k = 0$ and $\|\widehat{\mathbf{u}}_{k+1}\|_2 = 0$, the special choice (2.21m) proves the lemma. If $\lambda_k = 0$ and $\|\widehat{\mathbf{u}}_{k+1}\|_2 > 0$, then Eqs. (2.21k) and (2.24) prove the claimed result. All that remains is to consider the case when $\mathbf{u}_{k+1}$ is computed by Eq. (2.21i). This equation implies that for $1 \leq j \leq k$

$$\mathbf{u}_{k+1} \cdot \mathbf{v}_j = \widehat{\mathbf{u}}_{k+1} \cdot \mathbf{v}_j / \overline{\varrho_k}$$

then Eq. (2.21c) or (2.21e) produces

$$= \begin{cases} \left(\mathbf{A}^H \mathbf{u}_1 - \mathbf{u}_1 \overline{\delta_1}\right) \cdot \mathbf{v}_j / \overline{\varrho_k}, & k = 1 \\ \left(\mathbf{A}^H \mathbf{u}_k - \mathbf{u}_k \overline{\delta_k} - \mathbf{u}_{k-1} \lambda_{k-1}\right) \cdot \mathbf{v}_j / \overline{\varrho_k}, & k > 1 \end{cases}$$

then the inductive hypothesis (2.22b) on $\mathbf{u}_i \cdot \mathbf{v}_j$ implies that

$$= \begin{cases} \mathbf{u}_k \cdot \mathbf{A}\mathbf{v}_j / \overline{\varrho_k}, & 1 \leq j \leq k-2 \\ (\mathbf{u}_k \cdot \mathbf{A}\mathbf{v}_{k-1} - \lambda_{k-1}) / \overline{\varrho_k}, & j = k-1 \\ (\mathbf{u}_k \cdot \mathbf{A}\mathbf{v}_k - \delta_k) / \overline{\varrho_k}, & j = k \end{cases}$$

then the inductive hypothesis (2.22c) gives us

$$= \begin{cases} \mathbf{u}_k \cdot (\mathbf{v}_1 \delta_1 + \mathbf{v}_2 \lambda_2) / \overline{\varrho_k}, & 1 = j \leq k-2 \\ \mathbf{u}_k \cdot \left(\mathbf{v}_{j-1} \varrho_j + \mathbf{v}_j \delta_j + \mathbf{v}_{j+1} \lambda_j\right) / \overline{\varrho_k}, & 1 < j \leq k-2 \\ (\mathbf{u}_2 \cdot [\mathbf{v}_1 \delta_1 + \mathbf{v}_2 \lambda_2] - \lambda_1) / \overline{\varrho_2}, & 1 = j = k-1 \\ (\mathbf{u}_k \cdot [\mathbf{v}_{k-2} \varrho_{k-2} + \mathbf{v}_{k-1} \delta_{k-1} + \mathbf{v}_k \lambda_{k-1}] - \lambda_{k-1}) / \overline{\varrho_k}, & 1 < j = k-1 \\ (\mathbf{u}_k \cdot \mathbf{A}\mathbf{v}_k - \delta_k) / \overline{\varrho_k}, & j = k \end{cases}$$

then the inductive hypothesis (2.22b) on $\mathbf{u}_i \cdot \mathbf{v}_j$ implies that

$$= \begin{cases} 0, & 1 \leq j \leq k-2 \\ (\lambda_{k-1} - \lambda_{k-1}) / \overline{\varrho_k}, & 1 \leq j = k-1 \\ (\mathbf{u}_K \cdot \mathbf{A}\mathbf{v}_k - \delta_k) / \overline{\varrho_k}, & 1 \leq j = k \end{cases}$$

and then we use Eq. (2.21a) to get

$$= \begin{cases} 0, & 1 \leq j \leq k-1 \\ (\delta_k - \delta_k) / \overline{\varrho_K}, & 1 \leq j = k \end{cases} = 0 \,.$$

This completes the proof of (2.32).

Lemma 2.5.10 *The vectors* $\mathbf{v}_{k+1}$ *and* $\mathbf{u}_{k+1}$ *computed by the Lanczos Biorthogonalization Algorithm 2.5.6 satisfy*

$$\mathbf{u}_{k+1} \cdot \mathbf{v}_{k+1} = 1 \,. \tag{2.33}$$

Proof There are three cases. The claim is obviously true if $\mathbf{v}_{k+1}$ is determined by special choice (2.21l) and $\mathbf{u}_{k+1}$ is determined by special choice (2.21m). In the second case, $\mathbf{v}_{k+1}$ is given by the special choice (2.21j) and $\mathbf{u}_{k+1}$ is given by (2.21k). In this case, we have

$$\mathbf{u}_{k+1} \cdot \mathbf{v}_{k+1} = \left(\widehat{\mathbf{u}}_{k+1} / \left[\widehat{\mathbf{u}}_{k+1} \cdot \mathbf{v}_{k+1}\right]\right) \cdot \mathbf{v}_{k+1} = 1$$

because the denominator in this expression is real. In the remaining case, $\mathbf{v}_{k+1}$ is computed by Eq. (2.21g) and $\mathbf{u}_{k+1}$ is computed by Eq. (2.21i). Then Eq. (2.21h) implies that

$$\mathbf{u}_{k+1} \cdot \mathbf{v}_{k+1} = \left(\widehat{\mathbf{u}}_{k+1} / \overline{\varrho_k}\right) \mathbf{v}_{k+1} = 1 \; .$$

Lemma 2.5.11 *The vector* $\mathbf{v}_{k+1}$ *computed by the Lanczos Biorthogonalization Algorithm 2.5.6 satisfies*

$$\mathbf{A}\mathbf{v}_k = \begin{cases} \mathbf{v}_1 \delta_1 + \mathbf{v}_2 \lambda_1, & 1 = k \\ \mathbf{v}_{k-1} \varrho_{k-1} + \mathbf{v}_k \delta_k + \mathbf{v}_{k+1} \lambda_k, & 1 < k \end{cases} . \tag{2.34}$$

Proof Equation (2.21b) or (2.21d) for $\widehat{\mathbf{v}}_{k+1}$ implies that

$$\begin{aligned} \mathbf{A}\mathbf{v}_k &= \begin{cases} \widehat{\mathbf{v}}_2 + \mathbf{v}_1 \delta_1, & k = 1 \\ \widehat{\mathbf{v}}_{k+1} + \mathbf{v}_k \delta_k + \mathbf{v}_{k-1} \varrho_{k-1}, & k > 1 \end{cases} \\ &= \begin{cases} \mathbf{v}_2 \lambda_1 + \mathbf{v}_1 \delta_1, & k = 1 \text{ and } \lambda_1 > 0 \\ \mathbf{v}_{k+1} \lambda_k + \mathbf{v}_k \delta_k + \mathbf{v}_{k-1} \varrho_{k-1}, & k > 1 \text{ and } \lambda_k > 0 \\ \mathbf{v}_1 \delta_1, & k = 1 \text{ and } \lambda_1 = 0 \\ \mathbf{v}_k \delta_k + \mathbf{v}_{k-1} \varrho_{k-1}, & k > 1 \text{ and } \lambda_k = 0 \end{cases} \\ &= \begin{cases} \mathbf{v}_2 \lambda_1 + \mathbf{v}_1 \delta_1, & k = 1 \\ \mathbf{v}_{k+1} \lambda_k + \mathbf{v}_k \delta_k + \mathbf{v}_{k-1} \varrho_{k-1}, & k > 1 \end{cases} . \end{aligned}$$

Lemma 2.5.12 *The vector* $\mathbf{u}_{k+1}$ *computed by the Lanczos Biorthogonalization Algorithm 2.5.6 satisfies*

$$\mathbf{A}^H \mathbf{u}_k = \begin{cases} \mathbf{u}_1 \overline{\delta_1} + \mathbf{u}_2 \overline{\varrho_1}, & 1 = k \\ \mathbf{u}_{k-1} \lambda_{k-1} + \mathbf{u}_k \overline{\delta_k} + \mathbf{u}_{k+1} \overline{\varrho_k}, & 1 < k \end{cases} . \tag{2.35}$$

Proof Equation (2.21c) or (2.21e) for $\widehat{\mathbf{u}}_{k+1}$ implies that

$$\mathbf{A}^H \mathbf{u}_k = \begin{cases} \widehat{\mathbf{u}}_2 + \mathbf{u}_1 \overline{\delta_1}, & k = 1 \\ \widehat{\mathbf{u}}_{k+1} + \mathbf{u}_k \overline{\delta_k} + \mathbf{u}_{k-1} \lambda_{k-1}, & k > 1 \end{cases}$$

$$= \begin{cases} \mathbf{u}_2\overline{\varrho_1} + \mathbf{u}_1\overline{\delta_1}, & k = 1 \text{ and } \varrho_1 \neq 0 \\ \mathbf{u}_{k+1}\overline{\varrho_k} + \mathbf{u}_k\overline{\delta_k} + \mathbf{u}_{k-1}\lambda_{k-1}, & k > 1 \text{ and } \varrho_k \neq 0 \\ \mathbf{u}_1\overline{\delta_1}, & k = 1 \text{ and } \varrho_1 = 0 \\ \mathbf{u}_k\overline{\delta_k} + \mathbf{u}_{k-1}\lambda_{k-1}, & k > 1 \text{ and } \varrho_k = 0 \end{cases}$$

$$= \begin{cases} \mathbf{u}_2\overline{\varrho_1} + \mathbf{u}_1\overline{\delta_1}, & k = 1 \\ \mathbf{u}_{k+1}\overline{\varrho_k} + \mathbf{u}_k\overline{\delta_k} + \mathbf{u}_{k-1}\lambda_{k-1}, & k > 1 \end{cases} .$$

Theorem 2.5.6 (Lanczos Biorthogonalization) *Let* $\mathbf{A}$ *be an* $m \times m$ *matrix. Choose an* m*-vector* $\mathbf{v}_1$ *with* $\|\mathbf{v}_1\|_2 = 1$*, and choose an* m*-vector* $\mathbf{u}_1$ *with* $\mathbf{u}_1 \cdot \mathbf{v}_1 = 1$*. If the Lanczos Biorthogonalization Algorithm 2.5.6 does not stop before completing step* $k = m$*, then there are an* $m \times m$ *tridiagonal matrix* $\mathbf{T}$*, a nonsingular* $m \times m$ *matrix* $\mathbf{V}$ *with* $\mathbf{V}\mathbf{e}_1 = \mathbf{v}_1$ *and a nonsingular* $m \times m$ *matrix* $\mathbf{U}$ *with* $\mathbf{U}\mathbf{e}_1 = \mathbf{u}_1$ *so that*

$$\mathbf{A}\mathbf{V} = \mathbf{V}\mathbf{T}\ ,\ \mathbf{U}^H\mathbf{A} = \mathbf{T}\mathbf{U}^H\ ,\ \mathbf{U}^H\mathbf{V} = \mathbf{I} \text{ and } \|\mathbf{V}\mathbf{e}_j\|_2 = 1 \text{ for all } 1 \leq j \leq m\ .$$

Furthermore, the subdiagonal entries of $\mathbf{T}$ *are all real and nonnegative.*

Proof Let us write

$$\mathbf{T} = \begin{bmatrix} \delta_1 & \varrho_1 & & \\ \lambda_1 & \delta_2 & \ddots & \\ & \ddots & \ddots & \varrho_{m-1} \\ & & \lambda_{m-1} & \delta_m \end{bmatrix} .$$

The claimed results are equivalent to the assumptions (2.22) holding true for $1 \leq k \leq m$. For $k = 1$, the first two of these conditions (2.22a) and (2.22b) simplify to the assumptions that $\|\mathbf{v}_1\| = 1$ and $\mathbf{u}_1 \cdot \mathbf{v}_1 = 1$. The remaining conditions are vacuous when $k = 1$. We will prove that the claims (2.22) are true for $1 \leq k \leq m$ by induction.

Suppose that the inductive hypotheses (2.22) are true for indices at most k. Lemma 2.5.7 proves that inductive hypothesis (2.22a) is valid for $k + 1$. Lemmas 2.5.8, 2.5.9 and 2.5.10 prove that the inductive hypothesis (2.22b) holds for $k + 1$. Lemma 2.5.11 shows that the inductive hypothesis (2.22c) is true for $k + 1$, and Lemma 2.5.12 proves the inductive hypothesis (2.22d) for $k + 1$. In the Lanczos biorthogonalization algorithm, λ_k is either taken to be zero, or computed by Eq. (2.21f), so λ_k is real and nonnegative. Thus the inductive hypotheses are all valid for $k + 1$, completing the inductive proof.

The equation $\mathbf{U}^H\mathbf{V} = \mathbf{I}$ implies that both $\mathbf{U}$ and $\mathbf{V}$ are nonsingular. Since Lemma 2.5.5 shows that $\widehat{\mathbf{v}}_{m+1} = \mathbf{0}$ and $\widehat{\mathbf{u}}_{m+1} = \mathbf{0}$. As a result, $\lambda_m = 0$ and $\varrho_m = 0$, allowing $\mathbf{T}$ to be $m \times m$ tridiagonal.

Please note that the Lanczos Biorthogonalization Algorithm 2.5.6 can stop before completing the last step because of an unfortunate choice of the initial vectors $\mathbf{v}_1$ and $\mathbf{u}_1$. For example, if $\mathbf{v}_1$ is not an eigenvector of $\mathbf{A}$, then we can choose $\mathbf{u}_1$ so that

$\mathbf{u}_1 \cdot \mathbf{v}_1 = 1$ but $\mathbf{u}_1 \cdot \mathbf{A}\mathbf{v}_1 = 0$ and $\mathbf{u}_1 \cdot \mathbf{A}^2\mathbf{v}_1 = 0$. Then Eq. (2.21a) will compute $\delta_1 = \mathbf{u}_1 \cdot \mathbf{A}\mathbf{v}_1 = 0$. Afterward, Eq. (2.21b) will find that $\widehat{\mathbf{v}}_2 = \mathbf{A}\mathbf{v}_1$ and Eq. (2.21c) will produce $\widehat{\mathbf{u}}_2 = \mathbf{A}^H\mathbf{u}_1$. These will imply that

$$\widehat{\mathbf{u}}_2 \cdot \widehat{\mathbf{v}}_2 = \mathbf{u}_1 \cdot \mathbf{A}^2\mathbf{v}_1 = 0 ,$$

and the algorithm will stop. It is possible to avoid this unfortunate situation by employing **look-ahead**; for more information, see Brezinski et al. [27] or Parlett et al. [140].

Nevertheless, the Lanczos biorthogonalization algorithm will be related to the biconjugate gradient algorithm in Sect. 2.5.5.2, and used as part of the QMR algorithm in Sect. 2.6.3.

Exercise 2.5.5 Perform the Lanczos biorthogonalization algorithm on the matrix

$$\mathbf{A} = \begin{bmatrix} 9 & -1 & 8 & -9 \\ 6 & -1 & 5 & -5 \\ -5 & 1 & -4 & 5 \\ 4 & 0 & 5 & -4 \end{bmatrix} .$$

Take both $\mathbf{v}_1$ and $\mathbf{u}_1$ to be the first axis vector.

Exercise 2.5.6 Program the Lanczos biorthogonalization algorithm for a sparse matrix $\mathbf{A}$, and apply it to matrix **E20R0100** from MatrixMarket. Choose $\mathbf{v}_1$ and $\mathbf{u}_1$ to have random values between zero and one, then scale these vectors so that $\|\mathbf{v}_1\|_2 = 1$ and $\mathbf{u}_1 \cdot \mathbf{v}_1 = 1$. Then perform the Lanczos biorthogonal process. Compute $\|\mathbf{AV} - \mathbf{VT}\|_\infty$, $\|\mathbf{U}^H\mathbf{A} - \mathbf{TU}^H\|_\infty$ and $\|\mathbf{U}^H\mathbf{V} - \mathbf{I}\|_\infty$ and comment on the results.

2.5.5.2 BiCG

The **biconjugate gradient algorithm**, originally developed by Fletcher [67], is related to the Lanczos bi-orthogonalization process (2.5.6). Given an $m \times m$ matrix $\mathbf{A}$, and m-vectors $\mathbf{b}$ and $\mathbf{x}_0$, we define the initial residual to be $\mathbf{r}_0 = \mathbf{b} - \mathbf{A}\mathbf{x}_0$ and choose an m-vector $\mathbf{s}_0$ so that $\mathbf{s}_0 \cdot \mathbf{r}_0 \neq 0$. Then we compute sequences $\{\gamma_j\}_{j=0}^m$, $\{\mathbf{x}_j\}_{j=0}^m$, $\{\mathbf{p}_j\}_{j=0}^m$, $\{\mathbf{q}_j\}_{j=0}^m$, $\{\mathbf{r}_j\}_{j=0}^m$ and $\{\mathbf{s}_j\}_{j=0}^m$ by the following

Algorithm 2.5.7 (Biconjugate Gradients)

$$\mathbf{p}_0 = \mathbf{r}_0$$

$$\mathbf{q}_0 = \mathbf{s}_0$$

$$\gamma_0 = \mathbf{s}_0 \cdot \mathbf{r}_0$$

for $0 \le j < m$

$$\mathbf{z}_j = \mathbf{A}\mathbf{p}_j$$

$$\varepsilon_j = \mathbf{q}_j \cdot \mathbf{z}_j$$

if $\varepsilon_j = 0$ then

$n = j$

break

$$\alpha_j = \gamma_j / \varepsilon_j$$

$$\mathbf{x}_{j+1} = \mathbf{x}_j + \mathbf{p}_j \alpha_j$$

$$\mathbf{r}_{j+1} = \mathbf{r}_j - \mathbf{z}_j \alpha_j$$

$$\mathbf{s}_{j+1} = \mathbf{s}_j - \mathbf{A}^H \mathbf{q}_j \overline{\alpha_j}$$

$$\gamma_{j+1} = \mathbf{s}_{j+1} \cdot \mathbf{r}_{j+1}$$

if $\gamma_{j+1} = 0$ then

$n = j$

break

$$\beta_j = \gamma_{j+1} / \gamma_j$$

$$\mathbf{p}_{j+1} = \mathbf{r}_{j+1} + \mathbf{p}_j \beta_j$$

$$\mathbf{q}_{j+1} = \mathbf{s}_{j+1} + \mathbf{q}_j \overline{\beta_j}$$

This algorithm requires two matrix-vector multiplications, one by $\mathbf{A}$ and the other by $\mathbf{A}^H$. It also requires the storage of six m-vectors, namely $\mathbf{p}_j$, $\mathbf{q}_j$, $\mathbf{r}_j$, $\mathbf{s}_j$, $\mathbf{x}_j$ and $\mathbf{z}_j$.

The biconjugate gradient algorithm (2.5.7) has the following interesting consequences.

Lemma 2.5.13 *Suppose that* $\mathbf{A}$ *is an* $m \times m$ *matrix, and assume that both* $\mathbf{b}$ *and* $\mathbf{x}_0$ *are* m*-vectors. Define the initial residual to be* $\mathbf{r}_0 = \mathbf{b} - \mathbf{A}\mathbf{x}_0$*, and choose an* m*-vector* $\mathbf{s}_0$ *so that* $\mathbf{s}_0 \cdot \mathbf{r}_0 \neq 0$*. Compute the sequences* $\{\gamma_j\}_{j=0}^m$, $\{\mathbf{x}_j\}_{j=0}^m$, $\{\mathbf{r}_j\}_{j=0}^m$, $\{\mathbf{s}_j\}_{j=0}^m$, $\{\mathbf{p}_j\}_{j=0}^m$ *and* $\{\mathbf{q}_j\}_{j=0}^m$ *by the biconjugate gradient algorithm* (2.5.7)*. Then for all* $0 \le j \le m$ *we have* $\mathbf{r}_j = \mathbf{b} - \mathbf{A}\mathbf{x}_j$*. Furthermore, for all* $0 \le i \neq j \le m$ *we have* $\mathbf{s}_i \cdot \mathbf{r}_j = 0$ *and* $\mathbf{q}_i \cdot \mathbf{A}\mathbf{p}_j = 0$.

Proof We can prove that $\mathbf{r}_j = \mathbf{b} - \mathbf{A}\mathbf{x}_j$ by induction. By assumption, this claim is true for $j = 0$. Assume that the claim is true for $j = n$. Then

$$\mathbf{r}_{n+1} = \mathbf{r}_n - \mathbf{z}_n\alpha_n = (\mathbf{b} - \mathbf{A}\mathbf{x}_n) - \mathbf{A}\mathbf{p}_n\alpha_n = \mathbf{b} - \mathbf{A}(\mathbf{x}_n + \mathbf{p}_n\alpha_n) = \mathbf{b} - \mathbf{A}\mathbf{x}_{n+1}$$

Next, we will prove the second and third claims together by induction. First, we verify the claims for $n = 1$:

$$\mathbf{s}_0 \cdot \mathbf{r}_1 = \mathbf{s}_0 \cdot [\mathbf{r}_0 - \mathbf{z}_0\alpha_0] = \gamma_0 - \mathbf{q}_0 \cdot \mathbf{z}_0\alpha_0 = 0 ,$$

$$\mathbf{s}_1 \cdot \mathbf{r}_0 = \left[\mathbf{s}_0 - \mathbf{A}^H\mathbf{q}_0\overline{\alpha_0}\right] \cdot \mathbf{r}_0 = \gamma_0 - \mathbf{q}_0 \cdot \mathbf{A}\mathbf{r}_0\alpha_0 = \gamma_0 - \mathbf{q}_0 \cdot \mathbf{z}_0\alpha_0 = 0 ,$$

$$\mathbf{q}_0 \cdot \mathbf{A}\mathbf{p}_1 = \frac{1}{\alpha_0}[\mathbf{s}_0 - \mathbf{s}_1] \cdot [\mathbf{r}_1 + \mathbf{p}_0\beta_0] = \frac{1}{\alpha_0}[\mathbf{s}_0 \cdot \mathbf{r}_1 + \mathbf{s}_0 \cdot \mathbf{p}_0\beta_0 - \mathbf{s}_1 \cdot \mathbf{r}_1 - \mathbf{s}_1 \cdot \mathbf{p}_0\beta_0]$$

$$= \frac{1}{\alpha_0}[\gamma_0\beta_0 - \gamma_1 - \mathbf{s}_1 \cdot \mathbf{r}_0\beta_0] = \frac{1}{\alpha_0}[\gamma_0\beta_0 - \gamma_1] = 0 , \text{ and}$$

$$\mathbf{q}_1 \cdot \mathbf{z}_0 = \left[\mathbf{s}_1 + \mathbf{q}_0\overline{\beta_0}\right] \cdot [\mathbf{r}_0 - \mathbf{r}_1]\frac{1}{\alpha_0} = \frac{1}{\alpha_0}[\mathbf{s}_1 \cdot \mathbf{r}_0 - \mathbf{s}_1 \cdot \mathbf{r}_1 + \beta_0\mathbf{q}_0 \cdot \mathbf{r}_0 - \beta_0\mathbf{q}_0 \cdot \mathbf{r}_1]$$

$$= \frac{1}{\alpha_0}[-\gamma_1 + \beta_0\mathbf{s}_0 \cdot \mathbf{r}_0 - \beta_0\mathbf{q}_0 \cdot (\mathbf{r}_0 - \mathbf{z}_0\alpha_0)]$$

$$= \frac{1}{\alpha_0}[-\gamma_1 + \beta_0\gamma_0 - \beta_0(\mathbf{s}_0 \cdot \mathbf{r}_0 - \mathbf{q}_0 \cdot \mathbf{z}_0\alpha_0)] = -\frac{\beta_0}{\alpha_0}(\gamma_0 - \mathbf{q}_0 \cdot \mathbf{z}_0\alpha_0) = 0 .$$

Inductively, we will assume that the second and third claims are true for n, and verify the claims for $n + 1$. Then

$$\mathbf{s}_n \cdot \mathbf{r}_{n+1} = \mathbf{s}_n \cdot [\mathbf{r}_n - \mathbf{z}_n\alpha_n] = \gamma_n - \left[\mathbf{q}_n - \mathbf{q}_{n-1}\overline{\beta_{n-1}}\right] \cdot \mathbf{z}_n\alpha_n$$

$$= \gamma_n - \mathbf{q}_n \cdot \mathbf{z}_n\alpha_n + \beta_{n-1}\alpha_n\mathbf{q}_{n-1} \cdot \mathbf{z}_n = \gamma_n - \mathbf{q}_n \cdot \mathbf{z}_n\alpha_n = 0 ,$$

$$\mathbf{s}_{n+1} \cdot \mathbf{r}_n = \left[\mathbf{s}_n - \mathbf{A}^H\mathbf{q}_n\overline{\alpha_n}\right] \cdot \mathbf{r}_n = \gamma_n - \alpha_n\mathbf{q}_n \cdot \mathbf{A}\mathbf{r}_n$$

$$= \gamma_n - \alpha_n\mathbf{q}_n \cdot \mathbf{A}[\mathbf{p}_n - \mathbf{p}_{n-1}\beta_{n-1}] = \gamma_n - \alpha_n\mathbf{q}_n \cdot \mathbf{z}_n + \alpha_n\beta_{n-1}\mathbf{q}_n \cdot \mathbf{A}\mathbf{p}_{n-1}$$

$$= \gamma_n - \alpha_n\mathbf{q}_n \cdot \mathbf{z}_n = 0 ,$$

$$\mathbf{q}_n \cdot \mathbf{A}\mathbf{p}_{n+1} = \mathbf{q}_n \cdot \mathbf{A}[\mathbf{r}_{n+1} + \mathbf{p}_n\beta_n] = \frac{1}{\alpha_n}[\mathbf{s}_n - \mathbf{s}_{n+1}] \cdot \mathbf{r}_{n+1} + \mathbf{q}_n \cdot \mathbf{z}_n\beta_n$$

$$= -\frac{\gamma_{n+1}}{\alpha_n} + \frac{\gamma_n}{\alpha_n}\beta_n = 0 , \text{ and}$$

$$\mathbf{q}_{n+1} \cdot \mathbf{A}\mathbf{p}_n = \left[\mathbf{s}_{n+1} + \mathbf{q}_n\overline{\beta_n}\right] \cdot \mathbf{A}\mathbf{p}_n = \mathbf{s}_{n+1} \cdot \mathbf{z}_n + \beta_n\mathbf{q}_n \cdot \mathbf{z}_n$$

$$= \mathbf{s}_{n+1} \cdot [\mathbf{r}_n - \mathbf{r}_{n+1}]\frac{1}{\alpha_n} + \beta_n\frac{\gamma_n}{\alpha_n} = -\frac{\gamma_{n+1}}{\alpha_n} + \frac{\beta_n\gamma_n}{\alpha_n} = 0 ,$$

and for $0 \leq j \leq n-1$

$$\mathbf{s}_j \cdot \mathbf{r}_{n+1} = \mathbf{s}_j \cdot [\mathbf{r}_n - \mathbf{z}_n \alpha_n] = -\mathbf{s}_j \cdot \mathbf{z}_n \alpha_n = \left[\mathbf{q}_{j-1}\overline{\beta_{j-1}} - \mathbf{q}_j\right] \cdot \mathbf{z}_n \alpha_n = 0 \,,$$

$$\mathbf{s}_{n+1} \cdot \mathbf{r}_j = \left[\mathbf{s}_n - \mathbf{A}^H \mathbf{q}_n \overline{\alpha_n}\right] \cdot \mathbf{r}_j = -\alpha_n \mathbf{q}_n \cdot \mathbf{A}\mathbf{r}_j = -\alpha_n \mathbf{q}_n \cdot \mathbf{A}\left[\mathbf{p}_j - \mathbf{p}_{j-1}\beta_{j-1}\right]$$
$$= -\alpha_n \mathbf{q}_n \cdot \mathbf{z}_j + \alpha_n \beta_{j-1} \mathbf{q}_n \cdot \mathbf{z}_{j-1} = 0 \,,$$

$$\mathbf{q}_j \cdot \mathbf{A}\mathbf{p}_{n+1} = \mathbf{q}_j \cdot \mathbf{A}\left[\mathbf{r}_{n+1} + \mathbf{p}_n \beta_n\right] = \frac{1}{\alpha_j}\left[\mathbf{s}_j - \mathbf{s}_{j+1}\right] \cdot \mathbf{r}_{n+1} + \mathbf{q}_j \cdot \mathbf{z}_n \beta_n = 0 \quad \text{and}$$

$$\mathbf{q}_{n+1} \cdot \mathbf{z}_j = \left[\mathbf{s}_{n+1} + \mathbf{q}_n \overline{\beta_n}\right] \cdot \mathbf{z}_j = \mathbf{s}_{n+1} \cdot \mathbf{z}_j + \beta_n \mathbf{q}_n \cdot \mathbf{z}_j = \mathbf{s}_{n+1} \cdot \left[\mathbf{r}_j - \mathbf{r}_{j+1}\right] \frac{1}{\alpha_j} = 0 \,.$$

These results prove the claim for $n+1$.
This lemma implies that $\mathbf{s}_{n+1} \perp \mathbf{r}_0, \ldots, \mathbf{r}_n$. It follows that if $\mathbf{s}_{n+1} \cdot \mathbf{r}_{n+1} = \gamma_{n+1} \neq 0$ then $\mathbf{s}_{n+1} \notin \text{span}\{\mathbf{s}_0, \ldots, \mathbf{s}_n\}$. A similar statement holds for the residual vectors. Thus the spans of $\{\mathbf{s}_j\}_{j=0}^n$ and $\{\mathbf{r}_j\}_{j=0}^n$ expand as n increases, until the entire space of m vectors is spanned or the algorithm terminates with $\gamma_n = 0$.

A C^{++} implementation of biconjugate gradients can be found in file bicg.h. The Fortran 77 implementation this algorithm is available in files BiCG.f and BiCGREVCOM.f at `netlib.org`.

2.5.5.3 CGS

One unfortunate aspect of the Biconjugate Gradient Algorithm 2.5.7 is the need to compute matrix-vector products involving both $\mathbf{A}$ and $\mathbf{A}^H$. By studying the computations in this algorithm carefully, we can design a related algorithm that avoids multiplication by $\mathbf{A}^H$.

The vectors $\mathbf{r}_j$ and $\mathbf{p}_j$ lie in a Krylov subspace involving $\mathbf{A}$. This implies that there are polynomials ϱ_j and π_j of degree at most j so that

$$\mathbf{r}_j = \varrho_j(\mathbf{A})\mathbf{r}_0 \quad \text{and} \quad \mathbf{p}_j = \pi_j(\mathbf{A})\mathbf{r}_0 \,.$$

Similarly,

$$\mathbf{s}_j = \overline{\varrho_j}(\mathbf{A}^H)\mathbf{s}_0 \quad \text{and} \quad \mathbf{q}_j = \overline{\pi_j}(\mathbf{A}^H)\mathbf{s}_0 \,.$$

The crucial observation, due to Sonneveld [161], is that the coefficients in these polynomials depend only on the polynomials ϱ_j^2 and π_j^2. In particular, note that

$$\gamma_j = \mathbf{s}_j \cdot \mathbf{r}_j = \left[\overline{\varrho_j}\left(\mathbf{A}^H\right)\mathbf{s}_0\right] \cdot \left[\varrho_j(\mathbf{A})\mathbf{r}_0\right] = \mathbf{s}_0 \cdot \varrho_j(\mathbf{A})^2 \mathbf{r}_0 \quad \text{and}$$

$$\alpha_j = \frac{\mathbf{s}_j \cdot \mathbf{r}_j}{\mathbf{q}_j \cdot \mathbf{A}\mathbf{p}_j} = \frac{\left[\overline{\varrho_j}\left(\mathbf{A}^H\right)\mathbf{s}_0\right] \cdot \left[\varrho_j(\mathbf{A})\mathbf{r}_0\right]}{\left[\overline{\pi_j}\left(\mathbf{A}^H \mathbf{s}_0\right)\right] \cdot \mathbf{A}\pi_j(\mathbf{A})\mathbf{r}_0} = \frac{\mathbf{s}_0 \cdot \varrho_j(\mathbf{A})^2 \mathbf{r}_0}{\mathbf{s}_0 \cdot \mathbf{A}\pi_j(\mathbf{A})^2 \mathbf{r}_0} \,.$$

Recall that the biconjugate gradient algorithm computes

$$\begin{aligned} \mathbf{r}_{j+1} = \mathbf{r}_j - \mathbf{A}\mathbf{p}_j\alpha_j \quad &\Longrightarrow \quad \varrho_{j+1}(\zeta) = \varrho_j(\zeta) - \zeta\pi_j(\zeta)\alpha_j \;, \\ \mathbf{p}_{j+1} = \mathbf{r}_{j+1} + \mathbf{p}_j\beta_j \quad &\Longrightarrow \quad \pi_{j+1}(\zeta) = \varrho_{j+1}(\zeta) + \pi_j(\zeta)\beta_j \;. \end{aligned}$$

Of course, we have $\varrho_0(\zeta) = 1 = \pi_0(\zeta)$. We would like to develop recurrences for ϱ_j^2 and π_j^2.

If we define the polynomials

$$\omega_j(\zeta) = \varrho_{j+1}(\zeta)\pi_j(\zeta) \;,$$

then we obtain

$$\pi_{j+1}(\zeta)^2 = \left[\varrho_{j+1}(\zeta) + \pi_j(\zeta)\beta_j\right]^2 = \varrho_{j+1}(\zeta)2 + 2\varrho_{j+1}(\zeta)\pi_j(\zeta)\beta_j + \pi_j(\zeta)^2\beta_j^2 \;.$$

We notice that

$$\pi_j(\zeta)\varrho_j(\zeta) = \left[\varrho_j(\zeta) + \pi_{j-1}(\zeta)\beta_{j-1}\right]\varrho_j(\zeta) = \varrho_j(\zeta)^2 + \omega_{j-1}(\zeta)\beta_{j-1} \;,$$

from which we obtain

$$\begin{aligned} \varrho_{j+1}(\zeta)^2 &= \left[\varrho_j(\zeta) - \zeta\pi_j(\zeta)\alpha_j\right]^2 = \varrho_j(\zeta)^2 - 2\zeta\pi_j(\zeta)\varrho_j(\zeta)\alpha_j + \zeta^2\pi_j(\zeta)^2\alpha_j^2 \\ &= \varrho_j(\zeta)^2 - 2\zeta\left[\varrho_j(\zeta)^2 + \omega_{j-1}(\zeta)\beta_{j-1}\right]\alpha_j + \zeta^2\pi_j(\zeta)^2\alpha_j^2 \;. \end{aligned}$$

Finally,

$$\begin{aligned} \omega_j(\zeta) &= \left[\varrho_j(\zeta) - \zeta\pi_j(\zeta)\alpha_j\right]\pi_j(\zeta) = \varrho_j(\zeta)\left[\varrho_j(\zeta) + \pi_{j-1}(\zeta)\beta_{j-1}\right] - \zeta\pi_j(\zeta)^2\alpha_j \\ &= \varrho_j(\zeta)^2 + \omega_{j-1}(\zeta)\beta_{j-1} - \zeta\pi_j(\zeta)^2\alpha_j \;. \end{aligned}$$

These polynomials suggest that we define the vectors

$$\begin{aligned} \widehat{\mathbf{r}}_j &= \varrho_j(\mathbf{A})^2\mathbf{r}_0 = \varrho_j(\mathbf{A})\mathbf{r}_j \;, \\ \widehat{\mathbf{p}}_j &= \pi_j(\mathbf{A})^2\mathbf{r}_0 = \pi_j(\mathbf{A})\mathbf{p}_j \quad \text{and} \\ \widehat{\mathbf{w}}_j &= \omega_j(\mathbf{A})\mathbf{r}_0 \;. \end{aligned}$$

Then

$$\begin{aligned} \gamma_j &= \mathbf{s}_j \cdot \mathbf{r}_j = \mathbf{s}_0\varrho_j(\mathbf{A})^2\mathbf{r}_0 = \mathbf{s}_0 \cdot \widehat{\mathbf{r}}_j \quad \text{and} \\ \alpha_j &= \frac{\gamma_j}{\mathbf{q}_j \cdot \mathbf{A}\mathbf{p}_j} = \frac{\gamma_j}{\mathbf{s}_0 \cdot \mathbf{A}\pi_j(\mathbf{A})^2\mathbf{r}_0} = \frac{\gamma_j}{\mathbf{s}_0 \cdot \mathbf{A}\widehat{\mathbf{p}}_j} \;. \end{aligned}$$

If we also define the auxiliary vectors

$$\widehat{\mathbf{u}}_j = \widehat{\mathbf{r}}_j + \widehat{\mathbf{w}}_{j-1}\beta_{j-1} \;,$$

then we obtain the following recurrences:

$$\widehat{\mathbf{w}}_j = \varrho_j(\mathbf{A})^2\mathbf{r}_0 + \omega_{j-1}(\mathbf{A})\mathbf{r}_0\beta_{j-1} - \mathbf{A}\pi_j(\mathbf{A})^2\mathbf{r}_0\alpha_j$$
$$= \widehat{\mathbf{r}}_j + \widehat{\mathbf{w}}_{j-1}\beta_{j-1} - \mathbf{A}\widehat{\mathbf{p}}_j\alpha_j = \widehat{\mathbf{u}}_j - \mathbf{A}\widehat{\mathbf{p}}_j\alpha_j ,$$
$$\widehat{\mathbf{r}}_{j+1} = \varrho_{j+1}(\mathbf{A})^2\mathbf{r}_0 = \varrho_j(\mathbf{A})^2\mathbf{r}_0 - 2A\left[\varrho_j(\mathbf{A})^2 + \omega_{j-1}(\mathbf{A})\beta_{j-1}\right]\mathbf{r}_0\alpha_j + \mathbf{A}^2\pi_j(\mathbf{A})^2\mathbf{r}_0\alpha_j^2$$
$$= \widehat{\mathbf{r}}_j - \mathbf{A}\left[\widehat{\mathbf{r}}_j 2 + \widehat{\mathbf{w}}_{j-1}2\beta_{j-1} - \mathbf{A}\widehat{\mathbf{r}}_j\alpha_j\right]\alpha_j = \widehat{\mathbf{r}}_j - \mathbf{A}\left[\widehat{\mathbf{u}}_j + \widehat{\mathbf{w}}_j\right]\alpha_j \quad \text{and}$$
$$\widehat{\mathbf{p}}_{j+1} = \pi_{j+1}(\mathbf{A})^2\mathbf{r}_0 = \varrho_{j+1}(\mathbf{A})^2\mathbf{r}_0 + \omega_j(\mathbf{A})\mathbf{r}_0 2\beta_j + \pi_j(\mathbf{A})^2\mathbf{r}_0\beta_j^2$$
$$= \widehat{\mathbf{r}}_{j+1} + \widehat{\mathbf{w}}_j 2\beta_j + \widehat{\mathbf{p}}_j\beta_j^2 = \widehat{\mathbf{u}}_{j+1} + \left[\widehat{\mathbf{w}}_j + \widehat{\mathbf{p}}_j\beta_j\right]\beta_j .$$

These equations lead to the following

Algorithm 2.5.8 (Conjugate Gradient Squared)

$\mathbf{p}_0 = \mathbf{u}_0 = \mathbf{r}_0$

$\gamma_0 = \mathbf{s}_0 \cdot \mathbf{r}_0$

for $0 \le j < m$

  $\mathbf{z}_j = \mathbf{A}\mathbf{p}_j$

  $\varepsilon_j = \mathbf{s}_0 \cdot \mathbf{z}_j$

  if $\varepsilon_j = 0$ then

    $n = j$

    break

  $\alpha_j = \gamma_j / \varepsilon_j$

  $\mathbf{w}_j = \mathbf{u}_j - \mathbf{z}_j\alpha_j$

  $\mathbf{t}_j = \mathbf{u}_j + \mathbf{w}_j$

  $\mathbf{x}_{j+1} = \mathbf{x}_j + \mathbf{t}_j\alpha_j$

  $\mathbf{r}_{j+1} = \mathbf{r}_j - \mathbf{A}\mathbf{t}_j\alpha_j$

  $\gamma_{j+1} = \mathbf{s}_0 \cdot \mathbf{r}_{j+1}$

  if $\gamma_{j+1} = 0$ then

    $n = j$

    break

  $\beta_j = \gamma_{j+1} / \gamma_j$

  $\mathbf{u}_{j+1} = \mathbf{r}_{j+1} + \mathbf{w}_j\beta_j$

  $\mathbf{p}_{j+1} = \mathbf{u}_{j+1} + \left(\mathbf{w}_j + \mathbf{p}_j\beta_j\right)\beta_j .$

This algorithm requires two matrix-vector multiplications per iteration, both by **A**. It also requires storage of nine m-vectors.

A C^{++} implementation of conjugate gradient squared is available in file cgs.h. The corresponding Fortran 77 implementation of this program is available found in files CGS.f and CGSREVCOM.f at `netlib.org`.

2.5.5.4 BiCGSTAB

The conjugate gradient squared algorithm suffers from substantial rounding errors, due to use of the squares of the polynomials from the biconjugate gradient algorithm. Instead of taking the residuals to be $\widehat{\mathbf{r}}_j = \varrho_j(\mathbf{A})^2\widehat{\mathbf{r}}_0$, van der Vorst [176] suggested that we take

$$\widetilde{\mathbf{r}}_j = \psi_j(\mathbf{A})\varrho_j(\mathbf{A})\widetilde{\mathbf{r}}_0 ,$$

where $\psi_j(\zeta)$ is given recursively in the form

$$\psi_{j+1}(\zeta) = (1 - \kappa_j\zeta)\psi_j(\zeta) ,$$

and the scalar κ_j is chosen to achieve a steepest descent step.

The new polynomial ψ_j requires the development of new recurrences:

$$\begin{aligned}
\psi_{j+1}(\zeta)\varrho_{j+1}(\zeta) &= (1 - \zeta\kappa_j)\psi_j(\zeta)\left[\varrho_j(\zeta) - \zeta\pi_j(\zeta)\alpha_j\right] \\
&= (1 - \zeta\kappa_j)\left[\psi_j(\zeta)\varrho_j(\zeta) - \zeta\psi_j(\zeta)\pi_j(\zeta)\alpha_j\right] \quad \text{and} \\
\psi_j(\zeta)\pi_j(\zeta) &= \psi_j(\zeta)\left[\varrho_j(\zeta) + \pi_{j-1}(\zeta)\beta_{j-1}\right] \\
&= \psi_j(\zeta)\varrho_j(\zeta) - (1 - \zeta\kappa_{j-1})\psi_{j-1}(\zeta)\pi_{j-1}(\zeta)\beta_{j-1} .
\end{aligned}$$

If we formally write

$$\widetilde{\mathbf{p}}_j = \psi_j(\mathbf{A})\pi_j(\mathbf{A})\widetilde{\mathbf{r}}_0 ,$$

then we obtain the vector recurrences

$$\begin{aligned}
\widetilde{\mathbf{r}}_{j+1} &= \psi_{j+1}(\mathbf{A})\varrho_{j+1}(\mathbf{A})\widetilde{\mathbf{r}}_0 = (\mathbf{I} - \mathbf{A}\kappa_j)\left[\widetilde{\mathbf{r}}_j - \mathbf{A}\widetilde{\mathbf{p}}_j\alpha_j\right] \quad \text{and} \\
\widetilde{\mathbf{p}}_{j+1} &= \psi_{j+1}(\mathbf{A})\pi_{j+1}(\mathbf{A})\widetilde{\mathbf{r}}_0 = \widetilde{\mathbf{r}}_{j+1} + (\mathbf{I} - \mathbf{A}\kappa_j)\widetilde{\mathbf{p}}_j\beta_j .
\end{aligned}$$

The computation of the scalars in these recurrences is a bit more intricate. The basic idea uses the orthogonality of the vectors in Lemma 2.5.13. Let us formally write monomial expansions for the polynomials

$$\varrho_j(\zeta) = \sum_{i=0}^{j} \varrho_{ij}\zeta^i , \quad \pi_j(\zeta) = \sum_{i=0}^{j} \pi_{ij}\zeta^i \quad \text{and} \quad \psi_j(\zeta) = \sum_{i=0}^{j} \psi_{ij}\zeta^i .$$

Since $\varrho_{j+1}(\zeta) = \varrho_j(\zeta) - \zeta\pi_j(\zeta)\alpha_j$, it follows that the coefficients of ζ^{j+1} in this equation satisfy

$$\varrho_{j+1,j+1} = -\alpha_j\pi_{j,j} .$$

Because $\pi_{j+1}(\zeta) = \varrho_{j+1}(\zeta) + \pi_j(\zeta)\beta_j j$, it follows that

$$\pi_{j+1,j+1} = \varrho_{j+1,j+1} .$$

Finally, since $\psi_{j+1}(\zeta) = (1 - \zeta\kappa_j)\psi_j(\zeta)$, we must have

$$\psi_{j+1,j+1} = -\kappa_j\psi_{j,j} .$$

Since Lemma 2.5.13 proved that $\mathbf{s}_j \cdot \mathbf{r}_i = 0$ for all $0 \leq i < j$, it follows that $\mathbf{s}_j \cdot \varrho_i(\mathbf{A})\mathbf{r}_0 = 0$ for all $0 \leq i < j$. This in turn implies that $\mathbf{s}_j \cdot \mathbf{A}^i\mathbf{r}_0 = 0$ for all $0 \leq i < j$. Consequently,

$$\gamma_j = \mathbf{s}_j \cdot \mathbf{r}_j = \mathbf{s}_j \cdot \varrho_j(\mathbf{A})\mathbf{r}_0 = \mathbf{s}_j \cdot \sum_{i=0}^{j} \varrho_{i,j}\mathbf{A}^i\mathbf{r}_0 = \mathbf{s}_j \cdot \mathbf{A}^j\mathbf{r}_0\varrho_{j,j} = \mathbf{s}_j \cdot \left[\sum_{i=0}^{j} \psi_{ij}\mathbf{A}^i\mathbf{r}_0\right] \frac{\varrho_{j,j}}{\psi_{j,j}}$$
$$= \mathbf{s}_j \cdot \psi_j(\mathbf{A})\mathbf{r}_0 \frac{\varrho_{j,j}}{\psi_{j,j}} = \mathbf{s}_0 \cdot \varrho_j(\mathbf{A})\psi_j(\mathbf{A})\mathbf{r}_0 \frac{\varrho_{j,j}}{\psi_{j,j}} = \mathbf{s}_0 \cdot \widetilde{\mathbf{r}}_j \frac{\varrho_{j,j}}{\psi_{j,j}} .$$

Lemma 2.5.13 also proved that $\mathbf{q}_j \cdot \mathbf{A}\mathbf{p}_i = 0$ for all $0 \leq i < j$. This implies that $\mathbf{q}_j \cdot \mathbf{A}\pi_i(\mathbf{A})\mathbf{r}_0 = 0$ for all $0 \leq i < j$, and thus that $\mathbf{q}_j \cdot \mathbf{A}\mathbf{A}^i\mathbf{r}_0 = 0$ for all $0 \leq i < j$. Thus

$$\varepsilon_j = \mathbf{q}_j \cdot \mathbf{A}\mathbf{p}_j = \mathbf{q}_j \cdot \mathbf{A}\pi(\mathbf{A})\mathbf{r}_0 = \mathbf{q}_j \cdot \mathbf{A} \sum_{i=0}^{j} \pi_{i,j}\mathbf{A}^i\mathbf{r}_0 = \mathbf{q}_j \cdot \mathbf{A}\mathbf{A}^j\mathbf{r}_0\pi_{j,j}$$
$$= \mathbf{q}_j \cdot \mathbf{A}\left[\sum_{i=0}^{j} \psi_{ij}\mathbf{A}^i\mathbf{r}_0\right] \frac{\pi_{j,j}}{\psi_{j,j}} = \mathbf{q}_j \cdot \mathbf{A}\psi_j(\mathbf{A})\mathbf{r}_0 \frac{\pi_{j,j}}{\psi_{j,j}} = \mathbf{s}_0 \cdot \varrho_j(\mathbf{A})\mathbf{A}\psi_j(\mathbf{A})\mathbf{r}_0 \frac{\pi_{j,j}}{\psi_{j,j}}$$
$$= \mathbf{s}_0 \cdot \mathbf{A}\widetilde{\mathbf{p}}_j \frac{\pi_{j,j}}{\psi_{j,j}} .$$

The computations of α_j and $\beta_j j$ follow easily from these results.

It remains to compute κ_j. If we define

$$\widetilde{\mathbf{f}}_j = \widetilde{\mathbf{r}}_j - \mathbf{A}\widetilde{\mathbf{p}}\alpha_j ,$$

then we can write

$$\widetilde{\mathbf{r}}_{j+1} = (\mathbf{I} - \mathbf{A}\kappa_j)\left[\widetilde{\mathbf{r}}_j - \mathbf{A}\widetilde{\mathbf{p}}_j\alpha_j\right] = (\mathbf{I} - \mathbf{A}\kappa_j)\widetilde{\mathbf{f}}_j .$$

We choose a real scalar κ_j to minimize

$$\|\widetilde{\mathbf{r}}_{j+1}\|^2 = \widetilde{\mathbf{r}}_j \cdot \left(\mathbf{I} - \mathbf{A}^H \kappa_j\right) \left(\mathbf{I} - \mathbf{A}\kappa_j\right) \widetilde{\mathbf{r}}_j = \|\widetilde{\mathbf{f}}_j\|^2 - 2\Re\left(\widetilde{\mathbf{f}}_j \cdot \mathbf{A}\widetilde{\mathbf{f}}_j\right) + \|\mathbf{A}\widetilde{\mathbf{f}}_j\|^2 \kappa_j^2 .$$

This choice gives us

$$\kappa_j = \frac{\Re\left(\widetilde{\mathbf{f}}_j \cdot \mathbf{A}\widetilde{\mathbf{f}}_j\right)}{\|\mathbf{A}\widetilde{\mathbf{f}}_j\|^2} .$$

These equations lead to the following

Algorithm 2.5.9 (BiCGSTAB)

$\widetilde{\mathbf{p}}_0 = \widetilde{\mathbf{r}}_0$

$\widetilde{\gamma}_0 = \mathbf{s}_0 \cdot \widetilde{\mathbf{r}}_0$

for $0 \le j < m$

  $\widetilde{\mathbf{z}}_j = \mathbf{A}\widetilde{\mathbf{p}}_j$

  $\varepsilon_j = \mathbf{s}_0 \cdot \widetilde{\mathbf{z}}_j$

  if $\varepsilon_j = 0$ then

    $n = j$

    break

  $\alpha_j = \gamma_j / \varepsilon_j$

  $\widetilde{\mathbf{f}}_j = \widetilde{\mathbf{r}}_j - \widetilde{\mathbf{z}}_j \alpha_j$

  $\widetilde{\mathbf{t}}_j = \mathbf{A}\widetilde{\mathbf{f}}_j$

  $\kappa_j = \Re\left(\widetilde{\mathbf{f}}_j \cdot \widetilde{\mathbf{t}}_j\right) / \left\|\widetilde{\mathbf{t}}_j\right\|^2$

  $\widetilde{\mathbf{x}}_{j+1} = \widetilde{\mathbf{x}}_j + \widetilde{\mathbf{p}}_j \alpha_j + \widetilde{\mathbf{f}}_j \kappa_j$

  $\widetilde{\mathbf{r}}_{j+1} = \widetilde{\mathbf{f}}_j - \widetilde{\mathbf{t}}_j \kappa_j$

  $\widetilde{\gamma}_{j+1} = \mathbf{s}_0 \cdot \widetilde{\mathbf{r}}_{j+1}$

  if $\widetilde{\gamma}_{j+1} = 0$ then

    $n = j$

    break

  $\beta_j = \left(\widetilde{\gamma}_{j+1} \alpha_j\right) / \left(\widetilde{\gamma}_j \kappa_j\right)$

  $\widetilde{\mathbf{p}}_{j+1} = \widetilde{\mathbf{r}}_{j+1} + \left(\widetilde{\mathbf{p}}_j - \widetilde{\mathbf{z}}_j \kappa_j\right) \beta_j j$

This algorithm requires two matrix-vector multiplications per step, both by $\mathbf{A}$. It also requires storage of seven m-vectors.

A C^{++} implementation of BiCGSTAB can be found in file bicgstab.h. The corresponding Fortran 77 implementation is available in files BiCGSTAB.f and BiCGSTABREVCOM.f at `netlib.org`.

2.5.5.5 Recommendations

For Hermitian positive linear systems, the conjugate gradient algorithm is preferable to the other algorithms in this section. For general nonsingular linear systems, it is difficult to give a definitive recommendation. A paper by Nachtigal et al. [131] provided examples in which the conjugate gradient squared algorithm either vastly out-performed or under-performed the conjugate gradient algorithm applied to $\mathbf{A}\mathbf{A}^H\mathbf{y} = \mathbf{b}$. A discussion in Greenbaum [88, p. 92ff] draws similar conclusions, but generally favors the conjugate gradient squared and BiCGSTAB algorithms over the others.

Exercise 2.5.7 Get a copy of matrix **E20R0100** from MatrixMarket. This is a sparse nonsymmetric matrix, and should be represented in computer memory without storing the zero entries. If this matrix is denoted by $\mathbf{A}$, let $\mathbf{x}$ be the vector of ones, and $\mathbf{b} = \mathbf{A}\mathbf{x}$. Choose the initial guess $\widetilde{\mathbf{x}}$ for biconjugate gradients to have random entries chosen from a uniform probability distribution on $(0, 1)$. Program the biconjugate gradient iteration to solve $\mathbf{A}\mathbf{x} = \mathbf{b}$, and plot the logarithm base 10 of the error $\|\widetilde{\mathbf{x}} - \mathbf{x}\|_\infty$ versus iteration number. At what order of error does the iteration stop making progress? Relate this error size to the condition number of this matrix, which is estimated to be on the order of 10^{10}. Describe how you have implemented convergence criteria for this iteration.

2.6 Minimum Residual Methods

In Sect. 2.5, we studied gradient methods for linear systems. For Hermitian positive linear systems, the conjugate gradient algorithm minimized $\|\mathbf{b} - \mathbf{A}\mathbf{x}\|_{\mathbf{A}^{-1}}^2$ over subspaces of strictly increasing dimension. Extensions to non-Hermitian or non-positive linear systems were developed by generalizing the conjugacy conditions in the conjugate gradient algorithm.

In this section, we will examine iterative methods based on minimizing $\|\mathbf{b} - \mathbf{A}\mathbf{x}\|_{\mathbf{M}}^2$ for some Hermitian positive matrix $\mathbf{M}$. The resulting algorithms will be more complicated than conjugate gradients.

2.6.1 Orthomin

Suppose that $\mathbf{b}$ is an m-vector and that $\mathbf{A}$ is an $m \times m$ matrix. Note that we do not assume that $\mathbf{A}$ is Hermitian. Given an Hermitian positive matrix $\mathbf{M}$ of the same size as $\mathbf{A}$, we could solve $\mathbf{A}\mathbf{x} = \mathbf{b}$ by minimizing

$$\phi(\mathbf{x}) \equiv \frac{1}{2} \|\mathbf{b} - \mathbf{A}\mathbf{x}\|_{\mathbf{M}}^2 = \frac{1}{2}(\mathbf{b} - \mathbf{A}\mathbf{x})^H \mathbf{M}(\mathbf{b} - \mathbf{A}\mathbf{x}) .$$

Given an initial guess $\widetilde{\mathbf{x}}_0$, our minimization process will take the form of a recurrence

$$\widetilde{\mathbf{x}}_{n+1} = \widetilde{\mathbf{x}}_n + \sum_{j=0}^{n} \mathbf{p}_j \alpha_{j,n} ,$$

where the coefficients $\alpha_{j,n}$ are chosen so that the vectors $\mathbf{A}\mathbf{p}_j$ are linearly independent and the matrix of inner products $[\mathbf{A}\mathbf{p}_i, \mathbf{A}\mathbf{p}_j]_{\mathbf{M}}$ is nonsingular. The basic idea is contained in the following lemma.

Lemma 2.6.1 *Suppose that* $\mathbf{b}$ *is an m-vector,* $\mathbf{A}$ *and* $\mathbf{M}$ *are* $m \times m$ *matrices,* $\mathbf{P} = [\mathbf{p}_0, \ldots, \mathbf{p}_{n-1}]$ *is an* $m \times n$ *matrix, and* $\mathbf{M}$ *is Hermitian positive. Define the* $n \times n$ ***Gram matrix*** $\mathbf{G}$ *by*

$$\mathbf{G} = (\mathbf{A}\mathbf{P})^H \mathbf{M}(\mathbf{A}\mathbf{P}) .$$

1. $\mathbf{G}$ *is Hermitian.*
2. $\mathbf{G}$ *is positive if and only if the columns of* $\mathbf{A}\mathbf{P}$ *are linearly independent.*
3. *Given an m-vector* $\widetilde{\mathbf{x}}$*, if* $\mathbf{v} = \widetilde{\mathbf{x}} + \mathbf{P}\mathbf{a}$ *minimizes*

$$\phi(\mathbf{v}) = \frac{1}{2}(\mathbf{b} - \mathbf{A}\mathbf{v})^H \mathbf{M}(\mathbf{b} - \mathbf{A}\mathbf{v})$$

over all n-vectors $\mathbf{a}$*, then* $\mathbf{a}$ *solves*

$$\mathbf{G}\mathbf{a} = \mathbf{g} \equiv (\mathbf{A}\mathbf{P})^H \mathbf{M}(\mathbf{b} - \mathbf{A}\widetilde{\mathbf{x}}) . \quad (2.36)$$

4. *If the columns of* $\mathbf{A}\mathbf{P}$ *are linearly independent and* $\mathbf{a}$ *solves* $\mathbf{G}\mathbf{a} = \mathbf{g}$*, then* $\mathbf{v} = \widetilde{\mathbf{x}} + \mathbf{P}\mathbf{a}$ *uniquely minimizes* ϕ *over all possible values for* $\mathbf{a}$*.*
5. *If* $\mathbf{v} = \widetilde{\mathbf{x}} + \mathbf{P}\mathbf{a}$ *minimizes* $\phi(\mathbf{v})$ *over all possible* $\mathbf{a}$*, then the residual* $\mathbf{r} = \mathbf{b} - \mathbf{A}\mathbf{v}$ *satisfies*

$$(\mathbf{A}\mathbf{P}) \cdot \mathbf{M}\mathbf{r} = 0 . \quad (2.37)$$

Proof W prove each part in turn.

1. Since $\mathbf{G} = \mathbf{P}^H \mathbf{A}^H \mathbf{M}\mathbf{A}\mathbf{P}$ and $\mathbf{M}$ is Hermitian, $\mathbf{G}$ is Hermitian.

2. $\Longleftarrow$: Suppose that the columns of **AP** are linearly independent. If **a** is a n-vector, then since **M** is positive,

$$\mathbf{a} \cdot \mathbf{G}\mathbf{a} = (\mathbf{A}\mathbf{P}\mathbf{a}) \cdot \mathbf{M}(\mathbf{A}\mathbf{P}\mathbf{a}) \geq 0 \; .$$

If $\mathbf{a} \cdot \mathbf{G}\mathbf{a} = 0$, then the positiveness of **M** implies that $\mathbf{A}\mathbf{P}\mathbf{a} = \mathbf{0}$; since the columns of **AP** are linearly independent, we must have that $\mathbf{a} = \mathbf{0}$. This proves that **G** is positive.

$\Longrightarrow$: If the columns of **AP** are not linearly independent, then there exists $\mathbf{a} \neq \mathbf{0}$ so that $\mathbf{A}\mathbf{P}\mathbf{a} = \mathbf{0}$. Then $\mathbf{G}\mathbf{a} = (\mathbf{A}\mathbf{P})^H\mathbf{M}(\mathbf{A}\mathbf{P}\mathbf{a}) = \mathbf{0}$, so **G** is not positive.
3. If $\mathbf{v} = \widetilde{\mathbf{x}} + \mathbf{P}\mathbf{a}$ minimizes $\phi(\mathbf{v})$ over all choices of **a**, then the first-order conditions for the minimum of ϕ imply that

$$\mathbf{0} = \nabla_{\mathbf{a}}\phi\,(\widetilde{\mathbf{x}} + \mathbf{P}\mathbf{a}) = \nabla_{\mathbf{a}}\left[\frac{1}{2}(\mathbf{A}\widetilde{\mathbf{x}} + \mathbf{A}\mathbf{P}\mathbf{a} - \mathbf{b}) \cdot \mathbf{M}(\mathbf{A}\widetilde{\mathbf{x}} + \mathbf{A}\mathbf{P}\mathbf{a} - \mathbf{b})\right]$$
$$= (\mathbf{A}\mathbf{P})^H\mathbf{M}(\mathbf{A}\widetilde{\mathbf{x}} - \mathbf{b} + \mathbf{A}\mathbf{P}\mathbf{a}) = \mathbf{G}\mathbf{a} - \mathbf{g} \; .$$

4. If the columns of **AP** are linearly independent, then **G** is nonsingular, and $\mathbf{G}\mathbf{a} = \mathbf{g}$ has a unique solution **a**. Since $\mathbf{G} = \nabla_{\mathbf{a}}\nabla_{\mathbf{a}}{}^{\top}\phi$ is positive, **a** is the unique minimizer of $\frac{1}{2}\,|||\mathbf{A}(\widetilde{\mathbf{x}} + \mathbf{P}\mathbf{a}) - \mathbf{b}|||_{\mathbf{M}}^2$. This is equivalent to saying that $\mathbf{v} = \widetilde{\mathbf{x}} + \mathbf{P}\mathbf{a}$ is the unique minimizer of $\phi(\mathbf{v})$ over all choices of **a**.
5. If $\mathbf{v} = \widetilde{\mathbf{x}} + \mathbf{P}\mathbf{a}$ minimizes $\phi(\mathbf{v})$ over all choices of **a**, then we have already shown that

$$\mathbf{0} = (\mathbf{A}\mathbf{P})^H\mathbf{M}(\mathbf{A}\widetilde{\mathbf{x}} - \mathbf{b} + \mathbf{A}\mathbf{P}\mathbf{a}) = -(\mathbf{A}\mathbf{P})^H\mathbf{M}\mathbf{r} \; .$$

The next lemma will show us how to generate the **p**-vectors for our minimization process.

Lemma 2.6.2 *Suppose that* **A** *is a real $m \times m$ matrix, and that the real $m \times m$ matrix* **M** *is symmetric and positive. Suppose that the real $m \times n$ matrix* **P** *is such that* $(\mathbf{A}\mathbf{P})^H\mathbf{M}(\mathbf{A}\mathbf{P})$ *is diagonal and nonsingular. Given a real m-vector* **r**, *define the real n-vector* **q** *by*

$$(\mathbf{A}\mathbf{P})^H\mathbf{M}(\mathbf{A}\mathbf{P})\mathbf{q} = (\mathbf{A}\mathbf{P})^H\mathbf{M}(\mathbf{A}\mathbf{r}) \; .$$

If

$$\mathbf{p} = \mathbf{r} - \mathbf{P}\mathbf{q} \; , \tag{2.38}$$

then

$$(\mathbf{A}\mathbf{P})^H\mathbf{M}\mathbf{A}\mathbf{p} = \mathbf{0} \tag{2.39}$$

and $(\mathbf{A}[\mathbf{P},\ \mathbf{p}])^H\mathbf{M}(\mathbf{A}[\mathbf{P},\ \mathbf{p}])$ *is diagonal.*

Proof First, the definition (2.38) of $\mathbf{p}$ shows that

$$(\mathbf{AP})^H\mathbf{M}(\mathbf{Ap}) = (\mathbf{AP})^H\mathbf{M}(\mathbf{Ar}) - (\mathbf{AP})^H\mathbf{M}(\mathbf{APq}) = \mathbf{0} \, .$$

Since $(\mathbf{AP})^H\mathbf{M}(\mathbf{AP})$ is diagonal, so is

$$(\mathbf{A}[\mathbf{P},\ \mathbf{p}])^H\mathbf{M}(\mathbf{A}[\mathbf{P},\ \mathbf{p}]) = \begin{bmatrix} (\mathbf{AP})^H\mathbf{M}(\mathbf{AP}) & (\mathbf{AP})^H\mathbf{M}(\mathbf{Ap}) \\ (\mathbf{Ap})^H\mathbf{M}(\mathbf{AP}) & (\mathbf{Ap})^H\mathbf{M}(\mathbf{Ap}) \end{bmatrix}$$

$$= \begin{bmatrix} (\mathbf{AP})^H\mathbf{M}(\mathbf{AP}) & \mathbf{0} \\ \mathbf{0} & (\mathbf{Ap})^H\mathbf{M}(\mathbf{Ap}) \end{bmatrix} .$$

As a result, we obtain the following

Algorithm 2.6.1 (General Minimum Residual)

given $\widetilde{\mathbf{x}}_0$

$\mathbf{r}_0 = \mathbf{b} - \mathbf{A}\widetilde{\mathbf{x}}_0$

choose $\mathbf{p}_0 \neq \mathbf{0}$

for $n = 0, 1, \ldots$ until convergence

let $\mathbf{P} = [\mathbf{p}_0, \ldots, \mathbf{p}_{n-1}]$ (2.40a)

solve $(\mathbf{AP})^H\mathbf{M}(\mathbf{AP})\mathbf{a} = (\mathbf{AP})^H\mathbf{M}\mathbf{r}_{n-1}$ for the n-vector $\mathbf{a}$ (2.40b)

$\mathbf{d} = \mathbf{Pa}$

$\mathbf{s} = \mathbf{Ad}$

$\widetilde{\mathbf{x}}_n = \widetilde{\mathbf{x}}_{n-1} + \mathbf{d}$ (2.40c)

$\mathbf{r}_n = \mathbf{r}_{n-1} - \mathbf{s}$ (2.40d)

solve $(\mathbf{AP})^H\mathbf{M}(\mathbf{AP})\mathbf{q} = (\mathbf{AP})^H\mathbf{MA}\mathbf{r}_n$ for the n -vector $\mathbf{q}$ (2.40e)

$\mathbf{p}_n = \mathbf{r}_n - \mathbf{Pq}$. (2.40f)

This algorithm is not very efficient, and should not be used. We need to find ways to simplify the calculation of $\mathbf{a}$, $\widetilde{\mathbf{x}}_{n+1}$ and $\mathbf{r}_{n+1}$. We will see that $\mathbf{a}$ solves a diagonal system in which all but one of the entries of the right-hand side are zero, and $\mathbf{q}$ solves a diagonal system in which all of the entries of the right-hand side may be nonzero.

Lemma 2.6.3 ([7, Lemma 12.1]) *Suppose that* $\mathbf{A}$ *is a real* $m \times m$ *matrix, and that the real* $m \times m$ *matrix* $\mathbf{M}$ *is symmetric positive. Given real m-vectors* $\mathbf{b}$ *and* $\widetilde{\mathbf{x}}_0$*, suppose that we compute the solution vectors* $\widetilde{\mathbf{x}}_0, \ldots, \widetilde{\mathbf{x}}_n$*, the residual vectors* $\mathbf{r}_0, \ldots, \mathbf{r}_n$ *and the search directions* $\mathbf{p}_0, \ldots, \mathbf{p}_n$ *by the recurrences in* (2.6.1). *For* *each $n \geq 1$, let*

$$\mathbf{P} = [\mathbf{p}_0, \ldots, \mathbf{p}_{n-1}] \quad \textit{and} \quad \mathbf{R} = [\mathbf{r}_0, \ldots, \mathbf{r}_{n-1}] \ .$$

Finally, suppose that the matrix

$$\mathbf{G} \equiv (\mathbf{AP})^H \mathbf{M} (\mathbf{AP})$$

is nonsingular for all $n \geq 1$.

1. *For $n \geq 1$, $\widetilde{\mathbf{x}}_n = \widetilde{\mathbf{x}}_{n-1} + \mathbf{Pa}$ minimizes $\phi(\widetilde{\mathbf{x}}_n) \equiv \frac{1}{2}(\mathbf{b} - \mathbf{A}\widetilde{\mathbf{x}}_n)\mathbf{M}(\mathbf{b} - \mathbf{A}\widetilde{\mathbf{x}}_n)$ over all real n-vectors $\mathbf{a}$;*
2. *$\mathbf{r}_n = \mathbf{b} - \mathbf{A}\widetilde{\mathbf{x}}_n$ and $(\mathbf{AP})^H \mathbf{M}\mathbf{r}_n = \mathbf{0}$;*
3. *$(\mathbf{AP})^H \mathbf{M}(\mathbf{A}\mathbf{p}_n) = \mathbf{0}$;*
4. *For all $n \geq 1$, $\mathbf{G}$ is diagonal;*
5. *If $\mathbf{p}_0 = \mathbf{r}_0$ then for all $n \geq 1$ there is a real $n \times n$ upper triangular matrix $\mathbf{Q}$ with diagonal entries all equal to 1 so that $\mathbf{R} = \mathbf{PQ}$;*
6. *If there exists a real $n \times n$ matrix $\mathbf{Q}$ so that $\mathbf{R} = \mathbf{PQ}$, then $(\mathbf{AR})^H \mathbf{M}\mathbf{r}_n = \mathbf{0}$; and*
7. *$(\mathbf{A}[\mathbf{P}, \mathbf{p}_n])^H \mathbf{M}\mathbf{r}_n = (\mathbf{A}[\mathbf{P}, \mathbf{r}_n])^H \mathbf{M}\mathbf{r}_n$.*

Proof The first claim was proved in part (iii) of Lemma 2.6.1.

Let us prove the second claim. Note that $\mathbf{r}_0 = \mathbf{b} - \mathbf{A}\widetilde{\mathbf{x}}_0$, so the first part of the second claim is true for $n = 0$. Inductively, assume that it is true for $n - 1$. Then Eqs. (2.40c) and (2.40d) updating $\widetilde{\mathbf{x}}$ and $\mathbf{r}$ imply that

$$\mathbf{r}_n = \mathbf{r}_{n-1} - \mathbf{APa} = \mathbf{b} - \mathbf{A}\mathbf{x}_{n-1} - \mathbf{APa} = \mathbf{b} - \mathbf{A}(\mathbf{x}_{n-1} + \mathbf{Pa}) = \mathbf{b} - \mathbf{A}\mathbf{x}_n \ .$$

Finally, we note that Eq. (2.40c) and the definition (2.40b) of $\mathbf{a}$ imply that

$$(\mathbf{AP})^H \mathbf{M}\mathbf{r}_n = (\mathbf{AP})^H \mathbf{M}\mathbf{r}_{n-1} - (\mathbf{AP})^H \mathbf{M}(\mathbf{AP})\mathbf{a} = \mathbf{0} \ .$$

This completes the proof of the second claim.

To prove the third claim, we use Eq. (2.40f) for updating $\mathbf{p}$ and the definition (2.40e) of $\mathbf{q}$:

$$(\mathbf{AP})^H \mathbf{MA}\mathbf{p}_n = (\mathbf{AP})^H \mathbf{MA}[\mathbf{r}_n - \mathbf{Pq}] = \mathbf{0} \ .$$

We will prove the fourth claim by induction. For $n = 1$, the 1×1 matrix $\mathbf{p}_0 \cdot \mathbf{A}^H \mathbf{MA}\mathbf{p}_0$ is obviously diagonal. Assume that

$$\mathbf{G} \equiv (\mathbf{AP})^H \mathbf{M}(\mathbf{AP})$$

is diagonal when $\mathbf{P}$ has n columns. Then for $n+1$ the third claim implies that the corresponding matrix with $n+1$ columns is

$$(\mathbf{A}\,[\mathbf{P}\,,\;\mathbf{p}_n])^H\mathbf{M}\,(\mathbf{A}\,[\mathbf{P}\,,\;\mathbf{p}_n]) = \begin{bmatrix} (\mathbf{AP})^H\mathbf{M}(\mathbf{AP}) & (\mathbf{AP})^H\mathbf{M}\,(\mathbf{Ap}_n) \\ (\mathbf{Ap}_n)^H\mathbf{M}(\mathbf{AP}) & (\mathbf{Ap}_n)^H\mathbf{M}\,(\mathbf{Ap}_n) \end{bmatrix}$$

$$= \begin{bmatrix} \mathbf{G} & \mathbf{0} \\ \mathbf{0} & (\mathbf{Ap}_n)^H\mathbf{M}\,(\mathbf{Ap}_n) \end{bmatrix} .$$

We turn now to the fifth claim. If $\mathbf{p}_0 = \mathbf{r}_0$, then the claim is obviously true for $n = 1$. Assume that for n we have $\mathbf{R} = \mathbf{PQ}$ where $\mathbf{Q}$ is upper triangular with ones on the diagonal. Then Eq. (2.40f) implies that the corresponding matrix for $n+1$ is

$$[\mathbf{R}\,,\;\mathbf{r}_n] = [\mathbf{PQ}\,,\;\mathbf{Pq}+\mathbf{p}_n] = [\mathbf{P}\,,\;\mathbf{p}_n]\begin{bmatrix} \mathbf{Q} & \mathbf{q} \\ \mathbf{0} & 1 \end{bmatrix} .$$

Next, we prove the sixth claim. Equation (2.40d) for updating $\mathbf{r}$ and Eq. (2.40b) for defining $\mathbf{a}$ imply that

$$(\mathbf{AR})^H\mathbf{Mr}_n = \mathbf{Q}^H(\mathbf{AP})^H\mathbf{Mr}_n = \mathbf{Q}^H\left[(\mathbf{AP})^H\mathbf{Mr}_{n-1} - (\mathbf{AP})^H\mathbf{M}(\mathbf{AP})\mathbf{a}\right]$$
$$= \mathbf{0} .$$

Let us prove the final claim. Using Eq. (2.40f) for updating $\mathbf{p}$ and the second claim, we obtain

$$(\mathbf{A}\,[\mathbf{P}\,,\;\mathbf{p}_n])^H\mathbf{Mr}_n = [\mathbf{AP}\,,\;\mathbf{Ar}_n - \mathbf{APq}]^H\mathbf{Mr}_n$$
$$= (\mathbf{A}\,[\mathbf{P}\,,\;\mathbf{r}_n])^H\mathbf{Mr}_n - \mathbf{q}\cdot(\mathbf{AP})^H\mathbf{Mr}_n$$
$$= (\mathbf{A}\,[\mathbf{P}\,,\;\mathbf{r}_n])^H\mathbf{Mr}_n - \mathbf{q}\cdot\mathbf{0} = (\mathbf{A}\,[\mathbf{P}\,,\;\mathbf{r}_n])^H\mathbf{Mr}_n .$$

Next, we will prove a condition on the matrix M that will guarantee that the iterative method will succeed.

Lemma 2.6.4 ([7, Theorem 12.2]) *Suppose that* $\mathbf{A}$ *is an* $m \times m$ *matrix, and the* $m \times m$ *matrix* $\mathbf{M}$ *is Hermitian positive. Given m-vectors* $\mathbf{b}$ *and* $\widetilde{\mathbf{x}}_0$*, suppose that we compute the solution vectors* $\widetilde{\mathbf{x}}_0, \ldots, \widetilde{\mathbf{x}}_n$*, the residual vectors* $\mathbf{r}_0, \ldots, \mathbf{r}_n$ *and the search directions* $\mathbf{p}_0, \ldots, \mathbf{p}_n$ *by the recurrences in* (2.6.1)*. For each* $n \geq 1$*, let* $\mathbf{P} = [\mathbf{p}_0, \ldots, \mathbf{p}_{n-1}]$*. Finally, suppose that* $\mathbf{MA}+\mathbf{A}^H\mathbf{M}$ *is Hermitian and positive and that, for* $0 \leq i \leq n$, $\mathbf{r}_i \neq 0$*. Then the vectors* $\mathbf{Ap}_0, \ldots, \mathbf{Ap}_n$ *are linearly independent and the matrix* $(\mathbf{AP})^H\mathbf{M}(\mathbf{AP})$ *is nonsingular.*

Proof We will prove the lemma by contradiction. Suppose without loss of generality that n is the smallest index such that $(\mathbf{AP})^H\mathbf{M}(\mathbf{AP})$ is singular. Then the second claim of Lemma 2.6.1 shows that the set $\{\mathbf{Ap}_j\}_{j=0}^n$ is linearly dependent and

$\{\mathbf{Ap}_j\}_{j=0}^{n-1}$ is linearly independent. Thus there exists a real n-vector $\mathbf{c}$ so that

$$\mathbf{Ap}_n = (\mathbf{AP})\mathbf{c} \; .$$

Since recurrence (2.40f) for updating the search direction states that

$$\mathbf{r}_n = \mathbf{p}_n + \mathbf{Pq} \; ,$$

we have

$$\mathbf{Ar}_n = (\mathbf{AP})(\mathbf{q} + \mathbf{c}) \; .$$

It follows from the second claim of Lemma 2.6.3 that

$$[\mathbf{Ar}_n, \mathbf{r}_n]_{\mathbf{M}} = [(\mathbf{AP})(\mathbf{q}+\mathbf{c}), \mathbf{r}_n]_{\mathbf{M}} = (\mathbf{q}+\mathbf{c}) \cdot (\mathbf{AP})^H \mathbf{Mr}_n = \mathbf{0} \; .$$

This implies that

$$\mathbf{0} = [\mathbf{Ar}_n, \mathbf{r}_n]_{\mathbf{M}} + [\mathbf{r}_n, \mathbf{Ar}_n]_{\mathbf{M}} = \mathbf{r}_n \cdot \left[\mathbf{A}^H\mathbf{M} + \mathbf{MA}\right] \mathbf{r}_n \; .$$

Since $\mathbf{r}_n \neq 0$ and $\mathbf{A}^H\mathbf{M} + \mathbf{MA}$ is Hermitian and positive, we have a contradiction.

Note that the conjugacy of the search directions (2.39) implies that the matrix

$$\mathbf{G} = (\mathbf{AP})^H \mathbf{M} (\mathbf{AP}) \; ,$$

which is used in (2.36) to determine the coefficients $\mathbf{a}$, is diagonal. As a result, the minimum residual algorithm simplifies to

Algorithm 2.6.2 (MINRES)

$$\begin{aligned}
&\mathbf{r}_0 = \mathbf{b} - \mathbf{A}\widetilde{\mathbf{x}}_0 \\
&\mathbf{p}_0 = \mathbf{r}_0 \\
&\text{for } n = 1, \ldots \text{ until convergence} \\
&\quad \mathbf{z}_{n-1} = \mathbf{Ap}_{n-1} \\
&\quad \text{compute } \delta_{n-1} = [\mathbf{z}_{n-1}, \mathbf{z}_{n-1}]_{\mathbf{M}} \\
&\quad \alpha = [\mathbf{z}_{n-1}, \mathbf{r}_{n-1}]_{\mathbf{M}} / \delta_{n-1} \\
&\quad \widetilde{\mathbf{x}}_n = \widetilde{\mathbf{x}}_{n-1} + \mathbf{p}_{n-1}\alpha \\
&\quad \mathbf{r}_n = \mathbf{r}_{n-1} - \mathbf{z}_{n-1}\alpha \\
&\quad \mathbf{t}_n = \mathbf{Ar}_n \\
&\quad \text{for } 0 \le j < n \quad \omega_j = \left[\mathbf{Ap}_j, \mathbf{t}_n\right]_{\mathbf{M}} / \delta_j \\
&\quad \mathbf{p}_n = \mathbf{r}_n - \sum_{j=0}^{n-1} \mathbf{p}_j \omega_j \; .
\end{aligned} \tag{2.41}$$

This algorithm at step n requires storage for a total of $2n+5$ vectors, namely vectors $\mathbf{p}_j$ for $0 \le j < n$, vectors $\mathbf{Ap}_j$ for $0 \le j < n$, $\widetilde{\mathbf{x}}_n$, $\mathbf{r}_n$ and $\mathbf{Ar}_n$.

A C++ class to implement the minimum residual iteration can be found in class `MinResSolver` in files Solver.H and Solver.C. The constructor for this class needs to be passed a reference to a `SolverPreconditioner` to perform preconditioning as well.

The following lemma describes the convergence of the minimum residual algorithm.

Lemma 2.6.5 *Suppose that the hypotheses of Lemma 2.6.4 are satisfied, and that* $\mathbf{r}_{n-1} \neq 0$. *Let* $\mathbf{B} \equiv \mathbf{M}^{1/2}\mathbf{A}\mathbf{M}^{-1/2}$. *If* $\lambda_{\min}$ *is the smallest eigenvalue of its matrix argument, let*

$$\xi \equiv \lambda_{\min}\left(\frac{1}{2}\left[\mathbf{B}+\mathbf{B}^H\right]\right)\lambda_{\min}\left(\frac{1}{2}\left[\mathbf{B}^{-1}+\mathbf{B}^{-H}\right]\right) .$$

Then

$$\|\mathbf{r}_n\|_{\mathbf{M}}^2 \leq (1-\xi)\,\|\mathbf{r}_{n-1}\|_{\mathbf{M}}^2 .$$

If $\xi < \frac{1}{2}$, *then we have the better estimate*

$$\|\mathbf{r}_n\|_{\mathbf{M}}^2 \leq \left(1-\frac{\xi}{1-\xi}\right)\|\mathbf{r}_{n-1}\|_{\mathbf{M}}^2 .$$

Proof See [7, Theorem 12.5].

Example 2.6.1 ([7, p. 520]) Suppose that we want to solve

$$\mathbf{A}'\mathbf{x} = \mathbf{b}' .$$

The **orthomin** iteration is the minimum residual method that chooses $\mathbf{M} = \mathbf{I}$. Lemma 2.6.4 shows that this algorithm converges if $\mathbf{A}'+\mathbf{A}'^H$ is positive. If $\mathbf{A}'+\mathbf{A}'^H$ is not positive, we can choose $\mathbf{Q}$ appropriately (e.g., $\mathbf{Q} = [\mathbf{A}']^H$) and apply the iteration to

$$\mathbf{A}\mathbf{x} \equiv \mathbf{Q}\mathbf{A}'\mathbf{x} = \mathbf{Q}\mathbf{b}' \equiv \mathbf{b} .$$

Exercise 2.6.1 Suppose that $\mathbf{A}$ is symmetric positive and let $\mathbf{M} = \mathbf{A}^{-1}$. Show that in this case the minimum residual algorithm reduces to the standard conjugate gradient algorithm.

Exercise 2.6.2 Suppose that $\mathbf{M} = [(\mathbf{A} + \mathbf{A}^\top)/2]^{-1}$ is symmetric and positive. Describe the resulting minimum residual method, due to Concus et al. [7, p. 527].

2.6.2 GMRES

In Sect. 2.6.1 we generated search directions by minimizing a quadratic form over finite-dimensional subspaces of increasing dimension. This led to the recurrence (2.38) for the search directions. In this section, we will require the search directions to be mutually orthogonal. Given an initial search direction $\mathbf{p}_1$, we will generate these search directions recursively within the span of the Krylov subspace

$$\mathscr{K}_k \equiv \left\langle \mathbf{A}^0\mathbf{p}_1, \ldots, \mathbf{A}^{k-1}\mathbf{p}_1 \right\rangle .$$

More specifically, for each $1 \leq k \leq m$ we will seek real orthonormal m-vectors $\mathbf{p}_1, \ldots, \mathbf{p}_k$ and a real $k \times k$ upper Hessenberg matrix $\mathbf{H}_k = \left[\eta_{i,j}\right]$ so that $\mathbf{P}_k = [\mathbf{p}_1, \ldots, \mathbf{p}_k]$ satisfies

$$\mathbf{A}\mathbf{P}_k = \mathbf{P}_k\mathbf{H}_k + \mathbf{p}_{k+1}\eta_{k+1,k}\mathbf{e}_k{}^\top . \tag{2.42}$$

Given an initial guess $\widetilde{\mathbf{x}}_1$ for the solution $\mathbf{x}$ to $\mathbf{A}\mathbf{x} = \mathbf{b}$, we will seek $\widetilde{\mathbf{x}}_{k+1} \in \widetilde{\mathbf{x}}_1 + \mathscr{R}(\mathbf{P}_k)$ to minimize $\|\mathbf{b} - \mathbf{A}\widetilde{\mathbf{x}}_{k+1}\|_2$.

For $k \leq m$, Eq. (2.42) and the orthonormality of $\mathbf{p}_1, \ldots, \mathbf{p}_k$ imply that for $1 \leq i \leq k$ we have

$$\mathbf{p}_i \cdot \mathbf{A}\mathbf{p}_k = \mathbf{p}_i \cdot \mathbf{A}\mathbf{P}_k\mathbf{e}_k = \mathbf{p}_i \cdot \left(\mathbf{P}_k\mathbf{H}_k + \mathbf{p}_{k+1}\eta_{k+1,k}\mathbf{e}_k{}^\top\right)\mathbf{e}_k = \left(\mathbf{e}_i{}^\top\mathbf{H}_k\right)\mathbf{e}_k = \eta_{i,k} ,$$

and that

$$\mathbf{p}_{k+1} \cdot \mathbf{A}\mathbf{p}_k = \mathbf{p}_{k+1} \cdot \mathbf{A}\mathbf{P}_k\mathbf{e}_k = \mathbf{p}_{k+1} \cdot \left(\mathbf{P}_k\mathbf{H}_k + \mathbf{p}_{k+1}\eta_{k+1,k}\mathbf{e}_k{}^\top\right)\mathbf{e}_k = \eta_{k+1,k} .$$

Fortunately, we can use the **modified Gram-Schmidt process** to compute the orthonormal vectors $\mathbf{p}_k$ and the entries of $\mathbf{H}_k$, and obtain the following

Algorithm 2.6.3 (Arnoldi)

given $\mathbf{p}_1$ with $\|\mathbf{p}_1\|_2 = 1$
for $1 \leq k < m$
  $\widehat{\mathbf{p}}_{k+1} = \mathbf{A}\mathbf{p}_k$
  for $1 \leq i \leq k$
    $\eta_{i,k} = \mathbf{p}_i \cdot \widehat{\mathbf{p}}_{k+1}$
    $\widehat{\mathbf{p}}_{k+1} = \widehat{\mathbf{p}}_{k+1} - \mathbf{p}_i\eta_{i,k}$
  $\eta_{k+1,k} = \left\|\widehat{\mathbf{p}}_{k+1}\right\|_2$
  if $\eta_{k+1,k} > 0$ then
    $\mathbf{p}_{k+1} = \widehat{\mathbf{p}}_{k+1}/\eta_{k+1,k}$
  else
    choose $\mathbf{p}_{k+1} \perp \{\mathbf{p}_1, \ldots, \mathbf{p}_k\}$ so that $\|\mathbf{p}_{k+1}\|_2 = 1$.

Recall that the modified Gram-Schmidt process was described in Sect. 6.8 in Chap. 6 of Volume I For more details regarding the implementation of the modified Gram-Schmidt factorization (and, in particular, how to deal with small values of $\eta_{j+1,j}$), see Daniel et al. [43]. A version of the Arnoldi process using Householder reflectors can be found in Saad [154, p. 150]. The Arnoldi algorithm requires one matrix-vector multiplication per step, and for s steps requires the storage of $s+1$ vectors. Fortunately, the vectors $\mathbf{A}\mathbf{p}_k$, $\widehat{\mathbf{p}}_{k+1}$ and $\mathbf{p}_{k+1}$ can share the same storage.

The next lemma summarizes the Arnoldi process.

Lemma 2.6.6 *Suppose that* $\mathbf{A}$ *is an* $m \times m$ *matrix* $\mathbf{A}$. *For any positive integer* $k \leq m$, *let the* $m \times k$ *matrix* $\mathbf{P}_k = [\mathbf{p}_1, \ldots, \mathbf{p}_k]$ *and the* $k \times k$ *upper Hessenberg matrix* $\mathbf{H}_k = [\eta_{i,j}]$ *be computed by the Arnoldi process (Algorithm 2.6.3). Then*

1. *for all* $k \in [1, m]$, ${\mathbf{P}_k}^H \mathbf{P}_k = \mathbf{I}$;
2. *for all* $k \in [1, m]$, $\mathscr{R}(\mathbf{P}_k) = \mathscr{K}_k \equiv \langle \mathbf{A}^0 \mathbf{p}_1, \ldots, \mathbf{A}^{k-1} \mathbf{p}_1 \rangle$;
3. *for all* $k \in [2, m]$, $\mathscr{R}(\mathbf{P}_k) = \langle \mathbf{p}_1, \mathbf{A}\mathbf{p}_1, \ldots \mathbf{A}\mathbf{p}_{k-1} \rangle$;
4. *for all* $k \in [1, m]$, $\mathbf{H}_k = {\mathbf{P}_k}^H \mathbf{A} \mathbf{P}_k$; *and*
5. *for all* $k \in [1, m)$, $\mathbf{A}\mathbf{P}_k = \mathbf{P}_k \mathbf{H}_k + \mathbf{p}_{k+1} \eta_{k+1,k} {\mathbf{e}_k}^\top$.

Proof We will prove the first claim by induction. Note that $\|\mathbf{p}_1\|_2 = 1$, so the claim is true for $k = 1$. Assume inductively that ${\mathbf{P}_k}^H \mathbf{P}_k = \mathbf{I}$. Then Eq. (6.20) in Chap. 6 of Volume I describing successive orthogonal projection implies that for all $1 \leq i \leq k < m$

$$
\begin{aligned}
{\mathbf{e}_i}^\top {\mathbf{P}_k}^H \mathbf{p}_{k+1} \eta_{k+1,k} &= \mathbf{p}_i \cdot \left(\mathbf{I} - \mathbf{p}_k [\mathbf{p}_k]^H\right) \ldots \left(\mathbf{I} - \mathbf{p}_1 [\mathbf{p}_1]^H\right) \mathbf{A}\mathbf{p}_k \\
&= \mathbf{p}_i \cdot \left(\mathbf{I} - \mathbf{p}_i [\mathbf{p}_i]^H\right) \ldots \left(\mathbf{I} - \mathbf{p}_1 [\mathbf{p}_1]^H\right) \mathbf{A}\mathbf{p}_k = \mathbf{0}\,.
\end{aligned}
$$

We now consider two cases. If $\eta_{k+1,k} \neq 0$ then this implies that ${\mathbf{P}_k}^H \mathbf{p}_{k+1} = \mathbf{0}$. Otherwise the Arnoldi Algorithm 2.6.3 chooses $\mathbf{p}_{k+1} \perp \mathscr{R}(\mathbf{P}_k)$, and we also have ${\mathbf{P}_k}^H \mathbf{p}_{k+1} = \mathbf{0}$. In either case,

$$
[\mathbf{P}_k, \mathbf{p}_{k+1}]^H [\mathbf{P}_k, \mathbf{p}_{k+1}] = \begin{bmatrix} {\mathbf{P}_k}^H \mathbf{P}_k & {\mathbf{P}_k}^H \mathbf{p}_{k+1} \\ (\mathbf{P}_k \mathbf{p}_{k+1})^H & \mathbf{p}_{k+1} \cdot \mathbf{p}_{k+1} \end{bmatrix} = \begin{bmatrix} \mathbf{I} & \mathbf{0} \\ \mathbf{0} & 1 \end{bmatrix}.
$$

This completes the inductive proof of the first claim.

We will also prove the second claim by induction. The claim is obviously true for $k = 1$. Assume that the claim is true for some positive integer $k < m$. The Arnoldi process (Algorithm 2.6.3) and Eq. (6.20) in Chap. 6 of Volume I describing successive orthogonal projection imply that

$$
\begin{aligned}
\mathbf{p}_{k+1} &= \left(\mathbf{I} - \mathbf{p}_k [\mathbf{p}_k]^H\right) \ldots \left(\mathbf{I} - \mathbf{p}_1 [\mathbf{p}_1]^H\right) \mathbf{A}\mathbf{p}_k \\
&\in \langle \mathbf{p}_1, \ldots, \mathbf{p}_k, \mathbf{A}\mathbf{p}_k \rangle\ .
\end{aligned}
$$

Since the inductive hypothesis shows that

$$\langle \mathbf{p}_1, \ldots, \mathbf{p}_k \rangle = \left\langle \mathbf{A}^0 \mathbf{p}_1, \ldots, \mathbf{A}^{k-1} \mathbf{p}_1 \right\rangle = \mathscr{K}_k ,$$

the second claim now follows easily from the linear independence of the vectors $\mathbf{p}_i$, established in the first claim.

The proof of the third claim is similar. We have

$$\mathbf{p}_2 = \left(\mathbf{I} - \mathbf{p}_1 [\mathbf{p}_1]^H \right) \mathbf{A} \mathbf{p}_1 \in \langle \mathbf{p}_1, \mathbf{A} \mathbf{p}_1 \rangle ,$$

so the third claim is true for $k = 2$. Assume inductively that for some $k < m$ we have

$$\mathscr{R}(\mathbf{P}_k) = \langle \mathbf{p}_1, \mathbf{A}\mathbf{p}_1, \ldots, \mathbf{A}\mathbf{p}_{k-1} \rangle .$$

Since the proof of the second claim showed that

$$\mathbf{p}_{k+1} \in \langle \mathbf{p}_1, \ldots, \mathbf{p}_k, \mathbf{A}\mathbf{p}_k \rangle ,$$

we see that

$$\mathscr{R}([\mathbf{P}_k, \mathbf{p}_{k+1}]) = \mathscr{R}(\mathbf{P}_k) \cup \langle \mathbf{p}_{k+1} \rangle = \langle \mathbf{p}_1, \mathbf{A}\mathbf{p}_1, \ldots, \mathbf{A}\mathbf{p}_k \rangle .$$

We will prove the fourth claim by induction. Since $\mathbf{p}_1 \cdot \mathbf{A}\mathbf{p}_1$ is a 1×1 matrix, this matrix is obviously upper Hessenberg. Inductively, we assume that for some $k < m$ the matrix ${\mathbf{P}_k}^H \mathbf{A} \mathbf{P}_k = \mathbf{H}_k$ is upper Hessenberg. Note that the first claim implies that $\langle \mathbf{p}_1, \ldots, \mathbf{p}_k \rangle \perp \mathbf{p}_{k+1}$, and the third claim implies that

$$\langle \mathbf{A}\mathbf{p}_1, \ldots, \mathbf{A}\mathbf{p}_{k-1} \rangle \subset \langle \mathbf{p}_1, \mathbf{A}\mathbf{p}_1, \ldots, \mathbf{A}\mathbf{p}_{k-1} \rangle = \mathscr{R}(\mathbf{P}_k) \perp \mathbf{p}_{k+1} .$$

As a result,

$$[\mathbf{p}_{k+1}]^H \mathbf{A} \mathbf{P}_k = \mathbf{p}_{k+1} \cdot \mathbf{A}\mathbf{p}_k {\mathbf{e}_k}^\top .$$

This implies that

$$[\mathbf{P}_k, \mathbf{p}_{k+1}]^H \mathbf{A} [\mathbf{P}_k, \mathbf{p}_{k+1}] = \begin{bmatrix} {\mathbf{P}_k}^H \mathbf{A} \mathbf{P}_k & {\mathbf{P}_k}^H \mathbf{A} \mathbf{p}_{k+1} \\ (\mathbf{p}_{k+1})^H \mathbf{A} \mathbf{P}_k & \mathbf{p}_{k+1} \cdot \mathbf{A}\mathbf{p}_{k+1} \end{bmatrix}$$

$$= \begin{bmatrix} {\mathbf{P}_k}^H \mathbf{A} \mathbf{P}_k & {\mathbf{P}_k}^H \mathbf{A} \mathbf{p}_{k+1} \\ \mathbf{p}_{k+1} \cdot \mathbf{A}\mathbf{p}_k {\mathbf{e}_k}^\top & \mathbf{p}_{k+1} \cdot \mathbf{A}\mathbf{p}_{k+1} \end{bmatrix}$$

is upper Hessenberg.

We will prove the final claim by induction. First, we will show that this claim is true for $k = 1$. Note that the Arnoldi process (Algorithm 2.6.3) implies that

$$\mathbf{p}_2\eta_{2,1} = \left(\mathbf{I} - \mathbf{p}_1[\mathbf{p}_1]^H\right)\mathbf{A}\mathbf{p}_1 = \mathbf{A}\mathbf{p}_1 - \mathbf{p}_1\mathbf{p}_1 \cdot \mathbf{A}\mathbf{p}_1 \ .$$

This can be rewritten as

$$\mathbf{A}\mathbf{p}_1 = \mathbf{p}_1\mathbf{p}_1 \cdot \mathbf{A}\mathbf{p}_1 + \mathbf{p}_2\eta_{2,1} = \mathbf{P}_1\mathbf{H}_1 + \mathbf{p}_2\eta_{2,1} \ .$$

This proves the claim for $k = 1$. Inductively, we assume that for $k < m$ we have

$$\mathbf{A}\mathbf{P}_k = \mathbf{P}_k\mathbf{H}_k + \mathbf{p}_{k+1}\eta_{k+1,k}\mathbf{e}_k{}^\top \ .$$

The Arnoldi process (Algorithm 2.6.3) and the fourth claim imply that

$$\mathbf{p}_{k+2}\eta_{k+2,k+1} = \left(\mathbf{I} - \mathbf{P}_{k+1}\mathbf{P}_{k+1}{}^H\right)\mathbf{A}\mathbf{p}_{k+1} = \mathbf{A}\mathbf{p}_{k+1} - \mathbf{P}_{k+1}\mathbf{P}_{k+1}{}^H\mathbf{A}\mathbf{p}_{k+1}$$
$$= \mathbf{A}\mathbf{p}_{k+1} - \mathbf{P}_{k+1}\mathbf{H}_{k+1}\mathbf{e}_{k+1} = \mathbf{A}\mathbf{p}_{k+1} - \left[\mathbf{P}_k \ \mathbf{p}_{k+1}\right]\begin{bmatrix}\mathbf{H}_{k+1}\mathbf{e}_{k+1} \\ \eta_{k+1,k+1}\end{bmatrix} .$$

As a result, we have

$$\begin{aligned}\mathbf{A}\mathbf{P}_{k+1} &= \mathbf{A}\left[\mathbf{P}_k, \mathbf{p}_{k+1}\right] \\ &= \left[\mathbf{P}_k\mathbf{H}_k + \mathbf{p}_{k+1}\eta_{k+1,k}\mathbf{e}_k{}^\top, \mathbf{P}_k\mathbf{H}_{k+1}\mathbf{e}_{k+1} + \mathbf{p}_{k+1}\eta_{k+1,k+1} + \mathbf{p}_{k+2}\eta_{k+2,k+1}\right] \\ &= \left[\mathbf{P}_k, \mathbf{p}_{k+1}\right]\begin{bmatrix}\mathbf{H}_k & \mathbf{H}_{k+1}\mathbf{e}_{k+1} \\ \eta_{k+1,k}\mathbf{e}_k{}^\top & \eta_{k+1,k+1}\end{bmatrix} + \mathbf{p}_{k+2}\eta_{k+2,k+1}\mathbf{e}_{k+1}{}^\top \\ &= \mathbf{P}_{k+1}\mathbf{H}_{k+1} + \mathbf{p}_{k+2}\eta_{k+2,k+1}\mathbf{e}_{k+1}{}^\top \ .\end{aligned}$$

In particular, the Arnoldi process (Algorithm 2.6.3), shows that $\widehat{\mathbf{p}}_{m+1} = (\mathbf{I} - \mathbf{P}_m\mathbf{P}_m{}^H)\mathbf{A}\mathbf{p}_m$ is orthogonal to m orthonormal vectors, and must be zero. It follows that $\mathbf{A}\mathbf{P}_m = \mathbf{P}_m\mathbf{H}_m$.

Given an initial guess $\widetilde{\mathbf{x}}_0$ and the search vectors provided by the columns of the matrix $\mathbf{P}_k$, we will seek a real k-vector $\mathbf{y}_k$ to minimize

$$\|\mathbf{b} - \mathbf{A}\widetilde{\mathbf{x}}_{k+1}\|_2 = \|\mathbf{b} - \mathbf{A}\left[\widetilde{\mathbf{x}}_1 + \mathbf{P}_k\mathbf{y}_k\right]\|_2 = \|[\mathbf{b} - \mathbf{A}\widetilde{\mathbf{x}}_1] - \mathbf{A}\mathbf{P}_k\mathbf{y}_k\|_2 \ .$$

The next lemma discusses how we can perform this minimization.

Lemma 2.6.7 *Suppose that* $\mathbf{A}$ *is an* $m \times m$ *matrix,* $\mathbf{b}$ *is a* m*-vector, and the* m*-vector* $\widetilde{\mathbf{x}}_1$ *is such that* $\mathbf{r}_1 = \mathbf{b} - \mathbf{A}\widetilde{\mathbf{x}}_0 \neq \mathbf{0}$. *Let* $\mathbf{p}_1 = \mathbf{r}_1/\|\mathbf{r}_1\|_2$, *and suppose that* $\{\mathbf{p}_1, \ldots, \mathbf{p}_k\}$ *are mutually orthogonal vectors generated by the Arnoldi process (Algorithm 2.6.3).*

Define $\mathbf{P}_k = [\mathbf{p}_1, \ldots, \mathbf{p}_k]$*, and let* $\mathbf{H}_k = \mathbf{P}_k{}^H \mathbf{A} \mathbf{P}_k$*. Then*

1.

$$\min_{k\text{-vectors } \mathbf{y}} \|\mathbf{r}_1 - \mathbf{A}\mathbf{P}\mathbf{y}\|_2 = \min_{k\text{-vectors } \mathbf{y}} \left\| \begin{bmatrix} \mathbf{H}_k \\ \eta_{k+1,k} \mathbf{e}_k{}^\top \end{bmatrix} \mathbf{y} - \begin{bmatrix} \mathbf{e}_1 \\ 0 \end{bmatrix} \|\mathbf{r}_1\|_2 \right\|_2 ;$$

2. if

$$\mathbf{Q}_{k+1} \begin{bmatrix} \mathbf{H}_k \\ \eta_{k+1,k} \mathbf{e}_k{}^\top \end{bmatrix} = \begin{bmatrix} \mathbf{R}_k \\ \mathbf{0} \end{bmatrix} ,$$

where the $(k+1) \times (k+1)$ *matrix* $\mathbf{Q}_{k+1}$ *is unitary and the* $k \times k$ *matrix* $\mathbf{R}_k$ *is right triangular, then*

$$\min_{k\text{-vectors } \mathbf{y}} \|\mathbf{r}_1 - \mathbf{A}\mathbf{P}_k\mathbf{y}\|_2 = \min_{k\text{-vectors } \mathbf{y}} \left\| \mathbf{Q}_{k+1} \begin{bmatrix} \mathbf{e}_1 \\ 0 \end{bmatrix} \|\mathbf{r}_1\|_2 - \begin{bmatrix} \mathbf{R}_k \\ \mathbf{0} \end{bmatrix} \mathbf{y} \right\|_2 ;$$

3. if $\mathbf{R}_k$ *is nonsingular, then*

$$\min_{k\text{-vectors } \mathbf{y}} \|\mathbf{r}_1 - \mathbf{A}\mathbf{P}\mathbf{y}\|_2 = \left| \mathbf{e}_{k+1}{}^\top \mathbf{Q}_{k+1} \mathbf{e}_1 \right| \|\mathbf{r}_1\|_2 ;$$

4. if $\mathbf{R}_k$ *is nonsingular,* $\mathbf{y}_k$ *minimizes* $\|\mathbf{r}_1 - \mathbf{A}\mathbf{P}\mathbf{y}_k\|_2$ *and* $\widetilde{\mathbf{x}}_{k+1} = \widetilde{\mathbf{x}}_1 + \mathbf{P}_k \mathbf{y}_k$*, then*

$$\mathbf{b} - \mathbf{A}\widetilde{\mathbf{x}}_{k+1} = [\mathbf{P}_k, \mathbf{p}_{k+1}] \, \mathbf{Q}_{k+1}{}^H \mathbf{e}_{k+1} \mathbf{e}_{k+1}{}^\top \mathbf{Q}_{k+1} \mathbf{e}_1 \, \|\mathbf{r}_1\|_2 .$$

Proof Let us prove the first claim. Recall from Lemma 2.6.6 that

$$\mathbf{A}\mathbf{P}_k = \mathbf{P}_k \mathbf{H}_k + \mathbf{p}_{k+1} \eta_{k+1,k} \mathbf{e}_k{}^\top .$$

As a result,

$$\begin{aligned} \mathbf{r}_1 - \mathbf{A}\mathbf{P}_k\mathbf{y} &= \mathbf{r}_1 - \left(\mathbf{P}_k \mathbf{H}_k + \mathbf{p}_{k+1} \eta_{k+1,k} \mathbf{e}_k{}^\top \right) \mathbf{y} \\ &= [\mathbf{P}_k, \mathbf{p}_{k+1}] \left(\begin{bmatrix} \mathbf{e}_1 \\ 0 \end{bmatrix} \|\mathbf{r}_1\|_2 - \begin{bmatrix} \mathbf{H}_k \\ \eta_{k+1,k} \mathbf{e}_k{}^\top \end{bmatrix} \mathbf{y} \right) . \end{aligned} \quad (2.43)$$

Since Lemma 2.6.6 also showed that $[\mathbf{P}_k, \mathbf{p}_{k+1}]$ has orthonormal columns, the first claim follows easily.

Now we turn to the second claim. Using the first claim and the assumption that $\mathbf{Q}_{k+1}$ is unitary, we obtain

$$\|\mathbf{r}_1 - \mathbf{A}\mathbf{P}_k\mathbf{y}\|_2 = \left\| \begin{bmatrix} \mathbf{e}_1 \\ 0 \end{bmatrix} \|\mathbf{r}_1\|_2 - \begin{bmatrix} \mathbf{H}_k \\ \eta_{k+1,k}{\mathbf{e}_k}^\top \end{bmatrix} \mathbf{y} \right\|_2$$

$$= \left\| \begin{bmatrix} \mathbf{e}_1 \\ 0 \end{bmatrix} \|\mathbf{r}_1\|_2 - {\mathbf{Q}_{k+1}}^H \begin{bmatrix} \mathbf{R}_k \\ \mathbf{0} \end{bmatrix} \mathbf{y} \right\|_2 = \left\| \mathbf{Q}_{k+1}\mathbf{e}_1 \|\mathbf{r}_1\|_2 - \begin{bmatrix} \mathbf{R}_k \\ \mathbf{0} \end{bmatrix} \mathbf{y} \right\|_2 .$$

Let us prove the third claim. If $\mathbf{R}_k$ is nonsingular, then the second result shows that $\|\mathbf{r}_1 - \mathbf{A}\mathbf{P}_k\mathbf{y}\|_2$ is minimized by solving $\mathbf{R}_k\mathbf{y} = [\mathbf{I}, \mathbf{0}]\mathbf{Q}_{k+1}\mathbf{e}_1\|\mathbf{r}_1\|_2$. With this choice of $\mathbf{y}$ we get the claimed result.

All that remains is the final claim. If $\mathbf{y}_k$ is chosen to minimize $\|\mathbf{r}_1 - \mathbf{A}\mathbf{P}_k\mathbf{y}_k\|_2$, then Eq. (2.43) shows that the residual is

$$\mathbf{b} - \mathbf{A}\widetilde{\mathbf{x}}_{k+1} = \mathbf{r}_1 - \mathbf{A}\mathbf{P}_k\mathbf{y}_k = [\mathbf{P}_k, \mathbf{p}_{k+1}] \left(\begin{bmatrix} \mathbf{e}_1 \\ 0 \end{bmatrix} \|\mathbf{r}_1\|_2 - \begin{bmatrix} \mathbf{H}_k \\ \eta_{k+1,k}{\mathbf{e}_k}^\top \end{bmatrix} \mathbf{y}_k \right)$$

$$= [\mathbf{P}_k, \mathbf{p}_{k+1}]\, {\mathbf{Q}_{k+1}}^H \mathbf{Q}_{k+1} \left(\begin{bmatrix} \mathbf{e}_1 \\ 0 \end{bmatrix} \|\mathbf{r}_1\|_2 - \begin{bmatrix} \mathbf{H}_k \\ \eta_{k+1,k}{\mathbf{e}_k}^\top \end{bmatrix} \mathbf{y}_k \right)$$

$$= [\mathbf{P}_k, \mathbf{p}_{k+1}]\, {\mathbf{Q}_{k+1}}^H \left(\mathbf{Q}_{k+1}\mathbf{e}_1 \|\mathbf{r}_1\|_2 - \begin{bmatrix} \mathbf{R}_k \\ \mathbf{0}^\top \end{bmatrix} \mathbf{y}_k \right)$$

$$= [\mathbf{P}_k, \mathbf{p}_{k+1}]\, {\mathbf{Q}_{k+1}}^H \mathbf{e}_{k+1}{\mathbf{e}_{k+1}}^\top \mathbf{Q}_{k+1}\mathbf{e}_1 \|\mathbf{r}_1\|_2 .$$

Let us describe how to implement the QR factorization of the upper Hessenberg matrix. Consider the first step, in which $\mathbf{H}_1 = [\eta_{1,1}]$ is 1×1. Using Algorithm 6.9.1 in Chap. 6 of Volume I we compute a plane rotation

$$\mathbf{G}_{12} = \begin{bmatrix} \gamma_{12} & \sigma_{12} \\ -\overline{\sigma_{12}} & \gamma_{12} \end{bmatrix}$$

with γ_{12} real, so that

$$\mathbf{G}_{12} \begin{bmatrix} \eta_{1,1} \\ \eta_{2,1} \end{bmatrix} = \begin{bmatrix} \varrho_{1,1} \\ 0 \end{bmatrix} .$$

The 1×1 matrix $\mathbf{R}_1 = [\varrho_{1,1}]$ is obviously right-triangular. Let us define

$$\widehat{\mathbf{f}}_1 = \mathbf{e}_1 \|\mathbf{r}_1\|_2 \text{ and } \widehat{\mathbf{f}}_2 = \mathbf{G}_{1,2}\widehat{\mathbf{f}}_1 .$$

If we choose the 1-vector $\mathbf{y}_1$ to solve

$$\varrho_{1,1}\mathbf{y}_1 = \mathbf{R}_1\mathbf{y}_1 = \mathbf{e}_1 \cdot \mathbf{G}_{1,2}\mathbf{e}_1 \|\mathbf{r}_1\|_2 = \mathbf{e}_1 \cdot \widehat{\mathbf{f}}_2 ,$$

then Lemma 2.6.7 shows that we obtain

$$\|\mathbf{r}_1 - \mathbf{A}\mathbf{P}_1\mathbf{y}_1\|_2 = \left\| \begin{bmatrix} \mathbf{R}_1 \\ 0 \end{bmatrix} \mathbf{y}_1 - \widehat{\mathbf{f}}_2 \right\|_2 = \left| \mathbf{e}_2{}^\top \widehat{\mathbf{f}}_2 \right| = |\sigma_{1,2}| \, \|\mathbf{r}_1\|_2 \; .$$

Inductively, suppose that we have computed plane rotations $\mathbf{G}_{j,j+1}$ for $1 \leq j \leq k$ and formally represent their product by the $(k+1) \times (k+1)$ matrix

$$\mathbf{G}_{k,k+1} \ldots \mathbf{G}_{1,2} \equiv \mathbf{Q}_{k+1} = \begin{bmatrix} \widehat{\mathbf{Q}}_k & \mathbf{q}_k \\ \mathbf{q}_k{}^H & \omega_k \end{bmatrix} .$$

We assume that this product of plane rotations has been designed to factor

$$\begin{bmatrix} \widehat{\mathbf{Q}}_k & \mathbf{q}_k \\ \mathbf{q}_k{}^H & \omega_k \end{bmatrix} \begin{bmatrix} \mathbf{H}_k \\ \eta_{k+1,k} \mathbf{e}_k{}^\top \end{bmatrix} = \begin{bmatrix} \mathbf{R}_k \\ \mathbf{0}^\top \end{bmatrix}$$

where $\mathbf{R}_k$ is $k \times k$ right-triangular. To perform the next step, we begin by partitioning the $(k+1) \times (k+1)$ upper Hessenberg matrix

$$\mathbf{H}_{k+1} = \begin{bmatrix} \mathbf{H}_k & \mathbf{h}_k \\ \eta_{k+1,k} \mathbf{e}_k{}^\top & \eta_{k+1,k+1} \end{bmatrix} .$$

Then we compute

$$\begin{bmatrix} \mathbf{r}_k \\ \zeta_{k+1} \end{bmatrix} = \begin{bmatrix} \widehat{\mathbf{Q}}_k & \mathbf{q}_k \\ \mathbf{q}_k{}^H & \omega_k \end{bmatrix} \begin{bmatrix} \mathbf{h}_k \\ \eta_{k+1,k+1} \end{bmatrix}$$

Afterward, we choose a plane rotation

$$\mathbf{G}_{k+1,k+2} = \begin{bmatrix} \gamma_{k+1,k+2} & \sigma_{k+1,k+2} \\ -\overline{\sigma_{k+1,k+2}} & \gamma_{k+1,k+2} \end{bmatrix}$$

with real $\gamma_{k+1,k+2}$ so that

$$\begin{bmatrix} \gamma_{k+1,k+2} & \sigma_{k+1,k+2} \\ -\overline{\sigma_{k+1,k+2}} & \gamma_{k+1,k+2} \end{bmatrix} \begin{bmatrix} \zeta_{k+1} \\ \eta_{k+2,k+1} \end{bmatrix} = \begin{bmatrix} \varrho_{k+1,k+1} \\ 0 \end{bmatrix} .$$

As a result,

$$
\mathbf{Q}_{k+2}\begin{bmatrix}\mathbf{H}_{k+1}\\ \eta_{k+2,k+1}\mathbf{e}_{k+1}{}^\top\end{bmatrix} = \mathbf{G}_{k+1,k+2}\left(\mathbf{G}_{k,k+1}\ldots\mathbf{G}_{1,2}\right)\begin{bmatrix}\mathbf{H}_{k+1}\\ \eta_{k+2,k+1}\mathbf{e}_{k+1}{}^\top\end{bmatrix}
$$

$$
= \begin{bmatrix}\mathbf{I} & \mathbf{0} & \mathbf{0}\\ \mathbf{0}^\top & \gamma_{k+1,k+2} & \sigma_{k+1,k+2}\\ \mathbf{0}^\top & -\overline{\sigma_{k+1,k+2}} & \gamma_{k+1,k+2}\end{bmatrix}\begin{bmatrix}\widehat{\mathbf{Q}}_k & \mathbf{q}_k & \mathbf{0}\\ \mathbf{q}_k{}^H & \omega_k & 0\\ \mathbf{0}^\top & 0 & 1\end{bmatrix}\begin{bmatrix}\mathbf{H}_k & \mathbf{h}_k\\ \eta_{k+1,k}\mathbf{e}_k{}^\top & \eta_{k+1,k+1}\\ \mathbf{0}^\top & \eta_{k+2,k+1}\end{bmatrix}
$$

$$
= \begin{bmatrix}\mathbf{R}_k & \mathbf{r}_k\\ \mathbf{0}^\top & \varrho_{k+1,k+1}\\ \mathbf{0}^\top & 0\end{bmatrix}
$$

is right-trapezoidal. As part of our induction, we assume that we previously determined

$$
\widehat{\mathbf{f}}_{k+1} = \mathbf{Q}_{k+1}\widehat{\mathbf{f}}_1 = \mathbf{Q}_{k+1}\mathbf{e}_1\left\|\mathbf{r}_1\right\|_2 ,
$$

which we partitioned

$$
\widehat{\mathbf{f}}_{k+1} = \begin{bmatrix}\mathbf{f}_k\\ \widehat{\phi}_{k+1}\end{bmatrix} .
$$

Then we also have

$$
\widehat{\mathbf{f}}_{k+2} = \mathbf{G}_{k+1,k+2}\widehat{\mathbf{f}}_{k+1} = \begin{bmatrix}\mathbf{I} & \mathbf{0} & \mathbf{0}\\ \mathbf{0}^\top & \gamma_{k+1,k+2} & \sigma_{k+1,k+2}\\ \mathbf{0}^\top & -\overline{\sigma_{k+1,k+2}} & \gamma_{k+1,k+2}\end{bmatrix}\begin{bmatrix}\mathbf{f}_k\\ \widehat{\phi}_{k+1}\\ 0\end{bmatrix}
$$

$$
= \begin{bmatrix}\mathbf{f}_k\\ \gamma_{k+1,k+2}\widehat{\phi}_{k+1}\\ -\overline{\sigma_{k+1,k+2}}\widehat{\phi}_{k+1}\end{bmatrix} = \begin{bmatrix}\mathbf{f}_{k+1}\\ \widehat{\phi}_{k+2}\end{bmatrix} .
$$

The previous analysis leads immediately to the following [155]:

Algorithm 2.6.4 (GMRES)

$$
\begin{aligned}
&\mathbf{r}_1 = \mathbf{b} - \mathbf{A}\widetilde{\mathbf{x}}_1 \\
&\mathbf{p}_1 = \mathbf{r}_1 / \|\mathbf{r}_1\|_2 \\
&\mathbf{f}_1 = \mathbf{e}_1 \|\mathbf{r}_1\|_2 \\
&\text{for } 1 \le k \le s \\
&\quad \widehat{\mathbf{p}}_{k+1} = \mathbf{A}\mathbf{p}_k \\
&\quad \text{for } 0 \le j \le k \\
&\quad\quad \eta_{j,k} = \mathbf{p}_j \cdot \widehat{\mathbf{p}}_{k+1} \\
&\quad\quad \widehat{\mathbf{p}}_{k+)} = \widehat{\mathbf{p}}_{k+1} - \mathbf{p}_j \eta_{j,k} \\
&\quad \eta_{k+1,k} = \left\|\widehat{\mathbf{p}}_{k+1}\right\|_2 \\
&\quad \mathbf{p}_{k+1} = \widehat{\mathbf{p}}_{k+1} / \eta_{k+1,k} \\
&\quad \text{for } 1 \le j < k \\
&\quad \zeta_1 = \eta_{1,k} \\
&\quad \text{for } 1 \le j < k \\
&\quad\quad \begin{bmatrix} \varrho_{j,k} \\ \zeta_{j+1} \end{bmatrix} = \mathbf{G}_{j,j+1} \begin{bmatrix} \zeta_j \\ \eta_{j+1,k} \end{bmatrix} \\
&\quad \text{find } \mathbf{G}_{k,k+1} \text{ so that } \mathbf{G}_{k,k+1} \begin{bmatrix} \zeta_k \\ \eta_{k+1,k} \end{bmatrix} = \begin{bmatrix} \varrho_{k,k} \\ 0 \end{bmatrix} \\
&\quad \begin{bmatrix} \phi_k \\ \widehat{\phi}_{k+1} \end{bmatrix} = \mathbf{G}_{k,k+1} \begin{bmatrix} \widehat{\phi}_k \\ \mathbf{0} \end{bmatrix} \\
&\quad \text{if } |\widehat{\phi}_{k+1}| \le \varepsilon \text{ break} \\
&\text{back-solve } \mathbf{R}_s \mathbf{y}_s = \mathbf{f}_s \\
&\widetilde{\mathbf{x}}_s = \widetilde{\mathbf{x}}_1 + \mathbf{P}_s \mathbf{y}_s \\
&\mathbf{r}_s = \mathbf{P}_s \mathbf{Q}_s{}^H \mathbf{e}_s \widehat{\phi}_{s+1}
\end{aligned}
$$

This algorithm requires one matrix-vector multiplication per step, and a total of $s(s+1)/2$ orthogonal projections for s steps. It also requires the storage of $s+3$ m-vectors to complete s steps, the storage of s plane rotations, the storage of a right-triangular matrix $\mathbf{R}$, and the storage of one $(s+1)$-vector, namely $[\mathbf{f}_s, \widehat{\phi}_{s+1}]$.

The residual computation at the end of this algorithm was proved in the fourth claim of Lemma 2.6.7. It is useful for restarting the algorithm after some finite number of steps. Computing the residual in this form may involve less work than forming $\mathbf{r}_k = \mathbf{b} - \mathbf{A}\widetilde{\mathbf{x}}_k$ directly, if k is less than the average number of nonzero entries in a row of $\mathbf{A}$. In practice, k should not be taken to be too large, because the vectors $\mathbf{p}_j$ tend to lose numerical orthogonality. More details of the algorithm can be found in Saad and Schultz [155]. Saad [154, p. 160] also presents a version of the GMRES algorithm that uses Householder orthogonalization instead of the Gram-Schmidt process.

Here are some results regarding the convergence of GMRES.

Lemma 2.6.8 ([155]) *Suppose that* $\mathbf{A} = \mathbf{X}\Lambda\mathbf{X}^{-1}$ *where* Λ *is diagonal with diagonal entries* λ_i *for* $1 \leq i \leq m$*. Then the residual at the* $(k+1)$*th step of GMRES satisfies*

$$\|\mathbf{r}_{k+1}\|_2 \leq \kappa(\mathbf{X})\, \|\mathbf{r}_0\|_2 \min_{\substack{q \in \mathbb{P}_k \\ q(0)=1}} \max_{1 \leq i \leq m} |q(\lambda_i)| \ .$$

Corollary 2.6.1 *[155] Suppose that the real* $m \times m$ *matrix* $\mathbf{A}$ *is diagonalizable and that its symmetric part* $\mathbf{M} = \left(\mathbf{A} + \mathbf{A}^H\right)/2$ *is positive. Let* $0 < \alpha$ *be the smallest eigenvalue of* $\mathbf{M}$ *and let* β *be the largest eigenvalue of* $\mathbf{A}^H\mathbf{A}$*. Then for all* $1 \leq k \leq m$*, the residual at the kth step of GMRES satisfies*

$$\|\mathbf{r}_k\|_2 \leq \left[1 - \frac{\alpha}{\beta}\right]^{k/2} \|\mathbf{r}_0\|_2 \ .$$

Saad and Schultz [155] also provide an estimate for matrices $\mathbf{A}$ that have some eigenvalues with negative real parts.

A C^{++} implementation of GMRES can be found in gmres.h. The corresponding Fortran 77 implementation is available in files GMRES.f and GMRESREVCOM.f at `netlib.org`.

Exercise 2.6.3 Suppose that we want to solve $\mathbf{A}\mathbf{x} = \mathbf{b}$, where $\mathbf{A}$ is nonsingular. Show that we can transform this to a system $\widetilde{\mathbf{A}}\widetilde{\mathbf{x}} = \widetilde{\mathbf{b}}$, where, for $\nu > 0$ sufficiently large, $\widetilde{\mathbf{A}} = \mathbf{A} + \nu\mathbf{A}^H\mathbf{A}$ is positive.

Exercise 2.6.4 Consider the **GCR algorithm** [61] for solving $\mathbf{A}\mathbf{x} = \mathbf{b}$:

$$\begin{aligned}
&\mathbf{r}_0 = \mathbf{A}\widetilde{\mathbf{x}}_0 - \mathbf{b} \\
&\mathbf{p}_0 = -\mathbf{r}_0 \\
&\text{for } n = 0, 1, \ldots \text{ until convergence} \\
&\quad \alpha_n = -(\mathbf{r}_n)^\top \mathbf{A}\mathbf{p}_n / (\mathbf{A}\mathbf{p}_n)^\top (\mathbf{A}\mathbf{p}_n) \\
&\quad \widetilde{\mathbf{x}}_{n+1} = \widetilde{\mathbf{x}}_n + \mathbf{p}_n \alpha_n \\
&\quad \mathbf{r}_{n+1} = \mathbf{r}_n + \mathbf{A}\mathbf{p}_n \alpha_n \\
&\quad \mathbf{p}_{n+1} = -\mathbf{r}_{n+1} + \textstyle\sum_{j=0}^{n} \mathbf{p}_j \beta_{j,n} \ ,
\end{aligned} \tag{2.44}$$

where $\beta_{j,n}$ is chosen so that $(\mathbf{A}\mathbf{p}_{n+1})^\top \mathbf{A}\mathbf{p}_n = 0$.

1. What is the objective function that this algorithm is trying to minimize?
2. What is the conjugacy condition for the search directions?
3. Is this a minimum residual or an orthogonal error method?
4. Show that if $\mathbf{A} = \begin{bmatrix} 0 & 1 \\ -1 & 0 \end{bmatrix}$ and $\mathbf{b} = \begin{bmatrix} 1 \\ 1 \end{bmatrix}$, then the algorithm will break down before finding the solution.

Exercise 2.6.5 Show that when $\mathbf{A}$ is symmetric, the Arnoldi process generates a tridiagonal matrix $\mathbf{H}$. Compare this to the Lanczos process in Algorithm 1.3.13.

Exercise 2.6.6 Get a copy of matrix **E20R0100** from MatrixMarket. This is a sparse nonsymmetric matrix, and should be represented in computer memory without storing the zero entries. If this matrix is denoted by $\mathbf{A}$, let $\mathbf{x}$ be the vector of ones, and $\mathbf{b} = \mathbf{A}\mathbf{x}$. Choose the initial guess $\widetilde{\mathbf{x}}$ for GMRES to have random entries chosen from a uniform probability distribution on $(0, 1)$. Program the GMRES iteration to solve $\mathbf{A}\mathbf{x} = \mathbf{b}$, and plot the logarithm base 10 of the error $\|\widetilde{\mathbf{x}}-\mathbf{x}\|_\infty$ versus iteration number. At what order of error does the iteration stop making progress? Also describe how the convergence criteria for this iteration were implemented.

2.6.3 QMR

The **quasi-minimum residual (QMR)** algorithm, developed by Freund and Nachtigal [74], is similar to the GMRES algorithm. Whereas GMRES is based on the Arnoldi Algorithm 2.6.3 for recursively reducing a general matrix to upper Hessenberg form, the QMR algorithm is based on the Lanczos biorthogonalization procedure (Algorithm 2.5.6) for recursively reducing a matrix to tridiagonal form.

Another distinction between these two algorithms lies in their minimization objective. If the columns of $\mathbf{P}_k = [\mathbf{p}_1, \ldots, \mathbf{p}_k]$ are the orthonormal vectors generated by the Arnoldi process, the GMRES algorithm chooses approximate solutions of the form $\widetilde{\mathbf{x}}_{k+1} = \widetilde{\mathbf{x}}_1 + \mathbf{P}_k\mathbf{y}_k$ and minimizes

$$\|\mathbf{b} - \mathbf{A}\widetilde{\mathbf{x}}_{k+1}\|_2 = \|[\mathbf{b} - \mathbf{A}\widetilde{\mathbf{x}}_1] - \mathbf{A}\mathbf{P}_k\mathbf{y}_k\|_2 \ .$$

On the other hand, if the columns of $\mathbf{V}_k = [\mathbf{v}_1, \ldots, \mathbf{v}_k]$ are the unit vectors generated by the Lanczos biorthogonalization procedure are not orthonormal, the QMR algorithm chooses approximate solutions of the form $\widetilde{\mathbf{x}}_{k+1} = \widetilde{\mathbf{x}}_1 + \mathbf{V}_k\mathbf{y}_k$. QMR begins by choosing

$$\lambda_0 = \|\mathbf{r}_1\|_2 \text{ and } \mathbf{v}_1 = \mathbf{r}_1/\lambda_0 \ .$$

Afterward, Eq. (2.34) implies that the residual vector is

$$\mathbf{b} - \mathbf{A}\widetilde{\mathbf{x}}_{k+1} = [\mathbf{b} - \mathbf{A}\widetilde{\mathbf{x}}_1] - \mathbf{A}\mathbf{V}_k\mathbf{y}_k = \mathbf{V}_k\mathbf{e}_1\lambda_0 - \left(\mathbf{V}_k\mathbf{T}_k - \mathbf{v}_{k+1}\lambda_k{\mathbf{e}_k}^\top\right)\mathbf{y}_k$$
$$= \left[\mathbf{V}_k \ \mathbf{v}_{k+1}\right]\begin{bmatrix}\mathbf{e}_1 \\ 0\end{bmatrix}\lambda_0 - \begin{bmatrix}\mathbf{T}_k \\ \lambda_k{\mathbf{e}_k}^\top\end{bmatrix}\mathbf{y}_k \ .$$

As a result,

$$\|\mathbf{b} - \mathbf{A}\widetilde{\mathbf{x}}_{k+1}\|_2 \leq \left\| \begin{bmatrix} \mathbf{V}_k & \mathbf{v}_{k+1} \end{bmatrix} \right\|_2 \left\| \begin{bmatrix} \mathbf{e}_1 \\ 0 \end{bmatrix} \lambda_0 - \begin{bmatrix} \mathbf{T}_k \\ \lambda_k \mathbf{e}_k{}^\top \end{bmatrix} \mathbf{y}_k \right\|_2 .$$

The QMR algorithm chooses $\mathbf{y}_k$ to minimize

$$\left\| \begin{bmatrix} \mathbf{e}_1 \\ 0 \end{bmatrix} \lambda_0 - \begin{bmatrix} \mathbf{T}_k \\ \lambda_k \mathbf{e}_k{}^\top \end{bmatrix} \mathbf{y}_k \right\|_2 .$$

This minimization problem is solved by a QR factorization using plane rotations, much as in the GMRES algorithm.

Recall that the QR factorization in GMRES was applied to an upper Hessenberg matrix, and the factorization resulted in a full right-triangular matrix. In contrast, the QR factorization in QMR is applied to a tridiagonal matrix, so the factorization results in a right-triangular matrix with a diagonal and at most two super-diagonals. Pictorially, we have the following steps in this QR factorization:

$$\mathbf{H}_5 = \begin{bmatrix} \times & \times & & & \\ \times & \times & \times & & \\ & \times & \times & \times & \\ & & \times & \times & \times \\ & & & \times & \times \end{bmatrix} \overrightarrow{\mathbf{G}_{12}} \begin{bmatrix} * & * & + & & \\ 0 & * & * & & \\ & \times & \times & \times & \\ & & \times & \times & \times \\ & & & \times & \times \end{bmatrix} \overrightarrow{\mathbf{G}_{23}} \begin{bmatrix} * & * & + & & \\ 0 & * & * & + & \\ & 0 & * & * & \\ & & \times & \times & \times \\ & & & \times & \times \end{bmatrix}$$

$$\overrightarrow{\mathbf{G}_{34}} \begin{bmatrix} * & * & + & & \\ 0 & * & * & + & \\ & 0 & * & * & + \\ & & 0 & * & * \\ & & & \times & \times \end{bmatrix} \overrightarrow{\mathbf{G}_{45}} \begin{bmatrix} * & * & + & & \\ 0 & * & * & + & \\ & 0 & * & * & + \\ & & 0 & * & * \\ & & & 0 & * \end{bmatrix}$$

These symbolic results demonstrate that only the last two plane rotations are needed, in order to introduce a new column of $\mathbf{T}_{k+1}$ into this QR factorization, This simplifies the storage for both the right-triangular matrix $\mathbf{R}$ and the plane rotations. We can combine these ideas in the following

Algorithm 2.6.5 (QMR)

$$
\begin{aligned}
&\mathbf{r}_1 = \mathbf{b} - \mathbf{A}\widetilde{\mathbf{x}}_1 \\
&\lambda_0 = \|\mathbf{r}_1\|_2 \\
&\mathbf{v}_1 = \mathbf{r}_1/\lambda_0 \\
&\text{choose } \mathbf{u}_1 \text{ with } \mathbf{u}_1 \cdot \mathbf{v}_1 = 1 \\
&\mathbf{f}_1 = \mathbf{e}_1\lambda_0 \\
&\text{for } 1 \le k \le s \\
&\quad \mathbf{z}_k = \mathbf{A}\mathbf{v}_k \\
&\quad \delta_k = \mathbf{u}_k \cdot \mathbf{z}_k \\
&\quad \text{if } k = 1 \\
&\qquad \widehat{\mathbf{v}}_2 = \mathbf{z}_1 - \mathbf{v}_1\delta_1 \\
&\qquad \widehat{\mathbf{u}}_2 = \mathbf{A}^H\mathbf{u}_1 - \mathbf{u}_1\overline{\delta_1} \\
&\quad \text{else} \\
&\qquad \widehat{\mathbf{v}}_{k+1} = \mathbf{z}_k - \mathbf{v}_k\delta_k - \mathbf{v}_{k-1}\varrho_{k-1} \\
&\qquad \widehat{\mathbf{u}}_{k+1} = \mathbf{A}^H\mathbf{u}_k - \mathbf{u}_k\overline{\delta_k} - \mathbf{u}_{k-1}\lambda_{k-1} \\
&\quad \lambda_k = \|\widehat{\mathbf{v}}_{k+1}\|_2 \\
&\quad \text{if } \lambda_k = 0 \text{ stop} \\
&\quad \mathbf{v}_{k+1} = \widehat{\mathbf{v}}_{k+1}/\lambda_k \\
&\quad \varrho_k = \widehat{\mathbf{u}}_{k+1} \cdot \mathbf{v}_{k+1} \\
&\quad \text{if } \varrho_k = 0 \text{ stop} \\
&\quad \mathbf{u}_{k+1} = \widehat{\mathbf{u}}_{k+1}/\overline{\varrho_k} \\
&\quad \text{if } k > 2 \\
&\qquad \begin{bmatrix} \varrho_{k-2,k} \\ \zeta_{k-1} \end{bmatrix} = \mathbf{G}_{k-2,k-1} \begin{bmatrix} 0 \\ \varrho_{k-1} \end{bmatrix} \\
&\qquad \begin{bmatrix} \varrho_{k-1,k} \\ \zeta_k \end{bmatrix} = \mathbf{G}_{k-1,k} \begin{bmatrix} \zeta_{k-1} \\ \delta_k \end{bmatrix} \\
&\quad \text{else if } k = 2 \\
&\qquad \begin{bmatrix} \varrho_{k-1,k} \\ \zeta_k \end{bmatrix} = \mathbf{G}_{k-1,k} \begin{bmatrix} \varrho_{k-1} \\ \delta_k \end{bmatrix} \\
&\quad \text{else } \zeta_k = \delta_k \\
&\quad \text{find } \mathbf{G}_{k,k+1} \text{ so that } \mathbf{G}_{k,k+1} \begin{bmatrix} \zeta_k \\ \lambda_k \end{bmatrix} = \begin{bmatrix} \varrho_{k,k} \\ 0 \end{bmatrix} \\
&\quad \begin{bmatrix} \phi_k \\ \widehat{\phi}_{k+1} \end{bmatrix} = \mathbf{G}_{k,k+1} \begin{bmatrix} \widehat{\phi}_k \\ \mathbf{0} \end{bmatrix} \\
&\quad \text{if } |\widehat{\phi}_{k+1}| \le \varepsilon \text{ break} \\
&\text{back-solve } \mathbf{R}_s\mathbf{y}_s = \mathbf{f}_s \\
&\widetilde{\mathbf{x}}_s = \widetilde{\mathbf{x}}_1 + \mathbf{P}_s\mathbf{y}_s
\end{aligned}
$$

This algorithm requires two matrix-vector multiplications per step, one by $\mathbf{A}$ and the other by $\mathbf{A}^H$. It also requires the storage of six m-vectors, namely $\mathbf{v}_{k-1}$, $\mathbf{v}_k$, $\mathbf{v}_{k+1}$, $\mathbf{u}_{k-1}$, $\mathbf{u}_k$ and $\mathbf{u}_{k+1}$. The vectors $\mathbf{z}_k = \mathbf{A}\mathbf{v}_k$ and $\widehat{\mathbf{v}}_{k+1}$ can be stored in $\mathbf{v}_k$, and the vectors $\mathbf{A}^H\mathbf{u}_k$ and $\widehat{\mathbf{u}}_{k+1}$ can be stored in $\mathbf{u}_k$. The algorithm also requires the storage of a right tridiagonal matrix $\mathbf{R}$ and an s-vector $\mathbf{f}_s$, to complete s steps.

Readers can find error bounds for the QMR algorithm in Freund and Nachtigal [74]. These bounds indicate that QMR should converge at about the same rate as GMRES. Barrett et al. [10, p. 21] report that QMR typically converges more smoothly than biconjugate gradients.

A C^{++} implementation of QMR can be found in file qmr.h. The corresponding Fortran 77 implementation of QMR is available in files QMR.f and QMRREVCOM.f at `netlib.org`.

Exercise 2.6.7 Suppose that the sparse matrix $\mathbf{A}$ is represented in computer memory via the Yale sparse matrix format, described in Sect. 2.2. How would you use this storage scheme to compute $\mathbf{A}^H\mathbf{x}$?

Exercise 2.6.8 Get a copy of matrix **E20R0100** from MatrixMarket. This is a sparse nonsymmetric matrix, and should be represented in computer memory without storing the zero entries. If this matrix is denoted by $\mathbf{A}$, let $\mathbf{x}$ be the vector of ones, and $\mathbf{b} = \mathbf{A}\mathbf{x}$. Choose the initial guess $\widetilde{\mathbf{x}}$ for QMR to have random entries chosen from a uniform probability distribution on $(0, 1)$. Program the QMR iteration to solve $\mathbf{A}\mathbf{x} = \mathbf{b}$, and plot the logarithm base 10 of the error $\|\widetilde{\mathbf{x}} - \mathbf{x}\|_\infty$ versus iteration number. At what order of error does the iteration stop making progress? Also describe how the convergence criteria for this iteration were implemented.

2.7 Multigrid

The multigrid method is a very fast and effective method for solving certain symmetric positive linear systems. The initial idea is due to Fedorenko [63], but the popularity of the method is due to Brandt [25] and Hackbusch [89].

Currently, there are two basic flavors of multigrid. **Algebraic multigrid** views the problem in terms of matrices and vectors; a description of this approach can be found in Ruge and Stüben [153] and [167]. **Variational multigrid** considers linear equations that arise from finite element discretizations of elliptic partial differential equations; a description of this approach can be found in Bramble [24] or Mandel et al. [122], and in many recent finite element books, such as Braess [21] or Brenner and Scott [26]. Algebraic multigrid has not yet developed a comprehensive convergence theory, while variational multigrid requires several key results from finite element methods to prove convergence.

In this section, we will present the multigrid algorithm in algebraic form. We will also state a convergence theorem in which the key assumptions appear in algebraic form.

2.7.1 V-Cycle

Suppose that we want to solve $\mathbf{A}_f\mathbf{x}_f = \mathbf{b}_f$ on some fine grid, developed for the discretization of a partial differential equation. We will assume that $\mathbf{A}_f$ is Hermitian and positive. If possible, we would like to approximate the solution of this equation by solving a related equation $\mathbf{A}_c\mathbf{x}_c = \mathbf{b}_c$ on a coarser grid. Suppose that for each fine grid we are given a **prolongation** $\mathbf{P}_{cf}$ that maps vectors in the range of $\mathbf{A}_c$ to vectors in the domain of $\mathbf{A}_f$. We assume that the coarse matrix $\mathbf{A}_c$ is determined from the fine matrix by

$$\mathbf{A}_c = \mathbf{P}_{cf}{}^H\mathbf{A}_f\mathbf{P}_{cf} \; .$$

Also suppose that on each grid we are given a **smoother** $\mathbf{S}_f$ with good iterative improvement properties for use with $\mathbf{A}_f$.

The multigrid V-cycle algorithm is typically begun on the finest level. Each recursive call of the algorithm receives a residual vector on entry, and returns a change to the solution on exit. The algorithm takes the following recursive form:

Algorithm 2.7.1 (Multigrid V-Cycle)

$$\begin{aligned}
&\text{if there is no finer grid} && \\
&\quad \mathbf{r}_{f0} = \mathbf{A}_f\widetilde{\mathbf{x}_f}^{(0)} - \mathbf{b}_f && = \mathbf{A}_f\left(\widetilde{\mathbf{x}_f}^{(0)} - \mathbf{x}_f\right) \\
&\text{if there is a coarser grid} && \\
&\quad \mathbf{r}_f = \mathbf{r}_f^{(0)} && \\
&\quad \mathbf{d}_f^{(1)} = \mathbf{S}_f\mathbf{r}_f && \Longrightarrow \widetilde{\mathbf{x}_f}^{(1)} = \widetilde{\mathbf{x}_f}^{(0)} - \mathbf{d}_f^{(1)} \\
& && \Longrightarrow \widetilde{\mathbf{x}_f}^{(1)} - \mathbf{x}_f = \left(\mathbf{I} - \mathbf{S}_f\mathbf{A}_f\right)\left(\widetilde{\mathbf{x}_f}^{(0)} - \mathbf{x}_f\right) \\
&\quad \mathbf{r}_f = \mathbf{r}_f^{(0)} - \mathbf{A}_f\mathbf{d}_f^{(1)} && \equiv \mathbf{r}_f^{(1)} = \mathbf{A}_f\left(\widetilde{\mathbf{x}_f}^{(1)} - \mathbf{x}_f\right) \\
&\quad \mathbf{r}_c^{(0)} = \mathbf{P}_{cf}{}^H\mathbf{r}_f && = \mathbf{P}_{cf}{}^H\mathbf{A}_f\left(\widetilde{\mathbf{x}_f}^{(1)} - \mathbf{x}_f\right) \\
&\quad \text{call coarser multigrid with } \mathbf{r}_c^{(0)} && \Longrightarrow d_c = \mathbf{V}_c\mathbf{P}_{cf}{}^H\mathbf{A}_f\left(\widetilde{\mathbf{x}_f}^{(1)} - \mathbf{x}_f\right) \\
&\quad \mathbf{d}_f^{(2)} = \mathbf{d}_f^{(1)} + \mathbf{P}_{cf}d_c && \widetilde{\mathbf{x}_f}^{(2)} \equiv \widetilde{\mathbf{x}_f}^{(1)} - \mathbf{d}_f^{(2)} \\
& && \Longrightarrow \widetilde{\mathbf{x}_f}^{(2)} - \mathbf{x}_f = \left(\mathbf{I} - \mathbf{P}_{cf}\mathbf{V}_c\mathbf{P}_{cf}{}^H\mathbf{A}_f\right)\left(\widetilde{\mathbf{x}_f}^{(1)} - \mathbf{x}_f\right) \\
&\quad \mathbf{r}_f = \mathbf{r}_f^{(0)} - \mathbf{A}_f\mathbf{d}_f^{(2)} && \equiv \mathbf{r}_f^{(2)} = \mathbf{A}_f\left(\widetilde{\mathbf{x}_f}^{(2)} - \mathbf{x}_f\right) \\
&\quad \mathbf{d}_f^{(3)} = \mathbf{d}_f^{(2)} + \mathbf{S}_f{}^H\mathbf{r}_f && \Longrightarrow \widetilde{\mathbf{x}_f}^{(3)} = \widetilde{\mathbf{x}_f}^{(2)} - \mathbf{S}_f{}^H\mathbf{r}_f^{(2)} \\
& && \Longrightarrow \widetilde{\mathbf{x}_f}^{(3)} - \mathbf{x}_f = \left(\mathbf{I} - \mathbf{S}_f{}^H\mathbf{A}_f\right)\left(\widetilde{\mathbf{x}_f}^{(2)} - \mathbf{x}_f\right) \\
&\text{else} && \\
&\quad \text{solve } \mathbf{A}_c\mathbf{d}_c = \mathbf{r}_c^{(0)} && \Longrightarrow \mathbf{d}_c = \mathbf{A}_c^{-1}\mathbf{P}_{cf}{}^H\mathbf{r}_f \\
&\text{if there is no finer level then } \widetilde{\mathbf{x}_f} = \widetilde{\mathbf{x}_f}^{(0)} - \mathbf{d}_f^{(3)} &&
\end{aligned} \tag{2.45}$$

We will discuss choices of the prolongation $\mathbf{P}_{cf}$ in Sect. 2.7.5. The smoother could be given by Richardson's iteration (see Sect. 2.4.2), a relaxed Jacobi iteration (see Sect. 2.4.3), or a Gauss-Seidel iteration (see Sect. 2.4.4). The general requirements on the smoother are described in Theorem 2.7.1.

Note that our multigrid notation is not standard. It is more common to use integers to indicate levels of refinement, with the coarsest level typically having index 0 or 1, and finer levels having greater integer indices. We hope that the use of f and c subscripts makes the identification of fine and coarse levels more apparent, and the recursive implementation of the algorithm simpler for the reader.

Briggs [28]. discusses some heuristic notions that suggest why the multigrid iteration should be effective. However, the dependence on Fourier analysis in that discussion is a bit too restrictive for our purposes. Instead, we will prove Theorem 2.7.1, which has been adapted from Braess [21]. This theorem establishes a recurrence for the error reduction ratios on the separate levels of refinement, and proves a fixed upper bound for the error reduction ratios, independent of the number of equations. The essential features of the proof are that the effect of the coarse grid correction (the recursive call of multigrid) is to project the error onto the nullspace of ${\mathbf{P}_{cf}}^H \mathbf{A}_f$ along the range of $\mathbf{P}_{cf}$, and that the pre-smoother, which is used in the computation of $\mathbf{d}_f^{(1)}$, reduces errors in this nullspace. The post-smoother plays no role in this convergence theorem; rather it serves primarily to guarantee symmetry of the multigrid operator.

In our discussion of multigrid, we will make the following assumptions.

Assumption 2.7.1 Suppose that $m_f > m_c$ are integers, that the real $m_f \times m_c$ prolongation matrix $\mathbf{P}_{cf}$ has rank m_c, and that the $m_f \times m_f$ matrix $\mathbf{I}_f$ and the $m_c \times m_c$ matrix $\mathbf{I}_c$ are identity matrices. We assume that the real $m_f \times m_f$ matrix $\mathbf{A}_f$ is symmetric and positive. We assume that $\mathbf{A}_c = {\mathbf{P}_{cf}}^H \mathbf{A}_f \mathbf{P}_{cf}$, and we define

$$\mathbf{E}_{fc} = \mathbf{A}_c^{-1} {\mathbf{P}_{cf}}^H \mathbf{A}_f \; . \tag{2.46}$$

Finally, we assume that $\mathbf{S}_f$ is a real $m_f \times m_f$ matrix.

2.7.2 W-Cycle

A more general form of the multigrid iteration involves what is generally called a **W-cycle**. This involves repeated calls to the pre- and post-smoothers, and to the coarse grid, as follows:

Algorithm 2.7.2 (Multigrid W-Cycle)

$$
\begin{aligned}
&\mathbf{r}_f^{(1,0)} = \mathbf{r}_f^{(0)} \\
&\mathbf{d}_f^{(1,0)} = \mathbf{0} \\
&\text{for } 1 \le \ell \le n \\
&\quad \mathbf{d}_f^{(1,\ell)} = \mathbf{d}_f^{(1,\ell-1)} + \mathbf{S}_f \mathbf{r}_f^{(1,\ell-1)} \\
&\quad \mathbf{r}_f^{(1,\ell)} = \mathbf{r}_f^{(0)} - \mathbf{A}_f \mathbf{d}_f^{(1,\ell)} \\
&\mathbf{r}_c^{(0)} = \mathbf{P}_{cf}{}^\top \mathbf{r}_f^{(1,n)} \\
&\mathbf{d}_c^{(1)} = \mathbf{M}_c \mathbf{r}_c^{(0)} \\
&\text{for } 2 \le i \le p \\
&\quad \mathbf{r}_c^{(i-1)} = \mathbf{r}_c^{(0)} - \mathbf{A}_c \mathbf{d}_c^{(i-1)} \\
&\quad \mathbf{d}_c^{(i)} = \mathbf{d}_c^{(i-1)} + \mathbf{M}_c \mathbf{r}_c^{(i-1)} \\
&\mathbf{d}_f^{(2,0)} = \mathbf{d}_f^{(1,n)} + \mathbf{P}_{cf} \mathbf{d}_c^{(p)} \\
&\text{for } 1 \le \ell \le n \\
&\quad \mathbf{r}_f^{(2,\ell-1)} = \mathbf{r}_f^{(0)} - \mathbf{A}_f \mathbf{d}_f^{(2,\ell-1)} \\
&\quad \mathbf{d}_f^{(2,\ell)} = \mathbf{d}_f^{(2,\ell-1)} + \mathbf{S}_f{}^\top \mathbf{r}_f^{(2,\ell-1)} \\
&\mathbf{M}_f \mathbf{r}_f^{(0)} \equiv \mathbf{d}_f^{(2,n)} \; .
\end{aligned}
\tag{2.47}
$$

Of course, we take $p = 1$ whenever the coarse grid is the coarsest in the hierarchy; there is no point in repeating the coarsest grid correction, since $\mathbf{M}_c = \mathbf{A}_c^{-1}$ in this case. It is reasonable to ask if there is any advantage in performing multiple smoother iterations (i.e., $n > 1$), or in making multiple calls to the coarse grid computations (i.e., $p > 2$) on intermediate levels. Theorem 2.7.1 will provide bounds involving n and p that describe how these extra smoother iterations affect the convergence rate of the W-cycle algorithm.

2.7.3 Work Estimate

We will see in Theorem 2.7.1 that the multigrid algorithm reduces the error by a fixed ratio with each iteration, independent of the number of unknowns. It will follow that the number of iterations required to reduce the error by some given factor is *independent of the number of unknowns*. The next lemma shows that the work in each iteration of the multigrid W-cycle algorithm is proportional to the number of unknowns, provided that the number p of repetitions of the coarse grid correction is not too large. As a result, the total work required to solve a linear system via the multigrid algorithm is proportional to the number of unknowns.

Lemma 2.7.1 *Suppose that n, N and $m_0 < m_1 < \ldots < m_n = N$ are positive integers. Furthermore, assume that there exist integers $r > 1$ and $d \geq 1$ so that for all $0 < j \leq n$*

$$m_j \geq r^d m_{j-1} \; .$$

Assume that for each multigrid level $0 < j \leq n$, the application of the smoother requires at most $C_S m_j$ operations, that each residual computation requires at most $C_r m_j$ operations, and that each prolongation requires at most $C_P m_j$ operations. Suppose that we use $n \geq 1$ repetitions of the smoother and $p < r^d$ repetitions of the algorithm on the coarse grid. For $0 \leq j \leq n$ denote the work in one repetition of the multigrid W-cycle on level j by W_j. Then

$$W_n \leq m_n \left\{ \left[2n(C_S + C_r) + 2C_P + 1 + (p-1)(C_r + 1)r^{-d} \right] \frac{pr^{-d}}{1 - pr^{-d}} + \frac{W_0}{m_0} \right\} \; .$$

Proof By examining the W-Cycle Algorithm in 2.7.2, it is easy to see that for $n \geq j > 0$ the work in one repetition on the jth level is

$$\begin{aligned} W_j &\leq [2n(C_S + C_r) + 2C_P + 1] \, m_j + pW_{j-1} + (p-1)(C_r + 1)m_{j-1} \\ &\leq \left[2n(C_S + C_r) + 2C_P + 1 + (p-1)(C_r + 1)r^{-d} \right] m_j + pW_{j-1} \; . \end{aligned}$$

In order to simplify the expressions, we will write

$$C = 2n(C_S + C_r) + 2C_P + 1 + (p-1)(C_r + 1)r^{-d} \; ,$$

and obtain $W_j \leq Cm_j + pW_{j-1}$. We can solve this recurrence to see that

$$W_n \leq C \sum_{i=1}^{n} p^{n-i} m_i + p^n W_0 \; .$$

Note that

$$m_i = m_n \frac{m_i}{m_n} = m_n \prod_{\ell=i+1}^{n} \frac{m_{\ell-1}}{m_\ell} \leq m_n \prod_{\ell=i+1}^{n} r^{-d} = m_n r^{-d(n-i)} \; ,$$

and that

$$p^n \leq r^{nd} = \prod_{i=1}^{n} r^d \leq \prod_{i=1}^{n} \frac{m_i}{m_{i-1}} = \frac{m_n}{m_0} \; .$$

Combining our results, we obtain

$$W_n \leq Cm_n \sum_{i=1}^{n} \left(pr^{-d}\right)^{n-i} + m_n \frac{W_0}{m_0} = Cm_n pr^{-d} \frac{1-\left(pr^{-d}\right)^n}{1-pr^{-d}} + m_n \frac{W_0}{m_0}$$
$$\leq Cm_n \frac{pr^{-d}}{1-pr^{-d}} + m_n \frac{W_0}{m_0} \, .$$

Similar work estimates can be found in Braess [21, p. 257] or Brenner and Scott [26, p. 171].

Note that in some finite element computations it is common to bisect intervals, triangles and tetrahedra during mesh refinement. If so, then $r^d = 2$ in the previous lemma and we must take $p = 1$ to guarantee that the total work in the multigrid W-cycle is proportional to the number of unknowns, for an arbitrary number of levels in the multigrid algorithm. On the other hand, if 2D elements are subdivided into four children or 3D elements are subdivided into eight children, then the number p of coarse grid corrections can be taken to be as large as 3 or 7, respectively.

2.7.4 Convergence

The following theorem establishes conditions under which the multigrid W-cycle converges. It shows that repeated smoother steps reduce the contraction rate, through the factor b^{2m} in inequality (2.53), and in the denominator of the bounds in (2.54). The effect of repeated coarse grid corrections is determined implicitly through the recurrence (2.53) relating ϱ_f to ϱ_c.

There are three crucial assumptions in this theorem. Assumption (2.49) is basically the inductive hypothesis for the conclusion (2.53), and is easily satisfied at the beginning of the induction by solving the coarsest grid equations exactly. Proofs of the other two assumptions can be found in Braess [21, p. 251ff] or Trangenstein [173, p. 164].

Theorem 2.7.1 (Multigrid W-Cycle Convergence) *[21, p. 251] Suppose that Assumptions 2.7.1 are satisfied. Let λ_f be the largest eigenvalue of $\mathbf{A}_f$. Let $\mathbf{M}_c$ represent the action of the W-cycle iteration on the coarse grid; in particular, on the coarsest grid we have $\mathbf{M}_c = \mathbf{A}_c^{-1}$. Let $\mathbf{S}_f$ be some smoother matrix for $\mathbf{A}_f$. Assume that there is a constant $C_{(2.48)} \geq 1$ so that for all real, symmetric and positive $m \times m$ matrices $\mathbf{A}$ the Richardson iteration parameter μ is chosen so that*

$$\varrho(\mathbf{A}) \leq \mu \leq C_{(2.48)} \varrho(\mathbf{A}) \, . \tag{2.48}$$

Assume that there exists $\varrho_c \in [0, 1)$ so that for all A_f and all real m_c-vectors $\mathbf{v}_c$,

$$|||[\mathbf{I}_c - \mathbf{M}_c \mathbf{A}_c] \mathbf{v}_c|||_{\mathbf{A}_c} \leq \varrho_c \, |||\mathbf{v}_c|||_{\mathbf{A}_c} \, , \tag{2.49}$$

there exists $C_{(2.50)} > 0$ so that for all $\mathbf{A}_f$ and all real m_f-vectors $\mathbf{v}_f$

$$\left\|\left|\left(\mathbf{I}_f - \mathbf{P}_{c,f}\mathbf{A}_c^{-1}\mathbf{P}_{c,f}{}^\top\mathbf{A}_f\right)\mathbf{v}_f\right\|\right|_{\mathbf{A}_f} \leq C_{(2.50)}\|\mathbf{A}_f\mathbf{v}_f\|_2/\sqrt{\lambda_f} \quad \text{and} \tag{2.50}$$

there exists $C_{(2.48)} > 1$ so that for all $\mathbf{A}_f$ and all real nonzero m_f-vectors $\mathbf{v}_f$,

$$\left\|\left|\left(\mathbf{I}_f - \mathbf{S}_f\mathbf{A}_f\right)\mathbf{v}_f\right\|\right|_{\mathbf{A}_f} \leq \beta_R\left(\mathbf{v}_f\right)\left\|\left|\mathbf{v}_f\right\|\right|_{\mathbf{A}_f} , \tag{2.51}$$

where $\beta_R(\mathbf{w})$ is given by

$$\beta_R(\mathbf{w}) \equiv 1 - \frac{1}{C_{(2.48)}\varrho(\mathbf{A})}\frac{\|\mathbf{A}\mathbf{w}\|_2^2}{\|\!|\mathbf{w}|\!\|_{\mathbf{A}}^2} . \tag{2.52}$$

Given positive integers n and p, define the W-cycle matrix $\mathbf{M}_f$ by

$$\mathbf{I}_f - \mathbf{M}_f\mathbf{A}_f = \left(\mathbf{I}_f - \mathbf{S}_f{}^\top\mathbf{A}_f\right)^n\left(\mathbf{I}_f - \mathbf{P}_{c,f}\mathbf{M}_c\mathbf{A}_c\mathbf{P}_{c,f}{}^\top\mathbf{A}_f\right)^p\left(\mathbf{I}_f - \mathbf{S}_f\mathbf{A}_f\right)^n .$$

Then for all real nonzero m_f-vectors $\mathbf{v}_f$ and for all real $m_f \times m_f$ matrices $\mathbf{A}_f$ there exists $b \in [0, 1]$ so that

$$\begin{aligned}\left\|\left|\left[\mathbf{I}_f - \mathbf{M}_f\mathbf{A}_f\right]\mathbf{v}_f\right\|\right|_{\mathbf{A}_f}^2 &\leq \left\{(1-\varrho_c^{2p})\min\left\{1, C_{(2.50)}^2C_{(2.48)}\left[1-b\right]\right\} + \varrho_c^{2p}\right\}b^{2m}\left\|\left|\mathbf{v}_f\right\|\right|_{\mathbf{A}_f}^2 \\ &\equiv \varrho_f^2\left\|\left|\mathbf{v}_f\right\|\right|_{\mathbf{A}_f}^2 .\end{aligned} \tag{2.53}$$

Further,

$$\varrho_c \leq \sqrt{\frac{C_{(2.50)}^2C_{(2.48)}}{C_{(2.50)}^2C_{(2.48)} + 2n}} \implies \varrho_f \leq \sqrt{\frac{C_{(2.50)}^2C_{(2.48)}}{C_{(2.50)}^2C_{(2.48)} + 2n}} . \tag{2.54}$$

Note that on the coarsest level in the multigrid W-cycle we have $\mathbf{M}_c = \mathbf{A}_c^{-1}$, so the recurrence

$$\varrho_f^2 = \left\{(1-\varrho_c^{2p})\min\left\{1, C_{(2.50)}^2C_{(2.48)}\left[1-b\right]\right\} + \varrho_c^{2p}\right\}b^{2m} \tag{2.55}$$

begins with $\varrho_c = 0$ on the coarsest level.

The previous theorem shows that with each iteration the error in the multigrid W-cycle iteration is reduced by a constant factor, independent of the number of equations. In Lemma 2.7.1 we saw that the total work in one multigrid iteration is proportional to the number of unknowns. It follows that the total work required to solve a linear system by multigrid is proportional to the number of unknowns.

Suppose that the exact solution is known, and we are using it to measure the performance of the multigrid iteration. Under such circumstances, each multigrid iteration provides experimental values for quantities that appear in the hypotheses of Theorem 2.7.1. Let $\mathbf{x}_f$ be the true solution, $\widetilde{\mathbf{x}}_f^{(0)}$ be the initial solution guess and $\mathbf{r}_f^{(0)}$ be the residual corresponding to $\widetilde{\mathbf{x}}_f^{(0)}$. Also, let $\widetilde{\mathbf{x}}_f^{(1)}$ be the solution obtained after smoothing and let $\mathbf{r}_f^{(1)}$ be its corresponding residual. Then

$$\beta \equiv \frac{\left\|\left(\mathbf{I}_f - \mathbf{S}_f\mathbf{A}_f\right)\left(\widetilde{\mathbf{x}}_f^{(0)} - \mathbf{x}_f\right)\right\|_{\mathbf{A}_f}}{\left\|\widetilde{\mathbf{x}}_f^{(0)} - \mathbf{x}_f\right\|_{\mathbf{A}_f}} = \sqrt{\frac{\left(\mathbf{r}_f^{(1)}\right)^\top \left(\widetilde{\mathbf{x}}_f^{(1)} - \mathbf{x}_f\right)}{\mathbf{f}_f^{(0)\top}\left(\widetilde{\mathbf{x}}_f^{(0)} - \mathbf{x}_f\right)}}$$

provides an experimental lower value for $\beta_R\left(\widetilde{\mathbf{x}}_f^{(0)} - \mathbf{x}_f\right)$ in Eq. (2.52). We can estimate the spectral radius of $\mathbf{A}$ by an application of the Gerschgorin circle theorem, and estimate

$$C_{(2.48)} \le \frac{1}{1-\beta}\frac{\left\|\mathbf{r}_f^{(0)}\right\|_2^2}{\varrho(\mathbf{A}_f)\left\|\widetilde{\mathbf{x}}_f^{(0)} - \mathbf{x}_f\right\|_2} .$$

Similarly, if $\widetilde{\mathbf{x}}_f^{(2)}$ is the solution obtained after the coarse grid correction and $\mathbf{r}_f^{(2)}$ is its corresponding residual, then

$$\frac{\left\|\left[\mathbf{I}_f - \mathbf{P}_{c,f}\mathbf{V}_c\mathbf{P}_{c,f}{}^\top\mathbf{A}_f\right]\left(\widetilde{\mathbf{x}}_f^{(1)} - \mathbf{x}_f\right)\right\|_{\mathbf{A}_f}}{\|\mathbf{A}_f\left(\widetilde{\mathbf{x}}_f^{(1)} - \mathbf{x}_f\right)\|_2} = \frac{\sqrt{\left(\mathbf{r}_f^{(2)}\right)^\top\left(\widetilde{\mathbf{x}}_f^{(2)} - \mathbf{x}_f\right)}}{\left\|\mathbf{r}_f^{(1)}\right\|_2} .$$

We can use the Gerschgorin circle theorem to estimate the largest eigenvalue λ_f of $\mathbf{A}_f$, and then get a lower bound for $C_{(2.50)}$ as follows:

$$C_{(2.50)} \ge \frac{\sqrt{\left(\mathbf{r}_f^{(2)}\right)^\top\left(\widetilde{\mathbf{x}}_f^{(2)} - \mathbf{x}_f\right)}}{\left\|\mathbf{r}_f^{(1)}\right\|_2}\sqrt{\lambda_f} .$$

Finally, if $\widetilde{\mathbf{x}}_f$ is the approximate solution obtained after a complete multigrid step and $\mathbf{r}_f$ is its corresponding residual, then

$$\frac{\left\|\left[\mathbf{I}_f - \mathbf{M}_f\mathbf{A}_f\right]\left(\widetilde{\mathbf{x}}_f^{(0)} - \mathbf{x}_f\right)\right\|_{\mathbf{A}_f}}{\left\|\widetilde{\mathbf{x}}_f^{(0)} - \mathbf{x}_f\right\|_{\mathbf{A}_f}} = \sqrt{\frac{\mathbf{r}_f{}^\top\left(\widetilde{\mathbf{x}}_f - \mathbf{x}_f\right)}{\left(\mathbf{r}_f^{(0)}\right)^\top\left(\widetilde{\mathbf{x}}_f^{(0)} - \mathbf{x}_f\right)}} .$$

This ratio provides a lower bound for the factor ϱ_f in inequality (2.53).

2.7.5 Prolongation

In variational multigrid, the prolongation operator is determined by injection of coarse approximation functions into the space of finer approximation functions. For readers who have little familiarity with finite element methods, this statement would be impossible to implement in a computer program. To serve all readers, we will develop prolongations by other means.

In this section, we will describe an algebraic multigrid prolongation for matrices with positive diagonal entries and non-positive off-diagonal entries. More general algebraic multigrid prolongations are described in Sect. A.4.2 of Stuben [167].

Suppose that $\mathbf{A}$ is the matrix associated with some level of the multigrid algorithm, that the diagonal entries of $\mathbf{A}$ are all positive and the off-diagonal entries of $\mathbf{A}$ are all non-positive. Subdivide the fine grid indices into disjoint sets $\mathscr{C}$ and $\mathscr{F}$, where $\mathscr{C}$ corresponds to fine indices shared with the coarse grid. For each $i \in \mathscr{F}$ let

$$\mathscr{N}_i = \{j \neq i : \mathbf{A}_{ij} < 0\}$$

be the set of discretization stencil neighbors corresponding to negative off-diagonal entries in the ith row of $\mathbf{A}$. For some given $\tau \geq 1$, suppose that for all $i \in \mathscr{F}$ with $\mathscr{N}_i \cap \mathscr{C} \neq \emptyset$ we choose a non-empty set $\Pi_i \subset \mathscr{C} \cap \mathscr{N}_i$ such that

$$\sum_{j \in \mathscr{N}_i} |\mathbf{A}_{ij}| \leq \tau \sum_{j \in \Pi_i} |\mathbf{A}_{ij}| \, . \tag{2.56}$$

For problems in 1D, it is common to choose $\Pi_i = \mathscr{N}_i$. However, for problems in 2D, for which the coarsened multigrid matrix has a different sparsity pattern than the finest matrix, it may be convenient to choose Π_i to be the same set of neighbors as on the finest grid.

For $i \in \mathscr{C}$, define the prolongation operator to copy the coarse value to the same fine grid location. For $i \in \mathscr{F}$ with $\mathscr{N}_i \cap \mathscr{C} \neq \emptyset$, define the prolongation operator $\mathbf{P}$ by

$$(\mathbf{Px})_i = -\frac{\sum_{j \in \mathscr{N}_i} \mathbf{A}_{ij}}{\sum_{j \in \Pi_i} \mathbf{A}_{ij}} \frac{\sum_{j \in \Pi_i} \mathbf{A}_{ij} (\mathbf{Px})_j}{\mathbf{A}_{ii}} \, .$$

After these prolongation steps have been performed, consider $i \in \mathscr{F}$ such that $\mathscr{N}_i \cap \mathscr{C} = \emptyset$. If there is a subset $\Pi_i^f \subset \mathscr{N}_i$ such that $(\mathbf{Px})_j$ has been defined for all $j \in \Pi_i^f$ and such that (2.56) is satisfied, define

$$(\mathbf{Px})_i = -\frac{\sum_{j \in \mathscr{N}_i} \mathbf{A}_{ij}}{\sum_{j \in \Pi_i^f} \mathbf{A}_{ij}} \frac{\sum_{j \in \Pi_i^f} \mathbf{A}_{ij} (\mathbf{Px})_j}{\mathbf{A}_{ii}} \, .$$

We assume that this process of indirect interpolation can be continued until the prolongation is defined at all $i \in \mathscr{F}$.

Readers may find examples of various multigrid prolongation operators, both algebraic and variational, and in one or two dimensions, in Trangenstein [173, p. 168ff]. These are discussed in terms of discretizations of partial differential equations, which is beyond the scope of this book.

A program to perform the algebraic multigrid algorithm for the 1D Laplace equation can be found in Level1D.H and Level1D.C. This program uses a C++ class `Level` to contain the data arrays on the hierarchy of grids, and to perform the multigrid operations using recursion. In particular, note that the member function `Level::setup` shows how to compute the coarse grid matrix directly as $\mathbf{P}^\top \mathbf{A}_f \mathbf{P}$ in work proportional to the number of unknowns. The procedures `Level::prolong` and `Level::restrict` are chosen to be **adjoints** of each other, so that the overall multigrid cycle is symmetric. The form of this adjoint condition is

$$\sum_i \mathbf{f}_i \, [\mathbf{P}\mathbf{c}]_i = \sum_I [\mathbf{R}\mathbf{f}]_I \, \mathbf{c}_I$$

where $\mathbf{f}$ is an arbitrary fine vector, $\mathbf{c}$ is an arbitrary coarse vector, i is a fine index, I is a coarse index, $\mathbf{P}$ is the prolongation matrix and $\mathbf{R}$ is the restriction matrix. This transpose interrelationship between the prolongation and restriction allows the multigrid V-cycle to be used as a preconditioner for conjugate gradients. Similarly Level2D.H and Level2D.C contain code for two-dimensional grids. These two classes are designed for use with a main program to compare iterative methods, so the loop over multigrid iterations is outside the class.

Readers may also experiment with the JavaScript multigrid iteration program **multigrid1D.html** This program allows the user to select the number of grid cells, a differential equation to be solved, the smoother and the prolongation technique. The program plots various residuals and solution increments on the three finest levels of the grid hierarchy. Solution increments on the finest level are plotted against the exact solution increment, and the solution on the finest level is plotted against the true solution. The algorithm also plots $\log_{10} \left\|\widetilde{\mathbf{x}}^{(k)} - \mathbf{x}\right\|_{\mathbf{A}}$ versus the iteration number k. Reader can experiment with increasingly larger numbers of grid cells to see that the slope of this solution error plot is essentially independent of the number of grid cells.

2.7.6 Multigrid Debugging Techniques

There are a number of programming errors that can cause multigrid to fail, or to converge slowly. To remove these errors, it is useful to perform a variety of program tests. Most of these tests are implemented in the multigrid program files linked above.

1. Check that the fine matrix produces a symmetric linear operator. In other words, with the right-hand side set to zero, the computation of the residual given $\mathbf{x}_f$ should provide $\mathbf{A}_f\mathbf{x}_f$. Then for arbitrary values of $\mathbf{x}_f$ and $\mathbf{y}_f$ we should satisfy

$$\mathbf{y}_f{}^\top \mathbf{A}_f\mathbf{x}_f = \left[\mathbf{A}_f\mathbf{y}_f\right]^\top \mathbf{x}_f \,.$$

We can use a random number generator to select $\mathbf{x}_f$ and $\mathbf{y}_f$, apply the residual computation with $\mathbf{b}_f = 0$ to get $\mathbf{A}_f\mathbf{x}_f$ and $\mathbf{A}_f\mathbf{y}_f$, then form inner products to check the symmetry of $\mathbf{A}_f$. If the test fails, then the test should be repeated for $\mathbf{x}_f$ and $\mathbf{y}_f$ set equal to arbitrary axis vectors until the problem is isolated.
2. Check that the restriction is equal to the transpose of the prolongation. If $\mathbf{x}_f$ and $\mathbf{y}_c$ are arbitrary vectors, we should have

$$\mathbf{y}_c{}^\top \left[\mathbf{P}^\top \mathbf{x}_f\right] = [\mathbf{P}\mathbf{y}_c]^\top \mathbf{x}_f \,.$$

On a uniform grid with constant coefficients, the prolongation should produce averages of the coarse grid values at intermediate fine grid points, and the restriction should average fine grid values to produce coarse grid values.
3. Check that the coarse matrix is symmetric. This is similar to the symmetry test for $\mathbf{A}_f$. However, this test depends on the relationship between the prolongation and restriction, and on the code used to compute the coarse grid matrix from the fine grid matrix. For constant coefficients on uniform grids, we can often design the discretization so that the coarse grid matrix corresponds to the same difference scheme on the coarse grid.
4. Check that the pre-smoother and post-smoother are transposes of each other. If $\mathbf{x}_f$ and $\mathbf{y}_f$ are arbitrary vectors, we should have

$$\mathbf{y}_f{}^\top \left[\mathbf{S}_f\mathbf{x}_f\right] = \left[\mathbf{S}^\top \mathbf{y}_f\right]^\top \mathbf{x}_f \,.$$

We can apply the pre-smoother to $\mathbf{x}_f$ to get $\mathbf{S}_f\mathbf{x}_f$, and the post-smoother to $\mathbf{y}_f$ to get $\mathbf{S}^\top \mathbf{y}_f$. Then we can take appropriate inner products to perform the test.
5. Check that the coarse grid projection is a projection. Given a random vector $\mathbf{x}_c$, we want to check that

$$\left[\mathbf{I} - \mathbf{P}\left(\mathbf{P}^\top \mathbf{A}_f\mathbf{P}\right)^{-1}\mathbf{P}^\top \mathbf{A}_f\right]\mathbf{P}\mathbf{x}_c = 0 \,.$$

This test begins with a prolongation to compute $\mathbf{P}\mathbf{x}_c$, then with initial residual set to zero we perform the steps in the multigrid V-cycle that update the residual, restrict, recurse and prolong. Note that the subscript c here corresponds only to the coarsest level.

6. Check that the coarse grid projection is self-adjoint in the inner product generated by $\mathbf{A}_f$. Given arbitrary vectors $\mathbf{x}_f$ and $\mathbf{y}_f$, we compute the coarse grid projections

$$\mathbf{K}_f\mathbf{x}_f = \left[\mathbf{I} - \mathbf{P}\left(\mathbf{P}^\top\mathbf{A}_f\mathbf{P}\right)^{-1}\mathbf{P}^\top\mathbf{A}_f\right]\mathbf{x}_f$$

and $\mathbf{K}_f\mathbf{y}_f$. Then we check that

$$\left[\mathbf{K}_f\mathbf{x}_f, \mathbf{y}_f\right]_{\mathbf{A}_f} = \left(\mathbf{A}_f\mathbf{y}_f\right)^\top\left(\mathbf{K}_f\mathbf{x}_f\right) = \left(\mathbf{K}_f\mathbf{y}_f\right)^\top\left(\mathbf{A}_f\mathbf{x}_f\right) = \left[\mathbf{x}_f, \mathbf{K}_f\mathbf{y}_f\right]_{\mathbf{A}_f}$$

by computing appropriate inner projects.
7. Check the V-cycle is symmetric. If $\mathbf{r}_f$ and $\mathbf{s}_f$ are arbitrary, apply the multigrid V-cycle to compute the resulting corrections $\mathbf{d}_f = \mathbf{V}_f\mathbf{r}_f$ and $\mathbf{e}_f = \mathbf{V}_f\mathbf{s}_f$. Then compare the inner products $\mathbf{s}_f{}^\top\mathbf{d}_f$ and $\mathbf{r}^\top\mathbf{e}_f$.
8. Check that the post-smoother reduces the error in the matrix norm. Given an arbitrary value for $\mathbf{d}_f^{(2,0)}$, set the initial residual $\mathbf{r}_f^{(0)}$ to zero, compute the residual $\mathbf{r}_f^{(2,1)} = \mathbf{r}_f^{(0)} - \mathbf{A}_f\mathbf{d}_f^{(2,0)}$, and apply the post-smoother to get $\mathbf{d}_f^{(2,1)} = \mathbf{d}_f^{(2,0)} + \mathbf{S}_f\mathbf{r}_f^{(2,1)}$. Then check that

$$\left\|\mathbf{d}_f^{(2,1)}\right\|_{\mathbf{A}_f}^2 = \left[\mathbf{d}_f^{(2,1)}\right]^\top\mathbf{A}_f\mathbf{d}_f^{(2,1)} \le \left[\mathbf{d}_f^{(2,0)}\right]^\top\mathbf{A}_f\mathbf{d}_f^{(2,0)} = \left\|\mathbf{d}_f^{(2,0)}\right\|_{\mathbf{A}_f}^2 .$$

9. Check that the pre-smoother reduces the error in the matrix norm (the assumption in (2.51) of Theorem 2.7.1). Given an arbitrary value for $\mathbf{x}_f$, set the right-hand side $\mathbf{b}_f$ to zero, compute the initial residual $\mathbf{r}_f^{(1,0)} = \mathbf{r}_f^{(0)} = \mathbf{b}_f - \mathbf{A}_f\mathbf{x}_f$, apply the pre-smoother to get $\mathbf{d}_f^{(1,1)} = \mathbf{S}_f{}^\top\mathbf{r}_f^{(1,0)}$, update the residual to get $\mathbf{r}_f^{(1,1)} = \mathbf{r}_f^{(0)} - \mathbf{A}_f\mathbf{d}_f^{(1,1)}$, and compute the inner products $(\mathbf{x}_f - \mathbf{d}_f^{(1,1)})^\top\mathbf{r}_f^{(1,1)} = \left\|\left[\mathbf{I} - \mathbf{S}_f{}^\top\mathbf{A}_f\right]\mathbf{x}_f\right\|_{\mathbf{A}_f}^2$ and $\mathbf{x}_f{}^\top\mathbf{r}_f^{(0)} = \left\|\mathbf{x}_f\right\|_{\mathbf{A}_f}^2$. Estimate the largest eigenvalue λ_f of $\mathbf{A}_f$ using the Gerschgorin circle theorem, and compute

$$\beta(\mathbf{x}_f) = 1 - \frac{\left\|\mathbf{r}_f^{(0)}\right\|_2^2}{\lambda_f\mathbf{x}_f{}^\top\mathbf{r}_f^{(0)}} .$$

Then check that

$$\left\|\left[\mathbf{I} - \mathbf{S}_f{}^\top\mathbf{A}_f\right]\mathbf{x}_f\right\|_{\mathbf{A}_f} \le \beta(\mathbf{x}_f)\left\|\mathbf{x}_f\right\|_{\mathbf{A}_f} .$$

10. Check that the V-cycle reduces the error in the solution. If $\mathbf{x}_f$ is random, apply the multigrid V-cycle to initial residual $\mathbf{A}_f\mathbf{x}_f$ with $\mathbf{b}_f = 0$. The resulting vector

$$\mathbf{V}_f\mathbf{x}_f = \left(\mathbf{I} - \mathbf{S}_f{}^\top\mathbf{A}_f\right)\left(\mathbf{I} - \mathbf{P}\mathbf{V}_c\mathbf{P}^\top\mathbf{A}_f\right)\left(\mathbf{I} - \mathbf{S}_f\mathbf{A}_f\right)\mathbf{x}_f$$

should have components that are significantly smaller than $\mathbf{x}_f$. This could also be checked by taking $\mathbf{x}_f$ to be an arbitrary axis vector, and checking that $\mathbf{V}_f\mathbf{x}_f$ has entries that are small compared to one.

It is unfortunately common to produce a multigrid method that converges, but with increasingly more iterations as the size of the matrix increases. Often, this happens if the smoother is chosen to be the Jacobi iteration with relaxation factor $\gamma = 1$. In this case, the problem can be overcome by choosing γ to be the reciprocal of the maximum number of nonzero entries in any row of the matrix; better yet, the problem can be overcome by using a Gauss-Seidel iteration for the smoother.

Chapter 3
Nonlinear Systems

. . . [W]e must readily admit that for any computer algorithm there exist nonlinear functions (infinitely continuously differentiable, if you wish), perverse enough to defeat the algorithm. Therefore, all a user can be guaranteed from any algorithm applied to a nonlinear problem is the answer, 'An approximate solution to the problem is ___,' or, 'No approximate solution to the problem was found in the allotted time.'

J.E. Dennis Jr. and Robert B. Schnabel *[52, p. 16]*

If you optimize everything, you will always be unhappy.

Donald Knuth

Abstract This chapter is devoted to the problem of solving systems of nonlinear equations. The discussion begins with important notions from multi-variable calculus, and continues on to an analysis of the existence and uniqueness of zeros of nonlinear functions. The basic theory also covers perturbation analysis, and the notion of convergence rates of sequences. Newton's method is the first numerical method presented, and its convergence is analyzed, especially in the presence of rounding errors. Afterwards, the discussion turns to unconstrained minimization. Many numerical methods for these problems depend on descent directions, and successful numerical strategies must be judicious in the selection of step lengths and directions. These ideas lead to the development of global convergence strategies for finding unconstrained minima, often by trust regions. Quasi-Newton methods are then presented as efficient numerical techniques for solving systems of nonlinear equations or finding minima of nonlinear functionals. Iterative methods for solving systems of linear equations can be generalized to nonlinear systems through Krylov subspace methods. This chapter also discusses direct search methods and stochastic minimization.

Additional Material: The details of the computer programs referred in the text are available in the Springer website (http://extras.springer.com/2018/978-3-319-69107-7) for authorized users.

J.A. Trangenstein, *Scientific Computing*, Texts in Computational Science and Engineering 19, https://doi.org/10.1007/978-3-319-69107-7_3

3.1 Overview

Recall that we previously developed methods for solving systems of linear equations in Chap. 3 of Volume I and Chap. 2. The former of these two chapters discussed direct methods, based on matrix factorizations, while the latter chapter discussed iterative methods, presumably for large systems of linear equations. We have also discussed methods for finding zeros or minima of nonlinear functions of a single variable in Chap. 5 of Volume I. Our goal in this chapter is to combine these previous computational techniques with new ideas for solving two new and general problems. The first is to find a zero of a system of nonlinear equations, which can be written in the form

$$\mathbf{f}(\mathbf{z}) = 0 , \tag{3.1}$$

where $\mathbf{f}$ maps real n-vectors to real n-vectors. The second problem is to find a minimum of a nonlinear functional. This problem can be written in the form

$$\min \phi(\mathbf{x}) \tag{3.2}$$

where ϕ is a real-valued function of real n-vectors.

Systems of nonlinear equations (3.1) are more difficult to solve than either scalar nonlinear equations or systems of linear equations. In fact, it can be difficult just to guarantee the existence or uniqueness of a solution. Nevertheless, these problems are important in many practical applications. One computational approach in this chapter will be to find ways of approximating the solution of systems of nonlinear equations by a sequence of systems of linear equations. In order to improve the chances that an iteration converges, another computational approach in this chapter will construct search directions that depend on solving linear least squares problems. Those searches will be conducted along those directions in order to minimize certain nonlinear scalar functions. For the nonlinear minimization problem (3.2), a very successful computational approach will to convert this problem into a search for a zero of a system of nonlinear equations.

Here is a quick summary of the contents of this chapter. We will begin in Sect. 3.2 with some theory that will help us understand the conditions under which a system of nonlinear equations is suitable for scientific computing. Using calculus, we will develop conditions that guarantee the existence and uniqueness of a zero of a nonlinear function, or the local minimizer of a functional. In this same section, we will examine the sensitivity of such solutions to perturbations in the prescribed nonlinear functions. Section 3.3 will extend the discussion of rates of convergence to sequences of real vectors. In the remaining sections of this chapter, we will develop numerical methods to solve these problems and study the sensitivity of these methods to numerical perturbations. We will conclude the chapter will a brief summary of interesting test problems in Sect. 3.11, followed by a case study from chemistry in Sect. 3.12.

For more information about the material in this chapter, we recommend books by Dennis and Schnabel [52], Gill et al. [81], Kelley [107], Luenberger [120], Murray [130] and Ortega and Rheinboldt [133]. For software to solve systems of nonlinear equations, we recommend MINPACK, GSL (GNU Scientific Library) Multidimensional Root-Finding or Multidimensional Minimization, and PETSc (Portable, Extensible Toolkit for Scientific Computation). In MATLAB, we recommend commands fsolve, fminunc and fminsearch, or Tim Kelley's nsol. Scilab provides the command fsolve to solve a system of nonlinear equations, as well as the commands fminsearch and optim to find the minimizer of a nonlinear functional.

3.2 Calculus

Let us recall once again the steps in solving a scientific computing problem, as described in Sect. 1.3 in Chap. 1 of Volume I. The first step is to provide mathematical descriptions of the problems under consideration; these can be found in Eqs. (3.1) and (3.2). Next, we must determine conditions under which these problems have a solution, and the solution is unique. Afterward, we must determine how perturbations in the problem affect the solutions. These two goals are the impetus for the discussion in this section.

Our approach in Sects. 3.2.1–3.2.3 will be to use calculus to approximate the nonlinear functions by linear or quadratic models. Then we will find necessary and sufficient conditions for a local minimizer of a functional in Sect. 3.2.4. The first-order necessary condition for a local minimizer will require us to find the zero of a nonlinear function. The existence and uniqueness of the zero of a nonlinear function can be guaranteed in one of at least two ways, described in Sects. 3.2.5.1 and 3.2.5.2. We will end with perturbation analysis in Sect. 3.2.6.

3.2.1 Functionals

We begin with some notions of derivatives for scalar-valued functions of multiple variables.

Definition 3.2.1 A **functional** is a scalar-valued function of n-vectors, and a **real functional** is a real-valued function of real n-vectors. Suppose that ϕ is a continuous real functional, and let $\mathbf{s}$ be a nonzero real n-vector. Then ϕ has a **directional derivative** in direction $\mathbf{s}$ if and only if

$$\lim_{\varepsilon \downarrow 0} \frac{\phi(\mathbf{x} + \mathbf{s}\varepsilon) - \phi(\mathbf{x})}{\varepsilon}$$

exists. Let $\|\cdot\|$ be any norm on n-vectors. Then ϕ is **differentiable** at $\mathbf{x}$ if and only if there is an n-vector $\mathbf{g}$ so that

$$\lim_{\widetilde{\mathbf{x}}\to\mathbf{x}} \frac{|\phi(\widetilde{\mathbf{x}}) - \{\phi(\mathbf{x}) + \mathbf{g}\cdot(\widetilde{\mathbf{x}}-\mathbf{x})\}|}{\|\widetilde{\mathbf{x}}-\mathbf{x}\|} = 0 .$$

If ϕ is **differentiable** at $\mathbf{x}$, then $\mathbf{g}$ is called the **gradient** of ϕ at $\mathbf{x}$. In order to make the notation precise, we will write the gradient of ϕ at $\mathbf{x}$ as the vector $\mathbf{g}_\phi(\mathbf{x})$. Finally, if the gradient $\mathbf{g}_\phi(\mathbf{x})$ of a scalar-valued function ϕ is continuous at $\mathbf{x}$, then we say that ϕ is **continuously differentiable** at $\mathbf{x}$.

The definition of differentiability shows that if ϕ is continuously differentiable at $\mathbf{x}$, then the directional derivative of ϕ in direction $\mathbf{s}$ is $\mathbf{g}_\phi(\mathbf{x})\cdot\mathbf{s}$.

It is easy to see that the gradient of a differentiable function is unique. In particular, if $\mathbf{g}_1$ and $\mathbf{g}_2$ were two gradients of ϕ at $\mathbf{x}$, then for all n-vectors $\widetilde{\mathbf{x}}$

$$\begin{aligned}
&\lim_{\widetilde{\mathbf{x}}\to\mathbf{x}} \left|(\mathbf{g}_1-\mathbf{g}_2)\cdot\frac{\widetilde{\mathbf{x}}-\mathbf{x}}{\|\widetilde{\mathbf{x}}-\mathbf{x}\|}\right| \\
&\le \lim_{\widetilde{\mathbf{x}}\to\mathbf{x}} \frac{|\phi(\widetilde{\mathbf{x}}) - \{\phi(\mathbf{x}) + \mathbf{g}_1\cdot(\widetilde{\mathbf{x}}-\mathbf{x})\}|}{\|\widetilde{\mathbf{x}}-\mathbf{x}\|} + \lim_{\widetilde{\mathbf{x}}\to\mathbf{x}} \frac{|\phi(\widetilde{\mathbf{x}}) - \{\phi(\mathbf{x}) + \mathbf{g}_2\cdot(\widetilde{\mathbf{x}}-\mathbf{x})\}|}{\|\widetilde{\mathbf{x}}-\mathbf{x}\|} = 0 .
\end{aligned}$$

This inequality implies that $\mathbf{g}_1 - \mathbf{g}_2$ is orthogonal to all unit vectors, and therefore must be zero.

So far, we have avoided a discussion of partial derivatives. The following example will illustrate the kind of problems we must avoid.

Example 3.2.1 The functional

$$\phi(\mathbf{x}) = (\xi_1\xi_2)^{1/3}$$

is continuous everywhere with value $\phi(\mathbf{0}) = 0$. Its partial derivatives at the origin are

$$\begin{aligned}
\frac{\partial\phi}{\partial\xi_1}(\mathbf{0}) &= \lim_{\xi_1\to 0} \frac{\phi([\xi_1,0]) - \phi(\mathbf{0})}{\xi_1} = 0 \text{ and} \\
\frac{\partial\phi}{\partial\xi_2}(\mathbf{0}) &= \lim_{\xi_2\to 0} \frac{\phi([0,\xi_2]) - \phi(\mathbf{0})}{\xi_2} = 0 .
\end{aligned}$$

However, ϕ is not differentiable at the origin.

We will clarify the relationship between derivatives and partial derivatives in Sect. 3.2.2.

Here is a simple intermediate value result for real functionals.

Lemma 3.2.1 *Suppose that ϕ is a real functional, and assume that ϕ is continuously differentiable in some convex subset $\mathscr{D}$ of the set of all real n-vectors. Then* *for all* $\mathbf{x}$ *and* $\mathbf{x}+\mathbf{s} \in \mathscr{D}$*, then there is an n-vector* $\mathbf{y}$ *on the line between* $\mathbf{x}$ *and* $\mathbf{x}+\mathbf{s}$ *so that*

$$\phi(\mathbf{x}+\mathbf{s}) - \phi(\mathbf{x}) = \int_0^1 \mathbf{g}_\phi(\mathbf{x}+\mathbf{s}\tau) \cdot \mathbf{s} \, \mathrm{d}\tau = \mathbf{g}_\phi(\mathbf{y}) \cdot \mathbf{s} \, .$$

Proof Let

$$\gamma(\tau) \equiv \phi(\mathbf{x}+\mathbf{s}\tau) \, .$$

Then the derivative of this real-valued function of a real number is

$$\gamma'(\tau) = \mathbf{s} \cdot \mathbf{g}_\phi(\mathbf{x}+\mathbf{s}\tau) \, .$$

By the fundamental theorem of calculus 5.2.4 in Chap. 5 of Volume I and the intermediate value theorem 5.2.1 in Chap. 5 of Volume I, we see that

$$\gamma(1) - \gamma(0) = \int_0^1 \gamma'(\tau) \, \mathrm{d}\tau = \gamma'(\eta)$$

for some $\eta \in (0,1)$. By replacing γ and its derivatives with the equivalent functions involving ϕ, we get the desired result.

3.2.2 Vector-Valued Functions

Next, we consider vector-valued functions and their derivatives.

Definition 3.2.2 Suppose that $\mathbf{f}$ is a continuous function of real n-vectors, and takes values that are real m-vectors. Let $\mathbf{x}$ be an n-vector, and let $\mathbf{s}$ be a nonzero n-vector. Then $\mathbf{f}$ has a **directional derivative** at $\mathbf{x}$ in direction $\mathbf{s}$ if and only if

$$\lim_{\varepsilon \downarrow 0} \frac{\mathbf{f}(\mathbf{x}+\mathbf{s}\varepsilon) - \mathbf{f}(\mathbf{x})}{\varepsilon}$$

exists. Let $\|\cdot\|$ be a norm on real n-vectors. Then $\mathbf{f}$ is **differentiable** at $\mathbf{x}$ if and only if there is an $m \times n$ matrix $\mathbf{J}$ so that

$$\lim_{\widetilde{\mathbf{x}} \to \mathbf{x}} \frac{\|\mathbf{f}(\widetilde{\mathbf{x}}) - \{\mathbf{f}(\mathbf{x}) + \mathbf{J}(\widetilde{\mathbf{x}} - \mathbf{x})\}\|}{\|\widetilde{\mathbf{x}} - \mathbf{x}\|} = 0 \, .$$

If $\mathbf{f}$ is differentiable at $\mathbf{x}$, then we use the notation $\mathbf{J}_\mathbf{f}(\mathbf{x})$ for the **Jacobian** matrix of $\mathbf{f}$ at $\mathbf{x}$. Finally, $\mathbf{f}$ is **continuously differentiable** at $\mathbf{x}$ if and only if $\mathbf{J}_\mathbf{f}(\mathbf{x})$ is continuous at $\mathbf{x}$.

Some authors (e.g., Ortega and Rheinboldt [133, p. 59]) provide an alternative weaker definition of a derivative. Our definition corresponds to what those authors would call the **Fréchet derivative**.

It is common to write derivatives of functions in terms of partial derivatives. Example 3.2.1 shows that we must be cautious in doing so. However, the following result, which can be found in Dieudonné [55, p. 167], is more optimistic.

Lemma 3.2.2 *Suppose that if* $\mathbf{f}$ *is a function of n-vectors, and takes m-vectors as values. Let* $\mathbf{f}$ *be continuous in some open set* $\mathscr{D}$. *Then* $\mathbf{f}$ *is continuously differentiable in* $\mathscr{D}$ *if and only* $\mathbf{f}$ *is differentiable at each* $\mathbf{x} \in \mathscr{D}$ *with respect to each component of* $\mathbf{x}$, *and the derivatives of these restricted functions are continuous.*
In such a case, if $\mathbf{f}$ has component functions $\phi_i(\mathbf{x})$ then we can write

$$\mathbf{J_f}(\mathbf{x}) = \left[\frac{\partial \phi_i}{\partial \mathbf{x}_j}\right] .$$

The following lemma generalizes Lemma 3.2.1.

Lemma 3.2.3 *Suppose that* $\mathscr{D}$ *is an open subset of the set of all real n-vectors. Let* $\mathbf{f}$ *be continuously differentiable in* $\mathscr{D}$, *and let* $\mathbf{f}$ *take m-vectors as values. Then for all* $\mathbf{x}$ *and* $\mathbf{x} + \mathbf{s} \in \mathscr{D}$

$$\mathbf{f}(\mathbf{x} + \mathbf{s}) - \mathbf{f}(\mathbf{x}) = \int_0^1 \mathbf{J_f}(\mathbf{x} + \mathbf{s}\tau)\mathbf{s} \, d\tau .$$

Proof Let

$$\mathbf{p}(\tau) = \mathbf{f}(\mathbf{x} + \mathbf{s}\tau) .$$

Then $\mathbf{p}$ is continuously differentiable on $(0, 1)$, and the fundamental theorem of calculus 5.2.4 in Chap. 5 of Volume I applied to the components of $\mathbf{p}$ gives us

$$\mathbf{f}(\mathbf{x} + \mathbf{s}) - \mathbf{f}(\mathbf{x}) = \mathbf{p}(1) - \mathbf{p}(0) = \int_0^1 \mathbf{p}'(\tau) \, d\tau = \int_0^1 \mathbf{J_f}(\mathbf{x} + \mathbf{s}\tau)\mathbf{s} \, d\tau .$$

The next lemma is a generalization of the triangle inequality.

Lemma 3.2.4 *Let* $\mathscr{D}$ *be an open convex subset of the set of all real n-vectors. Assume that* $\mathbf{f}$ *is continuous on* $\mathscr{D}$ *and takes real m-vectors as values. Then for all norms* $\|\cdot\|$ *on real m-vectors and for all* $\mathbf{x}, \mathbf{x} + \mathbf{s} \in \mathscr{D}$, *we have*

$$\left\| \int_0^1 \mathbf{f}(\mathbf{x} + \mathbf{s}\tau) \, d\tau \right\| \le \int_0^1 \|\mathbf{f}(\mathbf{x} + \mathbf{s}\tau)\| \, d\tau .$$

Proof Since $\mathbf{f}$ is continuous, it has a Riemann integral. We can approximate the integral by a uniform Riemann sum and apply the triangle inequality. The result follows by taking the limit as the number of terms in the Riemann sum approaches infinity.

The following theorem should also be familiar.

Theorem 3.2.1 (Chain Rule) *Suppose that $\mathscr{D}_1$ is an open subset of real n-vectors, and $\mathscr{D}_2$ is an open subset of real k-vectors. Let $\mathbf{f}_1$ be a continuous function that maps $\mathscr{D}_1$ into $\mathscr{D}_2$, and suppose that $\mathbf{f}_1$ is differentiable at $\mathbf{u} \in \mathscr{D}_1$. Assume that $\mathbf{f}_2$ is a continuous function that maps $\mathscr{D}_2$ to real m-vectors, and that $\mathbf{f}_2$ is differentiable at $\mathbf{f}_1(\mathbf{u})$. Then $\mathbf{f}_2 \circ \mathbf{f}_1$ is differentiable at $\mathbf{u}$, and*

$$\mathbf{J}_{\mathbf{f}_2 \circ \mathbf{f}_1}(\mathbf{u}) = \mathbf{J}_{\mathbf{f}_2}(\mathbf{f}_1(\mathbf{u}))\mathbf{J}_{\mathbf{f}_1}(\mathbf{u}) . \tag{3.3}$$

Proof Since $\mathscr{D}_2$ is open, there exists $\delta_2 > 0$ so that

$$\|\mathbf{s}_2\| \leq \delta_2 \Longrightarrow \mathbf{f}_1(\mathbf{u}) + \mathbf{s}_2 \in \mathscr{D}_2 .$$

Since $\mathbf{f}_1$ is continuous in the open set $\mathscr{D}_1$, there exists $\delta_1 > 0$ so that

$$\|\mathbf{s}_1\| \leq \delta_1 \Longrightarrow \|\mathbf{f}_1(\mathbf{u}+\mathbf{s}_1) - \mathbf{f}_1(\mathbf{u})\| \leq \delta_2 ;$$

if $\mathbf{s}_1$ is chosen in this way, then $\mathbf{f}_1(\mathbf{u}+\mathbf{s}_1) \in \mathscr{D}_2$.

Since $\mathbf{f}_1$ is differentiable at $\mathbf{u} \in \mathscr{D}_1$, for any $\varepsilon_1' > 0$ there exists $\delta_1' > 0$ so that

$$\|\mathbf{s}_1'\| \leq \delta_1' \Longrightarrow \|\mathbf{f}_1(\mathbf{u}+\mathbf{s}_1) - \mathbf{f}_1(\mathbf{u}) - \mathbf{J}_{\mathbf{f}_1}(\mathbf{u})\| \leq \varepsilon_1' \|\mathbf{s}_1'\| .$$

Since $\mathbf{f}_2$ is differentiable at $\mathbf{f}_1(\mathbf{u}) \in \mathscr{D}_2$, for any $\varepsilon_2' > 0$ there exists $\delta_2' > 0$ so that

$$\|\mathbf{s}_2'\| \leq \delta_2' \Longrightarrow \|\mathbf{f}_2(\mathbf{f}_1(\mathbf{u}) + \mathbf{s}_2) - \mathbf{f}_2(\mathbf{f}_1(\mathbf{u})) - \mathbf{J}_{\mathbf{f}_2}(\mathbf{f}_1(\mathbf{u}))\mathbf{s}_2'\| \leq \varepsilon_2' \|\mathbf{s}_2'\| .$$

Given $\varepsilon > 0$, choose $\varepsilon_1' > 0$ so that

$$\varepsilon_1' \|\mathbf{J}_{\mathbf{f}_2}(\mathbf{f}_1(\mathbf{u}))\| \leq \varepsilon/2 .$$

Afterward, choose $\varepsilon_2' > 0$ so that

$$\varepsilon_2' \left(\varepsilon_1' + \|\mathbf{J}_{\mathbf{f}_1}(\mathbf{u})\|\right) \leq \varepsilon/2 .$$

If

$$\|\mathbf{s}_1\| \leq \min\left\{\delta_1 , \delta_1' , \frac{\delta_2'}{\varepsilon_1' + \|\mathbf{J}_{\mathbf{f}_1}(\mathbf{u})\|}\right\} ,$$

then

$$\|\mathbf{f}_1(\mathbf{u}+\mathbf{s}_1) - \mathbf{f}_1(\mathbf{u})\| \leq \left\{\|\mathbf{J}_{\mathbf{f}_1}(\mathbf{u})\mathbf{s}_1\| + \varepsilon_1'\|\mathbf{s}_1\|\right\} \leq \delta_2' ,$$

so

$$\|\mathbf{f}_2 \circ \mathbf{f}_1(\mathbf{u}+\mathbf{s}_1) - \mathbf{f}_2 \circ \mathbf{f}_1(\mathbf{u}) - \mathbf{J}_{\mathbf{f}_2}(\mathbf{f}_1(\mathbf{u}))\,\mathbf{J}_{\mathbf{f}_1}(\mathbf{u})\|$$
$$= \|\mathbf{f}_2(\mathbf{f}_1(\mathbf{u}+\mathbf{s}_1)) - \mathbf{f}_2(\mathbf{f}_1(\mathbf{u})) - \mathbf{J}_{\mathbf{f}_2}(\mathbf{f}_1(\mathbf{u}))\,\mathbf{J}_{\mathbf{f}_1}(\mathbf{u})\|$$
$$\leq \|\{\mathbf{f}_2(\mathbf{f}_1(\mathbf{u}+\mathbf{s}_1)) - \mathbf{f}_2(\mathbf{f}_1(\mathbf{u})) - \mathbf{J}_{\mathbf{f}_2}(\mathbf{f}_1(\mathbf{u}))\,[\mathbf{f}_1(\mathbf{u}+\mathbf{s}_1) - \mathbf{f}_1(\mathbf{u})]\}\|$$
$$+ \|\mathbf{J}_{\mathbf{f}_2}(\mathbf{f}_1(\mathbf{u}))\,\{\mathbf{f}_1(\mathbf{u}+\mathbf{s}_1) - \mathbf{f}_1(\mathbf{u}) - \mathbf{J}_{\mathbf{f}_1}(\mathbf{u})\mathbf{s}_1\}\|$$
$$\leq \varepsilon_2' \|\mathbf{f}_1(\mathbf{u}+\mathbf{s}_1) - \mathbf{f}_1(\mathbf{u})\| + \|\mathbf{J}_{\mathbf{f}_2}(\mathbf{f}_1(\mathbf{u}))\|\,\|\mathbf{f}_1(\mathbf{u}+\mathbf{s}_1) - \mathbf{f}_1(\mathbf{u}) - \mathbf{J}_{\mathbf{f}_1}(\mathbf{u})\mathbf{s}_1\|$$
$$\leq \varepsilon_2' \left\{\varepsilon_1'\|\mathbf{s}_1\| + \|\mathbf{J}_{\mathbf{f}_1}(\mathbf{u})\|\,\|\mathbf{s}_1\|\right\} + \|\mathbf{J}_{\mathbf{f}_2}(\mathbf{f}_1(\mathbf{u}))\|\,\varepsilon_1'\|\mathbf{s}_1\|$$
$$= \left\{\varepsilon_2'\left[\varepsilon_1' + \|\mathbf{J}_{\mathbf{f}_1}(\mathbf{u})\|\right] + \varepsilon_1'\|\mathbf{J}_{\mathbf{f}_2}(\mathbf{f}_1(\mathbf{u}))\|\right\}\|\mathbf{s}_1\| \leq \varepsilon\|\mathbf{s}_1\| \ .$$

This proves the claim.

Next, let us specialize the notion of continuity for vector-valued functions.

Definition 3.2.3 Suppose that $\mathscr{D}$ is an open subset of real n-vectors, and let $\mathbf{f}$ map $\mathscr{D}$ to real m-vectors. Then $\mathbf{f}$ is **Lipschitz continuous** at $\mathbf{u} \in \mathscr{D}$ if and only if there exists $\gamma_{\mathbf{u}} > 0$ such that for all $\widetilde{\mathbf{u}} \in \mathscr{D}$

$$\|\mathbf{f}(\widetilde{\mathbf{u}}) - \mathbf{f}(\mathbf{u})\| \leq \gamma_{\mathbf{u}}\|\widetilde{\mathbf{u}} - \mathbf{u}\| \ .$$

Also, $\mathbf{f}$ is **uniformly Lipschitz continuous** in $\mathscr{D}$ if and only if there exists $\gamma > 0$ such that for all $\mathbf{u}$ and $\widetilde{\mathbf{u}} \in \mathscr{D}$

$$\|\mathbf{f}(\widetilde{\mathbf{u}}) - \mathbf{f}(\mathbf{u})\| \leq \gamma\|\widetilde{\mathbf{u}} - \mathbf{u}\| \ .$$

We can use our new definition to bound the error in approximating a function by a tangent plane.

Lemma 3.2.5 *Suppose that $\mathscr{D}$ is an open and convex subset of real n-vectors, Assume that* $\mathbf{f}$ *map $\mathscr{D}$ to real m-vectors, and is continuously differentiable on $\mathscr{D}$. Let $\|\cdot\|$ represent a norm on either n-vectors or m-vectors. Suppose that* $\mathbf{J}_\mathbf{f}$ *be Lipschitz continuous at* $\mathbf{x} \in \mathscr{D}$ *with Lipschitz constant γ, related to the choice of norms on m-vectors and n-vectors. Then for all* $\mathbf{x}+\mathbf{s} \in \mathscr{D}$,

$$\|\mathbf{f}(\mathbf{x}+\mathbf{s}) - \mathbf{f}(\mathbf{x}) - \mathbf{J}_\mathbf{f}(\mathbf{x})\mathbf{s}\| \leq \frac{\gamma}{2}\|\mathbf{s}\|^2 \ .$$

Proof Lemma 3.2.3 shows that

$$\mathbf{f}(\mathbf{x}+\mathbf{s}) - \mathbf{f}(\mathbf{x}) - \mathbf{J}_\mathbf{f}(\mathbf{x})\mathbf{s} = \int_0^1 [\mathbf{J}_\mathbf{f}(\mathbf{x}+\mathbf{s}\tau) - \mathbf{J}_\mathbf{f}(\mathbf{x})]\,\mathbf{s}\,d\tau \ .$$

Now take norms of both sides, apply the triangle inequality and use Lipschitz continuity to get

$$\|\mathbf{f}(\mathbf{x}+\mathbf{s}) - \mathbf{f}(\mathbf{x}) - \mathbf{J_f}(\mathbf{x})\mathbf{s}\| \leq \int_0^1 \|\mathbf{J_f}(\mathbf{x}+\mathbf{s}\tau) - \mathbf{J_f}(\mathbf{x})\| \, \|\mathbf{s}\| \, d\tau$$

$$\leq \int_0^1 \{\gamma \|\mathbf{s}\| \tau\} \, \|\mathbf{s}\| \, d\tau = \frac{\gamma}{2} \|\mathbf{s}\|^2 \, .$$

As an easy consequence of this result, we get the following estimate for finite difference approximations to Jacobians.

Corollary 3.2.1 *Suppose that the hypotheses of Lemma 3.2.5 are satisfied. In addition, assume that the nonzero scalar ξ is such that $\mathbf{x}+\mathbf{e}_j\xi \in \mathscr{D}$ for all $1 \leq j \leq n$. Define the finite difference Jacobian columnwise by*

$$\widetilde{\mathbf{J}}_\mathbf{f}(\mathbf{x}, \xi)\mathbf{e}_j = \left[\mathbf{f}(\mathbf{x}+\mathbf{e}_j\xi) - \mathbf{f}(\mathbf{x})\right]/\xi \, .$$

Then for all $1 \leq j \leq n$

$$\left\| \left\{ \widetilde{\mathbf{J}}_\mathbf{f}(\mathbf{x}, \xi) - \mathbf{J_f}(\mathbf{x}) \right\} \mathbf{e}_j \right\| \leq \frac{\gamma}{2} |\xi| \, \|\mathbf{e}_j\|^2 \, .$$

In particular,

$$\left\| \widetilde{\mathbf{J}}_\mathbf{f}(\mathbf{x}, \xi) - \mathbf{J_f}(\mathbf{x}) \right\|_1 \leq \frac{\gamma}{2} |\xi| \, .$$

Proof The first claim follows immediately from Lemma 3.2.5 by choosing $\mathbf{s} = \mathbf{e}_j\xi$. The second claim then follows from the fact that $\|\mathbf{e}_j\|_1 = 1$ for all $1 \leq j \leq n$, and from Lemma 3.5.4 in Chap. 3 of Volume I which shows that the matrix 1-norm is the maximum 1-norm of a matrix column.

The next lemma will bound the error in approximating a function by a secant plane, and is proved in a fashion similar to Lemma 3.2.5.

Lemma 3.2.6 *Suppose that $\mathscr{D}$ is an open and convex subset of real n-vectors. Assume that $\|\cdot\|$ represents a norm on either n-vectors or m-vectors. Let $\mathbf{f}$ mapping $\mathscr{D}$ to real m-vectors be continuously differentiable on $\mathscr{D}$, and let its Jacobian $\mathbf{J_f}$ be uniformly Lipschitz continuous in $\mathscr{D}$ with Lipschitz constant γ with respect to the choices of norms on m-vectors and n-vectors. Then for all $\mathbf{x}$, $\mathbf{u}$ and $\mathbf{v} \in \mathscr{D}$, we have*

$$\|\mathbf{f}(\mathbf{v}) - \mathbf{f}(\mathbf{u}) - \mathbf{J_f}(\mathbf{x})(\mathbf{v}-\mathbf{u})\| \leq \frac{\gamma}{2} \{\|\mathbf{v}-\mathbf{x}\| + \|\mathbf{u}-\mathbf{x}\|\} \, \|\mathbf{v}-\mathbf{u}\| \, .$$

Proof Lemma 3.2.3 implies that

$$\|\mathbf{f}(\mathbf{v}) - \mathbf{f}(\mathbf{u}) - \mathbf{J_f}(\mathbf{x})(\mathbf{v}-\mathbf{u})\| = \left\| \int_0^1 \{\mathbf{J_f}(\mathbf{u}+[\mathbf{v}-\mathbf{u}]\tau) - \mathbf{J_f}(\mathbf{x})\} \{\mathbf{v}-\mathbf{u}\} \, d\tau \right\|$$

then Lemma 3.2.4 gives us

$$\leq \int_0^1 \|\mathbf{J_f}(\mathbf{u}+[\mathbf{v}-\mathbf{u}]\tau) - \mathbf{J_f}(\mathbf{x})\| \, \|\mathbf{v}-\mathbf{u}\| \, d\tau$$

then we use uniform Lipschitz continuity to get

$$\leq \int_0^1 \gamma \, \|\mathbf{u}+[\mathbf{v}-\mathbf{u}]\tau - \mathbf{x}\| \, \|\mathbf{v}-\mathbf{u}\| \, d\tau$$
$$= \int_0^1 \|(\mathbf{u}-\mathbf{x})(1-\tau) + (\mathbf{v}-\mathbf{x})\tau\| \, d\tau \|\mathbf{v}-\mathbf{u}\|$$

then the triangle inequality produces

$$\leq \gamma \int_0^1 \|\mathbf{u}-\mathbf{x}\|(1-\tau) + \|\mathbf{v}-\mathbf{x}\|\tau \, d\tau \|\mathbf{v}-\mathbf{u}\|$$
$$= \frac{\gamma}{2}\{\|\mathbf{u}-\mathbf{x}\| + \|\mathbf{v}-\mathbf{x}\|\} \, \|\mathbf{v}-\mathbf{u}\| \, .$$

3.2.3 Second-Order Derivatives

Next, we turn to second derivatives of functions.

Definition 3.2.4 Let the real functional ϕ be continuous at the real n-vector $\mathbf{x}$, and suppose that the gradient $\mathbf{g}_\phi$ of ϕ is also continuous at $\mathbf{x}$. Then ϕ is **twice differentiable** at $\mathbf{x}$ if and only if $\mathbf{g}_\phi$ is differentiable at $\mathbf{x}$. The **second derivative** of ϕ at $\mathbf{x}$ is the derivative of its gradient $\mathbf{g}_\phi(\mathbf{x})$ and is called the **Hessian** of ϕ. We will use the notation

$$\mathbf{H}_\phi(\mathbf{x}) = \mathbf{J}_{\mathbf{g}_\phi}(\mathbf{x})$$

for the Hessian of ϕ at $\mathbf{x}$.

The next example shows that we must be cautious in working with second-order partial derivatives.

Example 3.2.2 Let $\mathbf{x} = [\xi_1, \xi_2]$ and

$$\phi(\mathbf{x}) = \frac{\xi_1\xi_2(\xi_1^2 - \xi_2^2)}{\xi_1^2 + \xi_2^2}$$

for $\mathbf{x} \neq \mathbf{0}$. Define $\phi(\mathbf{x}) = 0$. It is not hard to see that ϕ is continuous at the origin. For $\mathbf{x} \neq \mathbf{0}$, its gradient is

$$\mathbf{g}_\phi(\mathbf{x}) = \begin{bmatrix} \xi_2 \left(3\xi_1^2 - \xi_2^2\right)\left(\xi_1^2 + \xi_2^2\right) - 2\xi_1^2\xi_2\left(\xi_1^2 - \xi_2^2\right) \\ -\xi_1 \left(3\xi_2^2 - \xi_1^2\right)\left(\xi_1^2 + \xi_2^2\right) + 2\xi_2^2\xi_1\left(\xi_2^2 - \xi_1^2\right) \end{bmatrix} \frac{1}{\left(\xi_1^2 + \xi_2^2\right)^2}$$

This gradient is also continuous at the origin. We can evaluate components of the gradient to see that

$$\frac{\partial \phi}{\partial \xi_1}(0, \xi_2) = -\xi_2 \text{ and } \frac{\partial \phi}{\partial \xi_2}(\xi_1, 0) = \xi_1 \; .$$

It follows that

$$\frac{\partial}{\partial \xi_2}\left(\frac{\partial \phi}{\partial \xi_1}\right)(0,0) = -1 \text{ and } \frac{\partial}{\partial \xi_1}\left(\frac{\partial \phi}{\partial \xi_2}\right)(0,0) = 1 \; .$$

Dieudonné [55, p. 176] shows that if a function is twice differentiable at a point, then its first-order partial derivatives are differentiable at that point, and its mixed second-order derivatives are equal at that point. Under such circumstances, the Hessian matrix is symmetric.

The following lemma will give us a technique for approximating second derivatives of functionals by difference quotients.

Lemma 3.2.7 *Suppose that the real functional ϕ is twice continuously differentiable at $\mathbf{x}$. Let $\mathbf{g}_\phi(\mathbf{x})$ be the gradient of ϕ evaluated at $\mathbf{x}$, and let $\mathbf{H}_\phi(\mathbf{x})$ be the Hessian of ϕ at $\mathbf{x}$. Then*

$$\lim_{\varepsilon \to 0} \frac{\mathbf{g}_\phi(\mathbf{x} + \mathbf{s}\varepsilon) - \mathbf{g}_\phi(\mathbf{x})}{\varepsilon} = \mathbf{H}_\phi(\mathbf{x})\mathbf{s} \; .$$

Proof Define the functions

$$\mathbf{y}(\varepsilon) = \mathbf{x} + \mathbf{s}\varepsilon \text{ and } \mathbf{f}(\varepsilon) = \mathbf{g}_\phi(\mathbf{x} + \mathbf{s}\varepsilon) = \mathbf{g}_\phi \circ \mathbf{y}(\varepsilon) \; .$$

Then the chain rule (3.3) implies that

$$\mathbf{J_f}(\varepsilon) = \mathbf{J}_{\mathbf{g}_\phi}(\mathbf{y}(\varepsilon))\, \mathbf{J_y}(\varepsilon) = \mathbf{H}_\phi(\mathbf{y}(\varepsilon))\, \mathbf{s} \; .$$

The claimed result follows by taking the limit as $\varepsilon \to 0$.

The next lemma provides an estimate for the error in approximating a functional by a tangent line.

Lemma 3.2.8 *Suppose that the real functional ϕ is twice continuously differentiable on the closed line segment between the real n-vectors $\mathbf{x}$ and $\mathbf{x} + \mathbf{s}$. Let $\| \cdot \|$*

represent either a norm on n-vectors, or the corresponding subordinate norm on $n \times n$ matrices. Then

$$\phi(\mathbf{x}+\mathbf{s}) - \phi(\mathbf{x}) - \mathbf{g}_\phi(\mathbf{x}) \cdot \mathbf{s} = \int_0^1 \int_0^\tau \mathbf{s} \cdot \mathbf{H}_\phi(\mathbf{x}+\mathbf{s}\sigma)\mathbf{s} \, d\sigma \, d\tau = \frac{1}{2}\mathbf{s} \cdot \mathbf{H}_\phi(\mathbf{x}+\mathbf{s}\tau)\mathbf{s}$$

for some $\tau \in (0,1)$. Furthermore,

$$\left|\phi(\mathbf{x}+\mathbf{s}) - \phi(\mathbf{x}) - \mathbf{g}_\phi(\mathbf{x}) \cdot \mathbf{s}\right| \le \frac{1}{2} \max_{0 \le \tau \le 1} \|\mathbf{H}_\phi(\mathbf{x}+\mathbf{s}\tau)\| \|\mathbf{s}\|^2 .$$

Proof Lemma 3.2.1 implies that

$$\phi(\mathbf{x}+\mathbf{s}) - \phi(\mathbf{x}) - \mathbf{s} \cdot \mathbf{g}_\phi(\mathbf{x}) = \int_0^1 \left[\mathbf{g}_\phi(\mathbf{x}+\mathbf{s}\tau) - \mathbf{g}_\phi(\mathbf{x})\right] \cdot \mathbf{s} \, d\tau .$$

This suggests that we define the functional

$$\gamma(\tau) \equiv \mathbf{g}_\phi(\mathbf{x}+\mathbf{s}\tau) \cdot \mathbf{s} .$$

Then the fundamental theorem of calculus 5.2.4 in Chap. 5 of Volume I implies that

$$\gamma(\tau) - \gamma(0) = \int_0^\tau \gamma'(\sigma) \, d\sigma = \int_0^\tau \mathbf{s} \cdot \mathbf{H}_\phi(\mathbf{x}+\mathbf{s}\sigma)\mathbf{s} \, d\sigma .$$

Substituting this into the first equation leads to

$$\begin{aligned}
\phi(\mathbf{x}+\mathbf{s}) - \phi(\mathbf{x}) - \mathbf{s}^\top \mathbf{g}_\phi(\mathbf{x}) &= \int_0^1 \int_0^\tau \mathbf{s} \cdot \mathbf{H}_\phi(\mathbf{x}+\mathbf{s}\sigma)\mathbf{s} \, d\sigma \, d\tau \\
&= \int_0^1 \int_\sigma^1 \mathbf{s} \cdot \mathbf{H}_\phi(\mathbf{x}+\mathbf{s}\sigma)\mathbf{s} \, d\tau \, d\sigma = \int_0^1 \mathbf{s} \cdot \mathbf{H}_\phi(\mathbf{x}+\mathbf{s}\sigma)\mathbf{s}(1-\sigma) \, d\sigma \\
&= \mathbf{s} \cdot \mathbf{H}_\phi(\mathbf{x}+\mathbf{s}\tau)\mathbf{s} \int_0^1 (1-\sigma) \, d\sigma = \frac{1}{2}\mathbf{s} \cdot \mathbf{H}_\phi(\mathbf{x}+\mathbf{s}\tau)\mathbf{s}
\end{aligned}$$

for some $\tau \in (0,1)$. This is true because $\mathbf{s} \cdot \mathbf{H}_\phi(\mathbf{x}+\mathbf{s}\tau)\mathbf{s}$ is a continuous function of τ, and $1-\sigma$ is positive for $\sigma \in (0,1)$. We also have

$$\begin{aligned}
&\left|\phi(\mathbf{x}+\mathbf{s}) - \phi(\mathbf{x}) - \mathbf{g}_\phi(\mathbf{x}) \cdot \mathbf{s}\right| = \left|\int_0^1 \int_0^\tau \mathbf{s} \cdot \mathbf{H}_\phi(\mathbf{x}+\mathbf{s}\sigma)\mathbf{s} \, d\sigma \, d\tau\right| \\
&\le \int_0^1 \int_\sigma^1 \left|\mathbf{s} \cdot \mathbf{H}_\phi(\mathbf{x}+\mathbf{s}\sigma)\mathbf{s}\right| \, d\tau \, d\sigma \le \max_{0 \le \sigma \le 1} \|\mathbf{H}_\phi(\mathbf{x}+\mathbf{s}\sigma)\| \|\mathbf{s}\|^2 \int_0^1 \int_\sigma^1 d\tau \, d\sigma \\
&= \frac{1}{2} \max_{0 \le \sigma \le 1} \|\mathbf{H}_\phi(\mathbf{x}+\mathbf{s}\sigma)\| \|\mathbf{s}\|_2^2 .
\end{aligned}$$

Next, we will bound the error in a quadratic interpolant to a real functional.

Lemma 3.2.9 *Suppose that $\mathscr{D}$ is an open and convex subset of real n-vectors. Let the real functional ϕ be twice continuously differentiable in $\mathscr{D}$. Also let $\|\cdot\|$ represent either a norm on n-vectors, or the corresponding subordinate norm on $n \times n$ matrices. Assume that $\mathbf{H}_\phi$ is Lipschitz continuous at $\mathbf{x} \in \mathscr{D}$ with Lipschitz constant γ with respect to the choice of norms. Then for all $\mathbf{x}+\mathbf{s} \in \mathscr{D}$,*

$$\left|\phi(\mathbf{x}+\mathbf{s}) - \phi(\mathbf{x}) - \mathbf{g}_\phi(\mathbf{x}) \cdot \mathbf{s} - \frac{1}{2}\mathbf{s} \cdot \mathbf{H}_\phi(\mathbf{x})\mathbf{s}\right| \leq \frac{\gamma}{6}\|\mathbf{s}\|^3 .$$

Proof Lemma 3.2.1 implies that

$$\begin{aligned}
&\phi(\mathbf{x}+\mathbf{s}) - \phi(\mathbf{x}) - \mathbf{g}_\phi(\mathbf{x}) \cdot \mathbf{s} - \frac{1}{2}\mathbf{s} \cdot \mathbf{H}_\phi(\mathbf{x})\mathbf{s} \\
&= \int_0^1 \mathbf{g}_\phi(\mathbf{x}+\mathbf{s}\tau) \cdot \mathbf{s} - \mathbf{g}_\phi(\mathbf{x}) \cdot \mathbf{s} - \frac{1}{2}\mathbf{s} \cdot \mathbf{H}_\phi(\mathbf{x})\mathbf{s} \, \mathrm{d}\tau
\end{aligned}$$

then Lemma 3.2.3 yields

$$= \int_0^1 \int_0^\tau \mathbf{s} \cdot \mathbf{H}_\phi(\mathbf{x}+\mathbf{s}\sigma)\mathbf{s} - \mathbf{s} \cdot \mathbf{H}_\phi(\mathbf{x})\mathbf{s} \, \mathrm{d}\sigma \mathrm{d}\tau .$$

Now we take absolute values of both sides and apply Lipschitz continuity:

$$\begin{aligned}
&\left|\phi(\mathbf{x}+\mathbf{s}) - \phi(\mathbf{x}) - \mathbf{g}_\phi(\mathbf{x}) \cdot \mathbf{s} - \frac{1}{2}\mathbf{s} \cdot \mathbf{H}_\phi(\mathbf{x})\mathbf{s}\right| \\
&\leq \int_0^1 \int_0^\tau \left|\mathbf{s} \cdot \mathbf{H}_\phi(\mathbf{x}+\mathbf{s}\sigma)\mathbf{s} - \mathbf{s} \cdot \mathbf{H}_\phi(\mathbf{x})\mathbf{s}\right| \mathrm{d}\sigma \mathrm{d}\tau . \\
&\leq \int_0^1 \int_0^\tau \|\mathbf{H}_\phi(\mathbf{x}+\mathbf{s}\sigma) - \mathbf{H}_\phi(\mathbf{x})\| \|\mathbf{s}\|^2 \mathrm{d}\sigma \mathrm{d}\tau \leq \int_0^1 \int_0^\tau \gamma \|\mathbf{s}\|^3 \sigma \, \mathrm{d}\sigma \, \mathrm{d}\tau = \frac{\gamma}{6}\|\mathbf{s}\|^3 .
\end{aligned}$$

3.2.4 Local Minima

Our next goal is to determine conditions that characterize the minimum of a real functional. First, let us specify what we are seeking.

Definition 3.2.5 Let $\mathscr{D}$ be an open subset of real n-vectors, and assume that the real functional ϕ is continuous on $\mathscr{D}$. Then $\mathbf{z}$ is a **local minimizer** for ϕ if and only if there is an open subset $\mathscr{D}' \subset \mathscr{D}$ such that $\mathbf{z} \in \mathscr{D}'$ and for all $\mathbf{x} \in \mathscr{D}'$ we have $\phi(\mathbf{z}) \leq \phi(\mathbf{x})$. Also, $\mathbf{z}$ is a **strict local minimizer** for ϕ if and only if there is an open subset $\mathscr{D}' \subset \mathscr{D}$ such that $\mathbf{z} \in \mathscr{D}'$ and for all $\mathbf{x} \in \mathscr{D}'$ with $\mathbf{x} \neq \mathbf{z}$ we have $\phi(\mathbf{z}) < \phi(\mathbf{x})$.

Next, we will show that the minimization problem is related to finding a zero of a system of nonlinear equations.

Lemma 3.2.10 (First-Order Necessary Conditions for a Local Minimum) *Suppose that $\mathscr{D}$ is an open and convex set of real n-vectors, and that the real functional ϕ is continuously differentiable on $\mathscr{D}$. If $\mathbf{z}$ is a local minimizer of ϕ, then the gradient of ϕ satisfies*

$$\mathbf{g}_\phi(\mathbf{z}) = \mathbf{0} \,.$$

Proof We will prove the claim by contradiction. If $\mathbf{g}_\phi(\mathbf{x}) \neq \mathbf{0}$, then we can choose an n-vector $\mathbf{s}$ so that

$$\mathbf{g}_\phi(\mathbf{x}) \cdot \mathbf{s} < 0 \,.$$

Since $\mathbf{g}_\phi$ is continuous at $\mathbf{x}$, there exists $\delta > 0$ so that for all $0 < \|\mathbf{s}\|\tau < \delta$ we have

$$\left\|\mathbf{g}_\phi(\mathbf{x} + \mathbf{s}\tau) - \mathbf{g}_\phi(\mathbf{x})\right\| < \frac{-\mathbf{g}_\phi(\mathbf{x}) \cdot \mathbf{s}}{2\|\mathbf{s}\|} \,.$$

It follows from Lemma 3.2.3 that

$$\begin{aligned}
\phi(\mathbf{x} + \mathbf{s}\tau) - \phi(\mathbf{x}) &= \int_0^\tau \mathbf{g}_\phi(\mathbf{x} + \mathbf{s}\sigma) \cdot \mathbf{s} \, d\sigma \\
&= \int_0^\tau \left[\mathbf{g}_\phi(\mathbf{x} + \mathbf{s}\sigma) - \mathbf{g}_\phi(\mathbf{x})\right] \cdot \mathbf{s} \, d\sigma + \mathbf{g}_\phi(\mathbf{x}) \cdot \mathbf{s}\tau
\end{aligned}$$

then the Cauchy inequality (3.15) in Chap. 3 of Volume I implies that

$$\begin{aligned}
&\leq \int_0^\tau \|\mathbf{g}_\phi(\mathbf{x} + \mathbf{s}\sigma) - \mathbf{g}_\phi(\mathbf{x})\| \|\mathbf{s}\| \, d\sigma + \mathbf{g}_\phi(\mathbf{x}) \cdot \mathbf{s}\tau < \frac{-\mathbf{g}_\phi(\mathbf{x}) \cdot \mathbf{s}}{2\|\mathbf{s}\|} \|\mathbf{s}\|\tau + \mathbf{g}_\phi(\mathbf{x}) \cdot \mathbf{s}\tau \\
&= \frac{1}{2} \mathbf{g}_\phi(\mathbf{x}) \cdot \mathbf{s}\tau < 0 \,.
\end{aligned}$$

This shows that $\mathbf{x}$ cannot be a local minimizer of ϕ.

The first-order condition in Lemma 3.2.10 is also satisfied by a local maximum. Thus, we will need additional conditions to distinguish minima from maxima.

Lemma 3.2.11 (Second-Order Necessary Conditions for a Local Minimum) *Suppose that $\mathscr{D}$ is an open and convex set of real n-vectors, and that the real functional ϕ is twice continuously differentiable on $\mathscr{D}$. If $\mathbf{z}$ is a local minimizer of ϕ, then $\mathbf{g}_\phi(\mathbf{z}) = \mathbf{0}$ and $\mathbf{H}_\phi(\mathbf{z})$ is nonnegative.*

Proof If $\mathbf{z}$ is a local minimizer of ϕ, then Lemma 3.2.10 shows that $\mathbf{g}_\phi(\mathbf{z}) = \mathbf{0}$. We will prove by contradiction that $\mathbf{H}_\phi(\mathbf{z})$ is nonnegative.

Suppose that $\mathbf{z} + \mathbf{s}\varepsilon \in \mathscr{D}$. Then the proofs of Lemmas 3.2.9 and 3.2.10 show that

$$\phi(\mathbf{z} + \mathbf{s}\varepsilon) - \phi(\mathbf{z}) = \varepsilon^2 \mathbf{s} \cdot \left\{ \frac{1}{2} \mathbf{H}_\phi(\mathbf{z}) + \int_0^1 \int_0^\tau \mathbf{H}_\phi(\mathbf{z} + \mathbf{s}\varepsilon\sigma) - \mathbf{H}_\phi(\mathbf{z}) \, d\sigma \, d\tau \right\} \mathbf{s} \; .$$

If $\mathbf{H}_\phi(\mathbf{z})$ is not nonnegative, then there is an eigenvector $\mathbf{s}$ of $\mathbf{H}_\phi(\mathbf{z})$ with eigenvalue $\lambda < 0$. We can scale $\mathbf{s}$ so that $\mathbf{z} + \mathbf{s}\varepsilon \in \mathscr{D}$ for all $\varepsilon \in (0, 1)$. Under these circumstances, we have

$$\phi(\mathbf{z}+\mathbf{s}\varepsilon) - \phi(\mathbf{z}) = \varepsilon^2 \|\mathbf{s}\|_2^2 \left\{ \frac{\lambda}{2} + \frac{\mathbf{s}}{\|\mathbf{s}\|_2} \cdot \int_0^1 \int_0^\tau \mathbf{H}_\phi(\mathbf{z} + \mathbf{s}\varepsilon\sigma) - \mathbf{H}_\phi(\mathbf{z}) \, d\sigma \, d\tau \frac{\mathbf{s}}{\|\mathbf{s}\|_2} \right\} .$$

Since λ is negative and $\mathbf{H}_\phi$ is continuous at $\mathbf{z}$, we can choose $\varepsilon > 0$ so that

$$\frac{\lambda}{2} + \frac{\mathbf{s}^\top}{\|\mathbf{s}\|_2} \int_0^1 \int_0^\tau \mathbf{H}_\phi(\mathbf{z} + \mathbf{s}\varepsilon\sigma) - \mathbf{H}_\phi(\mathbf{z}) \, d\sigma \, d\tau \frac{\mathbf{s}}{\|\mathbf{s}\|_2} < 0 \; .$$

It follows that

$$\phi(\mathbf{z} + \mathbf{s}\varepsilon) - \phi(\mathbf{z}) < 0 \; ,$$

which contradicts the assumption that $\mathbf{z}$ is a local minimizer of ϕ.

Next, let us describe conditions that guarantee the existence of a local minimum.

Lemma 3.2.12 (Sufficient Conditions for a Local Minimum) *Suppose that $\mathscr{D}$ is an open and convex set of n-vectors. Let the real functional ϕ be twice continuously differentiable in $\mathscr{D}$. If $\mathbf{z} \in \mathscr{D}$ is such that $\mathbf{g}_\phi(\mathbf{z}) = \mathbf{0}$ and the Hessian $\mathbf{H}_\phi(\mathbf{z})$ is positive, then $\mathbf{z}$ is a strict local minimizer of ϕ.*

Proof For small $\mathbf{s}$ we have $\mathbf{z} + \mathbf{s} \in \mathscr{D}$. Since $\mathbf{g}_\phi(\mathbf{z}) = \mathbf{0}$, the proof of Lemma 3.2.9 shows that

$$\phi(\mathbf{z} + \mathbf{s}) - \phi(\mathbf{z}) = \int_0^1 \mathbf{s} \cdot \mathbf{H}_\phi(\mathbf{z} + \mathbf{s}\tau)\mathbf{s} \, d\tau \; .$$

If the Hessian matrix $\mathbf{H}_\phi$ is positive, then the right-hand side of this equation is positive, proving that $\mathbf{z}$ is a strict local minimizer of ϕ.

We remark that if $\mathbf{g}_\phi(\mathbf{z}) = \mathbf{0}$ and $\mathbf{H}_\phi(\mathbf{z})$ is nonnegative, we cannot conclude that $\mathbf{z}$ is a local minimizer of ϕ. For example $\phi(\zeta) = \zeta^3$ is such that $\phi'(0) = 0$ and $\phi''(0) = 0$, but $\zeta = 0$ is not a local minimizer of ϕ.

Exercise 3.2.1 Let $\phi(\mathbf{x}) = 2\xi_1^2 + \xi_2^4 - 4\xi_1\xi_2$. Find the three points that satisfy the first-order necessary conditions for a local minimizer of ϕ, and determine which of these three points satisfy the second-order necessary conditions for a local minimizer.

Exercise 3.2.2 Let $\phi(\mathbf{x}) = (\xi_2 - \xi_1^2)(\xi_2 - 2\xi_1^2)$.

1. Show that $\mathbf{x} = \mathbf{0}$ satisfies the first-order and second-order necessary conditions for a local minimizer of ϕ.
2. Show that along any line through the origin, ϕ has a minimum at the origin.
3. Show that along the curve $\xi_2 = 3/2\xi_1^2$ we have $\phi(\mathbf{x}) < 0$ for $\mathbf{x} \neq \mathbf{0}$.
4. Conclude that $\mathbf{x} = \mathbf{0}$ is a saddle point of ϕ. Why doesn't this contradict the first-order and second-order necessary conditions for a minimum?

3.2.5 Existence and Uniqueness

In Sect. 3.2.4, we developed conditions that characterize a local minimizer of a functional. We saw that at a local minimizer, the gradient of the functional is zero and the Hessian is nonnegative. In this section, we would like to develop conditions that guarantee the existence of a zero of a nonlinear function. Our new results will be more difficult to obtain than the existence theorems for scalar nonlinear equations in Sect. 5.2.1 in Chap. 5 of Volume I, or the uniqueness results in Sect. 5.2.2 in Chap. 5 of Volume I.

3.2.5.1 Fixed Points

If $\mathbf{f}$ is a function mapping real n-vectors to real n-vectors, we can consider the function

$$\mathbf{s}(\mathbf{x}) \equiv \mathbf{x} - \mathbf{f}(\mathbf{x}) .$$

Then any zero $\mathbf{z}$ of $\mathbf{f}$ is a **fixed point** of $\mathbf{s}$, meaning that

$$\mathbf{s}(\mathbf{z}) = \mathbf{z} .$$

The following theorem guarantees the existence of a fixed point.

Theorem 3.2.2 (Contractive Mapping) *Suppose that $\mathscr{D}$ is a closed subset of real n-vectors, and that $\mathbf{s} : \mathscr{D} \to \mathscr{D}$. Let $\|\cdot\|$ be any norm on n-vectors. If there is a constant $\alpha \in [0, 1)$ such that for all $\mathbf{x}$ and $\widetilde{\mathbf{x}} \in \mathscr{D}$,*

$$\|\mathbf{s}(\widetilde{\mathbf{x}}) - \mathbf{s}(\mathbf{x})\| \leq \alpha \|\widetilde{\mathbf{x}} - \mathbf{x}\| , \tag{3.4}$$

then there exists a unique $\mathbf{z} \in \mathscr{D}$ such that $\mathbf{s}(\mathbf{z}) = \mathbf{z}$.

Proof It is easy to show that a fixed point must be unique. If $\mathbf{z}$ and $\mathbf{z}'$ are two fixed points, then

$$\|\mathbf{z}' - \mathbf{z}\| = \|\mathbf{s}(\mathbf{z}') - \mathbf{s}(\mathbf{z})\| \le \alpha \|\mathbf{z}' - \mathbf{z}\|$$

This can be rewritten in the form

$$(1-\alpha)\|\mathbf{z}' - \mathbf{z}\| \le 0 \,.$$

Since $1-\alpha > 0$, we conclude that $\|\mathbf{z}' - \mathbf{z}\| \le 0$, which implies that $\mathbf{z}' = \mathbf{z}$.

Next, we will prove that a fixed point exists. This proof will actually involve the construction of an algorithm for finding the fixed point.

Given $\mathbf{x}_0 \in \mathscr{D}$, for $k \ge 0$ define the sequence

$$\mathbf{x}_{k+1} = \mathbf{s}(\mathbf{x}_k) \,.$$

Since $\mathbf{s} : \mathscr{D} \to \mathscr{D}$, $\{\mathbf{x}_k\}_{k=0}^{\infty} \subset \mathscr{D}$. Next, inequality (3.4) implies that for $k \ge 1$

$$\|\mathbf{x}_{k+1} - \mathbf{x}_k\| = \|\mathbf{s}(\mathbf{x}_k) - \mathbf{s}(\mathbf{x}_{k-1})\| \le \alpha \|\mathbf{x}_k - \mathbf{x}_{k-1}\| \,.$$

We can solve this recursive inequality to find that

$$\|\mathbf{x}_{k+1} - \mathbf{x}_k\| \le \alpha^k \|\mathbf{x1} - \mathbf{x0}\| \,.$$

As a result,

$$\sum_{k=0}^{\infty} \|\mathbf{x}_{k+1} - \mathbf{x}_k\| \le \sum_{k=0}^{\infty} \alpha^k \|\mathbf{x1} - \mathbf{x0}\| = \frac{1}{1-\alpha}\|\mathbf{x1} - \mathbf{x0}\| \,.$$

This inequality shows that the sum on the left converges, which in turn implies that $\mathbf{x}_{k+1} - \mathbf{x}_k \to \mathbf{0}$ as $k \to \infty$.

Given any $\varepsilon > 0$ we can choose $k \ge 0$ so that for all $m \ge k$ we have

$$\|\mathbf{x}_{m+1} - \mathbf{x}_m\| \le \varepsilon(1-\alpha) \,.$$

This implies that for all $k \le \ell < m$ we have

$$\|\mathbf{x}_m - \mathbf{x}_\ell\| = \left\| \sum_{j=\ell}^{m-1} \mathbf{x}_{j+1} - \mathbf{x}_j \right\| \le \sum_{j=\ell}^{m-1} \left\|\mathbf{x}_{j+1} - \mathbf{x}_j\right\| \le \sum_{j=\ell}^{m-1} \alpha^{j-\ell} \|\mathbf{x}_{\ell+1} - \mathbf{x}_\ell\|$$
$$\le \frac{1}{1-\alpha} \|\mathbf{x}_{\ell+1} - \mathbf{x}_\ell\| \le \varepsilon \,.$$

Thus $\{\mathbf{x}_k\}_{k=0}^{\infty}$ is a Cauchy sequence. Since $\mathscr{D}$ is closed and the set of all real n-vectors is complete (see Kreyszig [114, p. 33] or Royden [152]), the sequence $\{\mathbf{x}_k\}_{k=0}^{\infty}$ converges to some point $\mathbf{z} \in \mathscr{D}$. In other words, for all $\varepsilon > 0$ there is an $m > 0$ such that for all $k > m$, $|\mathbf{z} - \mathbf{x}_k| < \varepsilon$. Then

$$\|\mathbf{s}(\mathbf{z}) - \mathbf{z}\| \le \|\mathbf{s}(\mathbf{z}) - \mathbf{s}(\mathbf{x}_k)\| + \|\mathbf{s}(\mathbf{x}_k) - \mathbf{z}\| \le \alpha\|\mathbf{z} - \mathbf{x}_k\| + \|\mathbf{x}_{k+1} - \mathbf{z}\| \le (1+\alpha)\varepsilon \; .$$

Since ε was arbitrary, we conclude that $\mathbf{s}(\mathbf{z}) = \mathbf{z}$.

Next, we will discuss a second way to guarantee the existence and uniqueness of a fixed point.

Definition 3.2.6 Let $\|\cdot\|$ be any norm on n-vectors. Suppose that $\mathbf{s}$ maps n-vectors to n-vectors, and that $\mathbf{s}$ is differentiable for $\|\mathbf{x} - \mathbf{x}_0\| \le r$. Let $0 < \varrho < r$. Assume that σ maps real scalars to real scalars, and that σ is differentiable on the closed interval $[\xi_0 \; , \; \xi_0 + \varrho]$. Then σ **majorizes s at** $(\xi_0, \mathbf{x}_0)$ **with respect to** (ϱ, r) if and only if

1. $\|\mathbf{s}(\mathbf{x}_0) - \mathbf{x}_0\| \le \sigma(\xi_0) - \xi_0$ and
2. $\|\mathbf{x} - \mathbf{x}_0\| \le \xi - \xi_0 \le \varrho$ implies that $\|\mathbf{J_s}(\mathbf{x})\| \le \sigma'(\xi)$.

The following lemma, due to Kantorovich and Akilov [103, p. 697], provides another existence result for fixed points.

Lemma 3.2.13 *Let ξ_0 be a real scalar, $\mathbf{x}_0$ be a real n-vector, and $\|\cdot\|$ be any norm on real n-vectors. Suppose that* $\mathbf{s}$ *maps real n-vectors to real n-vectors, and that there exists $r > 0$ so that* $\mathbf{s}$ *is continuously differentiable for all* $\mathbf{x}$ *satisfying $\|\mathbf{x} - \mathbf{x}_0\| \le r$. Assume that σ maps real scalars to real scalars, and that there exists $\varrho \in (0, r]$ so that σ is continuously differentiable for all $\xi \in [\xi_0, \xi_0 + \varrho]$. Let σ majorize* $\mathbf{s}$ *at $(\xi_0, \mathbf{x}_0)$ with respect to (ϱ, r), and assume that there exists $\zeta \in [\xi_0 \; , \; \xi_0 + \varrho]$ so that*

$$\zeta = \sigma(\zeta) \; .$$

Then there exists an n-vector $\mathbf{z}$ *so that* $\mathbf{s}(\mathbf{z}) = \mathbf{z}$ *and* $\|\mathbf{z} - \mathbf{x}_0\| \le \zeta - \xi_0$.
Note that since $\|\mathbf{z} - \mathbf{x}_0\| \le \zeta - \xi_0$ will be proved to be true for any fixed point of σ in the interval $[\xi_0 \; , \; \xi_0 + \varrho]$, this inequality must be true for the smallest fixed point of σ in this interval.

Proof Define $\{\xi_k\}_{k=0}^{\infty}$ recursively by

$$\xi_{k+1} = \sigma(\xi_k) \text{ for } k \ge 0 \; .$$

We will prove inductively that

$$\xi_k \le \zeta \text{ and} \tag{3.5a}$$

$$\xi_k \le \xi_{k+1} \tag{3.5b}$$

for all $k \geq 0$. The description of ζ shows that (3.5a) is true for $k = 0$. Since σ majorizes $\mathbf{s}$ at $(\xi_0, \mathbf{x}_0)$ with respect to (ϱ, r), we see that

$$\xi_1 - \xi_0 = \sigma(\xi_0) - \xi_0 \geq \|\mathbf{s}(\mathbf{x}_0) - \mathbf{x}_0\| \geq 0 \,,$$

This proves (3.5b) for $k = 0$. Inductively, assume that (3.5a) and (3.5b) are true for $k - 1 \geq 0$. Since σ majorizes $\mathbf{s}$, for any $\xi \in [\xi_0 \,,\, \xi_0 + \varrho]$ and any $\mathbf{x}$ satisfying $\|\mathbf{x} - \mathbf{x}_0\| \leq \xi - \xi_0$ we have

$$\sigma'(\xi) \geq \|\mathbf{J_s}(\mathbf{x})\| \geq 0 \,.$$

This implies that σ is nondecreasing on $[\xi_0 \,,\, \xi_0 + \varrho]$. Consequently,

$$\xi_k = \sigma(\xi_{k-1}) \leq \sigma(\zeta) = \zeta \leq \varrho \,.$$

This proves (3.5a) for k, and that $\sigma(\xi_k)$ is defined. Since σ is nondecreasing on $[\xi_0 \,,\, \xi_0 + \varrho]$, we also have

$$\xi_{k+1} = \sigma(\xi_k) \geq \sigma(\xi_{k-1}) = \xi_k \,.$$

This completes the inductive proof of (3.5).

Since $\{\xi_k\}_{k=0}^{\infty}$ is nondecreasing and bounded above by ζ, we conclude that there exists $\xi \in [\xi_0 \,,\, \zeta]$ so that

$$\xi_k \uparrow \xi \,.$$

Since σ is continuous, we conclude that

$$\xi = \lim_{k \to \infty} \xi_k = \lim_{k \to \infty} \sigma(\xi_{k-1}) = \sigma(\xi) \,.$$

Thus ξ is a fixed point of σ.

Let the sequence $\{\mathbf{x}_k\}_{k=0}^{\infty}$ be defined recursively by

$$\mathbf{x}_{k+1} = \mathbf{s}(\mathbf{x}_k) \text{ for } k \geq 0 \,,$$

We claim that

$$\|\mathbf{x}_k - \mathbf{x}_0\| \leq \zeta - \xi_0 \text{ and} \tag{3.6a}$$

$$\|\mathbf{x}_{k+1} - \mathbf{x}_k\| \leq \xi_{k+1} - \xi_k \tag{3.6b}$$

for all $k \geq 0$. Inequality (3.6a) is obviously true for $k = 0$. Since σ majorizes $\mathbf{s}$,

$$\|\mathbf{x}_1 - \mathbf{x}_0\| = \|\mathbf{s}(\mathbf{x}_0) - \mathbf{x}_0\| \leq \sigma(\xi_0) - \xi_0 = \xi_1 - \xi_0 \,,$$

and (3.6b) is also true for $k = 0$. In particular, $\|\mathbf{x}_1 - \mathbf{x}_0\| \leq \varrho \leq r$, so $\mathbf{s}(\mathbf{x}_1)$ is defined. Inductively, assume that inequalities (3.6) are true for all indices less than or equal to $k - 1 \geq 0$. First, we note that

$$\|\mathbf{x}_k - \mathbf{x}_0\| = \left\| \sum_{n=0}^{k-1} \mathbf{x}_{n+1} - \mathbf{x}_n \right\| \leq \sum_{n=0}^{k-1} \|\mathbf{x}_{n+1} - \mathbf{x}_n\| \leq \sum_{n=0}^{k-1} (\xi_{n+1} - \xi_n)$$
$$= \xi_k - \xi_0 \leq \zeta - \xi_0 \leq \varrho \; .$$

This proves (3.6a), and shows that $\mathbf{s}(\mathbf{x}_k)$ is defined. We also have

$$\mathbf{x}_{k+1} - \mathbf{x}_k = \mathbf{s}(\mathbf{x}_k) - \mathbf{s}(\mathbf{x}_{k-1})$$

then Lemma 3.2.3 implies that

$$= \int_0^1 \mathbf{J_s}(\mathbf{x}_{k-1} + [\mathbf{x}_k - \mathbf{x}_{k-1}]\tau)[\mathbf{x}_k - \mathbf{x}_{k-1}] \, \mathrm{d}\tau \; .$$

Let us define

$$\mathbf{x}(\tau) = \mathbf{x}_{k-1} + [\mathbf{x}_k - \mathbf{x}_{k-1}]\tau \text{ and } \xi(\tau) = \xi_{k-1} + [\xi_k - \xi_{k-1}]\tau$$

for all $\tau \in [0, 1]$. Then

$$\|\mathbf{x}(\tau) - \mathbf{x}_0\| = \|(\mathbf{x}_k - \mathbf{x}_0)\tau + (\mathbf{x}_{k-1} - \mathbf{x}_0)(1 - \tau)\|$$
$$\leq (\xi_k - \xi_0)\tau + (\xi_{k-1} - \xi_0)(1 - \tau) = \xi(\tau) - \xi_0 \; .$$

Since $\|\mathbf{x}_k - \mathbf{x}_0\| \leq \zeta - \xi_0$, $\|\mathbf{x}_{k-1} - \mathbf{x}_0\| \leq \zeta - \xi_0$ and the ball of radius $\zeta - \xi_0$ is convex, we must have $\|\mathbf{x}(\tau) - \mathbf{x}_0\| \leq \zeta - \xi_0$ for all $\tau \in [0, 1]$. Since σ majorizes $\mathbf{s}$,

$$\|\mathbf{x}_{k+1} - \mathbf{x}_k\| \leq \int_0^1 \|\mathbf{J_s}(\mathbf{x}(\tau)) \, \mathrm{d}\tau\| \|\mathbf{x}_k - \mathbf{x}_{k-1}\|$$
$$\leq \int_0^1 \sigma'(\xi(\tau)) \, \mathrm{d}\tau (\xi_k - \xi_{k-1}) = \sigma(\xi_k) - \sigma(\xi_{k-1}) = \xi_{k+1} - \xi_k \; .$$

This proves (3.6b) for k, and completes the inductive proof of inequalities (3.6).

Since $\xi_k \uparrow \xi$ as $k \to \infty$, given any $\varepsilon > 0$ we can choose $k \geq 0$ so that for all $\ell > 0$ we have

$$\xi_{k+\ell} - \xi_k < \varepsilon \; .$$

Then inequality (3.6b) implies that

$$\|\mathbf{x}_{k+\ell} - \mathbf{x}_k\| < \sum_{n=k}^{k+\ell-1} \|\mathbf{x}_{n+1} - \mathbf{x}_n\| \le \sum_{n=k}^{k+\ell-1} (\xi_{n+1} - \xi_n) = \xi_{k+\ell} - \xi_k < \varepsilon \; .$$

This implies that $\{\mathbf{x}_k\|_{k=0}^{\infty}$ is a Cauchy sequence. It is well-known (see Kreyszig [114, p. 33] or Royden [152]) that this implies that $\{\mathbf{x}_k\}_{k=0}^{\infty}$ converges to an n-vector $\mathbf{z}$. Since the set $\{\mathbf{x} : \|\mathbf{x} - \mathbf{x}_0\| \le \zeta - \xi_0\}$ is closed, we conclude that

$$\|\mathbf{z} - \mathbf{x}_0\| \le \zeta - \xi_0 \; .$$

Since $\mathbf{s}$ is continuous for all $\mathbf{x}$ satisfying $\|\mathbf{x} - \mathbf{x}_0\| \le r$ and $\zeta \le \varrho \le r$,

$$\mathbf{z} = \lim_{k \to \infty} \mathbf{x}_k = \lim_{k \to \infty} \mathbf{s}(\mathbf{x}_k) = \mathbf{s}(\mathbf{z}) \; .$$

Lemma 3.2.13 proves the existence of a fixed point of a nonlinear function subject to majorization, but does not prove uniqueness. For that, we need two additional assumptions, as suggested by Kantorovich and Akilov [103, p. 700].

Corollary 3.2.2 *In addition to the assumptions of Lemma 3.2.13, suppose that* $\sigma(\xi_0 + \varrho) \le \xi_0 + \varrho$ *and that there is a unique* $\zeta \in [\xi_0 \, , \; \xi_0 + \varrho]$ *so that* $\sigma(\zeta) = \zeta$. *Then there is a unique n-vector* $\mathbf{z}$ *such that* $\mathbf{s}(\mathbf{z}) = \mathbf{z}$ *and* $\|\mathbf{z} - \mathbf{x}_0\| \le r$; *in fact, the fixed point* $\mathbf{z}$ *satisfies* $\|\mathbf{z} - \mathbf{x}_0\| \le \zeta - \xi_0$.

Proof Let $\eta_0 = \xi_0 + \varrho$, and define $\{\eta_k\}_{k=0}^{\infty}$ recursively by

$$\eta_{k+1} = \sigma(\eta_k)$$

for $k \ge 0$. We will prove inductively that

$$\eta_k \ge \zeta \text{ and} \tag{3.7a}$$

$$\eta_k \ge \eta_{k+1} \tag{3.7b}$$

for $k \ge 0$. The definition of ζ shows that (3.7a) holds for $k = 0$. The first additional assumption of this corollary implies that

$$\eta_0 - \eta_1 = (\xi_0 + \varrho) - \sigma(\xi_0 + \varrho) \ge 0 \; ,$$

so (3.7b) is satisfied for $k = 0$. Inductively, we assume that inequalities (3.7) are satisfied for $k - 1 \ge 0$. Since σ is nondecreasing,

$$\eta_k = \sigma(\eta_{k-1}) \ge \sigma(\zeta) = \zeta \; ,$$

and

$$\eta_k = \sigma(\eta_{k-1}) \geq \sigma(\eta_{k-1}) = \eta_k ,$$

so inequalities (3.7) are both satisfied for k.

Since $\{\eta_k\}_{k=0}^{\infty}$ is nonincreasing and bounded below by ζ, this sequence has a limit $\eta \geq \zeta$. Since σ is continuous,

$$\eta = \lim_{k\to\infty} \eta_k = \lim_{k\to\infty} \sigma(\eta_k) = \sigma(\eta) .$$

Since σ has a unique fixed point in $[\xi_0, \xi_0 + \varrho]$, we conclude that $\eta_k \downarrow \zeta$.

Given any n-vector $\mathbf{y}_0$ with $\|\mathbf{y}_0 - \mathbf{x}_0\| \leq r$, recursively define $\{\mathbf{y}_k\}_{k=0}^{\infty}$ by

$$\mathbf{y}_{k+1} = \mathbf{s}(\mathbf{y}_k)$$

for all $k \geq 0$. We will prove inductively that

$$\|\mathbf{y}_k - \mathbf{x}_k\| \leq \eta_k - \xi_k \text{ and} \tag{3.8a}$$

$$\|\mathbf{y}_k - \mathbf{x}_0\| \leq \eta_k - \xi_0 \leq \varrho \leq r \tag{3.8b}$$

for all $k \geq 1$. Note that

$$\|\mathbf{y}_1 - \mathbf{x}_1\| = \|\mathbf{s}(\mathbf{y}_0) - \mathbf{s}(\mathbf{x}_0)\| = \left\| \int_0^1 \mathbf{J_s}(\mathbf{x}_0 + [\mathbf{y}_0 - \mathbf{x}_0]\tau)[\mathbf{y}_0 - \mathbf{x}_0] \, d\tau \right\|$$

$$\leq \int_0^1 \sigma'(\xi_0 + [\eta_0 - \xi_0]\tau) \, d\tau (\eta_0 - \xi_0) = \sigma(\eta_0) - \sigma(\xi_0) = \eta_1 - \xi_1 .$$

This inequality is satisfied because σ majorizes $\mathbf{s}$, and the integral on the right represents the largest possible value that can be obtained by majorization. This proves that (3.8a) is satisfied for $k = 1$. The triangle inequality also implies that

$$\|\mathbf{y}_1 - \mathbf{x}_0\| \leq \|\mathbf{y}_1 - \mathbf{x}_1\| + \|\mathbf{x}_1 - \mathbf{x}_0\| \leq (\eta_1 - \xi_1) + (\xi_1 - \xi_0) = \eta_1 - \xi_0 \leq \eta_0 - \xi_1 = \varrho \leq r ,$$

so (3.8b) is satisfied for $k = 1$, and $\mathbf{s}(\mathbf{y}_1)$ is defined. Inductively, assume that inequalities (3.8) are satisfied for k. Note that

$$\mathbf{y}_{k+1} - \mathbf{x}_{k+1} = \mathbf{s}(\mathbf{y}_k) - \mathbf{s}(\mathbf{x}_k) = \int_0^1 \mathbf{J_s}(\mathbf{x}_k + [\mathbf{y}_k - \mathbf{x}_k]\tau) [\mathbf{y}_k - \mathbf{x}_k] \, d\tau .$$

This suggests that we define

$$\mathbf{w}(\tau) = \mathbf{x}_k + [\mathbf{y}_k - \mathbf{x}_k]\tau = \mathbf{y}_k\tau + \mathbf{x}_k(1 - \tau) \text{ and } \omega(\tau) = \xi_k + [\eta_k - \xi_k]\tau .$$

Then

$$\|\mathbf{w}(\tau) - \mathbf{x}_0\| = \|(\mathbf{y}_k - \mathbf{x}_0)\tau + (\mathbf{x}_k - \mathbf{x}_0)(1 - \tau)\|$$
$$\le \|\mathbf{y}_k - \mathbf{x}_0\|\tau + \|\mathbf{x}_k - \mathbf{x}_0\|(1 - \tau) \le (\eta_k - \xi_0)\tau + (\xi_k - \xi_0)(1 - \tau) = \omega(\tau) - \xi_0$$

Since σ majorizes $\mathbf{s}$,

$$\|\mathbf{y}_{k+1} - \mathbf{x}_{k+1}\| \le \int_0^1 \|\mathbf{J}_\mathbf{s}(\mathbf{w}(\tau))\| \, d\tau \, \|\mathbf{y}_k - \mathbf{x}_k\|$$
$$\le \int_0^1 \sigma'(\omega(\tau)) \, d\tau (\eta_k - \xi_k) = \sigma(\eta_k) - \sigma(\xi_k) = \eta_{k+1} - \xi_{k+1} \,.$$

This proves (3.8a) for $k + 1$. We also have

$$\|\mathbf{y}_{k+1} - \mathbf{x}_0\| \le \|\mathbf{y}_{k+1} - \mathbf{x}_{k+1}\| + \|\mathbf{x}_{k+1} - \mathbf{x}_0\|$$
$$\le (\eta_{k+1} - \xi_{k+1}) + (\xi_{k+1} - \xi_0) = \eta_{k+1} - \xi_0 \le \eta_0 - \xi_0 = \varrho \le r \,,$$

so (3.8b) is satisfied for $k + 1$. This completes the inductive proof of inequalities (3.8).

Since $\eta_k - \xi_k \downarrow 0$, we conclude that

$$\lim_{k\to\infty} (\mathbf{y}_k - \mathbf{x}_k) = \mathbf{0} \,.$$

Since $\{\mathbf{x}_k\|_{k=0}^{\infty}$ converges to $\mathbf{z}$, so does $\{\mathbf{y}_k\|_{k=0}^{\infty}$. In particular, if $\mathbf{y}_0$ were a fixed point of $\mathbf{s}$, we would have $\mathbf{y}_k = \mathbf{y}_0$ for all $k \ge 0$, and $\mathbf{z} = \lim_{k\to\infty} \mathbf{y}_k = \mathbf{y}_0$.

Exercise 3.2.3 Consider the function

$$\mathbf{f}(\mathbf{x}) = \begin{bmatrix} \xi_1^2 + \xi_2^2 \\ e^{\xi_1 + \xi_2} - 1 \end{bmatrix}$$

1. Show that $\mathbf{z} = \mathbf{0}$ is the only zero of $\mathbf{f}$.
2. Can you find an open region containing $\mathbf{z}$ in which the function

$$\mathbf{s}(\mathbf{x}) = \mathbf{x} - \mathbf{J}_\mathbf{f}(\mathbf{x})^{-1}\mathbf{f}(\mathbf{x})$$

is contractive?

3.2.5.2 Zeros

In Sect. 3.2.5.1 we found several conditions that guarantee the existence and uniqueness of fixed points of nonlinear vector-valued functions. In this section, we would like to use the natural correspondence between fixed points and zeros of functions to establish conditions that guarantee the existence and uniqueness of zeros of nonlinear vector-valued functions. The following very important theorem is taken from Kantorovich and Akilov [103, p. 705].

Theorem 3.2.3 (Kantorovich) *Let $\mathbf{z}_0$ be a real n-vector, and let $\|\cdot\|$ represent either a norm on real n-vectors or the corresponding subordinate norm on real $n \times n$ matrices. Suppose that $\mathbf{f}$ maps real n-vectors to n-vectors, and is continuously differentiable for $\|\mathbf{x} - \mathbf{z}_0\| < r$. Further, suppose that the Jacobian $\mathbf{J_f}(\mathbf{z}_0)$ is nonsingular, and that $\mathbf{J_f}$ is uniformly Lipschitz continuous for $\|\mathbf{x} - \mathbf{z}_0\| < r$, with Lipschitz constant γ. Assume that $\mathbf{f}(\mathbf{z}_0)$ is sufficiently small, in the sense that*

$$\alpha \equiv \gamma \left\|\mathbf{J_f}(\mathbf{z}_0)^{-1}\right\| \left\|\mathbf{J_f}(\mathbf{z}_0)^{-1}\mathbf{f}(\mathbf{z}_0)\right\| < \frac{1}{2} .$$

Finally, assume that the region of Lipschitz continuity is sufficiently large, meaning that

$$r \geq \frac{1 - \sqrt{1 - 2\alpha}}{\gamma \left\|\mathbf{J_f}(\mathbf{z}_0)^{-1}\right\|} \equiv \zeta_0 .$$

Then there is a n-vector $\mathbf{z}$ so that $\|\mathbf{z} - \mathbf{z}_0\| \leq \zeta_0$ and $\mathbf{f}(\mathbf{z}) = \mathbf{0}$. Next, define

$$\zeta_1 = \frac{1 + \sqrt{1 - 2\alpha}}{\gamma \left\|\mathbf{J_f}(\mathbf{z}_0)^{-1}\right\|} .$$

If either $\alpha = 1/2$, or $\alpha < 1/2$ and $r < \zeta_1$, then $\mathbf{z}$ is the unique zero of f satisfying $\|\mathbf{z} - \mathbf{z}_0\| \leq r$.

Proof Define the function

$$\phi(\xi) = \frac{\gamma}{2} \left\|\mathbf{J_f}(\mathbf{z}_0)^{-1}\right\| \xi^2 - \xi + \left\|\mathbf{J_f}(\mathbf{z}_0)^{-1}\mathbf{f}(\mathbf{z}_0)\right\| .$$

Note that ζ_0 and ζ_1 are the roots of this quadratic. Furthermore, the quadratic ϕ was designed with the following interpolation properties:

$$\phi(0) = \left\|\mathbf{J_f}(\mathbf{z}_0)^{-1}\mathbf{f}(\mathbf{z}_0)\right\| \geq 0 ,$$
$$\phi'(0) = -1 \text{ and}$$
$$\phi''(\xi) = \gamma \left\|\mathbf{J_f}(\mathbf{z}_0)^{-1}\right\| \text{ for all } \xi .$$

Next, define

$$\mathbf{s}(\mathbf{x}) = \mathbf{x} - \mathbf{J_f}(\mathbf{z}_0)^{-1}\mathbf{f}(\mathbf{x}) \text{ and}$$
$$\sigma(\xi) = \xi - \frac{\phi(\xi)}{\phi'(0)} .$$

Note that $\mathbf{s}(\mathbf{z}) = \mathbf{z}$ if and only if $\mathbf{f}(\mathbf{z}) = \mathbf{0}$, and $\sigma(\zeta) = \zeta$ if and only if $\phi(\zeta) = 0$. We also remark that

$$\sigma'(\xi) = 1 - \frac{\phi'(\xi)}{\phi'(0)} = 1 + \left\{\gamma \left\| \mathbf{J_f}(\mathbf{z}_0)^{-1} \right\| \xi - 1\right\} - \gamma \left\| \mathbf{J_f}(\mathbf{z}_0)^{-1} \right\| \xi$$

Let us show that σ majorizes $\mathbf{s}$ at $(0, \mathbf{z}_0)$ with respect to (r, r). Since

$$\mathbf{J_s}(\mathbf{x}) = \mathbf{I} - \mathbf{J_f}(\mathbf{z}_0)^{-1}\mathbf{J_f}(\mathbf{x}) ,$$

it follows that $\mathbf{s}$ is continuously differentiable for $\|\mathbf{x}-\mathbf{z}_0\| \leq r$. Since ϕ is a quadratic, so is σ, and thus the latter is differentiable on the closed interval $[0, r]$. Now

$$\|\mathbf{s}(\mathbf{z}_0) - \mathbf{z}_0\| = \left\|\mathbf{J_f}(\mathbf{z}_0)^{-1}\mathbf{f}(\mathbf{z}_0)\right\| = -\frac{\phi(0)}{\phi'(0)} = \sigma(0)$$

and whenever $\|\mathbf{x} - \mathbf{z}_0\| \leq \xi$ we have

$$\begin{aligned}\|\mathbf{J_s}(\mathbf{x})\| &= \left\|\mathbf{I} - \mathbf{J_f}(\mathbf{z}_0)^{-1}\mathbf{J_f}(\mathbf{x})\right\| = \left\|\mathbf{J_f}(\mathbf{z}_0)^{-1}\left\{\mathbf{J_f}(\mathbf{x}) - \mathbf{J_f}(\mathbf{z}_0)\right\}\right\| \\ &\leq \gamma \left\|\mathbf{J_f}(\mathbf{z}_0)^{-1}\right\| \|\mathbf{x} - \mathbf{z}_0\| \leq \gamma \left\|\mathbf{J_f}(\mathbf{z}_0)^{-1}\right\| \xi = \sigma'(\xi) .\end{aligned}$$

At this point, we have demonstrated that σ majorizes $\mathbf{s}$.

Lemma 3.2.13 now implies that there is an n-vector $\mathbf{z}$ so that $\mathbf{s}(\mathbf{z}) = \mathbf{z}$ and $\|\mathbf{z} - \mathbf{z}_0\| \leq \zeta_1$. Since fixed points of $\mathbf{s}$ are zeros of $\mathbf{f}$, we have proved the first claim in Theorem 3.2.3.

Consider the case when $\alpha = 1/2$. Then $\sigma(\zeta_1) = \zeta_1$ is the unique fixed point of σ on $[0, \zeta_1]$. Since $r \geq \zeta_0 = \zeta_1$, Corollary 3.2.2 proves the second claim in Theorem 3.2.3 for this case.

Alternatively, consider the case when $\alpha < 1/2$ and $r < \zeta_1$. Since $\phi(\xi) < 0$ for $\zeta_0 < \xi < \zeta_1$, we see that

$$\sigma(r) = r + \phi(r) < r .$$

Because ζ_0 is the unique zero of ϕ in $[0, r]$, Corollary 3.2.2 again proves the second claim of Theorem 3.2.3 in this case.

Exercise 3.2.4 Consider the function

$$\mathbf{f}(\mathbf{x}) = \begin{bmatrix} 2(\xi_2 - \xi_1^2) \\ 1 - \xi_1 \end{bmatrix} .$$

1. Find the Jacobian $\mathbf{J_f}(\mathbf{x})$ and its inverse.
2. Show that the Jacobian is Lipschitz continuous, and find its Lipschitz constant.
3. Determine the 2-norm of $\mathbf{J_f}(\mathbf{x})^{-1}$ as a function of $\mathbf{x}$. (See Lemma 1.5.5 for information about how to determine the 2-norm of a square matrix.)

4. Evaluate $\|\mathbf{J_f}(\mathbf{x})^{-1}\mathbf{f}(\mathbf{x})\|_2$ as a function of $\mathbf{x}$.
5. Determine an expression for the Kantorovich inequality that guarantees convergence to a zero of $\mathbf{f}$. (See Theorem 3.2.3.)
6. Show that the function

$$\mathbf{p}(\mathbf{x}) = \mathbf{x} - \mathbf{J_f}(\mathbf{x})^{-1}\mathbf{f}(\mathbf{x})$$

is contractive for all $\mathbf{x}$ with $\xi_1 \in (1/2, 3/2)$.
7. Use Lemma 3.2.15 to bound the error in an approximation to the zero of $\mathbf{f}$ by the value of $\mathbf{f}$.

3.2.6 Perturbation Analysis

At this point, we have proved the Kantorovich Theorem 3.2.3, which provides conditions under which the general nonlinear system of Eqs. (3.1) has a unique solution. In Lemmas 3.2.11 and 3.2.12, we have also developed necessary and sufficient conditions for the minimization problem (3.2). Those necessary and sufficient conditions relate relate the minimization problem to the root-finding problem. This motivates our next goal, which is to bound perturbations in the zero of a nonlinear system due to perturbations in the data.

We will begin by using Corollary 3.6.1 in Chap. 3 of Volume I to bound perturbations in Jacobians.

Lemma 3.2.14 *Suppose that $\mathscr{D}$ is an open subset of real n-vectors, and that both $\mathbf{z}$ and $\widetilde{\mathbf{z}} \in \mathscr{D}$. Assume that $\|\cdot\|$ represents either a norm on real n-vectors, or the corresponding subordinate norm on real $n \times n$ matrices. Let $\mathbf{f}$ be a continuously differentiable mapping from $\mathscr{D}$ to real n-vectors, with $\mathbf{f}(\mathbf{z}) = \mathbf{0}$. Also assume that the Jacobian $\mathbf{J_f}$ is Lipschitz continuous at $\mathbf{z}$ with constant γ, and assume that $\mathbf{J_f}(\mathbf{z})$ is nonsingular. If*

$$\gamma \left\|\mathbf{J_f}(\mathbf{z})^{-1}\right\| \|\widetilde{\mathbf{z}} - \mathbf{z}\| < 1 , \tag{3.9}$$

then $\mathbf{J_f}(\widetilde{\mathbf{z}})$ is nonsingular,

$$\left\|\mathbf{J_f}(\widetilde{\mathbf{z}})^{-1}\right\| \leq \frac{\left\|\mathbf{J_f}(\mathbf{z})^{-1}\right\|}{1 - \gamma \|\mathbf{J_f}(\mathbf{z})^{-1}\| \|\widetilde{\mathbf{z}} - \mathbf{z}\|} ,$$

and

$$\left\|\mathbf{J_f}(\widetilde{\mathbf{z}})^{-1} - \mathbf{J_f}(\mathbf{z})^{-1}\right\| \leq \frac{\gamma \left\|\mathbf{J_f}(\mathbf{z})^{-1}\right\|^2 \|\widetilde{\mathbf{z}} - \mathbf{z}\|}{1 - \gamma \|\mathbf{J_f}(\mathbf{z})^{-1}\| \|\widetilde{\mathbf{z}} - \mathbf{z}\|} .$$

Proof Together, inequality (3.9) and Corollary 3.6.1 in Chap. 3 of Volume I show that $\mathbf{J_f}(\widetilde{\mathbf{z}})$ is nonsingular. Also, inequality (3.25) in Chap. 3 of Volume I and Lipschitz continuity prove that

$$\left\|\mathbf{J_f}(\widetilde{\mathbf{z}})^{-1}\right\| \leq \frac{\left\|\mathbf{J_f}(\mathbf{z})^{-1}\right\|}{1-\|\mathbf{J_f}(\mathbf{z})^{-1}(\mathbf{J_f}(\widetilde{\mathbf{z}})-\mathbf{J_f}(\mathbf{z}))\|} \leq \frac{\left\|\mathbf{J_f}(\mathbf{z})^{-1}\right\|}{1-\gamma\|\mathbf{J_f}(\mathbf{z})^{-1}\|\|\widetilde{\mathbf{z}}-\mathbf{z}\|} .$$

Finally, inequality (3.26) in Chap. 3 of Volume I and Lipschitz continuity give us

$$\left\|\mathbf{J_f}(\widetilde{\mathbf{z}})^{-1}-\mathbf{J_f}(\mathbf{z})^{-1}\right\| \leq \frac{\left\|\mathbf{J_f}(\mathbf{z})^{-1}\right\|\left\|\mathbf{J_f}(\mathbf{z})^{-1}(\mathbf{J_f}(\widetilde{\mathbf{z}})-\mathbf{J_f}(\mathbf{z}))\right\|}{1-\|\mathbf{J_f}(\mathbf{z})^{-1}(\mathbf{J_f}(\widetilde{\mathbf{z}})-\mathbf{J_f}(\mathbf{z}))\|}$$

$$\leq \frac{\gamma\left\|\mathbf{J_f}(\mathbf{z})^{-1}\right\|^2\|\widetilde{\mathbf{z}}-\mathbf{z}\|}{1-\gamma\|\mathbf{J_f}(\mathbf{z})^{-1}\|\|\widetilde{\mathbf{z}}-\mathbf{z}\|} .$$

One useful form of the perturbation problem for nonlinear systems relates to a general iteration to find a zero of a nonlinear function. Suppose that we stop an iteration to find a zero of $\mathbf{f}$ when we have found an approximate solution $\widetilde{\mathbf{z}}$ so that $\|\mathbf{f}(\widetilde{\mathbf{z}})\|$ is small. If $\mathbf{z}$ is the true zero, We would like to know how to estimate the error in $\widetilde{\mathbf{z}}$ as an approximation to $\mathbf{z}$. The information we seek can be found in the next lemma.

Lemma 3.2.15 *Suppose that $\mathscr{D}$ is an open convex subset of real n-vectors, and suppose that both $\mathbf{z}$ and $\widetilde{\mathbf{z}} \in \mathscr{D}$. Assume that $\|\cdot\|$ represents either a norm on real n-vectors, or the corresponding subordinate norm on real $n \times n$ matrices. Let $\mathbf{f}$ be a continuously differentiable mapping from $\mathscr{D}$ to real n-vectors, with $\mathbf{f}(\mathbf{z}) = \mathbf{0}$ and $\mathbf{J_f}(\mathbf{z})$ nonsingular.*

1. *Suppose that the perturbation in the Jacobian $\mathbf{J_f}$ is small along the line segment between $\mathbf{z}$ and $\widetilde{\mathbf{z}}$, meaning that there exists $\alpha \in [0,1)$ so that for all $\tau \in (0,1)$*

$$\left\|\mathbf{J_f}(\mathbf{z})^{-1}\{\mathbf{J_f}(\mathbf{z}+[\widetilde{\mathbf{z}}-\mathbf{z}]\tau)-\mathbf{J_f}(\mathbf{z})\}\right\| \leq \alpha .$$

   Then

$$\|\widetilde{\mathbf{z}}-\mathbf{z}\| \leq \frac{\left\|\mathbf{J_f}(\mathbf{z})^{-1}\right\|}{1-\alpha}\|\mathbf{f}(\widetilde{\mathbf{z}})\| .$$

2. *Suppose that the Jacobian $\mathbf{J_f}$ is uniformly Lipschitz continuous in $\mathscr{D}$ with Lipschitz constant γ, and $\mathbf{f}(\widetilde{\mathbf{z}})$ is sufficiently small, meaning that*

$$4\gamma\left\|\mathbf{J_f}(\mathbf{z})^{-1}\right\|^2\|\mathbf{f}(\widetilde{\mathbf{z}})\| < 1 .$$

   Then

$$\|\widetilde{\mathbf{z}}-\mathbf{z}\| \leq \frac{2\left\|\mathbf{J_f}(\mathbf{z})^{-1}\right\|}{1+\sqrt{1-4\gamma\|\mathbf{J_f}(\mathbf{z})^{-1}\|^2\|\mathbf{f}(\widetilde{\mathbf{z}})\|}}\|\mathbf{f}(\widetilde{\mathbf{z}})\| .$$

Proof Inequality (3.25) in Chap. 3 of Volume I shows that for all $\tau \in [0, 1]$,

$$\left\| \mathbf{J_f}\left(\mathbf{z} + \{\widetilde{\mathbf{z}} - \mathbf{z}\}\,\tau\right)^{-1} \right\| \leq \frac{\|\mathbf{J_f}(\mathbf{z})^{-1}\|}{1 - \|\mathbf{J_f}(\mathbf{z})^{-1}\,\{\mathbf{J_f}\left(\mathbf{z} + [\widetilde{\mathbf{z}} - \mathbf{z}]\,\tau\right) - \mathbf{J_f}(\mathbf{z})\}\|} \leq \frac{\|\mathbf{J_f}(\mathbf{z})^{-1}\|}{1 - \alpha} \,.$$

As a result,

$$\begin{aligned}\|\widetilde{\mathbf{z}} - \mathbf{z}\| &= \left\| \left[\int_0^1 \mathbf{J_f}\left(\mathbf{z} + \{\widetilde{\mathbf{z}} - \mathbf{z}\}\,\tau\right) \mathrm{d}\tau \right]^{-1} \mathbf{f}(\widetilde{\mathbf{z}}) \right\| \\ &\leq \int_0^1 \left\| \mathbf{J_f}\left(\mathbf{z} + \{\widetilde{\mathbf{z}} - \mathbf{z}\}\,\tau\right)^{-1} \right\| \mathrm{d}\tau \, \|\mathbf{f}(\widetilde{\mathbf{z}})\| \leq \frac{\|\mathbf{J_f}(\mathbf{z})^{-1}\|}{1 - \alpha} \|\mathbf{f}(\widetilde{\mathbf{z}})\| \,.\end{aligned}$$

This proves the first claim.

If we assume that $\mathbf{J_f}$ is Lipschitz continuous with constant γ in $\mathscr{D}$, then we find that

$$\begin{aligned}\|\widetilde{\mathbf{z}} - \mathbf{z}\| &\leq \int_0^1 \left\| \mathbf{J_f}\left(\mathbf{z} + \{\widetilde{\mathbf{z}} - \mathbf{z}\}\,\tau\right)^{-1} \right\| \mathrm{d}\tau \, \|\mathbf{f}(\widetilde{\mathbf{z}})\| \\ &\leq \int_0^1 \frac{\|\mathbf{J_f}(\mathbf{z})^{-1}\|}{1 - \|\mathbf{J_f}(\mathbf{z})^{-1}\,\{\mathbf{J_f}\left(\mathbf{z} + [\widetilde{\mathbf{z}} - \mathbf{z}]\,\tau\right) - \mathbf{J_f}(\mathbf{z})\}\|} \mathrm{d}\tau \, \|\mathbf{f}(\widetilde{\mathbf{z}})\| \\ &\leq \int_0^1 \frac{\|\mathbf{J_f}(\mathbf{z})^{-1}\|}{1 - \gamma\, \|\mathbf{J_f}(\mathbf{z})^{-1}\|\, \|\widetilde{\mathbf{z}} - \mathbf{z}\|\, \tau} \mathrm{d}\tau \, \|\mathbf{f}(\widetilde{\mathbf{z}})\| \\ &\leq \frac{\|\mathbf{J_f}(\mathbf{z})^{-1}\|}{1 - \gamma\, \|\mathbf{J_f}(\mathbf{z})^{-1}\|\, \|\widetilde{\mathbf{z}} - \mathbf{z}\|} \|\mathbf{f}(\widetilde{\mathbf{z}})\| \,.\end{aligned}$$

We can solve this inequality for $\|\widetilde{\mathbf{z}} - \mathbf{z}\|$ to prove the second claim.

We can also consider the effects of a perturbation on minimization problems. The following lemma will bound the error in an approximate minimizer of a real functional in terms of the error in the objective and the inverse of the Hessian at the true minimizer. Neither the minimum value of the objective function or the Hessian at the true minimizer will be known during the execution of a minimization algorithm. However, rapidly converging minimization algorithms will give us accurate estimates for the values we need.

Lemma 3.2.16 *Suppose that $\mathscr{D}$ is an open convex set of real n-vectors. Assume that $\|\cdot\|$ represents either a norm on n-vectors, or the corresponding subordinate norm on $n \times n$ matrices. Let the real functional ϕ be twice continuously differentiable on $\mathscr{D}$. Let $\mathbf{z} \in \mathscr{D}$ be a local minimizer for ϕ in $\mathscr{D}$, and let the Hessian $\mathbf{H}_\phi(\mathbf{z})$ be positive. Assume that we are given n-vector $\widetilde{\mathbf{z}} \in \mathscr{D}$, for which there exists $\varepsilon > 0$ so that for all $\tau \in (0, 1)$*

$$\left\| \mathbf{H}_\phi\left(\mathbf{z} + \{\widetilde{\mathbf{z}} - \mathbf{z}\}\,\tau\right) - \mathbf{H}_\phi(\mathbf{z}) \right\| \leq \varepsilon < \frac{1}{\|\mathbf{H}_\phi(\mathbf{z})^{-1}\|} \,.$$

Then

$$\|\widetilde{\mathbf{z}}-\mathbf{z}\| \leq \sqrt{\frac{2\left[\phi\left(\widetilde{\mathbf{z}}\right)-\phi(\mathbf{z})\right]}{1/\|\mathbf{H}_\phi(\mathbf{z})^{-1}\|-\varepsilon}} .$$

Proof The proof of Lemma 3.2.8 shows that

$$\phi\left(\widetilde{\mathbf{z}}\right)=\phi(\mathbf{z})+\int_0^1\int_0^\tau\{\widetilde{\mathbf{z}}-\mathbf{z}\}\cdot\mathbf{H}_\phi\left(\mathbf{x}+\{\widetilde{\mathbf{z}}-\mathbf{z}\}\sigma\right)\{\widetilde{\mathbf{z}}-\mathbf{z}\}\;\mathrm{d}\sigma\;\mathrm{d}\tau .$$

Since $\mathbf{H}_\phi(\mathbf{z})$ is symmetric positive, our discussion in Sect. 1.7.3 shows that it has a symmetric positive square root $\mathbf{S}=\sqrt{\mathbf{H}_\phi(\mathbf{z})}$. Then

$$[\widetilde{\mathbf{z}}-\mathbf{z}]\,\mathbf{H}_\phi(\mathbf{z})\,[\widetilde{\mathbf{z}}-\mathbf{z}]=\|\mathbf{S}\,[\widetilde{\mathbf{z}}-\mathbf{z}]\|^2\geq\left\|\mathbf{S}^{-1}\right\|^{-2}\|\widetilde{\mathbf{z}}-\mathbf{z}\|^2=\frac{\|\widetilde{\mathbf{z}}-\mathbf{z}\|^2}{\left\|\mathbf{H}_\phi(\mathbf{z})^{-1}\right\|} ,$$

because Lemma 1.3.3 shows that the 2-norm of a symmetric positive matrix is its largest eigenvalue. It follows that

$$\begin{aligned}\phi\left(\widetilde{\mathbf{z}}\right)-\phi(\mathbf{z})&=\int_0^1\int_0^\tau[\widetilde{\mathbf{z}}-\mathbf{z}]\cdot\mathbf{H}_\phi\left(\mathbf{z}+\{\widetilde{\mathbf{z}}-\mathbf{z}\}\sigma\right)[\widetilde{\mathbf{z}}-\mathbf{z}]\;\mathrm{d}\sigma\;\mathrm{d}\tau\\&=\frac{1}{2}[\widetilde{\mathbf{z}}-\mathbf{z}]\cdot\mathbf{H}_\phi(\mathbf{z})[\widetilde{\mathbf{z}}-\mathbf{z}]+[\widetilde{\mathbf{z}}-\mathbf{z}]\cdot\left[\int_0^1\int_0^\tau\mathbf{H}_\phi\left(\mathbf{z}+\{\widetilde{\mathbf{z}}-\mathbf{z}\}\sigma\right)-\mathbf{H}_\phi(\mathbf{z})\;\mathrm{d}\sigma\;\mathrm{d}\tau\right][\widetilde{\mathbf{z}}-\mathbf{z}]\\&\geq\frac{\|\widetilde{\mathbf{z}}-\mathbf{z}\|^2}{2\left\|\mathbf{H}_\phi(\mathbf{z})^{-1}\right\|}-\frac{\varepsilon}{2}\|\widetilde{\mathbf{z}}-\mathbf{z}\|^2 .\end{aligned}$$

We can solve for $\|\widetilde{\mathbf{z}}-\mathbf{z}\|^2$ to get

$$\|\widetilde{\mathbf{z}}-\mathbf{z}\|^2\leq\frac{2}{1/\|\mathbf{H}_\phi(\mathbf{z})^{-1}\|-\varepsilon}\left[\phi\left(\widetilde{\mathbf{z}}\right)-\phi(\mathbf{z})\right] .$$

This inequality implies the claimed result.

Under normal circumstances, we would expect that rounding errors in the computer evaluation of the objective function would be on the order of machine precision. Lemma 3.2.16 implies that if we use the change in the objective function to measure the progress of a minimization algorithm, then we expect that for well-conditioned problems (namely, those in which $\|\mathbf{H}_\phi(\mathbf{z})^{-1}\|\approx 1$) that a local minimizer can be determined only to an absolute accuracy on the order of the *square root of machine precision*. On the other hand, if we use the norm of the gradient $\mathbf{g}_\phi$ to measure the progress of the algorithm, then whenever $\left\|\mathbf{g}_\phi\left(\widetilde{\mathbf{z}}\right)\right\|$ is small Lemma 3.2.15 shows that we approximately have

$$\|\widetilde{\mathbf{z}}-\mathbf{z}\|\leq 2\left\|\mathbf{H}_\phi(\mathbf{z})^{-1}\right\|\,\|\mathbf{g}_\phi\left(\widetilde{\mathbf{z}}\right)\| .$$

Presumably, rounding errors in the evaluation of the gradient would also be on the order of machine precision. Both Lemmas 3.2.15 and 3.2.16 indicate that minimization problems are poorly conditioned whenever the matrix of second derivatives has a small eigenvalue. But measuring the norm of the gradient of the objective for well-conditioned problems can help us to determine the solution more accurately than measuring the change in the objective function.

Example 3.2.3 Let

$$\phi(\mathbf{x}) = (\xi_1 - 1)^2 + 10^{-6}(\xi_2 - 1)^2 + 1 .$$

The obvious global minimizer of ϕ is

$$\mathbf{z} = \begin{bmatrix} 1 \\ 1 \end{bmatrix} .$$

If $\mathbf{d} = [\delta_1, \delta_2]$, then

$$\phi(\mathbf{z} + \mathbf{d}) = \delta_1^2 + 10^6 \delta_2^2 .$$

If we require that $\phi(\mathbf{z} + \mathbf{d}) - \phi(\mathbf{z}) < \varepsilon$, then our test on objective values will be satisfied whenever

$$|\delta_2| < 10^{-3}\sqrt{\varepsilon} ;$$

if $\varepsilon = 10^{-16}$ is roughly double precision rounding error, then $|\delta_2| < 10^{-5}$.

However,

$$\mathbf{g}_\phi(\mathbf{x}) = \begin{bmatrix} 2(\xi_1 - 1) \\ 2 \times 10^6(\xi_2 - 1) \end{bmatrix} 2 .$$

Thus

$$\left\| \mathbf{g}_\phi(\mathbf{z} + \mathbf{d}) \right\|_2 = 2\sqrt{\delta_1^2 + 10^{12}\delta_2^2} .$$

If we require that $\|\mathbf{g}_\phi(\mathbf{z} + \mathbf{d})\|_2 < \varepsilon$, then we have

$$|\delta_2| < 0.5 \times 10^{-6}\varepsilon .$$

If $\varepsilon = 10^{-16}$, then $|\delta_2| < 0.5 \times 10^{-10}$. In this case, testing the gradient for accuracy of an approximate minimum leads to significantly greater accuracy in the solution.

3.3 Convergence Rates

In discussing numerical methods for solving systems of nonlinear equations, we will use the following important concepts to describe the relative performance of the methods.

Definition 3.3.1 Let $\mathbf{x}$ be an n-vector, let $\{\mathbf{x}_k\}_{k=0}^{\infty}$ be a sequence of n-vectors and let $\|\cdot\|$ be a norm on n-vectors. Then $\{\mathbf{x}_k\}_{k=0}^{\infty}$ **converges** to $\mathbf{x}$ if and only if for all $\varepsilon > 0$ there exists a positive integer K so that for all $k \geq K$

$$\|\mathbf{x}_k - \mathbf{x}\| \leq \varepsilon \ .$$

The sequence $\{\mathbf{x}_k\}_{k=0}^{\infty}$ is **linearly convergent** to $\mathbf{x}$ if and only if there exists an integer $K > 0$ and an error reduction factor $c \in [0, 1)$ so that for all $k > K$

$$\|\mathbf{x}_{k+1} - \mathbf{x}\| \leq c\|\mathbf{x}_k - \mathbf{x}\| \ .$$

The sequence $\{\mathbf{x}_k\}_{k=0}^{\infty}$ is **superlinearly convergent** to $\mathbf{x}$ if and only if there is a sequence $\{c_k\}$ of positive real numbers converging to zero such that

$$\|\mathbf{x}_{k+1} - \mathbf{x}\| \leq c_k\|\mathbf{x}_k - \mathbf{x}\| \ .$$

A sequence $\{\mathbf{x}_k\}$ is **quadratically convergent** if and only if there is a limit $\mathbf{x}$, an integer $K > 0$ and a scalar $c \geq 0$ such that for all $k > K$

$$\|\mathbf{x}_{k+1} - \mathbf{x}\| \leq c\|\mathbf{x}_k - \mathbf{x}\|^2 \ .$$

The sequence $\{\mathbf{x}_k\}_{k=0}^{\infty}$ is **convergent with rate** r to $\mathbf{x}$ if and only if there is a scalar $r \geq 1$, an integer $K > 0$ and a scalar $c \geq 0$ such that for all $k > K$

$$\|\mathbf{x}_{k+1} - \mathbf{x}\| \leq c\|\mathbf{x}_k - \mathbf{x}\|^r \ .$$

The reader should note that these definitions are similar to those in Definition 5.4.1 in Chap. 5 of Volume I regarding sequences of scalars. We also remind the reader that Example 5.4.2 in Chap. 5 of Volume I contains several examples of sequences that converge at different rates.

In order to discover the rate of convergence for a sequence, we can graph $-\log(\|\mathbf{x}_{k+1} - \mathbf{x}\|_2)$ versus $-\log(\|\mathbf{x}_k - \mathbf{x}\|_2)$, as in Fig. 3.1 (a). A linearly convergent sequence would produce a plot that tends toward a line with slope 1 that intersects the vertical axis below the origin. The corresponding plot of a quadratically convergent sequence would have limiting slope of 2.

In order to illustrate the relative behavior of convergent sequences, it is useful to plot $\log(\|\mathbf{x}_k - \mathbf{x}\|_2)$ versus $\log(k)$, as in Fig. 3.1 (b). If we perform roughly the same amount of work per iteration, we would expect the total work in the iteration to be proportional to k. In such a plot, we expect the graph of a

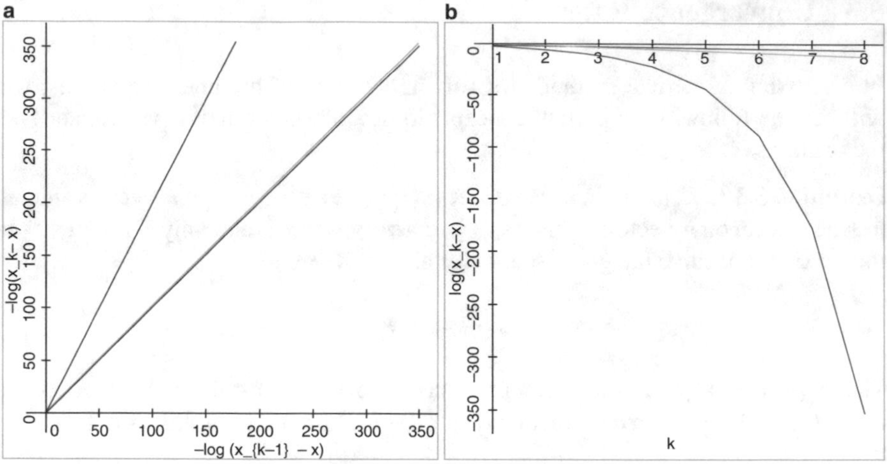

Fig. 3.1 Rates of convergence: red = linear $(1 + 2^{-k})$, green = superlinear $(1 + 2/k!)$, blue = quadratic $(1 + 2^{-2^k})$. (**a**) $-\log(\|\mathbf{x}_{k+1} - \mathbf{x}\|_2)$ versus $-\log(\|\mathbf{x}_k - \mathbf{x}\|_2)$. (**b**) $\log(\|\mathbf{x}_{k+1} - \mathbf{x}\|_2)$ versus $\log(k)$

linearly convergent sequence to tend toward a straight line with positive slope. A quadratically convergent sequence would curve downward exponentially fast.

Exercise 3.3.1 What is the convergence rate of $\{\mathbf{x}_k\}_{k=0}^{\infty}$ if

$$\mathbf{x}_k = \begin{bmatrix} 2^{-k} \\ 2/k! \end{bmatrix} ?$$

3.4 Newton's Method

Let us review our progress in applying scientific computing to solving systems of nonlinear equations and minimizing functionals. So far, we have examined conditions that guarantee the existence and uniqueness of local minima of nonlinear functionals, and zeros of nonlinear functions. We have also studied the sensitivity of these problems to perturbation. The next step in scientific computing for these problems is to develop numerical methods.

We will begin with Newton's method, because it is familiar to most readers, and easy to understand. Our first task will be to describe the method carefully in Sect. 3.4.1. Afterward, we will analyze its convergence behavior in Sect. 3.4.2. We will see that Newton's method for solving $\mathbf{f(z)} = \mathbf{0}$ requires both the evaluation of the Jacobian matrix $\mathbf{J_f(x)}$, and the solution of linear systems involving this Jacobian matrix. In Sect. 3.4.3, we will learn how to approximate the Jacobian by finite differences, in order to reduce the programming complexity of evaluating the

Jacobian. Our analysis of rounding errors in Sect. 3.4.4 will show that Newton's method with finite difference approximations to Jacobians are more sensitive to rounding errors than analytical Jacobians. These rounding errors will affect the maximum attainable accuracy of Newton's methods, as we will discover in Sect. 3.4.5. For very large systems of nonlinear equations, it may be attractive to solve the linear systems that arise in Newton's method by iteration. This leads to two nested iterations, and the inner iteration for the linear system solver does not need to be solved accurately until the true solution is close. We will discuss this issue in Sect. 3.4.6. At the end of this section, we will provide suggestions for software implementations of Newton's method and its variants.

3.4.1 Local Model

Recall that we have previously discussed Newton's method for nonlinear scalar equations in Sect. 5.4 in Chap. 5 of Volume I. In presenting Newton's method for systems, however, we will emphasize the role of developing a **local model** for the nonlinear equation. Given a guess $\mathbf{z}_k$ for the zero $\mathbf{z}$ of a continuously differentiable vector-valued function $\mathbf{f}$ of n variables, we approximate $\mathbf{f}$ by its tangent plane

$$\widetilde{\mathbf{f}}(\mathbf{x}) \equiv \mathbf{f}(\mathbf{z}_k) + \mathbf{J}_{\mathbf{f}}(\mathbf{z}_k)(\mathbf{x} - \mathbf{z}_k) \; .$$

If this local model is an accurate approximation to the original function $\mathbf{f}$, then we can approximate the zero $\mathbf{z}$ of $\mathbf{f}$ by the zero of the local model. Thus our new approximation is

$$\mathbf{z}_{k+1} = \mathbf{z}_k + \mathbf{s}_k \text{ where } \mathbf{J}_{\mathbf{f}}(\mathbf{z}_k)\mathbf{s}_k = -\mathbf{f}(\mathbf{z}_k)$$

Readers may experiment with the JavaScript **Newton's method** program. This program plots a vector field of function values for the function

$$\mathbf{f}(\mathbf{x}) = \begin{bmatrix} 2(\xi_2 - \xi_1^2) \\ 1 - \xi_1 \end{bmatrix} ,$$

and the Kantorovich convergence region described in Theorem 3.2.3. It also plots a path of Newton steps for this function. Readers may select an initial guess for Newton's method and see the resulting path of the iteration.

3.4.2 Convergence Behavior

The following theorem provides conditions under which Newton's method converges, provided that a solution exists.

Theorem 3.4.1 (Local Convergence for Newton's Method) *Suppose that $\mathscr{D}$ is a convex and open subset of real n-vectors. Let the mapping* $\mathbf{f}$ *from $\mathscr{D}$ to real n-vectors be continuously differentiable on $\mathscr{D}$. Also let $\|\cdot\|$ represent either a norm on n-vectors or the corresponding subordinate norm on $n \times n$ matrices. Assume that there exists $\mathbf{z} \in \mathscr{D}$ such that $\mathbf{f}(\mathbf{z}) = \mathbf{0}$. Suppose that the Jacobian $\mathbf{J_f}(\mathbf{z})$ is nonsingular, and that there are constants $\gamma > 0$ and $\varrho > 0$ such that $\|\mathbf{x} - \mathbf{z}\| < \varrho$ implies $\mathbf{x} \in \mathscr{D}$ and*

$$\|\mathbf{J_f}(\mathbf{x}) - \mathbf{J_f}(\mathbf{z})\| \le \gamma \|\mathbf{x} - \mathbf{z}\| \,.$$

Then there exists $\varepsilon > 0$ so that for all initial guesses $\mathbf{z}_0$ satisfying $\|\mathbf{z}_0 - \mathbf{z}\| < \varepsilon$, Newton's method converges quadratically to $\mathbf{z}$.

Proof Let

$$\varepsilon = \min\left\{\varrho \,,\, \frac{1}{2\gamma \, \|\mathbf{J_f}(\mathbf{z})^{-1}\|}\right\} \,,$$

and assume that $\|\mathbf{z}_k - \mathbf{z}\| \le \varepsilon$. Note that

$$\begin{aligned} \left\|\mathbf{J_f}(\mathbf{z})^{-1}(\mathbf{J_f}(\mathbf{z}_k) - \mathbf{J_f}(\mathbf{z}))\right\| &\le \left\|\mathbf{J_f}(\mathbf{z})^{-1}\right\| \, \|\mathbf{J_f}(\mathbf{z}_k) - \mathbf{J_f}(\mathbf{z})\| \\ &\le \gamma \left\|\mathbf{J_f}(\mathbf{z})^{-1}\right\| \, \|\mathbf{z}_k - \mathbf{z}\| \le \gamma\varepsilon \left\|\mathbf{J_f}(\mathbf{z})^{-1}\right\| \le \frac{1}{2} \,. \end{aligned}$$

Then Lemma 3.2.14 shows that $\mathbf{J_f}(\mathbf{z}_k)$ is nonsingular and

$$\left\|\mathbf{J_f}(\mathbf{z}_k)^{-1}\right\| \le \frac{\left\|\mathbf{J_f}(\mathbf{z})^{-1}\right\|}{1 - \|\mathbf{J_f}(\mathbf{z})^{-1}(\mathbf{J_f}(\mathbf{z}_k) - \mathbf{J_f}(\mathbf{z}))\|} \le 2\left\|\mathbf{J_f}(\mathbf{z})^{-1}\right\| \,.$$

Thus $\mathbf{z}_{k+1}$ is defined, and

$$\mathbf{z}_{k+1} - \mathbf{z} = \mathbf{z}_k - \mathbf{z} - \mathbf{J_f}(\mathbf{z}_k)^{-1}\mathbf{f}(\mathbf{z}_k) = \mathbf{J_f}(\mathbf{z}_k)^{-1}\left[\mathbf{f}(\mathbf{z}) - \mathbf{f}(\mathbf{z}_k) - \mathbf{J_f}(\mathbf{z}_k)(\mathbf{z} - \mathbf{z}_k)\right] \,.$$

Taking norms, Lemma 3.2.5 leads to

$$\begin{aligned} \|\mathbf{z}_{k+1} - \mathbf{z}\| &\le \left\|\mathbf{J_f}(\mathbf{z}_k)^{-1}\right\| \, \|\mathbf{f}(\mathbf{z}) - \mathbf{f}(\mathbf{z}_k) - \mathbf{J_f}(\mathbf{z}_k)(\mathbf{z} - \mathbf{z}_k)\| \\ &\le \left\|\mathbf{J_f}(\mathbf{z}_k)^{-1}\right\| \frac{\gamma}{2}\|\mathbf{z}_k - \mathbf{z}\|^2 \le \gamma \left\|\mathbf{J_f}(\mathbf{z})^{-1}\right\| \, \|\mathbf{z}_k - \mathbf{z}\|^2 \,. \end{aligned} \tag{3.10}$$

Since

$$\gamma \left\|\mathbf{J_f}(\mathbf{z})^{-1}\right\| \, \|\mathbf{z}_k - \mathbf{z}\| \le \frac{1}{2} \,,$$

we see that $\|\mathbf{z}_{k+1} - \mathbf{z}\| \leq \frac{1}{2}\|\mathbf{z}_k - \mathbf{z}\|$. In particular, this implies that $\mathbf{z}_{k+1} \in \mathscr{D}$, so Newton's iteration can be continued. This inequality also shows that the sequence converges, and inequality (3.10) shows that the convergence is quadratic.
We remark that the Kantorovich Theorem 3.2.3 can be used to determine another radius ε of local convergence for Newton's method.

One great advantage of Newton's method is its quadratic rate of convergence: when it converges, it converges rapidly. Two disadvantages are that it requires programming of the Jacobian matrix $\mathbf{J_f}$, and requires the solution of a system of linear equations with each iteration. Another disadvantage is that Newton's method is only locally convergent. It is also possible that the Jacobian matrix could be singular or poorly conditioned at some step of the iteration.

3.4.3 *Finite Difference Jacobian*

It is natural to construct modifications of Newton's method that attempt to overcome some of the difficulties. For example, we could avoid programming the Jacobian matrix for $\mathbf{f}$ by approximating the derivatives with difference quotients. In particular, we could approximate each jth column of the Jacobian at $\mathbf{z}_k$ by

$$\widetilde{\mathbf{J}}_{\mathbf{f}}(\mathbf{z}_k; \zeta_k)\mathbf{e}_j = \left[\mathbf{f}(\mathbf{z}_k + \mathbf{e}_j\zeta_k) - \mathbf{f}_i(\mathbf{z}_k)\right]/\zeta_k \tag{3.11a}$$

where ζ_k is a nonzero scalar for each k. Then we could try a modified Newton's method of the form

$$\widetilde{\mathbf{J}}_{\mathbf{f}}(\mathbf{z}_k\,,\ \zeta_k)\mathbf{s}_k = -\mathbf{f}(\mathbf{z}_k)\ , \tag{3.11b}$$

$$\mathbf{z}_{k+1} = \mathbf{z}_k + \mathbf{s}_k\ . \tag{3.11c}$$

The following theorem, which can be found in Dennis and Schnabel [52, p. 95], describes the local convergence of this method.

Theorem 3.4.2 (Local Convergence of Newton with Finite Difference Jacobian) *Suppose that $\mathscr{D}$ is a convex and open subset of real n-vectors. Assume that the mapping* $\mathbf{f}$ *from $\mathscr{D}$ to real n-vectors is continuously differentiable on $\mathscr{D}$. Suppose that there exists $\mathbf{z} \in \mathscr{D}$ such that $\mathbf{f}(\mathbf{z}) = \mathbf{0}$, and that $\mathbf{J_f}(\mathbf{z})$ is nonsingular. Assume that there are constants $\gamma > 0$ and $\varrho > 0$ such that $\|\mathbf{x} - \mathbf{z}\|_1 < \varrho$ implies $\mathbf{x} \in \mathscr{D}$ and*

$$\|\mathbf{J_f}(\mathbf{x}) - \mathbf{J_f}(\mathbf{z})\|_1 \leq \gamma\|\mathbf{x} - \mathbf{z}\|_1\ .$$

Then there exist $\varepsilon > 0$ and $\zeta > 0$ such that for any sequence $\{\zeta_k\}_{k=0}^{\infty}$ with $|\zeta_k| \leq \zeta$ for all k, and any $\mathbf{z}_0$ satisfying $\|\mathbf{z}_0 - \mathbf{z}\|_1 < \varepsilon$, the sequence $\{\mathbf{z}_k\}_{k=0}^{\infty}$ defined by (3.11) *is well-defined and converges (at least) linearly to* $\mathbf{z}$*. If* $\zeta_k \to 0$ *as* $k \to \infty$*, then* $\{\mathbf{z}_k\}_{kj=0}^{\infty}$ *converges (at least) superlinearly. If there exists a constant* $c > 0$ *such that*

$$|\zeta_k| \le c\|\mathbf{z}_k - \mathbf{z}\|_1 \ ,$$

then $\{\mathbf{z}_k\}_{k=0}^{\infty}$ *converges quadratically.*
An alternative condition for quadratic convergence is $|\zeta_k| \le c\|\mathbf{f}(\mathbf{z}_k)\|_1$.

Proof Suppose that $\mathbf{x} \in \mathscr{D}$, and that δ is sufficiently small so that $\mathbf{x} + \mathbf{e}_j\delta \in \mathscr{D}$ for all $1 \le j \le n$. Then the triangle inequality implies that

$$\begin{aligned}
&\left\|\mathbf{J_f}(\mathbf{z})^{-1}\left\{\widetilde{\mathbf{J}}_\mathbf{f}(\mathbf{x},\delta) - \mathbf{J_f}(\mathbf{z})\right\}\right\|_1 \\
&= \left\|\mathbf{J_f}(\mathbf{z})^{-1}\left\{\left[\widetilde{\mathbf{J}}_\mathbf{f}(\mathbf{x},\delta) - \mathbf{J_f}(\mathbf{x})\right] + \left[\mathbf{J_f}(\mathbf{x}) - \mathbf{J_f}(\mathbf{z})\right]\right\}\right\|_1 \\
&\le \left\|\mathbf{J_f}(\mathbf{z})^{-1}\right\|_1 \left\{\left\|\widetilde{\mathbf{J}}_\mathbf{f}(\mathbf{x},\delta) - \mathbf{J_f}(\mathbf{x})\right\|_1 + \|\mathbf{J_f}(\mathbf{x}) - \mathbf{J_f}(\mathbf{z})\|_1\right\}
\end{aligned}$$

and then Lipschitz continuity and Corollary 3.2.1 give us

$$\le \gamma \left\|\mathbf{J_f}(\mathbf{z})^{-1}\right\|_1 \left\{\frac{|\delta|}{2} + \|\mathbf{x} - \mathbf{z}\|_1\right\} \ . \tag{3.12}$$

If $\widetilde{\mathbf{J}}_\mathbf{f}(\mathbf{x},\delta)$ is nonsingular, these same assumptions on $\mathbf{x}$ and δ imply that

$$\begin{aligned}
&\left\|\mathbf{x} - \widetilde{\mathbf{J}}_\mathbf{f}(\mathbf{x},\delta)^{-1}\mathbf{f}(\mathbf{x}) - \mathbf{z}\right\|_1 = \left\|\widetilde{\mathbf{J}}_\mathbf{f}(\mathbf{x},\delta)^{-1}\left\{\mathbf{f}(\mathbf{z}) - \mathbf{f}(\mathbf{x}) - \widetilde{\mathbf{J}}_\mathbf{f}(\mathbf{x},\delta)(\mathbf{x} - \mathbf{z})\right\}\right\|_1 \\
&= \left\|\widetilde{\mathbf{J}}_\mathbf{f}(\mathbf{x},\delta)^{-1}\left\{\left[\mathbf{f}(\mathbf{z}) - \mathbf{f}(\mathbf{x}) - \mathbf{J_f}(\mathbf{x})(\mathbf{z} - \mathbf{x})\right] + \left[\mathbf{J_f}(\mathbf{x}) - \widetilde{\mathbf{J}}_\mathbf{f}(\mathbf{x},\delta)\right](\mathbf{z} - \mathbf{x})\right\}\right\|_1
\end{aligned}$$

then Lemma 3.2.5 and Corollary 3.2.1 yield

$$\le \frac{\gamma}{2}\left\|\widetilde{\mathbf{J}}_\mathbf{f}(\mathbf{x},\delta)^{-1}\right\|_1 \left\{\|\mathbf{x} - \mathbf{z}\|_1 + |\delta|\right\} \|\mathbf{x} - \mathbf{z}\|_1 \ . \tag{3.13}$$

Choose $\varepsilon > 0$ so that

$$\varepsilon < \min\left\{\varrho \ , \ \frac{1}{2\gamma \left\|\mathbf{J_f}(\mathbf{z})^{-1}\right\|_1}\right\} \ , \tag{3.14}$$

and then choose $\zeta > 0$ so that

$$\zeta < \min\left\{\varrho - \varepsilon \ , \ \frac{1}{2\gamma \left\|\mathbf{J_f}(\mathbf{z})^{-1}\right\|_1} - \varepsilon\right\} \ . \tag{3.15}$$

We will prove inductively that

$$\|\mathbf{z}_k - \mathbf{z}\|_1 < 2^{-k}\varepsilon \text{ and } \|\mathbf{z}_k + \mathbf{e}_j\zeta_k - \mathbf{z}\|_1 < 2^{-k}\varepsilon + \zeta \text{ for all } 1 \leq j \leq n \tag{3.16a}$$

[illegible] (3.16b)

$$\widetilde{\mathbf{J}}_\mathbf{f}(\mathbf{z}_k, \zeta_k) \text{ is nonsingular and } \left\|\widetilde{\mathbf{J}}_\mathbf{f}(\mathbf{z}_k, \zeta_k)^{-1}\right\|_1 \leq 2 \left\|\mathbf{J}_\mathbf{f}(\mathbf{z})^{-1}\right\|_1 \text{ and} \tag{3.16c}$$

$$\|\mathbf{z}_{k+1} - \mathbf{z}\|_1 \leq \frac{\gamma}{2} \left\|\widetilde{\mathbf{J}}_\mathbf{f}(\mathbf{z}_k\,,\ \zeta_k)^{-1}\right\|_1 \{\|\mathbf{z}_k - \mathbf{z}\|_1 + |\zeta_k|\} \|\mathbf{z}_k - \mathbf{z}\|_1 \tag{3.16d}$$

for all $k \geq 0$.

First, we will show that the inductive hypotheses (3.16) are satisfied for $k = 0$. Theorem 3.4.2 assumes that $\|\mathbf{z}_0 - \mathbf{z}\|_1 < \varepsilon$. Thus for all $1 \leq j \leq n$ we have

$$\|\mathbf{z}_0 + \mathbf{e}_j\zeta_0 - \mathbf{z}\|_1 \leq \|\mathbf{z}_0 - \mathbf{z}\|_1 + |\zeta_0| \leq \varepsilon + \zeta \,,$$

which proves inductive hypothesis (3.16a) $k = 0$. Because

$$\|\mathbf{z}_0 - \mathbf{z}\|_1 < \varepsilon < \varrho \,,$$

we conclude that $\mathbf{z}_0 \in \mathscr{D}$. Since $\zeta < \varrho - \varepsilon$, it follows that for all $1 \leq j \leq n$

$$\|\mathbf{z}_0 + \mathbf{e}_j\zeta_0 - \mathbf{z}\|_1 \leq \varepsilon + \zeta < \varrho \,,$$

so $\mathbf{z}_0 + \mathbf{e}_j\zeta_0 \in \mathscr{D}$. These prove inductive hypothesis (3.16b) for $k = 0$. Since inequality (3.12) implies that

$$\begin{aligned}\left\|\mathbf{J}_\mathbf{f}(\mathbf{z})^{-1}\left\{\widetilde{\mathbf{J}}_\mathbf{f}(\mathbf{z}_0, \zeta_0) - \mathbf{J}_\mathbf{f}(\mathbf{z})\right\}\right\|_1 &\leq \gamma \left\|\mathbf{J}_\mathbf{f}(\mathbf{z})^{-1}\right\|_1 \left\{\frac{|\zeta_0}{2} + \|\mathbf{z}_0 - \mathbf{z}\|_1\right\} \\ &< \gamma \left\|\mathbf{J}_\mathbf{f}(\mathbf{z})^{-1}\right\|_1 \{\zeta + \varepsilon\} < \frac{1}{2} \,,\end{aligned}$$

Corollary 3.6.1 in Chap. 3 of Volume I implies that $\widetilde{\mathbf{J}}_\mathbf{f}(\mathbf{z}_0, \zeta_0)$ is nonsingular and

$$\left\|\widetilde{\mathbf{J}}_\mathbf{f}(\mathbf{z}_0, \zeta_0)^{-1}\right\|_1 \leq \frac{\left\|\mathbf{J}_\mathbf{f}(\mathbf{z})^{-1}\right\|_1}{1 - \left\|\mathbf{J}_\mathbf{f}(\mathbf{z})^{-1}\left\{\widetilde{\mathbf{J}}_\mathbf{f}(\mathbf{z}_0, \zeta_0) - \mathbf{J}_\mathbf{f}(\mathbf{z})\right\}\right\|_1} < 2 \left\|\mathbf{J}_\mathbf{f}(\mathbf{z})^{-1}\right\|_1 \,.$$

This proves inductive hypothesis (3.16c) for $k = 0$. Also, inequality (3.13) implies inequality (3.16d) for $k = 0$. Thus we have proved that the inductive hypotheses (3.16) are satisfied for $k = 0$.

Inductively, assume that inequalities (3.16) are satisfied for $k - 1 \geq 0$. Then inequality (3.16d) for $k - 1$ implies that

$$\|\mathbf{z}_k - \mathbf{z}\|_1 \leq \frac{\gamma}{2} \left\|\widetilde{\mathbf{J}}_\mathbf{f}(\mathbf{z}_{k-1}\,,\ \zeta_{k-1})^{-1}\right\|_1 \{|\zeta_{k-1}| + \|\mathbf{z}_{k-1} - \mathbf{z}\|_1\} \|\mathbf{z}_{k-1} - \mathbf{z}\|_1$$

and then inequalities (3.16a) and (3.16c) for $k-1$ give us

$$< \gamma \left\|\mathbf{J_f}(\mathbf{z})^{-1}\right\|_1 \left\{\zeta + 2^{-(k-1)}\varepsilon\right\} \|\mathbf{z}_{k-1} - \mathbf{z}\|_1 \leq \gamma \left\|\mathbf{J_f}(\mathbf{z})^{-1}\right\|_1 \{\zeta + \varepsilon\} \|\mathbf{z}_{k-1} - \mathbf{z}\|_1$$

$$< \gamma \left\|\mathbf{J_f}(\mathbf{z})^{-1}\right\|_1 \frac{1}{2\gamma \left\|\mathbf{J_f}(\mathbf{z})^{-1}\right\|_1} \|\mathbf{z}_{k-1} - \mathbf{z}\|_1 = \frac{1}{2}\|\mathbf{z}_{k-1} - \mathbf{z}\|_1 \leq \frac{1}{2} 2^{-(k-1)}\varepsilon = 2^{-k}\varepsilon \; .$$

Furthermore,

$$\|\mathbf{z}_k + \mathbf{e}_j\zeta_k - \mathbf{z}\|_1 \leq \|\mathbf{z}_k - \mathbf{z}\|_1 + |\zeta_k| \leq 2^{-k}\varepsilon + \zeta \; ,$$

so inductive hypothesis (3.16a) is satisfied for k. Because $2^{-k}\varepsilon < \varrho$, we conclude that $\mathbf{z}_k \in \mathscr{D}$. Since $\zeta_k \leq \zeta < \varrho - \varepsilon$ we see that for all $1 \leq j \leq n$

$$\|\mathbf{z}_k + \mathbf{e}_j\zeta_k - \mathbf{z}\|_1 \leq 2^{-k}\varepsilon + \zeta < \varrho \; ,$$

so $\mathbf{z}_k + \mathbf{e}_j\zeta_k \in \mathscr{D}$. These prove inductive hypothesis (3.16b) for k. Inequality (3.12) implies that

$$\left\|\mathbf{J_f}(\mathbf{z})^{-1}\left\{\widetilde{\mathbf{J}}_\mathbf{f}(\mathbf{z}_{k-1}, \zeta_{k-1}) - \mathbf{J_f}(\mathbf{z})\right\}\right\|_1 \leq \gamma \left\|\mathbf{J_f}(\mathbf{z})^{-1}\right\|_1 \left\{\frac{|\zeta_k|}{2} + \|\mathbf{z}_k - \mathbf{z}\|_2\right\}$$

$$< \gamma \left\|\mathbf{J_f}(\mathbf{z})^{-1}\right\|_1 \left\{\frac{\zeta}{2} + 2^{-k}\varepsilon\right\} \leq \frac{\gamma\zeta}{2} \left\|\mathbf{J_f}(\mathbf{z})^{-1}\right\|_1 (\zeta + \varepsilon) < \frac{1}{2} \; ,$$

so Corollary 3.6.1 in Chap. 3 of Volume I implies that $\widetilde{\mathbf{J}}_\mathbf{f}(\mathbf{z}_{k-1}, \zeta_{k-1})$ is nonsingular and

$$\left\|\widetilde{\mathbf{J}}_\mathbf{f}(\mathbf{z}_{k-1}, \zeta_{k-1})^{-1}\right\|_1 \leq \frac{\left\|\mathbf{J_f}(\mathbf{z})^{-1}\right\|_1}{1 - \left\|\mathbf{J_f}(\mathbf{z})^{-1}\left\{\widetilde{\mathbf{J}}_\mathbf{f}(\mathbf{z}_{k-1}, \zeta_{k-1}) - \mathbf{J_f}(\mathbf{z})\right\}\right\|_1} < 2\left\|\mathbf{J_f}(\mathbf{z})^{-1}\right\|_1 \; .$$

This proves inductive hypothesis (3.16c) for k. Also, inequality (3.13) implies inductive hypothesis (3.16d) for k. This completes the inductive proof of (3.16).

Since

$$|\zeta_k| < \min\left\{\varrho \, , \, \frac{1}{2\gamma \left\|\mathbf{J_f}(\mathbf{z})^{-1}\right\|_1}\right\} - \varepsilon$$

for all $k \geq 0$, inequality (3.16d) shows that

$$\|\mathbf{z}_{k+1} - \mathbf{z}\|_1 \leq \frac{\gamma}{2} \left\|\widetilde{\mathbf{J}}_\mathbf{f}(\mathbf{z}_k \, , \, \zeta_k)^{-1}\right\|_1 \{\|\mathbf{z}_k - \mathbf{z}\|_1 + |\zeta_k|\} \|\mathbf{z}_k - \mathbf{z}\|_1$$

then inequality (3.16a) yields

$$\leq \frac{\gamma}{2} \left\|\widetilde{\mathbf{J}}_\mathbf{f}(\mathbf{z}_k \, , \, \zeta_k)^{-1}\right\|_1 \{2^{-k}\varepsilon + \zeta\} \|\mathbf{z}_k - \mathbf{z}\|_1$$

$$\leq \frac{\gamma}{2} \left\|\widetilde{\mathbf{J}}_\mathbf{f}(\mathbf{z}_k \, , \, \zeta_k)^{-1}\right\|_1 \{\varepsilon + \zeta\} \|\mathbf{z}_k - \mathbf{z}\|_1$$

then the definition (3.15) of ζ produces

$$< \frac{1}{2} \|\mathbf{z}_k - \mathbf{z}\|_1 \ .$$

This proves linear convergence.

If $\zeta_k \to 0$ as $k \to \infty$, then inequality (3.16d) shows that

$$\|\mathbf{z}_{k+1} - \mathbf{z}\|_1 \leq \frac{\gamma}{2} \left\| \widetilde{\mathbf{J}_\mathbf{f}}(\mathbf{z}_k \ , \ \zeta_k)^{-1} \right\|_1 \{\|\mathbf{z}_k - \mathbf{z}\|_1 + |\zeta_k|\} \, \|\mathbf{z}_k - \mathbf{z}\|_1 \ .$$

Since the coefficient of $\|\mathbf{z}_k - \mathbf{z}\|_1$ tends to zero as $k \to \infty$, we conclude that $\mathbf{z}_k \to \mathbf{z}$ superlinearly.

If $|\zeta_k| \leq c\|\mathbf{z}_k - \mathbf{z}\|_1$, then inequality (3.16d) shows that

$$\begin{aligned} \|\mathbf{z}_{k+1} - \mathbf{z}\|_1 &\leq \frac{\gamma}{2} \left\| \widetilde{\mathbf{J}_\mathbf{f}}(\mathbf{z}_k \ , \ \zeta_k)^{-1} \right\|_1 \{\|\mathbf{z}_k - \mathbf{z}\|_1 + |\zeta_k|\} \, \|\mathbf{z}_k - \mathbf{z}\|_1 \\ &\leq \frac{\gamma(1+c)}{2} \left\| \widetilde{\mathbf{J}_\mathbf{f}}(\mathbf{z}_k \ , \ \zeta_k)^{-1} \right\|_1 \|\mathbf{z}_k - \mathbf{z}\|_1^2 \ , \end{aligned}$$

so $\mathbf{z}_k \to \mathbf{z}$ quadratically.

3.4.4 Rounding Errors

Both of the Theorems 3.4.1 and 3.4.2, which describe convergence of Newton's method with analytical and finite difference Jacobians, assume that all computations use exact arithmetic. The next theorem analyzes the effects of rounding errors made in the evaluation of the function and the Jacobian, the solution of the linear system for the Newton step, and in updating the approximate solution. It shows that strict quadratic convergence is unexpected for Newton's method conducted in floating point arithmetic.

Lemma 3.4.1 *Suppose that $\mathscr{D}$ is an open convex subset of the set of all n-vectors, and that $\mathbf{f}$ is a continuously differentiable function on $\mathscr{D}$, with values given by n-vectors. Let $\| \cdot \|$ represent a norm on n-vectors, or the subordinate norm on $n \times n$ matrices. Assume that there is a point $\mathbf{z} \in \mathscr{D}$ such that $\mathbf{f}(\mathbf{z}) = \mathbf{0}$. Suppose that the Jacobian $\mathbf{J}_\mathbf{f}$ is nonsingular and Lipschitz continuous at $\mathbf{z}_k \in \mathscr{D}$, with Lipschitz constant γ. Assume that a numerical implementation of Newton's method computes*

$$\widetilde{\mathbf{f}}_k = \mathbf{f}(\mathbf{z}_k) + \mathbf{d}_k \ , \ \widetilde{\mathbf{J}}_k = \mathbf{J}_\mathbf{f}(\mathbf{z}_k) + \mathbf{E}_k \ , \ \left(\widetilde{\mathbf{J}}_k + \mathbf{F}_k\right) \widetilde{\mathbf{s}}_k = -\widetilde{\mathbf{f}}_k \ \textit{and} \ \widetilde{\mathbf{z}}_{k+1} = \mathbf{z}_k + \widetilde{\mathbf{s}}_k + \mathbf{c}_k \ .$$

Also assume that

$$\left\| \mathbf{J}_\mathbf{f}(\mathbf{z}_k)^{-1} \right\| \|\mathbf{E}_k + \mathbf{F}_k\| < 1 \ .$$

Then the subsequent Newton iterate $\widetilde{\mathbf{z}}_{k+1}$ *exists and*

$$\|\widetilde{\mathbf{z}}_{k+1} - \mathbf{z}\| \leq \alpha \|\mathbf{z}_k - \mathbf{z}\|^2 + (1 - \beta)\|\mathbf{z}_k - \mathbf{z}\| + \delta \tag{3.17a}$$

where

$$\alpha = \frac{\gamma}{2} \frac{\left\| \mathbf{J_f}(\mathbf{z}_k)^{-1} \right\|}{1 - \|\mathbf{J_f}(\mathbf{z}_k)^{-1} [\mathbf{E}_k + \mathbf{F}_k]\|} , \tag{3.17b}$$

$$1 - \beta = \frac{\left\| \mathbf{J_f}(\mathbf{z}_k)^{-1} [\mathbf{E}_k + \mathbf{F}_k] \right\|}{1 - \|\mathbf{J_f}(\mathbf{z}_k)^{-1} [\mathbf{E}_k + \mathbf{F}_k]\|} \kappa \left(\mathbf{J_f}(\mathbf{z}_k) \right) \text{ and} \tag{3.17c}$$

$$\delta = \frac{\left\| \mathbf{J_f}(\mathbf{z}_k)^{-1} \right\|}{1 - \|\mathbf{J_f}(\mathbf{z}_k)^{-1} [\mathbf{E}_k + \mathbf{F}_k]\|} \|\mathbf{d}_k\| + \|\mathbf{c}_k\| . \tag{3.17d}$$

The form of the definition for β will be convenient in our analysis of maximum attainable accuracy in Sect. 3.4.5. If the perturbations $\mathbf{E}_k$ and $\mathbf{F}_k$ are sufficiently small, then we will have $1 - \beta < 1$ and thus $\beta > 0$.

Proof Corollary 3.6.1 in Chap. 3 of Volume I implies that $\mathbf{J_f}(\mathbf{z}_k) + \mathbf{E}_k + \mathbf{F}_k$ is invertible, as well as

$$\left\| \{\mathbf{J_f}(\mathbf{z}_k) + \mathbf{E}_k + \mathbf{F}_k\}^{-1} \right\| \leq \frac{\left\| \mathbf{J_f}(\mathbf{z}_k)^{-1} \right\|}{1 - \|\mathbf{J_f}(\mathbf{z}_k)^{-1} \{\mathbf{E}_k + \mathbf{F}_k\}\|} .$$

and

$$\left\| \{\mathbf{J_f}(\mathbf{z}_k) + \mathbf{E}_k + \mathbf{F}_k\}^{-1} - \mathbf{J_f}(\mathbf{z}_k)^{-1} \right\| \leq \frac{\left\| \mathbf{J_f}(\mathbf{z}_k)^{-1} \right\| \left\| \mathbf{J_f}(\mathbf{z}_k)^{-1} \{\mathbf{E}_k + \mathbf{F}_k\} \right\|}{1 - \|\mathbf{J_f}(\mathbf{z}_k)^{-1} \{\mathbf{E}_k + \mathbf{F}_k\}\|} .$$

The rounding error assumptions imply that

$$\begin{aligned}
\widetilde{\mathbf{z}}_{k+1} - \mathbf{z} &= \mathbf{z}_k - \mathbf{z} - [\mathbf{J_f}(\mathbf{z}_k) + \mathbf{E}_k + \mathbf{F}_k]^{-1} [\mathbf{f}(\mathbf{z}_k) + \mathbf{d}_k] + \mathbf{c}_k \\
&= \mathbf{z}_k - \mathbf{z} - \mathbf{J_f}(\mathbf{z}_k)^{-1}\mathbf{f}(\mathbf{z}_k) + \left\{ \mathbf{J_f}(\mathbf{z}_k)^{-1} - [\mathbf{J_f}(\mathbf{z}_k) + \mathbf{E}_k + \mathbf{F}_k]^{-1} \right\} \mathbf{f}(\mathbf{z}_k) \\
&- [\mathbf{J_f}(\mathbf{z}_k) + \mathbf{E}_k + \mathbf{F}_k]^{-1} \mathbf{d}_k + \mathbf{c}_k \\
&= -\mathbf{J_f}(\mathbf{z}_k)^{-1} \{\mathbf{f}(\mathbf{z}_k) - \mathbf{f}(\mathbf{z}) - \mathbf{J_f}(\mathbf{z}_k)[\mathbf{z}_k - \mathbf{z}]\} \\
&+ \left\{ \mathbf{J_f}(\mathbf{z}_k)^{-1} - [\mathbf{J_f}(\mathbf{z}_k) + \mathbf{E}_k + \mathbf{F}_k]^{-1} \right\} [\mathbf{f}(\mathbf{z}_k) - \mathbf{f}(\mathbf{z})] \\
&- [\mathbf{J_f}(\mathbf{z}_k) + \mathbf{E}_k + \mathbf{F}_k]^{-1} \mathbf{d}_k + \mathbf{c}_k .
\end{aligned}$$

We can take norms of both sides to get

$$\begin{aligned}\|\widetilde{\mathbf{z}}_{k+1} - \mathbf{z}\| &\leq \left\|\mathbf{J_f}(\mathbf{z}_k)^{-1}\right\| \|\mathbf{f}(\mathbf{z}_k) - \mathbf{f}(\mathbf{z}) - \mathbf{J_f}(\mathbf{z}_k)\,(\mathbf{z}_k - \mathbf{z})\| \\ &+ \left\|[\mathbf{J_f}(\mathbf{z}_k) + \mathbf{E}_k + \mathbf{F}_k]^{-1} - \mathbf{J_f}(\mathbf{z}_k)^{-1}\right\| \|\mathbf{f}(\mathbf{z}_k) - \mathbf{f}(\mathbf{z})\| \\ &+ \left\|[\mathbf{J_f}(\mathbf{z}_k) + \mathbf{E}_k + \mathbf{F}_k]^{-1}\right\| \|\mathbf{d}_k\| + \|\mathbf{c}_k\|\end{aligned}$$

then we use Corollary 3.6.1 in Chap. 3 of Volume I to obtain

$$\begin{aligned}&\leq \left\|\mathbf{J_f}(\mathbf{z}_k)^{-1}\right\| \|\mathbf{f}(\mathbf{z}_k) - \mathbf{f}(\mathbf{z}) - \mathbf{J_f}(\mathbf{z}_k)\,(\mathbf{z}_k - \mathbf{z})\| \\ &+ \frac{\left\|\mathbf{J_f}(\mathbf{z}_k)^{-1}\right\| \left\|\mathbf{J_f}(\mathbf{z}_k)^{-1}[\mathbf{E}_k + \mathbf{F}_k]\right\|}{1 - \|\mathbf{J_f}(\mathbf{z}_k)^{-1}[\mathbf{E}_k + \mathbf{F}_k]\|} \|\mathbf{f}(\mathbf{z}_k) - \mathbf{f}(\mathbf{z})\| \\ &+ \frac{\left\|\mathbf{J_f}(\mathbf{z}_k)^{-1}\right\|}{1 - \|\mathbf{J_f}(\mathbf{z}_k)^{-1}[\mathbf{E}_k + \mathbf{F}_k]\|} \|\mathbf{d}_k\| + \|\mathbf{c}_k\|\end{aligned}$$

then Lemmas 3.2.5 and 3.2.3 give us

$$\begin{aligned}&\leq \left\|\mathbf{J_f}(\mathbf{z}_k)^{-1}\right\| \frac{\gamma}{2} \|\mathbf{z}_k - \mathbf{z}\|^2 \\ &+ \frac{\left\|\mathbf{J_f}(\mathbf{z}_k)^{-1}\right\| \left\|\mathbf{J_f}(\mathbf{z}_k)^{-1}[\mathbf{E}_k + \mathbf{F}_k]\right\|}{1 - \|\mathbf{J_f}(\mathbf{z}_k)^{-1}[\mathbf{E}_k + \mathbf{F}_k]\|} \int_0^1 \|\mathbf{J_f}\,(\mathbf{z}_k + [\mathbf{z} - \mathbf{z}_k]\,\tau)\;\mathrm{d}\tau\| \,\|\mathbf{z}_k - \mathbf{z}\| \\ &+ \frac{\left\|\mathbf{J_f}(\mathbf{z}_k)^{-1}\right\|}{1 - \|\mathbf{J_f}(\mathbf{z}_k)^{-1}[\mathbf{E}_k + \mathbf{F}_k]\|} \|\mathbf{d}_k\| + \|\mathbf{c}_k\| \,.\end{aligned}$$

Since

$$\begin{aligned}&\left\|\int_0^1 \mathbf{J_f}(\mathbf{z} + [\mathbf{z}_k - \mathbf{z}]\tau)\;\mathrm{d}\tau\right\| = \left\|\mathbf{J_f}(\mathbf{z}_k) + \int_0^1 \mathbf{J_f}\,(\mathbf{z}_k + [\mathbf{z} - \mathbf{z}_k](1 - \tau)) - \mathbf{J_f}(\mathbf{z}_k)\;\mathrm{d}\tau\right\| \\ &\leq \|\mathbf{J_f}(\mathbf{z}_k)\| + \int_0^1 \gamma\|\mathbf{z}_k - \mathbf{z}\|(1 - \tau)\;\mathrm{d}\tau = \|\mathbf{J_f}(\mathbf{z}_k)\| + \frac{\gamma}{2}\|\mathbf{z}_k - \mathbf{z}\| \,.\end{aligned}$$

The claimed results now follow easily.

Let us examine the implications of Lemma 3.4.1. If we evaluate the Jacobian analytically, then the numerical Jacobian typically satisfies

$$\left\|\widetilde{\mathbf{J}}_k - \mathbf{J_f}(\mathbf{z}_k)\right\|_1 = \|\,\mathrm{fl}\,(\mathbf{J_f}(\mathbf{z}_k)) - \mathbf{J_f}(\mathbf{z}_k)\|_1 \leq \|\mathbf{J_f}(\mathbf{z}_k)\|_1\,\varepsilon \,,$$

where ε is on the order of machine precision. On the other hand, if we use finite differences to approximate the Jacobian, then the numerical Jacobian has jth column

$$\widetilde{\mathbf{J}}_k\mathbf{e}_j = \mathrm{fl}\left(\left[\mathbf{f}(\mathbf{z}_k + \mathbf{e}_j\zeta) - \mathbf{f}(\mathbf{z}_k)\right]/\zeta\right) = \left[(\mathbf{I} + \mathbf{D}_+)\mathbf{f}(\mathbf{z}_k + \mathbf{e}_j\zeta) - (\mathbf{I} + \mathbf{D})\mathbf{f}(\mathbf{z}_k)\right](1 + \varepsilon_\div)/\zeta \,,$$

so

$$\left\|\widetilde{\mathbf{J}}_k\mathbf{e}_j - \mathbf{J}_\mathbf{f}(\mathbf{z}_k)\mathbf{e}_j\right\|_1 \leq \left\|\left[\mathbf{f}(\mathbf{z}_k + \mathbf{e}_j\zeta) - \mathbf{f}(\mathbf{z}_k) - \mathbf{J}_\mathbf{f}(\mathbf{z}_k)\mathbf{e}_j\zeta\right]/\zeta\right\|_1$$
$$+ \left\|\mathbf{D}_+\mathbf{f}(\mathbf{z}_k + \mathbf{e}_j\zeta)/\zeta\right\|_1 + \|\mathbf{D}\mathbf{f}(\mathbf{z}_k)/\zeta\|_1 + \varepsilon\left\|(\mathbf{I}+\mathbf{D}_+)\mathbf{f}(\mathbf{z}_k + \mathbf{e}_j\zeta)/\zeta\right\|_1 + \varepsilon\,\|(\mathbf{I}+\mathbf{D})\mathbf{f}(\mathbf{z}_k)/\zeta\|_1$$

then Lemma 3.2.5 and Lemma 3.8.1 in Chap. 3 of Volume I give us

$$\leq \frac{\gamma\zeta}{2} + \tau_\mathbf{f}(\mathbf{z}_k + \mathbf{e}_j\zeta)\frac{\varepsilon}{\zeta} + \tau_\mathbf{f}(\mathbf{z}_k)\frac{\varepsilon}{\zeta} + \tau_\mathbf{f}(\mathbf{z}_k + \mathbf{e}_j\zeta)\frac{\varepsilon}{\zeta(1-\varepsilon)} + \tau_\mathbf{f}(\mathbf{z}_k)\frac{\varepsilon}{\zeta(1-\varepsilon)}$$
$$\leq \frac{\gamma\zeta}{2} + \tau_\mathbf{f}(\mathbf{z}_k + \mathbf{e}_j\zeta)\frac{2\varepsilon}{\zeta(1-\varepsilon)} + \tau_\mathbf{f}(\mathbf{z}_k)\frac{2\varepsilon}{\zeta(1-\varepsilon)} \,.$$

Here $\tau_\mathbf{f}(\mathbf{x})$ represents the size of the terms involved in computing $\mathbf{f}$ at $\mathbf{x}$, and $\mathbf{D}_+$ and $\mathbf{D}$ are diagonal matrices with diagonal entries on the order of the machine precision ε. In summary, the numerical Jacobian satisfies

$$\widetilde{\mathbf{J}}_k = \mathbf{J}_\mathbf{f}(\mathbf{z}_k) + \mathbf{E}_k \,,$$

where

$$\|\mathbf{E}_k\|_1 \approx \begin{cases} \varepsilon\|\mathbf{J}_\mathbf{f}(\mathbf{z}_k)\|_1, & \text{analytical Jacobian} \\ \frac{\gamma\zeta}{2} + \frac{2\varepsilon}{\zeta(1-\varepsilon)}\left[\tau_\mathbf{f}(\mathbf{z}_k) + \max_{1\leq j\leq n}\left\{\tau_\mathbf{f}(\mathbf{z}_k + \mathbf{e}_j\zeta)\right\}\right], & \text{finite difference Jacobian} \end{cases} . \tag{3.18}$$

For the finite difference Jacobian, the term involving the Jacobian Lipschitz constant γ represents the error due to approximating the Jacobian by finite differences, and the second term estimates the rounding error.

Next, numerical evaluation of the Newton step produces

$$\left(\widetilde{\mathbf{J}}_k + \mathbf{F}_k\right)\widetilde{\mathbf{s}}_k = -\mathbf{f}(\mathbf{z}_k) + \mathbf{d}_k \text{ where } \|\mathbf{d}_k\| \leq \tau_\mathbf{f}(\mathbf{z}_k)\varepsilon \,.$$

If we use Gaussian factorization to solve the linear system, then the Wilkinson's Backward Error Analysis Theorem 3.8.2 in Chap. 3 of Volume I shows that

$$\|\mathbf{F}_k\|_2 \approx \frac{g_n n^2 \varepsilon}{1 - g_n n^2 \varepsilon}\left\|\widetilde{\mathbf{J}}_k\right\|_2 \leq \frac{g_n n^2 \varepsilon}{1 - g_n n^2 \varepsilon}\left(\|\mathbf{J}_\mathbf{f}(\mathbf{z}_k)\|_2 + \|\mathbf{E}_k\|_2\right) \,.$$

where g_n is the growth rate for pivots in Gaussian factorization. Finally, we update

$$\widetilde{\mathbf{x}}_{k+1} = \text{fl}(\mathbf{z}_k + \widetilde{\mathbf{s}}_k) = \mathbf{z}_k + \widetilde{\mathbf{s}}_k + \mathbf{c}_k$$

where

$$\|\mathbf{c}_k\| \leq \|\mathbf{z}_k + \widetilde{\mathbf{s}}_k\|\varepsilon \,.$$

It may be helpful to remember the forms of these rounding error terms as we discuss the next topic.

3.4.5 Maximum Attainable Accuracy

In order to choose reasonable termination criteria for a Newton iteration, we would like to understand the circumstances surrounding the maximum attainable accuracy of this method. The following lemma provides conditions under which we can guarantee that the error in the Newton iterates will decrease from one iteration to the next. If we cannot decrease the error in the solution because of the presence of rounding errors, then we have reached the **maximum attainable accuracy** of the algorithm.

Lemma 3.4.2 *Suppose that $\mathscr{D}$ is an open convex subset of real n-vectors, and that the mapping* $\mathbf{f}$ *from $\mathscr{D}$ to real n-vectors is continuously differentiable. Let $\|\cdot\|$ represent either a norm on n-vectors, or the corresponding subordinate norm on $n \times n$ matrices. Assume that there is a point $\mathbf{z} \in \mathscr{D}$ such that $\mathbf{f}(\mathbf{z}) = \mathbf{0}$. Suppose that the Jacobian $\mathbf{J_f}$ is nonsingular and Lipschitz continuous at $\mathbf{z}_k \in \mathscr{D}$, with Lipschitz constant γ. Assume that a numerical implementation of Newton's method computes an approximate function value $\widetilde{\mathbf{f}}_k = \mathbf{f}(\mathbf{z}_k) + \mathbf{d}_k$, an approximate Jacobian $\widetilde{\mathbf{J}}_k = \mathbf{J_f}(\mathbf{z}_k) + \mathbf{E}_k$, an approximate Newton step $\widetilde{\mathbf{s}}_k$ satisfying $\left(\widetilde{\mathbf{J}}_k + \mathbf{F}_k\right)\widetilde{\mathbf{s}}_k = -\widetilde{\mathbf{f}}_k$ and a subsequent Newton iterate $\widetilde{\mathbf{z}}_{k+1} = \mathbf{z}_k + \widetilde{\mathbf{s}}_k + \mathbf{c}_k$. Also assume that*

$$\left\|\mathbf{J_f}(\mathbf{z}_k)^{-1}\left[\mathbf{E}_k + \mathbf{F}_k\right]\right\| < \frac{1}{1 + \kappa\left(\mathbf{J_f}(\mathbf{z}_k)\right)} .$$

Define α, β and δ as in equations (3.17b) *through* (3.17d), *and assume that the rounding errors $\mathbf{c}_k$ and $\mathbf{d}_k$ are relatively small, meaning that*

$$\beta^2 - 4\alpha\delta > 0 .$$

If

$$\frac{2\delta}{\beta + \sqrt{\beta^2 - 4\alpha\delta}} < \|\mathbf{z}_k - \mathbf{z}\| < \frac{\beta + \sqrt{\beta^2 - 4\alpha\delta}}{2\alpha} , \tag{3.19}$$

then the subsequent Newton iterate $\widetilde{\mathbf{z}}_{k+1}$ exists and $\|\widetilde{\mathbf{z}}_{k+1} - \mathbf{z}\| < \|\widetilde{\mathbf{z}}_k - \mathbf{z}\|$.

Proof Since the assumptions of this lemma are stronger than the assumptions of Lemma 3.4.1, the latter shows that $\widetilde{\mathbf{z}}_{k+1}$ exists and satisfies

$$\|\widetilde{\mathbf{z}}_{k+1} - \mathbf{z}\| \le \alpha\|\mathbf{z}_k - \mathbf{z}\|^2 + (1 - \beta)\|\mathbf{z}_k - \mathbf{z}\| + \delta .$$

It is straightforward to check that $\min\{\alpha, \beta, \delta\} > 0$. Finally, if (3.19) is satisfied, then

$$\alpha\|\mathbf{z}_k - \mathbf{z}\|^2 - \beta\|\mathbf{z}_k - \mathbf{z}\| + \delta$$
$$= \alpha\left[\|\mathbf{z}_k - \mathbf{z}\| - \frac{2\delta}{\beta + \sqrt{\beta^2 - 4\alpha\delta}}\right]\left[\|\mathbf{z}_k - \mathbf{z}\| - \frac{\beta + \sqrt{\beta^2 - 4\alpha\delta}}{2\alpha}\right] < 0 .$$

This implies that

$$\|\widetilde{\mathbf{z}}_{k+1} - \mathbf{z}\| \le \alpha\|\mathbf{z}_k - \mathbf{z}\|^2 + (1-\beta)\|\mathbf{z}_k - \mathbf{z}\| + \delta < \|\mathbf{z}_k - \mathbf{z}\| .$$

Next, we would like to discuss the conditions under which the inequalities (3.19) are satisfied, either for direct evaluation of the Jacobian matrix, or for approximation of this matrix by finite differences. In this discussion, we will use the rounding error estimates from the end of Sect. 3.4.4.

3.4.5.1 Analytical Jacobian

If we are able to evaluate the Jacobian analytically, then the rounding errors in evaluating the Jacobian and solving a linear system involving the approximate Jacobian satisfy

$$\|\mathbf{E}_k\| \le \|\mathbf{J_f}(\mathbf{z}_k)\|\varepsilon \text{ and}$$
$$\|\mathbf{F}_k\| \le \frac{g_n n^2 \varepsilon}{1 - g_n n^2 \varepsilon}\left(\|\mathbf{J_f}(\mathbf{z}_k)\| + \|\mathbf{E}_k\|\right) .$$

These terms are on the order of the machine precision ε, so in most cases we will have

$$\left\|\mathbf{J_f}(\mathbf{z}_k)^{-1}\left[\mathbf{E}_k + \mathbf{F}_k\right]\right\| \ll 1 .$$

As a result, Eqs. (3.17b)–(3.17d) give us

$$\alpha \approx \frac{\gamma}{2}\|\mathbf{J_f}(\mathbf{z}_k)^{-1}\|$$
$$\beta \approx 1 \text{ and}$$
$$\delta \approx \|\mathbf{J_f}(\mathbf{z}_k)^{-1}\|\|\mathbf{d}_k\| + \|\mathbf{c}_k\| \le \varepsilon\left(\|\mathbf{J_f}(\mathbf{z}_k)^{-1}\|\tau_{\mathbf{f}}(\mathbf{z}_k) + \|\mathbf{z}_k + \widetilde{\mathbf{s}}_k\|\right) .$$

Consequently, the discriminant $\beta^2 - 4\alpha\delta$ is essentially one. Inequality (3.19) now shows that the error in Newton's method can decrease provided that

$$\delta \approx \varepsilon\left\{\|\mathbf{J_f}(\mathbf{z}_k)^{-1}\|\tau_{\mathbf{f}}(\mathbf{z}_k) + \|\mathbf{z}_k + \widetilde{\mathbf{s}}_k\|\right\} < \|\mathbf{z}_k - \mathbf{z}\| < \frac{2}{\gamma\|\mathbf{J_f}(\mathbf{z}_k)^{-1}\|} \approx \frac{1}{\alpha} .$$

This right-hand bound on $\mathbf{z}_k - \mathbf{z}$ is like a Kantorovich restriction on the initial guess, and the left-hand bound represents the **maximum attainable accuracy**.

3.4.5.2 Finite Differences

If we approximate the Jacobian by finite differences, then inequality (3.18) shows that we cannot guarantee that $\|\mathbf{E}_k\|_1$ is small. In fact, we expect this error to become large if the size ζ of the finite difference increment approaches machine precision ε.

Recall the inequalities in the collection (3.17). The condition that $\beta > 0$ is equivalent to

$$\left\|\mathbf{J_f}(\mathbf{z}_k)^{-1}\left[\mathbf{E}_k + \mathbf{F}_k\right]\right\|_1 < \frac{1}{1 + \kappa_1\left(\mathbf{J_f}(\mathbf{z}_k)\right)} .$$

This will be satisfied if

$$\|\mathbf{E}_k\|_1 + \|\mathbf{F}_k\|_1 < \frac{1}{\|\mathbf{J_f}(\mathbf{z}_k)^{-1}\|_1} \frac{1}{1 + \kappa_1\left(\mathbf{J_f}(\mathbf{z}_k)\right)} .$$

We can use the bound on $\|\mathbf{F}_k\|$ from Gaussian factorization to obtain

$$\|\mathbf{E}_k\|_1 + \frac{g_n n^2 \varepsilon}{1 - g_n n^2 \varepsilon}\left\{\|\mathbf{J_f}(\mathbf{z}_k)\|_1 + \|\mathbf{E}_k\|_1\right\} < \frac{1}{\|\mathbf{J_f}(\mathbf{z}_k)^{-1}\|_1} \frac{1}{1 + \kappa_1\left(\mathbf{J_f}(\mathbf{z}_k)\right)} .$$

Here we have assumed that the number of nonlinear equations n is small enough that $g_n n^2 \varepsilon < 1$, where g_n is the growth factor in Gaussian factorization and ε is machine precision. Using inequality (3.18) to bound $\mathbf{E}_k$ for a finite difference Jacobian approximation leads to

$$\begin{aligned}
&\frac{\gamma\zeta}{2} + \frac{2\varepsilon}{\zeta(1-\varepsilon)}\left[\tau_\mathbf{f}(\mathbf{z}_k) + \max_{1 \le j \le n}\left\{\tau_\mathbf{f}(\mathbf{z}_k + \mathbf{e}_j\zeta)\right\}\right] \\
&< \frac{1}{\|\mathbf{J_f}(\mathbf{z}_k)^{-1}\|_1} \frac{1 - g_n n^2 \varepsilon}{1 + \kappa_1\left(\mathbf{J_f}(\mathbf{z}_k)\right)} - g_n n^2 \varepsilon \|\mathbf{J_f}(\mathbf{z}_k)\|_1 .
\end{aligned}$$

If we ignore terms involving ε compared to terms of order one, we approximately have

$$\frac{\gamma\zeta}{2} + \frac{2\varepsilon}{\zeta}\left[\tau_\mathbf{f}(\mathbf{z}_k) + \max_{1 \le j \le n}\left\{\tau_\mathbf{f}(\mathbf{z}_k + \mathbf{e}_j\zeta)\right\}\right] < \frac{1}{\|\mathbf{J_f}(\mathbf{z}_k)^{-1}\|_1} \frac{1}{1 + \kappa_1\left(\mathbf{J_f}(\mathbf{z}_k)\right)} .$$

This is a quadratic inequality in ζ, with approximate solution (again ignoring terms that are small in ε)

$$\begin{aligned}
&2\varepsilon\left[\tau_\mathbf{f}(\mathbf{z}_k) + \max_{1 \le j \le n}\left\{\tau_\mathbf{f}(\mathbf{z}_k + \mathbf{e}_j\zeta)\right\}\right]\left\|\mathbf{J_f}(\mathbf{z}_k)^{-1}\right\|_1 \left(1 + \kappa_1\left(\mathbf{J_f}(\mathbf{z}_k)\right)\right) < \zeta \\
&< \frac{2}{\gamma\|\mathbf{J_f}(\mathbf{z}_k)^{-1}\|_1 \left(1 + \kappa_1\left(\mathbf{J_f}(\mathbf{z}_k)\right)\right)} .
\end{aligned} \tag{3.20}$$

In other words, we do not expect $\beta > 0$, and thus cannot expect a decrease in the error in the computed Newton $\mathbf{z}_k$, whenever the finite difference increment ζ violates these inequalities. Unless the Jacobian $\mathbf{J_f}(\mathbf{z}_k)$ is poorly conditioned, there should be a lot of room here to choose the finite difference increment ζ.

In order to discuss the **maximum attainable accuracy** for Newton's method with finite difference approximation of the Jacobian, we will consider two cases. First, suppose that

$$\varepsilon \ll \zeta \ll 1 \, .$$

Then Eqs. (3.17b)–(3.17d) give us

$$\begin{aligned}
\alpha &\approx \frac{\gamma}{2} \left\| \mathbf{J_f}(\mathbf{z}_k)^{-1} \right\|_1 \\
\beta &\approx 1 \text{ and} \\
\delta &\approx \left\| \mathbf{J_f}(\mathbf{z}_k)^{-1} \right\|_1 \tau_{\mathbf{f}}(\mathbf{z}_k)\varepsilon + \| \mathbf{z}_k + \widetilde{\mathbf{s}}_k \|_1 \varepsilon \, .
\end{aligned}$$

Under these circumstances, inequality (3.19) for the solution of the quadratic inequality controlling the error in the solution gives us

$$\left\| \mathbf{J_f}(\mathbf{z}_k)^{-1} \right\|_1 \tau_{\mathbf{f}}(\mathbf{z}_k)\varepsilon + \| \mathbf{z}_k + \widetilde{\mathbf{s}}_k \|_1 \varepsilon < \| \mathbf{z}_k - \mathbf{z} \|_1 < \frac{2}{\gamma \, \| \mathbf{J_f}(\mathbf{z}_k)^{-1} \|_1} \, .$$

These bounds are the same as those we found for analytical evaluation of the Jacobian.

On the other hand, suppose that we choose ζ to be a multiple of the lower bound in inequality (3.20):

$$\zeta = \mu \varepsilon \left\{ 1 + \kappa \left(\mathbf{J_f}(\mathbf{z}_k) \right) \right\} \left\| \mathbf{J_f}(\mathbf{z}_k)^{-1} \right\|_1 \tau_{\mathbf{f}}(\mathbf{z}_k) \, .$$

Here we assume that $\mu \geq 2$. Then we have

$$\begin{aligned}
\| \mathbf{E}_k \|_1 &\approx \frac{2}{\mu \left\{ 1 + \kappa \left(\mathbf{J_f}(\mathbf{z}_k) \right) \right\} \| \mathbf{J_f}(\mathbf{z}_k)^{-1} \|_1} \\
\alpha &\approx \frac{\gamma \mu \left\{ 1 + \kappa \left(\mathbf{J_f}(\mathbf{z}_k) \right) \right\}}{2\mu \left\{ 1 + \kappa \left(\mathbf{J_f}(\mathbf{z}_k) \right) \right\} - 4} \left\| \mathbf{J_f}(\mathbf{z}_k)^{-1} \right\|_1 , \\
\beta &\approx \frac{\mu \left\{ 1 + \kappa \left(\mathbf{J_f}(\mathbf{z}_k) \right) \right\}}{\mu \left\{ 1 + \kappa \left(\mathbf{J_f}(\mathbf{z}_k) \right) \right\} - 2} \text{ and} \\
\delta &\approx \frac{\mu \left\{ 1 + \kappa \left(\mathbf{J_f}(\mathbf{z}_k) \right) \right\}}{\mu \left\{ 1 + \kappa \left(\mathbf{J_f}(\mathbf{z}_k) \right) \right\} - 2} \left\| \mathbf{J_f}(\mathbf{z}_k)^{-1} \right\|_1 \tau_{\mathbf{f}}(\mathbf{z}_k)\varepsilon + \| \mathbf{z}_k + \widetilde{\mathbf{s}}_k \|_1 \varepsilon \, .
\end{aligned}$$

Under these circumstances, inequality (3.19) for the solution of the quadratic inequality controlling the error in the solution gives us

$$\left\|\mathbf{J_f}(\mathbf{z}_k)^{-1}\right\|_1 \tau_{\mathbf{f}}(\mathbf{z}_k)\varepsilon + \left[1 - \frac{2}{\mu(1+\kappa_1(\mathbf{J_f}(\mathbf{z}_k)))}\right] \|\mathbf{z}_k + \widetilde{\mathbf{s}}_k\|_1 \varepsilon \leq \|\mathbf{z}_k - \mathbf{z}\|_1 \leq \frac{1}{\gamma \left\|\mathbf{J_f}(\mathbf{z}_k)^{-1}\right\|_1}$$

In short, choosing the finite difference Jacobian increment ζ too large produces poor Jacobian approximations, and choosing it too small produces overwhelming rounding errors. A reasonable compromise is to take the increment ζ to be the geometric mean of the bounds in inequalities (3.20), which would give us

$$\zeta = \sqrt{\frac{2\varepsilon}{\gamma} \tau_{\mathbf{f}}(\mathbf{z}_k)} \; .$$

The difficulty with this choice is that the Lipschitz constant γ is not typically known, especially when direct computation of the Jacobian member functions is unavailable. Dennis and Schnabel [52, p. 97] suggest that in computing $\mathbf{J_f}(\mathbf{z}_k)\mathbf{e}_j$, we should use the increment

$$\zeta_j = \max\{|\mathbf{e}_j \cdot \mathbf{z}_k| \; , \; \tau_{\mathbf{f}}(\mathbf{z}_k)\} \text{sign}(\mathbf{e}_j \cdot \mathbf{z}_k) \sqrt{\tau_{\mathbf{f}}(\mathbf{z}_k)\varepsilon} \; .$$

In practice, there is no noticeable difference in performance between using difference approximations computed in this way and actual derivatives. Note that each step of the finite-difference version of Newton's method requires $n+1$ function evaluations, plus the solution of an $n \times n$ system of linear equations. However, we avoid programming the evaluation of the matrix of derivatives of $\mathbf{f}$, and the risk of an error in developing that code.

3.4.6 Iterative Linear Solvers

For large systems of nonlinear equations, it is natural to use an iterative method to solve the linear system for the Newton step. This means that we have two nested iterations: an outer iteration over Newton steps and an inner iteration for the linear solver.

When $\mathbf{f}(\mathbf{z}_k)$ is large, we have two issues to resolve. First, if we are far from the solution of the nonlinear system of equations, then there is not much need to solve the linear system for the Newton step $\mathbf{s}_k$ to high accuracy. Dembo et al. [47] have suggested some possible strategies, provided that we guarantee convergence of the iteration:

1. If the number of iterations in the linear system solver is chosen so that the residual in the linear system has norm bounded by some positive fraction times the norm of $\mathbf{f}(\mathbf{z}_k)$, then the approximate Newton iteration will be linearly convergent.

2. If the linear system residual has norm $\|\mathbf{r}^{(k)}\| \leq \eta_k \|\mathbf{f}(\mathbf{z}_k)\|_2$ for some sequence $\eta_k \to 0$ as $k \to \infty$, then the approximate Newton iteration will be superlinearly convergent.
3. If the norm of the linear system residual is bounded by a positive constant times $\|\mathbf{f}(\mathbf{z}_k)\|^2$, then the approximate Newton iteration will be quadratically convergent.

Note that these authors have imposed additional assumptions regarding the continuity of the Jacobian of **f**.

3.4.7 Software

There are several software packages for employ Newton's method to find zeros of systems of nonlinear equations. In the GSL (GNU Scientific Library), readers should consider using gsl_multiroot_fdfsolver_newton to perform Newton's method with analytical evaluation of the Jacobian, or gsl_multiroot_fsolver_dnewton to perform Newton's method with finite difference approximation of the Jacobian. The MINPACK package provides the routine hybrj1 to solve a system of nonlinear equations by evaluating Jacobians analytically, and the routine hybrd to solve a system of nonlinear equations by using Newton's method with the finite difference approximation of the Jacobian. Both `hybrj1` and `hybrd` are written in `FORTRAN`, and both use global convergence strategies to converge from arbitrary initial guesses. C or C^{++} programmers might also consider SNES in the PETSC toolkit.

Although MATLAB does not provide a procedure of its own for solving systems of nonlinear equations, Tim Kelley [107] has written the MATLAB package nsol. Kelley's nonlinear system solver uses finite differences to compute the Jacobian.

Exercise 3.4.1 Determine the rate of convergence of Newton's method when solving each of the following equations:

1. $f(x) = x^2$.
2. $f(x) = x^3$.
3. $f(x) = x + x^3$.

Exercise 3.4.2 For each of the functions $f_1(x) = x$, $f_2(x) = x^2 + x$ and $f_3(x) = e^x - 1$, answer the following questions:

1. What is the zero $\mathbf{z}$ and what is $f'(\mathbf{z})$?
2. What is a Lipschitz constant for f' in the interval $[-a, a]$?
3. What region of convergence does the Kantorovich theorem predict for Newton's method?
4. In what interval $b < 0 < c$ is Newton's method convergent?

Exercise 3.4.3 Using your answers to the previous exercise, consider applying Newton's method to

$$\mathbf{f}(\mathbf{x}) = \begin{bmatrix} \mathbf{x}_1 \\ \mathbf{x}_2^2 + \mathbf{x}_2 \\ e^{\mathbf{x}_3} - 1 \end{bmatrix} .$$

1. What is the zero $\mathbf{z}$ and what is $\mathbf{J})\mathbf{f}(\mathbf{z})$?
2. What is a Lipschitz constant on $\mathbf{J})\mathbf{f}(\mathbf{x})$ in a neighborhood of radius a around $\mathbf{z}$?
3. What region of convergence does the Kantorovich theorem predict?
4. What would be the region of convergence if $\mathbf{e}_3^\top \mathbf{x}_0 = 0$?
5. What would be the region of convergence if $\mathbf{e}_3^\top \mathbf{x}_0 = 0 = \mathbf{e}_2^\top \mathbf{x}_0$?

Exercise 3.4.4 Suppose that

$$\mathbf{f}(\mathbf{x}) = \begin{bmatrix} \mathbf{x}_1^2 + \mathbf{x}_2^2 - 2 \\ \frac{e}{\mathbf{x}_1} + \mathbf{x}_2^3 - 2 \end{bmatrix} .$$

Program Newton's method with analytical and finite difference derivatives to solve this problem. Given the initial condition $\mathbf{x}^\top = [2\ 3]$, run your code and compare the two methods for order of accuracy and efficiency:

1. Plot $\log \|\mathbf{x}_k - \mathbf{z}\|$ versus k for each method.
2. Add a loop to repeat each iteration step 10^6 times and plot the log of the error versus computer time.

Exercise 3.4.5 Recall that in Sect. 2.3.1.4 in Chap. 2 of Volume I we used Newton's method to compute the reciprocal of a number using only multiplication and addition. Consider the function $\mathbf{F}(\mathbf{X}) = \mathbf{I} - \mathbf{A}\mathbf{X}$. At each iteration, $\mathbf{X}$ is an approximation to $\mathbf{A}^{-1}$. The Newton iteration is $\mathbf{X} \leftarrow \mathbf{X} + \mathbf{A}^{-1}(\mathbf{I} - \mathbf{A}\mathbf{X})$, which can be approximated by the iteration $\mathbf{X} \leftarrow \mathbf{X} + \mathbf{X}(\mathbf{I} - \mathbf{A}\mathbf{X})$.

1. If we define the error $E(\mathbf{X}) = \mathbf{A}^{-1} - \mathbf{X}$, show that $\mathbf{F}(\mathbf{X}_{k+1}) = F(\mathbf{X}_k)^2$ and $E_{k+1} = E_k \mathbf{A} E_k$, and prove that the algorithm converges quadratically.
2. Determine the cost per iteration of this Newton method to find the inverse of a matrix. How many iterations would be needed to exceed the cost of determining the inverse by Gaussian factorization?
3. Write a computer program to compute the inverse using this iteration. Take $\mathbf{A}^\top / (\|\mathbf{A}\|_1 \|\mathbf{A}\|_\infty)$ as your initial guess.

Exercise 3.4.6 Analyze the iteration

$$\mathbf{S}_0 = \mathbf{A}$$

$$\text{for } 0 \le k \ \mathbf{S}_{k+1} = (\mathbf{S}_k + \mathbf{S}_k^{-1}\mathbf{A})/2$$

for computing the square root of an invertible matrix **A**. This iteration was presented in Sect. 1.7.3.

3.5 Unconstrained Minimization

Next, we will turn our attention to the related problem of finding a minimum of a nonlinear functional. We assume that we are given a real-valued function ϕ of n real variables, and we want to find a local minimizer $\mathbf{z}$ of ϕ. In other words, we want $\mathbf{z}$ to be such that there is a $\varrho > 0$ so that for all n-vectors $\mathbf{x}$ with $\|\mathbf{x} - \mathbf{z}\| < \varrho$ we have $\phi(\mathbf{z}) \leq \phi(\mathbf{x})$.

Recall that we developed necessary and sufficient conditions for the existence of a local minimizer in Sect. 3.2.4. In particular, we found that $\mathbf{g}_\phi(\mathbf{z}) = \mathbf{0}$ is a necessary condition for a local minimizer. This implies that we can use methods designed to solve nonlinear systems for finding local minima. We also saw that if $\mathbf{g}_\phi(\mathbf{z}) = \mathbf{0}$ and $\mathbf{H}_\phi(\mathbf{z})$ is positive, then $\mathbf{z}$ is a local minimizer of ϕ. This implies that minimization problems have extra properties that we might want to exploit in a numerical method.

3.5.1 Descent Directions

Let us begin by constructing a local model for ϕ. If ϕ is continuously differentiable, then we can approximate

$$\phi(\mathbf{x}) \approx \widetilde{\phi}(\mathbf{x}) \equiv \phi(\mathbf{z}_k) + \mathbf{g}_\phi(\mathbf{z}_k) \cdot (\mathbf{x} - \mathbf{z}_k) ,$$

where $\mathbf{g}_\phi$ is the gradient of ϕ. Unfortunately, this local model has no minimum. However, if $\mathbf{g}_\phi(\mathbf{z}_k) \neq 0$, this local model does suggest that in order to decrease ϕ as rapidly as possible, we could choose our new guess $\mathbf{z}_{k+1}$ to be in the direction of **steepest descent**:

$$\mathbf{z}_{k+1} = \mathbf{z}_k + \mathbf{s}_k \lambda \text{ where } \mathbf{s}_k = -\mathbf{g}_\phi(\mathbf{z}_k) .$$

One difficulty is that this local model does not provide a good estimate of step length λ.

We can, however, generalize the choice of step direction with the following definition.

Definition 3.5.1 Let the real functional ϕ be continuously differentiable, with gradient $\mathbf{g}_\phi$. Then for any point $\mathbf{x}$ where $\mathbf{g}_\phi(\mathbf{x})$ is defined, a vector $\mathbf{s}$ satisfying

$$\mathbf{g}_\phi(\mathbf{x}) \cdot \mathbf{s} < 0$$

is called a **descent direction** for ϕ at $\mathbf{x}$. A **critical point** of ϕ is a point $\mathbf{x}$ where

$$\mathbf{g}_\phi(\mathbf{x}) = \mathbf{0} .$$

The next lemma demonstrates why descent directions are useful.

Lemma 3.5.1 *Suppose that $\mathscr{D}$ is an open subset of n-vectors, and that the real functional ϕ is continuously differentiable on $\mathscr{D}$. If $\mathbf{x} \in \mathscr{D}$, $\mathbf{g}_\phi(\mathbf{x}) \neq \mathbf{0}$ and $\mathbf{s}$ is a descent direction for ϕ at $\mathbf{x}$, then there is a step length $\lambda > 0$ so that $\phi(\mathbf{x} + \mathbf{s}\lambda) < \phi(\mathbf{x})$.*

Proof Since $\mathscr{D}$ is open and $\mathbf{x} \in \mathscr{D}$, there is a scalar $\varrho > 0$ so that for all $\sigma \in [0, \varrho)$ we have $\mathbf{x} + \mathbf{s}\sigma \in \mathscr{D}$. Let

$$\varepsilon = \min\left\{\varrho \ , \ -\frac{\mathbf{g}_\phi(\mathbf{x}) \cdot \mathbf{s}}{2\|\mathbf{s}\|_2}\right\} .$$

Since $\mathbf{s}$ is a descent direction for ϕ at $\mathbf{x}$, we must have that $\varepsilon > 0$. Since $\mathbf{g}_\phi$ is continuous at $\mathbf{x}$, there exists $\delta > 0$ so that for all $\lambda \in (0 \ , \ \delta)$ we have

$$\left\|\mathbf{g}_\phi(\mathbf{x} + \mathbf{s}\lambda) - \mathbf{g}_\phi(\mathbf{x})\right\|_2 < \varepsilon .$$

Then Lemma 3.2.1 implies that

$$\phi(\mathbf{x} + \mathbf{s}\lambda) = \phi(\mathbf{x}) + \mathbf{g}_\phi(\mathbf{x}) \cdot \mathbf{s}\lambda + \int_0^1 \left[\mathbf{g}_\phi(\mathbf{x} + \mathbf{s}\lambda\tau) - \mathbf{g}_\phi(\mathbf{x})\right] \cdot \mathbf{s}\lambda \, d\tau$$

$$\leq \phi(\mathbf{x}) + \mathbf{g}_\phi(\mathbf{x}) \cdot \mathbf{s}\lambda + \int_0^1 \left\|\mathbf{g}_\phi(\mathbf{x} + \mathbf{s}\lambda\tau) - \mathbf{g}_\phi(\mathbf{x})\right\|_2 \|\mathbf{s}\|_2 \lambda \, d\tau$$

$$< \phi(\mathbf{x}) + \mathbf{g}_\phi(\mathbf{x}) \cdot \mathbf{s}\lambda + \varepsilon\|\mathbf{s}\|_2\lambda \leq \phi(\mathbf{x}) + \frac{1}{2}\mathbf{g}_\phi(\mathbf{x}) \cdot \mathbf{s}\lambda < \phi(\mathbf{x}) .$$

3.5.2 Newton's Method

For nonlinear minimization problems, it is more common to increase the complexity of the local model to include second-order derivatives. Our revised local model is

$$\widetilde{\phi}(\mathbf{x}) \equiv \phi(\mathbf{x}_k) + \mathbf{g}_{\phi(\mathbf{x}_k)} \cdot (\mathbf{x} - \mathbf{x}_k) + \frac{1}{2}(\mathbf{x} - \mathbf{x}_k) \cdot \mathbf{H}_\phi(\mathbf{x}_k)(\mathbf{x} - \mathbf{x}_k) .$$

Recall that for twice continuously differentiable functions ϕ, the Hessian matrix is always symmetric. If the Hessian matrix $\mathbf{H}_\phi(\mathbf{x}_k)$ is also nonsingular, then the critical point of the local model is of the form $\mathbf{x}_k + \mathbf{s}_k$, where $\mathbf{s}_k$ solves

$$\mathbf{H}_\phi(\mathbf{x}_k)\mathbf{s}_k = -\mathbf{g}_\phi(\mathbf{x}_k) . \tag{3.21}$$

If the Hessian matrix is also positive, then this critical point is the global minimizer of the local model $\widetilde{\phi}$. Choosing the next guess for a minimum of the original objective ϕ to be the critical point of the local model $\widetilde{\phi}$ is **Newton's method** for unconstrained minimization.

Note that Newton's method generates a descent direction whenever $\mathbf{H}_\phi(\mathbf{x}_k)$ is positive definite. On the other hand, if $\mathbf{H}_\phi(\mathbf{x}_k)$ has any negative eigenvalues, then step vectors $\mathbf{s}_k$ satisfying (3.21) are not necessarily descent directions.

3.5.2.1 Finite Differences

In order to perform Newton's method for unconstrained minimization, we need to evaluate both the gradient and the Hessian at the current approximation to the local minimizer. If we provide routines to compute both of these analytically, we must take care to avoid errors that could destroy the performance of the algorithm Generally, it is far easier to provide routines for analytical evaluation of the objective function, and possibly its gradient, but not for the Hessian.

If the gradient $\mathbf{g}_\phi(\mathbf{x})$ is available analytically, then there is a straightforward approximation to the Hessian matrix $\mathbf{H}_\phi(\mathbf{x})$ by difference quotients. First, we compute entries of the square matrix $\mathbf{H}$ by

$$\mathbf{H}_{ij} = \frac{\mathbf{e}_i \cdot \mathbf{g}_\phi(\mathbf{x} + \mathbf{e}_j \xi_j) - \mathbf{e}_i \cdot \mathbf{g}_\phi(\mathbf{x})}{\xi_j}$$

where ξ_j is computed as described near the end of Sect. 3.4.5.2. Then we approximate the Hessian $\mathbf{H}_\phi(\mathbf{x})$ by

$$\mathbf{H}_\phi(\mathbf{x}) \approx \left[\mathbf{H} + \mathbf{H}^\top\right]/2 \, .$$

The rounding errors due to this approximation are similar to the errors discussed in Sect. 3.4.3, so we will not discuss them further.

If the gradient $\mathbf{g}_\phi(\mathbf{x})$ is not available analytically, then we need to approximate both $\mathbf{g}_\phi(\mathbf{x})$ and $\mathbf{H}_\phi(\mathbf{x})$ by finite differences. The gradient could be approximated by first-order differences, similar to the approach in Sect. 3.4.3, for approximating the Jacobian of a nonlinear function. However, to approximate the Hessian matrix we need to form a second-order difference. For example,

$$\mathbf{e}_i \cdot \mathbf{H}_\phi(\mathbf{x})\mathbf{e}_j \approx \left[\frac{\phi(\mathbf{x} + \mathbf{e}_i\xi_i + \mathbf{e}_j\xi_j) - \phi(\mathbf{x} + \mathbf{e}_i\xi_i)}{\xi_j} - \frac{\phi(\mathbf{x} + \mathbf{e}_j\xi_j) - \phi(\mathbf{x})}{\xi_j}\right] / \xi_i$$

necessarily produces a symmetric Hessian approximation that has errors due both to the approximation and to rounding errors. It can be shown that

$$\left|\mathbf{e}_i{}^\top \mathbf{H}_\phi(\mathbf{x})\mathbf{e}_j - \left[\frac{\phi(\mathbf{x} + \mathbf{e}_i\xi_i + \mathbf{e}_j\xi_j) - \phi(\mathbf{x} + \mathbf{e}_i\xi_i)}{\xi_j} - \frac{\phi(\mathbf{x} + \mathbf{e}_j\xi_j) - \phi(\mathbf{x})}{\xi_j}\right] / \xi_i\right|$$
$$\leq \frac{5}{3}\gamma \max\{\xi_i, \xi_j\} + \frac{4\tau_\phi\varepsilon}{\xi_i\xi_j} \, ,$$

where γ is a Lipschitz constant for $\mathbf{g}_\phi$. Additional discussion of rounding errors in finite differences can be found in Sect. 2.2.2 in Chap. 2 of Volume III.

This estimate suggests that we choose the finite difference increments to equilibrate the approximation and rounding errors, by choosing the finite difference increment ξ to be on the order of the cube root of machine precision:

$$\xi_j = \max\{|\mathbf{x}_j|\ ,\ \tau_\mathbf{x}\}\text{sign}(\mathbf{x}_j)(\tau_\phi\varepsilon)^{1/3}\ .$$

This approach is sufficient for computing the Hessian by itself.

However, if the iteration converges at least linearly then we expect the gradient $\mathbf{g}_\phi(\mathbf{x})$ to be on the order of $\|\mathbf{x}-\mathbf{z}\|$ and therefore tend to zero as $\mathbf{x}$ tends to $\mathbf{z}$. This situation requires special accuracy from a finite difference approximation. In order to preserve quadratic convergence in the absence of rounding errors, we would require ξ_j to be on the order of $\|\mathbf{x}-\mathbf{z}\|^2$ for a forward-difference approximation to $\mathbf{g}_\phi$, such as

$$\mathbf{e}_j \cdot \mathbf{g}_\phi(\mathbf{x}) \approx \frac{\phi(\mathbf{x}+\mathbf{e}_j\xi_j)-\phi(\mathbf{x})}{\xi_j}\ ,$$

and on the order of $\|\mathbf{x}-\mathbf{z}\|$ for a centered difference approximation, such as

$$\mathbf{e}_j \cdot \mathbf{g}_\phi(\mathbf{x}) \approx \frac{\phi(\mathbf{x}+\mathbf{e}_j\xi_j)-\phi(\mathbf{x}-\mathbf{e}_j\xi_j)}{2\xi_j}\ .$$

It may not be easy to estimate the error $\|\mathbf{x}-\mathbf{z}\|$ under such circumstances, because we would probably have to use the finite difference approximation to the gradient to estimate $\|\mathbf{x}-\mathbf{z}\|$. The treatment of convergence criteria under such circumstances is somewhat delicate.

3.5.2.2 Maximum Attainable Accuracy

Let us examine the **maximum attainable accuracy** of Newton's method for minimization of a functional. If we use analytical evaluation of the gradient and Hessian, then the discussion in Sect. 3.4.5.1 will apply. If we use analytical evaluation of the gradient and finite difference approximation of the Hessian, then the discussion in Sect. 3.4.5.2 is appropriate. The remaining case occurs when finite differences are used to approximate both the gradient and the Hessian of the objective function.

We will assume that all rounding errors are bounded by a positive scalar ε. In particular, we will use the objective function floating point evaluations

$$\begin{aligned}
\widetilde{\phi_k} &= \phi(\mathbf{z}_k)(1+\delta_k)\ ,\\
\widetilde{\phi_{jk}} &= \phi(\mathbf{z}_k+\mathbf{e}_j\zeta_j)(1+\delta_{jk}) \text{ and}\\
\widetilde{\phi_{ijk}} &= \phi(\mathbf{z}_k+\mathbf{e}_i\zeta_i+\mathbf{e}_j\zeta_j)(1+\delta_{ijk}) \text{ where}\\
&|\delta_k|\ ,\ |\delta_{jk}|\ ,\ |\delta_{ijk}| \le \varepsilon\ .
\end{aligned}$$

We will also assume that the Hessian $\mathbf{H}_\phi$ is Lipschitz continuous at $\mathbf{z}_k$; in other words, there exists $\gamma > 0$ and $\varrho > 0$ so that for all $\|\mathbf{s}\| < \varrho$ we have

$$\left\|\mathbf{H}_\phi(\mathbf{z}_k + \mathbf{s}) - \mathbf{H}_\phi(\mathbf{z}_k)\right\| \leq \gamma \|\mathbf{s}\| \; .$$

It follows that

$$\left\|\mathbf{g}_\phi(\mathbf{z}_k + \mathbf{s}) - \mathbf{g}_\phi(\mathbf{z}_k)\right\| = \left\|\mathbf{H}_\phi(\mathbf{z}_k)\mathbf{s} + \int_0^1 \left\{\mathbf{H}_\phi(\mathbf{z}_k + \mathbf{s}\tau) - \mathbf{H}_\phi(\mathbf{z}_k)\right\} \mathbf{s} \, d\tau\right\|$$
$$\leq \left\|\mathbf{H}_\phi(\mathbf{z}_k)\right\| \; \|\mathbf{s}\| + \int_0^1 \left\|\mathbf{H}_\phi(\mathbf{z}_k + \mathbf{s}\tau) - \mathbf{H}_\phi(\mathbf{z}_k)\right\| \; \|\mathbf{s}\| \, d\tau = \left\|\mathbf{H}_\phi(\mathbf{z}_k)\right\| \; \|\mathbf{s}\| + \frac{1}{2}\gamma\|\mathbf{s}\|^2$$
$$\leq \left(\left\|\mathbf{H}_\phi(\mathbf{z}_k)\right\| + \frac{\gamma\varrho}{2}\right) \|\mathbf{s}\| \; .$$

For simplicity, we will assume that all finite differences use the same increment $\zeta > 0$. Finally, we will assume that the norm is chosen so that

$$\|\mathbf{e}_i\| = 1 \text{ and } \|\mathbf{e}_i\mathbf{e}_j{}^\top\| = 1$$

for all $1 \leq i,j \leq n$. These are true for the vector 1, 2 and ∞ norms.

The error in the approximation of the gradient by finite differences is

$$\mathbf{d}_k = \sum_{i=1}^n \mathbf{e}_i \frac{\widetilde{\phi}_{ik} - \widetilde{\phi}_k}{\zeta}(1 + \varepsilon_{ik-})(1 + \varepsilon_{ik\div}) - \mathbf{g}_\phi(\mathbf{z}_k)$$
$$= \sum_{i=1}^n \mathbf{e}_i \left\{ \frac{\phi(\mathbf{z}_k + \mathbf{e}_i\zeta) - \phi(\mathbf{z}_k)}{\zeta} - \mathbf{g}_\phi(\mathbf{z}_k) \cdot \mathbf{e}_i \right.$$
$$+\phi(\mathbf{z}_k + \mathbf{e}_i\zeta)\frac{(1 + \delta_{ik})(1 + \varepsilon_{ik-})(1 + \varepsilon_{ik\div}) - 1}{\zeta}$$
$$\left. -\phi(\mathbf{z}_k)\frac{(1 + \delta_k)(1 + \varepsilon_{ik-})(1 + \varepsilon_{ik\div}) - 1}{\zeta} \right\}$$

and Lemma 3.2.8 implies that

$$= \sum_{i=1}^n \mathbf{e}_i \left\{ \int_0^1 \int_0^\tau \mathbf{e}_i \cdot \mathbf{H}_\phi(\mathbf{z}_k + \mathbf{e}_i\zeta\sigma)\mathbf{e}_i\zeta \, d\sigma \, d\tau \right.$$
$$+\phi(\mathbf{z}_k + \mathbf{e}_i\zeta)\frac{(1 + \delta_{ik})(1 + \varepsilon_{ik-})(1 + \varepsilon_{ik\div}) - 1}{\zeta}$$
$$\left. -\phi(\mathbf{z}_k)\frac{(1 + \delta_k)(1 + \varepsilon_{ik-})(1 + \varepsilon_{ik\div}) - 1}{\zeta} \right\} \; .$$

We can use inequality (3.33) in Chap. 3 of Volume I to conclude that

$$\|\mathbf{d}_k\| \leq n \left\{ \frac{\zeta}{2}\|\mathbf{H}_\phi(\mathbf{z}_k)\| + \frac{\gamma\zeta^2}{6} \right.$$

$$\left. + \left(|\phi(\mathbf{z}_k)|\frac{1}{\zeta} + \|\mathbf{g}_\phi(\mathbf{z}_k)\| + \|\mathbf{H}_\phi(\mathbf{z}_k)\|\frac{\zeta}{2} + \frac{\gamma\zeta^2}{6} \right) \frac{3\varepsilon}{1-\varepsilon} + |\phi(\mathbf{z}_k)|\frac{1}{\zeta}\frac{3\varepsilon}{1-\varepsilon} \right\}$$

$$= |\phi(\mathbf{z}_k)|\frac{n}{\zeta}\frac{6\varepsilon}{1-\varepsilon} + \|\mathbf{g}_\phi(\mathbf{z}_k)\|\frac{3n\varepsilon}{1-\varepsilon} + \|\mathbf{H}_\phi(\mathbf{z}_k)\|\frac{n\zeta}{2}\left(1 + \frac{3\varepsilon}{1-\varepsilon}\right) + \gamma\frac{n\zeta^2}{6}\left(1 + \frac{3\varepsilon}{1-\varepsilon}\right) .$$

We also have the following formula for the rounding error in the approximation of the Hessian by finite differences:

$$\mathbf{E}_k = \sum_{i=1}^n \sum_{j=1}^n \mathbf{e}_i\mathbf{e}_j{}^\top \left\{ \left[\widetilde{\phi}_{ijk} - \widetilde{\phi}_{ik}\right](1 + \varepsilon_{iijk-}) \right.$$

$$\left. - \left[\widetilde{\phi}_{jk} - \widetilde{\phi}_k\right](1 + \varepsilon_{jk-}) \right\} \frac{(1 + \varepsilon_{ijk-})(1 + \varepsilon_{ij\times})(1 + \varepsilon_{ijk\div})}{\zeta^2} - \mathbf{H}_\phi(\mathbf{z}_k)$$

$$= \sum_{i=1}^n \sum_{j=1}^n \mathbf{e}_i\mathbf{e}_j{}^\top \left\{ \frac{\phi(\mathbf{z}_k + \mathbf{e}_i\zeta + \mathbf{e}_j\zeta) - \phi(\mathbf{z}_k + \mathbf{e}_i\zeta) - \phi(\mathbf{z}_k + \mathbf{e}_j\zeta) + \phi(\mathbf{z}_k)}{\zeta^2} - \mathbf{e}_i \cdot \mathbf{H}_\phi(\mathbf{z}_k)\mathbf{e}_j \right.$$

$$+ \phi(\mathbf{z}_k + \mathbf{e}_i\zeta + \mathbf{e}_j\zeta)\frac{(1 + \delta_{ijk})(1 + \varepsilon_{iijk-})(1 + \varepsilon_{ijk-})(1 + \varepsilon_{ij\times})(1 + \varepsilon_{ijk\div}) - 1}{\zeta^2}$$

$$- \phi(\mathbf{z}_k + \mathbf{e}_i\zeta)\frac{(1 + \delta_{ik})(1 + \varepsilon_{iijk-})(1 + \varepsilon_{ijk-})(1 + \varepsilon_{ij\times})(1 + \varepsilon_{ijk\div}) - 1}{\zeta^2}$$

$$- \phi(\mathbf{z}_k + \mathbf{e}_j\zeta)\frac{(1 + \delta_{jk})(1 + \varepsilon_{jk-})(1 + \varepsilon_{ijk-})(1 + \varepsilon_{ij\times})(1 + \varepsilon_{ijk\div}) - 1}{\zeta^2}$$

$$\left. + \phi(\mathbf{z}_k)\frac{(1 + \delta_k)(1 + \varepsilon_{jk-})(1 + \varepsilon_{ijk-})(1 + \varepsilon_{ij\times})(1 + \varepsilon_{ijk\div}) - 1}{\zeta^2} \right\}$$

and we can use Lemma 3.2.8 to see that

$$= \sum_{i=1}^n \sum_{j=1}^n \mathbf{e}_i\mathbf{e}_j{}^\top \left\{ \int_0^1 \int_0^\tau (\mathbf{e}_i + \mathbf{e}_j) \cdot \left\{\mathbf{H}_\phi(\mathbf{z}_k + \mathbf{e}_i\zeta\sigma + \mathbf{e}_j\zeta\sigma) - \mathbf{H}_\phi(\mathbf{z}_k)\right\}(\mathbf{e}_i + \mathbf{e}_j)\, d\sigma\, d\tau \right.$$

$$- \int_0^1 \int_0^\tau \mathbf{e}_i \cdot \left\{\mathbf{H}_\phi(\mathbf{z}_k + \mathbf{e}_i\zeta\sigma) - \mathbf{H}_\phi(\mathbf{z}_k)\right\}\mathbf{e}_i\, d\sigma\, d\tau$$

$$- \int_0^1 \int_0^\tau \mathbf{e}_j \cdot \left\{\mathbf{H}_\phi(\mathbf{z}_k + \mathbf{e}_j\zeta\sigma) - \mathbf{H}_\phi(\mathbf{z}_k)\right\}\mathbf{e}_j\, d\sigma\, d\tau$$

$$+ \phi(\mathbf{z}_k + \mathbf{e}_i\zeta + \mathbf{e}_j\zeta)\frac{(1 + \delta_{ijk})(1 + \varepsilon_{iijk-})(1 + \varepsilon_{ijk-})(1 + \varepsilon_{ij\times})(1 + \varepsilon_{ijk\div}) - 1}{\zeta^2}$$

$$
- \phi(\mathbf{z}_k + \mathbf{e}_i\zeta)\frac{(1+\delta_{ik})(1+\varepsilon_{iijk-})(1+\varepsilon_{ijk-})(1+\varepsilon_{ij\times})(1+\varepsilon_{ijk\div}) - 1}{\zeta^2}
$$

$$
- \phi(\mathbf{z}_k + \mathbf{e}_j\zeta)\frac{(1+\delta_{jk})(1+\varepsilon_{jk-})(1+\varepsilon_{ijk-})(1+\varepsilon_{ij\times})(1+\varepsilon_{ijk\div}) - 1}{\zeta^2}
$$

$$
+\phi(\mathbf{z}_k)\frac{(1+\delta_k)(1+\varepsilon_{jk-})(1+\varepsilon_{ijk-})(1+\varepsilon_{ij\times})(1+\varepsilon_{ijk\div}) - 1}{\zeta^2}\Bigg\} .
$$

We can take norms and use inequality (3.33) in Chap. 3 in Volume I to get

$$
\|\mathbf{E}_k\| \leq n^2 \left\{ \frac{8\gamma\zeta}{6} + \frac{\gamma\zeta}{6} + \frac{\gamma\zeta}{6} + \left[\frac{|\phi(\mathbf{z}_k)|}{\zeta^2} + \frac{2\|\mathbf{g}_\phi(\mathbf{z}_k)\|}{\zeta} + 2\|\mathbf{H}_\phi(\mathbf{z}_k)\| + \frac{4\gamma\zeta}{3} \right] \frac{5\varepsilon}{1-2\varepsilon} \right.
$$

$$
\left. +2\left[\frac{|\phi(\mathbf{z}_k)|}{\zeta^2} + \frac{\|\mathbf{g}_\phi(\mathbf{z}_k)\|}{\zeta} + \|\mathbf{H}_\phi(\mathbf{z}_k)\|\frac{1}{2} + \frac{\gamma\zeta}{6} \right] \frac{5\varepsilon}{1-2\varepsilon} + |\phi(\mathbf{z}_k)|\frac{5\varepsilon}{1-2\varepsilon} \right\}
$$

$$
= |\phi(\mathbf{z}_k)|\frac{4n^2}{\zeta^2}\frac{5\varepsilon}{1-2\varepsilon} + \|\mathbf{g}_\phi(\mathbf{z}_k)\|\frac{4n^2}{\zeta}\frac{5\varepsilon}{1-2\varepsilon}
$$

$$
+ \|\mathbf{H}_\phi(\mathbf{z}_k)\|3n^2\frac{5\varepsilon}{1-2\varepsilon} + \gamma\frac{5\zeta n^2}{3}\left(1 + \frac{5\varepsilon}{1-2\varepsilon}\right) .
$$

Next, we would like to use Lemma 3.4.2 to determine the **maximum attainable accuracy** with Newton's method for unconstrained minimization, with derivatives approximated by difference quotients. Assuming that our approximation to the Hessian $\mathbf{H}_\phi(\mathbf{z}_k)$ is positive, we can use a Cholesky factorization to solve the Newton equations. The rounding errors in this factorization should correspond to a matrix $\mathbf{F}_k$ where

$$
\|\mathbf{F}_k\|_1 \leq \frac{n^2\varepsilon}{1-n^2\varepsilon}\left\{\|\mathbf{H}_\phi(\mathbf{z}_k)\|_1 + \|\mathbf{E}_k\|_1\right\} ,
$$

since our discussion in Sect. 3.13.3.6 in Chap. 3 of Volume I showed that there is essentially no pivot growth in a Cholesky factorization. In order to satisfy the assumption

$$
\left\|\mathbf{J_f}(\mathbf{z}_k)^{-1}\left[\mathbf{E}_k + \mathbf{F}_k\right]\right\| < \frac{1}{1+\kappa\left(\mathbf{J_f}\left(\mathbf{z}_k\right)\right)}
$$

of Lemma 3.4.2, we will require that

$$
\|\mathbf{E}_k\|_1 \leq \frac{1}{\left\|\mathbf{H}_\phi(\mathbf{z}_k)^{-1}\right\|_1}\frac{1-n^2\varepsilon}{1+\kappa_1\left(\mathbf{H}_\phi(\mathbf{z}_k)\right)} - n^2\varepsilon\|\mathbf{H}_\phi(\mathbf{z}_k)\|_1 .
$$

If we ignore terms involving ε compared to terms of order one, we approximately require

$$|\phi(\mathbf{z}_k)|\frac{20n^2\varepsilon}{\zeta^2} + \|\mathbf{g}_\phi(\mathbf{z}_k)\|_1\frac{20n^2\varepsilon}{\zeta} + \|\mathbf{H}_\phi(\mathbf{z}_k)\|_1 15n^2\varepsilon + \gamma\frac{5\zeta n^2}{3}$$
$$\leq \frac{1}{\left\|\mathbf{H}_\phi(\mathbf{z}_k)^{-1}\right\|_1}\frac{1}{1+\kappa_1\left(\mathbf{H}_\phi(\mathbf{z}_k)\right)} .$$

This is a cubic inequality, and difficult to solve. However, the inequality will be satisfied if each term on the left is at most one fourth the value on the right. Such an assumption gives us

$$|\phi(\mathbf{z}_k)|\frac{20n^2\varepsilon}{\zeta^2} \leq \frac{1}{4}\frac{1}{\left\|\mathbf{H}_\phi(\mathbf{z}_k)^{-1}\right\|_1}\frac{1}{1+\kappa_1\left(\mathbf{H}_\phi(\mathbf{z}_k)\right)}$$
$$\Longleftrightarrow \zeta \geq \sqrt{80n^2\varepsilon|\phi(\mathbf{z}_k)|\left\|\mathbf{H}_\phi(\mathbf{z}_k)^{-1}\right\|_1\left(1+\kappa_1\left(\mathbf{H}_\phi(\mathbf{z}_k)\right)\right)} ,$$
$$\|\mathbf{g}_\phi(\mathbf{z}_k)\|_1\frac{20n^2\varepsilon}{\zeta} \leq \frac{1}{4}\frac{1}{\left\|\mathbf{H}_\phi(\mathbf{z}_k)^{-1}\right\|_1}\frac{1}{1+\kappa_1\left(\mathbf{H}_\phi(\mathbf{z}_k)\right)}$$
$$\Longleftrightarrow \zeta \geq 80n^2\varepsilon\|\mathbf{g}_\phi(\mathbf{z}_k)\|_1\left\|\mathbf{H}_\phi(\mathbf{z}_k)^{-1}\right\|_1\left(1+\kappa_1\left(\mathbf{H}_\phi(\mathbf{z}_k)\right)\right) ,$$
$$\|\mathbf{H}_\phi(\mathbf{z}_k)\|_1 15n^2\varepsilon \leq \frac{1}{4}\frac{1}{\left\|\mathbf{H}_\phi(\mathbf{z}_k)^{-1}\right\|_1}\frac{1}{1+\kappa_1\left(\mathbf{H}_\phi(\mathbf{z}_k)\right)}$$
$$\Longleftrightarrow 60n^2\varepsilon\kappa_1\left(\mathbf{H}_\phi(\mathbf{z}_k)\right)\left(1+\kappa_1\left(\mathbf{H}_\phi(\mathbf{z}_k)\right)\right) \leq 1 \text{ and}$$
$$\gamma\frac{5\zeta n^2}{3} \leq \frac{1}{4}\frac{1}{\left\|\mathbf{H}_\phi(\mathbf{z}_k)^{-1}\right\|_1}\frac{1}{1+\kappa_1\left(\mathbf{H}_\phi(\mathbf{z}_k)\right)}$$
$$\Longleftrightarrow \zeta \leq \frac{3}{20n^2\gamma}\frac{1}{\left\|\mathbf{H}_\phi(\mathbf{z}_k)^{-1}\right\|_1}\frac{1}{1+\kappa_1\left(\mathbf{H}_\phi(\mathbf{z}_k)\right)} .$$

Now we are ready to discuss the maximum attainable accuracy for Newton's method applied to unconstrained minimization while using finite difference approximation of all derivatives. We will assume that the finite difference increment satisfies

$$\sqrt{\varepsilon} \ll \zeta \ll 1 ,$$

Then the Eqs. (3.17) give us the approximations

$$\alpha \approx \frac{\gamma}{2}\left\|\mathbf{H}_\phi(\mathbf{z}_k)^{-1}\right\|_1 , \ \beta \approx 1 \text{ and } \delta \approx \kappa_1\left(\mathbf{H}_\phi(\mathbf{z}_k)\right)\frac{n\zeta}{2} .$$

As a result, the maximum attainable accuracy condition takes the form

$$\delta \approx \kappa_1\left(\mathbf{H}_\phi(\mathbf{z}_k)\right)\frac{n\zeta}{2} < \|\mathbf{z}_k - \mathbf{z}\|_1 < \frac{1}{\alpha} \approx \frac{2}{\gamma\left\|\mathbf{H}_\phi(\mathbf{z}_k)^{-1}\right\|_1} .$$

In summary, we should take $\zeta > O(\sqrt{\varepsilon})$ to guarantee that the numerical Hessian is nonsingular, but the closer we take ζ to be on the order of $\sqrt{\varepsilon}$, the better the maximum attainable accuracy can be.

Readers may experiment with the JavaScript **Newton minimization** program. This program plots contours for the modified Rosenbrock function

$$\mathbf{f}(\mathbf{x}) = \pi + \left\| \begin{bmatrix} 10(\xi_2 - \xi_1^2) \\ 1 - \xi_1 \end{bmatrix} \right\|_2^2 .$$

Users may click in this plot to select a starting point for a Newton iteration. Users may also select the number of iterations to perform, or the finite difference increment ζ used to approximate derivatives. The program will plot $\log\left(\|\mathbf{g}_\phi(\mathbf{z}_k)\|_1\right)$ versus k for three versions of Newton's method. The first uses analytically computed derivatives, while the second uses analytical gradients and finite difference Hessians. The third approximates both gradients and Hessians by finite differences. Users may verify that very small finite difference increments lead to convergence problems for finite difference gradients and Hessians.

Exercise 3.5.1 Let $\phi(\mathbf{x}) = \frac{1}{2}(\xi_1^2 - \xi_2)^2 + \frac{1}{2}(1 - \xi_1)^2$. Find its minimum $\mathbf{z}$. Then compute one step of Newton's method with initial guess $\mathbf{z}_0 = [2, 2]^\top$. Compare $\phi(\mathbf{z}_0)$ and $\phi(\mathbf{z}_1)$.

Exercise 3.5.2 Let $\phi(\mathbf{x}) = -\|\mathbf{x}\|_2^2$. Show that Newton's method will not converge to the maximum of ϕ if $\mathbf{z}_0 \neq \mathbf{z}$.

Exercise 3.5.3 Let $\phi(\mathbf{x}) = \xi_1^2 - \xi_2^2$. Show that Newton's method will not converge to the saddle point of ϕ unless $\mathbf{e}_2^\top \mathbf{z}_0 = 0$.

Exercise 3.5.4 Consider

$$\phi(\mathbf{x}) = \mathbf{e}^{\xi_1}(4\xi_1^2 + 2\xi_2^2 + 4\xi_1\xi_2 + 2\xi_2 + 1) ,$$

which has minimum $\mathbf{z} = [1, -2]^\top$.

1. Consider the starting point $\mathbf{z}_0 = [-4, 3]^\top$. Compute the steepest descent direction $-\mathbf{f}(\mathbf{z}_0)$ and find the step length λ that minimizes $\phi(\mathbf{z}_0 - \mathbf{f}(\mathbf{z}_0)\lambda)$. Show that $\mathbf{H}_\phi(\mathbf{z}_0)$ is not positive. Compute the safeguarded Cholesky factorization $\mathbf{L}\mathbf{L}^\top = \mathbf{H}_\phi(\mathbf{z}_k) + \mathbf{I}\mu$ by hand and use it to compute the modified Newton step.
2. Consider the starting point $\mathbf{z}_0 = [-3, 2]^\top$. Compute the gradient $\mathbf{f}(\mathbf{z}_0)$ and Hessian $\mathbf{H}_\phi(\mathbf{z}_0)$. Why are the steepest descent directions and the Newton step direction useless? What information could we obtain from $\mathbf{H}_\phi(\mathbf{z}_0)$ to determine a descent direction? (Hint: consider the eigenvectors of $\mathbf{H}_\phi(\mathbf{z}_0)$.)

3.5.3 *Step Selection*

Let us recall the unconstrained minimization problem. We are given a real functional ϕ, and we want to find an n-vector $\mathbf{z}$ for which there exists $\varrho > 0$ such that for all $\mathbf{x}$ satisfying $\|\mathbf{x} - \mathbf{z}\| < \varrho$ we have $\phi(\mathbf{z}) \leq \phi(\mathbf{x})$. Also recall that we developed the notion of a descent direction for this problem in Sect. 3.5.1. In this section, we want to consider reasonable alternatives for selecting both the step direction and the step length, since these will determine the convergence rate and possibly guarantee global convergence.

3.5.3.1 Step Direction

So far, we have seen two possibilities for the step direction. These are the **steepest descent direction**

$$\mathbf{s}_k^{(G)} = -\mathbf{g}_\phi(\mathbf{z}_k) ,$$

and the **Newton direction**, determined by solving

$$\mathbf{H}_\phi(\mathbf{z}_k)\mathbf{s}_k^{(N)} = -\mathbf{g}_\phi(\mathbf{z}_k)$$

whenever the Hessian $\mathbf{H}_\phi(\mathbf{z}_k)$ is positive-definite.

Recall that we discussed steepest descent methods for quadratic objective functions in Sect. 2.5.2. Those results are more generally applicable. Steepest descent is globally convergent to either a local minimizer or a saddle point of ϕ; convergence to saddle points occurs only if the initial guess is selected very carefully. Convergence to a critical point follows from Lemma 3.5.1, which shows that we can always decrease the value of a continuously differentiable objective if its gradient at a given point is nonzero. Consequently, any bounded sequence of steepest descent steps employing line searches (see Sect. 3.5.3.4 below) must converge to a zero of the gradient. However, if steepest descent is applied to general nonlinear functions, then the iteration generally converges linearly.

In summary, there are several difficulties with the steepest descent direction. First of all, it requires a strategy to select a step length. Second, even if we select the step length to be that value of λ that minimizes $\phi(\mathbf{z}_k - \mathbf{g}_\phi(\mathbf{z}_k)\lambda)$ for $\lambda > 0$, the convergence rate is at best linear. On the other hand, steepest descent has certain advantages. First, it only requires the gradient of ϕ, and not the Hessian. Another advantage of steepest descent is its global convergence.

3.5.3.2 Modified Newton Method

Since steepest descent does not require the evaluation of the Hessian, it is hard to determine if the limit point of a steepest descent iteration is a local minimizer or a saddle point. With Newton's method, this extra information is available at additional cost. Unfortunately, Newton's method for unconstrained minimization is only locally convergent. However, when Newton's method converges, it typically converges quadratically.

At each step of Newton's method, we need to solve the system of linear equations

$$\mathbf{H}_\phi(\mathbf{z}_k)\mathbf{s}_k = -\mathbf{g}_\phi(\mathbf{z}_k) \ .$$

Recall that in Sect. 3.13.3 in Chap. 3 of Volume I, we discussed the use of a Cholesky factorization to solve symmetric positive systems of linear equations. Thus, in the midst of unconstrained minimization it seems reasonable to attempt to factor

$$\mathbf{H}_\phi(\mathbf{z}_k) = \mathbf{L}\mathbf{L}^\top$$

where $\mathbf{L}$ is left-triangular. If this factorization process discovers that $\mathbf{H}_\phi(\mathbf{z}_k)$ is not positive, then we can choose a modified step direction $\mathbf{s}_k$ by solving

$$\left[\mathbf{H}_\phi(\mathbf{z}_k) + \mathbf{I}\mu\right]\mathbf{s}_k = -\mathbf{g}_\phi(\mathbf{z}_k) \ . \tag{3.22}$$

Note that for μ sufficiently large, the step $\mathbf{s}_k$ defined by (3.22) must be a descent direction.

If μ is chosen very large, then $\mathbf{s}_k$ tends toward a small scalar multiple of the steepest descent direction. The smallest possible μ is the negative of the smallest eigenvalue of $\mathbf{H}_\phi(\mathbf{z}_k)$, which is somewhat expensive to evaluate. Of course, the most negative eigenvalue can be bounded below by the Gerschgorin Circle Theorem 1.2.2. However, this can still provide too large a value for μ.

Instead, the Cholesky factorization can be modified to watch for negative diagonal values. A sufficient correction can be determined to prevent taking the square root of a negative number and producing large entries in the Cholesky factorization. Gill and Murray [78] suggest that we use a safeguarded Cholesky factorization to determine μ. This idea is available in Dennis and Schnabel [52, p. 318] as

Algorithm 3.5.1 (choldecomp)

$\varepsilon =$ machine precision , $\varepsilon_2 = \sqrt{\varepsilon}$, $\delta = 0$, $\lambda = 0$

for $0 \le i \le n$

  $\delta = \max\{\delta, |\mathbf{H}_{i,i}|\}$

  for $i < j \le n$, $\lambda = \max\{\lambda, |\mathbf{H}_{j,i}|\}$

$\overline{\lambda} = \sqrt{\max\{\delta \ , \ \lambda/\sqrt{n^2-1} \ , \ \varepsilon\}}$, $\underline{\lambda} = \overline{\lambda}\sqrt{\varepsilon_2}$

$\lambda_2 = \varepsilon_2\overline{\lambda}$, $\mu = 0$

for $1 \le k \le n$

$\mathbf{L}_{k,k} = \mathbf{H}_{k,k}$

for $1 \le j < k$, $\mathbf{L}_{k,k} = \mathbf{L}_{k,k} - \mathbf{L}_{k,j}^2$

$\underline{\lambda}_k = 0$

for $k < i \le n$

$\mathbf{L}_{i,k} = \mathbf{H}_{k,i}$

for $1 \le j < k$, $\mathbf{L}_{i,k} = \mathbf{L}_{i,k} - \mathbf{L}_{i,j}\mathbf{L}_{k,j}$

$\underline{\lambda}_k = \max\{\underline{\lambda}_k \, , \, |\mathbf{L}_{i,k}|\}$

$\underline{\lambda}_k = \max\{\underline{\lambda}_k/\overline{\lambda} \, , \, \underline{\lambda}\}$

if $\mathbf{L}_{k,k} > \underline{\lambda}_k^2$ then $\mathbf{L}_{k,k} = \sqrt{\mathbf{L}_{k,k}}$

else

$\underline{\lambda}_k = \max\{\underline{\lambda}_k \, , \, \lambda_2\}$, $\mu = \max\{\mu, (\underline{\lambda}_k)^2 - \mathbf{L}_{k,k}\}$, $\mathbf{L}_{k,k} = \underline{\lambda}_k$

for $j < i \le n$, $\mathbf{L}_{i,k} = \mathbf{L}_{i,k}/\mathbf{L}_{k,k}$

At the end of this algorithm, we have

$$\mathbf{L}\mathbf{L}^\top = \mathbf{H}_\phi(\mathbf{z}_k) + \mathbf{D}$$

where $\mathbf{D}$ is a nonnegative diagonal matrix with diagonal entries bounded above by the output value of μ. Also

$$\mathbf{L}_{i,i} \ge \underline{\lambda} \text{ for all } 1 \le i \le n \text{ and } |\mathbf{L}_{i,j}| \le \overline{\lambda} \text{ for all } 1 \le j < i \le n \, .$$

If $\mu > 0$, we may subsequently want to compute the Cholesky factorization of $\mathbf{H} + \mathbf{I}\mu$. In order to do so, we must provide separate storage for the Cholesky factor $\mathbf{L}$ and the Hessian $\mathbf{H}_\phi(\mathbf{z}_k)$. For example, the Hessian could be stored on and above the diagonal of a square matrix, while $\mathbf{L}$ could be stored below the diagonal with its diagonal entries in separate storage. Interested readers may find this approach in routine `modelhess` of Dennis and Schnabel [52, p. 315].

3.5.3.3 Negative Curvature

In order to develop a robust minimization program, we need to consider strategies to use when the Hessian of the objective function is not "sufficiently" positive. In Algorithm 3.5.1, we developed a safeguarded Cholesky factorization of the Hessian that could be used to compute a descent direction. An alternative is to compute a **direction of negative curvature**, and use it as a descent direction.

Suppose that we compute the eigenvectors and eigenvalues of the Hessian:

$$\mathbf{H}_\phi(\mathbf{x}) = \mathbf{X}\Lambda\mathbf{X}^\top \, .$$

Here the Spectral Theorem 1.3.1 guarantees that $\boldsymbol{\Lambda}$ can be chosen to be diagonal and real, and that $\mathbf{X}$ can be chosen to be orthogonal. Then a descent direction can be computed as

$$\mathbf{s} = -\mathbf{X}|\boldsymbol{\Lambda}|^{-1}\mathbf{X}^\top \mathbf{g}_\phi \ .$$

For the diagonal matrix $\boldsymbol{\Lambda} = \text{diag}(\lambda_1, \ldots, \lambda_n)$, its absolute value is

$$|\boldsymbol{\Lambda}| = \text{diag}(|\lambda_1|, \ldots, |\lambda_n|) \ .$$

This approach has the advantage of choosing the Newton step direction whenever $\mathbf{H}_\phi$ is positive, but has the disadvantage of requiring an expensive computation of the eigenvalues and eigenvectors of $\mathbf{H}_\phi$.

An alternative is to compute the symmetric indefinite factorization (see Theorem 3.13.1 in Chap. 3 of Volume I)

$$\mathbf{H}_\phi = \mathbf{M}\mathbf{D}\mathbf{M}^\top \ .$$

Here $\mathbf{M}$ is unit lower triangular and $\mathbf{D}$ is block diagonal with either 1×1 or 2×2 diagonal blocks. It is a simple matter to compute the eigenvalues and eigenvectors of $\mathbf{D}$ to get $|\mathbf{D}|$, and then to compute a descent direction by solving

$$\mathbf{M}|\mathbf{D}|\mathbf{M}^\top \mathbf{s} = -\mathbf{g}_\phi$$

for the descent direction $\mathbf{s}$.

For more information on directions of negative curvature for minimization problems, see Moré and Sorensen [127].

3.5.3.4 Line Searches

So far, we have seen several alternatives for descent directions in minimization: steepest descent (in Sect. 3.5.3), Newton's method (in Sect. 3.5.2), modified Newton steps computed by the safeguarded Cholesky factorization (in Sect. 3.5.3.2), and directions of negative curvature (in Sect. 3.5.3.3). Each of these step directions needs to be combined with a strategy for selecting a step length. Recall that we examined the performance of the steepest descent method with exact line searches for minimizing quadratic functions in Sect. 2.5.2; the resulting method is linearly and globally convergent. We have also seen in Theorem 3.4.1 that Newton's method without a step selection strategy produces a method with quadratic convergence rate, but may be only locally convergent.

Any step selection strategy needs to guarantee a sufficient decrease in the objective function, where "sufficient" is limited by what is both possible and cost-effective. Efficient step selection strategies will want to avoid evaluating the objective function too many times. This is because it is likely to be more productive to look in a new descent direction, rather than to obtain the last bit of decrease in some search direction, Finally, if our step direction with unit step length (such as the Newton step) produces superlinear local convergence without a step selection strategy, then we want any step selection strategy combined with this step direction to choose unit step lengths as the minimum of the objective is approached.

It is easy to construct examples of bad step length strategies.

Example 3.5.1 Suppose we want to find the minimum of $\phi(x) = x^2$. Let our initial guess be $z_0 = 2$ and that at step k our descent step is $s_k = (-1)^k$. If our step length is chosen to be

$$\lambda_k = 2 + 3/2^{k+1} ,$$

then it is easy to see that

$$z_k = (-1)^k \left(1 + 2^{-k}\right) .$$

As a result, $\phi(z_k)$ decreases to 1, which is not a local minimum. What is even worse, $\{z_k\}_{k=0}^{\infty}$ does not converge. The difficulty with the step length strategy in this example is that we do not obtain sufficient decrease in the objective function value with each step, relative to the decrease predicted by the tangent line:

$$\frac{\phi(z_k) - \phi(z_k + s_k\lambda_k)}{-\phi'(z_k)s_k\lambda_k} = \frac{z_k + z_{k+1}}{2z_k} = \frac{2^{-k-2}}{1 + 2^{-k}} \to 0 .$$

Example 3.5.2 Here is a different kind of example of a bad step length strategy. Again suppose that $\phi(x) = x^2$, and that our initial guess is $z_0 = 2$. Let our descent step always be $s_k = -1$. If our step length is $\lambda_k = 2^{-k-1}$, then

$$z_k = 1 + 2^{-k} .$$

As a result, $\phi(z_k)$ decreases to 1, but z_k converges to a point that is not a local minimizer of ϕ. The difficulty with the step length strategy in this example is that $\|s_k\lambda_k\|$ is so small that it becomes the same as the decrease predicted by the tangent line:

$$\frac{\phi(z_k) - \phi(z_k + s_k\lambda_k)}{-\phi'(z_k)s_k\lambda_k} = \frac{z_k + z_{k+1}}{2z_k} = \frac{2^{-k-2}}{1 + 2^{-k}} \to 1 .$$

At the current iterate $\mathbf{z}_k$, the purpose of a line search is to select a step length λ_k to obtain sufficient decrease in the objective along the search direction, without taking an unacceptably small step. In Sect. 5.7.4 in Chap. 5 of Volume I, this principle lead us to examine the function

$$\psi(\lambda) = \phi(\mathbf{z}_k + \mathbf{s}_k\lambda) ,$$

from which we developed the **Goldstein-Armijo Principle** (5.49) in Chap. 5 of Volume I. Note that in the current situation

$$\psi'(0) = \mathbf{g}_\phi(\mathbf{z}_k) \cdot \mathbf{s}_k \ .$$

This is negative if $\mathbf{s}_k$ is a descent direction. Recall that if we are given $0 < \alpha < 1$, the Goldstein-Armijo principle (5.49) in Chap. 5 of Volume I chooses $\lambda \in (0, \infty)$ such that

$$\psi(0) - \psi(\lambda) \geq -\alpha \psi'(0)\lambda \ . \tag{3.23a}$$

This condition requires that the decrease in the objective function is significant relative to the decrease predicted by the tangent line. Given $\alpha < \beta < 1$, we can avoid small step lengths by also requiring that

$$\psi(0) - \psi(\lambda) \leq -\beta \psi'(0)\lambda \ . \tag{3.23b}$$

Typical values are $\alpha = 10^{-4}$ and $\beta = 0.5$.

The next theorem provides circumstances under which the Goldstein-Armijo principle guarantees convergence of the minimization algorithm. This is a generalization of Wolfe's Theorem 5.7.3 in Chap. 5 of Volume I.

Theorem 3.5.1 (Convergence with the Goldstein-Armijo Principle) *Suppose that $\mathscr{D}$ is an open subset of real n-vectors, and that the real functional ϕ is continuously differentiable on $\mathscr{D}$. Assume that its gradient $\mathbf{g}_\phi$ is Lipschitz continuous on $\mathscr{D}$ with constant γ; in other words, for all n-vectors $\mathbf{x}$ and $\mathbf{y}$ in $\mathscr{D}$ we have*

$$\|\mathbf{g}_\phi(\mathbf{x}) - \mathbf{g}_\phi(\mathbf{y})\|_2 \leq \gamma \|\mathbf{x} - \mathbf{y}\|_2 \ .$$

Also suppose that ϕ is bounded below on $\mathscr{D}$. Finally, assume that $\{\mathbf{z}_k\}_{k=0}^{\infty} \subset \mathscr{D}$, $\{\mathbf{s}_k\}_{k=0}^{\infty} \subset \mathscr{D}$ and $\{\lambda_k\}_{k=0}^{\infty}$ are sequences such that

1. *for each k, either $\mathbf{g}_\phi(\mathbf{z}_k) = \mathbf{0}$ and $\lambda_k = 0$, or $\mathbf{s}_k$ is $\lambda_k > 0$ is such that*

$$\mathbf{z}_{k+1} = \mathbf{z}_k + \mathbf{s}_k \lambda_k \ ,$$

2. *and for each k, whenever $\mathbf{g}_\phi(\mathbf{z}_k) \neq \mathbf{0}$, the step direction $\mathbf{s}_k$ and the step length λ_k satisfy the Goldstein-Armijo principle* (3.23).

Then either there exists an iterate k such that $\mathbf{g}_\phi(\mathbf{z}_\ell) = \mathbf{0}$ for all $\ell \geq k$, or

$$\lim_{k \to \infty} \frac{\mathbf{g}_\phi(\mathbf{z}_k) \cdot \mathbf{s}_k}{\|\mathbf{s}_k\|^2} = 0 \ .$$

Proof If the gradient $\mathbf{g}_\phi$ is zero at some iterate, then $\lambda_k = 0$ so $\mathbf{z}_{k+1} = \mathbf{z}_k$. As a result, the gradient will be zero for all subsequent iterations. For the remainder of the proof, we will assume that $\mathbf{g}_\phi$ is nonzero for all iterates. To simplify the discussion, we will also assume that the step directions $\mathbf{s}_k$ are normalized so that $\|\mathbf{s}_k\|_2 = 1$.

Let us define

$$\sigma_k \equiv \mathbf{g}_\phi(\mathbf{z}_k) \cdot \mathbf{s}_k$$

and note that $\sigma_k < 0$ for all k. Note that

$$\phi(\mathbf{z}_j) - \phi(\mathbf{z}_0) = \sum_{k=0}^{j-1} \{\phi(\mathbf{z}_{k+1}) - \phi(\mathbf{z}_k)\} \leq \alpha \sum_{k=0}^{j-1} \mathbf{g}_\phi(\mathbf{z}_k) \cdot \mathbf{s}_k \lambda_k = \alpha \sum_{k=0}^{j-1} \sigma_k \lambda_k < 0 \; .$$

The final sum cannot be infinite because ϕ is bounded below. Therefore the sum converges, which implies that $\sigma_k \lambda_k \to 0$ as $k \to \infty$. All that remains is to show that $\sigma_k \to 0$.

Since the Goldstein-Armijo principle (3.23b) requires that

$$\mathbf{g}_\phi(\mathbf{z}_{k+1}) \cdot \mathbf{s}_k \geq \beta \mathbf{g}_\phi(\mathbf{z}_k) \cdot \mathbf{s}_k \; ,$$

it follows that

$$\left[\mathbf{g}_\phi(\mathbf{z}_{k+1}) - \mathbf{g}_\phi(\mathbf{z}_k)\right] \cdot \mathbf{s}_k \geq -(1-\beta)\mathbf{g}_\phi(\mathbf{z}_k) \cdot \mathbf{s}_k > 0 \; .$$

Then Lipschitz continuity of the gradient of ϕ and the definition of σ_k imply that

$$0 < -(1-\beta)\sigma_k \leq \|\mathbf{g}_\phi(\mathbf{z}_{k+1}) - \mathbf{g}_\phi(\mathbf{z}_k)\|_2 \leq \gamma \|\mathbf{z}_{k+1} - \mathbf{z}_k\|_2 = \gamma \lambda_k \; .$$

Thus

$$\lambda_k \geq -\frac{1-\beta}{\gamma} \sigma_k > 0 \; .$$

If we multiply this inequality by σ_k and take the limit as $k \to \infty$, we obtain

$$0 = \lim_{k\to\infty} \sigma_k \lambda_k \leq -\frac{1-\beta}{\gamma} \lim_{k\to\infty} \sigma_k^2 \leq 0 \; .$$

This result shows that $\sigma_k \to 0$ and completes the proof of the theorem.

Several comments are in order here. The theorem says that if ϕ is bounded below and sufficiently continuous, then eventually the gradient of ϕ is orthogonal to the step direction. If the step direction is chosen properly (for example, in the direction of steepest descent or along a Newton direction with a positive Hessian), then the gradient of ϕ must tend to zero.

In practice, we may not know if ϕ is bounded below before beginning the minimization algorithm. In such a case, $\mathbf{g}_\phi(\mathbf{z}_k) \cdot \mathbf{s}_k$ may not tend to zero. If it does not, then the proof shows that $\phi(\mathbf{z}_k) \to -\infty$ as $k \to \infty$.

The next lemma shows that for some choices of the step direction, eventually the choice of the step length is easy.

Theorem 3.5.2 *Suppose that $\mathscr{D}$ is an open convex subset of real n-vectors. Assume that the real functional ϕ is twice continuously differentiable in $\mathscr{D}$, and that ϕ is bounded below on $\mathscr{D}$. Let the Hessian $\mathbf{H}_\phi$ of ϕ be uniformly Lipschitz continuous in $\mathscr{D}$ with constant γ. Assume that the sequence $\{\mathbf{z}_k\}_{k=0}^\infty$ is of the form*

$$\mathbf{z}_{k+1} = \mathbf{z}_k + \mathbf{s}_k \lambda_k \; ,$$

where each n-vector $\mathbf{s}_k$ is a descent direction for ϕ at $\mathbf{z}_k$, and each λ_k satisfies the Goldstein-Armijo principle (3.23) *with $\alpha < \min\{\beta, 1/2\}$. Suppose that $\{\mathbf{z}_k\}_{k=0}^\infty$ converges to $\mathbf{z} \in \mathscr{D}$, and that the Hessian $\mathbf{H}_\phi(\mathbf{z})$ is positive. Finally, assume that the sequence $\{\mathbf{s}_k\}_{k=0}^\infty$ of step directions tends to the Newton direction:*

$$\lim_{k\to\infty} \frac{\|\mathbf{g}_\phi(\mathbf{z}_k) + \mathbf{H}_\phi(\mathbf{z}_k)\mathbf{s}_k\|_2}{\|\mathbf{s}_k\|_2} = 0 \; .$$

Then there is a step index $k_0 \geq 0$ such that for all $k \geq k_0$ the step length $\lambda_k = 1$ satisfies the Goldstein-Armijo principle.

Proof First, we will show that $\|\mathbf{s}_k\|_2 \to 0$. Define

$$\varrho_k \equiv \frac{\|\mathbf{g}_\phi(\mathbf{z}_k) + \mathbf{H}_\phi(\mathbf{z}_k)\mathbf{s}_k\|_2}{\|\mathbf{s}_k\|_2} \; .$$

Since $\mathbf{H}_\phi$ is continuous in $\mathscr{D}$ and $\{\mathbf{z}_k\}_{k=0}^\infty$ converges to $\mathbf{z}$, there is a step index $k_1 \geq 0$ such that for all $k \geq k_1$

$$\frac{1}{2}\left\|\mathbf{H}_\phi^{-1}(\mathbf{z})\right\|_2 \leq \|\mathbf{H}_\phi^{-1}(\mathbf{z}_k)\| \leq 2\left\|\mathbf{H}_\phi^{-1}(\mathbf{z})\right\|_2 \; .$$

Since $\{\varrho_k\}_{k=0}^\infty$ is assumed to converge to zero, there is a step index $k_2 \geq k_1$ such that for all $k \geq k_2$,

$$\varrho_k < \frac{1}{4\left\|\mathbf{H}_\phi^{-1}(\mathbf{z})\right\|_2} \; .$$

It follows that for $k \geq k_2$,

$$\begin{aligned} -\frac{\mathbf{g}_\phi(\mathbf{z}_k)\cdot\mathbf{s}_k}{\|\mathbf{s}_k\|_2} &= \frac{\mathbf{s}_k\cdot\mathbf{H}_\phi(\mathbf{z}_k)\mathbf{s}_k}{\|\mathbf{s}_k\|_2} - \frac{\mathbf{s}_k\cdot\left[\mathbf{g}_\phi(\mathbf{z}_k)+\mathbf{H}_\phi(\mathbf{z}_k)\mathbf{s}_k\right]}{\|\mathbf{s}_k\|_2} \\ &\geq \left(\frac{1}{2\left\|\mathbf{H}_\phi^{-1}(\mathbf{z})\right\|_2} - \varrho_k\right)\|\mathbf{s}_k\|_2 \geq \frac{\|\mathbf{s}_k\|_2}{4\left\|\mathbf{H}_\phi^{-1}(\mathbf{z})\right\|_2} \; . \end{aligned} \tag{3.24}$$

Since Theorem 3.5.1 shows that $\mathbf{s}_k\cdot\mathbf{g}_\phi(\mathbf{z}_k)/\|\mathbf{s}_k\|_2 \to 0$, we conclude that $\|\mathbf{s}_k\|_2 \to 0$.

Our next goal is to show that for sufficiently large k the first Goldstein-Armijo condition (3.23a) is satisfied. Notice that the intermediate-value result (3.2.8) implies that for some $\mathbf{y}_k$ along the line from $\mathbf{z}_k$ to $\mathbf{z}_k + \mathbf{s}_k$ we have

$$\phi(\mathbf{x} + \mathbf{s}_k) - \phi(\mathbf{z}_k) = \mathbf{g}_\phi(\mathbf{z}_k) \cdot \mathbf{s}_k + \frac{1}{2}\mathbf{s}_k \cdot \mathbf{H}_\phi(\mathbf{y}_k)\mathbf{s}_k \; .$$

It follows from the definition of ϱ_k and the Lipschitz continuity of $\mathbf{H}_\phi$ that

$$\begin{aligned}
&\phi(\mathbf{z}_k + \mathbf{s}_k) - \phi(\mathbf{z}_k) - \frac{1}{2}\mathbf{g}_\phi(\mathbf{z}_k) \cdot \mathbf{s}_k = \frac{1}{2}\mathbf{s}_k \cdot \left\{\mathbf{g}_\phi(\mathbf{z}_k) + \mathbf{H}_\phi(\mathbf{y}_k)\mathbf{s}_k\right\} \\
&= \frac{1}{2}\mathbf{s}_k \cdot \left\{\mathbf{g}_\phi(\mathbf{z}_k) + \mathbf{H}_\phi(\mathbf{z}_k)\mathbf{s}_k\right\} + \frac{1}{2}\mathbf{s}_k \cdot \left\{\mathbf{H}_\phi(\mathbf{y}_k) - \mathbf{H}_\phi(\mathbf{z}_k)\right\} \mathbf{s}_k \\
&\leq \frac{1}{2}\left(\varrho_k + \gamma \|\mathbf{s}_k\|_2\right) \|\mathbf{s}_k\|^2 \; . \qquad (3.25)
\end{aligned}$$

Since ϱ_k is assumed to converge to zero and since we have shown that $\mathbf{s}_k$ converges to zero, there is a $k_3 \geq k_2$ such that for all $k \geq k_3$

$$\varrho_k + \gamma \|\mathbf{s}_k\|_2 \leq \frac{\min\{\beta, 1 - 2\alpha\}}{2\left\|\mathbf{H}_\phi^{-1}(\mathbf{z})\right\|_2} \; .$$

Then (3.24) and (3.25) imply that

$$\begin{aligned}
&\phi(\mathbf{z}_k + \mathbf{s}_k) - \phi(\mathbf{z}_k) \leq \frac{1}{2}\mathbf{g}_\phi(\mathbf{z}_k) \cdot \mathbf{s}_k + \frac{(1 - 2\alpha)\|\mathbf{s}_k\|_2^2}{8\left\|\mathbf{H}_\phi^{-1}(\mathbf{z})\right\|_2} \leq \frac{1}{2}[1 - (1 - 2\alpha)]\mathbf{g}_\phi(\mathbf{z}_k) \cdot \mathbf{s}_k \\
&= \alpha \mathbf{g}_\phi(\mathbf{z}_k) \cdot \mathbf{s}_k \; .
\end{aligned}$$

This proves that (3.23a) is satisfied for $k \geq k_3$.

Next, we will prove that the second Goldstein-Armijo condition is satisfied for $k \geq k_3$. We can use the mean-value theorem to show that for some $\mathbf{w}_k$ on the line from $\mathbf{z}_k$ to $\mathbf{z}_k + \mathbf{s}_k$ we have

$$\begin{aligned}
\mathbf{s}_k \cdot \mathbf{g}_\phi(\mathbf{z}_k + \mathbf{s}_k) &= \mathbf{s}_k \cdot \left[\mathbf{g}_\phi(\mathbf{z}_k) + \mathbf{H}_\phi(\mathbf{w}_k)\mathbf{s}_k\right] \\
&= \mathbf{s}_k \cdot \left[\mathbf{g}_\phi(\mathbf{z}_k) + \mathbf{H}_\phi(\mathbf{z}_k)\mathbf{s}_k\right] + \mathbf{s}_k \cdot \left[\mathbf{H}_\phi(\mathbf{w}_k) - \mathbf{H}_\phi(\mathbf{z}_k)\right] \mathbf{s}_k \; .
\end{aligned}$$

Then Lipschitz continuity of $\mathbf{H}_\phi$ and inequality (3.24) imply that

$$\left|\mathbf{s}_k \cdot \mathbf{g}_\phi(\mathbf{z}_k + \mathbf{s}_k)\right| \leq \varrho_k \|\mathbf{s}_k\|_2^2 + \gamma \|\mathbf{s}_k\|_2^2 \leq \frac{\beta \|\mathbf{s}_k\|_2^2}{4\left\|\mathbf{H}_\phi^{-1}(\mathbf{z})\right\|_2} \leq -\beta \mathbf{s}_k \cdot \mathbf{g}_\phi(\mathbf{z}_k) \; .$$

This proves that the second condition (3.23b) of the Goldstein-Armijo principle is satisfied for $k \geq k_3$.

Let us define

$$\psi(\lambda) = \phi(\mathbf{z}_k + \mathbf{s}_k\lambda) \;,\; \psi'(\lambda) = \mathbf{g}_\phi(\mathbf{z}_k + \mathbf{s}_k\lambda) \cdot \mathbf{s}_k \text{ and}$$
$$\varrho(\lambda) = \psi(\lambda) - \psi(0) - \psi'(0)\lambda \;.$$

Then the Goldstein-Armijo conditions (3.23) can be rewritten in the form

$$-(1-\beta)\psi'(0) \leq \varrho(\lambda) \leq -(1-\alpha)\lambda\psi'(0) \;.$$

Suppose that the first Goldstein-Armijo condition (3.23a) is violated for $\lambda = 1$. Then

$$\varrho(1) \equiv \psi(1) - \psi(0) - \psi'(0) > -(1-\alpha)\psi'(0) > 0 \;.$$

Note that the quadratic

$$\pi_2(\lambda) = (1-\lambda)[\psi(0) + \lambda\{\psi(0) + \psi'(0)\}] + \lambda^2\psi(1) = \lambda^2\varrho(1) + \lambda\psi'(0) + \psi(0) \tag{3.26}$$

is such that $\pi_2(0) = \psi(0)$, $\pi_2'(0) = \psi'(0)$ and $\pi_2(1) = \psi(1)$. Furthermore, $\pi_2''(\lambda) = 2\varrho(1) > 0$, so the minimum of π_2 occurs at

$$\lambda_2 = -\frac{\psi'(0)}{2\varrho(1)} = -\frac{\psi'(0)}{2\{\psi(1) - \psi(0) - \psi'(0)\}} \;.$$

If $\alpha \leq 1/2$, then

$$\lambda_2 = \frac{-\psi'(0)}{2\varrho(1)} < \frac{-\psi'(0)}{-2(1-\alpha)\psi'(0)} = \frac{1}{2(1-\alpha)} \leq 1 \;.$$

The second formula for λ_2 also shows that if $\psi(1) \geq \psi(0)$ then we must have $\lambda_2 \leq 1/2$. To prevent very small step lengths, we replace λ_2 with $1/10$ if the former is smaller. If $\varrho(\lambda_2) \leq -(1-\alpha)\lambda_2\psi'(0)$, then we have achieved sufficient decrease without choosing the step length to be too small; under these conditions, we stop our search for the step length.

This leads us to the case when we have found two trial step lengths λ_1 and λ_2 at which the first Goldstein-Armijo condition (3.23a) has been violated. Initially we choose $\lambda_1 = 1$, and generally we have $0 < \lambda_2 < \lambda_1 \leq 1$. Consider the cubic

$$\begin{aligned}\pi_3(\lambda) = \;& \frac{\lambda^2(\lambda-\lambda_2)}{\lambda_1^2(\lambda_1-\lambda_2)}\psi(\lambda_1) + \frac{\lambda^2(\lambda_1-\lambda)}{\lambda_2^2(\lambda_1-\lambda_2)}\psi(\lambda_2) \\ &+ \frac{(\lambda_1-\lambda)(\lambda_2-\lambda)}{\lambda_1\lambda_2}\left\{\psi(0) + \lambda\left[\psi'(0) + \frac{\lambda_1+\lambda_2}{\lambda_1\lambda_2}\psi(0)\right]\right\} \;.\end{aligned} \tag{3.27}$$

This cubic has been designed so that $\pi_3(0) = \psi(0)$, $\pi_3'(0) = \psi'(0)$, $\pi_3(\lambda_1) = \psi(\lambda_1)$ and $\pi_3(\lambda_2) = \psi(\lambda_2)$. We can also see that

$$\pi_3'(\lambda) = 3\frac{\varrho(\lambda_1)/\lambda_1^2 - \varrho(\lambda_2)/\lambda_2^2}{\lambda_1 - \lambda_2}\lambda^2 - 2\frac{\lambda_2\varrho(\lambda_1)/\lambda_1^2 - \lambda_1\varrho(\lambda_2)/\lambda_2^2}{\lambda_1 - \lambda_2}\lambda + \psi'(0)$$
$$\equiv a\lambda^2 + b\lambda + \psi'(0) \; .$$

If $a > 0$ then π_3' has two real zeros with opposite signs, and the positive zero of π_3' is the local minimizer of π_3. If $a < 0$ and π_3' has real zeros, then both zeros have the same sign and the smaller zero of π_3' is the local minimizer of π_3. If $a = 0$, then π_3 is a quadratic; if $b > 0$ this quadratic has a local minimum. If the local minimizer of π_3 lies in the interval $[\lambda_2/10, \lambda/2]$, then we can use the local minimizer of π_3 as our next step length; otherwise we use the appropriate endpoint of this interval.

Alternatively, suppose instead that the second Goldstein-Armijo condition (3.23b) is violated. Then

$$\psi(1) < \psi(0) + \beta\psi'(0) < \psi(0) \; ,$$

so the step length $\lambda = 1$ leads to a decrease in the objective. The problem may be that the step $\mathbf{s}_k$ is small, so we search for a larger step length. As before, we can find the minimum of the quadratic π_2 defined in (3.26). Note that if $\beta \geq 3/4$, then

$$\lambda_2 = \frac{-\psi'(0)}{2\varrho(1)} > \frac{-\psi'(0)}{-2(1-\beta)\psi'(0)} = \frac{1}{2(1-\beta)} \geq 2 \; .$$

To prevent very large step lengths, we replace λ_2 by its minimum with 10. If $\psi(\lambda_2) \geq \psi(0) + \beta\psi'(0)\lambda_2$, then we have satisfied the second Goldstein-Armijo condition, and we use whichever of the two step lengths 1 and λ_2 has provided the smaller objective value. Otherwise we have decreased the objective even more but the second Goldstein-Armijo condition is still violated, so we continue to search for a longer step length.

This leads us to the case when we have found two trial step lengths λ_1 and λ_2 at which the second Goldstein-Armijo condition (3.23b) was violated. Initially we choose $\lambda_1 = 1$, and generally we have $1 < \lambda_1 < \lambda_2$. If it exists, we can find the local minimizer λ_3 of the cubic (3.27). To prevent inefficient searches, we replace λ_3 by its maximum with $2\lambda_2$ and minimum with $10\lambda_2$. If $\psi(\lambda_3) \geq \psi(0) + \beta\psi'(0)\lambda_3$, then we have satisfied the second Goldstein-Armijo condition, and we use whichever of the two step lengths λ_2 and λ_3 has provided the smaller objective value. Otherwise we have decreased the objective even more but the second Goldstein-Armijo condition is still violated, so we continue to search for a longer step length.

These ideas can be summarized by the following

Algorithm 3.5.2 (Line Search)

$\psi(0) = \phi(\mathbf{z}_k)$ and $\psi'(0) = \mathbf{g}_\phi(\mathbf{z}_k) \cdot \mathbf{s}_k$ and $\tau = -(1-\alpha)\psi'(0)$

$\lambda = \lambda_2 = \lambda_1 = 1$ and $\sigma_2 = \sigma_1 = \phi(\mathbf{z}_k + \mathbf{s}_k) - \psi(0) - \psi'(0)$

if $\sigma_2 > \tau$ then

  while $\sigma_2 > \tau/\lambda_2$ and $\mathbf{z}_k + \mathbf{s}_k\lambda_2 \neq \mathbf{z}_k$

    if $\lambda_2 = 1$ then $\lambda_2 = \min\{1/2\,,\ \max\{1/10\,,\ -\psi'(0)/[2\sigma_1]\}\}$

    else

      $\triangle\lambda = \lambda_1 - \lambda_2$ and $a = 3(\sigma_1 - \sigma_2)/\triangle\lambda$

      if $a = 0$ then

        $b = -(\lambda_2\sigma_1 - \lambda_1\sigma_2)/\triangle\lambda$

        $\lambda_3 = (b = 0\ ?\ \lambda_2\ :\ -\psi'(0)/(2b))$

      else

        $b = -(\lambda_2\sigma_1 - \lambda_1\sigma_2)/(a\triangle\lambda)$ and $c = \psi'(0)/a$ and $d = b^2 - c$

        if $d \geq 0$ then

          $d = \sqrt{d}$ and $R = (b \geq 0\ ?\ b + d\ :\ b - d)$ and $r = (R \neq 0\ ?\ c/R\ :\ 0)$

          $\lambda_3 = (a > 0\ ?\ \max\{R, r\}\ :\ \min\{R, r\})$

        else $\lambda_3 = \lambda_2$

      $\lambda_1 = \lambda_2$ and $\lambda = \lambda_2 = \min\{\lambda_2/2\,,\ \max\{\lambda_2/10\,,\ \lambda_3\}\}$ and $\sigma_1 = \sigma_2$

    $\sigma_2 = \left\{\phi(\mathbf{z}_k + \mathbf{s}_k\lambda_2) - \psi(0) - \psi'(0)\lambda_2\right\}/\lambda_2^2$

else

  $\tau = -(1-\beta)\psi'(0)$

  while $\sigma_2 < \tau/\lambda_2$

    if $\lambda_2 = 1$ then $\lambda_2 = \min\{10\,,\ \max\{2\,,\ -\psi'(0)/[2\sigma_1]\}\}$

    else

      $\triangle\lambda = \lambda_1 - \lambda_2$ and $a = 3(\sigma_1 - \sigma_2)/\triangle\lambda$

      if $a = 0$ then

        $b = -(\lambda_2\sigma_1 - \lambda_1\sigma_2)/\triangle\lambda$

        $\lambda_3 = (b = 0\ ?\ \lambda_2\ :\ -\psi'(0)/(2b))$

else

$b = -(\lambda_2\sigma_1 - \lambda_1\sigma_2)/(a\Delta\lambda)$ and $c = \psi'(0)/a$ and $d = b^2 - c$

if $d > 0$ then

$d = \sqrt{d}$ and $R = (b \geq 0\,?\, b + d \,:\, b - d)$ and $r = (R \neq 0\,?\, c/R \,:\, 0)$

$\lambda_3 = (a > 0\,?\, \max\{R, r\} \,:\, \min\{R, r\})$

else$\lambda_3 = \lambda_2$

$\lambda_1 = \lambda_2$ and $\lambda_2 = \min\{10\lambda_2\,,\, \max\{2\lambda_2\,,\, \lambda_3\}\}$

$\lambda = (\phi(\mathbf{z}_k + \mathbf{s}_k\lambda_2) < \phi(\mathbf{z}_k + \mathbf{s}_k\lambda_1)\,?\, \lambda_2 \,:\, \lambda_1)$ and $\sigma_1 = \sigma_2$

$\sigma_2 = \left\{\phi(\mathbf{z}_k + \mathbf{s}_k\lambda_2) - \psi(0) - \psi'(0)\lambda_2\right\}/\lambda_2^2$

The back-tracking portion of this algorithm (the first `while` loop) is similar to algorithm A6.3.1 in Dennis and Schnabel [52, p. 325ff].

Let us make some comments regarding line searches. First, we remark that the performance of the overall minimization algorithm typically depends more on the initial choice of each step length, rather than the Goldstein-Armijo principle. Typically, some descent directions have preferred step lengths, such as $\lambda = 1$ for Newton's method. In fact, the Goldstein-Armijo principle does not guarantee a good choice for the step length. For example, if α is close to zero and β is close to one, then small step lengths related to small reductions in the objective will satisfy the Goldstein-Armijo principle (3.23).

Secondly, the performance of a minimization algorithm depends more on the reduction in the value of the objective achieved by each step direction and corresponding line search. Note that the second Goldstein-Armijo condition (3.23b) does not require a decrease in the objective value. Rather, the purpose of this test is to make sure that the step length is not too small. Furthermore, the parameter α in the first Goldstein-Armijo condition (3.23a) is typically chosen to be very small, so that a small decrease in the objective is accepted.

Thirdly, we remark that if the gradient $\mathbf{g}_\phi$ of ϕ is expensive to evaluate, then we could modify the second Goldstein-Armijo condition (3.23b) to take the form

$$\mathbf{g}_\phi(\mathbf{z}_k + \mathbf{s}_k\lambda) \cdot \mathbf{s}_k \geq \beta \mathbf{g}_\phi(\mathbf{z}_k) \cdot \mathbf{s}_k \;.$$

This condition says that the directional derivative along the descent has been increased sufficiently above its initial negative value, and prevents the selection of very small step lengths.

It is difficult to provide uniform decisions to some questions regarding line searches. For example, it is not clear that we should always start at $\lambda = 1$ and reduce the step length if necessary. Far from the local minimizer, it is possible that $\mathbf{s}_k$ might be chosen to be very small, or that the directional derivative $\mathbf{g}_\phi(\mathbf{z}_k) \cdot \mathbf{s}_k$ is very small. In such cases, it might be advantageous to choose the step length λ to be greater than one. Also, fairly exact line searches, with β very small, are sometimes useful and other times wasteful.

As an alternative to the line search strategy by Dennis and Schnabel in Algorithm 3.5.2, readers may consider employing the Nelder-Mead Algorithm 3.9.2 to the function $\psi(\lambda) = \phi(\mathbf{z}_k + \mathbf{s}_k\lambda)$. The Nelder-Mead algorithm does not involve polynomial interpolations that may fail to have useful local minima, and naturally allows for the search region to expand or contract as it moves left or right:

Algorithm 3.5.3 (Nelder-Mead Approximate Line Search)

$\phi_k = \phi(\mathbf{z}_k)$ and $\sigma_k = \mathbf{g}_\phi(\mathbf{z}_k) \cdot \mathbf{s}_k$

$\lambda_0 = 0\ ,\ \psi_0 = \phi_k\ ,\ \lambda_1 = 1$ and $\psi_1 = \psi(1)$

fail1 $= (\phi_k - \psi_1 > -\alpha\sigma_k\lambda_1)$ and fail2 $= (\phi_k - \psi_1 \leq -\beta\sigma_k\lambda_1)$

if (not fail1 and not fail2) then return $\lambda = \lambda_1$

if $\psi_1 < \psi_0$ then

  $\tau = \lambda_0\ ,\ \lambda_0 = \lambda_1\ ,\ \lambda_1 = \tau$

  $\tau = \psi_0\ ,\ \psi_0 = \psi_1\ ,\ \psi_1 = \tau$

while true

  $\overline{\lambda} = (\lambda_0 + \lambda_1)/2$

  $\lambda_r = 2\overline{\lambda} - \lambda_1$ and $\psi_r = \psi(\lambda_r)$

  if $\psi_r < \psi_0$ then

    $\lambda_e = 3\overline{\lambda} - 2\lambda_1$ and $\psi_e = \psi(\lambda_e)$

    if $\psi_e < \psi_r$ then ($\lambda_1 = \lambda_e$ and $\psi_1 = \psi_e$)

    else ($\lambda_1 = \lambda_r$ and $\psi_1 = \psi_r$)

  else

    shrink = false

    if $\psi_r < \psi_1$ then ($\lambda_{oc} = 1.5\overline{\lambda} - 0.5\lambda_1$ and $\psi_{oc} = \psi(\lambda_{oc})$)

      if $\psi_{oc} \leq \psi_r$ then ($\lambda_1 = \lambda_{oc}$ and $\psi_1 = \psi_{oc}$)

      else shrink = true

    else

      $\lambda_{ic} = 0.5(\overline{\lambda} + \lambda_1)$ and $\psi_{ic} = \psi(\lambda_{ic})$

      if $\psi_{ic} \leq \psi_r$ then ($\lambda_1 = \lambda_{ic}$ and $\psi_1 = \psi_{ic}$)

else shrink $=$ true

if shrink then ($\lambda_1 = 1.5\lambda_0 - 0.5\lambda_1$ and $\psi_1 = \psi(\lambda_1)$)

if (fail1 and $\phi_k - \psi_1 \leq -\alpha\sigma_k\lambda_1$) then return $\lambda = \lambda_1$

if $\psi_1 < \psi_0$ then

$\tau = \lambda_0$, $\lambda_0 = \lambda_1$, $\lambda_1 = \tau$

$\tau = \psi_0$, $\psi_0 = \psi_1$, $\psi_1 = \tau$

if (fail2 and ($\phi_k - \psi_0 > -\beta\sigma_k\lambda_0$ or $\lambda_0 \leq \lambda_1$) then return $\lambda = \lambda_0$

We use the value of λ computed by this line search algorithm for the step length with step $\mathbf{s}_k$ in the minimization algorithm.

Readers may experiment with the JavaScript **descent direction** program. This program plots contours of function values for the Rosenbrock function

$$\mathbf{f}(\mathbf{x}) = \frac{1}{2}\left\|\begin{bmatrix}10(\mathbf{x}_2 - \mathbf{x}_1^2)\\ 1 - \mathbf{x}_1\end{bmatrix}\right\|_2^2 .$$

It also plots the surface of function values, and objective function values along the steepest descent and Newton directions. Readers may select an initial guess and see the functions associated with line searches in Algorithm 3.5.2 for these two step directions.

In summary, line searches are very useful, but we may waste much computational effort if we search too long in a fixed direction. Often, it is more efficient to adjust the step direction while determining the step length. We will develop such an approach in Sect. 3.6.1.

Exercise 3.5.5 Let $\phi(\mathbf{x}) = 3\mathbf{x}_1^2 + 2\mathbf{x}_1\mathbf{x}_2 + \mathbf{x}_2^2$ and $\mathbf{x}_0{}^\top = (1, 1)$.

1. What is the steepest descent direction for ϕ?
2. Is $\mathbf{s}^\top = (1, -1)$ a descent direction?

Exercise 3.5.6 Show that if $\mathbf{A}$ is symmetric positive-definite and

$$\phi(\mathbf{x}) = \frac{1}{2}\mathbf{x}^\top\mathbf{A}\mathbf{x} - \mathbf{b}^\top\mathbf{x} ,$$

and $\mathbf{A}$ is then for any $\mathbf{x}$, any $\alpha \leq 1/2$ and any $\beta > 0$, the Newton step satisfies the Goldstein-Armijo principle (3.23).

Exercise 3.5.7 Let $\phi(\mathbf{x}) = \mathbf{x}_1^4 + \mathbf{x}_2^2$, let $\mathbf{z}_k{}^\top = (1, 1)$ and let the search direction $\mathbf{s}_k$ be determined by the Newton step. Find $\mathbf{x}_{k+1}$ if it is determined by

1. the Newton step;
2. the back-tracking algorithm; and
3. a "perfect" line search: $\mathbf{s}_k \cdot \mathbf{g}_\phi(\mathbf{z}_k + 1) = 0$.

Exercise 3.5.8 Consider **Rosenbrock's function**

$$\phi(\mathbf{x}) = 100(\mathbf{x}_2 - \mathbf{x}_1^2)^2 + (1 - \mathbf{x}_1)^2 .$$

The minimum of this function is $\phi(\mathbf{z}) = 0$ at $\mathbf{z}^\top = (1, 1)$. Consider algorithms to minimize this function starting at the initial guess $(-1.2, 1)$.

1. Program the steepest descent method with a line search to minimize ϕ. Describe why you chose your form of the line search.
2. Program Newton's method to minimize ϕ. Describe how you choose your step length.
3. Plot $\ln(\phi(\mathbf{z}_k) - \phi(\mathbf{z}))$ versus step number k and versus computer time. (You may have to repeat each step many thousands of times in order to get meaningful timings.)

Exercise 3.5.9 Let the $n \times n$ matrix $\mathbf{A}$ be nonsingular, and let

$$\widetilde{\phi}(\mathbf{x}) = \phi(\mathbf{A}\mathbf{x}) .$$

1. Use the chain rule (3.3) to show that

$$\mathbf{g}_{\widetilde{\phi}}(\mathbf{x}) = \mathbf{A}^\top \mathbf{g}_\phi(\mathbf{A}\mathbf{x}) ,$$

   and

$$\mathbf{H}_{\widetilde{\phi}}(\mathbf{x}) = \mathbf{A}^\top \mathbf{H}_\phi(\mathbf{A}\mathbf{x})\mathbf{A} .$$

2. Let the steepest descent steps in the two coordinate systems be $\mathbf{s}^{(G)}(\mathbf{x}) = -\mathbf{g}_\phi(\mathbf{x})$ and $\widetilde{\mathbf{s}}^{(G)}(\mathbf{x}) = -\mathbf{g}_{\widetilde{\phi}}(\mathbf{x})$. Show that

$$\mathbf{s}^{(G)}(\mathbf{A}\mathbf{x}) = \mathbf{A}^{-\top}\widetilde{\mathbf{s}}^{(G)}(\mathbf{x}) .$$

3. Let the Newton steps in the two coordinate systems be $\mathbf{s}^{(N)}(\mathbf{x}) == -\mathbf{H}_\phi(\mathbf{x})^{-1}\mathbf{g}_\phi(\mathbf{x})$ and $\widetilde{\mathbf{s}}^{(N)}(\mathbf{x}) = -\mathbf{H}_{\widetilde{\phi}}(\mathbf{x})^{-1}\mathbf{g}_{\widetilde{\phi}}(\mathbf{x})$. Show that

$$\mathbf{s}^{(N)}(\mathbf{A}\mathbf{x}) = \mathbf{A}\widetilde{\mathbf{s}}^{(N)}(\mathbf{x}) .$$

4. Conclude that the Newton step is invariant under coordinate changes of any kind (i.e. $\mathbf{A}$ nonsingular), but the steepest descent step is invariant only under orthogonal coordinate changes (i.e., $\mathbf{A}$ orthogonal).

3.6 Global Convergence

An algorithm is **locally convergent** to a point if and only if there is some ball around the point such that whenever we start the algorithm within that ball, the algorithm converges to that point. An algorithm is **globally convergent** if and only if no matter where the algorithm starts, it converges to a point with certain desired properties (e.g., a zero of a vector-valued function, or a minimum of a scalar-valued function). Because we often encounter problems for which we do not have good initial guesses, globally convergent algorithms can be essential to optimization software.

In Sect. 3.5.3.1, we saw that steepest descent combined with exact line searches is globally convergent either to a saddle point or to a local minimizer. In Theorem 3.5.1, we saw that a sequence of descent directions coupled, with the Goldstein-Armijo principle approximate line searches, is globally convergent to a zero of the gradient of a functional that is bounded below. These two minimization strategies involve selecting a particular descent direction, and then selecting a step length for that direction in order to achieve sufficient decrease in the objective. Our goal in this section will be to combine the selection of the descent direction with the selection of the step length. Hopefully this strategy will reduce the overall computational work.

3.6.1 Trust Regions

In this section, we will describe some clever ideas developed by M.J.D. Powell [143] to encourage convergence of iterative methods from arbitrary initial guesses. Powell's approach uses a hybridization of Newton steps and steepest descent steps, and combines step selection with step length determination. Detailed discussion of the method can be found in Dennis and Schnabel [52, p. 111ff], Gill et al. [81, p. 113ff] or Kelley [108, p. 50ff]. Fortran implementations of the method are available in routine `ns12` of the Harwell Software Library, routines `hybrj1` or `hybrd1` of MINPACK, or in TOMS 611. Users preferring a MATLAB implementation can consider ntrust.m. or ENTRUST.

3.6.1.1 A Special Constrained Minimization Problem

The key idea in Powell's global convergence strategy is to minimize a quadratic model for the objective over a ball of some given radius, called a **trust region**. If the radius is chosen properly, the local model is accurate and its constrained minimum is a good approximation to a local minimizer of the original objective. If the local model is not accurate, the trust region radius will need to be decreased. If we are far from the true local minimizer but the trust region computation has been making good progress, we may be able to increase the radius.

Let us make the description of trust regions more precise. Suppose that we are given a real functional ϕ that is twice continuously differentiable. Suppose that at some step k of the trust region algorithm, we have obtained a current approximation $\mathbf{z}_k$ to the minimum $\mathbf{z}$ of ϕ. Our local model is

$$\widetilde{\phi}(\mathbf{s}) = \phi(\mathbf{z}_k) + \mathbf{g}_\phi(\mathbf{z}_k) \cdot \mathbf{s} + \frac{1}{2}\mathbf{s} \cdot \mathbf{H}_\phi(\mathbf{z}_k)\mathbf{s} \, .$$

For each approximate zero $\mathbf{z}_k$, the goal of the trust region is to choose the new step $\mathbf{s}_k$ to minimize $\widetilde{\phi}(\mathbf{s})$ subject to the constraint $\|\mathbf{s}\|_2 \le \delta$. The following lemma describes the solution of this constrained minimization problem.

Lemma 3.6.1 *Let δ be a positive scalar, $\mathbf{g}$ a real n-vector, and $\mathbf{H}$ a real symmetric $n \times n$ matrix. Suppose that the n-vector $\mathbf{s}$ minimizes*

$$\widetilde{\phi}(\mathbf{s}) = \mathbf{g} \cdot \mathbf{s} + \frac{1}{2}\mathbf{s} \cdot \mathbf{H}\mathbf{s} \textit{ subject to } \mathbf{s} \cdot \mathbf{s} \le \delta^2 \, .$$

Then there is a scalar $\lambda \ge 0$ so that

$$(\mathbf{H} + \mathbf{I}\lambda)\mathbf{s} = -\mathbf{g} \, .$$

Proof If the minimum $\mathbf{s}$ of $\widetilde{\phi}$ occurs in the interior of the constraint set, then the gradient of $\widetilde{\phi}$ must be zero there:

$$\mathbf{0} = \mathbf{g}_{\widetilde{\phi}}(\mathbf{s}) = \mathbf{g} + \mathbf{H}\mathbf{s} \Longrightarrow \mathbf{H}\mathbf{s} = -\mathbf{g} \, .$$

In this case, the claim holds with $\lambda = 0$.

Otherwise, the minimum of $\widetilde{\phi}$ occurs on the boundary of the constrained region. In this case, the Kuhn-Tucker first-order necessary conditions in Theorem 4.3.2 show that the gradient of $\widetilde{\phi}$ (i.e. the direction of steepest ascent) must be in the direction of the inner normal to the constraint set. In other words,

$$\mathbf{g}_k + \mathbf{H}_k\mathbf{s} = -\mathbf{s}\lambda$$

for some Lagrange multiplier $\lambda \ge 0$. This is equivalent to the claimed result.

Next, we examine the length of the constrained step.

Lemma 3.6.2 *Let δ be a positive scalar, $\mathbf{g}$ a real n-vector, and $\mathbf{H}$ a real symmetric $n \times n$ matrix. Suppose that $\mathbf{H} + \mathbf{I}\lambda$ is nonnegative,*

$$\mathbf{s}(\lambda) = -(\mathbf{H} + \mathbf{I}\lambda)^{-1}\mathbf{g} \textit{ and } \sigma(\lambda) = \|\mathbf{s}(\lambda)\|_2^2 \, .$$

Then

$$\sigma'(\lambda) \le 0 \textit{ and } \sigma''(\lambda) \ge 0 \, .$$

Proof Since

$$(\mathbf{H} + \mathbf{I}\lambda)\mathbf{s}(\lambda) = -\mathbf{g} ,$$

we can differentiate to get

$$\mathbf{s}(\lambda) + (\mathbf{H} + \mathbf{I}\lambda)\mathbf{s}'(\lambda) = \mathbf{0} , \tag{3.28}$$

and then solve for $\mathbf{s}'(\lambda)$ to get

$$\mathbf{s}'(\lambda) = -(\mathbf{H} + \mathbf{I}\lambda)^{-1}\mathbf{s}(\lambda) .$$

We can differentiate (3.28) to get

$$2\mathbf{s}'(\lambda) + (\mathbf{H} + \mathbf{I}\lambda)\mathbf{s}''(\lambda) = \mathbf{0} ,$$

and solve to get

$$\mathbf{s}''(\lambda) = -2(\mathbf{H} + \mathbf{I}\lambda)^{-1}\mathbf{s}'(\lambda) = 2(\mathbf{H} + \mathbf{I}\lambda)^{-2}\mathbf{s}(\lambda) .$$

Thus

$$\sigma'(\lambda) = 2\mathbf{s}(\lambda) \cdot \mathbf{s}'(\lambda) = -2\mathbf{s}(\lambda) \cdot (\mathbf{H} + \mathbf{I}\lambda)^{-1}\mathbf{s}(\lambda) \leq 0 ,$$

and

$$\sigma''(\lambda) = 2\mathbf{s}'(\lambda) \cdot \mathbf{s}'(\lambda) + 2\mathbf{s}(\lambda) \cdot \mathbf{s}''(\lambda) = 2\|\mathbf{s}'(\lambda)\|_2^2 + 4\|(\mathbf{H} + \mathbf{I}\lambda)^{-1}\mathbf{s}(\lambda)\|_2^2 \geq 0 .$$

Next, we show that if λ is chosen to be the Lagrange multiplier in the constrained minimization problem of Lemma 3.6.1, then it can be used to produce a useful modification of the Hessian.

Lemma 3.6.3 *Let δ be a positive scalar, $\mathbf{g}$ a real n-vector, and $\mathbf{H}$ a real symmetric $n \times n$ matrix. Let $\lambda \geq 0$ be the Lagrange multiplier for the constrained minimization problem*

$$\textit{minimize } \widetilde{\phi}(\mathbf{s}) \equiv \mathbf{g} \cdot \mathbf{s} + \frac{1}{2}\mathbf{s} \cdot \mathbf{H}\mathbf{s} \textit{ subject to } \mathbf{s} \cdot \mathbf{s} \leq \delta ,$$

such that the minimizer $\mathbf{s}$ satisfies

$$(\mathbf{H} + \mathbf{I}\lambda)\mathbf{s} = -\mathbf{g} .$$

Then $\mathbf{H} + \mathbf{I}\lambda$ is nonnegative.

Proof If $\|\mathbf{s}\|_2 < \delta$, then the second-order condition for a local minimizer in Lemma 3.2.11 shows that $\mathbf{H}$ is nonnegative, and the proof of Lemma 3.6.1 shows that $\lambda = 0$. In this case, the claim is obvious.

Otherwise, we have $\|\mathbf{s}\|_2 = \delta$, and Lemma 3.6.1 shows that there is a nonnegative scalar λ so that

$$(\mathbf{H} + \mathbf{I}\lambda)\mathbf{s} = -\mathbf{g} \; .$$

Let $\mathbf{d} \perp \mathbf{s}$ with $\|\mathbf{d}\|_2 = \delta$. Then for all $\varepsilon > 0$ the Pythagorean Theorem 3.2.2 in Chap. 3 of Volume I shows that

$$\|\mathbf{s}\sqrt{1-\varepsilon^2} + \mathbf{d}\varepsilon\|_2^2 = \|\mathbf{s}\|_2^2(1-\varepsilon^2) + \|\mathbf{d}\|_2^2\varepsilon^2 = \delta^2 \; ,$$

so $\mathbf{s}\sqrt{1-\varepsilon^2} + \mathbf{d}\varepsilon$ is on the boundary of the constraint region. Then

$$\begin{aligned}
\widetilde{\phi}(\mathbf{s}\sqrt{1-\varepsilon^2} + \mathbf{d}\varepsilon) &= \mathbf{g} \cdot (\mathbf{s}\sqrt{1-\varepsilon^2} + \mathbf{d}\varepsilon) + \frac{1}{2}(\mathbf{s}\sqrt{1-\varepsilon^2} + \mathbf{d}\varepsilon) \cdot \mathbf{H}(\mathbf{s}\sqrt{1-\varepsilon^2} + \mathbf{d}\varepsilon) \\
&= \mathbf{g} \cdot \mathbf{s}\sqrt{1-\varepsilon^2}\frac{1-\varepsilon^2}{2}\mathbf{s} \cdot \mathbf{Hs} + \varepsilon\mathbf{g} \cdot \mathbf{d} + \varepsilon\sqrt{1-\varepsilon^2}\mathbf{s} \cdot \mathbf{Hd} + \frac{\varepsilon^2}{2}\mathbf{d} \cdot \mathbf{Hd} \\
&= \widetilde{\phi}(\mathbf{s}) - \frac{\varepsilon^2}{1+\sqrt{1-\varepsilon^2}}\mathbf{g} \cdot \mathbf{s} - \frac{\varepsilon^2}{2}\mathbf{s} \cdot \mathbf{Hs} + \varepsilon\mathbf{g} \cdot \mathbf{d} - \varepsilon\sqrt{1-\varepsilon^2}(\mathbf{g} + \mathbf{s}\lambda) \cdot \mathbf{d} + \frac{\varepsilon^2}{2}\mathbf{d} \cdot \mathbf{Hd} \\
&= \widetilde{\phi}(\mathbf{s}) + \frac{\varepsilon^3}{1+\sqrt{1-\varepsilon^2}}\mathbf{g} \cdot \mathbf{d} + \varepsilon^2\left\{-\frac{\mathbf{g} \cdot \mathbf{s}}{1+\sqrt{1-\varepsilon^2}} - \frac{\mathbf{s} \cdot \mathbf{Hs}}{2} + \frac{\mathbf{d} \cdot \mathbf{Hd}}{2}\right\} \\
&= \widetilde{\phi}(\mathbf{s}) + \varepsilon^2\left\{-\frac{\mathbf{g} \cdot \mathbf{s}}{1+\sqrt{1-\varepsilon^2}} + \frac{\mathbf{s} \cdot (\mathbf{g} + \mathbf{s}\lambda)}{2} + \frac{\mathbf{d} \cdot \mathbf{Hd}}{2}\right\} + \frac{\varepsilon^3}{1+\sqrt{1-\varepsilon^2}}\mathbf{g} \cdot \mathbf{d} \\
&= \widetilde{\phi}(\mathbf{s}) + \varepsilon^2\left\{\frac{\varepsilon^2}{2(1+\sqrt{1-\varepsilon^2})}\mathbf{g} \cdot \mathbf{s} + \frac{\lambda}{2}\delta^2 + \frac{\mathbf{d} \cdot \mathbf{Hd}}{2}\right\} + \frac{\varepsilon^3}{1+\sqrt{1-\varepsilon^2}}\mathbf{g} \cdot \mathbf{d} \\
&= \widetilde{\phi}(\mathbf{s}) + \frac{\varepsilon^2}{2}\mathbf{d} \cdot (\mathbf{H} + \mathbf{I}\lambda)\mathbf{d} + \frac{\varepsilon^3}{1+\sqrt{1-\varepsilon^2}}\mathbf{g} \cdot \mathbf{d} + \frac{\varepsilon^4}{2(1+\sqrt{1-\varepsilon^2})}\mathbf{g} \cdot \mathbf{s} \; .
\end{aligned}$$

This shows that if $\mathbf{H} + \mathbf{I}\lambda$ is not nonnegative, we can choose $\mathbf{d}$ and ε to reduce the value of the objective.

In summary, if $\mathbf{H}$ has a negative eigenvalue, then the minimum of the trust region local model $\widetilde{\phi}$ will occur on the boundary of the trust region. Further, the solution of the constrained least squares problem will choose a Lagrange multiplier such that $\mathbf{H} + \mathbf{I}\lambda$ is nonnegative. If the Lagrange multiplier can be chosen to be zero, then the step chosen by the trust region is the Newton step.

It is harder to relate the Lagrange multiplier to the radius δ of the trust region. Qualitatively, all we can say is that if δ is large and $\mathbf{H}$ is positive, then the trust region step should be the Newton step; if δ is small (compared to the Newton step size) then the trust region step should look more like steepest descent.

3.6.1.2 Hook Step

Lemma 3.6.3 showed that the constrained minimization step satisfies

$$\mathbf{s}(\lambda) = -(\mathbf{H} + \mathbf{I}\lambda)^{-1}\mathbf{g}$$

for some nonnegative scalar λ that solves the nonlinear equation

$$\sigma(\lambda) \equiv \|\mathbf{s}(\lambda)\|_2 - \delta = 0 \; . \tag{3.29}$$

The proof of Lemma 3.6.2 shows that

$$\sigma'(\lambda) = \frac{\mathbf{s}(\lambda) \cdot \mathbf{s}'(\lambda)}{\|\mathbf{s}(\lambda)\|_2} = -\frac{\mathbf{s}(\lambda) \cdot [\mathbf{H} + \mathbf{I}\lambda]\,\mathbf{s}(\lambda)}{\|\mathbf{s}(\lambda)\|_2} \, ,$$

The computation of $\sigma(\lambda)$ and $\sigma'(\lambda)$ can be performed as follows. Recall that Lemma 3.6.3 showed that at the true value of λ the matrix $\mathbf{H} + \mathbf{I}\lambda$ must be nonnegative. Given a guess for λ, we can use Algorithm 3.5.1 to find $\mu \geq \lambda$ so that

$$\mathbf{H} + \mathbf{I}\mu = \mathbf{L}\mathbf{L}^\top \, ,$$

and then we solve

$$\mathbf{L}\mathbf{L}^\top \mathbf{s} = -\mathbf{g}$$

for $\mathbf{s}$. Next, we solve

$$\mathbf{L}\mathbf{y} = \mathbf{s}$$

for $\mathbf{y}$, and then compute

$$\sigma'(\lambda) = -\frac{\|\mathbf{y}\|_2^2}{\|\mathbf{s}\|_2} \, .$$

These computations can be summarized by the following

Algorithm 3.6.1 (Hook Step)

given λ_0

for $0 \leq \ell$

use algorithm (3.5.1) to factor $\mathbf{H} + \mathbf{I}\mu_\ell = \mathbf{L}\mathbf{L}^\top$ where $\mu_\ell \geq \lambda_\ell$

solve $\mathbf{L}\mathbf{L}^\top \mathbf{s} = -\mathbf{g}$ for $\mathbf{s}$

solve $\mathbf{L}\mathbf{y} = \mathbf{s}$ for $\mathbf{y}$

$$\lambda_{\ell+1} = \lambda_\ell + \frac{\|\mathbf{s}\|_2}{\|\mathbf{y}\|_2}(\|\mathbf{s}\|_2 - \delta)$$

The problem with this algorithm is that Lemma 3.6.2 shows that whenever $\mathbf{H} + \mathbf{I}\lambda$ is nonnegative, then $\sigma'(\lambda) \leq 0$ and $\sigma''(\lambda) \geq 0$. In other words, σ is decreasing and convex. Consequently, Newton's method will always underestimate the true value of the Lagrange multiplier λ.

To overcome this problem, we can approximate the trust region constraint

$$\sigma(\lambda) \equiv \|\mathbf{s}(\lambda)\| - \delta = 0$$

by the local model

$$\widetilde{\sigma}(\lambda) \equiv \frac{\alpha}{\beta + \lambda} - \delta = 0 \; .$$

Given a value λ_ℓ for the Lagrange multiplier, we will choose the two parameters α and β so that

$$\widetilde{\sigma}(\lambda_\ell) = \sigma(\lambda_\ell) \; , \text{ and } \widetilde{\sigma}'(\lambda_\ell) = \sigma'(\lambda_\ell) \; .$$

The solutions are

$$\alpha = -\frac{\|\mathbf{s}(\lambda_\ell)\|_2^2}{\sigma'(\lambda_\ell)} \text{ and } \beta = -\frac{\|\mathbf{s}(\lambda_\ell)\|_2}{\sigma'(\lambda_\ell)} - \lambda_\ell \; .$$

Our next Lagrange multiplier is chosen to be the zero of $\widetilde{\sigma}$:

$$\lambda_{\ell+1} = \frac{\alpha}{\delta} - \beta = \lambda_\ell - \frac{\|\mathbf{s}(\lambda_\ell)\|}{\delta} \frac{\sigma(\lambda_\ell)}{\sigma'(\lambda_\ell)} \; .$$

When the current guess λ_ℓ for the Lagrange multiplier produces a step $\mathbf{s}(\lambda_\ell)$ that lies outside the trust region, then this local model produces a correction to the Lagrange multiplier that is larger than that produced by Newton's method to find a zero of σ. When the step lies inside the trust region, then this local model produces a smaller correction to λ than Newton's method. As λ_ℓ approaches the true value λ, the local model produces an iterate that approaches the Newton iterate. We can summarize the new approach by the

Algorithm 3.6.2 (Hook Step via Local Model)

given λ_0

for $0 \leq \ell$

use algorithm (3.5.1) to factor $\mathbf{H} + \mathbf{I}\mu_\ell = \mathbf{L}\mathbf{L}^\top$ where $\mu_\ell \geq \lambda_\ell$

solve $\mathbf{L}\mathbf{L}^\top\mathbf{s} = -\mathbf{g}$ for $\mathbf{s}$

solve $\mathbf{L}\mathbf{y} = \mathbf{s}$ for $\mathbf{y}$

$$\lambda_{\ell+1} = \lambda_\ell - \left(\frac{\|\mathbf{s}\|_2}{\|\mathbf{y}\|_2}\right)^2 \left(1 - \frac{\|\mathbf{s}\|_2}{\delta}\right)$$

Still, we need to specify how we choose the initial Lagrange multiplier λ_0. For the first trust region, we can take $\lambda_0 = 0$. For subsequent trust regions, λ_0 can be obtained by using the factorization of $\mathbf{H} + \mathbf{I}\lambda$ from the previous trust region to compute $\sigma(\lambda)$ and $\sigma'(\lambda)$ for the new trust region radius δ.

Next, we remark that it is not necessary to solve Eq. (3.29) accurately for λ. Typically, we terminate the iteration for the Lagrange multiplier λ when $0.75 \leq \|\mathbf{s}(\lambda)\|_2/\delta \leq 1.5$.

As we saw in Sect. 5.6.1 in Chap. 5 of Volume I, it is useful to bracket the solution of a nonlinear scalar equation in order to enforce global convergence. Suppose that Algorithm 3.5.1 computes $\mu \in [0, \lambda)$ so that $\mathbf{H} + \mathbf{I}\mu$ is positive. Then the desired Lagrange multiplier λ satisfies

$$\delta = \left\|(\mathbf{H} + \mathbf{I}\lambda)^{-1}\mathbf{g}\right\|_2 = \left\|(\mathbf{H} + \mathbf{I}\mu] + \mathbf{I}[\lambda - \mu])^{-1}\mathbf{g}\right\|_2 \leq \|\mathbf{g}\|_2/(\lambda - \mu) \,.$$

This implies that

$$\mu \leq \lambda < \mu + \frac{\|\mathbf{g}\|_2}{\delta} \,.$$

Note that when we are close to the minimum of the objective, then $\mathbf{H}$ is positive and $\mu = 0$.

An implementation of the Hook step can be found in Dennis and Schnabel [52, p. 332ff]. Readers may also experiment with the JavaScript **hook step** program. This program plots contours of function values for the Rosenbrock function

$$\mathbf{f}(\mathbf{x}) = \frac{1}{2}\left\|\begin{bmatrix} 10(\mathbf{x}_2 - \mathbf{x}_1^2) \\ 1 - \mathbf{x}_1 \end{bmatrix}\right\|_2^2 \,.$$

Users may select a point in this graph, after which the program will plot the curve of points $\mathbf{x} - (\mathbf{H}_\phi(\mathbf{x}) + \mathbf{I}\mu)^{-1}\mathbf{g}_\phi(\mathbf{x})$ for which $\mathbf{H}_\phi(\mathbf{x}) + \mathbf{I}\mu$ is positive.

3.6.1.3 Double Dogleg Step

In Sect. 3.6.1.1, we learned how to determine the optimal solution to a constrained minimization problem involving a quadratic objective and an upper bound on the size of the step length. The solution of this problem showed that the optimal step is a modification of the Newton step, in which the Hessian is augmented by a scalar multiple of the identity matrix. We examined a way to choose this scalar so that the constraint is approximately satisfied; this lead to the so-called Hook step. Our next approach is to approximate the curve of Hook steps by a piecewise linear function, and find its intersection with the trust region.

In the development of this method, we will assume that we have a local model

$$\widetilde{\phi}(\mathbf{s}) = \mathbf{g} \cdot \mathbf{s} + \frac{1}{2}\mathbf{s} \cdot \mathbf{H}\mathbf{s}$$

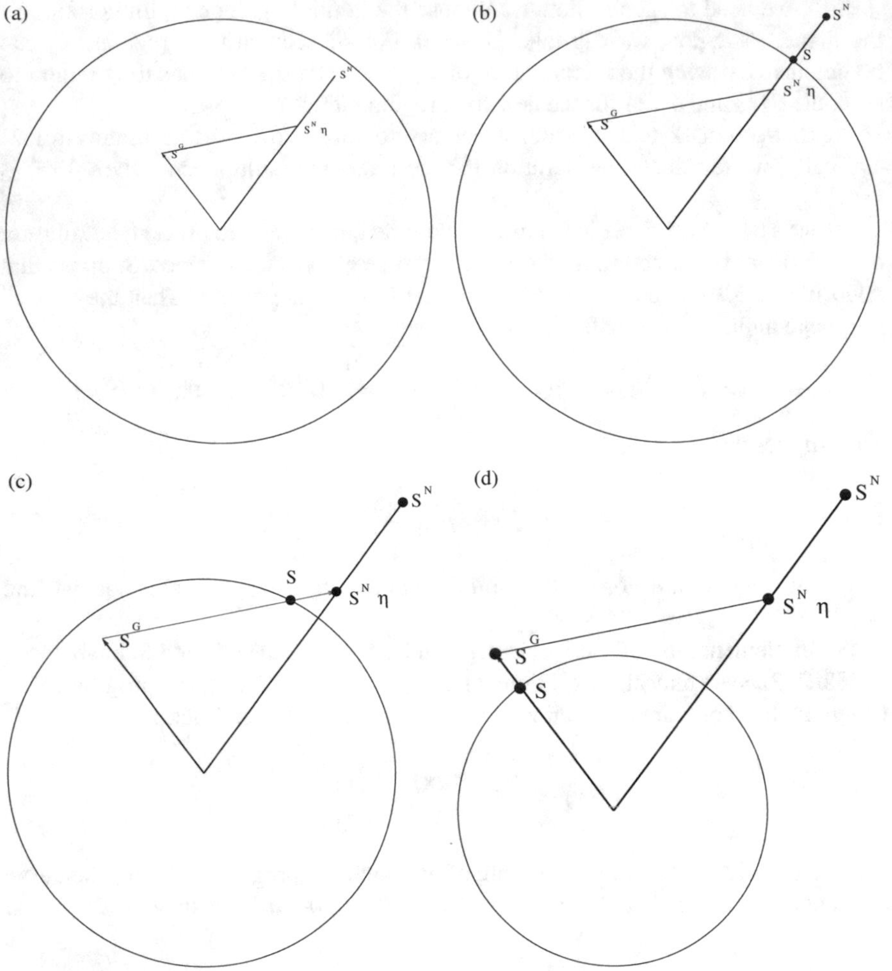

Fig. 3.2 Trust region step selection cases. (**a**) Newton step. (**b**) Scaled Newton step. (**c**) point between Cauchy step and scaled Newton step. (**d**) scaled Cauchy step

with positive matrix **H**. If necessary, **H** can be replaced by a safeguarded Cholesky factorization, using Algorithm 3.5.1.

There are three special points along the piecewise linear curve: The **Cauchy step** $\mathbf{s}^G$ is the minimum of the unconstrained local model $\widetilde{\phi}$ in the steepest descent direction. The **Newton step** $\mathbf{s}^N$ is the global minimizer of the unconstrained local model $\widetilde{\phi}$. Finally, the **scaled Newton step** $\mathbf{s}^N\eta$ is an appropriately chosen scalar multiple of the Newton point. The double dogleg strategy is illustrated in Fig. 3.2.

In order to compute the Newton step, we solve

$$\mathbf{H}\mathbf{s}^N = -\mathbf{g}$$

for $\mathbf{s}^N$. If $\|\mathbf{s}^N\| \leq \delta$, then the double dogleg algorithm uses this step to update the approximate minimum of the objective.

It is also easy to compute the Cauchy step. Along the steepest descent direction, the local model has the value

$$\widetilde{\phi}(-\mathbf{g}\lambda) = -\mathbf{g} \cdot \mathbf{g}\lambda + \mathbf{g} \cdot \mathbf{H}\mathbf{g}\lambda^2 .$$

The minimum occurs at

$$\mathbf{s}^G \equiv -\mathbf{g}\lambda^G \text{ where } \lambda^G = \frac{\mathbf{g} \cdot \mathbf{g}}{\mathbf{g} \cdot \mathbf{H}\mathbf{g}} . \tag{3.30}$$

If $\|\mathbf{s}^G\|_2 \geq \delta$, then the local model is decreasing from the center of the trust region to the Cauchy step. In this case, the double dogleg algorithm uses $\mathbf{s}^G$ to approximate the minimum of the objective.

This leaves the case when the Newton step lies outside the trust region but the Cauchy step lie inside. Under these circumstances, the double dogleg algorithm chooses a hybrid step. In order for this hybrid step to make sense, we need to guarantee that as we proceed along our double dogleg curve from the center of the trust region to the Cauchy step, then to the scaled Newton step and finally to the Newton step, both *the distance from the center of the trust region increases monotonically and the value of the local model decreases monotonically.*

Our first condition on the double dogleg curve will be guaranteed by the following lemma.

Lemma 3.6.4 *Suppose that* $\mathbf{g}$ *is a real nonzero n-vector, and* $\mathbf{H}$ *is a real symmetric positive n* $\times$ *n matrix. Define the Cauchy step*

$$\mathbf{s}^G = -\mathbf{g}\frac{\mathbf{g} \cdot \mathbf{g}}{\mathbf{g} \cdot \mathbf{H}\mathbf{g}}$$

and the Newton step

$$\mathbf{s}^N = -\mathbf{H}^{-1}\mathbf{g} .$$

Then

$$1 \geq \gamma \equiv \frac{\mathbf{g} \cdot \mathbf{s}^G}{\mathbf{g} \cdot \mathbf{s}^N} \tag{3.31}$$

and

$$\left\|\mathbf{s}^G\right\|_2 \leq \gamma \left\|\mathbf{s}^N\right\|_2 .$$

Proof Let

$$\mathbf{u} = \mathbf{H}^{1/2}\mathbf{g} \text{ and } \mathbf{w} = \mathbf{H}^{-1/2}\mathbf{g} .$$

The definitions of the Cauchy step and the Newton step imply that

$$\gamma = \frac{\mathbf{g} \cdot \mathbf{s}^G}{\mathbf{g} \cdot \mathbf{s}^N} = \frac{\mathbf{g} \cdot \mathbf{g}}{\mathbf{g} \cdot \mathbf{H}\mathbf{g}} \frac{\mathbf{g} \cdot \mathbf{g}}{\mathbf{g} \cdot \mathbf{H}^{-1}\mathbf{g}} = \frac{\mathbf{u} \cdot \mathbf{w}}{\mathbf{u} \cdot \mathbf{u}} \frac{\mathbf{u} \cdot \mathbf{w}}{\mathbf{w} \cdot \mathbf{w}}$$

then the Cauchy inequality (3.15) in Chap. 3 of Volume I gives us

$$\leq \frac{\|\mathbf{u}\|_2 \|\mathbf{w}\|_2}{\|\mathbf{u}\|_2^2} \frac{\|\mathbf{u}\|_2 \|\mathbf{w}\|_2}{\|\mathbf{w}\|_2^2} = 1 \, .$$

This proves the first claim.

To prove the second claim, we note that

$$\left\|\mathbf{s}^G\right\|_2 = \frac{\|\mathbf{g}\|_2^3}{\mathbf{g} \cdot \mathbf{H}\mathbf{g}}$$

then we use the Cauchy inequality (3.15) in Chap. 3 of Volume I to obtain

$$\leq \frac{\|\mathbf{g}\|_2^3}{\mathbf{g} \cdot \mathbf{H}\mathbf{g}} \frac{\|\mathbf{g}\|_2 \left\|\mathbf{H}^{-1}\mathbf{g}\right\|_2}{\mathbf{g} \cdot \mathbf{H}^{-1}\mathbf{g}} = \gamma \left\|\mathbf{H}^{-1}\mathbf{g}\right\|_2 = \gamma \left\|\mathbf{s}^N\right\|_2 \, .$$

To guarantee that the scaled Newton step is farther from the center of the trust region than the Cauchy step, we typically choose

$$\mathbf{s}^N \eta = -\mathbf{H}^{-1}\mathbf{g}\eta \text{ where } \eta = (1 + 4\gamma)/5 \, .$$

This is a weighted average of γ and 1, so $\eta \in (\gamma, 1)$. It follows that

$$\left\|\mathbf{s}^G\right\|_2 \leq \gamma \left\|\mathbf{s}^N\right\|_2 < \left\|\mathbf{s}^N \eta\right\|_2 < \left\|\mathbf{s}^N\right\|_2 \, .$$

The next lemma will guarantee that the local model decreases monotonically along the double dogleg curve.

Lemma 3.6.5 *Suppose that* $\mathbf{g}$ *is a real nonzero n-vector, and* $\mathbf{H}$ *is a real symmetric positive* $n \times n$ *matrix. Define the Cauchy step*

$$\mathbf{s}^G = -\mathbf{g} \frac{\mathbf{g} \cdot \mathbf{g}}{\mathbf{g} \cdot \mathbf{H}\mathbf{g}} \, ,$$

the Newton step

$$\mathbf{s}^N = -\mathbf{H}^{-1}\mathbf{g}$$

and the scaled Newton step $\mathbf{s}^N \eta$ *where*

$$\eta = 0.2 + 0.8\gamma \text{ and } \gamma = \frac{\mathbf{g} \cdot \mathbf{s}^G}{\mathbf{g} \cdot \mathbf{s}^N} \, .$$

Define the quadratic function

$$\widetilde{\phi}(\mathbf{s}) = \mathbf{g} \cdot \mathbf{s} + \frac{1}{2} \mathbf{s} \cdot \mathbf{H}_k \mathbf{s} .$$

Then $\widetilde{\phi}\left(\mathbf{s}^G \alpha\right)$, $\widetilde{\phi}\left(\mathbf{s}^N \alpha\right)$ *and* $\widetilde{\phi}\left(\mathbf{s}^G + \left[\mathbf{s}^N \eta - \mathbf{s}^G\right] \alpha\right)$ *are all decreasing functions of* α *for* $\alpha \in (0, 1)$.

Proof Let

$$\beta = \frac{\mathbf{g} \cdot \mathbf{g}}{\mathbf{g} \cdot \mathbf{H}\mathbf{g}} ,$$

so that $\mathbf{s}^G = -\mathbf{g}\beta$. Then note that the function

$$\psi_G(\alpha) \equiv \widetilde{\phi}\left(\mathbf{s}^G \alpha\right) = -\mathbf{g} \cdot \mathbf{g} \beta \alpha + \frac{\alpha^2 \beta^2}{2} \mathbf{g} \cdot \mathbf{H}\mathbf{g}$$

has derivative

$$\psi_G'(\alpha) = -\mathbf{g} \cdot \mathbf{g} \beta + \alpha \beta^2 \mathbf{g} \cdot \mathbf{H}\mathbf{g} = -(1-\alpha) \mathbf{g} \cdot \mathbf{g} \beta .$$

This is obviously negative for $\alpha \in (0, 1)$.

Next, we see that the function

$$\psi_N(\alpha) \equiv \widetilde{\phi}\left(\mathbf{s}^N \alpha\right) = -\mathbf{g} \cdot \mathbf{H}^{-1} \mathbf{g} \alpha + \frac{\alpha^2}{2} \left(\mathbf{H}^{-1}\mathbf{g}\right) \cdot \mathbf{H} \left(\mathbf{H}^{-1}\mathbf{g}\right) = \mathbf{g} \cdot \mathbf{H}^{-1} \mathbf{g}(-\alpha + \alpha^2/2)$$

has derivative

$$\psi_N'(\alpha) = -\mathbf{g} \cdot \mathbf{H}^{-1} \mathbf{g}(1-\alpha) .$$

Again, this is negative for $\alpha \in (0, 1)$.

Finally, we consider the function

$$\psi(\alpha) \equiv \widetilde{\phi}\left(\mathbf{s}^G + \mathbf{v}\alpha\right) \text{ where } \mathbf{v} \equiv \mathbf{s}^N \eta - \mathbf{s}^G .$$

This function has derivative

$$\psi'(\alpha) = \mathbf{g} \cdot \mathbf{v} + \mathbf{v} \cdot \mathbf{H}\left(\mathbf{s}^G + \mathbf{v}\alpha\right) ,$$

which is easily seen to be an increasing function of α. To make sure that this derivative is negative for all $\alpha \in [0, 1]$, we need only check that it is negative for $\alpha = 1$. Since $\mathbf{H}\mathbf{s}^N = -\mathbf{g}$, the directional derivative at the scaled Newton step is

$$\psi'(1) = \mathbf{v} \cdot \left[\mathbf{g} + \mathbf{H}\mathbf{s}^N \eta\right] = \mathbf{v} \cdot \mathbf{g}(1-\eta)$$

then we recall the definition of γ from Eq. (3.31) to get

$$= (1-\eta)(\eta-\gamma)\mathbf{s}^N \cdot \mathbf{g} = -(1-\eta)(\eta-\gamma)\mathbf{g}^\top \mathbf{H}^{-1}\mathbf{g} < 0 \,.$$

Consider the case where $\|\mathbf{s}^G\|_2 < \delta < \|\mathbf{s}^N\eta\|_2$. Lemma 3.6.5 shows that the smallest value of our quadratic objective along the double dogleg path within the trust region occurs at $\mathbf{s}^G + [\mathbf{s}^N\eta - \mathbf{s}^G]\nu$, where $\nu \in (0,1)$ is chosen so that $\|\mathbf{s}^G + [\mathbf{s}^N\eta - \mathbf{s}^G]\nu\|_2 = \delta$. This gives us a quadratic equation for ν with a single positive solution, and concludes our development of the double dogleg algorithm.

The single dogleg algorithm is available in the MINPACK Fortran routine dogleg or in the MATLAB routine ntrust. An implementation of the double dogleg algorithm can be found in Dennis and Schnabel [52, p. 336ff]. A similar implementation is given by the following

Algorithm 3.6.3 (Double Dogleg)

solve $\mathbf{H}\mathbf{s}^N = -\mathbf{g}$ for $\mathbf{s}^N$ and compute $\sigma^N = \|\mathbf{s}^N\|_2$

if $\sigma^N \le \delta$ then $\mathbf{s} = \mathbf{s}^N$

else

$\alpha = \mathbf{g}\cdot\mathbf{g}$ and $\beta = \mathbf{g}\cdot\mathbf{H}\mathbf{g}$ and $\sigma^G = \alpha\sqrt{\alpha}/\beta$

if $\sigma^G \ge \delta$ then $\mathbf{s} = \mathbf{s}^G\delta/\sigma^G$

else

$\varrho = \mathbf{g}\cdot\mathbf{s}^N$ and $\gamma = \alpha^2/(\beta\varrho)$ and $\eta = (1+4\gamma)/5$

if $\eta\sigma^N \le \delta$ then $\mathbf{s} = \mathbf{s}^N\delta/\sigma^N$

else

$\mathbf{v} = \mathbf{s}^N\eta - \mathbf{s}^G$ and $\nu = \|\mathbf{v}\|_2^2$ and $\omega = \mathbf{v}\cdot\mathbf{s}^G$

if $\omega < 0$

$$\tau = \frac{-\omega + \sqrt{\omega^2 + 4\nu\left(\delta-\sigma^G\right)\left(\delta+\sigma^G\right)}}{2\nu}$$

else

$$\tau = \frac{2\left(\delta-\sigma^G\right)\left(\delta+\sigma^G\right)}{\omega + \sqrt{\omega^2 + 4\nu\left(\delta-\sigma^G\right)\left(\delta+\sigma^G\right)}}$$

$\mathbf{s} = \mathbf{s}^G + \mathbf{v}\tau$

In practice the double dogleg step produces results that are almost as good as the hook step, but at lower cost per iteration.

Readers may experiment with the JavaScript **trust region** program. This program plots contours of function values for the Rosenbrock function

$$\phi(\mathbf{x}) = \frac{1}{2} \left\| \begin{bmatrix} 10(\mathbf{x}_2 - \mathbf{x}_1^2) \\ 1 - \mathbf{x}_1 \end{bmatrix} \right\|_2^2 .$$

Users may select a point in this graph, after which the program will draw the double dogleg path, and a trust region that contains the scaled Newton step in its interior.

3.6.1.4 Updating the Trust Region

It remains for us to describe how to select the initial trust region radius. The initial radius may be specified by an enlightened user. Alternatively, Powell [143] suggests that the initial trust region radius can be set to the length of the Cauchy step (3.30).

We also need to describe how to adjust the trust region radius during optimization. Suppose that at the kth step of the minimization process we are given an approximate minimum $\mathbf{z}_k$ and a trust region radius δ_k. We can use either the Hook step Algorithm 3.6.2 or the double dogleg Algorithm 3.6.3 to determine a step $\mathbf{s}_k$. We need to decide whether to accept this step, and whether to adjust the trust region radius.

Recall the Goldstein-Armijo condition for sufficient decrease (3.23a):

$$\phi(\mathbf{z}_k + \mathbf{s}_k) \leq \phi(\mathbf{z}_k) + \mathbf{g}_\phi(\mathbf{z}_k) \cdot \mathbf{s}_k \alpha .$$

The step $\mathbf{s}_k$ is acceptable if and only if it satisfies this inequality for some α between 0 and $\frac{1}{2}$. Typically, we take $\alpha = 10^{-4}$.

If the step is unacceptable, then for the next step we take the center of the trust region to be $\mathbf{z}_{k+1} = \mathbf{z}_k$. In order to guarantee that we use a different step during the next iteration, we also decrease the radius of the trust region. As in Algorithm 3.5.2, we can determine the minimum λ_2 of the quadratic that interpolates

$$\psi(\lambda) = \phi(\mathbf{z}_k + \mathbf{s}_k \lambda)$$

at $\lambda = 0$ and 1, and interpolates $\psi'(0)$. We take the new trust region radius to be

$$\delta_{k+1} = \delta_k \max\{\delta_k/10 \, , \, \min\{\delta_k/2 \, , \, \lambda_2\}\} .$$

This choice seeks to obtain the maximum decrease in the previous search direction on the next step, while making sure that the new trust region is decreased sufficiently but not too rapidly. In this case, the trust region radius is adjusted as in the line search Algorithm 3.5.2.

If the step is acceptable and the step was a Newton step, then we take the center of the new trust region to be $\mathbf{z}_{k+1} = \mathbf{z}_k + \mathbf{s}_k$. In this case, there is no need to adjust the trust region radius, since it is having no effect on the step selection.

If the step is acceptable but it is not a Newton step, then we consider the possibility of increasing the radius of the trust region without changing the center. Our goal is to avoid the cost of recomputing the gradient and Hessian in a case when the trust region radius has been restricting the step size.

There are two cases in which we will attempt to take a larger step with the same local model when $\mathbf{s}_k$ is not a Newton step. In the first case, the actual reduction in the objective was large, meaning that

$$\phi(\mathbf{z}_k + \mathbf{s}_k) \leq \phi(\mathbf{z}_k) + \mathbf{g}_k^\top \mathbf{s}_k \ .$$

In the second case, the actual reduction in the objective was very close to the predicted reduction, meaning that

$$\left| \phi(\mathbf{z}_k + \mathbf{s}_k) - \phi(\mathbf{z}_k) - \left\{ \mathbf{g}_\phi(\mathbf{z}_k) \cdot \mathbf{s}_k + \frac{1}{2} \mathbf{s}_k \cdot \mathbf{H}_\phi(\mathbf{z}_k) \mathbf{s}_k \right\} \right| \leq 0.1 \left| \phi(\mathbf{z}_k + \mathbf{s}_k) - \phi(\mathbf{z}_k) \right| .$$

The former case might occur if the objective has a direction of negative curvature, while the latter might occur if the trust region radius is very small. If either of these two circumstances occurs, then we take $\mathbf{z}_{k+1} = \mathbf{z}_k$ and $\delta_{k+1} = 2\delta_k$.

If the step is acceptable, is not a Newton step, and we do not attempt to double the trust region radius with the same local model, then we take the new center of the trust region to be $\mathbf{z}_{k+1} = \mathbf{z}_k + \mathbf{s}_k$ and adjust the trust region radius. If the actual reduction in the objective is less than three-fourths of the predicted reduction, i.e.

$$\phi(\mathbf{z}_k) - \phi(\mathbf{x}_{k+1}) < 0.75 \left[\mathbf{g}_\phi(\mathbf{z}_k) \cdot \mathbf{s}_k + \mathbf{s}_k \cdot \mathbf{H}_\phi(\mathbf{z}_k) \mathbf{s}_k \right] \ ,$$

then the local model has predicted the results reasonably well and we double the trust region radius. If the actual reduction is greater then 0.1 times the predicted reduction, then the local model has failed to predict the results and we halve the trust region radius. Otherwise, we leave the radius alone.

An algorithm for minimization via trust regions can be found in Dennis and Schnabel [52, p. 338ff]. A similar algorithm that summarizes our discussion in this section is available as

Algorithm 3.6.4 (Trust Region)

given $\mathbf{z}_0$ and α
$\phi_0 = \phi(\mathbf{z}_0)$ and $\mathbf{g} = \mathbf{g}_\phi(\mathbf{z}_0)$ and $\mathbf{H} = \mathbf{H}_\phi(\mathbf{z}_0)$
$\delta_0 = \|\mathbf{g}\|_2^3/(\mathbf{g} \cdot \mathbf{H}\mathbf{g})$
while not converged
  compute $\mathbf{s}_k$ using Algorithm 3.6.2 or 3.6.3
  $\sigma_k = \|\mathbf{s}_k\|_2$ and $\mathbf{z} = \mathbf{z}_k + \mathbf{s}_k$
  $\triangle\phi = \phi(\mathbf{z}) - \phi_k$ and $\varrho = \mathbf{g} \cdot \mathbf{s}_k$
  if $(\triangle\phi \geq \alpha\varrho)$ then /* step is unacceptable */
    $\mathbf{z}_{k+1} = \mathbf{z}_k$
    $\delta_{k+1} = \max\{\delta_k/10\ ,\ \min\{\delta_k/2\ ,\ -\varrho\sigma_k/(2[\triangle\phi - \varrho])\}\}$
  else /* step is acceptable */
    $\delta_{k+1} = \delta_k$
    if $\mathbf{s}_k$ is Newton step then /* trust region radius did not affect step */
      $\mathbf{z}_{k+1} = \mathbf{z}$ and $\mathbf{g} = \mathbf{g}_\phi(\mathbf{z})$ and $\mathbf{H} = \mathbf{H}_\phi(\mathbf{z})$
    else
      $\widetilde{\triangle\phi} = \varrho + \mathbf{s}_k \cdot \mathbf{H}\mathbf{s}_k/2$
      if $(\triangle\phi < \varrho$ or $|\widetilde{\triangle\phi} - \triangle\phi| < |\triangle\phi|/10)$ then /* try bigger radius at same point */
        $\mathbf{z}_{k+1} = \mathbf{z}_k$ and $\delta_{k+1} = 2\delta_k$
      else
        $\mathbf{z}_{k+1} = \mathbf{z}$ and $\mathbf{g} = \mathbf{g}_\phi(\mathbf{z})$ and $\mathbf{H} = \mathbf{H}_\phi(\mathbf{z})$
        if $(\triangle\phi \geq \widetilde{\triangle\phi}/10)$ then $\delta_{k+1} = \delta_k/2$
        else if $(\triangle\phi \leq 0.75\widetilde{\triangle\phi})$ then $\delta_{k+1} = 2\delta_k$

Exercise 3.6.1 Let $\phi(\mathbf{x}) = \xi_1^4 + \xi_1^2 + \xi_2^2$. Suppose that our current guess for the hook step is $\mathbf{z}_k^\top = [1, 1]$. Let the initial trust region radius be $\delta = 0.5$.

1. Compute $\mathbf{g}_\phi(\mathbf{z}_k)$ and $\mathbf{H}_\phi(\mathbf{z}_k)$.
2. Compute the Newton step $\mathbf{s}_k^N = -\mathbf{H}_\phi(\mathbf{z}_k)^{-1}\mathbf{g}_\phi(\mathbf{z}_k)$, and show that $\|\mathbf{s}_k^N\|_2 > 1$.
3. Take the initial Lagrange multiplier for the hook step to be zero, and perform two Newton iterations to approximate the Lagrange multiplier in the hook step.
4. Compute the reduction in the objective ϕ as a result of the hook step.

Exercise 3.6.2 Let ϕ and $\mathbf{z}_k$ be the same as in the previous exercise. Suppose that the trust region radius is $\delta = 3/4$.

1. Since the Newton step lies outside the trust region, compute the Cauchy step and show that it lies inside the trust region.
2. Since the Cauchy step lies inside the trust region, compute the scaled Newton step.
3. Compute the double dogleg step on the boundary of the trust region between the Cauchy step and the scaled Newton step.
4. Show that the double dogleg step is acceptable.
5. Compute the actual and predicted reductions in the objective.
6. Decide how to adjust the trust region radius.

3.6.2 Nonlinear Systems

Now that we have developed globally convergent methods for nonlinear minimization, it is natural to use them to develop globally convergent methods for finding zeros of nonlinear systems. Suppose that $\mathbf{f}$ maps n-vectors to n-vectors and $\mathbf{f}(\mathbf{z}) = \mathbf{0}$. Then $\mathbf{z}$ is a global minimizer of

$$\phi(\mathbf{x}) = \frac{1}{2}\|\mathbf{f}(\mathbf{x})\|_2^2 \,.$$

If $\mathbf{J_f}$ is the Jacobian of $\mathbf{f}$, then it is easy to see that the gradient of ϕ is

$$\mathbf{g}_\phi(x) = \mathbf{J_f}(\mathbf{x})^\top \mathbf{f}(\mathbf{x}) \,.$$

Thus, if $\mathbf{x}$ is not a zero of $\mathbf{f}$, then the Newton step

$$\mathbf{s}^N = -\mathbf{J_f}(\mathbf{x})^{-1}\mathbf{f}(\mathbf{x})$$

is a descent direction for ϕ, because

$$\mathbf{g}_\phi(\mathbf{x}) \cdot \mathbf{s}^N = \left[\mathbf{f}(\mathbf{x})^\top \mathbf{J_f}(\mathbf{x})\right]\left[-\mathbf{J_f}(\mathbf{x})^{-1}\mathbf{f}(\mathbf{x})\right] = -\mathbf{f}(\mathbf{x}) \cdot \mathbf{f}(\mathbf{x}) < 0 \,.$$

Since the second derivative of ϕ involves two partial derivatives of $\mathbf{f}$, we will approximate ϕ by the local model

$$\widetilde{\phi}(\mathbf{s}) = \frac{1}{2}\,\|\mathbf{f}(\mathbf{x}) + \mathbf{J_f}(\mathbf{x})\mathbf{s}\|^2 \,.$$

Obviously, the Newton step $\mathbf{s}^N$ gives us the global minimizer of the local model. The gradient of local model is

$$\mathbf{g}_{\widetilde{\phi}}(\mathbf{s}) = \mathbf{J_f}(\mathbf{x})^\top \left[\mathbf{J_f}(\mathbf{x})\mathbf{s} + \mathbf{f}(\mathbf{x})\right] \,,$$

and the Hessian is

$$\mathbf{H}_{\widetilde{\phi}}(\mathbf{s}) = \mathbf{J_f}(\mathbf{x})^\top \mathbf{J_f}(\mathbf{x}) \,.$$

This Hessian is always nonnegative, and typically positive.

If we apply a trust region technique to obtain global convergence of an iterative method for finding a minimum of ϕ, then we want to solve the constrained minimization problem

$$\min \widetilde{\phi}(\mathbf{s}) \text{ subject to } \|\mathbf{s}\|_2 \le \delta \,.$$

A hook step for this trust region takes the form

$$\left[\mathbf{J_f}^\top \mathbf{J_f} + \mathbf{I}\lambda\right] \mathbf{s} = \mathbf{J_f}^\top \mathbf{f} \,,$$

where λ is chosen so that $\|\mathbf{s}\|_2 \leq \delta$. In this case, it is useful to factor

$$\mathbf{J_f} = \mathbf{QR}$$

as in Sect. 6.7 or 6.8 in Chap. 6 in Volume I. Then the Newton step is determined by back-solving

$$\mathbf{R}\mathbf{s}^N = -\mathbf{Q}^\top \mathbf{f} \,,$$

the steepest descent step is given by

$$\mathbf{s}^{SD} = -R^\top \mathbf{Q}^\top \mathbf{f} \,,$$

the Cauchy step is

$$\mathbf{s}^G = \mathbf{s}^{SD}\alpha$$

where α minimizes

$$\psi(\alpha) = \widetilde{\phi}\left(-\mathbf{s}^{SD}\alpha\right) = \frac{1}{2}\left\|\mathbf{f} - \mathbf{J_f}\mathbf{J_f}^\top \mathbf{f}\alpha\right\|_2^2 \,.$$

This gives us

$$\alpha = \frac{\left\|\mathbf{J_f}^\top \mathbf{f}\right\|_2^2}{\left\|\mathbf{J_f}\mathbf{J_f}^\top \mathbf{f}\right\|_2^2} = \frac{\|\mathbf{s}^{SD}\|_2^2}{\|\mathbf{QR}\mathbf{s}^{SD}\|_2^2} = \frac{\|\mathbf{s}^{SD}\|_2^2}{\|\mathbf{R}\mathbf{s}^{SD}\|_2^2} \,.$$

Finally, the hook step satisfies

$$\left[\mathbf{R}^\top \mathbf{R} + \mathbf{I}\lambda\right] \mathbf{s} = \mathbf{R}^\top \mathbf{Q}^\top \mathbf{f}_k \,.$$

Note that the hook step requires an additional Cholesky factorization of $\mathbf{R}^\top\mathbf{R} + \mathbf{I}\lambda$ for each value of λ while enforcing $\|\mathbf{s}\|_2 \leq \delta$.

There are two possible difficulties with this approach. First of all, if $\mathbf{J_f}$ is nearly singular, then we cannot safely compute the Newton step. Thus, if the *QR* factorization of $\mathbf{J_f}$ indicates that its condition number is large (say, compared to the size of $\mathbf{J_f}$ divided by the square root of machine precision), then we can use the perturbed local model

$$\widetilde{\widetilde{\phi}}(\mathbf{s}) = \frac{1}{2}\|\mathbf{f}_k + \mathbf{J_f}\mathbf{s}\|_2^2 + \frac{1}{2}\mathbf{s}\cdot\mathbf{s}\left\|\mathbf{J_f}^\top \mathbf{J_f}\right\|_1 \sqrt{n\varepsilon} \,,$$

where n is the number of arguments to $\mathbf{f}$, and ε is machine precision.

The proof of the following theorem can be found in Moré and Trangenstein [128].

Theorem 3.6.1 (Global Convergence of Trust Regions) *Suppose that $\mathscr{D}$ is a convex open subset of real n-vectors, and let $\mathbf{z}_0 \in \mathscr{D}$. Assume that $\mathbf{f}$ is continuously differentiable in $\mathscr{D}$ and takes values that are real n-vectors. Define the level set*

$$L = \{\mathbf{x} \in \mathscr{D} : \|\mathbf{f}(\mathbf{x})\|_2 \leq \|\mathbf{f}(\mathbf{z}_0)\|_2\} .$$

Suppose that for some $\delta > 0$, the trust region

$$L_\delta = \{\text{real } n\text{-vectors } \mathbf{y} : \|\mathbf{y} - \mathbf{z}_0\| \leq \delta\}$$

is contained in $\mathscr{D}$. Assume that the Jacobian $\mathbf{J_f}$ is bounded on L_δ. Finally, let the sequence $\{\mathbf{z}_k\}_{k=0}^\infty$ be produced by trust regions using the double dogleg iteration. Then for any $\varepsilon > 0$ there is a $k > 0$ so that

$$\left\|\mathbf{J_f}(\mathbf{z}_k)^\top \mathbf{f}(\mathbf{z}_k)\right\|_2 < \varepsilon .$$

If $\mathbf{z}_k \to \mathbf{z} \in \mathscr{D}$, then

$$\mathbf{J_f}(\mathbf{z})^\top \mathbf{f}(\mathbf{z}) = \mathbf{0} .$$

Finally, if $\mathbf{z}_k \to \mathbf{z}$ and $\mathbf{J_f}(\mathbf{z})$ is nonsingular, then $\mathbf{f}(\mathbf{z}) = \mathbf{0}$ and $\{\mathbf{z}_k\}_{k=0}^\infty$ converges superlinearly to $\mathbf{z}$.

The proof shows that eventually every step chosen by the double dogleg iteration is a Newton step.

A Fortran implementation of a trust region algorithm for finding a zero of a system of nonlinear equations is available in the MINPACK routine hybrj. Readers may also experiment with the following JavaScript **trust region** program. This program displays iso-contours for $\|\mathbf{f}(\mathbf{x})\|_2^2/2$, where if $\mathbf{x}$ has components ξ_j then $\mathbf{f}$ is the Rosenbrock function

$$\mathbf{f}(\mathbf{x}) = \begin{bmatrix} 10(\xi_2 - \xi_1^2) \\ 1 - \xi_1 \end{bmatrix} .$$

Readers may select a starting point $\widetilde{\mathbf{x}}_0$ for the trust region algorithm. Afterward, the program will select a trust region radius, and choose among the Newton step, the Cauchy step, or the dogleg step to determine a provisional new approximate zero $\widetilde{\mathbf{x}}_{k+1}$ for $\mathbf{f}$.

Exercise 3.6.3 Show that the Newton step for the perturbed local model

$$\widetilde{\phi}(\mathbf{s}) = \frac{1}{2}\|\mathbf{f}(\mathbf{x}_k) + \mathbf{J_f}(\mathbf{x}_k)\mathbf{s}\|^2 + \frac{1}{2}\mathbf{s}\cdot\mathbf{s}\left\|\mathbf{J}_k{}^\top\mathbf{J}_k\right\|_1 \sqrt{n\varepsilon}$$

minimizes $\|\mathbf{J_f}(\mathbf{x}_k)\mathbf{s} + \mathbf{f}(\mathbf{x}_k)\|_2^2$ subject to $\|\mathbf{s}\|_2 \leq \delta$ for some δ. What is δ?

Exercise 3.6.4 Use the routine `hybrd1` from MINPACK to solve the following problems.

1.

$$\mathbf{f}_1(\mathbf{x}) = \begin{bmatrix} \xi_1 + \xi_2[\xi_2(5 - \xi_2) - 2] - 13 \\ \xi_1 + \xi_2[\xi_2(1 + \xi_2) - 14] - 29 \end{bmatrix}$$

with initial guess $\mathbf{x} = [15, -2]^\top$.

2.

$$\mathbf{f}_2(\mathbf{x}) = \begin{bmatrix} \xi_1^2 + \xi_2^2 + \xi_3^2 - 5 \\ \xi_1 + \xi_2 - 1 \\ \xi_1 + \xi_3 - 3 \end{bmatrix}$$

with initial guess $\mathbf{x} = [(1 + \sqrt{3}/2, (1 - \sqrt{3})/2, \sqrt{3}]^\top$.

3.

$$\mathbf{f}_3(\mathbf{x}) = \begin{bmatrix} \xi_1 + 10\xi_2 \\ \sqrt{5}(\xi_3 - \xi_4) \\ (\xi_2 - \xi_3)^2 \\ \sqrt{10}(\xi_1 - \xi_4)^2 \end{bmatrix}$$

with initial guess $\mathbf{x} = [1, 2, 1, 1]^\top$.

4.

$$\mathbf{f}_4(\mathbf{x}) = \begin{bmatrix} \xi_1 \\ 10\xi_1/(\xi_1 + 0.1) + 2\xi_2^2 \end{bmatrix}$$

with initial guess $\mathbf{x} = [1.8, 0]^\top$.

5.

$$\mathbf{f}_5(\mathbf{x}) = \begin{bmatrix} 10^4\xi_1\xi_2 - 1 \\ e^{-\xi_1} + e^{-\xi_2} - 1.0001 \end{bmatrix}$$

with initial guess $\mathbf{x} = [0, 1]^\top$.

Exercise 3.6.5 Obtain a copy of the Harwell Software Library routine ns12, and test it on the preceding problems.

3.7 Quasi-Newton Methods

As we saw in Sect. 3.4, Newton's method for finding a zero $\mathbf{z}$ of a nonlinear system $\mathbf{f}(\mathbf{z}) = \mathbf{0}$ requires us to evaluate the Jacobian matrix $\mathbf{J_f}$. In Sect. 3.4.3, we examined approximations of this Jacobian matrix by finite differences. Much of this work in evaluating or approximating the Jacobian may be wasted, since we only need good information about the inverse of the Jacobian acting in the direction of $\mathbf{f}$. Similarly, our discussion in Sect. 3.4 of Newton's method for unconstrained minimization required evaluation or approximation of a symmetric Hessian matrix.

In this section, we will examine techniques for performing a small rank perturbation of Jacobian or symmetric Hessian approximations, so that the result is sufficiently accurate along the path to the solution. The basic idea was originally proposed by Davidon [45], and later popularized by Fletcher and Powell [69]. This approach will require some interesting new ideas from linear algebra, as well as some interesting analysis of nonlinear systems.

3.7.1 Broyden's Method

Suppose that we want to solve $\mathbf{f}(\mathbf{z}) = \mathbf{0}$ without computing the derivative of $\mathbf{f}$. In the simple case when $\mathbf{f}$ is a real-valued function of a single real variable, we can consider the local model

$$\widetilde{\mathbf{f}}_k(\mathbf{x}) \equiv \mathbf{f}(\mathbf{z}_k) + \widetilde{\mathbf{J}}_k[\mathbf{x} - \mathbf{z}_k] \; .$$

In order to determine the approximate Jacobian $\widetilde{\mathbf{J}}_k$ uniquely, we can require that the local model agree with $\mathbf{f}$ at the previous iterate $\mathbf{z}_{k-1}$. This leads to the **secant method**, which we discussed for scalar nonlinear equations in Sect. 5.5 in Chap. 5 of Volume I. The model Jacobian in this simple case is given by

$$\widetilde{\mathbf{J}}_k = \frac{\mathbf{f}(\mathbf{z}_k) - \mathbf{f}(\mathbf{z}_{k-1})}{\mathbf{z}_k - \mathbf{z}_{k-1}} \; .$$

Next, consider the more general case when $\mathbf{f}$ maps real n-vectors to real n-vectors. Again, we will consider a local model of the form

$$\widetilde{\mathbf{f}}_k(\mathbf{x}) \equiv \mathbf{f}(\mathbf{z}_k) + \widetilde{\mathbf{J}}_k(\mathbf{x} - \mathbf{z}_k) \; .$$

Here $\widetilde{\mathbf{J}}_k$ is a real $n \times n$ matrix. In order to determine the Jacobian $\widetilde{\mathbf{J}}_k$ uniquely, we will require two conditions. First, we want the approximate Jacobian to satisfy the secant equation

$$\widetilde{\mathbf{J}}_k[\mathbf{z}_{k-1} - \mathbf{z}_k] = \mathbf{f}(\mathbf{z}_{k-1}) - \mathbf{f}(\mathbf{z}_k) \; . \qquad (3.32a)$$

Secondly, we want $\widetilde{\mathbf{J}}_k$ to minimize the change in the local model. This second requirement requires some elaboration. Let us write

$$\mathbf{s}_{k-1} = \mathbf{z}_k - \mathbf{z}_{k-1} \text{ and } \mathbf{y}_{k-1} = \mathbf{f}(\mathbf{z}_k) - \mathbf{f}(\mathbf{z}_{k-1})$$

Then for an arbitrary vector $\mathbf{x}$, the change in the local model is

$$\begin{aligned}\widetilde{\mathbf{f}}_k(\mathbf{x}) - \widetilde{\mathbf{f}}_{k-1}(\mathbf{x}) &= \left[\mathbf{f}(\mathbf{z}_k) + \widetilde{\mathbf{J}}_k(\mathbf{x} - \mathbf{z}_k)\right] - \left[\mathbf{f}(\mathbf{z}_{k-1}) + \widetilde{\mathbf{J}}_{k-1}(\mathbf{x} - \mathbf{z}_{k-1})\right] \\ &= \mathbf{y}_{k-1} - \widetilde{\mathbf{J}}_k\mathbf{s}_{k-1} + (\widetilde{\mathbf{J}}_k - \widetilde{\mathbf{J}}_{k-1})(\mathbf{x} - \mathbf{z}_{k-1}) = (\widetilde{\mathbf{J}}_k - \widetilde{\mathbf{J}}_{k-1})(\mathbf{x} - \mathbf{z}_{k-1}) \;.\end{aligned}$$

We can write

$$\mathbf{x} - \mathbf{z}_{k-1} = \mathbf{s}_{k-1}\alpha + \mathbf{t}i \text{ where } \mathbf{t} \perp \mathbf{s}_{k-1} \;.$$

Then the least change condition requires that we choose $\widetilde{\mathbf{J}}_k$ to solve

$$\begin{aligned}&\min \left\| \left(\widetilde{\mathbf{J}}_k - \widetilde{\mathbf{J}}_{k-1}\right) \mathbf{s}_{k-1}\alpha + \left(\widetilde{\mathbf{J}}_k - \widetilde{\mathbf{J}}_{k-1}\right) \mathbf{t} \right\|_2 \\ &\text{subject to } \|\mathbf{s}_{k-1}\alpha + \mathbf{t}\|_2 = 1 \;,\; \mathbf{t} \perp \mathbf{s}_{k-1} \text{ and } \widetilde{\mathbf{J}}_k\mathbf{s}_{k-1} = \mathbf{y}_{k-1} \;.\end{aligned} \tag{3.32b}$$

The solution of this minimization problem implies that

$$(\widetilde{\mathbf{J}}_k - \widetilde{\mathbf{J}}_{k-1})\mathbf{t} = \mathbf{0}$$

for all $\mathbf{t} \perp \mathbf{s}_{k-1}$. It follows that

$$\widetilde{\mathbf{J}}_k - \widetilde{\mathbf{J}}_{k-1} = \mathbf{u}\mathbf{s}_{k-1}{}^\top$$

for some real n-vector $\mathbf{u}$. The secant equation (3.32a) for $\widetilde{\mathbf{J}}_k$ requires that

$$\widetilde{\mathbf{J}}_k\mathbf{s}_{k-1} = \mathbf{y}_{k-1} \;,$$

so

$$(\widetilde{\mathbf{J}}_k - \widetilde{\mathbf{J}}_{k-1})\mathbf{s}_{k-1} = \mathbf{y}_{k-1} - \widetilde{\mathbf{J}}_{k-1}\mathbf{s}_{k-1} \;.$$

This determines $\mathbf{u}$. We conclude that

$$\widetilde{\mathbf{J}}_k = \widetilde{\mathbf{J}}_{k-1} + \left(\mathbf{y}_{k-1} - \widetilde{\mathbf{J}}_{k-1}\mathbf{s}_{k-1}\right) \frac{1}{\|\mathbf{s}_{k-1}\|_2^2}\mathbf{s}_k{}^\top \;.$$

This technique for approximating the Jacobian is due to Broyden [31], and leads to the following

Algorithm 3.7.1 (Broyden's Method)

$$
\begin{aligned}
&\text{given } \mathbf{z}_0 \text{ and } \widetilde{\mathbf{J}}_0 \\
&\text{for } 0 \le k \\
&\quad \text{solve } \widetilde{\mathbf{J}}_k \mathbf{s}_k = -\mathbf{f}(\mathbf{z}_k) \\
&\quad \mathbf{z}_{k+1} = \mathbf{z}_k + \mathbf{s}_k \\
&\quad \mathbf{y}_k = \mathbf{f}(\mathbf{z}_{k+1}) - \mathbf{f}(\mathbf{z}_k) \\
&\quad \widetilde{\mathbf{J}}_{k+1} = \widetilde{\mathbf{J}}_k + \left(\mathbf{y}_k - \widetilde{\mathbf{J}}_k \mathbf{s}_k\right) \frac{1}{\mathbf{s}_k \cdot \mathbf{s}_k} \mathbf{s}_k{}^\top
\end{aligned}
$$

Of course, Broyden's method introduces new questions. First of all, we would like to know if this method converges locally, and at what rate. We also want to know how to solve for $\mathbf{s}_k$. For example, if we have already factored $\widetilde{\mathbf{J}}_k$, can we factor $\widetilde{\mathbf{J}}_{k+1}$ easily?

3.7.2 *Local Convergence*

Now that we have developed Broyden's method as a generalization of the secant method, we would like to examine its convergence behavior. This convergence theory will develop in a series of lemmas, the first of which can be found in Broyden, Dennis and Moré [32].

Lemma 3.7.1 (Bounded Deterioration) *Suppose that $\mathscr{D}$ is an open and convex subset of real m-vectors. Let $\mathbf{f}$ map $\mathscr{D}$ to real m-vectors, and let f be continuously differentiable. Assume that there exists $\mathbf{z} \in \mathscr{D}$ so that $\mathbf{f}(\mathbf{z}) = \mathbf{0}$. Let $\| \cdot \|$ represent either the 2-norm or Frobenius norm on $m \times m$ matrices. Suppose that the Jacobian $\mathbf{J_f}$ of $\mathbf{f}$ is Lipschitz continuous at $\mathbf{z}$ with Lipschitz constant γ:*

$$\|\mathbf{J_f}(\mathbf{x}) - \mathbf{J_f}(\mathbf{z})\| \le \gamma \|\mathbf{x} - \mathbf{z}\|_2 \ .$$

Let $\widetilde{\mathbf{J}}_o$ be a nonsingular $m \times m$ matrix. Given $\mathbf{x} \in \mathscr{D}$ with $\mathbf{f}(\mathbf{x}) \neq \mathbf{0}$, let $\mathbf{s}$ solve

$$\widetilde{\mathbf{J}}_o \mathbf{s} = \mathbf{f}(\mathbf{x}) \ .$$

Also define

$$\mathbf{y} = \mathbf{f}(\mathbf{x} + \mathbf{s}) - \mathbf{f}(\mathbf{x}) \ , \ and\ \widetilde{\mathbf{J}}_n = \widetilde{\mathbf{J}}_o + [\mathbf{y} - \widetilde{\mathbf{J}}_o \mathbf{s}] \frac{1}{\mathbf{s} \cdot \mathbf{s}} \mathbf{s}^\top \ .$$

If $\mathbf{J_f}$ *is Lipschitz continuous in* $\mathscr{D}$ *with Lipschitz constant* γ*, then*

$$\left\| \widetilde{\mathbf{J}}_n - \mathbf{J_f}(\mathbf{x}+\mathbf{s}) \right\| \le \left\| \widetilde{\mathbf{J}}_o - \mathbf{J_f}(\mathbf{x}) \right\| + \frac{3\gamma}{2} \|\mathbf{s}\|_2 \, .$$

and

$$\left\| \widetilde{\mathbf{J}}_n - \mathbf{J_f}(\mathbf{z}) \right\| \le \left\| \widetilde{\mathbf{J}}_o - \mathbf{J_f}(\mathbf{z}) \right\| + \frac{\gamma}{2} \{ \|\mathbf{x}+\mathbf{s}-\mathbf{z}\|_2 + \|\mathbf{x}-\mathbf{z}\|_2 \} \, . \tag{3.33}$$

Proof The definition of $\mathbf{y}$ and $\mathbf{s}$, followed by Lemma 3.2.5, imply that

$$\|\mathbf{y} - \mathbf{J_f}(\mathbf{x})\mathbf{s}\|_2 = \|\mathbf{f}(\mathbf{x}+\mathbf{s}) - \mathbf{f}(\mathbf{x}) - \mathbf{J_f}(\mathbf{x})\mathbf{s}\|_2 \le \frac{\gamma}{2} \|\mathbf{s}\|_2^2 \, .$$

Also, the definition of $\widetilde{\mathbf{J}}_n$ implies that

$$\begin{aligned}
&\widetilde{\mathbf{J}}_n - \mathbf{J_f}(\mathbf{x}+\mathbf{s}) = \widetilde{\mathbf{J}}_o - \mathbf{J_f}(\mathbf{x}+\mathbf{s}) + [\mathbf{y} - \widetilde{\mathbf{J}}_o\mathbf{s}] \frac{1}{\mathbf{s}\cdot\mathbf{s}} \mathbf{s}^\top \\
&= \left[\widetilde{\mathbf{J}}_o - \mathbf{J_f}(\mathbf{x})\right] + [\mathbf{J_f}(\mathbf{x}) - \mathbf{J_f}(\mathbf{x}+\mathbf{s})] + \{\mathbf{y} - \mathbf{J_f}(\mathbf{x})\mathbf{s}\} \frac{1}{\mathbf{s}\cdot\mathbf{s}} \mathbf{s}^\top + \left[\mathbf{J_f}(\mathbf{x}) - \widetilde{\mathbf{J}}_o\right] \frac{1}{\mathbf{s}\cdot\mathbf{s}} \mathbf{s}^\top \\
&= \left[\widetilde{\mathbf{J}}_o - \mathbf{J_f}(\mathbf{x})\right] \left[\mathbf{I} - \mathbf{s}\frac{1}{\mathbf{s}\cdot\mathbf{s}}\mathbf{s}^\top\right] + [\mathbf{J_f}(\mathbf{x}) - \mathbf{J_f}(\mathbf{x}+\mathbf{s})] + \{\mathbf{y} - \mathbf{J_f}(\mathbf{x})\mathbf{s}\} \frac{1}{\mathbf{s}\cdot\mathbf{s}} \mathbf{s}^\top \, .
\end{aligned}$$

Since Definition 3.2.21 in Chap. 3 of Volume I shows that $\mathbf{I} - \mathbf{s}\mathbf{s}^\top/(\mathbf{s}\cdot\mathbf{s})$ is an orthogonal projector, Example 1.2.6 shows that its eigenvalues are either zero or one, and Lemma 1.3.3 shows that its 2-norm is 1. Then either Lemma 3.5.3 or 3.5.11 in Chap. 3 of Volume I gives us

$$\begin{aligned}
&\left\| \widetilde{\mathbf{J}}_n - \mathbf{J_f}(\mathbf{x}+\mathbf{s}) \right\| \\
&\le \left\| \widetilde{\mathbf{J}}_o - \mathbf{J_f}(\mathbf{x}) \right\| \left\| \mathbf{I} - \mathbf{s}\frac{1}{\mathbf{s}\cdot\mathbf{s}}\mathbf{s}^\top \right\|_2 + \|\mathbf{J_f}(\mathbf{x}) - \mathbf{J_f}(\mathbf{x}+\mathbf{s})\| + \|\mathbf{y} - \mathbf{J_f}(\mathbf{x})\mathbf{s}\|_2 \frac{1}{\|\mathbf{s}\|_2} \\
&\le \left\| \widetilde{\mathbf{J}}_o - \mathbf{J_f}(\mathbf{x}) \right\| + \gamma\|\mathbf{s}\|_2 + \frac{\gamma}{2}\|\mathbf{s}\|_2 \, .
\end{aligned}$$

This proves the first claim.

Similarly, the definition of $\mathbf{y}$ gives us

$$\|\mathbf{y} - \mathbf{J_f}(\mathbf{z})\mathbf{s}\|_2 = \|\mathbf{f}(\mathbf{x}+\mathbf{s}) - \mathbf{f}(\mathbf{x}) - \mathbf{J_f}(\mathbf{z})\mathbf{s}\|_2$$

and then Lemma 3.2.6 produces

$$\le \frac{\gamma}{2} \{ \|\mathbf{x}+\mathbf{s}-\mathbf{z}\|_2 + \|\mathbf{x}-\mathbf{z}\|_2 \} \|\mathbf{s}\|_2 \tag{3.34}$$

The definition of $\widetilde{\mathbf{J}}_n$ implies that

$$\begin{aligned}\widetilde{\mathbf{J}}_n - \mathbf{J_f}(\mathbf{z}) &= \widetilde{\mathbf{J}}_o - \mathbf{J_f}(\mathbf{z}) + (\mathbf{y} - \widetilde{\mathbf{J}}_o\mathbf{s})\frac{1}{\mathbf{s}\cdot\mathbf{s}}\mathbf{s}^\top \\ &= \left[\widetilde{\mathbf{J}}_o - \mathbf{J_f}(\mathbf{z})\right] + [\mathbf{y} - \mathbf{J_f}(\mathbf{z})\mathbf{s}]\,\frac{1}{\mathbf{s}\cdot\mathbf{s}}\mathbf{s}^\top + \left[\mathbf{J_f}(\mathbf{z}) - \widetilde{\mathbf{J}}_o\right]\mathbf{s}\frac{1}{\mathbf{s}\cdot\mathbf{s}}\mathbf{s}^\top \\ &= \left[\widetilde{\mathbf{J}}_o - \mathbf{J_f}(\mathbf{z})\right]\left[\mathbf{I} - \mathbf{s}\frac{1}{\mathbf{s}\cdot\mathbf{s}}\mathbf{s}^\top\right] + [\mathbf{y} - \mathbf{J_f}(\mathbf{z})\mathbf{s}]\,\frac{1}{\mathbf{s}\cdot\mathbf{s}}\mathbf{s}^\top\,. \end{aligned} \tag{3.35}$$

Then

$$\begin{aligned}\left\|\widetilde{\mathbf{J}}_n - \mathbf{J_f}(\mathbf{z})\right\| &\le \left\|\widetilde{\mathbf{J}}_o - \mathbf{J_f}(\mathbf{z})\right\| \left\|\mathbf{I} - \mathbf{s}\frac{1}{\mathbf{s}\cdot\mathbf{s}}\mathbf{s}^\top\right\|_2 + \frac{\|\mathbf{y} - \mathbf{J_f}(\mathbf{z})\mathbf{s}\|_2}{\|\mathbf{s}\|_2} \\ &\le \left\|\widetilde{\mathbf{J}}_o - \mathbf{J_f}(\mathbf{z})\right\| + \frac{\gamma}{2}\left\{\|\mathbf{x}+\mathbf{s}-\mathbf{z}\|_2 + \|\mathbf{x}-\mathbf{z}\|_2\right\}\,.\end{aligned}$$

This proves the second claim.

The following theorem is a special case of a more general result due to Broyden, Dennis and Moré [32]. This simplified form was taken from Dennis and Schnabel [52, p. 177].

Theorem 3.7.1 (Broyden's Local Convergence) *Suppose that $\mathscr{D}$ is a convex and open set of real n-vectors. Let $\mathbf{f}$ map $\mathscr{D}$ to real n-vectors, and assume that $\mathbf{f}$ is continuously differentiable. Assume that there exists $\mathbf{z} \in \mathscr{D}$ such that $\mathbf{f}(\mathbf{z}) = \mathbf{0}$, and that the Jacobian $\mathbf{J_f}(\mathbf{z})$ is nonsingular. Let $\|\cdot\|$ represent either the 2-norm or Frobenius norm on $n \times n$ matrices. Suppose that there are constants $\gamma > 0$ and $\varrho > 0$ such that $\|\mathbf{x} - \mathbf{z}\| < \varrho$ implies that $\mathbf{x} \in \mathscr{D}$ and that*

$$\|\mathbf{J_f}(\mathbf{x}) - \mathbf{J_f}(\mathbf{z})\| \le \gamma\|\mathbf{x} - \mathbf{z}\|_2\,.$$

Also assume that there is a constant $\delta > 0$ so that

$$\left\|\widetilde{\mathbf{J}}_0 - \mathbf{J_f}(\mathbf{z})\right\| \le \delta < \frac{1}{6\,\|\mathbf{J_f}(\mathbf{z})^{-1}\|}\,.$$

Let $\varepsilon \le \min\{\varrho, (2\delta)/(3\gamma)\}$, and let $\mathbf{z}0 \in \mathscr{D}$ satisfy

$$\|\mathbf{z}0 - \mathbf{z}\|_2 \le \varepsilon\,.$$

Let the sequences $\{\mathbf{z}_k\}_{k=0}^\infty$, $\{\mathbf{y}_k\}_{k=0}^\infty$, $\{\mathbf{s}_k\}_{k=0}^\infty$ and $\{\widetilde{\mathbf{J}}_k\}_{k=0}^\infty$ be generated by Broyden's method in Algorithm 3.7.1. Then the sequence $\{\mathbf{z}_k\}_{k=0}^\infty$ is well-defined and converges at least linearly to $\mathbf{z}$.

Proof We will show by induction that

$$\left\|\widetilde{\mathbf{J}}_k - \mathbf{J_f}(\mathbf{z})\right\| \le (2 - 2^{-k})\delta \tag{3.36}$$

and that

$$\|\mathbf{z}_k - \mathbf{z}\|_2 \leq 2^{-k}\varepsilon \;, \tag{3.37}$$

The assumptions of the theorem show that these two conditions are true for $k = 0$. Inductively, we will assume that both are true for $k - 1 \geq 0$, and prove that they are true for k. In this regard, we remark that whenever inequality (3.37) is true, we have

$$\|\mathbf{z}_k - \mathbf{z}\|_2 \leq 2^{-k}\varepsilon \leq \varepsilon \leq \varrho \;,$$

so $\mathbf{z}_k \in \mathscr{D}$.

First, we note that

$$\left\|\mathbf{J_f}(\mathbf{z})^{-1}\left(\widetilde{\mathbf{J}}_{k-1} - \mathbf{J_f}(\mathbf{z})\right)\right\| \leq \left\|\mathbf{J_f}(\mathbf{z})^{-1}\right\|\left\|\widetilde{\mathbf{J}}_{k-1} - \mathbf{J_f}(\mathbf{z})\right\| \leq \left\|\mathbf{J_f}(\mathbf{z})^{-1}\right\|(2 - 2^{1-k})\delta < \frac{1}{3}.$$

It follows from Lemma 3.6.1 in Chap. 3 of Volume I that $\widetilde{\mathbf{J}}_{k-1}$ is nonsingular, and

$$\left\|\widetilde{\mathbf{J}}_{k-1}^{-1}\right\| \leq \frac{\left\|\mathbf{J_f}(\mathbf{z})^{-1}\right\|}{1 - \left\|\mathbf{J_f}(\mathbf{z})^{-1}\left(\widetilde{\mathbf{J}}_{k-1} - \mathbf{J_f}(\mathbf{z})\right)\right\|} \leq \frac{\left\|\mathbf{J_f}(\mathbf{z})^{-1}\right\|}{1 - 1/3} = \frac{3}{2}\left\|\mathbf{J_f}(\mathbf{z})^{-1}\right\| \;. \tag{3.38}$$

Thus the hypotheses of Lemma 3.7.1 are satisfied, and inequality (3.33) implies that

$$\begin{aligned}\left\|\widetilde{\mathbf{J}}_k - \mathbf{J_f}(\mathbf{z})\right\| &\leq \left\|\widetilde{\mathbf{J}}_{k-1} - \mathbf{J_f}(\mathbf{z})\right\| + \frac{\gamma}{2}\left\{\|\mathbf{z}_k - \mathbf{z}\|_2 + \|\mathbf{z}_{k-1} - \mathbf{z}\|_2\right\} \\ &\leq \left(2 - 2^{1-k}\right)\delta + \frac{3\gamma}{4}\varepsilon 2^{1-k} < \left(2 - 2^{1-k}\right)\delta + 2^{-k}\delta = \left(2 - 2^{-k}\right)\delta \end{aligned} \tag{3.39}$$

This proves (3.36) for k.

Since $\widetilde{\mathbf{J}}_{k-1}$ is nonsingular, it follows that$\mathbf{s}_{k-1}$ is well-defined, and therefore $\mathbf{z}_k$ is well-defined. The definition of $\mathbf{z}_k$ implies that

$$\widetilde{\mathbf{J}}_{k-1}\left[\mathbf{z}_k - \mathbf{z}\right] = \widetilde{\mathbf{J}}_{k-1}\left[\mathbf{z}_{k-1} + \mathbf{s}_{k-1} - \mathbf{z}\right]$$

then the definition of $\mathbf{s}_{k-1}$ and the assumption that $\mathbf{f}(\mathbf{z}) = \mathbf{0}$ give us

$$\begin{aligned}&= \widetilde{\mathbf{J}}_{k-1}\left[\mathbf{z}_{k-1} - \mathbf{z}\right] - \left[\mathbf{f}(\mathbf{z}_{k-1}) - \mathbf{f}(\mathbf{z})\right] \\ &= \left[\widetilde{\mathbf{J}}_{k-1} - \mathbf{J_f}(\mathbf{z})\right]\left[\mathbf{z}_{k-1} - \mathbf{z}\right] - \left[\mathbf{f}(\mathbf{z}_{k-1}) - \mathbf{f}(\mathbf{z}) - \mathbf{J_f}(\mathbf{z})(\mathbf{z}_{k-1} - \mathbf{z})\right] \;. \end{aligned} \tag{3.40}$$

We can multiply both sides by $\widetilde{\mathbf{J}}_{k-1}^{-1}$ and take norms to see that

$$\begin{aligned}&\|\mathbf{z}_k - \mathbf{z}\|_2 \\ &\leq \left\|\widetilde{\mathbf{J}}_{k-1}^{-1}\right\|_2\left\{\left\|\widetilde{\mathbf{J}}_{k-1} - \mathbf{J_f}(\mathbf{z})\right\|\|\mathbf{z}_{k-1} - \mathbf{z}\|_2 + \|\mathbf{f}(\mathbf{z}_{k-1}) - \mathbf{f}(\mathbf{z}) - \mathbf{J_f}(\mathbf{z})(\mathbf{z}_{k-1} - \mathbf{z})\|_2\right\}\end{aligned}$$

then inequality (3.38), Lemma 3.2.5 and the inductive inequality (3.36) give us

$$< \frac{3}{2} \left\| \mathbf{J_f}(\mathbf{z})^{-1} \right\| \left\{ (2 - 2^{1-k})\delta + \frac{\gamma}{2} \left\| \mathbf{z}_{k-1} - \mathbf{z} \right\|_2 \right\} \left\| \mathbf{z}_{k-1} - \mathbf{z} \right\|_2$$

then we use the inductive hypothesis (3.37) to get

$$\begin{aligned}
&\le \frac{3}{2} \left\| \mathbf{J_f}(\mathbf{z})^{-1} \right\| \left\{ (2 - 2^{1-k})\delta + \gamma 2^{-k} \varepsilon \right\} \left\| \mathbf{z}_{k-1} - \mathbf{z} \right\|_2 \\
&\le \frac{3}{2} \left\| \mathbf{J_f}(\mathbf{z})^{-1} \right\| \left\{ 2 - 2^{1-k} + 2^{1-k}\delta/3 \right\} \delta \left\| \mathbf{z}_{k-1} - \mathbf{z} \right\|_2 < 3 \left\| \mathbf{J_f}(\mathbf{z})^{-1} \right\| \delta \left\| \mathbf{z}_{k-1} - \mathbf{z} \right\|_2 \\
&< \frac{1}{2} \left\| \mathbf{z}_{k-1} - \mathbf{z} \right\|_2 .
\end{aligned}$$

Combining this inequality with the inductive hypothesis (3.37) for $k - 1$ proves this inductive hypothesis for k. Linear convergence follows directly from the final inequality.

Next, we would like to determine the circumstances under which Broyden's method converges superlinearly. We make use of the following simple result.

Lemma 3.7.2 *If the sequence $\{\mathbf{z}_k\}_{k=0}^{\infty}$ of n-vectors converges superlinearly to $\mathbf{z}$ in some norm $\|\cdot\|$, then*

$$\lim_{k \to \infty} \frac{\|\mathbf{z}_{k+1} - \mathbf{z}_k\|}{\|\mathbf{z}_k - \mathbf{z}\|} = 1 .$$

Proof As usual, we write $\mathbf{s}_k = \mathbf{z}_{k+1} - \mathbf{z}_k$. Then

$$\left| \frac{\|\mathbf{s}_k\|}{\|\mathbf{z}_k - \mathbf{z}\|} - 1 \right| = \left| \frac{\|\mathbf{s}_k\| - \|\mathbf{z}_k - \mathbf{z}\|}{\|\mathbf{z}_k - \mathbf{z}\|} \right| \le \frac{\|\mathbf{s}_k + \mathbf{z}_k - \mathbf{z}\|}{\|\mathbf{z}_k - \mathbf{z}\|} = \frac{\|\mathbf{z}_{k+1} - \mathbf{z}\|}{\|\mathbf{z}_k - \mathbf{z}\|} .$$

Since superlinear convergence means that the right-hand side of this inequality tends to zero, the result follows.

By the way, this lemma implies that for superlinearly converging iterations, we can estimate the error in $\mathbf{z}_k$ by the size of the subsequent step $\mathbf{s}_k$.

The next theorem, which is due to Dennis and Moré [51], characterizes the circumstances under which we can expect Broyden's method to be superlinearly convergent.

Theorem 3.7.2 (Equivalent Conditions for Broyden Superlinear Convergence) *Let $\mathscr{D}$ be a convex and open set of real n-vectors. Assume that $\mathbf{f}$ maps $\mathscr{D}$ to real n-vectors, and that $\mathbf{f}$ is continuously differentiable. Suppose that the sequences $\{\mathbf{z}_k\}_{k=0}^{\infty}$, $\{\mathbf{y}_k\}_{k=0}^{\infty}$, $\{\mathbf{s}_k\}_{k=0}^{\infty}$ and $\{\widetilde{\mathbf{J}}_k\}_{k=0}^{\infty}$ are generated by Broyden's method (Algorithm 3.7.1). Assume that $\{\mathbf{z}_k\}_{k=0}^{\infty} \subset \mathscr{D}$ converges to $\mathbf{z} \in \mathscr{D}$, and for all k we have $\mathbf{z}_k \neq \mathbf{z}$. Suppose that the Jacobian $\mathbf{J_f}(\mathbf{z})$ is nonsingular. Let $\|\cdot\|$ represent either* *the 2-norm or Frobenius norm on $n \times n$ matrices. Finally, assume that there is a constant $\gamma > 0$ such that for all $\mathbf{x}, \mathbf{y} \in \mathscr{D}$*

$$\|\mathbf{J_f}(\mathbf{x}) - \mathbf{J_f}(\mathbf{y})\| \leq \gamma \|\mathbf{x} - \mathbf{y}\|_2 ,$$

Then $\{\mathbf{z}_k\}_{k=0}^{\infty}$ converges superlinearly to $\mathbf{z}$ and $\mathbf{f}(\mathbf{z}) = \mathbf{0}$ if and only if

$$\lim_{k \to \infty} \frac{\left\| \left\{ \widetilde{\mathbf{J}}_k - \mathbf{J_f}(\mathbf{z}) \right\} \mathbf{s}_k \right\|_2}{\|\mathbf{s}_k\|_2} = 0 . \tag{3.41}$$

Proof The definition of Broyden's method implies that

$$\mathbf{0} = \widetilde{\mathbf{J}}_k \mathbf{s}_k + \mathbf{f}(\mathbf{z}_k) = \left[\widetilde{\mathbf{J}}_k - \mathbf{J_f}(\mathbf{z}) \right] \mathbf{s}_k + \mathbf{f}(\mathbf{z}_k) + \mathbf{J_f}(\mathbf{z}) \mathbf{s}_k .$$

As a result,

$$-\mathbf{f}(\mathbf{z}_{k+1}) = \left[\widetilde{\mathbf{J}}_k - \mathbf{J_f}(\mathbf{z}) \right] \mathbf{s}_k + \left[-\mathbf{f}(\mathbf{z}_{k+1}) + \mathbf{f}(\mathbf{z}_k) + \mathbf{J_f}(\mathbf{z}) \mathbf{s}_k \right] . \tag{3.42}$$

We will begin by proving the reverse direction of the claimed equivalence. In other words, we will use Eq. (3.41) to show that $\mathbf{f}(\mathbf{z}) = \mathbf{0}$ and that $\mathbf{z}_k \to \mathbf{z}$ superlinearly. Note that Eq. (3.42) and Lemma 3.2.6 imply that

$$\frac{\|\mathbf{f}(\mathbf{z}_{k+1})\|_2}{\|\mathbf{s}_k\|_2} \leq \frac{\|\{\widetilde{\mathbf{J}}_k - \mathbf{J_f}(\mathbf{z})\}\mathbf{s}_k\|_2}{\|\mathbf{s}_k\|_2} + \frac{\|\mathbf{f}(\mathbf{z}_{k+1}) - \mathbf{f}(\mathbf{z}_k) - \mathbf{J_f}(\mathbf{z})\mathbf{s}_k\|_2}{\|\mathbf{s}_k\|_2}$$

$$\leq \frac{\|\{\widetilde{\mathbf{J}}_k - \mathbf{J_f}(\mathbf{z})\}\mathbf{s}_k\|_2}{\|\mathbf{s}_k\|_2} + \frac{\gamma}{2}(\|\mathbf{z}_k - \mathbf{z}\|_2 + \|\mathbf{z}_{k+1} - \mathbf{z}\|_2) .$$

Since $\{\mathbf{z}_k\}_{k=0}^{\infty}$ converges to $\mathbf{z}$, $\mathbf{z}_k - \mathbf{z} \to \mathbf{0}$ and $\mathbf{s}_k \to \mathbf{0}$. Then inequality (3.41) shows that the right-hand side of the previous inequality converges to zero. It follows that

$$\mathbf{f}(\mathbf{z}) = \lim_{k \to \infty} \mathbf{f}(\mathbf{z}_k) = \mathbf{0} .$$

Since $\mathbf{z}_k \to \mathbf{z}$ and $\mathbf{J_f} \neq \mathbf{0}$, there exists $k_0 \geq 0$ so that for all $k > k_0$ we have

$$\|\mathbf{z}_{k+1} - \mathbf{z}\|_2 \leq \|\mathbf{J_f}(\mathbf{z})\|_2 .$$

Since $\mathbf{J_f}(\mathbf{z})$ is nonsingular and $\mathbf{J_f}$ is Lipschitz continuous, Lemma 3.2.3 implies that for all $k > k_0$

$$\|\mathbf{f}(\mathbf{z}_{k+1})\|_2 = \|\mathbf{f}(\mathbf{z}_{k+1}) - \mathbf{f}(\mathbf{z})\|_2 = \left\| \int_0^1 \mathbf{J_f}(\mathbf{z} + [\mathbf{z}_{k+1} - \mathbf{z}]t)[\mathbf{z}_{k+1} - \mathbf{z}] \, dt \right\|_2$$

$$= \left\| \mathbf{J_f}(\mathbf{z})[\mathbf{z}_{k+1} - \mathbf{z}] + \int_0^1 \{\mathbf{J_f}(\mathbf{z} + [\mathbf{z}_{k+1} - \mathbf{z}]t) - \mathbf{J_f}(\mathbf{z})\} [\mathbf{z}_{k+1} - \mathbf{z}] \, dt \right\|_2$$

$$\geq \|\mathbf{J_f}(\mathbf{z})\| \|\mathbf{z}_{k+1} - \mathbf{z}\|_2 - \frac{\gamma}{2} \|\mathbf{z}_{k+1} - \mathbf{z}\|_2 \geq \frac{1}{2} \|\mathbf{J_f}(\mathbf{z})\| \|\mathbf{z}_{k+1} - \mathbf{z}\|_2 .$$

Then for $k > k_0$,

$$\frac{\|\mathbf{z}_{k+1} - \mathbf{z}\|_2}{\|\mathbf{z}_k - \mathbf{z}\|_2} \leq \frac{2}{\|\mathbf{J_f}(\mathbf{z})\|} \frac{\|\mathbf{f}(\mathbf{z}_{k+1})\|_2}{\|\mathbf{s}_k\|_2} \frac{\|\mathbf{s}_k\|_2}{\|\mathbf{z}_k - \mathbf{z}\|_2} \to \frac{2}{\|\mathbf{J_f}(\mathbf{z})\|}(0)(1) = 0 \; .$$

Thus $\{\mathbf{z}_k\}$ converges to $\mathbf{z}$ superlinearly.

Next, we will prove the forward direction of the claim. In other words, if $\{\mathbf{z}_k\}$ converges to $\mathbf{z}$ superlinearly and that $\mathbf{f}(\mathbf{z}) = \mathbf{0}$, then we will prove that Eq. (3.41) is satisfied. Since $\mathbf{J_f}(\mathbf{z})$ is nonzero and $\mathbf{z}_k \to \mathbf{z}$, there exists $k_1 > 0$ so that for all $k > k_1$ we have

$$\gamma \|\mathbf{z}_{k+1} - \mathbf{z}\|_2 \leq 2\|\mathbf{J_f}(\mathbf{z})\| \; .$$

Since $\mathbf{J_f}$ is Lipschitz continuous and $\mathbf{f}(\mathbf{z}) = \mathbf{0}$, Lemma 3.2.3 implies that for all $k > k_1$,

$$\begin{aligned}
\|\mathbf{f}(\mathbf{z}_{k+1})\|_2 &= \|\mathbf{f}(\mathbf{z}_{k+1}) - \mathbf{f}(\mathbf{z})\|_2 = \left\| \int_0^1 \mathbf{J_f}(\mathbf{z} + [\mathbf{z}_{k+1} - \mathbf{z}]t)[\mathbf{z}_{k+1} - \mathbf{z}] \, \mathrm{d}t \right\|_2 \\
&= \left\| \mathbf{J_f}(\mathbf{z})[\mathbf{z}_{k+1} - \mathbf{z}] + \int_0^1 \{\mathbf{J_f}(\mathbf{z} + [\mathbf{z}_{k+1} - \mathbf{z}]t) - \mathbf{J_f}(\mathbf{z})\} [\mathbf{z}_{k+1} - \mathbf{z}] \, \mathrm{d}t \right\|_2 \\
&\leq \|\mathbf{J_f}(\mathbf{z})\| \|\mathbf{z}_{k+1} - \mathbf{z}\|_2 + \frac{\gamma}{2}\|\mathbf{z}_{k+1} - \mathbf{z}\|_2^2 \leq 2\|\mathbf{J_f}(\mathbf{z})\| \|\mathbf{z}_{k+1} - \mathbf{z}\|_2 \; .
\end{aligned}$$

Since $\{\mathbf{z}_k\}$ converges superlinearly, Lemma 3.7.2 shows that

$$0 = \lim_{k\to\infty} \frac{2\|\mathbf{J_f}(\mathbf{z})\| \|\mathbf{z}_{k+1} - \mathbf{z}\|_2}{\|\mathbf{z}_k - \mathbf{z}\|_2} \geq \lim_{k\to\infty} \frac{\|\mathbf{f}(\mathbf{z}_{k+1})\|_2}{\|\mathbf{s}_k\|_2} \lim_{k\to\infty} \frac{\|\mathbf{s}_k\|_2}{\|\mathbf{z}_k - \mathbf{z}\|_2} = \lim_{k\to\infty} \frac{\|\mathbf{f}(\mathbf{z}_{k+1})\|_2}{\|\mathbf{s}_k\|_2} \; .$$

Next, Eq. (3.42) implies that

$$\begin{aligned}
\frac{\|\{\widetilde{\mathbf{J}}_k - \mathbf{J_f}(\mathbf{z})\}\mathbf{s}_k\|_2}{\|\mathbf{s}_k\|_2} &\leq \frac{\|\mathbf{f}(\mathbf{z}_{k+1})\|_2}{\|\mathbf{s}_k\|_2} + \frac{\|\mathbf{f}(\mathbf{z}_{k+1}) - \mathbf{f}(\mathbf{z}_k) - \mathbf{J_f}(\mathbf{z})\mathbf{s}_k\|_2}{\|\mathbf{s}_k\|_2} \\
&\leq \frac{\|\mathbf{f}(\mathbf{z}_{k+1})\|_2}{\|\mathbf{s}_k\|_2} + \frac{\gamma}{2}(\|\mathbf{z}_k - \mathbf{z}\|_2 + \|\mathbf{z}_{k+1} - \mathbf{z}\|_2) \; .
\end{aligned}$$

Since the right-hand side of this inequality tends to zero, Eq. (3.41) is proved and the theorem is complete.

There is an interesting interpretation of Eq. (3.41). Since $\mathbf{J_f}$ is Lipschitz continuous, we could change Eq. (3.41) to have the equivalent form

$$\lim_{k\to\infty} \frac{\left\|\{\widetilde{\mathbf{J}}_k - \mathbf{J_f}(\mathbf{z}_k)\} \mathbf{s}_k\right\|_2}{\|\mathbf{s}_k\|_2} = 0 \; .$$

Since Broyden's method defines $\mathbf{s}_k$ to solve

$$\widetilde{\mathbf{J}}_k \mathbf{s}_k = -\mathbf{f}(\mathbf{z}_k) ,$$

we obtain

$$\lim_{k \to \infty} \frac{\left\| \mathbf{J_f}(\mathbf{z}_k) \left\{ \mathbf{J_f}(\mathbf{z}_k)^{-1} \mathbf{f}(\mathbf{z}_k) + \mathbf{s}_k \right\} \right\|_2}{\|\mathbf{s}_k\|_2} = 0 .$$

This in turn is equivalent to requiring that $\mathbf{s}_k$ converge to the Newton step in both magnitude and direction.

Before proving the superlinear convergence of Broyden's method, we need to prove the following lemma.

Lemma 3.7.3 *Let* $\mathbf{s}$ *be a nonzero n-vector, and let and* $\mathbf{E}$ *be an* $n \times n$ *matrix. Then*

$$\left\| \mathbf{E} \left(\mathbf{I} - \frac{\mathbf{s}\mathbf{s}^H}{\mathbf{s} \cdot \mathbf{s}} \right) \right\|_F = \left[\|\mathbf{E}\|_F^2 - \left(\frac{\|\mathbf{E}\mathbf{s}\|_2}{\|\mathbf{s}\|_2} \right)^2 \right]^{1/2} \leq \|\mathbf{E}\|_F - \frac{1}{2\|\mathbf{E}\|_F} \left(\frac{\|\mathbf{E}\mathbf{s}\|_2}{\|\mathbf{s}\|_2} \right)^2 .$$

Proof Note that for any square matrix $\mathbf{A}$,

$$\|\mathbf{A}\|_F^2 = \operatorname{tr}\left(\mathbf{A}^H \mathbf{E} \right) .$$

In other words, $\operatorname{tr}(\mathbf{A}^H\mathbf{B})$ is an inner product on square matrices associated with the norm $\| \cdot \|_F$. We can use Lemma 1.2.6 to show that

$$\operatorname{tr}\left(\left\{ \mathbf{E} \left[\frac{\mathbf{s}\mathbf{s}^H}{\mathbf{s} \cdot \mathbf{s}} \right] \right\}^H \left\{ \mathbf{E} \left[\mathbf{I} - \frac{\mathbf{s}\mathbf{s}^H}{\mathbf{s} \cdot \mathbf{s}} \right] \right\} \right) = \operatorname{tr}\left(\mathbf{E} \left[\mathbf{I} - \frac{\mathbf{s}\mathbf{s}^H}{\mathbf{s} \cdot \mathbf{s}} \right] \frac{\mathbf{s}\mathbf{s}^H}{\mathbf{s} \cdot \mathbf{s}} \mathbf{E}^H \right) = 0 .$$

The Pythagorean Theorem 3.2.2 in Chap. 3 of Volume I shows that

$$\|\mathbf{E}\|_F^2 = \left\| \mathbf{E} \frac{\mathbf{s}\mathbf{s}^H}{\mathbf{s} \cdot \mathbf{s}} \right\|_F^2 + \left\| \mathbf{E} \left(\mathbf{I} - \frac{\mathbf{s}\mathbf{s}^H}{\mathbf{s} \cdot \mathbf{s}} \right) \right\|_F^2 .$$

Next, we note that for all n-vectors $\mathbf{u}$ and $\mathbf{w}$ we have

$$\left\| \mathbf{u}\mathbf{w}^H \right\|_F^2 = \operatorname{tr}\left(\mathbf{w}\mathbf{u}^H \mathbf{u}\mathbf{w}^H \right) = \mathbf{u}^H\mathbf{u} \operatorname{tr}\left(\mathbf{w}\mathbf{w}^H \right) = \|\mathbf{u}\|_2^2 \|\mathbf{w}\|_2^2 .$$

It follows that

$$\left\| \mathbf{E} \frac{\mathbf{s}\mathbf{s}^\top}{\mathbf{s} \cdot \mathbf{s}} \right\|_F^2 = \frac{\|\mathbf{E}\mathbf{s}\|_2}{\|\mathbf{s}\|_2} .$$

This proves the equality in the lemma. Next, we note that for any $0 \leq |\beta| \leq \alpha$ we have $\sqrt{\alpha^2 - \beta^2} \leq \alpha - \beta^2/2\alpha$. We choose $\alpha = \|\mathbf{E}\|_F$ and $\beta = \|\mathbf{E}\mathbf{s}\|_2/\|\mathbf{s}\|_2$ to complete the lemma.

Our final theorem in this section, which is also due to Dennis and Moré [51], proves the superlinear convergence of Broyden's method.

Theorem 3.7.3 (Superlinear Convergence of Broyden's Method) *Suppose that $\mathscr{D}$ is a convex and open set of real n-vectors. Let $\mathbf{f}$ map $\mathscr{D}$ to real n-vectors, and let $\mathbf{f}$ be continuously differentiable function on $\mathscr{D}$. Assume that there exists $\mathbf{z} \in \mathscr{D}$ such that $\mathbf{f}(\mathbf{z}) = \mathbf{0}$, and that the Jacobian $\mathbf{J_f}(\mathbf{z})$ is nonsingular. Suppose that there are constants $\gamma > 0$ and $\varrho > 0$ such that $\|\mathbf{x} - \mathbf{z}\| < \varrho$ implies $\mathbf{x} \in \mathscr{D}$ and*

$$\|\mathbf{J_f}(\mathbf{x}) - \mathbf{J_f}(\mathbf{z})\|_F \leq \gamma \|\mathbf{x} - \mathbf{z}\|_2 \; .$$

Assume that

$$\left\|\widetilde{\mathbf{J}}_0 - \mathbf{J_f}(\mathbf{z})\right\|_F \leq \delta < \frac{1}{6\|\mathbf{J_f}(\mathbf{z})^{-1}\|_F} \; ,$$

and that

$$\|\mathbf{z}0 - \mathbf{z}\|_2 \leq \varepsilon \leq \frac{2\delta}{3\gamma} \; .$$

Let the sequences $\{\mathbf{z}_k\}_{k=0}^{\infty}$, $\{\mathbf{y}_k\}_{k=0}^{\infty}$, $\{\mathbf{s}_k\}_{k=0}^{\infty}$ and $\{\widetilde{\mathbf{J}}_k\}_{k=0}^{\infty}$ be generated by Broyden's method (Algorithm 3.7.1). Then the sequence $\{\mathbf{z}_k\}_{k=0}^{\infty}$ is well-defined and converges superlinearly to $\mathbf{z}$.

Proof Theorem 3.7.1 proves that $\{\mathbf{z}_k\}_{k=0}^{\infty}$ is well-defined and converges at least linearly to $\mathbf{z}$. If there exists k so that $\mathbf{z}_k = \mathbf{z}$, then $\mathbf{s}_k = \mathbf{0}$ and we will have $\mathbf{z}_\ell = \mathbf{z}$ for all $\ell \geq k$. In this case the claim is obvious. In the remainder of the proof, we will assume that $\mathbf{z}_k \neq \mathbf{z}$ for all $k \geq 0$. As a result, Theorem 3.7.2, implies that we need only show that

$$\frac{\left\|\left\{\widetilde{\mathbf{J}}_k - \mathbf{J_f}(\mathbf{z})\right\} \mathbf{s}_k\right\|_2}{\|\mathbf{s}_k\|} \to 0 \; .$$

Lemma 1.2.6 shows that for all n-vectors $\mathbf{u}$ and $\mathbf{w}$ we have

$$\left\|\mathbf{u}\mathbf{w}^H\right\|_F^2 = \operatorname{tr}\left(\left[\mathbf{u}\mathbf{w}^H\right]^H \mathbf{u}\mathbf{w}^H\right) = \operatorname{tr}\left(\mathbf{w}\mathbf{u}^H\mathbf{u}\mathbf{w}^H\right) = \mathbf{u}\cdot\mathbf{u}\operatorname{tr}\left(\mathbf{w}\mathbf{w}^H\right)$$

$$= \mathbf{u}\cdot\mathbf{u}\mathbf{w}\cdot\mathbf{w} = \|\mathbf{u}\|_2^2\|\mathbf{w}\|_2^2 \; .$$

Consequently, we see that

$$\left\|\frac{\{\mathbf{y}_k - \mathbf{J_f}(\mathbf{z})\mathbf{s}_k\}\, \mathbf{s}_k{}^\top}{\mathbf{s}_k \cdot \mathbf{s}_k}\right\|_F = \frac{\|\mathbf{y}_k - \mathbf{J_f}(\mathbf{z})\mathbf{s}_k\|_2 \, \|\mathbf{s}_k\|_2}{\|\mathbf{s}_k\|_2^2}$$

then inequality (3.34) shows that

$$\leq \frac{\gamma}{2}\{\|\mathbf{z}_k - \mathbf{z}\|_2 + \|\mathbf{z}_{k+1} - \mathbf{z}\|_2\} . \tag{3.43}$$

Also, Lemma 3.7.3 shows that

$$\left\| \{\widetilde{\mathbf{J}}_k - \mathbf{J}_\mathbf{f}(\mathbf{z})\} \left\{ \mathbf{I} - \frac{\mathbf{s}_k {\mathbf{s}_k}^\top}{\mathbf{s}_k \cdot \mathbf{s}_k} \right\} \right\|_F$$
$$\leq \left\| \widetilde{\mathbf{J}}_k - \mathbf{J}_\mathbf{f}(\mathbf{z}) \right\|_F - \frac{1}{2\left\| \widetilde{\mathbf{J}}_k - \mathbf{J}_\mathbf{f}(\mathbf{z}) \right\|_F} \left(\frac{\left\| \{\widetilde{\mathbf{J}}_k - \mathbf{J}_\mathbf{f}(\mathbf{z})\}\mathbf{s}_k \right\|_2}{\|\mathbf{s}_k\|_2} \right)^2 . \tag{3.44}$$

Equation (3.35) under our current circumstances can be rewritten in the form

$$\widetilde{\mathbf{J}}_{k+1} - \mathbf{J}_\mathbf{f}(\mathbf{z}) = \left[\widetilde{\mathbf{J}}_k - \mathbf{J}_\mathbf{f}(\mathbf{z}) \right] \left[\mathbf{I} - \mathbf{s}_k \frac{1}{\mathbf{s}_k \cdot \mathbf{s}_k} {\mathbf{s}_k}^\top \right] + [\mathbf{y}_k - \mathbf{J}_\mathbf{f}(\mathbf{z})\mathbf{s}_k] \frac{1}{\mathbf{s}_k \cdot \mathbf{s}_k} {\mathbf{s}_k}^\top$$

By taking norms of both sides of this equation, we obtain

$$\left\| \widetilde{\mathbf{J}}_{k+1} - \mathbf{J}_\mathbf{f}(\mathbf{z}) \right\|_F \leq \left\| \{\widetilde{\mathbf{J}}_k - \mathbf{J}_\mathbf{f}(\mathbf{z})\} \left\{ \mathbf{I} - \frac{\mathbf{s}_k {\mathbf{s}_k}^\top}{\mathbf{s}_k \cdot \mathbf{s}_k} \right\} \right\|_F + \frac{\left\| \{\mathbf{y}_k - \mathbf{J}_\mathbf{f}(\mathbf{z})\mathbf{s}_k\} {\mathbf{s}_k}^\top \right\|_F}{\mathbf{s}_k \cdot \mathbf{s}_k} .$$

then inequalities (3.43) and (3.44) give us

$$\leq \left\| \widetilde{\mathbf{J}}_k - \mathbf{J}_\mathbf{f}(\mathbf{z}) \right\|_F - \frac{\left\| \{\widetilde{\mathbf{J}}_k - \mathbf{J}_\mathbf{f}(\mathbf{z})\}\mathbf{s}_k \right\|_2^2}{2\left\| \{\widetilde{\mathbf{J}}_k - \mathbf{J}_\mathbf{f}(\mathbf{z})\} \right\|_F \|\mathbf{s}_k\|_2^2} + \frac{\gamma}{2}\{\|\mathbf{z}_k - \mathbf{z}\|_2 + \|\mathbf{z}_{k+1} - \mathbf{z}\|_2\}$$

and since the proof of Theorem 3.7.1 showed that $\|\mathbf{z}_{k+1} - \mathbf{z}\|_2 \leq \frac{1}{2}\|\mathbf{z}_k - \mathbf{z}\|_2$, we get

$$\leq \left\| \widetilde{\mathbf{J}}_k - \mathbf{J}_\mathbf{f}(\mathbf{z}) \right\|_F - \frac{\left\| \{\widetilde{\mathbf{J}}_k - \mathbf{J}_\mathbf{f}(\mathbf{z})\}\mathbf{s}_k \right\|_2^2}{2\left\| \{\widetilde{\mathbf{J}}_k - \mathbf{J}_\mathbf{f}(\mathbf{z})\} \right\|_F \|\mathbf{s}_k\|_2^2} + \frac{3\gamma}{4}\|\mathbf{z}_k - \mathbf{z}\|_2 .$$

This inequality can be rewritten in the form

$$\frac{\left\| \{\widetilde{\mathbf{J}}_k - \mathbf{J}_\mathbf{f}(\mathbf{z})\}\mathbf{s}_k \right\|_2^2}{\|\mathbf{s}_k\|_2^2}$$
$$\leq 2\left\| \widetilde{\mathbf{J}}_k - \mathbf{J}_\mathbf{f}(\mathbf{z}) \right\|_F \left[-\left\| \widetilde{\mathbf{J}}_{k+1} - \mathbf{J}_\mathbf{f}(z) \right\|_F + \left\| \widetilde{\mathbf{J}}_k - \mathbf{J}_\mathbf{f}(\mathbf{z}) \right\|_F + \frac{3\gamma}{4}\|\mathbf{z}_k - \mathbf{z}\|_2 \right]$$

then we use inequality (3.39) to get

$$\leq 4\delta \left[-\left\| \widetilde{\mathbf{J}}_{k+1} - \mathbf{J}_\mathbf{f}(z) \right\|_F + \left\| \widetilde{\mathbf{J}}_k - \mathbf{J}_\mathbf{f}(\mathbf{z}) \right\|_F + \frac{3\gamma}{4}\|\mathbf{z}_k - \mathbf{z}\|_2 \right] \tag{3.45}$$

Since

$$\|\mathbf{z}_{k+1} - \mathbf{z}\|_2 \leq \frac{1}{2} \|\mathbf{z}_k - \mathbf{z}\|_2$$

for all k and $\|\mathbf{z}_0 - \mathbf{z}\| \leq \varepsilon$, we conclude that

$$\sum_{k=0}^{\infty} \|\mathbf{z}_k - \mathbf{z}\|_2 \leq 2\varepsilon \ .$$

We can sum (3.45) to obtain

$$\sum_{k=1}^{N} \frac{\left\|\left\{\widetilde{\mathbf{J}}_k - \mathbf{J}_\mathbf{f}(\mathbf{z})\right\} \mathbf{s}_k\right\|_2^2}{\|\mathbf{s}_k\|_2^2} \leq 4\delta \left[\left\|\widetilde{\mathbf{J}}_0 - \mathbf{J}_\mathbf{f}(\mathbf{z})\right\|_F - \left\|\widetilde{\mathbf{J}}_{N+1} - \mathbf{J}_\mathbf{f}(\mathbf{z})\right\|_F + \frac{3\gamma}{4} \sum_{k=0}^{N} \|\mathbf{z}_k - \mathbf{z}\|_2 \right]$$
$$\leq 4\delta \left[\|\mathbf{E}_0\|_F + \frac{3\gamma\varepsilon}{2} \right] \leq 4\delta \left(\delta + \frac{3\gamma\varepsilon}{2} \right) .$$

Since this inequality is true for any $N > 0$, the infinite sum converges. The convergence of the infinite sum on the left in turn implies that the terms in the sum converge to zero as $k \to \infty$. This proves the local superlinear convergence of Broyden's method.

3.7.3 Numerical Implementation

In Algorithm 3.7.1. for Broyden's method, the interesting numerical issues arise in solving

$$\widetilde{\mathbf{J}}_k \mathbf{s}_k = -\mathbf{f}(\mathbf{z}_k)$$

for $\mathbf{s}_k$, and in updating

$$\widetilde{\mathbf{J}}_{k+1} = \widetilde{\mathbf{J}}_k + \left(\mathbf{y}_k - \widetilde{\mathbf{J}}_k \mathbf{s}_k\right) \frac{1}{\mathbf{s}_k \cdot \mathbf{s}_k} {\mathbf{s}_k}^\top .$$

Given the inverse of $\widetilde{\mathbf{J}}_k$, we will learn how to compute the inverse of $\widetilde{\mathbf{J}}_{k+1}$. Then we could use this inverse to compute $\mathbf{s}_{k+1}$ in an order of n^2 operations. Even better, given a $\mathbf{QR}$ factorization of $\widetilde{\mathbf{J}}_k$, we will see how to compute the $\mathbf{QR}$ factorization of $\widetilde{\mathbf{J}}_{k+1}$ in an order of n^2 operations.

3.7.3.1 Inverse Updates

In order to maintain $\widetilde{\mathbf{J}}_k^{-1}$ and update it during the Quasi-Newton iteration, we will make use of the following interesting lemma.

Lemma 3.7.4 (Sherman-Morrison-Woodbury Formula) *Suppose that* **u** *and* **v** *are n-vectors, and that* **A** *is a nonsingular n×n matrix. Then* $\mathbf{A}+\mathbf{u}\mathbf{v}^H$ *is nonsingular if and only if*

$$1 + \mathbf{v}\cdot\mathbf{A}^{-1}\mathbf{u} \neq 0 \,.$$

Furthermore, if $\mathbf{A}+\mathbf{u}\mathbf{v}^H$ *is nonsingular, then*

$$\left(\mathbf{A}+\mathbf{u}\mathbf{v}^H\right)^{-1} = \mathbf{A}^{-1} - \mathbf{A}^{-1}\mathbf{u}\frac{1}{1+\mathbf{v}\cdot\mathbf{A}^{-1}\mathbf{u}}\mathbf{v}^H\mathbf{A}^{-1} \,. \tag{3.46}$$

Proof Suppose that $1 + \mathbf{v}\cdot\mathbf{A}^{-1}\mathbf{u} = 0$. Then $\mathbf{u} \neq 0$ and

$$\left(\mathbf{A}+\mathbf{u}\mathbf{v}^H\right)\mathbf{A}^{-1}\mathbf{u} = \mathbf{u}\left(1+\mathbf{v}\cdot\mathbf{A}^{-1}\mathbf{u}\right) = \mathbf{0} \,.$$

This proves that $\mathbf{A}+\mathbf{u}\mathbf{v}^H$ is singular, and establishes the forward direction of the first claim.

If $1 + \mathbf{v}\cdot\mathbf{A}^{-1}\mathbf{u} \neq 0$, then

$$\begin{aligned}
&\left(\mathbf{A}+\mathbf{u}\mathbf{v}^H\right)\left[\mathbf{A}^{-1} - \mathbf{A}^{-1}\mathbf{u}\frac{1}{1+\mathbf{v}\cdot\mathbf{A}^{-1}\mathbf{u}}\mathbf{v}^H\mathbf{A}^{-1}\right] \\
&= \mathbf{I} + \mathbf{u}\mathbf{v}^H\mathbf{A}^{-1} - \mathbf{u}\frac{1}{1+\mathbf{v}\cdot\mathbf{A}^{-1}\mathbf{u}}\mathbf{v}^H\mathbf{A}^{-1} - \mathbf{u}\mathbf{v}^H\mathbf{A}^{-1}\mathbf{u}\frac{1}{1+\mathbf{v}\cdot\mathbf{A}^{-1}\mathbf{u}}\mathbf{v}^H\mathbf{A}^{-1} \\
&= \mathbf{I} + \mathbf{u}\frac{1+\mathbf{v}\cdot\mathbf{A}^{-1}\mathbf{u} - 1 - \mathbf{v}\cdot\mathbf{A}^{-1}\mathbf{u}}{1+\mathbf{v}\cdot\mathbf{A}^{-1}\mathbf{u}}\mathbf{v}^H\mathbf{A}^{-1} = \mathbf{I} \,.
\end{aligned}$$

In this case, $\mathbf{A}+\mathbf{u}\mathbf{v}^H$ is nonsingular, proving the reverse direction of the first claim, as well as the second claim.

As a result of the lemma, we might implement Broyden's method as follows:

Algorithm 3.7.2 (Broyden's Method via Inverse Updates)

$$\begin{aligned}
\mathbf{s}_k &= -\widetilde{\mathbf{J}}_k^{-1}\mathbf{f}(\mathbf{z}_k) \\
\mathbf{z}_{k+1} &= \mathbf{z}_k + \mathbf{s}_k \\
\mathbf{y}_k &= \mathbf{f}(\mathbf{z}_{k+1}) - \mathbf{f}(\mathbf{z}k) \\
\mathbf{u}_k &= \widetilde{\mathbf{J}}_k^{-\top}\mathbf{s}_k \\
\delta_k &= \mathbf{u}_k\cdot\mathbf{y}_k \\
\mathbf{v}_k &= \widetilde{\mathbf{J}}_k^{-1}\mathbf{y}_k \\
\mathbf{w}_k &= \left[\mathbf{s}_k - \mathbf{v}_k\right]/\delta_k \\
\widetilde{\mathbf{J}}_{k+1}^{-1} &= \widetilde{\mathbf{J}}_k^{-1} + \mathbf{w}_k\mathbf{u}_k{}^\top \,.
\end{aligned}$$

If we are working with n-vectors, then we note that $\mathbf{s}_k$, $\mathbf{u}_k$, $\mathbf{v}_k$ and $\widetilde{\mathbf{J}}_{k+1}^{-1}$ can each be computed in an order of n^2 operations, while $\mathbf{z}_{k+1}$, δ_k and $\mathbf{w}_k$ can each be computed in an order of n operations. Also, the vector $\mathbf{y}_k$ can be computed with one new function evaluation. If we had approximated the new Jacobian by finite differences and then factored it to prepare for solving a linear system, we would have required $n+1$ new function evaluations and an order of n^3 operations for the matrix factorization.

3.7.3.2 Factorization Updates

The inverse update approach is not used in practice because it does not allow us to detect ill-conditioning of the Jacobian approximations $\widetilde{\mathbf{J}}_k$. A better approach is to compute a Householder LQ factorization of $\widetilde{\mathbf{J}}_k$ and update the factorization.

Suppose that we have factored

$$\widetilde{\mathbf{J}}_k = \mathsf{Q}_k \mathbf{R}_k$$

where Q_k is orthogonal and $\mathbf{R}_k$ is right-triangular. We will store Q_k as an $n \times n$ matrix, not as a product of Householder reflectors; thus this factorization proceeds in a different manner from the algorithms we developed in Sect. 6.7.4.1 in Chap. 6 of Volume I.

Since Broyden's method updates the approximate Jacobians by the equation

$$\widetilde{\mathbf{J}}_{k+1} = \widetilde{\mathbf{J}}_k + \left(\mathbf{y}_k - \widetilde{\mathbf{J}}_k \mathbf{s}_k\right) \frac{1}{\mathbf{s}_k \cdot \mathbf{s}_k} {\mathbf{s}_k}^\top ,$$

we have

$${\mathsf{Q}_k}^\top \widetilde{\mathbf{J}}_{k+1} = \mathbf{R}_k + {\mathsf{Q}_k}^\top \left(\mathbf{y}_k - \widetilde{\mathbf{J}}_k \mathbf{s}_k\right) \frac{1}{\mathbf{s}_k \cdot \mathbf{s}_k} {\mathbf{s}_k}^\top .$$

This suggests that we compute

$$\mathbf{u}_k \equiv {\mathsf{Q}_k}^\top \left(\mathbf{y}_k - \widetilde{\mathbf{J}}_k \mathbf{s}_k\right) = {\mathsf{Q}_k}^\top \mathbf{y}_k - \mathbf{R}_k \mathbf{s}_k \text{ and } \mathbf{w}_k = \mathbf{s}_k / \|\mathbf{s}_k\|_2^2 .$$

Then

$${\mathsf{Q}_k}^\top \widetilde{\mathbf{J}}_{k+1} = \mathbf{R}_k + \mathbf{u}_k {\mathbf{w}_k}^\top$$

is a rank-one modification of a right-triangular matrix. We will determine a product $\mathbf{G}_k$ of plane rotations so that

$${\mathbf{G}_k}^\top \left(\mathbf{R}_k + \mathbf{u}_k {\mathbf{w}_k}^\top\right) = \mathbf{R}_{k+1}$$

is left-triangular. Then

$$\widetilde{\mathbf{J}}_{k+1} = Q_k \left\{\mathbf{R}_k + \mathbf{u}_k {\mathbf{w}_k}^\top\right\} = \left(Q_k {\mathbf{G}_k}^\top\right)\left(\mathbf{G}_k \left\{\mathbf{R}_k + \mathbf{u}_k {\mathbf{w}_k}^\top\right\}\right) = Q_{k+1} \mathbf{R}_{k+1}$$

is a QR factorization of $\widetilde{\mathbf{J}}_{k+1}$.

In order to understand the following discussion carefully, the reader should review Sect. 6.9 in Chap. 6 of Volume I on plane rotations. In the beginning, the arrays have the form

$$\mathbf{R}_k + \mathbf{u}_k {\mathbf{w}_k}^\top = \begin{bmatrix} \times & \times & \times & \times & \times \\ & \times & \times & \times & \times \\ & & \times & \times & \times \\ & & & \times & \times \\ & & & & \times \end{bmatrix} + \begin{bmatrix} \times \\ \times \\ \times \\ \times \\ \times \end{bmatrix} \begin{bmatrix} \times & \times & \times & \times & \times \end{bmatrix} .$$

In the first step of the QR update, we choose a plane rotation in the last two entries to zero out the last entry of $\mathbf{u}_k$; this gives us

$$\mathbf{G}_{n-1,n}\left(\mathbf{R}_k + \mathbf{u}_k {\mathbf{w}_k}^\top\right) = \begin{bmatrix} \times & \times & \times & \times & \times \\ & \times & \times & \times & \times \\ & & \times & \times & \times \\ & & & \ast & \ast \\ & & & + & \ast \end{bmatrix} + \begin{bmatrix} \times \\ \times \\ \times \\ \ast \\ 0 \end{bmatrix} \begin{bmatrix} \times & \times & \times & \times & \times \end{bmatrix} .$$

We continue in this way, zeroing entries in $\mathbf{u}_k$ until we obtain the upper Hessenberg form

$$\mathbf{G}_{1,2} \ldots \mathbf{G}_{n-1,n}\left(\mathbf{R}_k + \mathbf{u}_k {\mathbf{w}_k}^\top\right) = \begin{bmatrix} \ast & \ast & \ast & \ast & \ast \\ + & \ast & \ast & \ast & \ast \\ & + & \ast & \ast & \ast \\ & & + & \ast & \ast \\ & & & + & \ast \end{bmatrix} + \begin{bmatrix} \ast \\ 0 \\ 0 \\ 0 \\ 0 \end{bmatrix} \begin{bmatrix} \times & \times & \times & \times & \times \end{bmatrix}$$

$$= \begin{bmatrix} \times & \times & \times & \times & \times \\ \times & \times & \times & \times & \times \\ & \times & \times & \times & \times \\ & & \times & \times & \times \\ & & & \times & \times \end{bmatrix} \equiv \mathbf{H}_{k+1} .$$

Next, we choose plane rotations to zero out the sub-diagonal entries. In the first step of this process we obtain

$$\widetilde{\mathbf{G}}_{1,2}\mathbf{H}_{k+1} = \begin{bmatrix} * & * & * & * & * \\ 0 & * & * & * & * \\ & \times & \times & \times & \times \\ & & \times & \times & \times \\ & & & \times & \times \end{bmatrix} .$$

We continue until we obtain

$$\widetilde{\mathbf{G}}_{n-1,n} \ldots \widetilde{\mathbf{G}}_{1,2}\mathbf{H}_{k+1} = \begin{bmatrix} * & * & * & * & * \\ 0 & * & * & * & * \\ & 0 & * & * & * \\ & & 0 & * & * \\ & & & 0 & * \end{bmatrix} = \mathbf{R}_{k+1} .$$

We also need to compute

$$\mathbf{Q}_{k+1} = \mathbf{Q}_k\mathbf{G}_{n-1,n}{}^\top \ldots \mathbf{G}_{1,2}{}^\top\widetilde{\mathbf{G}}_{1,2}{}^\top \ldots \widetilde{\mathbf{G}}_{n-1,n}{}^\top .$$

This product is performed while we determine the plane reflectors.

Readers may read more about updating matrix factorizations in the paper by Daniel et al. [43] or the book by Golub and van Loan [84, p. 334ff]. The MINPACK library contains code that implements factorization updates for Broyden's method. The initial QR factorization is performed by routine qrfac. Updates of QR factorizations are performed by routine r1updt.

Codes such as MINPACK employ some additional features to enhance the convergence of the iteration. For example, if Broyden's method produces a step direction that is not a descent direction for $\|\mathbf{f}(\mathbf{x})\|_2^2$, then the true Jacobian is computed and used to restart the Jacobian approximations. For additional details, it is best to examine the code for MINPACK routines hybrj1 or hybrd1.

3.7.4 Unconstrained Minimization

When we minimize a real functional ϕ via a Quasi-Newton method, we want an approximation $\widetilde{\mathbf{H}}_k$ to the Hessian such that

1. $\widetilde{\mathbf{H}}_{k+1}\mathbf{s}_k = \mathbf{y}_k \equiv \mathbf{g}_\phi(\mathbf{z}_{k+1}) - \mathbf{g}_\phi(\mathbf{z}_k)$, where $\mathbf{s}_k \equiv \mathbf{z}_{k+1} - \mathbf{z}_k$,
2. $\widetilde{\mathbf{H}}_{k+1}$ is symmetric and as close to $\widetilde{\mathbf{H}}_k$ as possible, and
3. $\widetilde{\mathbf{H}}_{k+1}$ is positive-definite.

Broyden's method is not satisfactory for use in unconstrained minimization, because it does not maintain symmetry of the Hessian.

The most popular choice is provided by the **Broyden-Fletcher-Goldfarb-Shanno (BFGS) method**. Although it can be formally written in the form

$$\widetilde{\mathbf{H}}_{k+1} = \widetilde{\mathbf{H}}_k + \mathbf{y}_k \frac{1}{\mathbf{y}_k{}^\top \mathbf{s}_k} \mathbf{y}_k{}^\top - \widetilde{\mathbf{H}}_k \mathbf{s}_k \frac{1}{\mathbf{s}_k{}^\top \widetilde{\mathbf{H}}_k \mathbf{s}_k} \mathbf{s}_k{}^\top \widetilde{\mathbf{H}}_k ,$$

it is typically implemented as an update of a Cholesky factorization. If $\widetilde{\mathbf{H}}_k = \mathbf{L}_k \mathbf{L}_k{}^\top$ is the Cholesky factorization of the previous Hessian approximation and $\mathbf{s}_k \cdot \mathbf{y}_k \geq 0$, we begin by computing

$$\mathbf{v}_k \equiv \mathbf{L}_k{}^\top \mathbf{s}_k \frac{\sqrt{\mathbf{s}_k \cdot \mathbf{y}_k}}{\left\| \mathbf{L}_k{}^\top \mathbf{s}_k \right\|_2} \text{ and } \mathbf{A}_k \equiv \mathbf{L}_k + (\mathbf{y}_k - \mathbf{L}_k \mathbf{v}_k) \frac{1}{\mathbf{v}_k \cdot \mathbf{v}_k} \mathbf{v}_k{}^\top .$$

Then we factor

$$\mathbf{A}_k = \mathbf{L}_{k+1} \mathbf{Q}_{k+1} .$$

This factorization can be accomplished by plane rotations in a manner similar to that in Sect. 3.7.3.2 for a rank-one modification of a right-triangular matrix.

Note that

$$\mathbf{v}_k \cdot \mathbf{v}_k = \mathbf{s}_k \cdot \mathbf{y}_k ,$$

so the process fails whenever $\mathbf{s}_k \cdot \mathbf{y}_k \leq 0$; otherwise, BFGS maintains positivity. This is because

$$\begin{aligned}
\mathbf{L}_{k+1} \mathbf{L}_{k+1}{}^\top &= \mathbf{A}_k \mathbf{A}_k{}^\top \\
&= \mathbf{L}_k \mathbf{L}^\top + (\mathbf{y}_k - \mathbf{L}_k \mathbf{v}_k) \frac{1}{\mathbf{v}_k \cdot \mathbf{v}_k} \mathbf{v}_k{}^\top \mathbf{L}_k{}^\top + \mathbf{L}_k{}^\top \mathbf{v}_k \frac{1}{\mathbf{v}_k \cdot \mathbf{v}_k} (\mathbf{y}_k - \mathbf{L}_k{}^\top \mathbf{v}_k)^\top \\
&+ (\mathbf{y}_k - \mathbf{L}_k{}^\top \mathbf{v}_k) \frac{1}{\mathbf{v}_k \cdot \mathbf{v}_k} (\mathbf{y}_k - \mathbf{L}_k{}^\top \mathbf{v}_k)^\top \\
&= \mathbf{L}_k \mathbf{L}^\top + \mathbf{y}_k \frac{1}{\mathbf{s}_k \cdot \mathbf{y}_k} \mathbf{y}_k{}^\top - \mathbf{L}_k \mathbf{v}_k \frac{1}{\mathbf{s}_k \cdot \mathbf{y}_k} \mathbf{v}_k{}^\top \mathbf{L}_k{}^\top \\
&= \mathbf{L}_k \mathbf{L}^\top + \mathbf{y}_k \frac{1}{\mathbf{s}_k \cdot \mathbf{y}_k} \mathbf{y}_k{}^\top - \mathbf{L}_k \mathbf{L}^\top \mathbf{s}_k \frac{1}{\mathbf{s}_k \cdot \mathbf{L}_k \mathbf{L}^\top \mathbf{s}_k} \mathbf{s}_k{}^\top \mathbf{L}_k \mathbf{L}^\top \\
&= \widetilde{\mathbf{H}}_k + \mathbf{y}_k \frac{1}{\mathbf{s}_k \cdot \mathbf{y}_k} \mathbf{y}_k{}^\top - \widetilde{\mathbf{H}}_k \mathbf{s}_k \frac{1}{\mathbf{s}_k \cdot \widetilde{\mathbf{H}}_k \mathbf{s}_k} \mathbf{s}_k{}^\top \widetilde{\mathbf{H}}_k = \widetilde{\mathbf{H}}_{k+1} .
\end{aligned}$$

A Fortran program by Schnabel, Koontz and Weiss to solve unconstrained minimization problems using trust regions, the BFGS quasi-Newton update and either the dogleg or hook step is called UNCMIN. MATLAB programmers can use function fminunc.

Exercise 3.7.1 Given a symmetric $n \times n$ matrix $\mathbf{H}_o$, as well as n-vectors $\mathbf{s}$ and $\mathbf{y}$, show that the unique solution $\mathbf{H}_n$ to the minimization problem

$$\min \|\mathbf{H}_n - \mathbf{H}_o\|_F \text{ subject to } \mathbf{H}_n\mathbf{s} = \mathbf{y} \text{ and } \mathbf{H}_n = {\mathbf{H}_n}^\top$$

is the **Powell-Symmetric-Broyden update**

$$\mathbf{H}_n = \mathbf{H}_o + \begin{bmatrix} \mathbf{y} - \mathbf{H}_o\mathbf{s} & \mathbf{s} \end{bmatrix} \begin{bmatrix} 0 & 1 \\ 1 & -\mathbf{s}^\top(\mathbf{y} - \mathbf{H}_o\mathbf{s})/\mathbf{s}\cdot\mathbf{s} \end{bmatrix} \begin{bmatrix} (\mathbf{y} - \mathbf{H}_o\mathbf{s})^\top \\ \mathbf{s}^\top \end{bmatrix} \frac{1}{\mathbf{s}\cdot\mathbf{s}} .$$

The Powell-Symmetric-Broyden update is unpopular because even when $\mathbf{H}_o$ is positive, it is possible that $\mathbf{H}_n$ is not.

Exercise 3.7.2 Given an invertible $n \times n$ matrix $\mathbf{L}_o$ as well as n-vectors $\mathbf{s}$ and $\mathbf{y}$ such that $\mathbf{y}\cdot\mathbf{s} > 0$, show that the $n \times n$ matrix $\mathbf{L}_n$ that solves the minimization problem

$$\min \left\| {\mathbf{L}_n}^{-\top} - {\mathbf{L}_o}^{-\top} \right\|_F \text{ subject to } {\mathbf{L}_n}^{-\top}{\mathbf{L}_n}^{\top}\mathbf{y} = \mathbf{s}$$

is given by

$${\mathbf{L}_n}^{-\top} = {\mathbf{L}_o}^{-\top} + \left[\mathbf{s} - {\mathbf{L}_o}^{-\top}\mathbf{w}\right] \frac{1}{\mathbf{w}\cdot\mathbf{w}}\mathbf{w}^\top \text{ where } \mathbf{w} = {\mathbf{L}_o}^{-1}\mathbf{y}\sqrt{\frac{\mathbf{y}\cdot\mathbf{s}}{\mathbf{y}\cdot{\mathbf{L}_o}^{-\top}\mathbf{L}_o^{-1}\mathbf{y}}} .$$

If $\mathbf{H}_o = \mathbf{L}_o\mathbf{L}^\top{}_o$, show that $\mathbf{H}_n = \mathbf{L}_n\mathbf{L}^\top{}_n$ is given by the **Davidon-Fletcher-Powell update**,

$$\mathbf{H}_n = \mathbf{H}_o + \frac{(\mathbf{y} - \mathbf{H}_o\mathbf{s})\mathbf{y}^\top + \mathbf{y}(\mathbf{y} - \mathbf{H}_o\mathbf{s})^\top}{\mathbf{y}\cdot\mathbf{s}} - \mathbf{y}\frac{\mathbf{y}\cdot(\mathbf{y} - \mathbf{H}_o\mathbf{s})}{(\mathbf{y}\cdot\mathbf{s})^2}\mathbf{y}^\top .$$

This method seems not to work as well in practice as the BFGS update.

3.8 Krylov Subspace Iterative Methods

So far in this chapter, all of our methods have been willing to store Jacobian or Hessian matrices, and to solve linear systems involving those matrices. For very large nonlinear systems, particularly those in which very few independent variables are involved in any individual nonlinear equation, these methods can involve substantial computer memory and computational work. In this section, we will adopt some ideas from Chap. 2 on iterative linear algebra to systems of nonlinear equations.

3.8.1 Conjugate Gradients

Luenberger [120, p. 182] describes how to modify the conjugate gradient algorithm of Sect. 2.5.3 to minimize a real-valued function of n real variables. His algorithm takes the form

Algorithm 3.8.1 (Nonlinear Conjugate Gradients)

until convergence do

$$\mathbf{g}_0 = \mathbf{g}_\phi(\mathbf{z}_0)$$

$$\mathbf{s}_0 = -\mathbf{g}_0$$

for $0 \leq k < n$

$$\mathbf{H}_k = \mathbf{H}_\phi(\mathbf{z}_k)$$

$$\mathbf{v}_k = \mathbf{H}_k \mathbf{s}_k$$

$$\alpha_k = -\frac{\mathbf{g}_k \cdot \mathbf{s}_k}{\mathbf{s}_k \cdot \mathbf{v}_k}$$

$$\mathbf{z}_{k+1} = \mathbf{z}_k + \mathbf{s}_k \alpha_k$$

$$\mathbf{g}_{k+1} = \mathbf{g}_\phi(\mathbf{z}_{k+1})$$

if $k < n-1$

$$\beta_k = \frac{\mathbf{g}_{k+1} \cdot \mathbf{H}_k \mathbf{s}_k}{\mathbf{s}_k \cdot \mathbf{v}_k}$$

$$\mathbf{s}_{k+1} = -\mathbf{g}_{k+1} + \mathbf{s}_k \beta_k$$

$$\mathbf{z}_0 = \mathbf{z}_n$$

Of course, the unfortunate feature of this algorithm is the evaluation of the Hessian $\mathbf{H}_k$ in the innermost loop, and the requirement that this matrix be positive.

An implementation of the nonlinear conjugate gradient algorithm is available as routine snescg in PETSc.

3.8.2 Fletcher-Reeves

The Fletcher-Reeves algorithm [70] implements the conjugate gradient algorithm without evaluating the Hessian. This algorithm for minimizing a real functional ϕ the following form:

Algorithm 3.8.2 (Fletcher-Reeves)

until convergence do

$\quad \mathbf{g}_0 = \mathbf{g}_\phi(\mathbf{x}_0)$

$\quad \mathbf{s}_0 = -\mathbf{g}_0$

$\quad$ for $0 \leq k < n$

$\quad\quad \alpha_k = \text{argmin}\ \phi(\mathbf{x}_k + \mathbf{s}_k\alpha)$ (i.e., perform line search)

$\quad\quad \mathbf{x}_{k+1} = \mathbf{x}_k + \mathbf{s}_k\alpha_k$

$\quad\quad \mathbf{g}_{k+1} = \mathbf{g}_\phi(\mathbf{x}_{k+1})$

$\quad\quad$ if $k < n - 1$

$$\beta_k = \frac{\mathbf{g}_{k+1} \cdot \mathbf{g}_{k+1}}{\mathbf{g}_k \cdot \mathbf{g}_k}$$

$\quad\quad\quad \mathbf{s}_{k+1} = -\mathbf{g}_{k+1} + \mathbf{s}_k\beta_k$

$\quad \mathbf{x}_0 = \mathbf{x}_n$

An implementation of the Fletcher-Reeves algorithm is available in C.

3.8.3 GMRES

Brown and Saad [30] have described a method for solving a system of nonlinear equations by a modification of the GMRES solver, which we previously described in Sect. 2.6.2. Readers may also read about this nonlinear solver in Kelley [107, p. 101ff] or Trangenstein [173, p. 156].

Suppose that we want to solve a system of nonlinear equations $\mathbf{f}(\mathbf{z}) = \mathbf{0}$ by means of a Newton-like iteration. We assume that we have a current guess $\mathbf{x}_k$ for the solution of this nonlinear system, and we want to find a step $\mathbf{s}_k$ satisfying $\mathbf{J}_\mathbf{f}(\mathbf{z}_k)\mathbf{s}_k = -\mathbf{f}\mathbf{z}_k$. The linear system for $\mathbf{s}_k$ can be solved approximately by a GMRES iteration. Recall that the GMRES iteration requires values of the Jacobian matrix times a given vector, in order to form the Krylov subspace approximations. The key idea is that matrix-vector products can be approximated by finite differences:

$$\mathbf{J}_{\mathbf{f}}(\mathbf{z}_k)\mathbf{w} \approx \left[\mathbf{f}(\mathbf{z}_k + \mathbf{w}\delta) - \mathbf{f}(\mathbf{z}_k)\right]/\delta \,,$$

where $\delta = \sqrt{\varepsilon}\|\mathbf{z}_k\|/\|\mathbf{w}\|$ and ε is machine precision.

Another issue is how long we should continue to build the Krylov expansion for the iterates, versus restarting the GMRES portion of the iterative process at the current value of the increment $\mathbf{s}_k$, or even restarting both the nonlinear iterative process and the GMRES process at some approximate solution $\mathbf{z}_k$. As the number of GMRES iterations becomes large, the storage and work for the iterative solution of the linear system increases, in order to handle the upper Hessenberg system for the search vectors. In practice, Brown and Saad suggest using at most five Krylov vectors in a typical iteration; then the GMRES iteration is restarted with initial guess given by the approximate solution increment from the previous GMRES computation.

We must also consider using a global convergence strategy for the iteration. First, let us consider developing a strategy for minimizing

$$\phi(\sigma) = \frac{1}{2}\left\|\mathbf{f}\left(\mathbf{z}_k + \widetilde{\mathbf{s}}_k\sigma\right)\right\|_2^2$$

where $\widetilde{\mathbf{s}}_k$ is the current approximate search direction. If $\widetilde{\mathbf{s}}_k$ approximates the solution to the Newton equation with residual

$$\mathbf{r} = \mathbf{J}_{\mathbf{f}}(\mathbf{z}_k)\widetilde{\mathbf{s}}_k + \mathbf{f}\mathbf{z}_k \,,$$

then

$$\psi'(0) = \mathbf{f}(\mathbf{z}_k) \cdot \mathbf{J}_{\mathbf{f}}(\mathbf{z}_k)\widetilde{\mathbf{s}}_k = -\mathbf{f}(\mathbf{z}_k) \cdot [\mathbf{f}(\mathbf{z}_k) - \mathbf{r}] \;.$$

Thus the approximate GMRES search direction $\widetilde{\mathbf{s}}_k$ will be a descent direction for ψ at $\mathbf{z}_k$ if and only if

$$\mathbf{f}(\mathbf{z}_k) \cdot \mathbf{r} < \|\mathbf{f}(\mathbf{z}_k)\|_2^2 \;.$$

The Cauchy inequality (3.15) in Chap. 3 of Volume I shows that this condition will be satisfied whenever $\|\mathbf{r}\|_2 < \|\mathbf{f}(\mathbf{z}_k)\|_2$. The following lemma due to Brown and Saad [29] provides fairly simple conditions under which the residual will satisfy this condition.

Lemma 3.8.1 ([29]) *Suppose that* $\mathbf{f}$ *maps real n-vectors to real n-vectors, and that* $\mathbf{f}$ *is differentiable at* $\mathbf{z}_k$ *with Jacobian* $\mathbf{J}_{\mathbf{f}}(\mathbf{z}_k)$*. Let* $\widetilde{\mathbf{s}}_{k,m}$ *be the solution increment determined by m iterations of the GMRES method with initial guess* $\widetilde{\mathbf{s}}_{k,0} = \mathbf{0}$*. Define*

$$\psi(\sigma) = \frac{1}{2}\left\|\mathbf{f}\left(\mathbf{z}_k + \widetilde{\mathbf{s}}_{k,m}\sigma\right)\right\|_2^2 \;.$$

Then

$$\psi'(0) = -\mathbf{f}(\mathbf{z}_k) \cdot \mathbf{J_f}(\mathbf{z}_k)\widetilde{\mathbf{s}}_{k,m} = \|\mathbf{J_f}(\mathbf{z}_k)\widetilde{\mathbf{s}}_{k,m} + \mathbf{f}(\mathbf{z}_k)\|_2^2 - \|\mathbf{f}^{(k)}\|_2^2 \; .$$

If in addition $\mathbf{J}(\mathbf{z}_k)$ *is nonsingular and* $\widetilde{\mathbf{s}}_{k,m} \neq \mathbf{0}$, *then* $\widetilde{\mathbf{s}}_{k,m}$ *is a descent direction.*

Proof Let $\mathbf{r}_0 = \mathbf{f}(\mathbf{z}_k)$ and form the associated Krylov subspace

$$\mathscr{K}_m = \left\langle \left[\mathbf{J}^{(k)}\right]^0 \mathbf{r}_0, \ldots, \left[\mathbf{J}^{(k)}\right]^{m-1} \mathbf{r}_0 \right\rangle \; .$$

Since the GMRES algorithm chooses $\widetilde{\mathbf{s}}_{k,m}$ so that

$$\|\mathbf{r}_m\|_2 \equiv \|\mathbf{J_f}(\mathbf{z}_k)\widetilde{\mathbf{s}}_{k,m} + \mathbf{f}(\mathbf{z}_k)\|_2 = \min_{\mathbf{s}\in\mathscr{K}_m} \|\mathbf{J_f}(\mathbf{z}_k)\mathbf{d} + \mathbf{f}(\mathbf{z}_k)\|_2 \; ,$$

it follows that

$$\mathbf{r}_m = \mathbf{J_f}\mathbf{z}_k\widetilde{\mathbf{s}}_{k,m} + \mathbf{f}(\mathbf{z}_k) \perp \mathscr{K}_m \; .$$

This implies that

$$\|\mathbf{r}_m\|_2^2 = \mathbf{r}_m \cdot [\mathbf{J_f}\mathbf{z}_k\widetilde{\mathbf{s}}_{k,m} + \mathbf{f}\mathbf{z}_k] = \mathbf{r}_m \cdot \mathbf{f}(\mathbf{z}_k) \; .$$

Consequently,

$$\begin{aligned}\psi'(0) = \mathbf{f}(\mathbf{z}_k) \cdot \mathbf{J_f}(\mathbf{z}_k)\widetilde{\mathbf{s}}_{k,m} &= -\left\|\mathbf{f}^{(k)}\right\|_2^2 + \mathbf{f}(\mathbf{z}_k) \cdot \mathbf{r}_m \\ &= \|\mathbf{r}_m\|_2^2 - \|\mathbf{f}(\mathbf{z}_k)\|_2^2 \; .\end{aligned}$$

If $\mathbf{J_f}\mathbf{z}_k$ is nonsingular and $\widetilde{\mathbf{s}}_{k,m} \neq \mathbf{0}$, the Pythagorean theorem implies that

$$\|\mathbf{f}(\mathbf{z}_k)\|_2^2 = \|\mathbf{J_f}(\mathbf{z}_k)\widetilde{\mathbf{s}}_{k,m} - \mathbf{r}_m\|_2^2 = \|\mathbf{J_f}(\mathbf{z}_k)\widetilde{\mathbf{s}}_{k,m}\|_2^2 + \|\mathbf{r}_m\|_2^2 > \|\mathbf{r}_m\|_2^2 \; ,$$

and the previous inequality shows that $\widetilde{\mathbf{s}}_{k,m}$ is a descent direction.

An implementation of the nonlinear GMRES algorithm is available as routine snesngmres in PETSc. This technique can also be found in `KINSOL`, which is part of the SUNDIALS package.

3.9 Direct Search Methods

Kelley [108, p. 135ff] and Swann [168] survey derivative-free methods for minimizing a real-valued function of multiple real variables. We will limit our discussion to a brief summary of some of the ideas.

3.9.1 Alternating Variable Method

One very simple method performs line searches in the independent variables.

Algorithm 3.9.1 (Alternating Variable Method)

$$
\begin{aligned}
&\text{until convergence do} \\
&\quad \text{for } 1 \leq i \leq n \\
&\qquad \mathbf{x}_i = \text{argmin } \phi(\mathbf{x} \pm \mathbf{e}_i \xi)
\end{aligned}
$$

This algorithm is also very slow and expensive. To see this algorithm in action, readers may experiment with the JavaScript **alternating variable** method. This program plots contours of function values for the Rosenbrock function

$$
\phi(\mathbf{x}) = \frac{1}{2} \left\| \begin{bmatrix} 10(\mathbf{x}_2 - \mathbf{x}_1^2) \\ 1 - \mathbf{x}_1 \end{bmatrix} \right\|_2^2 .
$$

Users may select a point in this graph, after which the program will perform one line search in each of the coordinate direction. Afterward, the reader may continue iteration to watch the progress of the algorithm.

3.9.2 Simplex Search

Spendley et al. [162] suggested an algorithm in which a multidimensional regular simplex is formed and pivoted about one of its faces until the minimum of some function is contained within. At that point, the simplex is contracted and the algorithm repeated, until the simplex is acceptably small.

A more successful variation of this idea is due to Nelder and Mead [132]. Unlike the Spendley-Hext-Himsworth algorithm, the Nelder-Mead algorithm does not require the simplex to be regular. A detailed description of the algorithm can be found in Kelley [108, p. 135ff]. Given n-vectors $\mathbf{x}_0, \ldots, \mathbf{x}_n$, this algorithm takes the following form.

Algorithm 3.9.2 (Nelder-Mead)

for $0 \leq i \leq n$, $\phi_i = \phi(\mathbf{x}_i)$

sort $\{\mathbf{x}_i\}_{i=0}^n$ so that $\phi_0 = \phi(\mathbf{x}_0) \leq \ldots \leq \phi_n = \phi(\mathbf{x}_n)$

while $\phi_n - \phi_0 > \tau$

$$\overline{\mathbf{x}} = \left(\sum_{i=0}^{n-1} \mathbf{x}_i \right) / n$$

$\mathbf{x}_r = 2\overline{\mathbf{x}} - \mathbf{x}_n$ and $\phi_r = \phi(\mathbf{x}_r)$ /* reflect */

if $\phi_r < \phi_0$

$\mathbf{x}_e = 3\overline{\mathbf{x}} - 2\mathbf{x}_n$ and $\phi_e = \phi(\mathbf{x}_e)$ /* expand */

if $\phi_e < \phi_r$ then $\{\mathbf{x}_n = \mathbf{x}_e$ and $\phi_n = \phi_e\}$

else $\mathbf{x}_n = \mathbf{x}_r$ and $\phi_n = \phi_r$

else if $\phi_r < \phi_{n-1}$ then $\{\mathbf{x}_n = \mathbf{x}_r$ and $\phi_n = \phi_r\}$

else

shrink = false

if $\phi_r < \phi_n$

$\mathbf{x}_{oc} = 1.5\overline{\mathbf{x}} - 0.5\mathbf{x}_n$ and $\phi_{oc} = \phi(\mathbf{x}_{oc})$ /* outer contraction */

if $\phi_{oc} \leq \phi_r$ then $\{\mathbf{x}_n = \mathbf{x}_{oc}$ and $\phi_n = \phi_{oc}\}$

else shrink = true

else

$\mathbf{x}_{ic} = 0.5\,(\overline{\mathbf{x}} + \mathbf{x}_n)$ and $\phi_{ic} = \phi(\mathbf{x}_{ic})$ /* inner contraction */

if $\phi_{ic} \leq \phi_r$ then $\{\mathbf{x}_n = \mathbf{x}_{ic}$ and $\phi_n = \phi_{ic}\}$

else shrink = true

if shrink then for $1 \leq i \leq n\,\{\mathbf{x}_i = 1.5\mathbf{x}_0 - 0.5\mathbf{x}_i$ and $\phi_i = \phi(\mathbf{x}_i)\}$

sort $\{\mathbf{x}_i\}_{i=0}^n$ so that $\phi_0 = \phi(\mathbf{x}_0) \leq \ldots \leq \phi_n = \phi(\mathbf{x}_n)$

The Nelder-Mead algorithm is not guaranteed to converge. For example, the Wikipedia Nelder-Mead method web page has an animated illustration of nonconvergence of this method.

Readers may experiment with the JavaScript **Nelder-Mead** program. This program plots contours of function values for the Rosenbrock function

$$\phi(\mathbf{x}) = \frac{1}{2} \left\| \begin{bmatrix} 10(\mathbf{x}_2 - \mathbf{x}_1^2) \\ 1 - \mathbf{x}_1 \end{bmatrix} \right\|_2^2 .$$

Users may select three points in this graph, after which the program will draw the corresponding initial simplex. Afterward, the reader may run the program to see subsequent simplices generated by the Nelder-Mead simplex method.

The Nelder-Mead algorithm is available from the GNU Scientific Library in routine gsl_multimin_fminimizer_nmsimplex. or two other variants. The NLopt library contains a variety of local derivative-free optimization methods, including the Nelder-Mead algorithm. The Nelder-Mead algorithm is also available in MATLAB as function fminsearch.

3.10 Stochastic Minimization

A **local minimizer** of a scalar-valued function was described in Definition 3.2.5. Such a point has function value no larger than the values at all points in some ball around that the local minimizer. The second-order necessary conditions in Lemma 3.2.11 show that if a real functional ϕ is twice continuously differentiable, then its local minimizer is a point $\mathbf{z}$ where the gradient $\mathbf{g}_\phi(\mathbf{z})$ is zero and the Hessian $\mathbf{H}_\phi(\mathbf{x})$ is nonnegative. All of our minimization algorithms in Sects. 3.5, 3.6, 3.7, 3.8 and 3.9 were designed to find local minima.

Unfortunately, there may be several local minima for a given objective function. In such cases, we may want to find the **global minimizer**. If $\mathscr{D}$ is the domain of ϕ, then $\mathbf{z}$ is a global minimizer of ϕ if and only if $\phi(\mathbf{z}) \leq \phi(\mathbf{x})$ for all $\mathbf{x} \in \mathscr{D}$. In order to determine if a point is a global minima, it may be necessary to compare its objective value to the objective values at all other local minima. If the function has many local minima, it may be very difficult to guarantee that all local minima are examined.

In this section, we will present two global minimization algorithms that depend on the selection of random points in the search for the global minimizer. The convergence of these algorithms will be expressed in terms of probabilities that the computed objective value is within a desired tolerance of the global minimizer.

3.10.1 Random Search

Random search in multiple dimensions is very similar to random search in one dimension, which was discussed in Sect. 5.7.7.1 in Chap. 5 of Volume I. Given some n-dimensional ball

$$B_r(\mathbf{c}) = \{\mathbf{x} : \|\mathbf{x} - \mathbf{c}\| < r\} ,$$

we select a set of independent points $\{\mathbf{x}_k\}_{k=1}^N \subset B_r(\mathbf{c})$ from a uniform probability distribution, and then evaluate the objective $\phi(\mathbf{x}_k)$ for each point. The global minimum of ϕ over $B_r(\mathbf{c})$ can be approximated by the minimum value of $\phi(\mathbf{x}_k)$.

Suppose that $\mathbf{z} \in B_r(\mathbf{c})$ is the unique point where ϕ achieves its minimum over $B_r(\mathbf{c})$. Then the probability that for a sequence of length n there is a point with absolute error of at most δ in the objective is

$$\begin{aligned}
&\mathscr{P}\{\text{ there exists } k \in [1,N] : \phi(\mathbf{x}_k) \leq \phi(\mathbf{z}) + \delta\} \\
&= 1 - \mathscr{P}\{\text{ for all } k \in [1,N]\ ,\ \phi(\mathbf{x}_k) > \phi(\mathbf{z}) + \delta\} \\
&= 1 - \mathscr{P}\{\mathbf{x} \in B_r(\mathbf{c}) : \phi(\mathbf{x}) > \phi(\mathbf{z}) + \delta\}^N \\
&= 1 - [1 - \mathscr{P}\{\mathbf{x} \in B_r(\mathbf{c}) : \phi(\mathbf{x}) \leq \phi(\mathbf{z}) + \delta\}]^N \\
&= 1 - \left[1 - \frac{\int_{\{\mathbf{x} \in B_r(\mathbf{c}) : \phi(x) - \phi(z) < \delta\}} \, d\mathbf{x}}{\int_{B_r(\mathbf{c})} \, d\mathbf{x}}\right]^N
\end{aligned}$$

next Lemma 3.2.9 gives us

$$\approx 1 - \left[1 - \frac{\int_{\{\mathbf{x} : (\mathbf{x}-\mathbf{z})^\top \mathbf{H}_\phi(\mathbf{z})(\mathbf{x}-\mathbf{z}) \leq 2\delta\}} \, d\mathbf{x}}{\int_{B_r(\mathbf{c})} \, d\mathbf{x}}\right]^N$$

and finally a change of variables in integration leads to

$$\begin{aligned}
&= 1 - \left[1 - \frac{1}{\sqrt{\det \mathbf{H}_\phi(\mathbf{z})}} \frac{\int_{B_{\sqrt{2\delta}}(\mathbf{z})} \, d\mathbf{x}}{\int_{B_r(\mathbf{c})} \, d\mathbf{x}}\right]^N = 1 - \left[1 - \frac{1}{\sqrt{\det \mathbf{H}_\phi(\mathbf{z})}} \left(\frac{\sqrt{2\delta}}{r}\right)^n\right]^N \\
&\approx 1 - \exp\left(-\frac{N}{\sqrt{\det \mathbf{H}_\phi(\mathbf{z})}} \left[\frac{\sqrt{2\delta}}{r}\right]^n\right) .
\end{aligned}$$

We would like to choose N sufficient large so that

$$\mathscr{P}\left\{\min_{1 \leq k \leq N} \phi(\mathbf{x}_k) \leq \phi(\mathbf{z}) + \delta\right\} \geq 1 - \varepsilon .$$

Using the previous approximation for the probability, we can solve for N to get

$$N \geq \left[\frac{\sqrt{2\delta}}{r}\right]^{-n} \sqrt{\det \mathbf{H}_\phi(\mathbf{z})} \log\left(\frac{1}{\varepsilon}\right) = \sqrt{\det\left[\mathbf{H}_\phi(\mathbf{z})\frac{r^2}{2\delta}\right]} \log\left(\frac{1}{\varepsilon}\right) .$$

This inequality describes how the cost of the algorithm depends on the allowable error δ in the objective and the probability ε of failure.

For more information regarding random search methods, see Zhigljavsky [190]. The program TOMS 667 by Aluffi-Pentini et al. [3] performs stochastic minimization. A program to perform stochastic minimization subject to constraints, due to Rabinowitz [147], is available in TOMS 744. The NLopt library contains a variety of global optimization algorithms. MATLAB programmers may be interested in its GlobalSearch class.

3.10.2 Simulated Annealing

We discussed the simulated annealing algorithm for minimization of a function of a single variable in Sect. 5.7.7.2 in Chap. 5 of Volume I. Similar ideas apply to minimization of functions of n variables. Suppose that we are given some rectangular region $\mathscr{D} = \{\mathbf{x} : \mathbf{a} \leq \mathbf{x} \leq \mathbf{b}\}$ in which we want to find the global minimizer of ϕ. We begin by selecting a point $\mathbf{x}_0 \in \mathscr{D}$. As the step index k increases, we want the expected step length $\|\mathbf{x}_{k+1} - \mathbf{x}_k\|_2$ to approach zero. Let the "temperature" function $T(k)$ be such that $T(k) \rightarrow 0$ as $k \rightarrow \infty$. For example, we could use $T(k) = 1/k$ or $T(k) = \mathrm{e}^{-k}$. Our objective function ϕ corresponds to the "energy," and its argument $\mathbf{x}$ corresponds to the "state." We can determine a new candidate state by selecting a component index i, choosing a random number $r_k \in [0, 1]$ and then computing

$$\mathbf{e}_i \cdot \mathbf{x}_{k+1} = \max\{\mathbf{a}_i\ ,\ \min\{\mathbf{b}_i\ ,\ \mathbf{e}_i \cdot \mathbf{x}_k + (\mathbf{b}_i - \mathbf{a}_i)(r_k - 1/2)\}\}\ .$$

Next, we need to define the "probability" $P(\mathbf{x}_k, \mathbf{x}_{k+1}, T(k)))$ of accepting a move from the current state $\mathbf{x}_k$ to the candidate state $\mathbf{x}_{k+1}$. This probability function $P(\mathbf{x}, \mathbf{y}, T)$ must be such that

$$\lim_{T \downarrow 0} P(\mathbf{x}, \mathbf{y}, T) = \begin{cases} 0, f(\mathbf{y}) \geq f(\mathbf{x}) \\ \gamma, f(\mathbf{y}) < f(\mathbf{x}) \end{cases}\ ,$$

where $\gamma > 0$. This implies that there is always some chance of accepting a move that increases energy, but the chance of doing so goes to zero as the temperature approaches zero. At temperatures above zero, the chance of increasing energy allows moves to states that may lead to other local minima with lower energy.

In practice, it is almost always difficult to guarantee that the integral of P over its state space is one, as would be required for a true probability function. However, the following strategy will at least guarantee that P takes values between zero and one. If Δ_ϕ is some estimate for the difference between the highest and lowest possible values for ϕ, then one possible choice of the function P is the following:

$$P(\mathbf{x}_k, \mathbf{x}_{k+1}, T(k)) = \begin{cases} 1\ , \phi(\mathbf{x}_{k+1}) < \phi(\mathbf{x}_k) \\ \exp\left(-\frac{\phi(\mathbf{x}_{k+1}) - \phi(\mathbf{x}_k)}{T(k)\Delta_f}\right)\ , \phi(\mathbf{x}_{k+1}) \geq \phi(\mathbf{x}_k) \end{cases}\ .$$

A convergence proof for simulated annealing can be found in Granville et al. [87]. Other useful discussions of simulated annealing can be found in Kirkpatrick et al. [110] and Johnson et al. [100].

The GNU Scientific Library (GSL) contains the routine gsl_siman_solve to perform simulated annealing. MATLAB programmers may use the function simulannealbnd.

3.11 Test Problems

Moré et al. [129] have collected a number of interesting test problems for finding zeros of nonlinear functions and unconstrained optimization. Here is a partial list of their test functions for nonlinear equations.

Example 3.11.1 (Rosenbrock Function)

$$\mathbf{f}(\mathbf{x}) = \begin{bmatrix} 10\left(\xi_2 - \xi_1^2\right) \\ 1 - \xi_1 \end{bmatrix} \text{ with initial guess } \mathbf{x} = \begin{bmatrix} -1.2 \\ 1 \end{bmatrix}$$

The zero is

$$\mathbf{z} = \begin{bmatrix} 1 \\ 1 \end{bmatrix}$$

Example 3.11.2 (Freudenstein and Roth Function)

$$\mathbf{f}(\mathbf{x}) = \begin{bmatrix} -13 + \xi_1 + ((5 - \xi_2)\xi_2 - 2)\xi_2 \\ -29 + \xi_1 + ((\xi_2 + 1)\xi_2 - 14)\xi_2 \end{bmatrix} \text{ with initial guess } \mathbf{x} = \begin{bmatrix} 1/2 \\ -2 \end{bmatrix}$$

The zero is

$$\mathbf{z} = \begin{bmatrix} 5 \\ 4 \end{bmatrix}$$

Example 3.11.3 (Powell Badly Scaled Function)

$$\mathbf{f}(\mathbf{x}) = \begin{bmatrix} 10^4\xi_1\xi_2 - 1 \\ \exp(-\xi_1) + \exp(-\xi_2) - 1.0001 \end{bmatrix} \text{ with initial guess } \mathbf{x} = \begin{bmatrix} 0 \\ 1 \end{bmatrix}$$

The zero is

$$\mathbf{z} \approx \begin{bmatrix} 1.098 \times 10^{-5} \\ 9.106 \end{bmatrix}$$

Example 3.11.4 (Helical Valley Function)

$$\mathbf{f}(\mathbf{x}) = \begin{bmatrix} 10\left[\xi_3 - 10\,\theta(\xi_1, \xi_2)\right] \\ 10\left[\sqrt{\xi_1^2 + \xi_2^2} - 1\right] \\ \xi_3 \end{bmatrix} \text{ with initial guess } \xi = \begin{bmatrix} -1 \\ 0 \\ 0 \end{bmatrix}$$

where

$$\theta(a, b) = \frac{1}{2\pi} \begin{cases} \arctan(b/a), \ a \geq 0 \\ \arctan(b/a) + \pi, \ a < 0 \end{cases}$$

The global minimizer is

$$\mathbf{z} = \begin{bmatrix} 1 \\ 0 \\ 0 \end{bmatrix}$$

Example 3.11.5 (Powell Singular Function)

$$\mathbf{f}(\mathbf{x}) = \begin{bmatrix} \xi_1 + 10\xi_2 \\ \sqrt{5}(\xi_3 - \xi_4) \\ (\xi_2 - 2\xi_3)^2 \\ \sqrt{10}(\xi_1 - \xi_4)^2 \end{bmatrix} \text{ with initial guess } \mathbf{x} = \begin{bmatrix} 3 \\ -1 \\ 0 \\ 1 \end{bmatrix}$$

The global minimizer is

$$\mathbf{z} = \begin{bmatrix} 0 \\ 0 \\ 0 \\ 0 \end{bmatrix}$$

3.12 Case Study: Lennard Jones Potential

After examining a number of algorithms for unconstrained minimization, we are ready to apply them to an interesting problem from chemistry. The **Lennard Jones potential** is often used in chemistry to model the potential energy of a collection of neutral atoms or molecules. It involves both an attractive term and a repulsive term. Let ϱ bee the distance at which the potential energy between two such atoms is zero, and let $-\varepsilon$ be the minimum energy for a pair of atoms. Then the energy of a collection of n atoms located at points $\mathbf{x}_k$ for $1 \leq k \leq n$ is

$$\phi\left(\{\mathbf{x}_k\}_{k=1}^n\right) = 4\varepsilon \sum_{i=1}^{n-1} \sum_{j=i+1}^{n} \left[\left(\frac{\varrho}{\|\mathbf{x}_i - \mathbf{x}_j\|_2}\right)^{12} - \left(\frac{\varrho}{\|\mathbf{x}_i - \mathbf{x}_j\|_2}\right)^{6}\right] .$$

Some preliminary work can be used to simplify this problem. Since only the distances between atoms is important, we can fix the first atom to be at the origin: $\mathbf{x}_1 = \mathbf{0}$. Next, we can rotate the collection of atoms around the first atom so that the second atom lies along the positive first coordinate axis: $\mathbf{x}_2 = \mathbf{e}_1 \xi_{2,1}$. If we have at least three atoms, we can rotate the collection of atoms around the first coordinate axis so that the third atom lies in the 1, 2-plane: $\mathbf{x}_3 = \mathbf{e}_1 \xi_{3,1} + \mathbf{e}_2 \xi_{3,2}$ These observations can be used to reduce the number of unknown parameters in the minimization problem.

3.12.1 Two Atoms

Suppose that we would like to minimize the energy for a pair of atoms. With our simplifications, this energy is

$$\phi\left(\xi_{2,1}\right) = 4\varepsilon\left[\left(\frac{\varrho}{\xi_{2,1}}\right)^{12} - \left(\frac{\varrho}{\xi_{2,1}}\right)^{6}\right] .$$

The derivative is

$$\phi'\left(\xi_{2,1}\right) = \frac{4\varepsilon}{\varrho}\left[-12\left(\frac{\varrho}{\xi_{2,1}}\right)^{13} + 6\left(\frac{\varrho}{\xi_{2,1}}\right)^{7}\right] = \frac{24\varepsilon}{\varrho}\left(\frac{\varrho}{\xi_{2,1}}\right)^{7}\left[-2\left(\frac{\varrho}{\xi_{2,1}}\right)^{6} + 1\right] ,$$

so the extreme points occur at $\xi_{2,1} = \varrho\sqrt[6]{2}$ or $\xi_{2,1} = \infty$. It is not hard to see that the former is the only local minimizer, with energy $\phi\left(\xi_{2,1}\right) = -\varepsilon$.

We used the GSL routine `gsl_multimin_fdfminimizer_vector_bfgs2` to compute the minimum energy for the Lennard Jones with two atoms. In 20 trials with initial guesses perturbed from the global minimizer by a random relative error on the order of 10, this algorithm found six points with nearly zero energy, and found the global minimizer 14 times. On the other hand, the GSL routine `gsl_multimin_fminimizer_nmsimplex2` performed more reliably. In 20 trials with initial guesses perturbed from the global minimizer by a random relative error on the order of 10 and initial step sizes on the order of one, this algorithm found the global minimizer every time. The code that performed these calculations is available as LennardJones.C.

Readers may also experiment with the JavaScript simulated annealing program for the **Lennard Jones potential.** Readers may select the number of atoms, and repeat the simulation with additional random observations. This simulated annealing algorithm gets one digit of accuracy in the energy and the position of the atoms with little work, but struggles to improve the accuracy afterward.

3.12.2 Three Atoms

For three atoms, the optimal location of the atoms is

$$\mathbf{x}_1 = \begin{bmatrix} 0 \\ 0 \\ 0 \end{bmatrix}, \ \mathbf{x}_2 = \begin{bmatrix} \sqrt[6]{2} \\ 0 \\ 0 \end{bmatrix} \varrho \text{ and } \mathbf{x}_3 = \begin{bmatrix} \sqrt[6]{2}/2 \\ \sqrt[6]{2}\sqrt{3/4} \\ 0 \end{bmatrix} \varrho \,,$$

with energy -3ε. In this configuration, the atoms are located at the vertices of an equilateral triangle, with each side of the triangle having the same length as the distance between a pair of atoms at lowest energy.

We used the GSL routine `gsl_multimin_fminimizer_nmsimplex2` to conduct 20 trials with initial guesses perturbed from the global minimum by a random relative error on the order of 10 and initial step sizes on the order of one. This algorithm found three points with energy $-\varepsilon$, and three other points with energy between $-\varepsilon$ and -3ε. In the other 14 cases, it found the global minimizer. Apparently, this minimization problem has multiple local minima.

3.12.3 Four Atoms

For four atoms, the optimal location of the atoms is

$$\mathbf{x}_1 = \begin{bmatrix} 0 \\ 0 \\ 0 \end{bmatrix}, \ \mathbf{x}_2 = \begin{bmatrix} \sqrt[6]{2} \\ 0 \\ 0 \end{bmatrix} \varrho \,, \ \mathbf{x}_3 = \begin{bmatrix} \sqrt[6]{2}/2 \\ \sqrt[6]{2}\sqrt{3/4} \\ 0 \end{bmatrix} \varrho \text{ and } \mathbf{x}_3 = \begin{bmatrix} \sqrt[6]{2}/2 \\ \sqrt[6]{2}\sqrt{1/12} \\ \sqrt[6]{2}\sqrt{2/3} \end{bmatrix} \varrho$$

with energy -6ε. In this configuration, the atoms are located at the vertices of a regular tetrahedron, with each edge of the tetrahedron having the same length as the distance between a pair of atoms at minimum energy.

3.12.4 Five Atoms

With five atoms, it is not obvious to non-chemists what the optimal locations of the atoms should be. One way to enhance the chances of computing the optimal location is to select the initial positions carefully. If we are using either a trust region algorithm or a simplex search, we can place the first four atoms at their optimal locations, and then place the fifth atom at a significant distance away from the other four. These initial positions will produce an energy slightly above -6ε. For stochastic optimization or the other optimization methods, we can place the first four atoms at their optimal locations, and place the fifth atom at the location of the fourth atom with the sign of its third component negated. This will produce a smaller search region for stochastic optimization than the previously described initial guess.

Simulated annealing can search for the global minimizer, but it converges slowly. We can use the best search point from simulated annealing as an initial guess for either a trust region algorithm or a simplex method. The code that performed these calculations is available as LennardJonesSA.C.

Chapter 4
Constrained Optimization

George Dantzig will go down in history as one of the founders and chief contributors to the field of mathematical programming, and as the creator of the simplex algorithm for linear programming, perhaps the most important algorithm developed in the 20th century.

Richard M. Karp *[105]*

Instead of freaking out about these constraints, embrace them. Let them guide you. Constraints drive innovation and force focus. Instead of trying to remove them, use them to your advantage.

Jason Fried and Heinemeier David Hansson *[75]*

Abstract This chapter discusses the solution of constrained optimization problems. The chapter begins with the linear programming problem, to minimize a linear functional subject to linear equality or inequality constraints. A standard forms of this problem is presented, its feasibility is analyzed, and the fundamental theorem of linear programming is proved. Afterwards, the simplex method is developed, together with starting techniques and implementation details. Then the discussion returns to duality theory, complementary slackness and perturbation theory. These lead to the dual simplex method and Karmarkar's algorithm. For nonlinear constrained optimization problems, the discussion begins with the Kuhn-Tucker theory, Lagrangians and nonlinear duality. Quadratic programming problems are analyzed, and numerical methods for their solution are developed. The chapter ends with general nonlinear constrained optimization problems and a variety of numerical techniques based on feasible searches, penalty methods, augmented Lagrangians and sequential quadratic programming.

Additional Material: The details of the computer programs referred in the text are available in the Springer website (http://extras.springer.com/2018/978-3-319-69107-7) for authorized users.

J.A. Trangenstein, *Scientific Computing*, Texts in Computational Science and Engineering 19, https://doi.org/10.1007/978-3-319-69107-7_4

4.1 Overview

In previous chapters, we have examined several important optimization problems. We learned how to solve least squares problems in Chap. 6 of Volume I. We also learned how to compute minimizers of nonlinear functions of a single variable in Sect. 5.7 in Chap. 5 of Volume I and minimizers of nonlinear functionals of multiple variables in Sects. 3.5 and 3.9. In the midst of these developments, we also took a little time to study some simple quadratic programming problems in Sects. 3.6.1.1 and 6.12 in Chap. 6 of Volume I. In the former we minimized a quadratic functional subject to either linear equality constraints, and in the latter we minimized subject to a bound on the 2-norm of the solution.

In this chapter, we will study optimization problems subject to constraints that may be linear or nonlinear, and may involve equations or inequalities. We will begin in Sect. 4.2 with linear programming, which involves a linear objective subject to linear equality or inequality constraints. In Sect. 4.3 we will develop the Kuhn-Tucker conditions for the constrained minimum of a functional. The Kuhn-Tucker theory will guide us in our study of quadratic programming problems in Sect. 6.12 in Chap. 6 of Volume I; these problems involve optimizing a quadratic objective subject to linear equality or inequality constraints. We will end the chapter with Sect. 4.5 on general nonlinear programming, in which we optimize a nonlinear objective subject to nonlinear equality or inequality constraints.

Our goal in this chapter is to develop effective algorithms to solve these problems. The development of these algorithms will require some interesting new mathematical theory, sometimes relating a constrained problem to a dual problem. We will also need to develop supporting ideas in numerical linear algebra, in order to solve linear systems that are perturbed by column exchanges in the coefficient matrix, or the addition/deletion of a row or column.

For more information about the material in this chapter, we recommend nonlinear optimization books by Avriel [5], Chvatal [36], Dantzig [44], Luenberger [120], Mangasarian [123], Schrijver [159] and Sierksma [160].

For linear programming problem software, we recommend GLPK (GNU Linear Programming Kit) or algorithms listed on the Linear Programming Software Survey. Several spreadsheet programs, such as Excel are able to solve linear programming problems. MATLAB users should become familiar with the command linprog. The GNU Octave project provides an interpreted programming language to solve linear programming problems, using the glpk function. GNU Octave is publicly available, and uses a programming syntax similar to MATLAB. Scilab provides the command karmarkar to solve linear programming problems.

Software for solving quadratic programming problems is available for a fee at Stanford Business Software Inc. MATLAB contains the quadprog command for solving quadratic programming problems. The GNU Octave project provides an interpreted programming language to solve quadratic programming problems by means of the qp function. Scilab provides the commands qld and qpsolve to solve quadratic programming problems.

For general nonlinear programming problem software, the reader can consult the Nonlinear Programming Software Survey. For nonlinear programming problems in MATLAB we recommend fmincon. The GNU Octave project provides an interpreted programming language to solve nonlinear programming problems by means of the sqp command. A number of other nonlinear programming problem algorithms are available in NLopt.

4.2 Linear Programming

There are various forms of the linear programming problem. We will begin with the following.

Definition 4.2.1 Suppose that m and n are positive integers, $\mathbf{A}$ is a real $m \times n$ matrix with rank $(\mathbf{A}) = m$, the nonnegative real m-vector $\mathbf{b}$ is in the range of $\mathbf{A}$, and $\mathbf{c}$ is a real n-vector. Then the **standard form** of the **linear programming problem** is to find a real n-vector $\mathbf{x}$ to solve

$$\min \mathbf{c}^\top \mathbf{x} \text{ subject to } \mathbf{A}\mathbf{x} = \mathbf{b} \text{ and } \mathbf{x} \geq \mathbf{0} \,. \tag{4.1}$$

A real n-vector $\mathbf{x}$ is **feasible** for (4.1) if and only if its satisfies the constraints $\mathbf{A}\mathbf{x} = \mathbf{b}$ and $\mathbf{x} \geq \mathbf{0}$.

Figure 4.1 shows the contours of two standard form objectives within the feasible region $\mathbf{e}^\top \mathbf{x} = 1$ and $\mathbf{x} \geq \mathbf{0}$. In the figure on the left, the objective uses vector

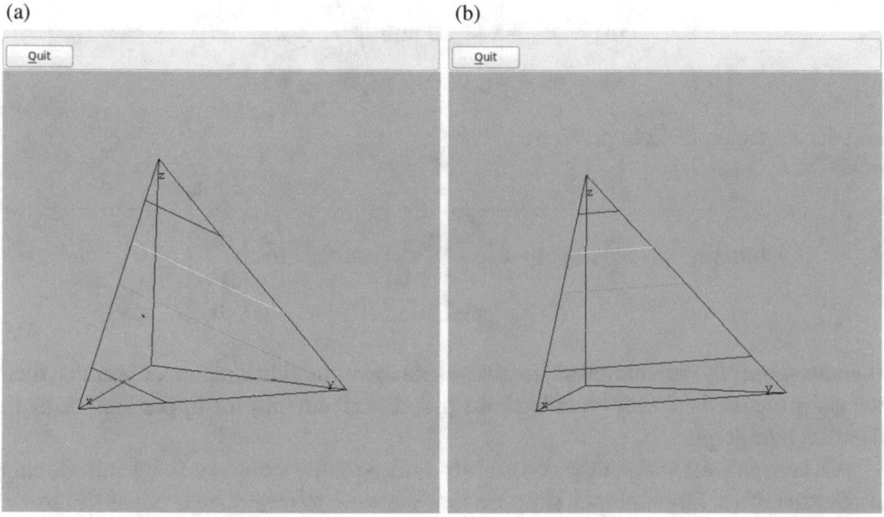

Fig. 4.1 Contours of $\mathbf{c}^\top \mathbf{x}$ for $\mathbf{e}^\top \mathbf{x} = 1$ and $\mathbf{x} \geq \mathbf{0}$. (**a**) $\mathbf{c}^\top = [-3, -2, -1]$. (**b**) $\mathbf{c}^\top = [-2, -2, -1]$

$\mathbf{c}^\top = [-3, , -2, -1]$, so the optimal solution is obviously $\mathbf{x} = [1, 0, 0]$. In the figure on the right, the objective uses vector $\mathbf{c}^\top = [-2, , -2, -1]$, and any vector of the form $\mathbf{x} = [\xi, 1 - \xi, 0]$ with $\xi \in [0, 1]$ is optimal. This figure shows that the optimal solution of a linear programming problem need not be unique. Both figures also indicate that optimal solutions occur at vertices of the feasible region. Both figures were generated by the C^{++} program lp_standard_form_objective.C.

4.2.1 Converting to Standard Form

General linear programming problems may involve a combination of equality and inequality constraints. Here are some techniques for converting general linear programming problems into standard form.

If the original objective involves a maximum, then replace the objective vector $\mathbf{c}$ with its negative. This suggestion uses the observation that

$$\max \mathbf{c}^\top \mathbf{x} = -\min (-\mathbf{c})^\top \mathbf{x} .$$

If the original constraints involve bounds on the solution vector, define a new solution vector for which these bounds correspond to nonnegative constraints. For example, suppose that we want to solve

$$\min \mathbf{c}_1{}^\top \mathbf{x}_1 + \mathbf{c}_2{}^\top \mathbf{x}_2 + \mathbf{c}_3{}^\top \mathbf{x}_3 \text{ subject to } \mathbf{x}_1 \geq \boldsymbol{\ell}_1 \text{ , } \mathbf{x}_2 \leq \mathbf{u}_2 \text{ and } \mathbf{x}_3 \text{ is unrestricted} .$$

We can define

$$\begin{aligned} \mathbf{p}_1 &= \mathbf{x}_1 - \boldsymbol{\ell}_1 \geq \mathbf{0} \text{ ,} \\ \mathbf{m}_2 &= \mathbf{u}_2 - \mathbf{x}_2 \geq \mathbf{0} \text{ and} \\ \mathbf{x}_3 &= \mathbf{p}_3 - \mathbf{m}_3 \text{ where } \mathbf{p}_3, \mathbf{m}_3 \geq \mathbf{0} \text{ ,} \end{aligned}$$

and derive the equivalent problem

$$\min \left[\mathbf{c}_1{}^\top, -\mathbf{c}_2{}^\top, \mathbf{c}_3{}^\top, -\mathbf{c}_3{}^\top\right] \begin{bmatrix} \mathbf{p}_1 \\ \mathbf{m}_2 \\ \mathbf{p}_3 \\ \mathbf{m}_3 \end{bmatrix} \text{ subject to } \begin{bmatrix} \mathbf{p}_1 \\ \mathbf{m}_2 \\ \mathbf{p}_3 \\ \mathbf{m}_3 \end{bmatrix} \geq \mathbf{0} .$$

If some subset of the unknowns involves both lower bounds and upper bounds, then we can process the lower bounds as we just described, and the upper bounds as in the next paragraph.

We can convert inequality constraints into equality constraints by introducing **slack variables**. For example, suppose that we want to minimize $\mathbf{c}^\top \mathbf{x}$ subject to

$$\mathbf{A}_1 \mathbf{x} = \mathbf{b}_1 \text{ , } \mathbf{A}_2 \mathbf{x} \leq \mathbf{b}_2 \text{ , } \mathbf{A}_3 \mathbf{x} \geq \mathbf{b}_3 \text{ and } \mathbf{x} \geq \mathbf{0} .$$

We define

$$\mathbf{s}_2 = \mathbf{b}_2 - \mathbf{A}_2\mathbf{x} \text{ and } \mathbf{s}_3 = \mathbf{A}_3\mathbf{x} - \mathbf{b}_3 \; .$$

Then it is equivalent to solve the following linear programming problem in standard form

$$\min \left[\mathbf{c}^\top, \mathbf{0}^\top, \mathbf{0}^\top\right] \begin{bmatrix} \mathbf{x} \\ \mathbf{s}_2 \\ \mathbf{s}_3 \end{bmatrix} \text{ subject to } \begin{bmatrix} \mathbf{A}_1 & \mathbf{0} & \mathbf{0} \\ \mathbf{A}_2 & \mathbf{I} & \mathbf{0} \\ \mathbf{A}_3 & \mathbf{0} & -\mathbf{I} \end{bmatrix} \begin{bmatrix} \mathbf{x} \\ \mathbf{s}_2 \\ \mathbf{s}_3 \end{bmatrix} = \begin{bmatrix} \mathbf{b}_1 \\ \mathbf{b}_2 \\ \mathbf{b}_3 \end{bmatrix} \text{ and } \begin{bmatrix} \mathbf{x} \\ \mathbf{s}_2 \\ \mathbf{s}_3 \end{bmatrix} \geq \mathbf{0} \; .$$

Once we have transformed a linear programming problem to involve only equality constraints, it is possible that the right-hand side $\mathbf{b}$ has negative entries. For each negative ith component of $\mathbf{b}$, we can multiply it and the corresponding ith row of $\mathbf{A}$ by negative one. The resulting linear programming problem will then involve equality constraints with $\mathbf{b} \geq \mathbf{0}$.

Finally, it is possible that $\text{rank}\,(\mathbf{A}) < m$. If $\mathbf{A} \neq \mathbf{0}$, the LR Theorem 3.7.1 in Chap. 3 of Volume I guarantees that we can find an integer r satisfying $1 \leq r \leq \min\{m, n\}$, permutation matrices $\mathbf{Q}$ and $\mathbf{P}$, an $m \times r$ unit left trapezoidal matrix $\mathbf{L}$, and an $r \times n$ right trapezoidal matrix $\mathbf{R}$ with nonzero diagonal entries so that

$$\mathbf{Q}^\top\mathbf{A}\mathbf{P} = \mathbf{L}\mathbf{R} \; .$$

We can partition

$$\mathbf{L} = \begin{bmatrix} \mathbf{L}_B \\ \mathbf{L}_N \end{bmatrix} \text{ and } \mathbf{b} = \begin{bmatrix} \mathbf{b}_B \\ \mathbf{b}_N \end{bmatrix}$$

where $\mathbf{L}_B$ is $r \times r$ unit left triangular and $\mathbf{b}_B$ is an r-vector. We can solve

$$\mathbf{L}_B\mathbf{y} = \mathbf{b}_B$$

for $\mathbf{y}$. Then $\mathbf{b} \in \mathscr{R}\,(\mathbf{A})$ if and only if $\mathbf{b}_N$ satisfies

$$\mathbf{L}_N\mathbf{y} = \mathbf{b}_N \; .$$

If $\mathbf{b} \notin \mathscr{R}\,(\mathbf{A})$, then the linear programming problem has no solution. Otherwise,

$$\mathbf{A}\mathbf{x} = \mathbf{b} \Longleftrightarrow \mathbf{Q}^\top\mathbf{A}\mathbf{P}\mathbf{P}^\top\mathbf{x} = \mathbf{Q}^\top\mathbf{b} \Longleftrightarrow \begin{bmatrix} \mathbf{L}_B \\ \mathbf{L}_N \end{bmatrix} \mathbf{R}\mathbf{P}^\top\mathbf{x} = \begin{bmatrix} \mathbf{b}_B \\ \mathbf{b}_N \end{bmatrix} \; .$$

The last $m - r$ equations in the ordering provided by the permutation matrix $\mathbf{Q}$ are redundant, and can be eliminated from the set of constraints. We are left with the equivalent constraints

$$\mathbf{L}_B\mathbf{R}\mathbf{P}^\top\mathbf{x} = \mathbf{b}_B \; .$$

This linear system involves an $r \times n$ matrix with rank r.

Exercise 4.2.1 Transform the following linear programming problems to take standard form.

1. Maximize $3\xi_1 - 2\xi_2$ subject to

$$\begin{array}{rrrrl} 5\xi_1 & +2\xi_2 & -3\xi_3 & +\xi_4 & \leq 7 \\ & 3\xi_2 & -4\xi_3 & & \leq 6 \\ \xi_1 & & +\xi_3 & -\xi_4 & \geq 11 \\ \xi_1, & \xi_2, & \xi_3, & \xi_4 & \geq 0 \end{array}$$

2. Maximize $\xi_1 + 2\xi_2 + 4\xi_3$ subject to

$$|4\xi_1 + 3\xi_2 - 7\xi_3| \leq \xi_1 + \xi_2 + \xi_3$$
$$\xi_1, \xi_2, \xi_3 \geq 0$$

Exercise 4.2.2 Suppose that we want to minimize $\mathbf{c}^\top \mathbf{x}$ subject to $\mathbf{A}\mathbf{x} = \mathbf{b}$ and $\mathbf{x} \leq \mathbf{u}$. Let $\mathbf{x}$ have components ξ_j, $\mathbf{u}$ have components υ_j and $\mathbf{p}$ have components π_j defined by

$$\pi_j = \max\{\upsilon_j - \xi_j, 0\} \text{ for } 1 \leq j \leq n .$$

Show that any feasible $\mathbf{x}$ can be written $\mathbf{x} = \mathbf{u} - \mathbf{p}$, and that the original problem is equivalent to minimizing $-\mathbf{c}^\top \mathbf{p}$ subject to $\mathbf{A}\mathbf{p} = \mathbf{A}\mathbf{u} - \mathbf{b}$ and $\mathbf{p} \geq \mathbf{0}$.

Exercise 4.2.3 Suppose that the objective in our linear programming problem is to minimize $\|\mathbf{x}\|_1$, where $\mathbf{x}$ has components ξ_j. Let the vectors $\mathbf{p}$ and $\mathbf{m}$ have components π_j and μ_j, respectively defined by

$$\pi_j = \max\{\xi_j, 0\} \text{ and } \mu_j = -\min\{\xi_j, 0\} .$$

Show that the original objective is equivalent to minimizing $\sum_{j=1}^n (\pi_j + \mu_j)$.

Exercise 4.2.4 Suppose that we want to solve a linear programming problem subject to the nonnegativity constraint $\mathbf{x} \geq \mathbf{0}$ with objective $\min \max_{1 \leq j \leq n} \xi_j$. Recursively define

$$\mu_0 = 0 \text{ and } \mu_j = \max\{\mu_{j-1}, \xi_j\} \text{ for } 1 \leq j \leq n .$$

Then define the scalars

$$\pi_j = \max\{\xi_j - \mu_{j-1}, 0\} \text{ and } \nu_j = -\min\{\xi_j - \mu_{j-1}, 0\} \text{ for } 1 \leq j \leq n .$$

Show that

$$\mu_j = \sum_{k=1}^{j} \pi_k \text{ for } 0 \leq j \leq n$$

and

$$\xi_j = \mu_j - \nu_j \text{ for } 1 < j \leq n \,.$$

Then show that the original objective can be written as $\sum_{k=1}^{n} \pi_k$.

4.2.2 *Feasibility of Standard Form*

In order to be able to solve the linear programming problem in standard form (see Definition 4.2.1), we need to be sure that this problem is feasible. Obviously, we don't want to waste time looking for a solution if none exists.

It is common to discuss feasibility for linear programming problems in terms of **theorems of the alternative**. For a system of linear equations, the following famous theorem describes this approach.

Theorem 4.2.1 (Fredholm Alternative) *Suppose that* $\mathbf{A}$ *is an* $m \times n$ *matrix and* $\mathbf{b}$ *is an m-vector. Then either there exists an n-vector* $\mathbf{x}$ *so that* $\mathbf{A}\mathbf{x} = \mathbf{b}$*, or there exists an m-vector* $\mathbf{y}$ *so that* $\mathbf{y}^H\mathbf{A} = \mathbf{0}$ *and* $\mathbf{y}^H\mathbf{b} < 0$*, but not both.*

Proof The Fundamental Theorem of Linear Algebra 3.2.3 in Chap. 3 of Volume I shows that there exists an n-vector $\mathbf{x}$ and a unique m-vector $\mathbf{r} \in \mathscr{N}\left(\mathbf{A}^H\right)$ so that

$$\mathbf{b} = \mathbf{A}\mathbf{x} + \mathbf{r} \,.$$

If $\mathbf{r} = \mathbf{0}$, then $\mathbf{A}\mathbf{x} = \mathbf{b}$, and for all m-vectors $\mathbf{y}$ so that $\mathbf{y}^H\mathbf{A} = \mathbf{0}$ we have

$$\mathbf{y}^H\mathbf{b} = \mathbf{y}^H\mathbf{A}\mathbf{x} = \mathbf{0}^H\mathbf{x} = 0 \,.$$

Thus the first alternative is satisfied and the second alternative is excluded in this case.

Otherwise $\mathbf{r} \neq \mathbf{0}$, so $\mathbf{A}\mathbf{x} \neq \mathbf{b}$ and

$$(-\mathbf{r})^H\mathbf{b} = -\mathbf{r}^H\left(\mathbf{A}\mathbf{x} + \mathbf{r}\right) = -\mathbf{0}^H\mathbf{x} - \mathbf{r}^H\mathbf{r} < 0 \,.$$

Thus the second alternative is satisfied and the first alternative is excluded in this case.

Geometrically, this theorem says that either $\mathbf{b}$ is in the range of $\mathbf{A}$, or there is a vector in the nullspace of $\mathbf{A}$ that has a negative inner product with $\mathbf{b}$. The left-hand graph in Fig. 4.2 shows an example of the second alternative in this theorem.

Recall that a vector $\mathbf{x}$ is feasible for the standard form of the linear programming problem if and only if $\mathbf{A}\mathbf{x} = \mathbf{b}$ and $\mathbf{x} \geq \mathbf{0}$. Feasibility of the standard form of the linear programming problem is described by the following famous result due to Farkas [62], which is also proved in Mangasarian [123, p. 21].

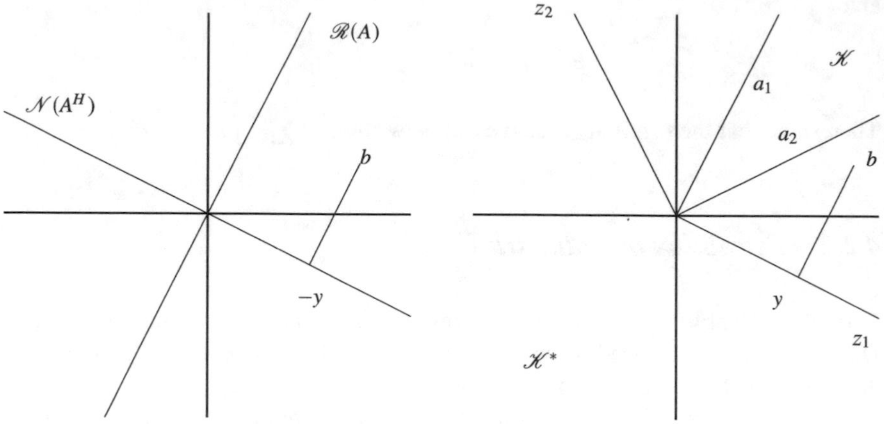

Fig. 4.2 Fredholm alternative (left) and Farkas theorem (right)

Theorem 4.2.2 (Farkas) *Suppose that* $\mathbf{A}$ *is a real* $m \times n$ *matrix and* $\mathbf{b}$ *is a real* m*-vector. Then, either there exists a real* n*-vector* $\mathbf{x}$ *so that* $\mathbf{A}\mathbf{x} = \mathbf{b}$ *and* $\mathbf{x} \geq \mathbf{0}$*, or there exists a real* m*-vector* $\mathbf{y}$ *so that* $\mathbf{y}^\top \mathbf{A} \leq \mathbf{0}$ *and* $\mathbf{y}^\top \mathbf{b} > 0$*, but not both.*
Geometrically, we can express Farkas' Theorem in terms of the **cones**

$$\mathscr{K} = \{\mathbf{A}\mathbf{x} : \mathbf{x} \geq \mathbf{0}\} \text{ and } \mathscr{K}^* = \left\{\mathbf{y} : \mathbf{y}^\top \mathbf{A} \leq \mathbf{0}\right\} .$$

Farkas' Theorem says that either $\mathbf{b} \in \mathscr{K}$ (i.e., the standard form of the linear programming problem is feasible), or there exists $\mathbf{y} \in \mathscr{K}^*$ so that $\mathbf{y}^\top \mathbf{b} > 0$, but not both. Figure 4.2 shows an example of the second alternative in this theorem. In this figure, $\mathbf{z}_1 \perp \mathbf{a}_1$, $\mathbf{z}_2 \perp \mathbf{a}_2$. The cone $\mathscr{K}$ consists of all nonnegative linear combinations of $\mathbf{a}_1$ and $\mathbf{a}_2$, while $\mathscr{K}^*$ consists of all vectors that have a nonpositive inner product with both $\mathbf{a}_1$ and $\mathbf{a}_2$. The vector $\mathbf{b}$ is outside the cone $\mathscr{K}$, and the vector $\mathbf{y} \in \mathscr{K}^*$ is such that $\mathbf{y}^\top \mathbf{b} > 0$.

Later, in Sect. 4.2.5 we will learn how to compute a feasible solution to a linear program in standard form, or determine that no feasible solution exists.

Exercise 4.2.5 Show that the set of feasible vectors for a linear program in standard form is convex. In other words, if $\mathbf{x}_1$ and $\mathbf{x}_2$ are feasible, show that $\mathbf{x}_1\tau + \mathbf{x}_2(1 - \tau)$ is feasible for any $\tau \in [0, 1]$.

4.2.3 Fundamental Theorem

Now that we know when linear programming problems in standard form have feasible solutions, we would like to reduce the number of feasible solutions that we need to examine. First, let us define what we are seeking.

Definition 4.2.2 Suppose that $\mathbf{A}$ is a real $m \times n$ matrix and $\mathbf{b}$ is a real m-vector. Then a real n-vector $\mathbf{x}$ is **basic feasible** for a programming problem in standard form (4.1) if and only if $\mathbf{x}$ is feasible (i.e., $\mathbf{A}\mathbf{x} = \mathbf{b}$ and $\mathbf{x} \geq \mathbf{0}$), and the columns of $\mathbf{A}$ corresponding to nonzero entries of $\mathbf{x}$ are linearly independent.
This definition leads us to the following very important result.

Theorem 4.2.3 (Fundamental Theorem of Linear Programming) *Suppose that* $\mathbf{A}$ *is a real* $m \times n$ *matrix with* rank $(\mathbf{A}) = m$. *Let* $\mathbf{b}$ *be a real* m*-vector, and* $\mathbf{c}$ *be a real* n*-vector. If the constraints* $\mathbf{A}\mathbf{x} = \mathbf{b}$ *and* $\mathbf{x} \geq \mathbf{0}$ *have a solution, then they have a basic feasible solution. Also, if the linear programming problem* $\min \mathbf{c}^\top \mathbf{x}$ *subject to* $\mathbf{A}\mathbf{x} = \mathbf{b}$ *and* $\mathbf{x} \geq \mathbf{0}$ *has an optimal feasible solution, then it has an optimal basic feasible solution.*

Proof We will prove the first claim. The proof of the second claim is very similar.

Suppose that $\mathbf{x}$ is feasible. Using a permutation matrix $\mathbf{P}$, we can reorder the entries of $\mathbf{x}$ and columns of $\mathbf{A}$ so that we can partition

$$\mathbf{A}\mathbf{P} = \begin{bmatrix} \mathbf{A}_B & \mathbf{A}_N \end{bmatrix} \text{ and } \mathbf{P}^\top \mathbf{x} = \begin{bmatrix} \mathbf{x}_B \\ \mathbf{0} \end{bmatrix} .$$

Here $\mathbf{x}_B > \mathbf{0}$ is a p-vector. If the columns of $\mathbf{A}_B$ are linearly independent, we are done. Otherwise, there exists a p-vector $\mathbf{z} \in \mathscr{N}(\mathbf{A}_B)$ with a positive component, say the ith. Then for all real scalars ε we have

$$\mathbf{b} = \mathbf{A}_B \mathbf{x}_B = \mathbf{A}_B \mathbf{x}_B - \mathbf{A}_B \mathbf{z}\varepsilon = \mathbf{A}_B(\mathbf{x}_B - \mathbf{z}\varepsilon) .$$

If $\mathbf{x}_B$ has components ξ_j and $\mathbf{z}_B$ has components ζ_j, choose

$$\varepsilon = \min_{1 \leq j \leq p} \left\{ \frac{\xi_j}{\zeta_j} : \zeta_j > 0 \right\} .$$

Then

$$\mathbf{x}'_B = \mathbf{x}_B - \mathbf{z}\varepsilon \geq \mathbf{0}$$

and this vector has ith component equal to zero. This gives us a new feasible vector with fewer nonzero components. We can reorder the columns of $\mathbf{A}$ to correspond to this new feasible vector, and continue until we achieve $\mathscr{N}(\mathbf{A}_B) = \{\mathbf{0}\}$.

Since $\mathbf{A}$ has n columns and any maximal linearly independent subset subset of its columns contains m vectors, there are at most $\binom{n}{m}$ basic solutions to $\mathbf{A}\mathbf{x} = \mathbf{b}$, some of which may not be nonnegative. At any rate, we need only consider a finite number of possible solutions to the standard form of the linear programming problem.

In view of the Fundamental Theorem of Linear Programming 4.2.3, we can rewrite the standard form of the linear programming problem as follows. Given a real $m \times n$ matrix $\mathbf{A}$ with rank $(\mathbf{A}) = m$, a nonnegative real m-vector $\mathbf{b} \in \mathscr{R}(\mathbf{A})$ and a real n-vector $\mathbf{c}$, we want to find an $n \times n$ permutation matrix $\mathbf{P}$ and a real m-vector $\mathbf{x}_B$ so that

1. we can partition

$$\mathbf{AP} = \begin{bmatrix}\mathbf{A}_B & \mathbf{A}_N\end{bmatrix} \text{ where } \mathbf{A}_B \text{ is nonsingular, and}$$
$$\mathbf{c}^\top\mathbf{P} = \begin{bmatrix}\mathbf{c}_B{}^\top & \mathbf{c}_N{}^\top\end{bmatrix} ,$$

2. the vector

$$\mathbf{x} = \mathbf{P}\begin{bmatrix}\mathbf{x}_B \\ \mathbf{0}\end{bmatrix}$$

is feasible, meaning that

$$\mathbf{b} = \mathbf{Ax} = \mathbf{A}_B\mathbf{x}_B \text{ and } 0 \le \mathbf{x}_B ,$$

3. and $\mathbf{x}$ is optimal, meaning that $\mathbf{x}$ minimizes

$$\mathbf{c}^\top\mathbf{x} = \mathbf{c}_B{}^\top\mathbf{x}_B .$$

Exercise 4.2.6 Consider the standard form linear programming problem with

$$\mathbf{A} = \begin{bmatrix}1 & 1 & -2 & 3 \\ -2 & 0 & 1 & 0\end{bmatrix} , \mathbf{b} = \begin{bmatrix}2 \\ 2\end{bmatrix} \text{ and } \mathbf{c}^\top = \begin{bmatrix}5 & 2 & 3 & 1\end{bmatrix} .$$

1. Draw $\mathbf{b}$ and the columns of $\mathbf{A}$ on a graph, then show geometrically that this problem has only two basic feasible solutions.
2. Compute the two basic feasible solutions.
3. Use the nonnegativity condition on $\mathbf{x}$ to show that the objective $\mathbf{c}^\top\mathbf{x}$ is bounded below.
4. Use the Fundamental Theorem of Linear Programming 4.2.3 to find the optimal solution of the linear programming problem.

4.2.4 Simplex Method

In this section, we will describe George Dantzig's [44], very successful **simplex method** for solving linear programming problems. Given a basic feasible vector for the standard form of the linear programming problem, the algorithm roughly consists of the following three steps. First, we choose a non-basic variable that allows us to reduce the objective; this choice implies a direction of change in the current solution in order to maintain the equality constraints. Next, we choose the largest possible value for this non-basic variable, in such a way that the new feasible vector is basic; this will imply that some previously basic variable will become non-basic. Finally, we re-order the basic variables first and non-basic variables last, and then continue the solution process.

Of course, this algorithm assumes that we know how to find an initial basic feasible starting point for the simplex algorithm. We will show how to use the simplex method to find a basic feasible starting point in Sect. 4.2.5.

4.2.4.1 How to Choose the Nonbasic Variable

Suppose that we are given a real $m \times n$ matrix $\mathbf{A}$ with $\operatorname{rank}(\mathbf{A}) = m$, a nonnegative real m-vector $\mathbf{b} \in \mathscr{R}(\mathbf{A})$, and a real n-vector $\mathbf{c}$ for the objective of the linear programming problem in standard form

$$\min \mathbf{c}^\top \mathbf{x} \text{ subject to } \mathbf{A}\mathbf{x} = \mathbf{b} \text{ and } \mathbf{x} \geq \mathbf{0} \,.$$

Also assume that we are given be a real n-vector $\mathbf{x}$ that is basic feasible for this linear programming problem. In other words, we can find a permutation matrix $\mathbf{P}$ so that

$$\mathbf{P}^\top \mathbf{x} = \begin{bmatrix} \mathbf{x}_B \\ \mathbf{0} \end{bmatrix} \,,\ \mathbf{A}\mathbf{P} = \begin{bmatrix} \mathbf{A}_B & \mathbf{A}_N \end{bmatrix} \,,\ \mathbf{A}_B\mathbf{x}_B = \mathbf{b} \text{ and } \mathbf{x}_B \geq \mathbf{0} \,,$$

and $\mathbf{A}_B$ is nonsingular. We call $\mathbf{A}_B$ the **basic matrix**. If $\mathbf{h}$ is a real m-vector, j_N is a nonbasic index and ε is a positive scalar, the perturbed vector

$$\mathbf{P}^\top \mathbf{x}_\varepsilon = \begin{bmatrix} \mathbf{x}_B - \mathbf{h}\varepsilon \\ \mathbf{e}_{j_N}\varepsilon \end{bmatrix} \,. \tag{4.2}$$

is feasible if and only if both

$$\mathbf{0} = \mathbf{A}(\mathbf{x}_\varepsilon - \mathbf{x}) = (\mathbf{A}\mathbf{P})\mathbf{P}^\top(\mathbf{x}_\varepsilon - \mathbf{x}) = \begin{bmatrix} \mathbf{A}_B & \mathbf{A}_N \end{bmatrix} \begin{bmatrix} -\mathbf{h} \\ \mathbf{e}_{j_N} \end{bmatrix} \varepsilon \tag{4.3a}$$

and

$$0 \leq \mathbf{P}^\top \mathbf{x}_\varepsilon = \begin{bmatrix} \mathbf{x}_B - \mathbf{h}\varepsilon \\ \mathbf{e}_{j_N}\varepsilon \end{bmatrix} \,. \tag{4.3b}$$

Condition (4.3a) implies that $\mathbf{h}$ solves the system of linear equations

$$\mathbf{A}_B\mathbf{h} = \mathbf{A}_N\mathbf{e}_{j_N} \,.$$

Since $\mathbf{A}_B$ is nonsingular, this equation determines $\mathbf{h}$ after we have determined the nonbasic index j_N.

In order to choose j_N, let us permute and partition

$$\mathbf{c}^\top \mathbf{P} = \left[\mathbf{c}_B{}^\top \ \mathbf{c}_N{}^\top \right] ,$$

where $\mathbf{c}_B$ is an m-vector. In the simplex method, *we choose the nonbasic index j_N to get the largest possible reduction in the objective*. The change in the objective is

$$\mathbf{c}^\top \mathbf{x}_\varepsilon - \mathbf{c}^\top \mathbf{x} = \mathbf{c}_B{}^\top (\mathbf{x}_B - \mathbf{h}\varepsilon) + \mathbf{c}_N{}^\top \mathbf{e}_{j_N} \varepsilon - \mathbf{c}_B{}^\top \mathbf{x}_B = \left(\mathbf{c}_N{}^\top - \mathbf{c}_B{}^\top \mathbf{A}_B^{-1} \mathbf{A}_N \right) \mathbf{e}_{j_N} \varepsilon \ .$$

This suggests that we compute the **cost reduction vector**

$$\mathbf{r}_N{}^\top = \mathbf{c}_N{}^\top - \mathbf{c}_B{}^\top \mathbf{A}_B^{-1} \mathbf{A}_N \tag{4.4}$$

and find its most negative entry j_N. *If $\mathbf{r}_N \geq \mathbf{0}$, then no reduction in the objective is possible, and the given solution vector $\mathbf{x}$ must be basic optimal.*

4.2.4.2 How to Choose the Nonbasic Variable Value

Suppose that $\mathbf{x}$ is basic feasible, $\mathbf{r}_N$ is the corresponding cost reduction vector defined in Eq. (4.4), and j_N is the index of the most negative entry of $\mathbf{r}_N$. In order for the new feasible vector $\mathbf{x}_\varepsilon$ defined in Eq. (4.2) to be feasible, Eq. (4.3b) requires that $\varepsilon \geq 0$ and $\mathbf{x}_B - \mathbf{h}\varepsilon \geq \mathbf{0}$. These two conditions imply that if $\mathbf{h}$ has any positive entries, then ε is restricted in size:

$$\varepsilon = \min_{1 \leq i \leq m} \left\{ \frac{\mathbf{e}_i{}^\top \mathbf{x}_B}{\mathbf{e}_i{}^\top \mathbf{h}} : \mathbf{e}_i{}^\top \mathbf{h} > 0 \right\} \ . \tag{4.5}$$

On the other hand, *if $\mathbf{h} \leq \mathbf{0}$ then the linear programming problem is unbounded.*

With ε chosen by Eq. (4.5), at least one component i_B of $\mathbf{x}_B - \mathbf{h}\varepsilon$ is zero. As a result,

$$\mathbf{x}_\varepsilon = \mathbf{P} \begin{bmatrix} \mathbf{x}_B - \mathbf{h}\varepsilon \\ \mathbf{e}_{j_N} \varepsilon \end{bmatrix}$$

is basic feasible.

We can interchange the new basic component i_B with the previously basic component j_N to determine a new permutation matrix $\mathbf{P}$. This interchange allows us to continue the algorithm as before.

4.2.4.3 Algorithm

We can summarize Dantzig's ideas with the following

Algorithm 4.2.1 (Simplex)

given permutation matrix $\mathbf{P}$ so that $\mathbf{A}\mathbf{P} = \begin{bmatrix}\mathbf{A}_B & \mathbf{A}_N\end{bmatrix}$ with $\mathbf{A}_B$ nonsingular

given $\mathbf{x}_B$ so that $\mathbf{A}_B\mathbf{x}_B = \mathbf{b}$ and $\mathbf{x}_B \geq \mathbf{0}$

partition $\mathbf{c}^\top\mathbf{P} = \begin{bmatrix}\mathbf{c}_B{}^\top & \mathbf{c}_N{}^\top\end{bmatrix}$

while true

  solve $\mathbf{y}_B{}^\top\mathbf{A}_B = \mathbf{c}_B{}^\top$ for $\mathbf{y}_B{}^\top$

  $\mathbf{r}_N{}^\top = \mathbf{c}_N{}^\top - \mathbf{y}_B{}^\top\mathbf{A}_N$

  if $\mathbf{r}_N{}^\top \geq \mathbf{0}^\top$ then (return optimal solution $\mathbf{x}$)

  find j_N so that $\mathbf{r}_N{}^\top\mathbf{e}_{j_N} = \min_j\{\mathbf{r}_N{}^\top\mathbf{e}_j\}$

  solve $\mathbf{A}_B\mathbf{h} = \mathbf{A}_N\mathbf{e}_{j_N}$ for $\mathbf{H}$

  if $\mathbf{h} \leq \mathbf{0}$ then return (since problem is unbounded)

  find i_B so that $\varepsilon = \dfrac{\mathbf{e}_{i_B}\cdot\mathbf{x}_B}{\mathbf{e}_{i_B}\cdot\mathbf{h}} = \min_i\left\{\dfrac{\mathbf{e}_i\cdot\mathbf{x}_B}{\mathbf{e}_i\cdot\mathbf{h}} : \mathbf{e}_i\cdot\mathbf{h} > 0\right\}$

  update $\mathbf{x} = \mathbf{P}\begin{bmatrix}\mathbf{x}_B - \mathbf{h}\varepsilon \\ \mathbf{e}_{j_N}\varepsilon\end{bmatrix}$

  update $\mathbf{P}$: interchange basic entry i_B with non-basic entry j_N

  update $\mathbf{A}_B$, $\mathbf{A}_N$: interchange column i_B of $\mathbf{A}_B$ with column j_N of $\mathbf{A}_N$

  update $\mathbf{x}_B$, $\mathbf{x}_N$: interchange component i_B of $\mathbf{x}_B$ with component j_N of $\mathbf{x}_N$

  update $\mathbf{c}_B$, $\mathbf{c}_N$: interchange component i_B of $\mathbf{c}_B$ with component j_N of $\mathbf{c}_N$

Generally, the simplex method is reasonably efficient. However, there are examples that demonstrate that the simplex method can be very computationally expensive.

Example 4.2.1 (Klee-Minty) The linear programming problem

$$\max \sum_{i=1}^{m} 2^{m-i} \mathbf{y}_i$$

$$\text{s.t.} \sum_{i=1}^{j-1} 2^{j+1-i} \mathbf{y}_i + \mathbf{y}_j \leq 5^i \text{ for } 1 \leq j \leq m \text{ and}$$

$$\mathbf{y} \geq \mathbf{0}$$

has a feasible region containing 2^m basic feasible points. Obviously, $\mathbf{y} = \mathbf{0}$ is feasible for this problem. This problem can be transformed to standard form. Klee and Minty [111] have shown that if the simplex method is begun from $\mathbf{y} = \mathbf{0}$, then it will visit all 2^m basic feasible points.

In Sect. 4.2.13, we will discuss an alternative algorithm that seems to work much better than the simplex method for very large linear programming problems.

4.2.5 *Initial Basic Feasible Guess*

When we described the simplex method in Sect. 4.2.4, we assumed that we were given an initial basic feasible vector $\mathbf{x}$. The following discussion will show how to find such a vector.

Suppose that we are given a real $m \times n$ matrix $\mathbf{A}$ with rank $(\mathbf{A}) = m$, and a nonnegative real m-vector $\mathbf{b} \in \mathscr{R}(\mathbf{A})$. We would like to find a real n-vector $\mathbf{x} \geq \mathbf{0}$ so that $\mathbf{A}\mathbf{x} = \mathbf{b}$, and such that the columns of $\mathbf{A}$ corresponding to the nonzero entries of $\mathbf{x}$ are linearly independent. Consider the **auxiliary problem**

$$\min \mathbf{e}^\top \mathbf{a} \text{ subject to } \begin{bmatrix} \mathbf{I} & \mathbf{A} \end{bmatrix} \begin{bmatrix} \mathbf{a} \\ \mathbf{x} \end{bmatrix} = \mathbf{b} \text{ and } \begin{bmatrix} \mathbf{a} \\ \mathbf{x} \end{bmatrix} \geq \mathbf{0} \,.$$

Here $\mathbf{e}$ is the vector of ones. Note that $\mathbf{a} = \mathbf{b}$ and $\mathbf{x} = \mathbf{0}$ is basic feasible for this auxiliary problem. Also note that the objective is bounded below by zero, so the auxiliary linear programming is not unbounded. This means that we can apply the simplex method to find an optimal solution of the auxiliary problem.

If some optimal solution to this auxiliary problem has $\mathbf{a} = \mathbf{0}$, then the corresponding vector $\mathbf{x}$ is feasible for the original standard form problem. Conversely, if the original standard form problem is feasible, then the auxiliary problem must have an optimal solution with $\mathbf{a} = \mathbf{0}$.

The contrapositive of this second statement is the following: if no optimal solution to the auxiliary problem has $\mathbf{a} = \mathbf{0}$, then the original linear programming problem is infeasible. In other words, if the optimal objective value for the auxiliary problem is positive, then the original problem is infeasible.

4.2.6 Simplex Tableau

Now that we understand the basic ideas in implementing the simplex method, we would like to describe how to implement these ideas. We will begin by describing an approach that is related to Gaussian factorization, which we described in Sect. 3.7 in Chap. 3 of Volume I. The new approach is commonly called **Gauss-Jordan elimination**. We will present this method in terms of a **tableau**.

Suppose that we are given a real $m \times n$ matrix $\mathbf{A}$ with rank $(\mathbf{A}) = m$, a real nonnegative m-vector $\mathbf{b} \in \mathscr{R}(\mathbf{A})$, and a real n-vector $\mathbf{c}$. We want to solve

$$\min \mathbf{c}^\top \mathbf{x} \qquad \mathbf{c}^\top = \begin{bmatrix} 1 & 3 & 2 \end{bmatrix}$$

$$\text{subject to } \mathbf{A}\mathbf{x} = \mathbf{b} \text{ and } \mathbf{x} \geq \mathbf{0} \qquad \mathbf{A} = \begin{bmatrix} 1 & 1 & 1 \end{bmatrix} \text{ and } \mathbf{b} = \begin{bmatrix} 1 \end{bmatrix} .$$

We will set the initial tableau to be as in Table 4.1.

Suppose that we know a basic feasible subset of the variables, with indices in the array $\mathbf{p}_B$, and the remaining variable indices in the array $\mathbf{p}_N$. It is convenient to reorder the basic feasible variables first, as in Table 4.2.

Until we are done, we repeat the following steps. First, we conduct Gauss-Jordan elimination, by performing elementary row operations until we obtain a tableau as in Table 4.3.

Next, we choose the nonbasic variable index j_N corresponding to the most negative entry in $\mathbf{r}_N{}^\top$; then $\mathbf{h}$ is the corresponding column of $\mathbf{H} = \mathbf{A}_B^{-1}\mathbf{A}_N$. We also choose the index i_B of the basic variable with smallest ratio $\mathbf{e}_i{}^\top \mathbf{x}_B / \mathbf{e}_i{}^\top \mathbf{h}$ where $\mathbf{e}_i{}^\top \mathbf{h} > 0$. Then we interchange basic variable i_B with non-basic variable j_N.

Table 4.1 Initial tableau for standard form

$\mathbf{p}^\top$			1	2	3	
$\mathbf{A}$	$\mathbf{b}$		1	1	1	1
$\mathbf{c}^\top$	0		1	3	2	0

Table 4.2 Initial tableau for standard form after ordering basic first

$\mathbf{p}_B{}^\top$	$\mathbf{p}_N{}^\top$			2	1	3	
$\mathbf{A}_B$	$\mathbf{A}_N$	$\mathbf{b}$		1	1	1	1
$\mathbf{c}_B{}^\top$	$\mathbf{c}_N{}^\top$	0		3	1	2	0

Table 4.3 Tableau for standard form after elimination

$\mathbf{p}_B{}^\top$	$\mathbf{p}_N{}^\top$			2	1	3	
$\mathbf{I}$	$\mathbf{H} = \mathbf{A}_B^{-1}\mathbf{A}_N$	$\mathbf{x}_B = \mathbf{A}_B^{-1}\mathbf{b}$		1	1	1	1
$\mathbf{0}^\top$	$\mathbf{r}_N{}^\top = \mathbf{c}_N{}^\top - \mathbf{c}_B{}^\top \mathbf{A}_B^{-1}\mathbf{A}_N$	$-\mathbf{c}_B{}^\top \mathbf{x}_B$		0	-2	-1	-3

The simplex tableau computations can be summarized by the following

Algorithm 4.2.2 (Simplex Method via Gauss-Jordan Elimination)

assume that the first m variables are basic
for $1 \leq k \leq m$
  for $k < j \leq n$ $(\mathbf{A}_{kj} = \mathbf{A}_{kj}/\mathbf{A}_{kk})$ /* BLAS routine dscal */
  $\mathbf{b}_k = \mathbf{b}_k/\mathbf{A}_{kk}$
  $\mathbf{A}_{kk} = 1$
  for $k < j \leq n$, for $1 \leq i < k$ $(\mathbf{A}_{ij} = \mathbf{A}_{ij} - \mathbf{A}_{ik}\mathbf{A}_{kj})$ /* BLAS routine dger */
  for $1 \leq i < k$ $(\mathbf{A}_{ik} = 0)$
  for $k < j \leq n$ for $k < i \leq m$ $(\mathbf{A}_{ij} = \mathbf{A}_{ij} - \mathbf{A}_{ik}\mathbf{A}_{kj})$ /* BLAS routine dger */
  for $k < i \leq m$ $(\mathbf{A}_{ik} = 0)$
  for $1 \leq i < k$ $(\mathbf{b}_i = \mathbf{b}_i - \mathbf{A}_{ik}\mathbf{b}_k)$ /* BLAS routine daxpy */
  for $k < i \leq m$ $(\mathbf{b}_i = \mathbf{b}_i - \mathbf{A}_{ik}\mathbf{b}_k)$ /* BLAS routine daxpy */
  for $k < j \leq n$ $(\mathbf{c}_j = \mathbf{c}_j - \mathbf{c}_k\mathbf{A}_{kj})$ /* BLAS routine daxpy */
while true
  find j_N so that $\mathbf{c}_{j_N} = \min_{m<j\leq n} \mathbf{c}_j$ /* BLAS routine idmin */
  if $\mathbf{c}_{j_N} \geq 0$ return optimal $\mathbf{x}_B$ /* stored in **b** */
  $\varepsilon = \infty$, $i_B = 0$
  for $1 \leq i \leq m$
    if $\mathbf{A}_{i,j_N} > 0$ then $\varepsilon = \min\{\varepsilon\ ,\ \mathbf{b}_i/\mathbf{A}_{i,j_N}\}$and $i_B = i$
  if $i_B = 0$ return unbounded
  for $1 \leq i \leq m$ $(\mathbf{A}_{i,i_B} \leftrightarrow \mathbf{A}_{i,j_N})$ /* BLAS routine dswap */
  for $i_B < j \leq n$ $(\mathbf{A}_{kj} = \mathbf{A}_{kj}/\mathbf{A}_{kk})$ /* BLAS routine dscal */
  $\mathbf{b}_{i_B} = \mathbf{b}_{i_B}/\mathbf{A}_{i_B i_B}$
  $\mathbf{A}_{kk} = 1$
  for $i_B < j \leq n$, for $1 \leq i < i_B$ $(\mathbf{A}_{ij} = \mathbf{A}_{ij} - \mathbf{A}_{ik}\mathbf{A}_{kj})$ /* BLAS routine dger */
  for $1 \leq i < i_B$ $\mathbf{A}_{ik} = 0$
  for $i_B < j \leq n$, for $i_B < i \leq m$ $(\mathbf{A}_{ij} = \mathbf{A}_{ij} - \mathbf{A}_{ik}\mathbf{A}_{kj})$ /* BLAS routine dger */
  for $i_B < i \leq m$ $\mathbf{A}_{ik} = 0$
  for $1 \leq i < i_B$ $(\mathbf{b}_i = \mathbf{b}_i - \mathbf{A}_{ik}\mathbf{b}_{i_B})$ /* BLAS routine daxpy */
  for $i_B < i \leq m$ $(\mathbf{b}_i = \mathbf{b}_i - \mathbf{A}_{ik}\mathbf{b}_{i_B})$ /* BLAS routine daxpy */
  for $i_B < j \leq n$ $(\mathbf{c}_j = \mathbf{c}_j - \mathbf{c}_{i_B}\mathbf{A}_{kj})$ /* BLAS routine daxpy */

Example 4.2.2 Suppose that we want to solve

$$\begin{array}{rrrrl}
\max\ 6\mathbf{x}_1 & +8\mathbf{x}_2 & +5\mathbf{x}_3 & +9\mathbf{x}_4 & \\
\text{s.t. } 2\mathbf{x}_1 & +\mathbf{x}_2 & +\mathbf{x}_3 & +3\mathbf{x}_4 & \leq 5 \\
\mathbf{x}_1 & +3\mathbf{x}_2 & +\mathbf{x}_3 & +2\mathbf{x}_4 & \leq 3 \\
\mathbf{x}_1, & \mathbf{x}_2, & \mathbf{x}_3, & \mathbf{x}_4 & \geq 0
\end{array}$$

Table 4.4 Example tableaus

$\mathbf{x}_1$	$\mathbf{x}_2$	$\mathbf{x}_3$	$\mathbf{x}_4$	$\mathbf{s}_1$	$\mathbf{s}_2$	
2	1	1	3	1	0	5
1	3	1	2	0	1	3
−6	−8	−5	−9	0	0	0

→

$\mathbf{s}_1$	$\mathbf{s}_2$	$\mathbf{x}_1$	$\mathbf{x}_2$	$\mathbf{x}_3$	$\mathbf{x}_4$		
1	0	2	1	1	3	5	
0	1	1	3	1	2	3	←
0	0	−6	−8	−5	−9	0	
					↑		

⟹

$\mathbf{s}_1$	$\mathbf{x}_4$	$\mathbf{x}_1$	$\mathbf{x}_2$	$\mathbf{x}_3$	$\mathbf{s}_2$	
1	3	2	1	1	0	5
0	2	1	3	1	1	3
0	−9	−6	−8	−5	0	0

⟹

$\mathbf{s}_1$	$\mathbf{x}_4$	$\mathbf{x}_1$	$\mathbf{x}_2$	$\mathbf{x}_3$	$\mathbf{s}_2$		
1	0	1/2	−7/2	−1/2	−3/2	1/2	←
0	1	1/2	3/2	1/2	1/2	3/2	
0	0	−3/2	11/2	−1/2	9/2	27/2	
		↑					

⟹

$\mathbf{s}_1$	$\mathbf{x}_4$	$\mathbf{x}_1$	$\mathbf{x}_2$	$\mathbf{x}_3$	$\mathbf{s}_2$	
1/2	0	1	−7/2	−1/2	−3/2	1/2
1/2	1	0	3/2	1/2	1/2	3/2
−3/2	0	0	11/2	−1/2	9/2	27/2

⟹

$\mathbf{x}_1$	$\mathbf{x}_4$	$\mathbf{s}_1$	$\mathbf{x}_2$	$\mathbf{x}_3$	$\mathbf{s}_2$		
1	0	1/2	2	−7	−1	−3	
0	1	1/2	−1	5	1	2	←
0	0	−3/2	3	−5	−2	0	
			↑				

⟹

$\mathbf{x}_1$	$\mathbf{x}_3$	$\mathbf{s}_1$	$\mathbf{x}_4$	$\mathbf{x}_3$	$\mathbf{s}_2$	
1	−7	2	0	−1	−3	1
9	5	−1	1	1	2	1
0	−5	3	0	−2	0	15

⟹

$\mathbf{x}_1$	$\mathbf{x}_2$	$\mathbf{s}_1$	$\mathbf{x}_4$	$\mathbf{x}_3$	$\mathbf{s}_2$		
1	0	3/5	7/5	2/5	−1/5	12/5	
0	1	−1/5	1/5	1/5	2/5	1/5	←
0	0	2	1	−1	2	16	
				↑			

⟹

$\mathbf{x}_1$	$\mathbf{x}_3$	$\mathbf{s}_1$	$\mathbf{x}_4$	$\mathbf{x}_3$	$\mathbf{s}_2$	
1	2/5	3/5	7/5	0	−1/5	12/5
0	1/5	−1/5	1/5	1	2/5	1/5
0	−1	2	1	0	2	16

⟹

$\mathbf{x}_1$	$\mathbf{x}_3$	$\mathbf{s}_1$	$\mathbf{x}_4$	$\mathbf{x}_2$	$\mathbf{s}_2$	
1	0	1	1	−2	−1	2
0	1	−1	1	5	2	1
0	0	1	2	1	4	17

We multiply the coefficients in the objective by minus one to convert the maximum objective to a minimum. Next, we introduce slack variables to convert the inequality constraints into equalities. Our initial tableau is shown at the beginning of Table 4.4. We reorder the slack variables first, since they are obviously basic feasible. The resulting tableau and the subsequent tableau steps are also shown in Table 4.4. No Gauss-Jordan elimination is needed at this point. Non-basic variable $\mathbf{x}_4$ corresponds to the most negative entry of the cost reduction vector $\mathbf{r}_N{}^\top$. The second row corresponds to the smallest ratio of an entry of $\mathbf{x}_B$ to a positive entry of $\mathbf{h}$. So, we interchange $\mathbf{x}_4$ and $\mathbf{s}_2$. Then we perform Gauss-Jordan elimination on the second column. The first row corresponds to the smallest ratio of an entry of $\mathbf{x}_B$ to a positive entry of $\mathbf{h}$. So, we interchange $\mathbf{x}_1$ and $\mathbf{s}_1$. We perform Gauss-Jordan elimination on the first column: The second row corresponds to the smallest ratio of an entry of $\mathbf{x}_B$ to a positive entry of $\mathbf{h}$. So, we interchange $\mathbf{x}_4$ and $\mathbf{x}_2$. We perform Gauss-Jordan elimination on the second column: The second row corresponds to the smallest ratio of an entry of $\mathbf{x}_B$ to a positive entry of $\mathbf{h}$. So, we interchange $\mathbf{x}_3$ and $\mathbf{x}_2$. We perform Gauss-Jordan elimination on the second column: In this tableau, the cost reduction vector has no negative entries. At this point, we conclude that the current solution is optimal:

$$\mathbf{x}_B = \begin{bmatrix} \mathbf{x}_1 \\ \mathbf{x}_3 \end{bmatrix} = \begin{bmatrix} 2 \\ 1 \end{bmatrix} .$$

Bartels [11, pp. 418–421] has performed a rounding error analysis of Gauss-Jordan elimination for the standard form of the simplex method. His analysis shows that the rounding errors are unbounded. As a result, Gauss-Jordan elimination is not numerically stable, and should be used with care or not at all.

4.2.7 Inverse Updates

The tableau in Sect. 4.2.6 is inefficient, because it computes all columns of $\mathbf{H} = \mathbf{A}_B^{-1}\mathbf{A}_N$, when only one column is needed. Alternatively, we could compute

$$\mathbf{y}_B{}^\top = \mathbf{c}_B{}^\top \mathbf{A}_B^{-1} ,$$

and then compute the cost reduction vector

$$\mathbf{r}_N{}^\top = \mathbf{c}_n{}^\top - \mathbf{y}_B{}^\top \mathbf{A}_N .$$

We could find the index j_N of the most negative entry of $\mathbf{r}_N$, and then compute

$$\mathbf{h} = \mathbf{A}_B^{-1}\left(\mathbf{A}_N \mathbf{e}_{j_N}\right) .$$

For this approach, we might consider updating $\mathbf{A}_B^{-1}$ by means of the Sherman-Morrison-Woodbury Formula (3.46). Since our new basic matrix is

$$\widetilde{\mathbf{A}}_B = \mathbf{A}_B + \left[\mathbf{A}_N \mathbf{e}_{j_N} - \mathbf{A}_B \mathbf{e}_{i_B}\right] \mathbf{e}_{i_B}{}^\top ,$$

the Sherman-Morrison-Woodbury Formula implies that

$$\widetilde{\mathbf{A}}_B^{-1} = \mathbf{A}_B^{-1} - \mathbf{A}_B^{-1}\left[\mathbf{A}_N \mathbf{e}_{j_N} - \mathbf{A}_B \mathbf{e}_{i_B}\right] \frac{1}{1 + \mathbf{e}_{i_B}{}^\top \mathbf{A}_B^{-1}\left[\mathbf{A}_N \mathbf{e}_{j_N} - \mathbf{A}_B \mathbf{e}_{i_B}\right]} \mathbf{e}_{i_B}{}^\top \mathbf{A}_B^{-1}$$

$$= \mathbf{A}_B^{-1} - \left[\mathbf{h} - \mathbf{e}_{i_B}\right] \frac{1}{\mathbf{e}_{i_B}{}^\top \mathbf{h}} \mathbf{e}_{i_B}{}^\top \mathbf{A}_B^{-1} .$$

The new matrix $\widetilde{\mathbf{A}}_B^{-1}$ can be used to determine the other vectors needed to complete a step in the simplex method.

4.2.8 *Factorization Updates*

Numerical stability is the major problem for both the simplex tableau via Gauss-Jordan elimination and the simplex method via inverse updates. Although columns of the coefficient array $\mathbf{A}$ are reordered during the simplex method, this pivoting is done to reduce the objective, and not to guarantee numerical stability in matrix operations.

Bartels and Golub [12] have suggested the use of Gaussian factorization with row pivoting to solve the linear systems that arise in the simplex method. As we saw in Sect. 3.7.3 in Chap. 3 of Volume I, this algorithm has reasonable numerical stability. The difficulty arises in updating the factorization during the simplex method. The following discussion will illustrate the problems.

Suppose that we have factored

$$\mathbf{Q}_B{}^\top \mathbf{A}_B = \mathbf{L}_B \mathbf{R}_B$$

where $\mathbf{A}_B$ is the basic part of the constraint coefficient matrix in our standard form of the linear programming problem, $\mathbf{Q}_B$ is an $m \times m$ permutation matrix, $\mathbf{L}_B$ is unit left triangular and $\mathbf{R}_B$ is right triangular with nonzero diagonal entries. During the simplex method, we will identify the non-basic column number j_N to become basic, and the basic column number i_B to become non-basic. We can partition

$$\mathbf{A}_B = \begin{bmatrix} \mathbf{A}_{B1} & \mathbf{a}_{i_B} & \mathbf{A}_{B2} \end{bmatrix}$$

and

$$\mathbf{L}_B{}^{-1} \mathbf{Q}_B{}^\top \mathbf{A}_B = \begin{bmatrix} \mathbf{R}_{B11} & \mathbf{r}_B & \mathbf{R}_{B12} \\ \mathbf{0} & \mathbf{e}_1 \varrho_B & \mathbf{R}_{B22} \end{bmatrix} ,$$

where $\mathbf{R}_{B11}$ is $i_B \times i_B$. In the process of computing the vector

$$\mathbf{h} = \mathbf{A}_B{}^{-1} \mathbf{A}_N \mathbf{e}_{j_N} = \mathbf{R}_B{}^{-1} \mathbf{L}_B{}^{-1} \mathbf{Q}_B{}^\top \mathbf{A}_N \mathbf{e}_{j_N}$$

in the Simplex Algorithm 4.2.1, we will have computed the following column of $\mathbf{R}$:

$$\begin{bmatrix} \mathbf{r}_{N1} \\ \mathbf{r}_{N2} \end{bmatrix} = \mathbf{r}_N = \mathbf{L}_B{}^{-1} \mathbf{Q}_B{}^\top \mathbf{A}_N \mathbf{e}_{j_N} .$$

(Please do not confuse this with the cost reduction vector.) If we delete the i_B-th column of $\mathbf{A}_B$ and insert the j_N-th column of $\mathbf{A}_N$ as the last column, then we have

$$\mathbf{L}_B{}^{-1} \mathbf{Q}_B{}^\top \mathbf{A}_{B'} \equiv \mathbf{L}_B{}^{-1} \mathbf{Q}_B{}^\top \begin{bmatrix} \mathbf{A}_{B1} & \mathbf{A}_{B2} & \mathbf{a}_{j_N} \end{bmatrix} = \begin{bmatrix} \mathbf{R}_{B11} & \mathbf{R}_{B12} & \mathbf{r}_{N1} \\ \mathbf{0} & \mathbf{R}_{B22} & \mathbf{r}_{N2} \end{bmatrix} = \mathbf{H}_{B'} .$$

This matrix is upper Hessenberg. We can use Gaussian factorization with partial pivoting to factor

$$\mathbf{Q}_{B'}^\top \mathbf{H}_{B'} = \mathbf{L}_{B'} \mathbf{R}_{B'}$$

where $\mathbf{Q}_{B'}$ is a permutation matrix, $\mathbf{L}_{B'}$ is unit left triangular and $\mathbf{R}_{B'}$ is right triangular with nonzero diagonal entries. Then the new basic matrix has the factorization

$$\mathbf{A}_{B'} = \mathbf{Q}_B \mathbf{L}_B \mathbf{Q}_{B'} \mathbf{L}_{B'} \mathbf{R}_{B'} \ .$$

Unfortunately, we cannot simplify this factorization into a standard Gaussian factorization with row pivoting. As a result, the Bartels and Golub algorithm chose to remember the sequence of row pivots and left-triangular factors for a sequence of simplex steps. If at some point the storage and computation becomes too burdensome, then Bartels and Golub would determine the Gaussian factorization of the basic matrix directly.

Bartels [11, pp. 423–431] shows that these Gaussian factorization updates for the simplex method are numerically stable. We also remark that Reid [149] and Saunders [157] have constructed numerically stable variants of the Bartels-Golub algorithm that can exploit sparsity of the constraint coefficient matrix $\mathbf{A}$.

As an alternative to Gaussian factorization, we might use the Gram-Schmidt Algorithm 6.8.1or 6.8.2 in Chap. 6 of Volume I to factor

$$\mathbf{A}_B = \mathbf{Q}_B \mathbf{R}_B \ ,$$

where $\mathbf{Q}_B$ is an orthogonal matrix and $\mathbf{R}_B$ is right-triangular with positive diagonal entries. Suppose that some simplex step chooses non-basic column number j_N to become basic, and basic column number i_B to become non-basic. We could partition

$$\mathbf{A}_B = \begin{bmatrix} \mathbf{A}_{B1} & \mathbf{a}_{i_B} & \mathbf{A}_{B2} \end{bmatrix}$$

and

$$\mathbf{Q}_B^\top \mathbf{A}_B = \begin{bmatrix} \mathbf{R}_{B11} & \mathbf{r}_B & \mathbf{R}_{B12} \\ \mathbf{0} & \mathbf{e}_1 \varrho_B & \mathbf{R}_{B22} \end{bmatrix}$$

where $\mathbf{R}_{B11}$ is $i_B \times i_B$. In the process of computing the vector

$$\mathbf{h} = \mathbf{A}_B^{-1} \mathbf{A}_N \mathbf{e}_{j_N} = \mathbf{R}_B^{-1} \mathbf{Q}_B^\top \mathbf{A}_N \mathbf{e}_{j_N}$$

in the Simplex Algorithm 4.2.1, we will have computed

$$\begin{bmatrix} \mathbf{r}_{N1} \\ \mathbf{r}_{N2} \end{bmatrix} = \mathbf{r}_N = \mathbf{Q}_B^\top \mathbf{A}_N \mathbf{e}_{j_N} \ .$$

If we delete the i_B-th column of $\mathbf{A}_B$ and insert the j_N-th column of $\mathbf{A}_N$ as the last column, then we have

$$\mathbf{Q}_B{}^\top \mathbf{A}_{B'} = \mathbf{Q}_B{}^\top \begin{bmatrix} \mathbf{A}_{B1} & \mathbf{A}_{B2} & \mathbf{a}_{j_N} \end{bmatrix} = \begin{bmatrix} \mathbf{R}_{B11} & \mathbf{R}_{B12} & \mathbf{r}_{N1} \\ \mathbf{0} & \mathbf{R}_{B22} & \mathbf{r}_{N2} \end{bmatrix} \equiv \mathbf{H}_{B'} \; .$$

This matrix is upper Hessenberg. As in Sect. 1.4.8.3, we can use a sequence of Given rotations to transform $\mathbf{H}_{B'}$ to right-triangular form:

$$\mathbf{G}_{B'}{}^\top \mathbf{H}_{B'} = \mathbf{R}_{B'} \; .$$

Then we have the factorization

$$\mathbf{A}_{B'} = \mathbf{Q}_B \mathbf{G}_{B'} \mathbf{R}_{B'} \equiv \mathbf{Q}_{B'} \mathbf{R}_{B'} \; .$$

This is a $\mathbf{Q}\mathbf{R}$ factorization of the new basic coefficient matrix. For more information regarding Gram-Schmidt factorization updates, see Daniel et al. [43].

The GNU Linear Programming Kit (GLPK) contains routine `glp_set_bfcp` to select how the basic coefficient matrix is factored. Users may choose a $\mathbf{L}\mathbf{R}$ factorization, a block $\mathbf{L}\mathbf{R}$ factorization, or a sparse factorization. Documentation for the GLPK is distributed with the software. MATLAB users should use the command linprog. This command has an option to perform the simplex method, but it is not clear how linear systems involving the basic matrix are solved.

Exercise 4.2.7 Describe how to transform the Klee-Minty linear program in Example 4.2.1 to standard form. Then solve the transformed linear program using your favorite linear programming problem solver for $m = 10$. Verify by hand that the computed solution is optimal.

Exercise 4.2.8 Read and enable CUTEst on your machine for your favorite linear program solver. Next, download the BOEING1 file from The NETLIB LP Test Problem Set. Then solve the problem and describe the steps you took to perform this exercise.

4.2.9 Duality

Next, we will consider two new forms of the linear programming problem that are intimately related. Eventually, we will see that either form will be infeasible if and only if the other is unbounded. Also, we will discover that whenever both are feasible then both are bounded and their optimal objective values must be equal.

Let us begin by defining the new forms of the linear programming problem.

Definition 4.2.3 Suppose that $\mathbf{A}$ is a real $m \times n$ matrix, $\mathbf{b}$ is a real m-vector and $\mathbf{c}$ is a real n-vector. Then the **standard min form of the linear programming problem** is

$$\min \mathbf{c}^\top \mathbf{x} \text{ subject to } \mathbf{A}\mathbf{x} \geq \mathbf{b} \text{ and } \mathbf{x} \geq \mathbf{0} \,. \tag{4.6a}$$

and the **standard max form of the linear programming problem** is

$$\max \mathbf{y}^\top \mathbf{b} \text{ subject to } \mathbf{y}^\top \mathbf{A} \leq \mathbf{c}^\top \text{ and } \mathbf{y} \geq \mathbf{0} \,. \tag{4.6b}$$

These problems are said to be **dual** to each other.

It is easy to convert the standard form (4.1) to the standard min form (4.6a). We begin by finding an $n \times n$ permutation matrix $\mathbf{P}$ so that

$$\mathbf{A}\mathbf{P} = \begin{bmatrix} \mathbf{A}_B & \mathbf{A}_N \end{bmatrix}$$

where $\mathbf{A}_B$ is nonsingular. This step is possible, because $\text{rank}\,(\mathbf{A}) = m$. Given any real $n - m$ vector $\mathbf{x}_N \geq \mathbf{0}$, the n-vector

$$\mathbf{x} = \mathbf{P} \begin{bmatrix} \mathbf{x}_B \\ \mathbf{x}_N \end{bmatrix}$$

satisfies the constraints of the standard form (4.1) of the linear programming problem if and only if

$$\mathbf{x}_B = \mathbf{A}_B^{-1} (\mathbf{b} - \mathbf{A}_N \mathbf{x}_N) \geq \mathbf{0} \,.$$

Let us partition

$$\mathbf{c}^\top \mathbf{P} = \begin{bmatrix} \mathbf{c}_B{}^\top & \mathbf{c}_N{}^\top \end{bmatrix} \,.$$

Then the objective for the standard form of the linear programming problem is

$$\begin{aligned} \mathbf{c}^\top \mathbf{x} &= \mathbf{c}_B{}^\top \mathbf{x}_B + \mathbf{c}_N{}^\top \mathbf{x}_N = \mathbf{c}_B{}^\top \mathbf{A}_B^{-1} (\mathbf{b} - \mathbf{A}_N \mathbf{x}_N) + \mathbf{c}_N{}^\top \mathbf{x}_N \\ &= \left(\mathbf{c}_N{}^\top - \mathbf{c}_B{}^\top \mathbf{A}_B^{-1} \mathbf{A}_N \right) \mathbf{x}_N + \mathbf{c}_B{}^\top \mathbf{A}_B^{-1} \mathbf{b} \,. \end{aligned}$$

Thus the standard form (4.1) of the linear programming problem is equivalent to the standard min linear programming problem

$$\begin{aligned} &\min \left(\mathbf{c}_N{}^\top - \mathbf{c}_B{}^\top \mathbf{A}_B^{-1} \mathbf{A}_N \right) \mathbf{x}_N \\ &\text{s.t.} - \mathbf{A}_B^{-1} \mathbf{A}_N \mathbf{x}_N \geq -\mathbf{A}_B^{-1} \mathbf{b} \text{ and } \mathbf{x}_N \geq \mathbf{0} \,. \end{aligned}$$

It is similarly easy to convert a standard form (4.1) to a standard max form (4.6b).

Note that if we convert an equality constraint

$$\mathbf{a}^\top \mathbf{x} = \beta$$

into the pair of inequality constraints

$$\begin{bmatrix} \mathbf{a}^\top \\ -\mathbf{a}^\top \end{bmatrix} \mathbf{x} \geq \begin{bmatrix} \beta \\ -\beta \end{bmatrix} ,$$

then we introduce a linearly dependent pair of rows into the constraint matrix. This can lead to some unfortunate decisions based on rounding errors in the simplex method. This manner of converting equality constraints into inequality constraints should be avoided.

4.2.9.1 Duals of Linear Programs

The Definition 4.2.3 of certain dual linear programming problems has implications for other linear programming problems as well.

Lemma 4.2.1 *The dual of the standard form of the linear programming problem*

$$\min \mathbf{c}^\top \mathbf{x} \textit{ subject to } \mathbf{A}\mathbf{x} = \mathbf{b} \textit{ and } \mathbf{x} \geq \mathbf{0} .$$

is

$$\max \mathbf{y}^\top \mathbf{b} \textit{ subject to } \mathbf{y}^\top \mathbf{A} \leq \mathbf{c}^\top . \tag{4.7}$$

Note that in the dual of the standard form, the variables $\mathbf{y}$ are unrestricted in sign.

Proof As we saw in the introduction to Sect. 4.2.9, the standard form of the linear programming problem is equivalent to the standard min problem

$$\begin{aligned} &\min \left(\mathbf{c}_N{}^\top - \mathbf{c}_B{}^\top \mathbf{A}_B^{-1} \mathbf{A}_N \right) \mathbf{x}_N + \mathbf{c}_B{}^\top \mathbf{A}_B^{1} \mathbf{b} \\ &\text{s.t.} - \mathbf{A}_B^{-1} \mathbf{A}_N \mathbf{x}_N \geq -\mathbf{A}_B^{-1} \mathbf{b} \text{ and } \mathbf{x}_N \geq \mathbf{0} . \end{aligned}$$

This has dual

$$\begin{aligned} &\max - \mathbf{y}_B{}^\top \mathbf{A}_B^{-1} \mathbf{b} + \mathbf{c}_B{}^\top \mathbf{A}_B^{1} \mathbf{b} \\ &\text{s.t.} - \mathbf{y}_B{}^\top \mathbf{A}_B^{-1} \mathbf{A}_N \leq \mathbf{c}_N{}^\top - \mathbf{c}_B{}^\top \mathbf{A}_B^{-1} \mathbf{A}_N \text{ and } \mathbf{y}_B \geq \mathbf{0} . \end{aligned}$$

If we define

$$\mathbf{y}^\top = \left(\mathbf{c}_B{}^\top - \mathbf{y}_B{}^\top \right) \mathbf{A}_B^{-1} ,$$

then

$$\mathbf{y}^\top \mathbf{A}_B = \mathbf{c}_B{}^\top - \mathbf{y}_B{}^\top \le \mathbf{c}_B{}^\top \text{ and}$$
$$\mathbf{y}^\top \mathbf{A}_N = \left(\mathbf{c}_B{}^\top - \mathbf{y}_B{}^\top\right) \mathbf{A}_B^{-1} \mathbf{A}_N \le \mathbf{c}_N{}^\top .$$

Thus $\mathbf{y}$ solves the linear programming problem (4.7).

The following lemma generalizes the forms of linear programming problems and their duals.

Lemma 4.2.2 *The dual of the* ***general mixed min form of the linear programming problem***

$$\begin{aligned} &\min \mathbf{c}_P{}^\top \mathbf{x}_P + \mathbf{c}_U{}^\top \mathbf{x}_U \\ &\text{s.t. } \mathbf{A}_{PP}\mathbf{x}_P + \mathbf{A}_{PU}\mathbf{x}_U \ge \mathbf{b}_P \\ &\quad \mathbf{A}_{UP}\mathbf{x}_P + \mathbf{A}_{UU}\mathbf{x}_U = \mathbf{b}_U \\ &\quad \mathbf{x}_P \ge \mathbf{0} \end{aligned}$$

is

$$\begin{aligned} &\max \mathbf{y}_P{}^\top \mathbf{b}_P + \mathbf{y}_U{}^\top \mathbf{b}_U \\ &\text{s.t. } \mathbf{y}_P{}^\top \mathbf{A}_{PP} + \mathbf{y}_U{}^\top \mathbf{A}_{UP} \le \mathbf{c}_P{}^\top \\ &\quad \mathbf{y}_P{}^\top \mathbf{A}_{PU} + \mathbf{y}_U{}^\top \mathbf{A}_{UU} = \mathbf{c}_U{}^\top \\ &\quad \mathbf{y}_P \ge \mathbf{0} \end{aligned}$$

Proof Following the ideas in Sect. 4.2.1, we can convert the general mixed min form of the linear programming problem to the standard form

$$\begin{aligned} &\min \mathbf{c}_P{}^\top \mathbf{x}_P + \mathbf{C}_U{}^\top \mathbf{p}_U - \mathbf{c}_U{}^\top \mathbf{m}_U \\ &\text{s.t.} \mathbf{A}_{PP}\mathbf{x}_P + \mathbf{A}_{PU}\mathbf{p}_U - \mathbf{A}_{PU}\mathbf{m}_U - \mathbf{s}_P = \mathbf{b}_P \\ &\mathbf{A}_{UP}\mathbf{x}_P + \mathbf{A}_{UU}\mathbf{p}_U - \mathbf{A}_{UU}\mathbf{m}_U = \mathbf{b}_U \\ &\mathbf{x}_P, \mathbf{p}_U, \mathbf{m}_U, \mathbf{s}_P \ge \mathbf{0} \end{aligned}$$

Lemma 4.2.1 shows that this has the dual

$$\begin{aligned} &\max \mathbf{y}_P{}^\top \mathbf{b}_P + \mathbf{y}_U{}^\top \mathbf{b}_U \\ &\text{s.t.} \mathbf{y}_P{}^\top \mathbf{A}_{PP} + \mathbf{y}_U{}^\top \mathbf{A}_{UP} \le \mathbf{c}_P{}^\top \\ &\mathbf{y}_P{}^\top \mathbf{A}_{PU} + \mathbf{y}_U{}^\top \mathbf{A}_{UU} \le \mathbf{c}_U{}^\top \\ &-\mathbf{y}_P{}^\top \mathbf{A}_{PU} - \mathbf{y}_U{}^\top \mathbf{A}_{UU} \le -\mathbf{c}_U{}^\top \\ &\mathbf{y}_P \ge \mathbf{0} \end{aligned}$$

This can be simplified to the claimed linear programming problem.

4.2.9.2 Theory

We began our discussion of linear programming problems by providing a mathematical description of the problem and then developing the simplex algorithm to solve the problem. However, we have not yet found conditions that guarantee that a solution exists, is unique, or depends continuously on the data.

In this section, we will prove a series of results that address the question of existence of a solution. These results will lead to the Duality Theorems 4.2.7 and 4.2.8, which relate the optimal objective values of the standard min and max problems (4.6a) and (4.6b). Uniqueness can be determined by examining the cost reduction vector at an optimal solution: if the cost reduction vector has any zero entries, then the optimal solution is not unique. We will leave verification of this easy fact as an exercise for the reader. Finally, we remark that we will discuss continuous dependence on the data in Sect. 4.2.11.

We will begin our proof of existence with the following lemma, which shows how the standard max and min linear programs can be used to provide bounds on the optimal objective values for each other.

Lemma 4.2.3 *Let* $\mathbf{A}$ *be a real* $m \times n$ *matrix,* $\mathbf{b}$ *be a real* m*-vector and* $\mathbf{c}$ *be a real* n*-vector. Suppose that the real* n*-vector* $\mathbf{x}$ *satisfies*

$$\mathbf{A}\mathbf{x} \geq \mathbf{b} \text{ and } \mathbf{x} \geq \mathbf{0} ,$$

and that the real m*-vector* $\mathbf{y}$ *satisfies*

$$\mathbf{y}^\top \mathbf{A} \leq \mathbf{c}^\top \text{ and } \mathbf{y} \geq \mathbf{0} .$$

Then

$$\mathbf{c}^\top \mathbf{x} \geq \mathbf{y}^\top \mathbf{b} .$$

Proof Let

$$\mathbf{s} = \mathbf{A}\mathbf{x} - \mathbf{b} \geq \mathbf{0} \text{ and } \mathbf{r}^\top = \mathbf{c}^\top - \mathbf{y}^\top \mathbf{A} \geq \mathbf{0} .$$

Since $\mathbf{r}$ and $\mathbf{x}$ are both nonnegative, their inner product is nonnegative. Similarly, the inner product of $\mathbf{y}$ and $\mathbf{s}$ is nonnegative. As a result,

$$0 \leq \mathbf{r}^\top \mathbf{x} + \mathbf{y}^\top \mathbf{s} = \left(\mathbf{c}^\top - \mathbf{y}^\top \mathbf{A}\right) \mathbf{x} + \mathbf{y}^\top (\mathbf{A}\mathbf{x} - \mathbf{b}) = \mathbf{c}^\top \mathbf{x} - \mathbf{y}^\top \mathbf{b} .$$

Lemma 4.2.3 implies the following easy observation.

Corollary 4.2.1 *If either the standard min or max form of the linear programming problem is unbounded, then the other form is infeasible.*

Next, we will delve a bit more deeply into the properties of vectors that provide equal objective values for the standard min and max problems.

Lemma 4.2.4 *Let* $\mathbf{A}$ *be a real* $m \times n$ *matrix,* $\mathbf{b}$ *be a real m-vector and* $\mathbf{c}$ *be a real n-vector. Suppose that the real n-vector* $\mathbf{x}$ *satisfies*

$$\mathbf{A}\mathbf{x} \geq \mathbf{b} \text{ and } \mathbf{x} \geq \mathbf{0} ,$$

and that the real m-vector $\mathbf{y}$ *satisfies*

$$\mathbf{y}^\top \mathbf{A} \leq \mathbf{c}^\top \text{ and } \mathbf{y} \geq \mathbf{0} .$$

If $\mathbf{c}^\top \mathbf{x} = \mathbf{y}^\top \mathbf{b}$*, then* $\mathbf{x}$ *is optimal for the standard min form of the linear programming problem and* $\mathbf{y}$ *is optimal for the standard max form.*

Proof Suppose that $\widetilde{\mathbf{x}}$ is feasible for the standard min form of the linear programming problem. Then Lemma 4.2.3 implies that

$$\mathbf{c}^\top \widetilde{\mathbf{x}} \geq \mathbf{y}^\top \mathbf{b} = \mathbf{c}^\top \mathbf{x} .$$

This implies that $\mathbf{x}$ is optimal. A similar argument shows that $\mathbf{y}$ is optimal for the max problem.

A proof of the following theorem can be found in Mangasarian [123, p. 29].

Theorem 4.2.4 (Motzkin Alternative) *Let* $\mathbf{F}$ *be a real* $m_1 \times n$ *matrix,* $\mathbf{G}$ *be a real* $m_2 \times n$ *matrix and* $\mathbf{H}$ *be a real* $m_3 \times n$ *matrix. Either there is a real n-vector* $\mathbf{x}$ *such that*

$$\mathbf{F}\mathbf{x} > \mathbf{0} \text{ and } \mathbf{G}\mathbf{x} \geq \mathbf{0} \text{ and } \mathbf{H}\mathbf{x} = \mathbf{0}$$

or there is a real m_1*-vector* $\mathbf{y_F}$*, a real* m_2*-vector* $\mathbf{y_G}$ *and a real* m_3*-vector* $\mathbf{y_H}$ *so that*

$$\mathbf{y_F}^\top \mathbf{F} + \mathbf{y_G}^\top \mathbf{G} + \mathbf{y_H}^\top \mathbf{H} = \mathbf{0} \; , \; \mathbf{y_F} \geq \mathbf{0} \; , \; \mathbf{y_F} \neq \mathbf{0} \text{ and } \mathbf{y_G} \geq \mathbf{0} \; ,$$

but not both.

We can use the Motzkin Alternative Theorem to prove the following.

Theorem 4.2.5 (Standard Min Alternative) *Let* $\mathbf{A}$ *be a real* $m \times n$ *matrix and* $\mathbf{b}$ *be a real m-vector. Either there is a real n-vector* $\mathbf{x}$ *so that*

$$\mathbf{A}\mathbf{x} \geq \mathbf{b} \text{ and } \mathbf{x} \geq \mathbf{0}$$

or there is a real m-vector $\mathbf{y}$ *so that*

$$\mathbf{y}^\top \mathbf{A} \leq \mathbf{0} \; , \; \mathbf{y} \geq \mathbf{0} \text{ and } \mathbf{y}^\top \mathbf{b} > 0 \; ,$$

but not both.

Proof First, we note that the second alternative

$$\mathbf{y}^\top \mathbf{A} \leq \mathbf{0} \text{ and } \mathbf{y} \geq \mathbf{0} \text{ and } \mathbf{y}^\top \mathbf{b} > 0$$

is equivalent to

$$\mathbf{b}^\top \mathbf{y} > 0 \text{ and } \begin{bmatrix} -\mathbf{A}^\top \\ \mathbf{I} \end{bmatrix} \mathbf{y} \geq \mathbf{0} . \tag{4.8}$$

This suggests that we define

$$\mathbf{F} = -\mathbf{b}^\top \text{ and } \mathbf{G} = \begin{bmatrix} -\mathbf{A}^\top \\ \mathbf{I} \end{bmatrix} .$$

We will let $\mathbf{H}$ have no rows. Then Motzkin's Theorem 4.2.4 applied to (4.8) implies that this is an exclusive alternative with the existence of a scalar $\xi_\mathbf{F}$, a real n-vector $\mathbf{x}_{\mathbf{G},1}$ and a real m-vector $\mathbf{x}_{G,2}$ so that

$$\xi_\mathbf{F} \mathbf{b}^\top + \left[{\mathbf{x}_{\mathbf{G},1}}^\top , {\mathbf{x}_{G,2}}^\top \right] \begin{bmatrix} -\mathbf{A}^\top \\ \mathbf{I} \end{bmatrix} = \mathbf{0} \, , \; \xi_\mathbf{F} > 0 \, , \; \mathbf{x}_{\mathbf{G},1} \geq \mathbf{0} \text{ and } \mathbf{x}_{\mathbf{G},2} \geq \mathbf{0} \, .$$

Let us define

$$\mathbf{x} = \mathbf{x}_{\mathbf{G},1} / \xi_\mathbf{F} \text{ and } \mathbf{s} = \mathbf{x}_{\mathbf{G},2} / \xi_\mathbf{F} \, .$$

Then we can divide the Motzkin alternative by $\xi_\mathbf{F}$ and take the transpose to get

$$\mathbf{b} - \mathbf{A}\mathbf{x} + \mathbf{s} = \mathbf{0} \, , \; \mathbf{x} \geq \mathbf{0} \text{ and } \mathbf{s} \geq \mathbf{0} \, .$$

Removing the slack variables $\mathbf{s}$ gives us the feasibility conditions for the standard min form of the linear programming problem.

Similarly, we can develop alternative conditions for the standard max problem.

Theorem 4.2.6 (Standard Max Alternative) *Let* $\mathbf{A}$ *be a real* $m \times n$ *matrix and* $\mathbf{c}$ *be a real n-vector. Either there is a real m-vector* $\mathbf{y}$ *so that*

$$\mathbf{y}^\top \mathbf{A} \leq \mathbf{c}^\top \textit{ and } \mathbf{y} \geq \mathbf{0}$$

or there is a real n-vector $\mathbf{x}$ *so that*

$$\mathbf{A}\mathbf{x} \geq \mathbf{0} \, , \; \mathbf{x} \geq \mathbf{0} \textit{ and } \mathbf{c}^\top \mathbf{x} < 0 \, ,$$

but not both.

Proof First, we note that the second alternative

$$\mathbf{A}\mathbf{x} \geq \mathbf{0} \, , \; \mathbf{x} \geq \mathbf{0} \text{ and } \mathbf{c}^\top \mathbf{x} < 0$$

is equivalent to

$$(-\mathbf{c})^\top \mathbf{x} > 0 \text{ and } \begin{bmatrix} \mathbf{A} \\ \mathbf{I} \end{bmatrix} \mathbf{x} \geq \mathbf{0} \,. \tag{4.9}$$

This suggests that we define

$$\mathbf{F} = -\mathbf{c}^\top \text{ and } \mathbf{G} = \begin{bmatrix} \mathbf{A} \\ \mathbf{I} \end{bmatrix} \,.$$

We will let $\mathbf{H}$ have no rows. Then Motzkin's Theorem 4.2.4 applied to (4.9) implies that this is an exclusive alternative with the existence of a scalar $\eta_\mathbf{F}$, a real m-vectors $\mathbf{x}_{\mathbf{G},1}$ and a real n-vector $\mathbf{x}_{G,2}$ so that

$$\eta_\mathbf{F}(-\mathbf{c})^\top + \left[\mathbf{y}_{\mathbf{G},1}{}^\top \; \mathbf{y}_{G,2}{}^\top \right] \begin{bmatrix} \mathbf{A} \\ \mathbf{I} \end{bmatrix} = \mathbf{0} \text{ and } \eta_\mathbf{F} > 0 \text{ and } \mathbf{y}_{\mathbf{G},1} \geq \mathbf{0} \text{ and } \mathbf{y}_{\mathbf{G},2} \geq \mathbf{0} \,.$$

Let us define

$$\mathbf{y} = \mathbf{y}_{\mathbf{G},1}/\eta_\mathbf{F} \text{ and}$$
$$\mathbf{r} = \mathbf{y}_{\mathbf{G},2}/\eta_\mathbf{F} \,.$$

Then we can divide the Motzkin alternative by $\eta_\mathbf{F}$ and transpose to get

$$-\mathbf{c}^\top + \mathbf{y}^\top \mathbf{A} + \mathbf{r}^\top = \mathbf{0} \text{ and } \mathbf{y} \geq \mathbf{0} \text{ and } \mathbf{r} \geq \mathbf{0} \,.$$

Removing the slack variables $\mathbf{r}$ gives us the feasibility conditions for the standard max form of the linear programming problem.

The following lemma provides our first proof of existence of solutions to the standard min and max problems.

Lemma 4.2.5 *Let* $\mathbf{A}$ *be a real* $m \times n$ *matrix,* $\mathbf{b}$ *be a real* m*-vector and* $\mathbf{c}$ *be a real* n*-vector. Suppose that there is an* n*-vector* $\mathbf{x}$ *so that the standard min linear programming problem is feasible:*

$$\mathbf{A}\mathbf{x} \geq \mathbf{b} \textit{ and } \mathbf{x} \geq \mathbf{0} \,,$$

and there is an m*-vector* $\mathbf{y}$ *so that the standard max linear programming problem is feasible:*

$$\mathbf{y}^\top \mathbf{A} \leq \mathbf{c}^\top \textit{ and } \mathbf{y} \geq \mathbf{0} \,.$$

Then both the standard min and max linear programming problems have optimal solutions, and at the optimal solutions the corresponding objectives are equal.

Proof Let

$$\mathscr{M} = \left\{ \begin{bmatrix} \widetilde{\mathbf{x}} \\ \widetilde{\mathbf{y}} \end{bmatrix} \geq \mathbf{0} : \begin{bmatrix} \mathbf{A} & \mathbf{0} \\ \mathbf{0} & -\mathbf{A}^\top \\ \mathbf{c}^\top & \mathbf{b}^\top \end{bmatrix} \begin{bmatrix} \widetilde{\mathbf{x}} \\ \widetilde{\mathbf{y}} \end{bmatrix} \geq \begin{bmatrix} \mathbf{b} \\ -\mathbf{c} \\ 0 \end{bmatrix} \right\} .$$

If $\begin{bmatrix} \widetilde{\mathbf{x}} \, \widetilde{\mathbf{y}} \end{bmatrix} \in \mathscr{M}$, then it is easy to see that $\widetilde{\mathbf{x}}$ is feasible for the standard min form, and that $\widetilde{\mathbf{y}}$ is feasible for the standard max form. Furthermore,

$$\mathbf{c}^\top \widetilde{\mathbf{x}} \leq \widetilde{\mathbf{y}}^\top \mathbf{b} ,$$

so Lemmas 4.2.3 and 4.2.4 imply that $\widetilde{\mathbf{x}}$ and $\widetilde{\mathbf{y}}$ are optimal. Let us show by contradiction that $\mathscr{M}$ is not empty.

Suppose that $\mathscr{M}$ is empty. By the Standard Min Problem Theorem of the Alternative 4.2.5, there is a vector $\begin{bmatrix} \mathbf{v} \; \mathbf{w} \; \zeta \end{bmatrix} \geq \mathbf{0}$ so that

$$\begin{bmatrix} \mathbf{v}^\top & \mathbf{w}^\top & \zeta \end{bmatrix} \begin{bmatrix} \mathbf{A} & \mathbf{0} \\ \mathbf{0} & -\mathbf{A}^\top \\ -\mathbf{c}^\top & \mathbf{b}^\top \end{bmatrix} \geq \mathbf{0}^\top \text{ and } \begin{bmatrix} \mathbf{v}^\top & \mathbf{w}^\top & \zeta \end{bmatrix} \begin{bmatrix} \mathbf{b} \\ -\mathbf{c} \\ 0 \end{bmatrix} > 0 .$$

Equivalently,

$$\mathbf{v}^\top \mathbf{A} \leq \mathbf{c}^\top , \; \mathbf{A}\mathbf{w} \geq \mathbf{b} \text{ and } \mathbf{v}^\top \mathbf{b} > \mathbf{c}^\top \mathbf{w} .$$

If $\zeta = 0$, then

$$\mathbf{v}^\top \mathbf{A} \leq \mathbf{0}^\top \text{ and } \mathbf{A}\mathbf{w} \geq \mathbf{0} .$$

Since $\mathbf{x}$ is feasible for the standard min problem and $\mathbf{v} \geq \mathbf{0}$, we must have

$$0 \geq \mathbf{v}^\top \mathbf{A}\mathbf{x} \geq \mathbf{v}^\top \mathbf{b} .$$

Similarly, since $\mathbf{y}$ is feasible for the standard max problem and $\mathbf{w} \geq \mathbf{0}$, we must have

$$0 \leq \mathbf{y}^\top \mathbf{A}\mathbf{w} \leq \mathbf{c}^\top \mathbf{w} .$$

This contradicts the previously established condition that $\mathbf{v}^\top \mathbf{b} > \mathbf{c}^\top \mathbf{w}$. Thus we must have $\zeta > 0$. This means that $\mathbf{w}/\zeta$ is feasible for the standard min problem and $\mathbf{v}/\zeta$ is feasible for the standard max problem. Then Lemma 4.2.3 implies that

$$\mathbf{c}^\top (\mathbf{w}/\zeta) \geq (\mathbf{v}/\zeta)^\top \mathbf{b} ,$$

which contradicts the condition that $\mathbf{v}^\top \mathbf{b} > \mathbf{c}^\top \mathbf{w}$. We conclude that $\mathscr{M} \neq \emptyset$.

We can summarize some previous results with the following important theorem.

Theorem 4.2.7 (Duality I) *Let* $\mathbf{A}$ *be a real* $m \times n$ *matrix,* $\mathbf{b}$ *be a real* m*-vector and* $\mathbf{c}$ *be a real* n*-vector. If* $\mathbf{x}$ *is feasible for the standard min form* (4.6a) *of the linear programming problem, and* $\mathbf{y}$ *is feasible for the standard max form* (4.6b) *of the linear programming problem, then* $\mathbf{x}$ *and* $\mathbf{y}$ *are optimal if and only if*

$$\mathbf{c}^\top \mathbf{x} = \mathbf{y}^\top \mathbf{b} .$$

Proof If $\mathbf{x}$ and $\mathbf{y}$ are optimal, then Lemma 4.2.5 shows that $\mathbf{c}^\top \mathbf{x} = \mathbf{y}^\top \mathbf{b}$.

Conversely, suppose that $\mathbf{x}$ and $\mathbf{y}$ are feasible with $\mathbf{c}^\top \mathbf{x} = \mathbf{y}^\top \mathbf{b}$. If $\widetilde{\mathbf{x}}$ is feasible for the standard min form and $\widetilde{\mathbf{y}}$ is feasible for the standard max form, then Lemma 4.2.3 implies that

$$\mathbf{c}^\top \widetilde{\mathbf{x}} \geq \mathbf{y}^\top \mathbf{b} = \mathbf{c}^\top \mathbf{x} \Longrightarrow \mathbf{x} \text{ optimal}$$
$$\widetilde{\mathbf{y}}^\top \mathbf{b} \leq \mathbf{c}^\top \mathbf{x} = \mathbf{y}^\top \mathbf{b} \Longrightarrow \mathbf{y} \text{ optimal} .$$

Next, we will examine conditions that prevent a linear programming problem from having a solution.

Lemma 4.2.6 *If either the standard min form or the standard max form of the linear programming problem is infeasible, then neither has an optimal solution.*

Proof Let $\mathbf{A}$ be a real $m \times n$ matrix, $\mathbf{b}$ be a real m-vector and $\mathbf{c}$ be a real n-vector. Suppose that the standard min form of the linear programming problem in Definition 4.2.3 is infeasible. By the Standard Min Problem Theorem of the Alternative 4.2.5, there exists $\mathbf{z} \geq \mathbf{0}$ so that

$$\mathbf{z}^\top \mathbf{A} \leq \mathbf{0} \text{ and } \mathbf{z}^\top \mathbf{b} > 0 .$$

If the standard max form of the problem is also infeasible, then we are done. Otherwise, there exists $\mathbf{y} \geq \mathbf{0}$ so that

$$\mathbf{y}^\top \mathbf{A} \leq \mathbf{c}^\top .$$

Then for all $\lambda > 0$,

$$\mathbf{y} + \mathbf{z}\lambda \geq \mathbf{0} \text{ and } (\mathbf{y} + \mathbf{z}\lambda)^\top \mathbf{A} \leq \mathbf{c}^\top ,$$

so $\mathbf{y} + \mathbf{z}\lambda$ is feasible for the standard max form. Since $\mathbf{z}^\top \mathbf{b} > 0$, we see that

$$(\mathbf{y} + \mathbf{z}\lambda)^\top \mathbf{b} = \mathbf{y}^\top \mathbf{b} + \lambda \mathbf{z}^\top \mathbf{b} \to \infty \text{ as } \lambda \to \infty .$$

Thus the standard max problem is unbounded, and therefore has no optimal solution.

If the standard max problem is infeasible, a similar proof shows that the standard min problem is either infeasible or unbounded.

The following is the contrapositive of Lemma 4.2.6.

Corollary 4.2.2 *If either the standard min form or the standard max form of the linear programming problem has an optimal solution, then both are feasible.*

Example 4.2.3 The standard min problem

$$\min \begin{bmatrix} -1 & 1 \end{bmatrix} \mathbf{x} \text{ subject to } \begin{bmatrix} 1 & 0 \\ 0 & -1 \end{bmatrix} \mathbf{x} \geq \begin{bmatrix} 0 \\ 1 \end{bmatrix} \text{ and } \mathbf{x} \geq \mathbf{0}$$

and its dual

$$\max \mathbf{y}^\top \begin{bmatrix} 0 \\ 1 \end{bmatrix} \text{ subject to } \mathbf{y}^\top \begin{bmatrix} 1 & 0 \\ 0 & -1 \end{bmatrix} \leq \begin{bmatrix} -1 & 1 \end{bmatrix} \text{ and } \mathbf{y} \geq \mathbf{0}$$

are both infeasible.

In the first Duality Theorem 4.2.7, we established one existence result, namely that if both the standard min and standard max forms of the linear programming problem are feasible, then both have optimal solutions with equal objectives. Next, we will provide a different way to guarantee existence of a solution to a linear programming problem.

Theorem 4.2.8 (Duality II) *If either the standard min form or the standard max form of the linear programming problem has an optimal solution, then so does the other problem, and the optimal values of the objective functions are equal.*

Proof Let $\mathbf{A}$ be a real $m \times n$ matrix, $\mathbf{b}$ be a real m-vector and $\mathbf{c}$ be a real n-vector. Suppose that the standard min form of the linear programming problem in Definition 4.2.3 has an optimal solution. Then Corollary 4.2.2 shows that the standard max form is feasible. Since both forms are feasible, Lemma 4.2.5 shows that there are optimal solutions $\mathbf{x}$ and $\mathbf{y}$, respectively, so that

$$\mathbf{c}^\top \mathbf{x} = \mathbf{y}^\top \mathbf{b} .$$

Since Lemma 4.2.3 shows that any feasible guesses $\widetilde{\mathbf{x}}$ and $\widetilde{\mathbf{y}}$ must satisfy

$$\mathbf{c}^\top \widetilde{\mathbf{x}} \geq \widetilde{\mathbf{y}}^\top \mathbf{b} ,$$

any optimal solutions must have equal objective values.

A similar proof shows that if the standard max form has an optimal solution, then so does the standard min form.

Exercise 4.2.9 Let

$$\mathbf{A} = \begin{bmatrix} 3 & 1 & -4 \\ -2 & 6 & 8 \\ 7 & -1 & -2 \\ 1 & 5 & -1 \end{bmatrix} , \ \mathbf{b} = \begin{bmatrix} 4 \\ 10 \\ -3 \\ 2 \end{bmatrix} \text{ and } \mathbf{c}^\top = \begin{bmatrix} 26 & 30 & 10 \end{bmatrix} .$$

1. Show that $\mathbf{y} = \begin{bmatrix} 54/5, 26/5, 0, 0 \end{bmatrix}$ is feasible for the standard max linear programming problem with the given arrays.
2. Compute the standard max objective $\mathbf{y} \cdot \mathbf{b}$.
3. Determine the standard min form of the linear programming problem for these arrays.
4. Show that $\mathbf{x} = \begin{bmatrix} 7/10, 19/10, 0 \end{bmatrix}$ is feasible for the standard min problem.
5. Compute $\mathbf{c} \cdot \mathbf{x}$.
6. What can you conclude about the existence of solutions to the standard min and max problems for these arrays?

4.2.10 Complementary Slackness

In Lemma 4.2.5, we showed that if both the standard min and the standard max forms of a linear programming problem are feasible, then both have optimal solutions. We can strengthen this statement into a specification of precisely which feasible solutions are optimal.

Theorem 4.2.9 (Weak Complementary Slackness) *Let* $\mathbf{A}$ *be a real* $m \times n$ *matrix,* $\mathbf{b}$ *be a real* m*-vector and* $\mathbf{c}$ *be a real* n*-vector. Suppose that* $\mathbf{x}$ *is feasible for the standard min form of the linear programming problem:*

$$\mathbf{A}\mathbf{x} \geq \mathbf{b} \text{ and } \mathbf{x} \geq \mathbf{0} \ ,$$

and that $\mathbf{y}$ *is feasible for the standard max form of the linear programming problem:*

$$\mathbf{y}^\top \mathbf{A} \leq \mathbf{c}^\top \text{ and } \mathbf{y} \geq \mathbf{0} \ .$$

Then $\mathbf{x}$ *and* $\mathbf{y}$ *are optimal for their respective problems if and only if*

$$\left(\mathbf{c}^\top - \mathbf{y}^\top \mathbf{A}\right) \mathbf{x} = 0 = \mathbf{y}^\top (\mathbf{A}\mathbf{x} - \mathbf{b}) \ . \tag{4.10}$$

Proof We will begin by proving the reverse direction of the claim. Suppose that $\mathbf{x}$ and $\mathbf{y}$ are feasible and complementary. Then Eq. (4.10) implies that

$$\mathbf{c}^\top \mathbf{x} = \mathbf{y}^\top \mathbf{A}\mathbf{x} = \mathbf{y}^\top \mathbf{b} \ .$$

Then the first Duality Theorem 4.2.7 shows that $\mathbf{x}$ and $\mathbf{y}$ are optimal.

Next, we will prove the forward direction of the claim. Suppose that $\mathbf{x}$ and $\mathbf{y}$ are optimal. Define

$$\mathbf{s} = \mathbf{A}\mathbf{x} - \mathbf{b} \geq \mathbf{0} \text{ and } \mathbf{r}^\top = \mathbf{c}^\top - \mathbf{y}^\top \mathbf{A} \geq \mathbf{0}^\top \ .$$

Then the first Duality Theorem 4.2.7 implies that

$$\begin{bmatrix}\mathbf{r}^\top & \mathbf{y}^\top\end{bmatrix}\begin{bmatrix}\mathbf{x}\\ \mathbf{s}\end{bmatrix} = \left(\mathbf{c}^\top - \mathbf{y}^\top\mathbf{A}\right)\mathbf{x} + \mathbf{y}^\top\left(\mathbf{A}\mathbf{x} - \mathbf{b}\right) = \mathbf{c}^\top\mathbf{x} - \mathbf{y}^\top\mathbf{b} = 0\ .$$

Since $\begin{bmatrix}\mathbf{r}^\top & \mathbf{y}^\top\end{bmatrix} \geq \mathbf{0}^\top$ and $\begin{bmatrix}\mathbf{x}^\top & \mathbf{s}^\top\end{bmatrix} \geq \mathbf{0}^\top$, all terms in the inner product must be zero:

$$0 = \mathbf{r}^\top\mathbf{x} = \left(\mathbf{c}^\top - \mathbf{y}^\top\mathbf{A}\right)\mathbf{x} \text{ and } 0 = \mathbf{y}^\top\mathbf{s} = \mathbf{y}^\top\left(\mathbf{A}\mathbf{x} - \mathbf{b}\right)\ .$$

Note that both sides of the weak **complementary slackness** equation (4.10) involve inner products of nonnegative vectors. This theorem implies that each individual product term in either one of the inner products is zero. In other words, whenever a constraint on an optimal solution is inactive, the corresponding entry of the optimal dual solution must be zero.

We can also provide a complementary slackness condition for the standard form of the linear programming problems (4.1) and its dual (4.7).

Corollary 4.2.3 *Let* $\mathbf{A}$ *be a real* $m \times n$ *matrix,* $\mathbf{b}$ *be a real* m*-vector and* $\mathbf{c}$ *be a real* n*-vector. Suppose that* $\mathbf{x}$ *is feasible for the standard form* (4.1) *of the linear programming problem:*

$$\mathbf{A}\mathbf{x} = \mathbf{b} \text{ and } \mathbf{x} \geq \mathbf{0}\ ,$$

and $\mathbf{y}$ *is feasible for the dual problem (see Lemma 4.2.1):*

$$\mathbf{y}^\top\mathbf{A} \leq \mathbf{c}^\top\ .$$

Then

$$\left(\mathbf{y}^\top\mathbf{A} - \mathbf{c}^\top\right)\mathbf{x} = \mathbf{0}\ . \tag{4.11}$$

Proof We can rewrite the constraints for $\mathbf{x}$ in the form

$$\begin{bmatrix}\mathbf{A}\\ -\mathbf{A}\end{bmatrix}\mathbf{x} \geq \begin{bmatrix}\mathbf{b}\\ -\mathbf{b}\end{bmatrix} \text{ and } \mathbf{x} \geq \mathbf{0}\ . \tag{4.12}$$

These are constraints for the standard min form of a certain linear programming problem. Suppose that $\mathbf{y}$ is feasible for the dual to the standard form. Since $\mathbf{y}$ is unrestricted in sign, we can write

$$\mathbf{y}^\top = \mathbf{y}_P{}^\top - \mathbf{y}_M{}^\top \text{ where } \mathbf{y}_P\ ,\ \mathbf{y}_M \geq \mathbf{0}$$

and see that

$$\begin{bmatrix}\mathbf{y}_P{}^\top, & \mathbf{y}_M{}^\top\end{bmatrix}\begin{bmatrix}\mathbf{A}\\ -\mathbf{A}\end{bmatrix} \leq \mathbf{c}^\top \text{ and } \begin{bmatrix}\mathbf{y}_P{}^\top, & \mathbf{y}_M{}^\top\end{bmatrix} \geq \mathbf{0}\ .$$

These constraints correspond to the dual of the standard min problem with constraints (4.12). Then the Weak Complementary Slackness Theorem 4.2.9 implies that $\mathbf{x}$ and $\left[\mathbf{y}_P{}^\top \; \mathbf{y}_M{}^\top\right]$ are optimal for their respective problems if and only if

$$0 = \left(\left[\mathbf{y}_P{}^\top \; \mathbf{y}_M{}^\top\right]\begin{bmatrix}\mathbf{A}\\-\mathbf{A}\end{bmatrix} - \mathbf{c}^\top\right)\mathbf{x} = \left(\mathbf{y}^\top\mathbf{A} - \mathbf{c}^\top\right)\mathbf{x}$$

$$= \left[\mathbf{y}_P{}^\top \; \mathbf{y}_M{}^\top\right]\left(\begin{bmatrix}\mathbf{A}\\-\mathbf{A}\end{bmatrix}\mathbf{x} - \begin{bmatrix}\mathbf{b}\\-\mathbf{b}\end{bmatrix}\right) = \left(\mathbf{y}_P{}^\top - \mathbf{y}_M{}^\top\right)(\mathbf{A}\mathbf{x} - \mathbf{b})$$

$$= \mathbf{y}^\top(\mathbf{A}\mathbf{x} - \mathbf{b}) \; .$$

Since feasibility implies that $\mathbf{A}\mathbf{x} = \mathbf{b}$, we omit the second half of these complementary slackness conditions.

We will also present a useful and stronger form of complementary slackness. This theorem will be useful in developing the dual simplex method in Sect. 4.2.12.

Theorem 4.2.10 (Strong Complementary Slackness) *Let* $\mathbf{A}$ *be a real* $m \times n$ *matrix,* $\mathbf{b}$ *be a real m-vector and* $\mathbf{c}$ *be a real n-vector. If the standard min and max forms* (4.6a) *and* (4.6b) *of the linear programming problem are both feasible, then there are optimal solutions* $\mathbf{x}$ *and* $\mathbf{y}$ *to the respective problems so that*

$$\mathbf{x}_j > 0 \text{ if and only if } \left(\mathbf{c}^\top - \mathbf{y}^\top\mathbf{A}\right)_j = 0 \text{ and}$$

$$\mathbf{y}_i > 0 \text{ if and only if } (\mathbf{A}\mathbf{x} - \mathbf{b})_i = 0 \; .$$

A proof of this theorem may be found in Dantzig [44], Goldman and Tucker [83], Schrijver [159] or Sierksma [160].

The Strong Complementary Slackness Theorem 4.2.10 leads to the following important observations about the standard max and min forms of the linear programming problem. Let us reorder the entries of $\mathbf{A}$, $\mathbf{b}$, $\mathbf{c}$, $\mathbf{x}$ and $\mathbf{y}$ so that active constraints are ordered before inactive constraints. This reordering will determine permutation matrices $\mathbf{P}$ and $\mathbf{Q}$, as described below. For the standard max form, we can partition

$$\mathbf{A}\mathbf{P} = \left[\mathbf{A}_B \; \mathbf{A}_N\right] \; , \; \mathbf{P}^\top\mathbf{x} = \begin{bmatrix}\mathbf{x}_B\\\mathbf{x}_N\end{bmatrix} \text{ and } \mathbf{c}^\top\mathbf{P} = \left[\mathbf{c}_B{}^\top \; \mathbf{c}_N{}^\top\right] \; ,$$

where

$$\mathbf{x}_B > \mathbf{0} \Longrightarrow \mathbf{y}^\top\mathbf{A}_B = \mathbf{c}_B{}^\top \; , \text{ and } \mathbf{y}^\top\mathbf{A}_N < \mathbf{c}_N{}^\top \Longrightarrow \mathbf{x}_N = \mathbf{0} \; .$$

The permutation matrix $\mathbf{Q}$ allows us to partition the constraints for the standard min form as follows:

$$\mathbf{Q}^\top\mathbf{A}\mathbf{P} = \begin{bmatrix}\mathbf{A}_{BB} & \mathbf{A}_{BN}\\\mathbf{A}_{NB} & \mathbf{A}_{NN}\end{bmatrix} \; , \; \mathbf{Q}^\top\mathbf{b} = \begin{bmatrix}\mathbf{b}_B\\\mathbf{b}_N\end{bmatrix} \text{ and } \mathbf{y}^\top\mathbf{Q} = \left[\mathbf{y}_B{}^\top \; \mathbf{y}_N{}^\top\right] \; .$$

Here

$$\mathbf{A}_{BB}\mathbf{x}_B = \mathbf{b}_B \iff \mathbf{y}_B > 0 \text{ and}$$
$$\mathbf{A}_{NB}\mathbf{x}_B > \mathbf{b}_N \iff \mathbf{y}_N = 0$$

This additional partitioning implies that

$$\mathbf{y}_B{}^\top \mathbf{A}_{BB} = \mathbf{c}_B{}^\top \iff \mathbf{x}_B > 0 \text{ , and } \mathbf{y}_B{}^\top \mathbf{A}_{BN} < \mathbf{c}_N{}^\top \iff \mathbf{x}_N = \mathbf{0} \,.$$

Thus if $\mathbf{A}_{BB}$ is nonsingular, the active constraints determine the optimal solutions $\mathbf{x}$ and $\mathbf{y}$ by the equations

$$\mathbf{A}_{BB}\mathbf{x}_B = \mathbf{b}_B \text{ and } \mathbf{y}_B{}^\top \mathbf{A}_{BB} = \mathbf{c}_B{}^\top \,.$$

As a check of optimality, we note that

$$\mathbf{c}^\top \mathbf{x} = \mathbf{c}_B{}^\top \mathbf{x}_B = \mathbf{y}_B{}^\top \mathbf{A}_{BB}\mathbf{x}_B = \mathbf{y}_B{}^\top \mathbf{b}_B = \mathbf{y}^\top \mathbf{b} \,.$$

Example 4.2.4 Let

$$\mathbf{A} = \begin{bmatrix} 8 & 4 & 4 \\ 6 & 2 & 3/2 \\ 1 & 3/2 & 1/2 \end{bmatrix} \,,\; \mathbf{b} = \begin{bmatrix} 60 \\ 30 \\ 20 \end{bmatrix} \text{ and } \mathbf{c}^\top = \begin{bmatrix} 48, \, 20, \, 8 \end{bmatrix} \,.$$

It is easy to see that

$$\mathbf{y}^\top = \begin{bmatrix} 2, \, 0, \, 8 \end{bmatrix}$$

is feasible for the standard max problem

$$\max \mathbf{y}^\top \mathbf{b} \text{ subject to } \mathbf{y}^\top \mathbf{A} \le \mathbf{c}^\top \text{ and } \mathbf{y} \ge \mathbf{0} \,.$$

If $\mathbf{y}$ is optimal for this problem, then we can use strong complementary slackness to determine the solution $\mathbf{x}$ to the standard min problem

$$\min \mathbf{c}^\top \mathbf{x} \text{ subject to } \mathbf{A}\mathbf{x} \ge \mathbf{b} \text{ and } \mathbf{x} \ge \mathbf{0} \,.$$

Since the first entry of

$$\mathbf{r}^\top = \mathbf{c}^\top - \mathbf{y}^\top \mathbf{A} = \begin{bmatrix} 24, \, 0, \, 0 \end{bmatrix}$$

is positive, we choose the permutation matrix $\mathbf{P}$ to interchange the first and third columns of $\mathbf{A}$. Since the second entry of $\mathbf{y}$ is zero, we choose the permutation matrix

$\mathbf{Q}$ to interchange the second and third rows of $\mathbf{A}$. Thus we have the permutation matrices

$$\mathbf{P} = \left[\mathbf{e}_3, \mathbf{e}_2, \mathbf{e}_1\right] \text{ and } \mathbf{Q} = \left[\mathbf{e}_1, \mathbf{e}_3, \mathbf{e}_2\right] .$$

Then

$$\mathbf{Q}^\top \mathbf{A} \mathbf{P} = \begin{bmatrix} \begin{bmatrix} 2 & 4 \\ 1/2 & 3/2 \end{bmatrix} & \begin{bmatrix} 8 \\ 1 \end{bmatrix} \\ \begin{bmatrix} 3/2 & 2 \end{bmatrix} & \begin{bmatrix} 6 \end{bmatrix} \end{bmatrix} = \begin{bmatrix} \mathbf{A}_{BB} & \mathbf{A}_{BN} \\ \mathbf{A}_{NB} & \mathbf{A}_{NN} \end{bmatrix} , \ \mathbf{Q}^\top \mathbf{b} = \begin{bmatrix} \begin{bmatrix} 60 \\ 20 \end{bmatrix} \\ \begin{bmatrix} 30 \end{bmatrix} \end{bmatrix} = \begin{bmatrix} \mathbf{b}_B \\ \mathbf{b}_N \end{bmatrix} ,$$

$$\mathbf{c}^\top \mathbf{P} = \left[[8, 20] \ [48] \right] = \left[\mathbf{c}_B{}^\top, \mathbf{c}_N{}^\top \right] \text{ and } \mathbf{y}^\top \mathbf{Q} = \left[[2, 8] \ [0] \right] = \left[\mathbf{y}_B{}^\top, \mathbf{y}_N{}^\top \right] .$$

Note that $\mathbf{y}_B$ solves $\mathbf{y}_B{}^\top \mathbf{A}_{BB} = \mathbf{c}_B{}^\top$, which is

$$[2, 8] \begin{bmatrix} 2 & 4 \\ 1/2 & 3/2 \end{bmatrix} = [8, 20] .$$

The dual solution has been reordered

$$\mathbf{P}^\top \mathbf{x} = \begin{bmatrix} \begin{bmatrix} \xi_3 \\ \xi_2 \end{bmatrix} \\ \begin{bmatrix} \xi_1 \end{bmatrix} \end{bmatrix} = \begin{bmatrix} \mathbf{x}_B \\ \mathbf{0} \end{bmatrix} ,$$

where $\mathbf{x}_B$ solves $\mathbf{A}_{BB} \mathbf{x}_B = \mathbf{b}_B$, which is

$$\begin{bmatrix} 2 & 4 \\ 1/2 & 3/2 \end{bmatrix} \begin{bmatrix} \xi_3 \\ \xi_2 \end{bmatrix} = \begin{bmatrix} 60 \\ 20 \end{bmatrix} .$$

This has solution

$$\mathbf{x}_B = \begin{bmatrix} 10 \\ 10 \end{bmatrix} .$$

In the original ordering of the variables, we have

$$\mathbf{x} = \begin{bmatrix} 0 \\ 10 \\ 10 \end{bmatrix} , \ \mathbf{s} = \mathbf{A}\mathbf{x} - \mathbf{b} = \begin{bmatrix} 0 \\ 5 \\ 0 \end{bmatrix} ,$$

$$\mathbf{y}^\top = \left[2, 0, 8\right] \text{ and } \mathbf{r}^\top = \mathbf{c}^\top - \mathbf{y}^\top \mathbf{A} = \left[24, 0, 0\right] .$$

Since $\mathbf{x} \geq \mathbf{0}$ and $\mathbf{s} \geq \mathbf{0}$, $\mathbf{x}$ is feasible for the standard min problem. Since $\mathbf{y} \geq \mathbf{0}$ and $\mathbf{r} \geq \mathbf{0}$, $\mathbf{y}$ is feasible for the standard max problem. By design, $\mathbf{x}$ and $\mathbf{r}$ are complementary, and $\mathbf{y}$ and $\mathbf{s}$ are complementary. The Weak Complementary

Slackness Theorem 4.2.9 implies that $\mathbf{x}$ and $\mathbf{y}$ are optimal for their respective linear programming problems. We can check optimality by verifying the first Duality Theorem 4.2.7, namely that

$$\mathbf{c}^\top \mathbf{x} = 280 = \mathbf{y}^\top \mathbf{b} .$$

Exercise 4.2.10 Let

$$\mathbf{A} = \begin{bmatrix} 1 & 0 & 1 \\ 0 & 1 & 1 \\ 1 & 1 & 2 \\ 1 & -1 & 0 \end{bmatrix} , \ \mathbf{b} = \begin{bmatrix} 2 \\ 2 \\ 0 \\ 0 \end{bmatrix} \text{ and } \mathbf{c}^\top = \begin{bmatrix} 1, \ 1, \ 3 \end{bmatrix} .$$

1. Show that $\mathbf{y}^\top = \begin{bmatrix} 1, \ 1, \ 0, \ 0 \end{bmatrix}$ is feasible for the standard max problem with these arrays, and that $\mathbf{x} = \mathbf{e}$ is feasible for the standard min problem.
2. Compute $\mathbf{s} = \mathbf{A}\mathbf{x} - \mathbf{b}$, and show that $\mathbf{s}$ and $\mathbf{y}$ are complementary.
3. Compute $\mathbf{r}^\top = \mathbf{c}^\top - \mathbf{y}^\top \mathbf{A}$, and show that $\mathbf{x}$ and $\mathbf{r}$ are not complementary.
4. Assuming that $\mathbf{y}$ is optimal for the standard max problem, use strong complementary slackness to determine the corresponding optimal solution of the standard min problem.

4.2.11 *Perturbation Analysis*

In Sect. 4.2.9, we developed some theory to determine when a linear programming problem has an optimal solution. Our discussion of complementary slackness in Sect. 4.2.10 helped us to specify these optimal solutions more precisely. We will use such ideas in this section to examine the effect of perturbations on the solutions of linear programming problems. Specifically, we want to know how the optimal solution will change if we modify an entry of $\mathbf{A}$, $\mathbf{b}$ or $\mathbf{c}$, or if we append another constraint, or insert another unknown.

In our discussion of perturbations, we will consider the standard min or max forms of the linear programming problem

$$\min \mathbf{c}^\top \mathbf{x} \text{ subject to } \mathbf{A}\mathbf{x} \geq \mathbf{b} \, , \ \mathbf{x} \geq \mathbf{0} \text{ and}$$
$$\max \mathbf{y}^\top \mathbf{b} \text{ subject to } \mathbf{y}^\top \mathbf{A} \leq \mathbf{c}^\top \, , \ \mathbf{y} \geq \mathbf{0} \, .$$

We will also make use of the following assumptions.

Assumption 4.2.1 Suppose that $\mathbf{A}$ is an $m \times n$ matrix with rank $(\mathbf{A}) > 0$, that $\mathbf{y}$ and $\mathbf{b}$ are m-vectors, and that $\mathbf{x}$ and $\mathbf{c}$ are n-vectors. Assume that we have found permutation matrices $\mathbf{P}$ and $\mathbf{Q}$ so that we may partition

$$\mathbf{Q}^\top \mathbf{A} \mathbf{P} = \begin{bmatrix} \mathbf{A}_{BB} & \mathbf{A}_{BN} \\ \mathbf{A}_{NB} & \mathbf{A}_{NN} \end{bmatrix} , \; \mathbf{P}^\top \mathbf{x} = \begin{bmatrix} \mathbf{x}_B \\ \mathbf{0} \end{bmatrix} , \; \mathbf{Q}^\top \mathbf{b} = \begin{bmatrix} \mathbf{b}_B \\ \mathbf{b}_N \end{bmatrix}$$
$$\mathbf{y}^\top \mathbf{Q} = \begin{bmatrix} \mathbf{y}_B{}^\top & \mathbf{0} \end{bmatrix} \text{ and } \mathbf{c}^\top \mathbf{P} = \begin{bmatrix} \mathbf{c}_B{}^\top & \mathbf{c}_N{}^\top \end{bmatrix} .$$

Here $\mathbf{A}_{BB}$ is a $k \times k$ nonsingular matrix with $k > 0$. Also suppose that $\mathbf{x}$ and $\mathbf{y}$ are complementary optimal solutions:

$$\mathbf{A}_{BB}\mathbf{x}_B = \mathbf{b}_B \; , \; \mathbf{x}_B \geq \mathbf{0} \text{ and } \mathbf{s}_N \equiv \mathbf{A}_{NB}\mathbf{x}_B - \mathbf{b}_N \geq \mathbf{0} \; ,$$
$$\mathbf{y}_B{}^\top \mathbf{A}_{BB} = \mathbf{c}_B{}^\top \; , \; \mathbf{y}_B{}^\top \geq \mathbf{0}^\top \text{ and } \mathbf{r}_N{}^\top \equiv \mathbf{c}_N{}^\top - \mathbf{y}_B{}^\top \mathbf{A}_{BN} \geq \mathbf{0} \; .$$

4.2.11.1 Perturbing $\mathbf{c}_N$ or $\mathbf{b}_N$

We will begin by considering perturbations to non-basic entries of $\mathbf{b}$ or $\mathbf{c}$.

Lemma 4.2.7 *Suppose that Assumptions 4.2.1 are satisfied. If*

$$\widetilde{\mathbf{c}}^\top = \begin{bmatrix} \mathbf{c}_B{}^\top & \widetilde{\mathbf{c}_N}{}^\top \end{bmatrix} \mathbf{P}^\top ,$$

where $\widetilde{\mathbf{c}_N}{}^\top \geq \mathbf{y}_B{}^\top \mathbf{A}_{BN}$, *then the vectors* $\mathbf{x}$ *and* $\mathbf{y}$ *are optimal for the perturbed standard min and max linear programming problem with coefficient arrays* $\mathbf{A}$, $\mathbf{b}$ *and* $\widetilde{\mathbf{c}}$. *Also, if*

$$\widetilde{\mathbf{b}} = \mathbf{Q} \begin{bmatrix} \mathbf{b}_B \\ \widetilde{\mathbf{b}_N} \end{bmatrix} ,$$

where $\widetilde{\mathbf{b}_N} \leq \mathbf{A}_{NB}\mathbf{x}_B$, *then the vectors* $\mathbf{x}$ *and* $\mathbf{y}$ *are optimal for the perturbed standard min and max linear programming problem with coefficient arrays* $\mathbf{A}$, $\widetilde{\mathbf{b}}$ *and* $\mathbf{c}$.

Proof Suppose that we change $\mathbf{c}_N$ to $\widetilde{\mathbf{c}_N}$. These entries correspond to non-basic variables $\mathbf{x}_N$ for the standard min problem, and to inactive constraints for the standard max problem. Complementary slackness (4.10) shows that the optimal solutions to the standard min and max problems will be unchanged if

$$\widetilde{\mathbf{r}_N}{}^\top = \widetilde{\mathbf{c}_N}{}^\top - \mathbf{y}_B{}^\top \mathbf{A}_{BN} \geq \mathbf{0} \text{ which is equivalent to } \widetilde{\mathbf{c}_N} \geq \mathbf{y}_B{}^\top \mathbf{A}_{BN} \; .$$

Similarly, entries of $\mathbf{b}_N$ correspond to non-basic variables $\mathbf{y}_N$ for the standard max problem, and to inactive constraints for the standard min problem. Complementary slackness (4.10) shows that the optimal solutions to the standard min and max problems will be unchanged if

$$\widetilde{\mathbf{s}_N} = \mathbf{A}_{NB}\mathbf{x}_B - \mathbf{b}_N \geq \mathbf{0} \text{ which is equivalent to } \mathbf{A}_{NB}\mathbf{x}_B \geq \mathbf{b}_N \; .$$

In other words, increasing any or all entries of $\mathbf{c}_N$ has no effect on the optimal solutions, and sufficiently small decreases to entries of $\mathbf{c}_N$ also have no effect. However, if a non-basic component j_N of $\mathbf{c}_N$ is decreased too much, meaning that for some component index j_N we have

$$\widetilde{\mathbf{c}_N}^\top \mathbf{e}_{j_N} < \mathbf{y}_B{}^\top \mathbf{A}_{BN} \mathbf{e}_{j_N} ,$$

then the set of basic variables for the standard min form of the problem is changed, so the solutions to both linear programs change. Similarly, decreasing entries of $\mathbf{b}_N$ has no effect on the optimal solutions, and sufficiently small increases have no effect. However, if a non-basic component i_N of $\mathbf{b}_N$ is increased too much, meaning that

$$\mathbf{e}_{i_N}{}^\top \widetilde{\mathbf{b}_N} > \mathbf{e}_{i_N}{}^\top \mathbf{A}_{NB} \mathbf{x}_B ,$$

then the set of basic variables for the standard max form of the problem is changed, so the solutions to both linear programs change.

4.2.11.2 Perturbing a Single Basic Entry of $\mathbf{c}_B$ or $\mathbf{b}_B$

Next, we will consider perturbations to single basic entries of **b** or **c**.

Lemma 4.2.8 *Suppose that Assumptions 4.2.1 are satisfied. Let*

$$\widetilde{\mathbf{c}_B}^\top = \mathbf{c}_B{}^\top + \gamma \mathbf{e}_i{}^\top , \ \mathbf{g}_B{}^\top = \mathbf{e}_i{}^\top \mathbf{A}_{BB}^{-1} \text{ and } \mathbf{g}_N{}^\top = \mathbf{g}_B{}^\top \mathbf{A}_{BN} .$$

If the scalar γ satisfies the constraints

$$-\mathbf{y}_B{}^\top \le \gamma \mathbf{g}_B{}^\top \text{ and } \gamma \mathbf{g}_N{}^\top \le \mathbf{r}_N{}^\top ,$$

then **x** *and*

$$\widetilde{\mathbf{y}}^\top = \left[\mathbf{y}_B{}^\top + \gamma \mathbf{g}_B{}^\top , \mathbf{0}^\top\right] \mathbf{Q}^\top$$

are optimal for the standard min and max problems, respectively, with coefficients **A**, **b** *and*

$$\widetilde{\mathbf{c}}^\top = \left[\widetilde{\mathbf{c}_B}^\top , \mathbf{c}_N{}^\top\right] \mathbf{P}^\top .$$

Similarly, we can let

$$\widetilde{\mathbf{b}_B} = \mathbf{b}_B + \mathbf{e}_j \beta , \ \mathbf{h}_B = \mathbf{A}_{BB}^{-1} \mathbf{e}_j \text{ and } \mathbf{h}_N = \mathbf{A}_{NB} \mathbf{A}_{BB}^{-1} \mathbf{e}_j .$$

If the scalar β satisfies the constraints

$$-\mathbf{x}_B \leq \mathbf{h}_B\beta \text{ and } -\mathbf{h}_N\beta \leq \mathbf{s}_N ,$$

then

$$\widetilde{\mathbf{x}} = \mathbf{P}\begin{bmatrix} \mathbf{x}_B + \mathbf{h}_B\beta \\ \mathbf{0} \end{bmatrix}$$

and $\mathbf{y}$ *are optimal for the standard min and max problems, respectively, with coefficients* $\mathbf{A}$, $\mathbf{c}$ *and*

$$\widetilde{\mathbf{b}} = \mathbf{Q}\begin{bmatrix} \widetilde{\mathbf{b}}_B \\ \mathbf{b}_N \end{bmatrix} .$$

Proof Suppose that we change entry i of $\mathbf{c}_B$. Weak complementary slackness shows that this perturbation leads to the perturbed values

$$\widetilde{\mathbf{c}_B}^\top = \mathbf{c}_B{}^\top + \gamma\mathbf{e}_i{}^\top ,$$
$$\widetilde{\mathbf{y}_B}^\top = \mathbf{y}_B{}^\top + \gamma\mathbf{e}_i{}^\top\mathbf{A}_{BB}^{-1} = \mathbf{y}_B{}^\top + \gamma\mathbf{g}_B{}^\top \text{ and}$$
$$\widetilde{\mathbf{r}_N}^\top = \mathbf{r}_N{}^\top - \gamma\mathbf{e}_i{}^\top\mathbf{A}_{BB}^{-1}\mathbf{A}_{BN} = \mathbf{r}_N{}^\top - \gamma\mathbf{g}_N{}^\top .$$

In order for the perturbed solution $\widetilde{\mathbf{y}}$ of the standard max problem to be optimal, we require

$$0 \leq \mathbf{y}_B{}^\top + \gamma\mathbf{g}_B{}^\top \text{ and } 0 \leq \mathbf{r}_N{}^\top - \gamma\mathbf{g}_N{}^\top .$$

Similarly, suppose that we change entry j of $\mathbf{b}_B$. Weak complementary slackness shows that this perturbation leads to the perturbed values

$$\widetilde{\mathbf{b}_B} = \mathbf{b}_B + \mathbf{e}_j\beta ,$$
$$\widetilde{\mathbf{x}_B} = \mathbf{x}_B + \mathbf{A}_{BB}^{-1}\mathbf{e}_j\beta = \mathbf{x}_B + \mathbf{h}_B\beta \text{ and}$$
$$\widetilde{\mathbf{s}_N} = \mathbf{s}_N + \mathbf{A}_{NB}\mathbf{A}_{BB}^{-1}\mathbf{e}_j\beta = \mathbf{s}_N + \mathbf{h}_N\beta .$$

In order for the perturbed solution $\widetilde{\mathbf{x}}$ to the standard min problem to be optimal, we require

$$0 \leq \mathbf{x}_B + \mathbf{h}_B\beta \text{ and } 0 \leq \mathbf{s}_N + \mathbf{h}_N\beta .$$

In summary, we require the perturbation γ to the ith component of $\mathbf{c}_B$ to satisfy

$$\lambda_{i,B} \equiv \max_i \left\{ -\frac{\mathbf{y}_B{}^\top \mathbf{e}_j}{\mathbf{g}_B{}^\top \mathbf{e}_j} : \mathbf{g}_B{}^\top \mathbf{e}_j > 0 \right\} \leq \gamma \leq \min_j \left\{ -\frac{\mathbf{y}_B{}^\top \mathbf{e}_j}{\mathbf{g}_B{}^\top \mathbf{e}_j} : \mathbf{g}_B{}^\top \mathbf{e}_j < 0 \right\} \equiv \upsilon_{i,B} \text{ and} \tag{4.13a}$$

$$\lambda_{i,N} \equiv \max_j \left\{ \frac{\mathbf{r}_N{}^\top \mathbf{e}_j}{\mathbf{g}_N{}^\top \mathbf{e}_j} : \mathbf{g}_N{}^\top \mathbf{e}_j < 0 \right\} \leq \gamma \leq \min_j \left\{ \frac{\mathbf{r}_N{}^\top \mathbf{e}_j}{\mathbf{g}_N{}^\top \mathbf{e}_j} : \mathbf{g}_N{}^\top \mathbf{e}_j > 0 \right\} \equiv \upsilon_{i,N} \ . \tag{4.13b}$$

Furthermore, the set of basic variables will be unchanged, $\mathbf{x}_B$ will be unchanged, and $\widetilde{\mathbf{y}}_B{}^\top = \mathbf{y}_B{}^\top + \gamma \mathbf{e}_i{}^\top \mathbf{A}_{BB}^{-1}$ will be the new vector of optimal basic variables for the perturbed standard max problem provided that

$$\max\left\{\lambda_{i,B}, \lambda_{i,N}\right\} \leq \left(\widetilde{\mathbf{c}}_B{}^\top - \mathbf{c}_B{}^\top\right) \mathbf{e}_i \leq \min\left\{\upsilon_{i,B}, \upsilon_{i,N}\right\} \ .$$

Similarly, we require that the perturbation β to the jth basic entry of $\mathbf{b}$ satisfy

$$\lambda_{j,B} \equiv \max_i \left\{ -\frac{\mathbf{e}_i{}^\top \mathbf{x}_B}{\mathbf{e}_i{}^\top \mathbf{h}_{j,B}} : \mathbf{e}_i{}^\top \mathbf{h}_{j,B} > 0 \right\} \leq \beta \leq \min_i \left\{ -\frac{\mathbf{e}_i{}^\top \mathbf{x}_B}{\mathbf{e}_i{}^\top \mathbf{h}_{i,B}} : \mathbf{e}_i{}^\top \mathbf{h}_{i,B} < 0 \right\} \equiv \upsilon_{j,B} \text{ and} \tag{4.14a}$$

$$\lambda_{j,N} \equiv \max_i \left\{ \frac{\mathbf{e}_i{}^\top \mathbf{s}_N}{\mathbf{e}_i{}^\top \mathbf{h}_{j,N}} : \mathbf{e}_i{}^\top \mathbf{h}_{j,N} < 0 \right\} \leq \beta \leq \min_i \left\{ \frac{\mathbf{e}_i{}^\top \mathbf{s}_N}{\mathbf{e}_i{}^\top \mathbf{h}_{j,N}} : \mathbf{e}_i{}^\top \mathbf{h}_{j,N} > 0 \right\} \equiv \upsilon_{i,N} \equiv \upsilon_{j,N} \ . \tag{4.14b}$$

Then the set of basic variables will be unchanged, $\mathbf{y}_B$ will be unchanged, and $\widetilde{\mathbf{x}}_B = \mathbf{x}_B + \mathbf{A}_{BB}^{-1}\mathbf{e}_j\varepsilon$ will be the vector of optimal basic variables for the perturbed standard min problem if

$$\max\left\{\lambda_{j,B}, \lambda_{j,N}\right\} \leq \mathbf{e}_j{}^\top \left(\widetilde{\mathbf{b}}_B - \mathbf{b}_B\right) \leq \min\left\{\upsilon_{j,B}, \upsilon_{j,N}\right\} \ .$$

4.2.11.3 Perturbing $\mathbf{c}_B$ or $\mathbf{b}_B$

Next, we will consider perturbations to multiple components of $\mathbf{b}_B$ or $\mathbf{c}_B$.

Lemma 4.2.9 *Suppose that Assumptions 4.2.1 are satisfied. Let* $\mathbf{e}$ *be the vector of ones, and let* $\mathbf{a}$ *be a nonnegative k-vector such that*

$$\mathbf{e}^\top \mathbf{a} \leq 1 \ .$$

For each $1 \leq i \leq k$, *define the upper and lower bounds as in inequalities* (4.13a) *and* (4.13b). *Also let* $\triangle \mathbf{c}_B$ *be a k-vector such that for all* $1 \leq i \leq k$

$$\max\{\lambda_{i,B} \ , \ \lambda_{i,N}\}\mathbf{a}_i \leq \mathbf{e}_i{}^\top \triangle \mathbf{c}_B \leq \min\{\upsilon_{i,B} \ , \ \upsilon_{i,N}\}\mathbf{a}_i \ , \tag{4.15}$$

Then $\mathbf{x}$ *and*

$$\widetilde{\mathbf{y}}^\top = \left[\mathbf{y}_B{}^\top + \triangle\mathbf{c}_B{}^\top \mathbf{A}_{BB}^{-1}, \mathbf{0}^\top\right]\mathbf{Q}$$

are optimal for the standard max and min problems with coefficients $\mathbf{A}$, $\mathbf{b}$ *and*

$$\widetilde{\mathbf{c}}^\top = \left[\mathbf{c_B}{}^\top + \triangle\mathbf{c}_B{}^\top, \mathbf{c}_N{}^\top\right]\mathbf{P}^\top .$$

Alternatively, for each $1 \leq i \leq k$*, define the upper and lower bounds as in inequalities* (4.14a) *and* (4.14b). *Also let* $\triangle\mathbf{b}_B$ *be a* k*-vector such that for all* $1 \leq i \leq k$

$$\max\{\lambda_{i,B}\ ,\ \lambda_{i,N}\}\mathbf{a}_i \leq \mathbf{e}_i \cdot \triangle\mathbf{b}_B \leq \min\{\upsilon_{i,B}\ ,\ \upsilon_{i,N}\}\mathbf{a}_i\ , \tag{4.16}$$

Then

$$\widetilde{\mathbf{x}} = \mathbf{P}\begin{bmatrix}\mathbf{x}_B + \mathbf{A}_{BB}^{-1}\triangle\mathbf{b}_B \\ \mathbf{0}\end{bmatrix}$$

and $\mathbf{y}$ *are optimal for the standard max and min problems with coefficients* $\mathbf{A}$, $\mathbf{c}$ *and*

$$\widetilde{\mathbf{b}} = \mathbf{Q}\begin{bmatrix}\mathbf{b_B} + \triangle\mathbf{b}_B \\ \mathbf{b}_N\end{bmatrix} .$$

Proof First, we will consider a perturbation to $\mathbf{c}_B$. Let

$$\begin{aligned}
\widetilde{\mathbf{c}}_B{}^\top &= \mathbf{c}_B{}^\top + \triangle\mathbf{c}_B{}^\top , \\
\widetilde{\mathbf{y}}_B{}^\top &= \widetilde{\mathbf{c}}_B{}^\top \mathbf{A}_{BB}^{-1} = \mathbf{y}_B{}^\top + \triangle\mathbf{c}_B{}^\top \mathbf{A}_{BB}^{-1} \text{ and} \\
\widetilde{\mathbf{r}}_N{}^\top &= \mathbf{c}_N{}^\top - \widetilde{\mathbf{c}}_B{}^\top \mathbf{A}_{BB}^{-1}\mathbf{A}_{BN} = \mathbf{r}_N{}^\top - \triangle\mathbf{c}_B{}^\top \mathbf{A}_{BB}^{-1}\mathbf{A}_{BN} .
\end{aligned}$$

From the Definition (4.13a) of $\lambda_{i,B}$ and $\upsilon_{i,B}$, we see that if $\mathbf{e}_i{}^\top \mathbf{A}_{BB}^{-1}\mathbf{e}_j > 0$ we have

$$-\mathbf{y}_B{}^\top \mathbf{e}_j\mathbf{a}_i \leq \mathbf{e}_i{}^\top \mathbf{A}_{BB}^{-1}\mathbf{e}_j\lambda_{i,B}\mathbf{a}_i \leq \mathbf{e}_i{}^\top \mathbf{A}_{BB}^{-1}\mathbf{e}_j\triangle\mathbf{c}_i$$

and if $\mathbf{e}_i{}^\top \mathbf{A}_{BB}^{-1}\mathbf{e}_j < 0$ we have

$$-\mathbf{y}_B{}^\top \mathbf{e}_j\mathbf{a}_i \leq \mathbf{e}_i{}^\top \mathbf{A}_{BB}^{-1}\mathbf{e}_j\upsilon_{i,B}\mathbf{a}_i \leq \mathbf{e}_i{}^\top \mathbf{A}_{BB}^{-1}\mathbf{e}_j\triangle\mathbf{c}_i .$$

These imply that for all $1 \leq j \leq k$

$$\triangle\mathbf{c}^\top \mathbf{A}_{BB}^{-1}\mathbf{e}_j = \sum_{i=1}^{k} \triangle\mathbf{c}_i\mathbf{e}_i{}^\top \mathbf{A}_{BB}^{-1}\mathbf{e}_j \geq -\mathbf{y}_B{}^\top \mathbf{e}_j \sum_{i=1}^{k} \mathbf{a}_i \geq -\mathbf{y}_B{}^\top \mathbf{e}_j .$$

From the Definition (4.13b) of $\lambda_{i,N}$ and $\upsilon_{i,N}$, we see that if $\mathbf{e}_i{}^\top \mathbf{A}_{BB}^{-1}\mathbf{A}_{BN}\mathbf{e}_j > 0$ we have

$$-\mathbf{r}_N{}^\top \mathbf{e}_j \mathbf{a}_i > \mathbf{e}_i{}^\top \mathbf{A}_{BB}^{-1}\mathbf{A}_{BN}\mathbf{e}_j \upsilon_{i,B}\mathbf{a}_i \geq \mathbf{e}_i{}^\top \mathbf{A}_{BB}^{-1}\mathbf{A}_{BN}\mathbf{e}_j \mathbf{d}_i$$

and if $\mathbf{e}_i{}^\top \mathbf{A}_{BB}^{-1}\mathbf{A}_{BN}\mathbf{e}_j < 0$ we have

$$\mathbf{r}_N{}^\top \mathbf{e}_j \mathbf{a}_i \geq \mathbf{e}_i{}^\top \mathbf{A}_{BB}^{-1}\mathbf{A}_{BN}\mathbf{e}_j \lambda_{i,N}\mathbf{a}_i \geq \mathbf{e}_i{}^\top \mathbf{A}_{BB}^{-1}\mathbf{A}_{BN}\mathbf{e}_j \mathbf{d}_i \; .$$

These imply that for all $1 \leq j \leq k$

$$\triangle\mathbf{c}^\top \mathbf{A}_{BB}^{-1}\mathbf{A}_{BN}\mathbf{e}_j = \sum_{i=1}^{k} \triangle\mathbf{c}_i \mathbf{e}_i{}^\top \mathbf{A}_{BB}^{-1}\mathbf{A}_{BN}\mathbf{e}_j \leq \mathbf{r}_N{}^\top \mathbf{e}_j \sum_{i=1}^{k} \mathbf{a}_i \leq \mathbf{r}_N{}^\top \mathbf{e}_j \; .$$

As a result,

$$\widetilde{\mathbf{y}}_B{}^\top = \mathbf{y}_B{}^\top + \triangle\mathbf{c}^\top \mathbf{A}_{BB}^{-1} \geq \mathbf{y}_B{}^\top - \mathbf{y}_B{}^\top = \mathbf{0} \; ,$$

and

$$\widetilde{\mathbf{r}}_N{}^\top = \mathbf{r}_N{}^\top - \triangle\mathbf{c}^\top \mathbf{A}_{BB}^{-1}\mathbf{A}_{BN} \geq \mathbf{r}_N{}^\top - \mathbf{r}_N{}^\top = \mathbf{0} \; .$$

Thus $\widetilde{\mathbf{y}}_B$ is feasible for the perturbed standard max problem. Since the constraints did not change for the standard min problem, $\mathbf{x}_B$ is the set of basic variables for a feasible solution to this problem. Since

$$\widetilde{\mathbf{y}}_B{}^\top \mathbf{b}_B = \widetilde{\mathbf{c}}_B{}^\top \mathbf{A}_{BB}^{-1}\mathbf{b}_B = \widetilde{\mathbf{c}}_B{}^\top \mathbf{x}_B \; ,$$

the max and min objective values are equal for these two feasible solutions. Thus both are optimal. In other words, if the perturbation $\mathbf{d}$ to $\mathbf{c}_B$ satisfies (4.15), then the set of basic variables is unchanged.

Perturbations to basic entries of $\mathbf{b}$ are handled in a similar fashion.
The vector $\mathbf{a}$ allows us to average the bounds on the perturbations, in order to decide how much influence we will allow each component of the perturbation to have.

The rate of change of the optimal objective value $\zeta \equiv \mathbf{c}^\top\mathbf{x} = \mathbf{y}^\top\mathbf{b}$ is

$$\frac{\partial \zeta}{\partial \mathbf{c}} = \mathbf{x}^\top \; ,$$

and the rate of change of the optimal solution to the standard max problem is

$$\frac{\partial \mathbf{y}_B}{\partial \mathbf{c}_B} = \mathbf{A}_{BB}{}^{-\top} \; .$$

If the perturbation $\triangle \mathbf{c}_B$ satisfies the conditions in Lemma 4.2.9, then the perturbation in the solution to the standard max problem satisfies

$$\frac{\|\widetilde{\mathbf{y}}_B - \mathbf{y}_B\|}{\|\mathbf{y}_B\|} = \frac{\|{\mathbf{A}_{BB}}^{-\top}(\widetilde{\mathbf{c}}_B - \mathbf{c}_B)\|}{\|{\mathbf{A}_{BB}}^{-\top}\mathbf{c}_B\|} \leq \frac{\|{\mathbf{A}_{BB}}^{-\top}\|\|\widetilde{\mathbf{c}}_B - \mathbf{c}_B\|}{\|\mathbf{c}_B\|/\|{\mathbf{A}_{BB}}^{\top}\|} = \kappa\left({\mathbf{A}_{BB}}^{\top}\right)\frac{\|\widetilde{\mathbf{c}}_B - \mathbf{c}_B\|}{\|\mathbf{c}_B\|}$$

for any vector norm and corresponding subordinate matrix norm. Similarly, the rate of change of the optimal objective value $\zeta \equiv \mathbf{c}^\top \mathbf{x} = \mathbf{y}^\top \mathbf{b}$ is

$$\frac{\partial \zeta}{\partial \mathbf{b}} = \mathbf{y}^\top ,$$

and the rate of change of the optimal solution to the standard max problem is

$$\frac{\partial \mathbf{x}_B}{\partial \mathbf{b}_B} = \mathbf{A}_{BB}^{-1} .$$

If the perturbation $\triangle \mathbf{b}_B$ satisfies the conditions in Lemma 4.2.9, then the perturbation in the solution to the standard max problem satisfies

$$\frac{\|\widetilde{\mathbf{x}}_B - \mathbf{x}_B\|}{\|\mathbf{x}_B\|} = \frac{\|\mathbf{A}_{BB}^{-1}(\widetilde{\mathbf{b}}_B - \mathbf{b}_B)\|}{\|\mathbf{A}_{BB}^{-1}\mathbf{b}_B\|} \leq \frac{\|\mathbf{A}_{BB}^{-1}\|\|\widetilde{\mathbf{b}}_B - \mathbf{b}_B\|}{\|\mathbf{b}_B\|/\|\mathbf{A}_{BB}\|} = \kappa\left(\mathbf{A}_{BB}\right)\frac{\|\widetilde{\mathbf{b}}_B - \mathbf{b}_B\|}{\|\mathbf{b}_B\|} .$$

4.2.11.4 Perturbing $\mathbf{A}_{NN}$

Since $\mathbf{A}_{NN}$ does not appear in the complementary slackness conditions, we can make arbitrary changes to this portion of $\mathbf{A}$ without changing the set of basic variables, or the optimal solutions to either the standard min or max problems.

4.2.11.5 Perturbing $\mathbf{A}_{BN}$ or $\mathbf{A}_{NB}$

The next lemma determines the effect of perturbations in $\mathbf{A}_{BN}$ or $\mathbf{A}_{NB}$.

Lemma 4.2.10 *Suppose that Assumptions 4.2.1 are satisfied. Let the matrix $\triangle \mathbf{A}_{NB}$ satisfy*

$$\|\triangle \mathbf{A}_{NB}\|_\infty \leq \frac{\min_i |{\mathbf{e}_i}^\top \mathbf{s}_N|}{\|\mathbf{x}_B\|_\infty} .$$

Then $\mathbf{x}$ and $\mathbf{y}$ are optimal for the standard max and min linear programming problems with coefficients

$$\widetilde{\mathbf{A}} = \mathbf{Q}\begin{bmatrix} \mathbf{A}_{BB} & \mathbf{A}_{BN} \\ \mathbf{A}_{NB} + \triangle \mathbf{A}_{NB} & \mathbf{A}_{NN} \end{bmatrix}\mathbf{P}^\top ,$$

b *and* **c**. *Similarly, if the matrix* $\triangle \mathbf{A}_{BN}$ *satisfies*

$$\|\triangle \mathbf{A}_{BN}\|_1 \leq \frac{\min_i |\mathbf{r}_N{}^\top \mathbf{e}_j|}{\|\mathbf{y}_B\|_\infty} ,$$

then **x** *and* **y** *are optimal for the standard max and min linear programming problems with coefficients* **b**, **c** *and*

$$\widetilde{\mathbf{A}} = \mathbf{Q} \begin{bmatrix} \mathbf{A}_{BB} & \mathbf{A}_{BN} + \triangle \mathbf{A}_{BN} \\ \mathbf{A}_{NB} & \mathbf{A}_{NN} \end{bmatrix} \mathbf{P}^\top .$$

Proof For the first matrix perturbation, the perturbed slack variables for the standard min problem are

$$\widetilde{\mathbf{s}}_N = (\mathbf{A}_{NB} + \triangle \mathbf{A}_{NB})\, \mathbf{x}_B - \mathbf{b}_N = \mathbf{s}_N + \triangle \mathbf{A}_{NB} \mathbf{x}_B .$$

Since

$$\|\widetilde{\mathbf{s}}_N - \mathbf{s}_N\|_\infty = \|\triangle \mathbf{A}_{NB} \mathbf{x}_B\|_\infty \leq \|\triangle \mathbf{A}_{NB}\|_\infty \|\mathbf{x}_B\|_\infty \leq \min_i |\mathbf{e}_i{}^\top \mathbf{s}_N| ,$$

it follows that $\widetilde{\mathbf{s}}_N \geq \mathbf{0}$. As a result, the current set of basic variables is unchanged, and the optimal solutions $\mathbf{x}_B$ and $\mathbf{y}_B$ are unchanged.

For the second matrix perturbation, the perturbed slack variables for the standard max problem are

$$\widetilde{\mathbf{r}}_N{}^\top = \mathbf{c}_N{}^\top - \mathbf{y}_B{}^\top (\mathbf{A}_{BN} + \triangle \mathbf{A}_{BN}) = \mathbf{r}_N{}^\top - \mathbf{y}_B{}^\top \triangle \mathbf{A}_{BN} .$$

Since

$$\|\widetilde{\mathbf{r}}_N - \mathbf{r}_N\|_\infty = \left\| \triangle \mathbf{A}_{BN}{}^\top \mathbf{y}_B \right\|_\infty \leq \left\| \triangle \mathbf{A}_{BN}{}^\top \right\|_\infty \|\mathbf{y}_B\|_\infty \leq \min_i |\mathbf{e}_i{}^\top \mathbf{r}_N| ,$$

it follows that $\widetilde{\mathbf{r}}_N \geq \mathbf{0}$. As a result, the current set of basic variables is unchanged, and the optimal solutions $\mathbf{x}_B$ and $\mathbf{y}_B$ are unchanged.

4.2.11.6 Perturbing $\mathbf{A}_{BB}$

We will end our perturbation analysis with the following lemma regarding perturbations in the portion of the coefficient matrix **A** corresponding to basic rows and columns.

Lemma 4.2.11 *Suppose that Assumptions 4.2.1 are satisfied. Let the matrix* $\triangle \mathbf{A}_{BB}$ *satisfy*

$$\|\mathbf{A}_{BB}^{-1} \triangle \mathbf{A}_{BB}\|_\infty < 1 \,, \tag{4.17a}$$

$$\|\triangle \mathbf{A}_{BB} \mathbf{A}_{BB}^{-1}\|_1 < 1 \,, \tag{4.17b}$$

$$\left\|\mathbf{A}_{BB}^{-1} \triangle \mathbf{A}_{BB}\right\|_\infty \leq \min \left\{ \frac{\min_j |\mathbf{e}_j{}^\top \mathbf{x}_B|}{\|\mathbf{x}_B\|_\infty + \min_j |\mathbf{e}_j{}^\top \mathbf{x}_B|} \,,\, \frac{\min_i |\mathbf{e}_i{}^\top \mathbf{s}_N|}{\|\mathbf{A}_{NB}\|_\infty \|\mathbf{x}_B\|_\infty + \min_i |\mathbf{e}_i{}^\top \mathbf{s}_N|} \right\} \text{ and} \tag{4.17c}$$

$$\left\|\triangle \mathbf{A}_{BB} \mathbf{A}_{BB}^{-1}\right\|_1 \leq \min \left\{ \frac{\min_i |\mathbf{y}_B{}^\top \mathbf{e}_i|}{\|\mathbf{y}_B\|_\infty + \min_i |\mathbf{y}_B{}^\top \mathbf{e}_i|} \,,\, \frac{\min_j |\mathbf{r}_N{}^\top \mathbf{e}_j|}{\|\mathbf{A}_{BN}\|_1 \|\mathbf{y}_B\|_\infty + \min_j |\mathbf{r}_N{}^\top \mathbf{e}_j|} \right\} . \tag{4.17d}$$

Then

$$\widetilde{\mathbf{x}} = \mathbf{P} \begin{bmatrix} (\mathbf{A}_{BB} + \triangle \mathbf{A}_{BB})^{-1} \mathbf{b}_B \\ \mathbf{0} \end{bmatrix} \text{ and } \widetilde{\mathbf{y}}^\top = \left[\mathbf{c}_B{}^\top (\mathbf{A}_{BB} + \triangle \mathbf{A}_{BB})^{-1} \,,\, \mathbf{0}^\top \right] \mathbf{Q}^\top$$

are optimal for the standard max and min linear programming problems with coefficients **b**, **c** *and*

$$\widetilde{\mathbf{A}} = \mathbf{Q} \begin{bmatrix} \mathbf{A}_{BB} + \triangle \mathbf{A}_{BB} & \mathbf{A}_{BN} \\ \mathbf{A}_{NB} & \mathbf{A}_{NN} \end{bmatrix} \mathbf{P}^\top .$$

Proof Inequality (4.17a) and Corollary 3.6.1 in Chap. 3 of Volume I show that $\mathbf{A}_{BB} + \triangle \mathbf{A}_{BB}$ is nonsingular. Then inequality (3.28) in Chap. 3 of Volume I is valid, and gives us

$$\|\widetilde{\mathbf{x}}_B - \mathbf{x}_B\|_\infty \leq \frac{\left\|\mathbf{A}_{BB}^{-1} \triangle \mathbf{A}_{BB}\right\|_\infty}{1 - \left\|\mathbf{A}_{BB}^{-1} \triangle \mathbf{A}_{BB}\right\|_\infty} \|\mathbf{x}_B\|_\infty$$

and afterward the first bound on the left of inequality (4.17c) yields

$$\leq \min_j \left| \mathbf{e}_j \cdot \mathbf{x}_B \right| \,.$$

This inequality implies that $\widetilde{\mathbf{x}}_B \geq \mathbf{0}$, so we conclude that $\widetilde{\mathbf{x}} \geq \mathbf{0}$. Similarly, inequality (4.17b) and inequality (3.28) in Chap. 3 of Volume I produce

$$\|\widetilde{\mathbf{y}}_B - \mathbf{y}_B\|_\infty \leq \frac{\left\|\mathbf{A}_{BB}{}^{-\top} \triangle \mathbf{A}_{BB}{}^\top\right\|_\infty}{1 - \left\|\mathbf{A}_{BB}{}^{-\top} \triangle \mathbf{A}_{BB}{}^\top\right\|_\infty} \|\mathbf{y}_B\|_\infty$$

then the first bound on the left of inequality (4.17d) yields

$$\leq \min_i |\mathbf{y}_B \cdot \mathbf{e}_i| \ .$$

This inequality implies that $\mathbf{y}_B \geq \mathbf{0}$, so we conclude that $\widetilde{\mathbf{y}} \geq \mathbf{0}$.

Next, we note that the perturbed non-basic slack variables for the standard min problem are

$$\widetilde{\mathbf{s}}_N = \mathbf{A}_{NB}\widetilde{\mathbf{x}}_B - \mathbf{b}_N \ .$$

It follows that

$$\|\widetilde{\mathbf{s}}_N - \mathbf{s}_N\|_\infty = \|\mathbf{A}_{NB}\left(\widetilde{\mathbf{x}}_B - \mathbf{x}_B\right)\|_\infty$$

then we use inequality (3.28) in Chap. 3 of Volume I to get

$$\leq \|\mathbf{A}_{NB}\|_\infty \frac{\left\|\mathbf{A}_{BB}^{-1}\triangle\mathbf{A}_{BB}\right\|_\infty}{1-\left\|\mathbf{A}_{BB}^{-1}\triangle\mathbf{A}_{BB}\right\|_\infty} \|\mathbf{x}_B\|_\infty$$

then the second bound on the left of inequality (4.17d) yields

$$\leq \min_i |\mathbf{e}_i \cdot \mathbf{s}_N| \ .$$

This inequality implies that $\widetilde{\mathbf{s}}_N \geq \mathbf{0}$, so we conclude that $\mathbf{s} \equiv \mathbf{A}\mathbf{x} - \mathbf{b} \geq \mathbf{0}$. The proof that $\mathbf{r}^\top = \mathbf{c}^\top - \mathbf{y}^\top\mathbf{A} \geq \mathbf{0}$ is similar.

Note that during the proof of this lemma, we showed that

$$\frac{\|\widetilde{\mathbf{x}}_B - \mathbf{x}_B\|_\infty}{\|\mathbf{x}_B\|_\infty} \leq \frac{\left\|\mathbf{A}_{BB}^{-1}\triangle\mathbf{A}_{BB}\right\|_\infty}{1-\left\|\mathbf{A}_{BB}^{-1}\triangle\mathbf{A}_{BB}\right\|_\infty}$$

and that

$$\frac{\|\widetilde{\mathbf{y}}_B - \mathbf{y}_B\|_\infty}{\|\mathbf{y}_B\|_\infty} \leq \frac{\left\|\mathbf{A}_{BB}{}^{-\top}\triangle\mathbf{A}_{BB}{}^\top\right\|_\infty}{1-\left\|\mathbf{A}_{BB}{}^{-\top}\triangle\mathbf{A}_{BB}{}^\top\right\|_\infty} \ .$$

4.2.11.7 Adding a New Constraint

Lemma 4.2.12 *Suppose that Assumptions 4.2.1 are satisfied. Given an m-vector* **a**, *partition*

$$\mathbf{Q}^\top\mathbf{a} = \begin{bmatrix}\mathbf{a}_B \\ \mathbf{a}_N\end{bmatrix} \ .$$

Also suppose that we are given a scalar γ *such that*

$$\varrho \equiv \gamma - \mathbf{y}^\top \mathbf{a} = \gamma - \mathbf{y}_B{}^\top \mathbf{a}_B \geq 0 \, .$$

Then

$$\widetilde{\mathbf{x}} = \begin{bmatrix} \mathbf{x} \\ 0 \end{bmatrix}$$

is optimal for the perturbed standard min linear programming problem with coefficient arrays

$$\widetilde{\mathbf{A}} = \begin{bmatrix} \mathbf{A}, \mathbf{a} \end{bmatrix} \; , \; \mathbf{b} \; and \; \widetilde{\mathbf{c}}^\top = \begin{bmatrix} \mathbf{c}^\top, \gamma \end{bmatrix} \, .$$

and $\mathbf{y}$ *is optimal for the perturbed standard max linear programming problem with these same coefficient arrays.*

Proof Note that $\widetilde{\mathbf{x}} \geq \mathbf{0}$, and

$$\mathbf{Q}^\top \left[\widetilde{\mathbf{A}} \widetilde{\mathbf{x}} - \widetilde{\mathbf{b}} \right] = \begin{bmatrix} \mathbf{A}_{BB} & \mathbf{A}_{BN} & \mathbf{a}_B \\ \mathbf{A}_{NB} & \mathbf{A}_{NN} & \mathbf{a}_N \end{bmatrix} \begin{bmatrix} \mathbf{x}_B \\ \mathbf{0} \\ 0 \end{bmatrix} - \begin{bmatrix} \mathbf{b}_B \\ \mathbf{b}_N \end{bmatrix} = \begin{bmatrix} \mathbf{0} \\ \mathbf{A}_{NB}\mathbf{x}_B - \mathbf{b}_N \end{bmatrix} \geq \mathbf{0} \, .$$

Thus $\widetilde{\mathbf{x}}$ is feasible for the perturbed standard min linear programming problem. The perturbed min objective is

$$\widetilde{\mathbf{c}}^\top \widetilde{\mathbf{x}} = \begin{bmatrix} \mathbf{c}_B{}^\top, \mathbf{c}_N{}^\top, \gamma \end{bmatrix} \begin{bmatrix} \mathbf{x}_B \\ \mathbf{0} \\ 0 \end{bmatrix} = \mathbf{c}_B{}^\top \mathbf{x}_B = \mathbf{c}_B{}^\top \mathbf{A}_{BB}^{-1} \mathbf{b}_B \, .$$

We are given that $\mathbf{y} \geq \mathbf{0}$. We also have

$$\begin{aligned} \left[\widetilde{\mathbf{c}}^\top - \mathbf{y}^\top \widetilde{\mathbf{A}} \right] \begin{bmatrix} \mathbf{P} & \\ & 1 \end{bmatrix} &= \begin{bmatrix} \mathbf{c}_B{}^\top, \mathbf{c}_N{}^\top, \gamma \end{bmatrix} - \begin{bmatrix} \mathbf{y}_B{}^\top, \mathbf{0}^\top \end{bmatrix} \begin{bmatrix} \mathbf{A}_{BB} & \mathbf{A}_{BN} & \mathbf{a}_B \\ \mathbf{A}_{NB} & \mathbf{A}_{NN} & \mathbf{a}_N \end{bmatrix} \\ &= \begin{bmatrix} \mathbf{c}_B{}^\top - \mathbf{y}_B{}^\top \mathbf{A}_{BB}, \mathbf{c}_N{}^\top - \mathbf{y}_B{}^\top \mathbf{A}_{BN}, \gamma - \mathbf{y}_B{}^\top \mathbf{a}_B \end{bmatrix} \\ &= \begin{bmatrix} \mathbf{0}^\top, \mathbf{c}_N{}^\top - \mathbf{y}_B{}^\top \mathbf{A}_{BN}, \gamma - \mathbf{y}_B{}^\top \mathbf{a}_B \end{bmatrix} \geq \mathbf{0}^\top \, . \end{aligned}$$

Thus $\mathbf{y}$ is feasible for the perturbed standard max linear programming problem. The perturbed max objective is

$$\mathbf{y}^\top \mathbf{b} = \begin{bmatrix} \mathbf{y}_B{}^\top, \mathbf{0}^\top \end{bmatrix} \begin{bmatrix} \mathbf{b}_B \\ \mathbf{b}_N \end{bmatrix} = \mathbf{y}_B{}^\top \mathbf{b}_B = \mathbf{c}_B{}^\top \mathbf{A}_{BB}^{-1} \mathbf{b}_B \, .$$

Since the proposed feasible vectors for the perturbed standard min and max linear programming problems have the same objective values, they are optimal.

Exercise 4.2.11 The following exercise has been adapted from Winston [188, p. 309]. A furniture manufacturer would like to maximize his income from producing a mixture of desks, tables and chairs. His linear programming problem involves the following coefficient arrays:

$$\mathbf{A} = \begin{bmatrix} 8 \overset{\text{feet lumber}}{\text{per desk}} & 4 \overset{\text{finish hours}}{\text{per desk}} & 2 \overset{\text{carpentry hours}}{\text{per desk}} \\ 6 \overset{\text{feet lumber}}{\text{per table}} & 2 \overset{\text{finish hours}}{\text{per table}} & 3/2 \overset{\text{carpentry hours}}{\text{per table}} \\ 1 \overset{\text{feet lumber}}{\text{per chair}} & 3/2 \overset{\text{finish hours}}{\text{per chair}} & 1/2 \overset{\text{carpentry hours}}{\text{per chair}} \end{bmatrix},$$

$$\mathbf{b} = \begin{bmatrix} 600 \overset{\text{dollars}}{\text{per desk}} \\ 300 \overset{\text{dollars}}{\text{per table}} \\ 200 \overset{\text{dollars}}{\text{per chair}} \end{bmatrix} \text{ and}$$

$$\mathbf{c}^\top = \left[48 \text{ feet lumber, } 20 \text{ finish hours, } 8 \text{ carpentry hours}\right] .$$

1. Determine for the manufacturer the array

$$\mathbf{y}^\top = \left[\eta_1 \text{ desks }, \eta_2 \text{ tables }, \eta_3 \text{ chairs}\right]$$

   of scheduled manufacturing, in order to maximize the his income $\mathbf{y}^\top\mathbf{b}$ subject to the constraints $\mathbf{y}^\top\mathbf{A} \leq \mathbf{c}^\top$ that manufacturing does not exceed the available supplies, and that $\mathbf{y}^\top \geq \mathbf{0}^\top$ i.e. manufacturing is nonnegative.
2. In the dual linear programming problem, an investor might wish to buy the furniture manufacturer's supplies as inexpensively as possible. Determine for the investor the array

$$\mathbf{x} = \begin{bmatrix} \xi_1 \overset{\text{dollars}}{\text{per feet lumber}} \\ \xi_2 \overset{\text{dollars}}{\text{per finish hour}} \\ \xi_3 \overset{\text{dollars}}{\text{per carpentry hour}} \end{bmatrix}$$

   in order to minimize the purchase price $\mathbf{c}^\top\mathbf{x}$ subject to the constraints $\mathbf{A}\mathbf{x} \geq \mathbf{b}$ that he pays enough to cover the selling price the manufacturer could receive for each product, and that $\mathbf{x} \geq \mathbf{0}$ i.e. the price the investor pays for each individual supply item is nonnegative.
3. Verify that the maximum value of the manufacturer's income is equal to the minimum purchase price for the investor.

4. If we reorder basic variables first, we get

$$\mathbf{Q}^\top \mathbf{A} \mathbf{P} = \left[\begin{array}{cc|c} 2 \ \overset{\text{carpentry hours}}{\text{per desk}} & 4 \ \overset{\text{finish hours}}{\text{per desk}} & 8 \ \overset{\text{feet lumber}}{\text{per desk}} \\ 1/2 \ \overset{\text{carpentry hours}}{\text{per chair}} & 3/2 \ \overset{\text{finish hours}}{\text{per chair}} & 1 \ \overset{\text{feet lumber}}{\text{per chair}} \\ \hline 3/2 \ \overset{\text{carpentry hours}}{\text{per table}} & 2 \ \overset{\text{finish hours}}{\text{per table}} & 6 \ \overset{\text{feet lumber}}{\text{per table}} \end{array} \right] = \begin{bmatrix} \mathbf{A}_{BB} & \mathbf{A}_{BN} \\ \mathbf{A}_{NB} & \mathbf{A}_{NN} \end{bmatrix}$$

$$\mathbf{Q}^\top \mathbf{b} = \left[\begin{array}{c} 600 \ \overset{\text{dollars}}{\text{per desk}} \\ 200 \ \overset{\text{dollars}}{\text{per chair}} \\ \hline 300 \ \overset{\text{dollars}}{\text{per table}} \end{array} \right] = \begin{bmatrix} \mathbf{b}_B \\ \mathbf{b}_N \end{bmatrix} \text{ and}$$

$$\mathbf{c}^\top \mathbf{P} = \left[8 \text{ carpentry hours}, \ 20 \text{ finish hours}, \middle| \, 48 \text{ feet lumber} \right] = \left[\mathbf{c}_B{}^\top, \ \mathbf{c}_N{}^\top \right] .$$

Find the total amount of lumber required for current production, and find a lower bound on the lumber supply that will allow the current production schedule to remain unchanged.
5. Find upper and lower bounds on the supply of finish hours so that it will remain optimal for the manufacturer to produce only tables and chairs.
6. How much would the selling price of a table have to increase in order for it to become advantageous for the manufacturer to produce tables?
7. For what range of selling prices for desks would the manufacturer continue to make just desks and chairs, assuming that the selling prices of tables and chairs remain the same?
8. Suppose that the manufacturer finds a way to make a table with just onc foot of lumber per table. Would he begin to produce tables?
9. Suppose that the manufacturer could find a way to make a table using only one finish hour. How would this affect his optimal production schedule?
10. Suppose that the manufacturer could find a way to make a desk using only 3 finish hours. How would this affect his optimal production schedule?

4.2.12 Dual Simplex Method

In Sect. 4.2.4, we described a computational method for solving the standard form of the linear programming problem. Our goal in this section is to describe similar techniques for solving the standard min or max forms of the linear programming problem.

4.2.12.1 Standard Min Problem

Let $\mathbf{A}$ be a real $m \times n$ matrix, $\mathbf{b}$ be a real m-vector and $\mathbf{c}$ be a real n-vector. Suppose that we want to solve

$$\min \mathbf{c}^\top \mathbf{x} \text{ subject to } \mathbf{A}\mathbf{x} \geq \mathbf{b} \text{ and } \mathbf{x} \geq \mathbf{0} .$$

We will begin by finding a nonnegative n-vector $\mathbf{x}$ so that

$$\mathbf{A}\mathbf{x} \geq \mathbf{b} .$$

There is a permutation matrix $\mathbf{Q_b}$ so that

$$\mathbf{Q_b}^\top \mathbf{b} = \begin{bmatrix} \mathbf{b}_P \\ \mathbf{b}_M \end{bmatrix} \text{ where } \mathbf{b}_P > \mathbf{0} \text{ and } \mathbf{b}_M \leq \mathbf{0} \text{ and}$$

$$\mathbf{Q_b}^\top \mathbf{A} = \begin{bmatrix} \mathbf{A}_P \\ \mathbf{A}_M \end{bmatrix} .$$

Consider the **auxiliary standard min problem**

$$\min \begin{bmatrix} \mathbf{e}^\top & \mathbf{0}^\top \end{bmatrix} \begin{bmatrix} \mathbf{a}_P \\ \mathbf{x} \end{bmatrix}$$

$$\text{s.t.} \begin{bmatrix} \mathbf{I} & \mathbf{A}_P \\ \mathbf{0} & \mathbf{A}_M \end{bmatrix} \begin{bmatrix} \mathbf{a}_P \\ \mathbf{x} \end{bmatrix} \geq \begin{bmatrix} \mathbf{b}_P \\ \mathbf{b}_M \end{bmatrix} , \begin{bmatrix} \mathbf{a}_P \\ \mathbf{x} \end{bmatrix} \geq \mathbf{0} .$$

Note that $\mathbf{a}_P = \mathbf{b}_P$ and $\mathbf{x} = \mathbf{0}$ is basic feasible for this standard min problem. We can use the dual simplex algorithm, which will be described next, to solve this auxiliary problem. If the optimal solution of the auxiliary problem has $\mathbf{a}_P = \mathbf{0}$, then the corresponding optimal $\mathbf{x}$ is feasible for the original standard min problem. If no optimal solution has $\mathbf{a}_P = \mathbf{0}$, then the original standard min problem is infeasible.

Next, let us describe the **dual simplex method** for the standard min problem. Assume that we have found a permutation matrix $\mathbf{P}$ so that

$$\mathbf{x} = \mathbf{P} \begin{bmatrix} \mathbf{x}_B \\ \mathbf{0} \end{bmatrix}$$

is basic feasible for the standard min form of the linear programming problem. Also assume that we have found a permutation matrix $\mathbf{Q}$, partitioned

$$\mathbf{Q}^\top \mathbf{A} \mathbf{P} = \begin{bmatrix} \mathbf{A}_{BB} & \mathbf{A}_{BN} \\ \mathbf{A}_{NB} & \mathbf{A}_{NN} \end{bmatrix} , \ \mathbf{Q}^\top \mathbf{b} = \begin{bmatrix} \mathbf{b}_B \\ \mathbf{b}_N \end{bmatrix} \text{ and } \mathbf{c}^\top \mathbf{P} = \begin{bmatrix} \mathbf{c}_B{}^\top & \mathbf{c}_N{}^\top \end{bmatrix} ,$$

solved

$$\mathbf{A}_{BB}\mathbf{x}_B = \mathbf{b}_B$$

to find $\mathbf{x}_B \geq \mathbf{0}$, and computed

$$\mathbf{s}_N = \mathbf{A}_{NB}\mathbf{x}_B - \mathbf{b}_N$$

to find that $\mathbf{s}_N \geq \mathbf{0}$.

To motivate the dual simplex method for the standard min problem, we will examine connections with the tableau for the problem in standard form. We can put the standard min problem in standard form by introducing slack variables:

$$\min \begin{bmatrix} \mathbf{c}_B{}^\top & \mathbf{c}_N{}^\top & \mathbf{0}^\top & \mathbf{0}^\top \end{bmatrix} \begin{bmatrix} \mathbf{x}_B \\ \mathbf{x}_N \\ \mathbf{s}_B \\ \mathbf{s}_N \end{bmatrix}$$

$$\text{subject to} \begin{bmatrix} \mathbf{A}_{BB} & \mathbf{A}_{BN} & -\mathbf{I} & \mathbf{0} \\ \mathbf{A}_{NB} & \mathbf{A}_{NN} & \mathbf{0} & -\mathbf{I} \end{bmatrix} \begin{bmatrix} \mathbf{x}_B \\ \mathbf{x}_N \\ \mathbf{s}_B \\ \mathbf{s}_N \end{bmatrix} = \begin{bmatrix} \mathbf{b}_B \\ \mathbf{b}_N \end{bmatrix} \text{ and } \begin{bmatrix} \mathbf{x}_B \\ \mathbf{x}_N \\ \mathbf{s}_B \\ \mathbf{s}_N \end{bmatrix} \geq \mathbf{0} \,.$$

We can multiply the set of basic constraints by $\mathbf{A}_{BB}^{-1}$ to get

$$\mathbf{x}_B + \mathbf{A}_{BB}^{-1}\mathbf{A}_{BN}\mathbf{x}_N - \mathbf{A}_{BB}^{-1}\mathbf{s}_B = \mathbf{A}_{BB}^{-1}\mathbf{b}_B \,.$$

Then we can solve for $\mathbf{x}_B$ and substitute the value into the nonbasic constraints to get

$$\left(\mathbf{A}_{NN} - \mathbf{A}_{NB}\mathbf{A}_{BB}^{-1}\mathbf{A}_{BN}\right)\mathbf{x}_N + \mathbf{A}_{NB}\mathbf{A}_{BB}^{-1}\mathbf{s}_B - \mathbf{s}_N = \mathbf{b}_N - \mathbf{A}_{NB}\mathbf{A}_{BB}^{-1}\mathbf{b}_B \,.$$

Similarly, the objective can be rewritten as

$$\mathbf{c}_B{}^\top\mathbf{x}_B + \mathbf{c}_N{}^\top\mathbf{x}_N = \left(\mathbf{c}_N{}^\top - \mathbf{c}_B{}^\top\mathbf{A}_{BB}^{-1}\mathbf{A}_{BN}\right)\mathbf{x}_N + \mathbf{c}_B{}^\top\mathbf{A}_{BB}^{-1}\mathbf{s}_B + \mathbf{c}_B{}^\top\mathbf{A}_{BB}^{-1}\mathbf{b}_B \,.$$

This leads to the simplex tableau in Table 4.5.

Table 4.5 Tableau for standard min problem

$\mathbf{x}_B{}^\top$	$\mathbf{s}_N{}^\top$	$\mathbf{s}_B{}^\top$	$\mathbf{x}_N{}^\top$	
$\mathbf{I}$	$\mathbf{0}$	$-\mathbf{A}_{BB}^{-1}$	$\mathbf{A}_{BB}^{-1}\mathbf{A}_{BN}$	$\mathbf{A}_{BB}^{-1}\mathbf{b}_B = \mathbf{x}_B$
$\mathbf{0}$	$\mathbf{I}$	$-\mathbf{A}_{NB}\mathbf{A}_{BB}^{-1}$	$\mathbf{A}_{NB}\mathbf{A}_{BB}^{-1}\mathbf{A}_{BN} - \mathbf{A}_{NN}$	$\mathbf{A}_{NB}\mathbf{x}_B - \mathbf{b}_N = \mathbf{s}_N$
$\mathbf{0}$	$\mathbf{0}$	$\mathbf{c}_B{}^\top\mathbf{A}_{BB}^{-1} = \mathbf{y}_B{}^\top$	$\mathbf{c}_N{}^\top - \mathbf{y}_B{}^\top\mathbf{A}_{BN} = \mathbf{r}_N{}^\top$	$-\mathbf{c}_B{}^\top\mathbf{A}_{BB}^{-1}\mathbf{b}_B$

These computations suggest that we solve

$$\mathbf{A}_{BB}{}^\top \mathbf{y}_B = \mathbf{c}_B$$

for $\mathbf{y}_B$, and compute

$$\mathbf{r}_N{}^\top = \mathbf{c}_N{}^\top - \mathbf{y}_B{}^\top \mathbf{A}_{BN} \, .$$

If $\mathbf{y}_B \geq \mathbf{0}$ and $\mathbf{r}_N \geq \mathbf{0}$, then the Weak Complementary Slackness Theorem 4.2.9 shows that $\mathbf{x}$ is optimal for the standard min problem and

$$\mathbf{y}^\top = \left[\mathbf{y}_B{}^\top \; \mathbf{0}\right] \mathbf{Q}^\top$$

is optimal for the dual standard max problem. Otherwise, we search for the index of the most negative entry of the vector $\left[\mathbf{y}_B{}^\top \; \mathbf{r}_N{}^\top\right]$.

Suppose that the most negative is component i_B of $\mathbf{y}_B$. Then we solve

$$\mathbf{A}_{BB}\mathbf{g}_B = -\mathbf{e}_{i_B}$$

and find component index j_B so that

$$\frac{\mathbf{e}_{j_B}{}^\top \mathbf{x}_B}{\mathbf{e}_{j_B}{}^\top \mathbf{g}_B} = \varepsilon_B \equiv \min \left\{ \frac{\mathbf{e}_j{}^\top \mathbf{x}_B}{\mathbf{e}_j{}^\top \mathbf{g}_B} : \mathbf{e}_j{}^\top \mathbf{g}_B > 0 \right\} \, .$$

We also compute

$$\mathbf{g}_N = \mathbf{A}_{NB}\mathbf{g}_B$$

and find component index i_N so that

$$\frac{\mathbf{e}_{i_N}{}^\top \mathbf{s}_N}{\mathbf{e}_{i_N}{}^\top \mathbf{g}_N} = \varepsilon_N \equiv \min \left\{ \frac{\mathbf{e}_i{}^\top \mathbf{s}_N}{\mathbf{e}_i{}^\top \mathbf{g}_N} : \mathbf{e}_i{}^\top \mathbf{g}_N > 0 \right\} \, .$$

If $\mathbf{g}_B \leq \mathbf{0}$ and $\mathbf{g}_N \leq \mathbf{0}$, then the standard min problem for $\mathbf{x}$ is unbounded. If $\varepsilon_B \leq \varepsilon_N$, then we exchange non-basic entry i_B of $\mathbf{s}_B$ for basic entry j_B of $\mathbf{x}_B$. In this case,

- column j_B and row i_B are removed from $\mathbf{A}_{BB}$,
- entry j_B of $\mathbf{x}_B$ becomes nonbasic (i.e., zero), and
- entry i_B of $\mathbf{s}_B$ becomes basic (the corresponding constraint becomes inactive).

Otherwise, we exchange nonbasic entry i_B of $\mathbf{s}_B$ for basic entry i_N of $\mathbf{s}_N$. In this case,

- row i_N of $\mathbf{A}_{NB}$ is exchanged with row i_B of $\mathbf{A}_{BB}$,
- entry i_N of $\mathbf{s}_N$ becomes nonbasic (the corresponding constraint becomes active), and
- entry i_B of $\mathbf{s}_B$ becomes basic (the corresponding constraint becomes inactive).

Otherwise, $\mathbf{r}_N$ has the most negative entry j_N. We solve

$$\mathbf{A}_{BB}\mathbf{h}_B = \mathbf{A}_{BN}\mathbf{e}_{j_N}$$

and find component index j_B so that

$$\frac{{\mathbf{e}_{j_B}}^\top \mathbf{x}_B}{{\mathbf{e}_{j_B}}^\top \mathbf{h}_B} = \varepsilon_B \equiv \min\left\{\frac{{\mathbf{e}_j}^\top \mathbf{x}_B}{{\mathbf{e}_j}^\top \mathbf{h}_B} : {\mathbf{e}_j}^\top \mathbf{h}_B > 0\right\} .$$

We also compute

$$\mathbf{h}_N = \mathbf{A}_{NB}\mathbf{h}_B - \mathbf{A}_{NN}\mathbf{e}_{j_N}$$

and find component index i_N so that

$$\frac{{\mathbf{e}_{i_N}}^\top \mathbf{s}_N}{{\mathbf{e}_{i_N}}^\top \mathbf{h}_N} = \varepsilon_N \equiv \min\left\{\frac{{\mathbf{e}_i}^\top \mathbf{s}_N}{{\mathbf{e}_i}^\top \mathbf{h}_N} : {\mathbf{e}_i}^\top \mathbf{h}_N > 0\right\} .$$

If $\mathbf{h}_B \leq \mathbf{0}$ and $\mathbf{h}_N \leq \mathbf{0}$, then the standard min problem for $\mathbf{x}$ is unbounded. If $\varepsilon_B \leq \varepsilon_N$, then we exchange non-basic entry j_N of $\mathbf{x}_N$ for basic entry j_B of $\mathbf{x}_B$. In this case,

- column j_N of $\mathbf{A}_{BN}$ (corresponding to a nonbasic entry of $\mathbf{x}_N$) is exchanged with column j_B of $\mathbf{A}_{BB}$ (which corresponds to a basic entry of $\mathbf{x}_B$),
- entry j_B of $\mathbf{x}_B$ becomes nonbasic (i.e., zero), and
- entry j_N of $\mathbf{x}_N$ becomes basic (and possibly nonzero).

Otherwise, we exchange nonbasic entry j_N of $\mathbf{x}_N$ for basic entry i_N of $\mathbf{s}_N$. In this case,

- row i_N of $\mathbf{A}_{NB}$ (a currently inactive constraint) and column j_N of $\mathbf{A}_{BN}$ (a currently inactive variable) and entry (i_N, j_N) of $\mathbf{A}_{NN}$ move into $\mathbf{A}_{BB}$,
- entry i_N of $\mathbf{s}_N$ becomes nonbasic (i.e., zero: the corresponding constraint becomes active), and
- entry j_N of $\mathbf{x}_N$ becomes basic (and possibly nonzero).

4.2.12.2 Standard Max Problem

The dual simplex method for the standard max linear programming problem is similar to the dual simplex method for the min problem. Again, let $\mathbf{A}$ be a real $m \times n$ matrix, $\mathbf{b}$ be a real m-vector and $\mathbf{c}$ be a real n-vector. Suppose that we want to solve

$$\max \mathbf{y}^\top \mathbf{b} \text{ subject to } \mathbf{y}^\top \mathbf{A} \leq \mathbf{c}^\top \text{ and } \mathbf{y} \geq \mathbf{0} .$$

We will begin by finding a nonnegative m-vector $\mathbf{y}$ so that

$$\mathbf{y}^\top \mathbf{A} \leq \mathbf{c}^\top .$$

First, we will find a permutation matrix $\mathbf{P}$ so that

$$\mathbf{c}^\top \mathbf{P} = \begin{bmatrix} \mathbf{c}_M{}^\top & \mathbf{c}_P{}^\top \end{bmatrix} \text{ where } \mathbf{c}_M < \mathbf{0} \text{ and } \mathbf{c}_P \geq \mathbf{0}$$
$$\mathbf{A}\mathbf{P} = \begin{bmatrix} \mathbf{A}_M & \mathbf{A}_P \end{bmatrix} .$$

Then we will consider the **auxiliary standard max problem**

$$\max \begin{bmatrix} \mathbf{a}_M{}^\top & \mathbf{y}^\top \end{bmatrix} \begin{bmatrix} -\mathbf{e} \\ \mathbf{0} \end{bmatrix}$$
$$\text{s.t.} \begin{bmatrix} \mathbf{a}_M{}^\top & \mathbf{y}^\top \end{bmatrix} \begin{bmatrix} -\mathbf{I} & \mathbf{0} \\ \mathbf{A}_M & \mathbf{A}_P \end{bmatrix} \leq \begin{bmatrix} \mathbf{c}_M{}^\top & \mathbf{c}_P{}^\top \end{bmatrix} , \begin{bmatrix} \mathbf{a}_M \\ \mathbf{y} \end{bmatrix} \geq \mathbf{0} .$$

Note that $\mathbf{a}_M = -\mathbf{c}_M$ and $\mathbf{y} = \mathbf{0}$ is basic feasible for this standard max problem. We can use the dual simplex algorithm (described next) to solve this auxiliary problem. If the optimal solution has $\mathbf{a}_M = \mathbf{0}$, then the corresponding optimal $\mathbf{y}$ is feasible for the original standard max problem. If no optimal solution has $\mathbf{a}_M = \mathbf{0}$, then the original standard max problem is infeasible.

Next, we will describe the **dual simplex method** for the standard max problem. Assume that we have found permutation matrices $\mathbf{Q}$ and $\mathbf{P}$ so that

$$\mathbf{y}^\top = \begin{bmatrix} \mathbf{y}_B{}^\top \\ \mathbf{y}_N{}^\top \end{bmatrix} \mathbf{Q}$$

is basic feasible for this linear programming problem. This means that we have partitioned

$$\mathbf{Q}^\top \mathbf{A}\mathbf{P} = \begin{bmatrix} \mathbf{A}_{BB} & \mathbf{A}_{BN} \\ \mathbf{A}_{NB} & \mathbf{A}_{NN} \end{bmatrix} , \ \mathbf{Q}^\top \mathbf{b} = \begin{bmatrix} \mathbf{b} \\ \mathbf{b}_N \end{bmatrix} \text{ and } \mathbf{c}^\top \mathbf{P} = \begin{bmatrix} \mathbf{c}_B{}^\top & \mathbf{c}_N{}^\top \end{bmatrix} ,$$

solved

$$\mathbf{y}_B{}^\top \mathbf{A}_{BB} = \mathbf{c}_B{}^\top$$

to find $\mathbf{y}_B \geq \mathbf{0}$, and computed

$$\mathbf{r}_N{}^\top = \mathbf{c}_N{}^\top - \mathbf{c}_B{}^\top \mathbf{A}_{BB}^{-1} \mathbf{A}_{BN}$$

to find that $\mathbf{r}_N \geq \mathbf{0}$.

To motivate the dual simplex method for the standard max problem, we will examine connections with the simplex tableau. We can put the standard max problem in standard form by introducing slack variables:

$$\min \begin{bmatrix} \mathbf{b}_B{}^\top & \mathbf{b}_N{}^\top & \mathbf{0}^\top & \mathbf{0}^\top \end{bmatrix} \begin{bmatrix} \mathbf{y}_B \\ \mathbf{y}_N \\ \mathbf{r}_B \\ \mathbf{r}_N \end{bmatrix}$$

$$\text{subject to } \begin{bmatrix} \mathbf{A}_{BB}{}^\top & \mathbf{A}_{NB}{}^\top & \mathbf{I} & \mathbf{0} \\ \mathbf{A}_{BN}{}^\top & \mathbf{A}_{NN}{}^\top & \mathbf{0} & \mathbf{I} \end{bmatrix} \begin{bmatrix} \mathbf{y}_B \\ \mathbf{y}_N \\ \mathbf{r}_B \\ \mathbf{r}_N \end{bmatrix} = \begin{bmatrix} \mathbf{c}_B \\ \mathbf{c}_N \end{bmatrix} \text{ and } \begin{bmatrix} \mathbf{y}_B \\ \mathbf{y}_N \\ \mathbf{r}_B \\ \mathbf{r}_N \end{bmatrix} \geq \mathbf{0} \, .$$

We can multiply the set of basic constraints by $\mathbf{A}_{BB}{}^{-\top}$ to get

$$\mathbf{y}_B + \mathbf{A}_{BB}{}^{-\top}\mathbf{A}_{NB}{}^\top\mathbf{y}_N - \mathbf{A}_{BB}{}^{-\top}\mathbf{r}_B = \mathbf{A}_{BB}{}^{-\top}\mathbf{c}_B \, .$$

Then we can solve for $\mathbf{y}_B$ and substitute the value into the nonbasic constraints to get

$$\left(\mathbf{A}_{NN}{}^\top - \mathbf{A}_{BN}{}^\top\mathbf{A}_{BB}{}^{-\top}\mathbf{A}_{NB}{}^\top\right)\mathbf{y}_N - \mathbf{A}_{BN}{}^\top\mathbf{A}_{BB}{}^{-\top}\mathbf{r}_B + \mathbf{r}_N = \mathbf{c}_N - \mathbf{A}_{BN}{}^\top\mathbf{A}_{BB}{}^{-\top}\mathbf{c}_B \, .$$

Similarly, the objective can be rewritten as

$$\mathbf{b}_B{}^\top\mathbf{y}_B + \mathbf{b}_N{}^\top\mathbf{y}_N = \left(\mathbf{b}_N{}^\top - \mathbf{b}_B{}^\top\mathbf{A}_{BB}{}^{-\top}\mathbf{A}_{NB}{}^\top\right)\mathbf{y}_N - \mathbf{b}_B{}^\top\mathbf{A}_{BB}{}^{-\top}\mathbf{r}_B + \mathbf{b}_B{}^\top\mathbf{A}_{BB}{}^{-\top}\mathbf{c}_B \, .$$

This leads to the simplex tableau in Table 4.6.

These computations suggest that we solve

$$\mathbf{A}_{BB}\mathbf{x}_B = \mathbf{b}_B$$

for $\mathbf{x}_B$, and compute

$$\mathbf{s}_N = \mathbf{A}_{NB}\mathbf{x}_B - \mathbf{b}_N \, .$$

Table 4.6 Tableau for standard max problem

$\mathbf{y}_B{}^\top$	$\mathbf{r}_N{}^\top$	$\mathbf{r}_B{}^\top$	$\mathbf{y}_N{}^\top$	
$\mathbf{I}$	$\mathbf{0}$	$\mathbf{A}_{BB}{}^{-\top}$	$\mathbf{A}_{BB}{}^{-\top}\mathbf{A}_{NB}{}^\top$	$\mathbf{A}_{BB}{}^{-\top}\mathbf{c}_B = \mathbf{y}_B$
$\mathbf{0}$	$\mathbf{I}$	$-\mathbf{A}_{BN}{}^\top\mathbf{A}_{BB}{}^{-\top}$	$\mathbf{A}_{NN}{}^\top - \mathbf{A}_{BN}{}^\top\mathbf{A}_{BB}{}^{-\top}\mathbf{A}_{NB}{}^\top$	$\mathbf{c}_N - \mathbf{A}_{BN}{}^\top\mathbf{A}_{BB}{}^{-\top}\mathbf{c}_B = \mathbf{r}_N$
$\mathbf{0}$	$\mathbf{0}$	$\mathbf{b}_B{}^\top\mathbf{A}_{BB}{}^{-\top} = \mathbf{x}_B{}^\top$	$\mathbf{b}_B{}^\top\mathbf{A}_{BB}{}^{-\top}\mathbf{A}_{NB}{}^\top - \mathbf{b}_N{}^\top = \mathbf{s}_N{}^\top$	$\mathbf{b}_B{}^\top\mathbf{A}_{BB}{}^{-\top}\mathbf{c}_B$

If $\mathbf{x}_B \geq \mathbf{0}$ and $\mathbf{s}_N \geq \mathbf{0}$, then the Weak Complementary Slackness Theorem 4.2.9 shows that $\mathbf{y}$ is optimal for the standard max problem and

$$\mathbf{x} = \mathbf{P}\begin{bmatrix}\mathbf{x}_B{}^\top & \mathbf{0}\end{bmatrix}$$

is optimal for the dual standard min problem. Otherwise, we search for the index of the most negative entry of the vector $\begin{bmatrix}\mathbf{x}_B{}^\top & \mathbf{s}_N{}^\top\end{bmatrix}$.

Suppose that the most negative is component j_B of $\mathbf{x}_B$. Then we solve

$$\mathbf{g}_B{}^\top \mathbf{A}_{BB} = \mathbf{e}_{j_B}{}^\top$$

and find component index i_B so that

$$\frac{\mathbf{y}_B{}^\top \mathbf{e}_{i_B}}{\mathbf{g}_B{}^\top \mathbf{e}_{i_B}} = \varepsilon_B \equiv \min\left\{\frac{\mathbf{y}_B{}^\top \mathbf{e}_i}{\mathbf{g}_B{}^\top \mathbf{e}_i} : \mathbf{b}_B{}^\top \mathbf{e}_i > 0\right\} .$$

We also compute

$$\mathbf{g}_N{}^\top = -\mathbf{g}_B{}^\top \mathbf{A}_{BN}$$

and find component index j_N so that

$$\frac{\mathbf{r}_N{}^\top \mathbf{e}_{j_N}}{\mathbf{g}_N{}^\top \mathbf{e}_{j_N}} = \varepsilon_N \equiv \min\left\{\frac{\mathbf{r}_N{}^\top \mathbf{e}_j}{\mathbf{g}_N{}^\top \mathbf{e}_j} : \mathbf{g}_N{}^\top \mathbf{e}_j > 0\right\} .$$

If $\mathbf{g}_B \leq \mathbf{0}$ and $\mathbf{g}_N \leq \mathbf{0}$, then the standard max problem for $\mathbf{y}$ is unbounded. If $\varepsilon_B \leq \varepsilon_N$, then we exchange non-basic entry j_B of $\mathbf{r}_B$ for basic entry i_B of $\mathbf{y}_B$. In this case,

- row i_B and column j_B are removed from $\mathbf{A}_{BB}$,
- entry i_B of $\mathbf{y}_B$ becomes nonbasic (i.ê., zero), and
- entry j_B of $\mathbf{r}_B$ becomes basic (and possibly nonzero: the corresponding constraint becomes inactive).

Otherwise, we exchange nonbasic entry j_B of $\mathbf{r}_B$ for basic entry j_N of $\mathbf{r}_N$. In this case,

- column j_N of $\mathbf{A}_{BN}$ (a currently inactive constraint) is exchanged with column j_B of $\mathbf{A}_{BB}$ (a currently active constraint).
- entry j_N of $\mathbf{r}_N$ becomes nonbasic (i.e., zero: the corresponding constraint becomes active), and
- entry j_B of $\mathbf{r}_B$ becomes basic (and possibly nonzero: the corresponding constraint becomes inactive).

Otherwise, $\mathbf{s}_N$ has the most negative entry i_N. We solve

$$\mathbf{h}_B{}^\top \mathbf{A}_{BB} = \mathbf{e}_{i_N}{}^\top \mathbf{A}_{NB}$$

for $\mathbf{h}_B{}^\top$ and find component index i_B so that

$$\frac{\mathbf{y}_B{}^\top \mathbf{e}_{i_B}}{\mathbf{h}_B{}^\top \mathbf{e}_{i_B}} = \varepsilon_B \equiv \min \left\{ \frac{\mathbf{y}_B{}^\top \mathbf{e}_i}{\mathbf{h}_B{}^\top \mathbf{e}_i} : \mathbf{h}_B{}^\top \mathbf{e}_i > 0 \right\} .$$

We also compute

$$\mathbf{h}_N{}^\top = \mathbf{e}_{i_N}{}^\top \mathbf{A}_{NN} - \mathbf{h}_B{}^\top \mathbf{A}_{BN}$$

and find component index j_N so that

$$\frac{\mathbf{r}_N{}^\top \mathbf{e}_{j_N}}{\mathbf{h}_N{}^\top \mathbf{e}_{j_N}} = \varepsilon_N \equiv \min \left\{ \frac{\mathbf{r}_N{}^\top \mathbf{e}_j}{\mathbf{h}_N{}^\top \mathbf{e}_j} : \mathbf{h}_N{}^\top \mathbf{e}_j > 0 \right\} .$$

If $\mathbf{h}_B \leq \mathbf{0}$ and $\mathbf{h}_N \leq \mathbf{0}$, then the standard max problem for $\mathbf{x}$ is unbounded. If $\varepsilon_B \leq \varepsilon_N$, then we exchange non-basic entry i_N of $\mathbf{y}_N$ for basic entry i_B of $\mathbf{y}_B$. In this case,

- row i_N of $\mathbf{A}_{NB}$ is exchanged with row i_B of $\mathbf{A}_{BB}$,
- entry i_B of $\mathbf{y}_B$ becomes nonbasic (i.e., zero), and
- entry i_N of $\mathbf{y}_N$ becomes basic (and possibly nonzero).

Otherwise, we exchange nonbasic entry i_N of $\mathbf{y}_N$ for basic entry j_N of $\mathbf{r}_N$. In this case,

- row i_N of $\mathbf{A}_{NB}$, column j_N of $\mathbf{A}_{BN}$ and entry (i_N, j_N) of $\mathbf{A}_{NN}$ move into $\mathbf{A}_{BB}$,
- entry i_N of $\mathbf{y}_N$ becomes basic (and possibly nonzero) and
- entry j_N of $\mathbf{r}_N$ becomes basic (i.e., zero).

Software for solving linear programming problems by the simplex method is available at Netlib TOMS 551, Netlib TOMS 552, Netlib AMPL and GLPK (GNU Linear Programming Kit). Within the GLPK, routine `glp_simplex` performs the dual simplex method. The GNU Octave project provides an interpreted programming language to solve linear programming problems, using the glpk function. GNU Octave is publicly available, and uses a programming syntax similar to MATLAB. MATLAB routine linprog can be used to solve linear programming problems via the simplex method.

4.2.13 Karmarkar's Algorithm

The simplex method operates by moving from one vertex to another along the boundary of the feasible region. An alternative approach, developed by Karmarkar [104], determines a sequence of points in the interior of the feasible region that approach the optimal solution of a linear programming problem. For very large problems, such interior point methods can converge much faster than the simplex method.

Let us describe this new **interior point method**. Suppose that we are given a real $m \times n$ matrix $\mathbf{A}$ with rank $(\mathbf{A}) = m$, a real m-vector $\mathbf{b}$ and a real n-vector $\mathbf{c}$. We want to solve the linear programming problem in standard form

$$\min \mathbf{c}^\top \mathbf{x} \text{ subject to } \mathbf{A}\mathbf{x} = \mathbf{b} \text{ and } \mathbf{x} \geq \mathbf{0} . \tag{4.18}$$

Also suppose that we are given $\widetilde{\mathbf{x}}$ so that

$$\mathbf{A}\widetilde{\mathbf{x}} = \mathbf{b} \text{ and } \widetilde{\mathbf{x}} > \mathbf{0} .$$

In other words, $\widetilde{\mathbf{x}}$ is a feasible interior point. We want to find an increment $\widetilde{\mathbf{z}}$ and step length $\lambda > 0$ so that

$$\mathbf{c}^\top \widetilde{\mathbf{z}} > 0 \ , \ \mathbf{A}\widetilde{\mathbf{z}} = \mathbf{0} \text{ and } \widetilde{\mathbf{z}}\lambda < \widetilde{\mathbf{x}} .$$

Then $\widetilde{\mathbf{x}} - \widetilde{\mathbf{z}}\lambda$ will be a feasible interior point with lower objective value for the linear programming problem (4.18).

Let $\mathbf{A}^\dagger$ be the pseudo-inverse of $\mathbf{A}$, which was defined in Theorem 6.4.1 in Chap. 6 of Volume I. Then Lemma 6.4.3 in Chap. 6 of Volume I shows that the orthogonal projector onto the nullspace $\mathscr{N}(\mathbf{A})$ is

$$\mathscr{P}_{\mathscr{N}(\mathbf{A})} = \mathbf{I} - \mathbf{A}^\dagger \mathbf{A} .$$

Consequently, the fundamental theorem of linear algebra 3.2.3 in Chap. 3 of Volume I implies that

$$\mathbf{c}^\top \mathscr{P}_{\mathscr{N}(\mathbf{A})}\mathbf{c} = \left\{ \mathscr{P}_{\mathscr{R}(\mathbf{A}^\top)}\mathbf{c} + \mathscr{P}_{\mathscr{N}(\mathbf{A})}\mathbf{c} \right\}^\top \mathscr{P}_{\mathscr{N}(\mathbf{A})}\mathbf{c} = \left\| \mathscr{P}_{\mathscr{N}(\mathbf{A})}\mathbf{c} \right\|_2^2 \geq 0 .$$

This means that whenever $\mathscr{P}_{\mathscr{N}(\mathbf{A})}\mathbf{c} \neq \mathbf{0}$, we can reduce the objective in the linear program (4.18) by choosing the direction of the increment to the solution to be

$$\widetilde{\mathbf{z}} = \mathscr{P}_{\mathscr{N}(\mathbf{A})}\mathbf{c} .$$

The step length λ is chosen so that $\widetilde{\mathbf{x}} - \widetilde{\mathbf{z}}\lambda > \mathbf{0}$. This suggests that we choose

$$\lambda = \alpha \min \left\{ \frac{\widetilde{\mathbf{x}}_j}{\widetilde{\mathbf{z}}_j} : \widetilde{\mathbf{z}}_j > 0 \right\} .$$

Here $\alpha \in (0, 1)$ is a safety factor; typically we take $\alpha = 0.96$.

In order to improve the performance of the algorithm in subsequent steps, Karmarkar performs a rescaling of the problem. If $\widetilde{\mathbf{x}}$ is a feasible interior point for the linear programming problem (4.18), define the diagonal matrix

$$\mathbf{D} = \operatorname{diag}(n\widetilde{\mathbf{x}}_1, \ldots, n\widetilde{\mathbf{x}}_n) .$$

Then

$$\mathbf{D}^{-1}\widetilde{\mathbf{x}} = \mathbf{e}\frac{1}{n} ,$$

so $\mathbf{e}^\top \mathbf{D}^{-1}\widetilde{\mathbf{x}} = 1$. In other words, the rescaling has moved us to the center of the simplex $\{\mathbf{x}' : \mathbf{e}^\top \mathbf{x}' = 1 \text{ and } \mathbf{x}' \geq \mathbf{0}\}$. In general, we will define

$$\mathbf{c}' = \mathbf{D}\mathbf{c} \text{ and}$$
$$\mathbf{A}' = \mathbf{A}\mathbf{D} .$$

We also perform the change of variables

$$\mathbf{x}' = \mathbf{D}^{-1}\mathbf{x} ,$$

and apply Karmarkar's algorithm to the rescaled linear programming problem

$$\min {\mathbf{c}'}^\top \mathbf{x}' \text{ subject to } \mathbf{A}'\mathbf{x}' = \mathbf{b} \text{ and } \mathbf{x}' \geq \mathbf{0} . \tag{4.19}$$

Since $\widetilde{\mathbf{x}}$ is a feasible interior point for (4.18), $\widetilde{\mathbf{x}}' = \mathbf{e}/n$ is a feasible interior point for the rescaled problem (4.19). If we increment $\widetilde{\mathbf{x}}'$ by the step $-\mathscr{P}_{\mathscr{N}(\mathbf{A}')}\mathbf{c}'\lambda'$, then the new approximate solution to the original problem is $\widetilde{\mathbf{x}} - \mathbf{D}\mathscr{P}_{\mathscr{N}(\mathbf{A}')}\mathbf{c}'\lambda'$. This gives us a new feasible interior point for the original problem (4.18), which we can use to repeat the rescaling step.

To compute the orthogonal projector onto $\mathscr{N}(\mathbf{A})$, we could factor

$$\mathbf{A} = \mathbf{L}\mathbf{Q}$$

where $\mathbf{L}$ is left-triangular and

$$\mathbf{Q}\mathbf{Q}^\top = \mathbf{I} .$$

Let

$$\mathbf{Q} = \left[\mathbf{q}_1, \ldots, \mathbf{q}_n\right] .$$

Then

$$\mathscr{P}_{\mathscr{N}(\mathbf{A})}\mathbf{c} = \mathbf{I} - \mathbf{A}^\top \left(\mathbf{A}\mathbf{A}^\top\right)^{-1} \mathbf{A} = \mathbf{I} - \mathbf{Q}^\top \mathbf{Q} = \left(\mathbf{I} - \mathbf{q}_n{\mathbf{q}_n}^\top\right) \ldots \left(\mathbf{I} - \mathbf{q}_1{\mathbf{q}_1}^\top\right) \mathbf{c} .$$

This means that we can compute $\mathscr{P}_{\mathscr{N}(\mathbf{A})}\mathbf{c}$ by successive orthogonal projection.

Our discussion in this section can be summarized by the following

Algorithm 4.2.3 (Karmarkar's)

given $m \times n$ matrix $\mathbf{A}$ with rank $(\mathbf{A}) = m$
given m vector $\mathbf{b}$ and n vector $\mathbf{c}$
given n vector $\mathbf{x} > \mathbf{0}$ so that $\mathbf{Ax} = \mathbf{b}$
given $\alpha \in (0, 1)$
factor $\mathbf{A} = \mathbf{LQ}$ where $\mathbf{L}$ is left triangular and $\mathbf{QQ}^\top = \mathbf{I}$
/* LAPACK routine dgelqf */
$\mathbf{z} = (\mathbf{I} - \mathbf{QQ}^\top)\mathbf{c}$
/* Algorithm 3.4.9 in Chap. 3 of Volume I */
$\lambda = \alpha \min\{\mathbf{x}_j/\mathbf{z}_j : \mathbf{z}_j > 0\}$
$\mathbf{x} = \mathbf{x} - \mathbf{z}\lambda$
until (converged)
  $\mathbf{D} = \operatorname{diag}(\mathbf{x})n$
  factor $\mathbf{AD} = \mathbf{LQ}$ where $\mathbf{L}$ is left triangular and $\mathbf{QQ}^\top = \mathbf{I}$
/* LAPACK routine dgelqf */
  $\mathbf{z} = (\mathbf{I} - \mathbf{QQ}^\top)\mathbf{Dc}$
/* Algorithm 3.4.9 in Chap. 3 of Volume I */
  $\lambda = \alpha \min\{1/(n\mathbf{z}_j) : \mathbf{z}_j > 0\}$
  $\mathbf{x} = \mathbf{D}(\mathbf{e}/n - \mathbf{z}\lambda)$

Example 4.2.5 Suppose that

$$\mathbf{A} = \begin{bmatrix} 1 & 1 & 1 & 0 \\ 1 & -1 & 0 & 1 \end{bmatrix}, \ \mathbf{b} = \begin{bmatrix} 4 \\ 6 \end{bmatrix} \text{ and}$$
$$\mathbf{c}^\top = \begin{bmatrix} 2 & -3 & 0 & 0 \end{bmatrix} .$$

For future reference, we remark that the optimal solution is

$$\mathbf{x} = \begin{bmatrix} 0 \\ 4 \\ 0 \\ 10 \end{bmatrix} .$$

However, our initial guess is

$$\widetilde{\mathbf{x}} = \begin{bmatrix} 1 \\ 1 \\ 2 \\ 6 \end{bmatrix} .$$

To begin Karmarkar's method, we factor

$$\mathbf{A} = \mathbf{L}\mathbf{Q} = \begin{bmatrix} \sqrt{3} & \\ 0 & \sqrt{3} \end{bmatrix} \begin{bmatrix} 1/\sqrt{3} & 1/\sqrt{3} & 1/\sqrt{3} & 0 \\ 1/\sqrt{3} & -1/\sqrt{3} & 0 & 1/\sqrt{3} \end{bmatrix} .$$

Then

$$\widetilde{\mathbf{z}} = \mathscr{P}_{\mathscr{N}(\mathbf{A})}\mathbf{c} = \left(\mathbf{I} - \mathbf{Q}^\top\mathbf{Q}\right)\mathbf{c} = \begin{bmatrix} 2/3 \\ -1 \\ 1/3 \\ -5/3 \end{bmatrix} ,$$

from which we compute

$$\lambda = 0.96 \ \min\left\{\frac{\widetilde{\mathbf{x}}_j}{\widetilde{\mathbf{z}}_j} : \widetilde{\mathbf{z}}_j > 0\right\} = 0.96 \ \min\left\{\frac{1}{2/3} , \frac{2}{1/3}\right\} = 1.44 .$$

Thus the first step of Karmarkar's algorithm gives us

$$\widetilde{\mathbf{x}} - \widetilde{\mathbf{z}}\lambda = \begin{bmatrix} 1 \\ 1 \\ 2 \\ 6 \end{bmatrix} - \begin{bmatrix} 2/3 \\ -1 \\ 1/3 \\ -5/3 \end{bmatrix} 1.44 = \begin{bmatrix} 0.04 \\ 2.44 \\ 1.52 \\ 8.40 \end{bmatrix} .$$

At this point, it is not obvious which components of $\mathbf{x}$ should be nonbasic.

In the second step, we define the diagonal matrix

$$\mathbf{D} = \begin{bmatrix} 0.04 & & & \\ & 2.44 & & \\ & & 1.52 & \\ & & & 8.40 \end{bmatrix} 4 = \begin{bmatrix} 0.16 & & & \\ & 9.76 & & \\ & & 6.08 & \\ & & & 33.60 \end{bmatrix} .$$

This allows us to rescale

$$\mathbf{c}'^\top = \mathbf{c}^\top\mathbf{D} = \begin{bmatrix} 2 & -3 & 0 & 0 \end{bmatrix} \begin{bmatrix} 0.16 & & & \\ & 9.76 & & \\ & & 6.08 & \\ & & & 33.60 \end{bmatrix} = \begin{bmatrix} 0.32 & -29.28 & 0 & 0 \end{bmatrix}$$

and

$$\mathbf{A}' = \mathbf{A}\mathbf{D} = \begin{bmatrix} 1 & 1 & 1 & 0 \\ 1 & -1 & 0 & 1 \end{bmatrix} \begin{bmatrix} 0.16 & & & \\ & 9.76 & & \\ & & 6.08 & \\ & & & 33.60 \end{bmatrix} = \begin{bmatrix} 0.16 & 9.76 & 6.08 & 0 \\ 0.16 & -9.76 & 0 & 33.6 \end{bmatrix} .$$

Then we factor

$$\mathbf{A}' = \begin{bmatrix} 11.500 & \\ -8.2811 & 33.9951 \end{bmatrix} \begin{bmatrix} 0.0139 & 0.8487 & 0.5287 & 0 \\ 0.0081 & -0.0804 & 0.1288 & 0.9884 \end{bmatrix} .$$

We use this factorization to compute

$$\widetilde{\mathbf{z}'} = \mathscr{P}_{\mathscr{N}(\mathbf{A}')}\mathbf{c}' = \begin{bmatrix} 0.6466 \\ -8.0045 \\ 12.8323 \\ -2.3282 \end{bmatrix} ,$$

and

$$\lambda' = 0.96 \min \left\{ \frac{0.25}{0.6466} , \frac{0.25}{12.8323} \right\} = 0.0187 .$$

Thus the second step of Karmarkar's algorithm gives us the approximate solution

$$\begin{aligned} \widetilde{\mathbf{x}} \leftarrow \mathbf{D}\left(\widetilde{\mathbf{x}'} - \widetilde{\mathbf{z}'}\lambda\right) &= \begin{bmatrix} 0.16 & & & \\ & 9.76 & & \\ & & 6.08 & \\ & & & 33.60 \end{bmatrix} \left(\begin{bmatrix} 0.25 \\ 0.25 \\ 0.25 \\ 0.25 \end{bmatrix} - \begin{bmatrix} 0.6466 \\ -8.0045 \\ 12.8323 \\ -2.3282 \end{bmatrix} 0.0187 \right) \\ &= \begin{bmatrix} 0.0381 \\ 3.9011 \\ 0.0608 \\ 9.8631 \end{bmatrix} . \end{aligned}$$

At this point, it is starting to look like the first and third components of $\mathbf{x}$ should be nonbasic.

Let us perform one more step. We define the diagonal matrix

$$\mathbf{D} = \begin{bmatrix} 0.0381 & & & \\ & 3.9011 & & \\ & & 0.0608 & \\ & & & 9.8631 \end{bmatrix} 4 = \begin{bmatrix} 0.1523 & & & \\ & 15.6045 & & \\ & & 0.2432 & \\ & & & 39.4523 \end{bmatrix} .$$

This allows us to rescale

$$\mathbf{c}'^{\top} = \mathbf{c}^{\top}\mathbf{D} = \begin{bmatrix} 2 & -3 & 0 & 0 \end{bmatrix} \begin{bmatrix} 0.1523 & & & \\ & 15.6045 & & \\ & & 0.2432 & \\ & & & 39.4523 \end{bmatrix} = \begin{bmatrix} 0.3045 & -46.8136 & 0 & 0 \end{bmatrix}$$

and

$$\mathbf{A}' = \mathbf{A}\mathbf{D} = \begin{bmatrix} 1 & 1 & 1 & 0 \\ 1 & -1 & 0 & 1 \end{bmatrix} \begin{bmatrix} 0.1523 & & & \\ & 15.6045 & & \\ & & 0.2432 & \\ & & & 39.4523 \end{bmatrix}$$

$$= \begin{bmatrix} 0.1523 & 15.6045 & 0.2432 & 0 \\ 0.1523 & -15.6045 & 0 & 39.4523 \end{bmatrix} .$$

Then we factor

$$\mathbf{A}' = \begin{bmatrix} 15.6072 & \\ -15.6004 & 39.4542 \end{bmatrix} \begin{bmatrix} 0.0098 & 0.9998 & 0.0156 & 0 \\ 0.0077 & -0.0002 & 0.0062 & 1.0000 \end{bmatrix} .$$

project

$$\widetilde{\mathbf{z}'} = \mathscr{P}_{\mathscr{N}(\mathbf{A}')}\mathbf{c}' = \begin{bmatrix} 0.7610 \\ -0.0188 \\ 0.7292 \\ -0.0104 \end{bmatrix} ,$$

and compute

$$\lambda' = 0.96 \min \left\{ \frac{0.25}{0.7610} , \frac{0.25}{0.7292} \right\} = 0.3154 .$$

Thus the third step of Karmarkar's algorithm gives us the approximate solution

$$\widetilde{\mathbf{x}} \leftarrow \mathbf{D}\left(\widetilde{\mathbf{x}'} - \widetilde{\mathbf{z}'}\lambda\right) = \begin{bmatrix} 0.1523 & & & \\ & 15.6045 & & \\ & & 0.2432 & \\ & & & 39.4523 \end{bmatrix} \left(\begin{bmatrix} 0.25 \\ 0.25 \\ 0.25 \\ 0.25 \end{bmatrix} - \begin{bmatrix} 0.7610 \\ -0.0188 \\ 0.7292 \\ -0.0104 \end{bmatrix} 0.3154 \right)$$

$$= \begin{bmatrix} 0.0015 \\ 3.9936 \\ 0.0049 \\ 9.9921 \end{bmatrix} .$$

At this point, it is pretty obvious that the first and third components of $\mathbf{x}$ should be nonbasic. This means that the true solution should solve

$$\begin{bmatrix} \mathbf{A}\mathbf{e}_2 & \mathbf{A}\mathbf{e}_4 \end{bmatrix} \begin{bmatrix} \mathbf{x}_2 \\ \mathbf{x}_4 \end{bmatrix} = \mathbf{b} ,$$

which in this case is

$$\begin{bmatrix} 1 & 0 \\ -1 & 1 \end{bmatrix} \begin{bmatrix} \mathbf{x}_2 \\ \mathbf{x}_4 \end{bmatrix} = \begin{bmatrix} 4 \\ 6 \end{bmatrix} .$$

The solution of this linear system gives us the optimal solution claimed at the beginning of the example.

The LQ factorization could be updated due to rescaling of the columns of $\mathbf{A}$, in order to avoid work on the order of mn^2 floating point operations to perform a new LQ of $\mathbf{AD}$. Karmarkar [104, p. 388] discusses a different approach in which $\left(\mathbf{A}\mathbf{A}^\top\right)^{-1}$ is updated by the Sherman-Morrison-Woodbury formula (3.46).

Karmarkar [104] shows that a variant of this algorithm converges to the machine representation of the solution of the linear programming problem in $O(n)$ steps. As a result, the linear programming problem can be solved by work that is proportional to a polynomial function of the matrix dimensions.

Next, we would like to discuss how to formulate a linear programming problem in standard form with optimal objective value equal to zero, and a known feasible interior point. These ideas are discussed in Karmarkar [104, p. 385ff].

Lemma 4.2.13 *Let* $\mathbf{A}$ *be a real* $m \times n$ *matrix with* $m \leq n$, $\mathbf{b}$ *be a real* m*-vector, and* $\mathbf{c}$ *be a real* n*-vector. Let* $\widetilde{\mathbf{y}} > \mathbf{0}$ *and* $\widetilde{\mathbf{s}} > \mathbf{0}$ *be* m*-vectors; also let* $\widetilde{\mathbf{x}} > \mathbf{0}$ *and* $\widetilde{\mathbf{r}} > \mathbf{0}$ *be* n*-vectors. Then* $\left[\widetilde{\mathbf{x}}^\top, \widetilde{\mathbf{s}}^\top, \widetilde{\mathbf{y}}^\top, \widetilde{\mathbf{r}}^\top, 1\right]^\top$ *is a feasible interior point for the standard form linear programming problem*

$$\min \begin{bmatrix} \mathbf{0}^\top & \mathbf{0}^\top & \mathbf{0}^\top & \mathbf{0}^\top & 1 \end{bmatrix} \begin{bmatrix} \mathbf{x} \\ \mathbf{s} \\ \mathbf{y} \\ \mathbf{r} \\ \lambda \end{bmatrix} \tag{4.20}$$

$$\text{s.t.} \begin{bmatrix} \mathbf{A} & -\mathbf{I} & \mathbf{0} & \mathbf{0} & \mathbf{b} - \mathbf{A}\widetilde{\mathbf{x}} + \widetilde{\mathbf{s}} \\ \mathbf{0} & \mathbf{0} & \mathbf{A}^\top & \mathbf{I} & \mathbf{c} - \mathbf{A}^\top\widetilde{\mathbf{y}} - \widetilde{\mathbf{r}} \\ \mathbf{c}^\top & \mathbf{0} & -\mathbf{b}^\top & \mathbf{0} & \widetilde{\mathbf{y}}^\top\mathbf{b} - \mathbf{c}^\top\widetilde{\mathbf{x}} \end{bmatrix} \begin{bmatrix} \mathbf{x} \\ \mathbf{x} \\ \mathbf{y} \\ \mathbf{r} \\ \lambda \end{bmatrix} = \begin{bmatrix} \mathbf{b} \\ \mathbf{c} \\ 0 \end{bmatrix} \text{ and } \begin{bmatrix} \mathbf{x} \\ \mathbf{x} \\ \mathbf{y} \\ \mathbf{r} \\ \lambda \end{bmatrix} \geq \mathbf{0} . \tag{4.21}$$

Furthermore, if $\mathbf{x}$ *is optimal for the standard min problem*

$$\min \mathbf{c}^\top \mathbf{x} \text{ subject to } \mathbf{A}\mathbf{x} \geq \mathbf{b} \text{ and } \mathbf{x} \geq \mathbf{0} \tag{4.22}$$

and $\mathbf{y}$ *is optimal for the standard max problem*

$$\min \mathbf{y}^\top \mathbf{b} \text{ subject to } \mathbf{y}^\top \mathbf{A} \leq \mathbf{c}^\top \text{ and } \mathbf{y} \geq \mathbf{0} , \tag{4.23}$$

then $\left[\mathbf{x},\, \mathbf{A}\mathbf{x}-\mathbf{b},\, \mathbf{y},\, \mathbf{c}-\mathbf{A}^\top\mathbf{y},\, 0\right]$ *is optimal for the standard form linear programming problem* (4.21). *Conversely, if* $\left[\mathbf{x},\, \mathbf{s},\, \mathbf{y},\, \mathbf{r},\, 0\right]$ *is optimal for* (4.21), *then* $\mathbf{x}$ *is optimal for* (4.22) *and* $\mathbf{y}$ *is optimal for* (4.23).

Proof It is easy to see that $\left[\widetilde{\mathbf{x}}\, \widetilde{\mathbf{s}}\, \widetilde{\mathbf{y}}\, \widetilde{\mathbf{r}}\; 1\right]$ is a feasible point for (4.21).

$$\begin{bmatrix} \mathbf{A} & -\mathbf{I} & \mathbf{0} & \mathbf{0} & \mathbf{b}-\mathbf{A}\widetilde{\mathbf{x}}+\widetilde{\mathbf{s}} \\ \mathbf{0} & \mathbf{0} & \mathbf{A}^\top & \mathbf{I} & \mathbf{c}-\mathbf{A}^\top\widetilde{\mathbf{y}}-\widetilde{\mathbf{r}} \\ \mathbf{c}^\top & \mathbf{0} & -\mathbf{b}^\top & \mathbf{0} & \widetilde{\mathbf{y}}^\top\mathbf{b}-\mathbf{c}^\top\widetilde{\mathbf{x}} \end{bmatrix} \begin{bmatrix} \widetilde{\mathbf{x}} \\ \widetilde{\mathbf{s}} \\ \widetilde{\mathbf{y}} \\ \widetilde{\mathbf{r}} \\ 1 \end{bmatrix} = \begin{bmatrix} \mathbf{A}\widetilde{\mathbf{x}}-\widetilde{\mathbf{s}}+(\mathbf{b}-\mathbf{A}\widetilde{\mathbf{x}}+\widetilde{\mathbf{s}}) \\ \mathbf{A}^\top\widetilde{\mathbf{y}}+\widetilde{\mathbf{r}}+(\mathbf{c}-\mathbf{A}^\top\widetilde{\mathbf{y}}-\widetilde{\mathbf{r}}) \\ \mathbf{c}^\top\widetilde{\mathbf{x}}-\mathbf{b}^\top\widetilde{\mathbf{y}}+(\widetilde{\mathbf{y}\mathbf{b}}-\mathbf{c}^\top\widetilde{\mathbf{x}}) \end{bmatrix} = \begin{bmatrix} \mathbf{b} \\ \mathbf{c} \\ 0 \end{bmatrix} .$$

Since all entries of this vector are positive, it is a feasible interior point for (4.21).

Next, suppose that $\mathbf{x}$ is optimal for the standard min form of the linear programming problem (4.22), and $\mathbf{y}$ is optimal for its dual (4.23). Then the Duality Theorem 4.2.7 implies that $\mathbf{c}^\top\mathbf{x} = \mathbf{y}^\top\mathbf{b}$. Furthermore,

$$\begin{bmatrix} \mathbf{A} & -\mathbf{I} & \mathbf{0} & \mathbf{0} & \mathbf{b}-\mathbf{A}\widetilde{\mathbf{x}}+\widetilde{\mathbf{s}} \\ \mathbf{0} & \mathbf{0} & \mathbf{A}^\top & \mathbf{I} & \mathbf{c}-\mathbf{A}^\top\widetilde{\mathbf{y}}-\widetilde{\mathbf{r}} \\ \mathbf{c}^\top & \mathbf{0} & -\mathbf{b}^\top & \mathbf{0} & \widetilde{\mathbf{y}}^\top\mathbf{b}-\mathbf{c}^\top\widetilde{\mathbf{x}} \end{bmatrix} \begin{bmatrix} \mathbf{x} \\ \mathbf{A}\mathbf{x}-\mathbf{b} \\ \mathbf{y} \\ \mathbf{c}-\mathbf{A}^\top\mathbf{y} \\ 0 \end{bmatrix} = \begin{bmatrix} \mathbf{A}\mathbf{x}-(\mathbf{A}\mathbf{x}-\mathbf{b}) \\ \mathbf{A}^\top\mathbf{y}+(\mathbf{c}-\mathbf{A}^\top\mathbf{y}) \\ \mathbf{c}^\top\mathbf{x}-\mathbf{b}^\top\mathbf{y} \end{bmatrix} = \begin{bmatrix} \mathbf{b} \\ \mathbf{c} \\ 0 \end{bmatrix} .$$

This shows that $\left[\mathbf{x}\,\mathbf{s}\,\mathbf{y}\,\mathbf{r}\;0\right]$ is feasible for (4.21). The constraints for the standard form problem (4.21) imply that any feasible point must have a nonnegative objective value. Since the feasible point $\left[\mathbf{x}\,\mathbf{s}\,\mathbf{y}\,\mathbf{r}\;0\right]$ has zero objective value, it must be optimal. Conversely, suppose that $\left[\mathbf{x}\,\mathbf{s}\,\mathbf{y}\,\mathbf{r}\;0\right]$ is optimal for the standard form problem (4.21). Then the constraints for (4.21) show that

$$\begin{aligned} &\mathbf{A}\mathbf{x}-\mathbf{s}=\mathbf{b} \Longrightarrow \mathbf{A}\mathbf{x} \geq \mathbf{b} , \\ &\mathbf{y}^\top\mathbf{A}+\mathbf{r}^\top = \mathbf{c}^\top \Longrightarrow \mathbf{y}^\top\mathbf{A} \leq \mathbf{c}^\top \text{ and} \\ &\mathbf{c}^\top\mathbf{x}-\mathbf{y}^\top\mathbf{b} = \mathbf{0} . \end{aligned}$$

The first condition implies that $\mathbf{x}$ is feasible for (4.22), the second condition implies that $\mathbf{y}$ is feasible for (4.23), and the third condition together with the Duality Theorem 4.2.7 implies that $\mathbf{x}$ and $\mathbf{y}$ are optimal.

Algorithm 4.2.3 has been superceded by an approach due to Mehrotra [125]. For more detailed information on interior point methods, see the survey article by Potra and Wright [141].

Software for solving linear programming problems by interior point methods is available at Netlib bpmpd.tar.gz. Also, the GLPK (GNU Linear Programming Kit) contains routine `glp_interior` to solve linear programming problems by interior point methods. By default, MATLAB routine linprog uses interior point methods to solve linear programming problems.

Interested readers may also execute a JavaScript program for **Karmarkar's method**. This program solves

$$\min[-3, -2, -1]\mathbf{x} \text{ subject to } \mathbf{e}^\top \mathbf{x} = 1 \text{ and } \mathbf{x} \geq \mathbf{0} .$$

The program chooses a random point in the interior of the feasible region, and then performs Karmarkar's algorithm to solve the problem.

4.3 Kuhn-Tucker Theory

In this section, we will develop the theory needed to characterize solutions of general nonlinear programming problems. This theory will show that the direction of steepest descent of the objective is orthogonal to the plane tangent to the constraints active at the minimizer. Consequently, our first task in Sect. 4.3.1 will be to describe these tangent planes mathematically. Afterward, in Sect. 4.3.2 we will find necessary and sufficient conditions for minimizers of nonlinear objectives subject to equality constraints. We will extend these results to inequality constraints in Sect. 4.3.3. We will close with a discussion of Lagrangians in Sect. 4.3.4 and then a discussion of nonlinear duality in Sect. 4.3.5.

Let us begin our discussion by describing the general form of the problem we wish to solve.

Definition 4.3.1 Suppose that $0 \leq p \leq n$ are integers, and q is a nonnegative integer. Let $\mathscr{D}$ be a closed subset of real n-vectors, and assume that the real functional ϕ is defined and continuous on $\mathscr{D}$. If $p > 0$, let the function $\mathbf{a}$ map $\mathscr{D}$ to real p-vectors. If $q > 0$, let the function $\mathbf{b}$ map $\mathscr{D}$ to real q-vectors. Assume that $\mathscr{D}$ contains the **feasible region**

$$\mathscr{F} = \{\mathbf{x} : \text{ if } p > 0 \text{ then } \mathbf{a}(\mathbf{x}) = \mathbf{0} \text{ , and if } q > 0 \text{ then } \mathbf{b}(\mathbf{x}) \leq \mathbf{0}\} .$$

Then the **nonlinear programming problem** is

$$\min \phi(\mathbf{x}) \text{ subject to } \mathbf{a}(\mathbf{x}) = \mathbf{0} \text{ and } \mathbf{b}(\mathbf{x}) \leq \mathbf{0} .$$

In other words, we seek a minimizer $\mathbf{z} \in \mathscr{F}$ so that for all $\widetilde{\mathbf{x}} \in \mathscr{F}$ we have $\phi(\mathbf{z}) \leq \phi(\widetilde{\mathbf{x}})$.

A point $\widetilde{\mathbf{x}} \in \mathscr{D}$ is **feasible** for the nonlinear programming problem if and only if $\mathbf{x} \in \mathscr{F}$. If $\widetilde{\mathbf{x}}$ is feasible, then the ith inequality constraint is **active** at $\widetilde{\mathbf{x}}$ if and only if there exists an index $1 \leq i \leq q$ so that

$$\mathbf{e}_i \cdot \mathbf{b}(\widetilde{\mathbf{x}}) = 0 .$$

4.3.1 Hypersurfaces and Tangent Planes

Now that we have described our general nonlinear programming problem mathematically, let us examine its constraints in more detail. The set of all points that satisfy some collection of equality constraints is the subject of our first definition.

Definition 4.3.2 Let $\mathscr{A}$ be an open subset of real n-vectors, and let the mapping $\mathbf{a}$ from $\mathscr{A}$ to real p-vectors be continuous. Then the set

$$\mathscr{S}_{\mathbf{a}} \equiv \{\mathbf{x} \in \mathscr{A} : \mathbf{a}(\mathbf{x}) = \mathbf{0}\} \tag{4.24}$$

is called a **hypersurface** in the set of all real n-vectors. The hypersurface $\mathscr{S}_{\mathbf{a}}$ is **smooth** if and only if $\mathbf{a}$ is continuously differentiable.
A hypersurface can be described mathematically as solution of some nonlinear equation as in the previous definition, or can be described geometrically through the ideas in the next definition.

Definition 4.3.3 Let $\mathscr{A}$ be an open subset of real n-vectors, and let the mapping $\mathbf{a}$ from $\mathscr{A}$ to real p-vectors be continuous, with corresponding hypersurface $\mathscr{S}_{\mathbf{a}}$. Assume that $\mathscr{I}$ is an open interval of real numbers, and that $\mathbf{c}$ maps $\mathscr{I}$ into $\mathscr{A}$. Then

$$\mathscr{C}_{\mathbf{c}} \equiv \{\mathbf{c}(\tau) : \tau \in \mathscr{I}\} \tag{4.25}$$

is a **curve on the hypersurface** $\mathscr{S}_{\mathbf{a}}$ if and only if for all $\tau \in \mathscr{I}$ we have

$$\mathbf{a}(\mathbf{c}(\tau)) = \mathbf{0} .$$

The curve $\mathscr{C}_{\mathbf{c}}$ is **differentiable** if and only if $\mathbf{c}(\tau)$ is differentiable on $\mathscr{I}$. If $\mathbf{x}_0 \in \mathscr{S}_{\mathbf{a}}$, then the curve $\mathscr{C}_{\mathbf{c}}$ **passes through $\mathbf{x}_0$ at $\boldsymbol{\tau}_0$** if and only if there exists $\tau_0 \in \mathscr{I}$ such that $\mathbf{c}(\tau_0) = \mathbf{x}_0$.

Eventually, we will characterize solutions of nonlinear programming problems by means of tangents to hypersurfaces. The next definition will make this notion precise, in a way that does not depend on a parametrization of the hypersurface.

Definition 4.3.4 Let $\mathscr{A}$ be an open subset of real n-vectors, and let the mapping $\mathbf{a}$ from $\mathscr{A}$ to real p-vectors be continuous, with corresponding hypersurface $\mathscr{S}_{\mathbf{a}}$. Then the **plane tangent to $\mathscr{S}_{\mathbf{a}}$ at $\mathbf{x}_0$** is

$$\begin{aligned}\mathscr{T}_{\mathscr{S}_{\mathbf{a}}}(\mathbf{x}_0) = \big\{&\mathbf{x}_0 + \mathbf{c}'(\tau_0) \;:\; \mathbf{x}_0 \in \mathscr{S}_{\mathbf{a}} \text{ and} \\ &\mathscr{C}_{\mathbf{c}} \text{ is a curve given by Eq. (4.25) on } \mathscr{S}_{\mathbf{a}} \text{ so that } \mathbf{c}(\tau_0) = \mathbf{x}_0\big\} .\end{aligned} \tag{4.26}$$

Because our nonlinear programming problems will describe hypersurfaces as the zero set of some nonlinear function $\mathbf{a}$, it will be useful to to describe their tangent planes in terms of $\mathbf{a}$. The next definition will assist this description.

Definition 4.3.5 Let $\mathscr{A}$ be an open subset of real n-vectors, and let the mapping $\mathbf{a}$ from $\mathscr{A}$ to real p-vectors be continuous. Assume that $\mathbf{x}_0 \in \mathscr{A}$ satisfies the constraint $\mathbf{a}(\mathbf{x}_0) = \mathbf{0}$. Then $\mathbf{x}_0$ is a **regular point** of this constraint if and only if the Jacobian matrix $\mathbf{J_a}(\mathbf{x}_0)$ satisfies

$$\text{rank}\,(\mathbf{J_a}(\mathbf{x}_0)) = p\ .$$

Of course, the status of a point $\mathbf{x}_0$ as a regular point for a constraint is dependent on the parametrization $\mathbf{a}$ of the constraint.

The reader will probably recall the following important result.

Theorem 4.3.1 (Implicit Function) *Suppose that n and p are two positive integers. Let $\mathscr{D}$ be an open subset of ordered pairs of real n-vectors and real p-vectors, and let $(\mathbf{x}_0, \mathbf{y}_0) \in \mathscr{D}$. Assume that $\mathbf{f}$ maps $\mathscr{D}$ to real p-vectors, that $\mathbf{f}(\mathbf{x}_0, \mathbf{y}_0) = \mathbf{0}$, that $\mathbf{f}$ is continuous at $(\mathbf{x}_0, \mathbf{y}_0)$, that $\partial\mathbf{f}/\partial\mathbf{y}$ exists in $\mathscr{D}$ and that $\partial\mathbf{f}/\partial\mathbf{y}(\mathbf{x}_0, \mathbf{y}_0)$ is nonsingular. Then there exists an open set $\mathscr{D}_{\mathbf{x}_0}$ of real n-vectors and an open set $\mathscr{D}_{\mathbf{y}_0}$ of real p-vectors with $\mathbf{x}_0 \in \mathscr{D}_{\mathbf{x}_0}$, $\mathbf{y}_0 \in \mathscr{D}_{\mathbf{y}_0}$ and $\mathscr{D}_{\mathbf{x}_0} \times \mathscr{D}_{\mathbf{y}_0} \subset \mathscr{D}$, and there exists a continuous function $\mathbf{g}$ mapping $\mathscr{D}_{\mathbf{x}_0}$ to $\mathscr{D}_{\mathbf{y}_0}$ so that for all $\mathbf{x} \in \mathscr{D}_{\mathbf{x}_0}$ we have*

$$\mathbf{f}\,(\mathbf{x}\ ,\ \mathbf{g}(\mathbf{x})) = \mathbf{0}\ .$$

Furthermore, if $\partial\mathbf{f}/\partial\mathbf{x}(\mathbf{x}_0, \mathbf{y}_0)$ exists, then $\mathbf{g}$ is differentiable at $\mathbf{x}_0$ and

$$\frac{\partial\mathbf{g}}{\partial\mathbf{x}}(\mathbf{x}_0) = -\left[\frac{\partial\mathbf{f}}{\partial\mathbf{y}}(\mathbf{x}_0, \mathbf{y}_0)\right]^{-1}\frac{\partial\mathbf{f}}{\partial\mathbf{x}}(\mathbf{x}_0, \mathbf{y}_0)\ .$$

A proof of the implicit function theorem can be found in Dieudonné [55, p. 265] or Ortega and Rheinboldt [133, p. 128].

The implicit function theorem allows us to prove the next lemma, which gives us a useful characterization of a plane tangent to a hypersurface.

Lemma 4.3.1 *Let $\mathscr{A}$ be an open subset of real n-vectors, and let the mapping $\mathbf{a}$ from $\mathscr{A}$ to real p-vectors be continuous and differentiable at $\mathbf{x}_0$. Suppose that $\mathscr{S}_\mathbf{a}$ is the hypersurface of points where $\mathbf{a}$ is zero. Assume that $\mathbf{x}_0$ is a regular point of the constraint $\mathbf{a}(\mathbf{x}_0) = \mathbf{0}$. Then the plane tangent to $\mathscr{S}_\mathbf{a}$ at $\mathbf{x}_0$ is*

$$\mathscr{M}_{\mathscr{S}_\mathbf{a}}(\mathbf{x}_0) \equiv \{\mathbf{x}_0 + \mathbf{t}\ :\ \mathbf{J_a}(\mathbf{x}_0)\mathbf{t} = \mathbf{0}\}\ .$$

Proof First, we will prove that $\mathscr{T}_{\mathscr{S}_\mathbf{a}}(\mathbf{x}_0) \subset \mathscr{M}_{\mathscr{S}_\mathbf{a}}(\mathbf{x}_0)$. Let $\mathscr{C}_\mathbf{c}$ be a differentiable curve on $\mathscr{S}_\mathbf{a}$ that passes through $\mathbf{x}_0$ at τ_0. Then there is an open interval $\mathscr{I}$ such that $\mathbf{c}$ maps $\mathscr{I}$ into $\mathscr{A}$, and $\mathbf{a} \circ \mathbf{c}(\tau) = \mathbf{0}$ for all $\tau \in \mathscr{I}$. Then the chain rule (3.3) implies that

$$\mathbf{0} = \frac{\mathrm{d}\mathbf{a} \circ \mathbf{c}}{\mathrm{d}\tau}(\tau_0) = \mathbf{J_a}(\mathbf{x}_0)\mathbf{c}'(\tau_0)\ .$$

Next, we will show that $\mathscr{M}_{\mathscr{S}_{\mathbf{a}}}(\mathbf{x}_0) \subset \mathscr{T}_{\mathscr{S}_{\mathbf{a}}}(\mathbf{x}_0)$. To do so, we will prove that for every real n-vector $\mathbf{t}$ satisfying $\mathbf{J_a}(\mathbf{x}_0)\mathbf{t} = \mathbf{0}$, there is a curve $\mathscr{C}_{\mathbf{c}}$ on $\mathscr{S}_{\mathbf{a}}$ that passes through $\mathbf{x}$ at $\tau = 0$, with $\mathbf{c}'(0) = \mathbf{t}$. Suppose that

$$\mathbf{J_a}(\mathbf{x}_0)\mathbf{t} = \mathbf{0} \,.$$

Consider the mapping

$$\mathbf{s}(\tau \;,\; \mathbf{u}) \equiv \mathbf{a}\left(\mathbf{x}_0 + \mathbf{t}\tau + \mathbf{J_a}(\mathbf{x}_0)^\top \mathbf{u}\right) \,.$$

Then

$$\mathbf{s}(0, \mathbf{0}) = \mathbf{a}(\mathbf{x}_0) = \mathbf{0}$$

and

$$\frac{\partial \mathbf{s}}{\partial \tau}(0, \mathbf{0}) = \mathbf{J_a}(\mathbf{x}_0)\mathbf{t} = \mathbf{0} \,.$$

Also, since $\mathbf{x}_0$ is a regular point of the constraint $\mathbf{a}(\mathbf{x}_0) = \mathbf{0}$, we see that

$$\frac{\partial \mathbf{s}}{\partial \mathbf{u}}(0, \mathbf{0}) = \mathbf{J_a}(\mathbf{x}_0)\mathbf{J_a}(\mathbf{x}_0)^\top$$

is nonsingular. The Implicit Function Theorem 4.3.1 implies that there is an open interval $(-\alpha, \alpha)$, an open set $\mathscr{R}_{\mathbf{0}}$ of real n-vectors containing the origin, and a differentiable function $\mathbf{u}$ mapping $(-\alpha, \alpha)$ into $\mathscr{R}_{\mathbf{0}}$ such that for all $\tau \in (-\alpha, \alpha)$ we have

$$\mathbf{0} = \mathbf{s}(\tau, \mathbf{u}(\tau)) = \mathbf{a}\left(\mathbf{x}_0 + \mathbf{t}\tau + \mathbf{J_a}(\mathbf{x}_0)^\top \mathbf{u}(\tau)\right) \,,$$

and that

$$\frac{\mathrm{d}\mathbf{u}}{\mathrm{d}\tau}(0) = -\left[\frac{\partial \mathbf{s}}{\partial \mathbf{u}}(0, \mathbf{0})\right]^{-1} \frac{\partial \mathbf{s}}{\partial \tau}(0, \mathbf{0}) = \mathbf{0} \,.$$

The definition of $\mathbf{u}(\tau)$ shows that

$$\mathbf{c}(\tau) = \mathbf{x}_0 + \mathbf{t}\tau + \mathbf{J_a}(\mathbf{x}_0)^\top \mathbf{u}(\tau)$$

defines a curve on $\mathscr{S}_{\mathbf{a}}$ that passes through $\mathbf{x}_0$ at $\tau_0 = 0$. To complete the proof, we note that

$$\mathbf{x}'(0) = \mathbf{t} + \mathbf{J_a}(\mathbf{x}_0)^\top \frac{\mathrm{d}\mathbf{u}}{\mathrm{d}\tau}(0) = \mathbf{t} \,.$$

4.3.2 Equality Constraints

Our next goal is to establish first- and second-order conditions that characterize the minimum of a nonlinear real functional subject to a system of equality constraints. Our work here will generalize our previous results regarding unconstrained minimization in Sect. 3.2.4.

First, we will show that at a constrained local minimum, the direction of steepest descent of the objective must be orthogonal to the plane that is tangent to the constraint hypersurface.

Lemma 4.3.2 *Suppose that $p < n$ are positive integers. Let $\mathscr{A}$ be an open subset of real n-vectors, and assume that $\mathbf{a}$ is a continuously differentiable mapping from $\mathscr{A}$ to real p-vectors. Let ϕ be a real functional that is continuously differentiable on the open set $\mathscr{D} \subset \mathscr{A}$. Suppose that $\mathbf{z} \in \mathscr{D}$ is a regular point of the constraint $\mathbf{a}(\mathbf{z}) = \mathbf{0}$. Finally, assume that $\mathbf{z}$ is a local minimum of ϕ restricted to the hypersurface $\mathscr{S}_{\mathbf{a}}$ defined by Eq. (4.24). Then there is a real p-vector $\boldsymbol{\ell}$ so that*

$$\mathbf{g}_\phi(\mathbf{z})^\top + \boldsymbol{\ell}^\top \mathbf{J}_\mathbf{a}(\mathbf{z}) = \mathbf{0}^\top . \tag{4.27}$$

Proof Suppose that the n-vector $\mathbf{t}$ satisfies $\mathbf{J}_\mathbf{a}(\mathbf{z})\mathbf{t} = \mathbf{0}$. Then the proof of Lemma 4.3.1 shows that there is an interval $(-\alpha, \alpha)$, there is an open subset $\mathscr{R}_\mathbf{c} \subset \mathscr{A}$, and there is a differentiable function $\mathbf{c}$ mapping $(-\alpha, \alpha)$ into $\mathscr{R}_\mathbf{c}$ such that $\mathbf{c}(0) = \mathbf{z}$, $\mathbf{c}'(0) = \mathbf{t}$ and

$$\mathbf{a}(\mathbf{c}(\tau)) = \mathbf{0}$$

for all $\tau \in (-\alpha, \alpha)$. Since 0 is a local minimum of $\phi \circ \mathbf{c}$, Lemma 5.7.2 in Chap. 5 of Volume I shows that

$$0 = \frac{\mathrm{d}\phi \circ \mathbf{c}}{\mathrm{d}\tau}(0)$$

then the chain rule (3.3) produces

$$= \mathbf{g}_\phi(\mathbf{z}) \cdot \mathbf{c}'(0) = \mathbf{g}_\phi(\mathbf{z}) \cdot \mathbf{t} .$$

Since $\mathbf{t}$ is arbitrary, this equation shows that $\mathbf{g}_\phi(\mathbf{z}) \perp \mathscr{N}(\mathbf{J}_\mathbf{a}(\mathbf{z}))$. Then the Fundamental Theorem of Linear Algebra 3.2.3 in Chap. 3 of Volume I shows that $\mathbf{g}_\phi(\mathbf{z}) \in \mathscr{R}\left(\mathbf{J}_\mathbf{a}(\mathbf{z})^\top\right)$. This is equivalent to the claimed result.

The components of the vector $\boldsymbol{\ell}$ in Eq. (4.27) are called **Lagrange multipliers**. These numbers provide the following useful service. In solving the nonlinear programming problem

$$\text{minimize } \phi(\mathbf{x}) \text{ subject to } \mathbf{a}(\mathbf{x}) = \mathbf{0} ,$$

we have n unknown components of the minimizer $\mathbf{z}$ and p unknown components of the Lagrange multipliers $\boldsymbol{\ell}$. Barring certain degeneracies, these $n+p$ unknowns are determined by the n tangent plane orthogonality equations

$$\mathbf{g}_\phi(\mathbf{z}) + \mathbf{J}_\mathbf{a}(\mathbf{z})^\top \boldsymbol{\ell} = \mathbf{0} ,$$

and the p constraint equations

$$\mathbf{a}(\mathbf{z}) = \mathbf{0} .$$

Example 4.3.1 Consider the quadratic programming problem

$$\text{minimize } \frac{1}{2}\mathbf{x}^\top \mathbf{F}\mathbf{x} - \mathbf{f}^\top \mathbf{x} \text{ subject to } \mathbf{H}\mathbf{x} - \mathbf{a} = \mathbf{0} .$$

Lemma 4.3.2 shows that the minimizer $\mathbf{z}$ and the associated Lagrange multipliers $\boldsymbol{\ell}$ satisfy the system of linear equations

$$\mathbf{F}\mathbf{z} - \mathbf{f} + \mathbf{H}^\top \boldsymbol{\ell} = \mathbf{0} \text{ and } \mathbf{H}\mathbf{z} - \mathbf{a} = \mathbf{0} .$$

These equations are the same as those determined previously by other means in Lemma 6.12.1 in Chap. 6 of Volume I.

Next, we will turn to second-order necessary conditions for a constrained minimum.

Lemma 4.3.3 *Suppose that $p < n$ are positive integers. Let $\mathscr{A}$ be an open subset of real n-vectors, and assume that $\mathbf{a}$ is a twice continuously differentiable mapping from $\mathscr{A}$ to real p-vectors. Let ϕ be a real functional that is twice continuously differentiable on the open set $\mathscr{D} \subset \mathscr{A}$. Suppose that $\mathbf{z} \in \mathscr{D}$ is a regular point of the constraint $\mathbf{a}(\mathbf{x}) = \mathbf{0}$, and that $\mathbf{z}$ is a local minimizer of ϕ restricted to the hypersurface $\mathscr{S}_\mathbf{a}$. Finally, let the real p-vector $\boldsymbol{\ell}$ of Lagrange multipliers satisfy (4.27). Then the matrix $\mathbf{H}_\phi(\mathbf{z}) + \sum_{i=1}^p \mathbf{H}_{\mathbf{a}_i}(\mathbf{z})\ell_i$ is nonnegative on the plane tangent to $\mathscr{S}_\mathbf{a}$ at $\mathbf{z}$.*

Proof Let $\mathbf{c}(\tau)$ be a twice continuously differentiable mapping such that $\mathscr{C}_\mathbf{c} = \{\mathbf{c}(\tau) : \tau \in (-\alpha, \alpha)\}$ is a curve in $\mathscr{S}_\mathbf{a}$ that passes through $\mathbf{z}$ at $\tau = 0$. Since $\mathbf{a} \circ \mathbf{c}(\tau) = \mathbf{0}$ for all $\tau \in (-\alpha, \alpha)$, the chain rule (3.3) implies that

$$\mathbf{0} = \frac{\mathrm{d}\mathbf{a} \circ \mathbf{c}}{\mathrm{d}\tau}(0) = \mathbf{J}_\mathbf{a}(\mathbf{z})\mathbf{c}'(0) = \mathbf{0} .$$

We can multiply by the Lagrange multipliers $\boldsymbol{\ell}$ to see that $\boldsymbol{\ell} \cdot \mathbf{a} \circ \mathbf{c}(\tau) = 0$ for all $\tau \in (-\alpha, \alpha)$, then we can differentiate twice to get

$$0 = \frac{\mathrm{d}^2 \boldsymbol{\ell} \cdot \mathbf{a} \circ \mathbf{c}}{\mathrm{d}\tau^2}(0) = \boldsymbol{\ell}^\top \mathbf{J}_\mathbf{a}(\mathbf{z}) \frac{\mathrm{d}^2 \mathbf{c}}{\mathrm{d}\tau^2}(0) + \mathbf{c}'(0)^\top \left\{ \sum_{i=1}^p \ell_i \mathbf{H}_{\mathbf{a}_i}(\mathbf{z}) \right\} \mathbf{c}'(0) .$$

Since $\phi \circ \mathbf{x}$ has a local minimum at $\tau_0 = 0$, Lemma 5.7.1 in Chap. 5 of Volume I implies that

$$0 < \frac{\mathrm{d}^2 \phi \circ \mathbf{c}}{\mathrm{d}\tau^2}(0) = \frac{\mathrm{d}^2 \phi \circ \mathbf{c}}{\mathrm{d}\tau^2}(0) + \frac{\mathrm{d}^2 \lambda \cdot \mathbf{a} \circ \mathbf{c}}{\mathrm{d}\tau^2}(0)$$

$$= \left\{ \mathbf{g}_\phi(\mathbf{z}) + \boldsymbol{\ell}^\top \mathbf{J_a}(\mathbf{z}) \right\} \mathbf{c}''(0) + \mathbf{c}'(0)^\top \left\{ \mathbf{H}_\phi(\mathbf{z}) + \sum_{i=1}^p \ell_i \mathbf{H}_{\mathbf{a}_i}(\mathbf{z}) \right\} \mathbf{c}'(0)$$

$$= \mathbf{c}'(0)^\top \left\{ \mathbf{H}_\phi(\mathbf{z}) + \sum_{i=1}^p \ell_i \mathbf{H}_{\mathbf{a}_i}(\mathbf{z}) \right\} \mathbf{c}'(0) \; .$$

Since $\mathbf{c}'(0)$ is arbitrary, the claimed result follows.

The next lemma provides sufficient conditions for a constrained minimum.

Lemma 4.3.4 *Suppose that $p < n$ are positive integers. Let $\mathscr{A}$ be an open subset of real n-vectors, and assume that $\mathbf{a}$ is a twice continuously differentiable mapping from $\mathscr{A}$ to real p-vectors. Let ϕ be a real functional that is twice continuously differentiable on the open set $\mathscr{D} \subset \mathscr{A}$. Suppose that $\mathbf{z} \in \mathscr{D}$ satisfies $\mathbf{a}(\mathbf{z}) = \mathbf{0}$, and that there is a real p-vector $\boldsymbol{\ell}$ so that*

$$\mathbf{g}_\phi(\mathbf{z})^\top + \boldsymbol{\ell}^\top \mathbf{J_a}(\mathbf{z}) = \mathbf{0}^\top \; .$$

Also suppose that the matrix $\mathbf{H}_\phi(\mathbf{z}) + \sum_{i=1}^p \ell_i \mathbf{H}_{\mathbf{a}_i}(\mathbf{z})$ is positive on the plane tangent to the constraint hypersurface

$$\mathscr{S}_\mathbf{a} = \{ \mathbf{x} \in \mathscr{A} \; : \; \mathbf{a}(\mathbf{x}) = \mathbf{0} \}$$

at $\mathbf{z}$. Then $\mathbf{z}$ is a strict local minimizer for ϕ on $\mathscr{S}_\mathbf{a}$.

Proof We will prove the claim by contradiction. If $\mathbf{z}$ is not a strict local minimizer of ϕ on $\mathscr{S}_\mathbf{a}$, then there is a sequence of points $\{\widetilde{\mathbf{z}}_k\}_{k=0}^\infty \subset \mathscr{S}_\mathbf{a}$ that converges to $\mathbf{z}$ and satisfies $\phi(\widetilde{\mathbf{z}}_k) \leq \phi(\mathbf{z})$ for all $k \geq 0$. Let

$$\mathbf{s}_k = \frac{\widetilde{\mathbf{z}}_k - \mathbf{z}}{\|\widetilde{\mathbf{z}}_k - \mathbf{z}\|_2} \text{ and } \delta_k = \|\widetilde{\mathbf{z}}_k - \mathbf{z}\|_2 \; .$$

Since the sequence $\{\mathbf{s}_k\}_{k=0}^\infty$ is bounded, the Bolzano-Weierstrass theorem guarantees that there is a convergent subsequence converging to some real unit n-vector $\mathbf{s}$. For notational simplicity, we will use the same notation for this subsequence as the original sequence. Lemma 3.2.3 shows that for all $k \geq 0$

$$\mathbf{0} = \frac{\mathbf{a}(\widetilde{\mathbf{z}}_k) - \mathbf{a}(\mathbf{z})}{\|\widetilde{\mathbf{z}}_k - \mathbf{z}\|_2} = \int_0^1 \mathbf{J_a}(\mathbf{z} + \mathbf{s}_k \delta_k \tau) \mathbf{s}_k \, \mathrm{d}\tau$$

In the limit as $k \to \infty$ we get

$$\mathbf{0} = \mathbf{J_a}(\mathbf{z})\mathbf{s} .$$

We can use Lemma 3.2.8 to show that for all $1 \leq i \leq p$ we have

$$0 = \mathbf{a}_i(\widetilde{\mathbf{z}}_k) = \mathbf{a}_i(\mathbf{z}) + \mathbf{g}_{\mathbf{a}_i}(\mathbf{z}) \cdot \mathbf{s}_k\delta_k + \int_0^1 \int_0^\tau (\mathbf{s}_k\delta_k)^\top \mathbf{H}_{\mathbf{a}_i}(\mathbf{z} + \mathbf{s}_k\delta_k\sigma)\mathbf{s}_k\delta_k \, d\sigma \, d\tau .$$

We can multiply by $\boldsymbol{\ell}_i$ and sum over $1 \leq i \leq p$ to get

$$0 = \boldsymbol{\ell}^\top \mathbf{a}(\mathbf{z}) + \boldsymbol{\ell}^\top \mathbf{J_a}(\mathbf{z})\mathbf{s}_k\delta_k + \int_0^1 \int_0^\tau (\mathbf{s}_k\delta_k)^\top \left\{ \sum_{i=1}^p \boldsymbol{\ell}_i \mathbf{H}_{\mathbf{a}_i}(\mathbf{z} + \mathbf{s}_k\delta_k\sigma) \right\} \mathbf{s}_k\delta_k \, d\sigma \, d\tau . \tag{4.28}$$

Lemma 3.2.8 also gives us

$$0 \geq \phi(\widetilde{\mathbf{z}}_k) - \phi(\mathbf{z}) = \mathbf{g}_\phi(\mathbf{z})^\top \mathbf{s}_k\delta_k + \int_0^1 \int_0^\tau (\mathbf{s}_k\delta_k)^\top \mathbf{H}_\phi(\mathbf{z} + \mathbf{s}_k\delta_k\sigma)\mathbf{s}_k\delta_k \, d\sigma \, d\tau \tag{4.29}$$

Adding (4.28) and (4.29) produces

$$0 \geq \boldsymbol{\ell}^\top \mathbf{a}(\mathbf{z}) + \left\{ \mathbf{g}_\phi(\mathbf{z})^\top + \boldsymbol{\ell}^\top \mathbf{J_a}(\mathbf{z}) \right\} \mathbf{s}_k\delta_k$$
$$+ (\mathbf{s}_k\delta_k)^\top \int_0^1 \int_0^\tau \left\{ \mathbf{H}_\phi(\mathbf{z} + \mathbf{s}_k\delta_k\sigma) + \sum_{i=1}^p \boldsymbol{\ell}_i \mathbf{H}_{\mathbf{a}_i}(\mathbf{z} + \mathbf{s}_k\delta_k\sigma) \right\} \, d\sigma \, d\tau \mathbf{s}_k\delta_k$$

then the constraint $\mathbf{a}(\mathbf{z}) = \mathbf{0}$ and the Eq. (4.27) defining the Lagrange multipliers give us

$$= (\mathbf{s}_k\delta_k)^\top \int_0^1 \int_0^\tau \left\{ \mathbf{H}_\phi(\mathbf{z} + \mathbf{s}_k\delta_k\sigma) + \sum_{i=1}^p \boldsymbol{\ell}_i \mathbf{H}_{\mathbf{a}_i}(\mathbf{z} + \mathbf{s}_k\delta_k\sigma) \right\} \, d\sigma \, d\tau \mathbf{s}_k\delta_k$$

We can divide by δ_k^2 and take the limit as $k \to \infty$ to get

$$0 \geq \frac{1}{2}\mathbf{s}^\top \left\{ \mathbf{H}_\phi(\mathbf{z}) + \sum_{i=1}^p \boldsymbol{\ell}_i \mathbf{H}_{\mathbf{a}_i}(\mathbf{z}) \right\} \mathbf{s} .$$

Since $\mathbf{s}$ is a unit vector and the matrix inside the braces is positive, we have contradicted the assumption that $\mathbf{z}$ is not a strict local minimum.

4.3.3 Kuhn-Tucker Theorems

Now we are ready to characterize solutions of the general nonlinear programming problem described in Definition 4.3.1. We will begin by revising the Definition 4.3.5 of a regular point for our new circumstances.

Definition 4.3.6 Let $p < n$ and q be nonnegative integers, and let $\mathscr{A}$ and $\mathscr{B}$ be non-disjoint open subsets of real n-vectors. Suppose that $\mathbf{x}_0 \in \mathscr{A} \cap \mathscr{B}$. Let $\mathbf{a}$ mapping $\mathscr{A}$ into real p-vectors be differentiable at $\mathbf{x}_0$, and $\mathbf{b}$ mapping $\mathscr{B}$ into real q-vectors be differentiable at $\mathbf{x}_0$. Assume that $\mathbf{a}(\mathbf{x}_0) = \mathbf{0}$. Let $\mathbf{P}$ be a $q \times q$ permutation matrix so that

$$\mathbf{P}^\top \mathbf{b}(\mathbf{x}_0) = \begin{bmatrix} \mathbf{b}_B(\mathbf{x}_0) \\ \mathbf{b}_N(\mathbf{x}_0) \end{bmatrix}$$

where $\mathbf{b}_B(\mathbf{x}_0) = \mathbf{0}$ if it has any components. Then $\mathbf{x}_0$ is a **regular point** of the constraints $\mathbf{a}(\mathbf{x}) = \mathbf{0}$ and $\mathbf{b}(\mathbf{x}) \leq \mathbf{0}$ if and only if the matrix $\begin{bmatrix} \mathbf{J_a}(\mathbf{x}_0) \\ \mathbf{J}_{\mathbf{b}_B}(\mathbf{x}_0) \end{bmatrix}$ has rank equal to the number of its rows.

With this definition, we are able to prove the following first-order necessary conditions for the solution of a nonlinear programming problem.

Theorem 4.3.2 (Kuhn-Tucker First-Order Necessary Conditions) *Let $0 \leq p < n$ and q be nonnegative integers, and let $\mathscr{D}, \mathscr{A}$ and $\mathscr{B}$ be non-disjoint open subsets of real n-vectors. Suppose that the mapping $\mathbf{a}$ from $\mathscr{A}$ to real p-vectors is continuously differentiable, that the mapping $\mathbf{b}$ from $\mathscr{B}$ to real q-vectors is continuously differentiable, and that the real functional ϕ is continuously differentiable on $\mathscr{D}$. Assume that $\mathbf{z} \in \mathscr{A} \cap \mathscr{B} \cap \mathscr{D}$ is a regular point of the constraints $\mathbf{a}(\mathbf{z}) = \mathbf{0}$ and $\mathbf{b}(\mathbf{z}) \leq \mathbf{0}$, and assume that $\mathbf{z}$ is a local minimum of ϕ subject to these constraints. Then there is a real p-vector $\boldsymbol{\ell}$ and a real q-vector $\mathbf{m} \geq \mathbf{0}$ so that*

$$\mathbf{g}_\phi(\mathbf{z})^\top + \boldsymbol{\ell}^\top \mathbf{J_a}(\mathbf{z}) + \mathbf{m}^\top \mathbf{J_b}(\mathbf{z}) = \mathbf{0}^\top \text{ and} \tag{4.30a}$$

$$\mathbf{m}^\top \mathbf{b}(\mathbf{z}) = 0 \,. \tag{4.30b}$$

Proof Define the $q \times q$ permutation matrix $\mathbf{P}$ so that

$$\mathbf{P}^\top \mathbf{b}(\mathbf{z}) = \begin{bmatrix} \mathbf{b}_B(\mathbf{z}) \\ \mathbf{b}_N(\mathbf{z}) \end{bmatrix}$$

where $\mathbf{b}_B(\mathbf{z}) = \mathbf{0}$ if the number q_B of its components is positive. Then the first-order necessary conditions (4.27) for the equality constraints $\mathbf{a}(\mathbf{z}) = \mathbf{0}$ and $\mathbf{b}_B(\mathbf{z}) = \mathbf{0}$ imply that there exists a real p-vector $\boldsymbol{\ell}$ and a real q_B-vector $\mathbf{m}_B$ so that

$$\mathbf{g_a}(\mathbf{z})^\top + \left[\boldsymbol{\ell}^\top, \mathbf{m}_B{}^\top\right] \begin{bmatrix} \mathbf{J_a}(\mathbf{z}) \\ \mathbf{J}_{\mathbf{b}_B}(\mathbf{z}) \end{bmatrix} = \mathbf{0}^\top \,.$$

Define the q-vector $\mathbf{m}$ by

$$\mathbf{m} = \mathbf{P} \begin{bmatrix} \mathbf{m}_B \\ \mathbf{0} \end{bmatrix} .$$

We will prove by contradiction that $\mathbf{m} \geq \mathbf{0}$.

Suppose that $\mathbf{m}$ has a negative component. Such a component must correspond to an entry of $\mathbf{m}_B$; in other words, it must correspond to an active inequality constraint. We can group all other active inequality constraints with the equality constraints, and treat the active inequality constraint corresponding to a negative entry of $\mathbf{m}_B$ as the only entry of $\mathbf{b}_B$.

Since $\mathbf{z}$ is a regular point of the constraints $\mathbf{a}(\mathbf{z}) = \mathbf{0}$ and $\mathbf{b}_B(\mathbf{z}) = \mathbf{0}$, there is a real n-vector $\mathbf{t}$ so that

$$\begin{bmatrix} \mathbf{J_a}(\mathbf{z}) \\ \mathbf{J}_{\mathbf{b}_B}(\mathbf{z}) \end{bmatrix} \mathbf{t} = \begin{bmatrix} \mathbf{0} \\ -1 \end{bmatrix} .$$

Let $\mathbf{c}(\tau)$ define a curve on the hypersurface

$$\mathscr{S}_{\mathbf{a}} = \{\mathbf{x} \ : ; \mathbf{a}(\mathbf{x}) = \mathbf{0}\}$$

and assume that $\mathbf{c}$ passes through $\mathbf{z}$ at $\tau_0 = 0$ with $\mathbf{c}'(0) = \mathbf{t}$. Note that

$$\frac{\mathrm{d}\mathbf{g}_{\mathbf{b}_B} \circ \mathbf{c}}{\mathrm{d}\tau}(0) = \mathbf{e}_1 \cdot \mathbf{J}_{\mathbf{b}_B}(\mathbf{z})\mathbf{c}'(0) = \mathbf{e}_1 \cdot \mathbf{J}_{\mathbf{b}_B}(\mathbf{z})\mathbf{t} = -1 ,$$

so for sufficiently small values of $\tau > 0$ we must have that $\mathbf{b}_B(\mathbf{c}(\tau) < \mathbf{0}$. But we also see that

$$\frac{\mathrm{d}\phi \circ \mathbf{c}}{\mathrm{d}\tau}(0) = \mathbf{g}_\phi(\mathbf{z}) \cdot \mathbf{c}'(0) = -\left[\boldsymbol{\ell}^\top, \, \mathbf{m}_B{}^\top\right] \begin{bmatrix} \mathbf{J_a}(\mathbf{z}) \\ \mathbf{J}_{\mathbf{b}_B}(\mathbf{z}) \end{bmatrix} \mathbf{t} = \mathbf{m}_B{}^\top \mathbf{e}_1 < 0 .$$

In summary, for sufficiently small values of $\tau > 0$ we have $\mathbf{a}(\mathbf{c}(\tau)) = \mathbf{0}$, $\mathbf{b}(\mathbf{c}(\tau)) < 0$ and $\phi(\mathbf{c}(\tau)) < \phi(\mathbf{c}(0)) = \phi(\mathbf{z})$. Thus the assumption that $\mathbf{m}$ has a negative component contradicts the hypothesis that $\mathbf{z}$ is a local minimum of ϕ subject to the constraints.

Here is a geometric interpretation of this theorem. First, suppose that $\mathbf{t}$ is a real n-vector in the plane tangent to the hypersurface of active constraints:

$$\begin{bmatrix} \mathbf{J_a}(\mathbf{z}) \\ \mathbf{J}_{\mathbf{b}_B}(\mathbf{z}) \end{bmatrix} \mathbf{t} = \begin{bmatrix} \mathbf{0} \\ \mathbf{0} \end{bmatrix} .$$

Then

$$\mathbf{g}_\phi(\mathbf{z}) \cdot \mathbf{t} = -\left[\boldsymbol{\ell}^\top , \mathbf{m}^\top\right] \begin{bmatrix} \mathbf{J}_\mathbf{a}(\mathbf{z}) \\ \mathbf{J}_{\mathbf{b}_B}(\mathbf{z}) \end{bmatrix} \mathbf{t} = 0 \ .$$

Geometrically, this says that the gradient of the objective is orthogonal to the plane tangent to the hypersurface given by the active constraints.

Next, suppose that $\mathbf{t}$ is a real n-vector that is orthogonal to the gradient of every equality constraint, and orthogonal to all but one of the active inequality constraints, with which it makes a positive inner product (which without loss of generality we will take to be one):

$$\begin{bmatrix} \mathbf{J}_\mathbf{a}(\mathbf{z}) \\ \mathbf{J}_{\mathbf{b}_B}(\mathbf{z}) \end{bmatrix} \mathbf{t} = \begin{bmatrix} \mathbf{0} \\ \mathbf{e}_j \end{bmatrix} \ .$$

Then

$$\mathbf{b}(\mathbf{z} + \mathbf{t}\varepsilon) \approx \mathbf{b}(\mathbf{z}) + \mathbf{J}_\mathbf{b}(\mathbf{z})\mathbf{t}\varepsilon = \begin{bmatrix} \mathbf{0} + \mathbf{J}_{\mathbf{b}_B}(\mathbf{z})\mathbf{t}\varepsilon \\ \mathbf{b}_N(\mathbf{z}) + \mathbf{J}_{\mathbf{b}_N}(\mathbf{z})\mathbf{t}\varepsilon \end{bmatrix} \approx \begin{bmatrix} \mathbf{e}_j\varepsilon \\ \mathbf{b}_N(\mathbf{z}) \end{bmatrix} ,$$

so $\mathbf{t}$ points out of the feasible region. Furthermore,

$$\mathbf{g}_\phi(\mathbf{z}) \cdot \mathbf{t} = -\left[\boldsymbol{\ell}^\top , \mathbf{m}^\top\right] \begin{bmatrix} \mathbf{J}_\mathbf{a}(\mathbf{z}) \\ \mathbf{J}_{\mathbf{b}_B}(\mathbf{z}) \end{bmatrix} \mathbf{t} = -\mathbf{m}^\top \mathbf{e}_j < 0 \ ,$$

so the rate of change of ϕ in the direction of $\mathbf{t}$ is negative. Geometrically speaking, the direction of steepest descent of the objective be orthogonal to the tangent plane of the active constraints and must point out of the feasible region.

Figure 4.3 shows contours of $\phi(\mathbf{x}) = -\mathbf{x}^\top\mathbf{x}$ for feasible $\mathbf{x}$. In the figure on the left, the feasible region consists of all vectors $\mathbf{x}$ so that

$$2\mathbf{x}_1^2 + \mathbf{x}_2^2 \leq 4 \ , \ \mathbf{x}_1^2 + 3\mathbf{x}_2^2 \leq 9 \text{ and } \mathbf{x} \geq \mathbf{0} \ .$$

This figure also shows the normals to the two constraints that are active at the optimal solution $\mathbf{x} = [\sqrt{3/5} \ , \ \sqrt{14/5}]^\top$, and the negative of the gradients of the objective. Note that $-\mathbf{g}_\phi(\mathbf{z})$ must be a nonnegative linear combination of the gradients of the active constraints. In the figure on the right, the feasible region consists of all vectors $\mathbf{x}$ satisfying

$$\mathbf{x}_1^3 + \mathbf{x}_2^3 \leq 1 \text{ and } \mathbf{x} \geq \mathbf{0} \ .$$

This figure also shows $-\mathbf{g}_\phi(\mathbf{z})$ at the optimal solution $\mathbf{z} = [2/\sqrt[3]{16} \ , \ 2/\sqrt[3]{16}]^\top$. This vector is obviously normal to the constraint at $\mathbf{z}$. These figures were generated by the C^{++} programs kuhn_tucker.C and kuhn_tucker2.C.

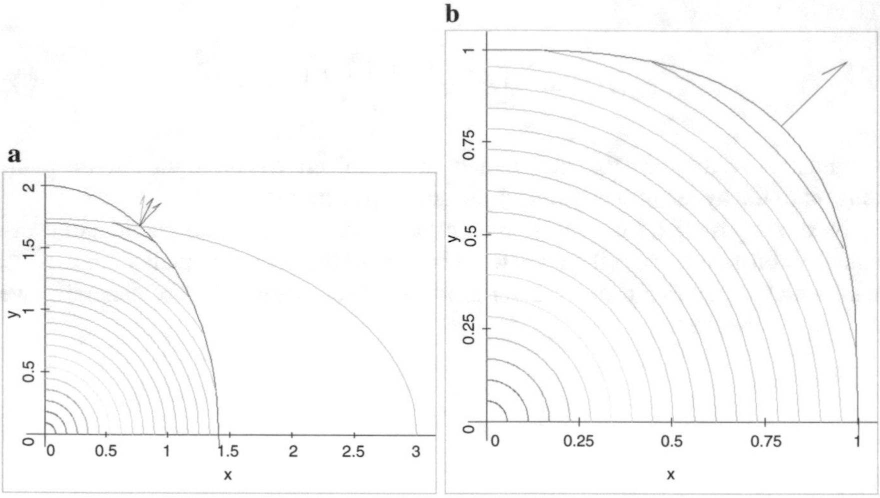

Fig. 4.3 Contours of $-\mathbf{x}_1^2 - \mathbf{x}_2^2$ for $\mathbf{x}$ feasible. (**a**) $\mathbf{x}_1^2 + 3\mathbf{x}_2^2 \leq 9$ and $\mathbf{x} \geq \mathbf{0}$. (**b**) $2\mathbf{x}_1^2+\mathbf{x}_2^2 \leq 4$, $\mathbf{x}_1^3+\mathbf{x}_2^3 \leq 1$ and $\mathbf{x} \geq \mathbf{0}$

Next, we note that the first-order necessary condition (4.30b) is a **complementarity** condition. Since $\mathbf{m} \geq \mathbf{0}$ and $\mathbf{b}(\mathbf{z}) \leq \mathbf{0}$, the condition $\mathbf{m}^\top \mathbf{b}(\mathbf{z}) = 0$ says that for each $1 \leq j \leq q$ either $\mathbf{m}_j = 0$ or $\mathbf{b}_j(\mathbf{z}) = 0$.

Note that there are $n + p + q_B$ unknowns in the nonlinear programming problem of Definition 4.3.1, namely the components of $\mathbf{z}$, $\boldsymbol{\ell}$ and the components of $\mathbf{m}$ corresponding to the active inequality constraints. The first-order necessary conditions (4.30a) give us n equations, the equality constraints $\mathbf{a}(\mathbf{z}) = \mathbf{0}$ provide p more equations, and the active inequality constraints $\mathbf{b}_B(\mathbf{z}) = \mathbf{0}$ provide the remaining q_B equations.

Next, we will establish second-order necessary conditions for a constrained minimum.

Theorem 4.3.3 (Kuhn-Tucker Second-Order Necessary Conditions) *Let* $0 \leq p < n$ *and* q *be nonnegative integers, and let* $\mathscr{D}, \mathscr{A}$ *and* $\mathscr{B}$ *be non-disjoint open subsets of real* n*-vectors. Suppose that the mapping* $\mathbf{a}$ *from* $\mathscr{A}$ *to real* p*-vectors is twice continuously differentiable, that the mapping* $\mathbf{b}$ *from* $\mathscr{B}$ *to real* q*-vectors is twice continuously differentiable, and that the real functional* ϕ *is twice continuously differentiable on* $\mathscr{D}$*. Assume that* $\mathbf{z} \in \mathscr{A} \cap \mathscr{B} \cap \mathscr{D}$ *is a regular point of the constraints* $\mathbf{a}(\mathbf{z}) = \mathbf{0}$ *and* $\mathbf{b}(\mathbf{z}) \leq \mathbf{0}$*, and assume that* $\mathbf{z}$ *is a local minimum of* ϕ *subject to these constraints. Then there is a real* p*-vector* $\boldsymbol{\ell}$ *and a real* q*-vector*

$\mathbf{m} \geq \mathbf{0}$ *so that Eqs.* (4.30) *are satisfied. Furthermore, the matrix*

$$\mathbf{H}_\phi(\mathbf{z}) + \sum_{i=1}^{p} \ell_i \mathbf{H}_{\mathbf{a}_i}(\mathbf{z}) + \sum_{j=1}^{q} \mathbf{m}_j \mathbf{H}_{\mathbf{b}_j}(\mathbf{z})$$

is nonnegative on the plane tangent to the hypersurface of active constraints at $\mathbf{z}$.

Proof The first claim follows immediately from Theorem 4.3.2. Lemma 4.3.3 shows that the matrix

$$\mathbf{H}_\phi(\mathbf{z}) + \sum_{i=1}^{p} \ell_i \mathbf{H}_{\mathbf{a}_i}(\mathbf{z}) + \sum_{\substack{1 \leq j \leq q \\ \mathbf{b}_j(\mathbf{z})=0}} \mathbf{m}_j \mathbf{H}_{\mathbf{b}_j}(\mathbf{z})$$

is nonnegative on the plane tangent to the hypersurface of active constraints at $\mathbf{z}$. Then the complementarity condition (4.30b) shows that this is the same as the matrix in the second claim of this theorem.

Finally, we will establish second-order sufficient conditions for a constrained minimum.

Theorem 4.3.4 (Kuhn-Tucker Second-Order Sufficient Conditions) *Let* $0 \leq p < n$ *and* q *be nonnegative integers, and let* $\mathscr{D}, \mathscr{A}$ *and* $\mathscr{B}$ *be non-disjoint open subsets of real n-vectors. Suppose that the mapping* $\mathbf{a}$ *from* $\mathscr{A}$ *to real p-vectors is twice continuously differentiable, that the mapping* $\mathbf{b}$ *from* $\mathscr{B}$ *to real q-vectors is twice continuously differentiable, and that the real functional* ϕ *is twice continuously differentiable on* $\mathscr{D}$. *Assume that* $\mathbf{z} \in \mathscr{A} \cap \mathscr{B} \cap \mathscr{D}$ *satisfies the constraints* $\mathbf{a}(\mathbf{z}) = \mathbf{0}$ *and* $\mathbf{b}(\mathbf{z}) \leq \mathbf{0}$. *Let there be a real p-vector* $\boldsymbol{\ell}$ *and a real q-vector* $\mathbf{m} \geq \mathbf{0}$ *so that*

$$\mathbf{g}_\phi(\mathbf{z})^\top + \boldsymbol{\ell}^\top \mathbf{J}_\mathbf{a}(\mathbf{z}) + \mathbf{m}^\top \mathbf{J}_\mathbf{b}(\mathbf{z}) = \mathbf{0}^\top \text{ and}$$
$$\mathbf{m}^\top \mathbf{b}(\mathbf{z}) = 0 \; .$$

Finally, assume that the matrix

$$\mathbf{H}_\phi(\mathbf{z}) + \sum_{i=1}^{p} \ell_i \mathbf{H}_{\mathbf{a}_i}(\mathbf{z}) + \sum_{j=1}^{q} \mathbf{m}_j \mathbf{H}_{\mathbf{b}_j}(\mathbf{z})$$

is positive on the subspace

$$\mathscr{M}_+ \equiv \left\{\mathbf{t} : \mathbf{J}_\mathbf{a}(\mathbf{z})\mathbf{t} = \mathbf{0} \text{ and } \mathbf{e}_j{}^\top \mathbf{J}_\mathbf{b}(\mathbf{z})\mathbf{t} = 0 \text{ for all } j \text{ such that } \mathbf{m}_j > 0\right\} \; .$$

Then $\mathbf{z}$ *is a strict local minimum of* $\phi(\mathbf{x})$ *subject to the constraints* $\mathbf{a}(\mathbf{x}) = \mathbf{0}$ *and* $\mathbf{b}(\mathbf{x}) \leq \mathbf{0}$.

Proof We will prove the claim by contradiction. As in the proof of Lemma 4.3.4, if $\mathbf{z}$ is not a strict local minimum of ϕ subject to the given constraints, then there is a sequence of feasible points $\{\widetilde{\mathbf{z}}_k\}_{k=0}^{\infty}$ that converges to $\mathbf{z}$ and satisfies $\phi(\widetilde{\mathbf{z}}_k) \leq \phi(\mathbf{z})$ for all $k \geq 0$. Let

$$\mathbf{s}_k = \frac{\widetilde{\mathbf{z}}_k - \mathbf{z}}{\|\widetilde{\mathbf{z}}_k - \mathbf{z}\|_2} \text{ and } \delta_k = \|\widetilde{\mathbf{z}}_k - \mathbf{z}\|_2 \ .$$

Since the sequence $\{\mathbf{s}_k\}_{k=0}^{\infty}$ is bounded, the Bolzano-Weierstrass theorem guarantees that there is a convergent subsequence converging to some real unit n-vector $\mathbf{s}$. For notational simplicity, we will use the same notation for this subsequence as the original sequence. The proof of Lemma 4.3.4, shows that

$$\mathbf{0} = \mathbf{J_a}(\mathbf{z})\mathbf{s} \ .$$

If $\mathbf{b}_j(\mathbf{z}) = 0$, then

$$0 \geq \frac{\mathbf{b}_j(\widetilde{\mathbf{z}}_k) - \mathbf{b}_j(\mathbf{z})}{\|\widetilde{\mathbf{z}}_k - \mathbf{z}\|_2} = \int_0^1 \mathbf{e}_j{}^\top \mathbf{J_b}(\mathbf{z} + \mathbf{s}_k\delta_k\tau)\mathbf{s}_k \, d\tau \ .$$

In the limit as $k \to \infty$ we get

$$0 \geq \mathbf{e}_j{}^\top \mathbf{J_b}(\mathbf{z})\mathbf{s} \ .$$

The complementarity condition $\mathbf{m} \cdot \mathbf{b}(\mathbf{z}) = 0$ now implies that whenever $\mathbf{m}_j > 0$ we must have $0 \geq \mathbf{e}_j{}^\top \mathbf{J_b}(\mathbf{z})\mathbf{s}$.

As in the proof of Lemma 4.3.4, for all $k \geq 0$ we have

$$0 = \boldsymbol{\ell}^\top \mathbf{a}(\mathbf{z}) + \boldsymbol{\ell}^\top \mathbf{J_a}(\mathbf{z})\mathbf{s}_k\delta_k + \int_0^1 \int_0^\tau (\mathbf{s}_k\delta_k)^\top \left\{ \sum_{i=1}^p \boldsymbol{\ell}_i \mathbf{H}_{\mathbf{a}_i}(\mathbf{z} + \mathbf{s}_k\delta_k\sigma) \right\} \mathbf{s}_k\delta_k \, d\sigma \, d\tau \ . \tag{4.31}$$

Similarly, we can use Lemma 3.2.8 to show that for all $1 \leq j \leq q$

$$0 \geq \mathbf{b}_j(\widetilde{\mathbf{z}}_k) = \mathbf{b}_j(\mathbf{z}) + \mathbf{g}_{\mathbf{b}_j}(\mathbf{z}) \cdot \mathbf{s}_k\delta_k + \int_0^1 \int_0^\tau (\mathbf{s}_k\delta_k)^\top \mathbf{H}_{\mathbf{b}_j}(\mathbf{z} + \mathbf{s}_k\delta_k\sigma)\mathbf{s}_k\delta_k \, d\sigma \, d\tau \ .$$

We can multiply by $\mathbf{m}_j$ and sum over $1 \leq j \leq q$ to get

$$0 \geq \mathbf{m}^\top \mathbf{b}(\mathbf{z}) + \mathbf{m}^\top \mathbf{J_b}(\mathbf{z})\mathbf{s}_k\delta_k + \int_0^1 \int_0^\tau (\mathbf{s}_k\delta_k)^\top \left\{ \sum_{j=1}^q \mathbf{m}_j \mathbf{H}_{\mathbf{b}_j}(\mathbf{z} + \mathbf{s}_k\delta_k\sigma) \right\} \mathbf{s}_k\delta_k \, d\sigma \, d\tau \ . \tag{4.32}$$

Lemma 3.2.8 also gives us

$$0 \geq \phi(\widetilde{\mathbf{z}}_k) - \phi(\mathbf{z}) = \mathbf{g}_\phi(\mathbf{z})^\top \mathbf{s}_k\delta_k + \int_0^1 \int_0^\tau (\mathbf{s}_k\delta_k)^\top \mathbf{H}_\phi(\mathbf{z} + \mathbf{s}_k\delta_k\sigma)\mathbf{s}_k\delta_k \, d\sigma \, d\tau \tag{4.33}$$

Adding (4.31)–(4.33) produces

$$0 \geq \boldsymbol{\ell}^\top \mathbf{a}(\mathbf{z}) + \mathbf{m}^\top \mathbf{b}(\mathbf{z}) + \left\{\mathbf{g}_\phi(\mathbf{z})^\top + \boldsymbol{\ell}^\top \mathbf{J}_\mathbf{a}(\mathbf{z}) + \mathbf{m}^\top \mathbf{J}_\mathbf{b}(\mathbf{z})\right\} \mathbf{s}_k\delta_k$$

$$+ (\mathbf{s}_k\delta_k)^\top \int_0^1 \int_0^\tau \left\{ \mathbf{H}_\phi(\mathbf{z} + \mathbf{s}_k\delta_k\sigma) + \sum_{i=1}^p \boldsymbol{\ell}_i \mathbf{H}_{\mathbf{a}_i}(\mathbf{z} + \mathbf{s}_k\delta_k\sigma) \right.$$

$$\left. + \sum_{j=1}^q \mathbf{m}_j \mathbf{H}_{\mathbf{b}_j}(\mathbf{z} + \mathbf{s}_k\delta_k\sigma) \right\} \, d\sigma \, d\tau \mathbf{s}_k\delta_k$$

then the constraint $\mathbf{a}(\mathbf{z}) = \mathbf{0}$, complementarity condition $\mathbf{m} \cdot \mathbf{b}(\mathbf{z}) = 0$, and Eq. (4.30a) defining the Lagrange multipliers give us

$$= (\mathbf{s}_k\delta_k)^\top \int_0^1 \int_0^\tau \left\{ \mathbf{H}_\phi(\mathbf{z} + \mathbf{s}_k\delta_k\sigma) + \sum_{i=1}^p \boldsymbol{\ell}_i \mathbf{H}_{\mathbf{a}_i}(\mathbf{z} + \mathbf{s}_k\delta_k\sigma) \right.$$

$$\left. + \sum_{j=1}^q \mathbf{m}_j \mathbf{H}_{\mathbf{b}_j}(\mathbf{z} + \mathbf{s}_k\delta_k\sigma) \right\} \, d\sigma \, d\tau \mathbf{s}_k\delta_k$$

We can divide by δ_k^2 and take the limit as $k \to \infty$ to get

$$0 \geq \frac{1}{2}\mathbf{s}^\top \left\{ \mathbf{H}_\phi(\mathbf{z}) + \sum_{i=1}^p \boldsymbol{\ell}_i \mathbf{H}_{\mathbf{a}_i}(\mathbf{z}) + \sum_{j=1}^q \mathbf{m}_j \mathbf{H}_{\mathbf{b}_j}(\mathbf{z}) \right\} \mathbf{s} \, . \tag{4.34}$$

Since $\mathbf{s}$ is a unit vector and the matrix inside the braces is positive, we have contradicted the assumption that $\mathbf{z}$ is not a strict local minimum.

If

$$0 = \mathbf{e}_j{}^\top \mathbf{J}_\mathbf{b}(\mathbf{z})\mathbf{s}$$

for all $1 \leq j \leq q$ such that $\mathbf{m}_j > 0$, then $\mathbf{s} \in \mathscr{M}_+$ and inequality (4.34) contradicts the hypothesis that the matrix in this inequality is positive on $\mathscr{M}_+$. In other words, we have contradicted the assumption that $\mathbf{z}$ is not a strict local minimum of ϕ subject to the constraints.

Otherwise, there exists at least one index j for which $\mathbf{m}_j > 0$ and $0 > \mathbf{e}_j{}^\top \mathbf{J}_\mathbf{b}(\mathbf{z})\mathbf{s}$. But for all $k \geq 0$ we must also have

$$0 \geq \frac{\phi(\widetilde{\mathbf{z}}_k) - \phi(\mathbf{z})}{\|\widetilde{\mathbf{z}}_k - \mathbf{z}\|_2} = \int_0^1 \mathbf{g}_\phi(\mathbf{z} + \mathbf{s}_k \delta_k \tau) \cdot \mathbf{s}_k \, d\tau \, .$$

In the limit as $k \to \infty$ we get

$$0 \geq \mathbf{g}_\phi(\mathbf{z}) \cdot \mathbf{s}$$

then Eq. (4.30a) defining the Lagrange multipliers gives us

$$= -\boldsymbol{\ell}^\top \mathbf{J}_\mathbf{a}(\mathbf{z})\mathbf{s} - \mathbf{m}^\top \mathbf{J}_\mathbf{b}(\mathbf{z})\mathbf{s} = -\mathbf{m}^\top \mathbf{J}_\mathbf{b}(\mathbf{z})\mathbf{s} > 0 \, .$$

This again contradicts the assumption that $\mathbf{z}$ is not a strict local minimum of ϕ subject to the constraints, and completes the proof.

Exercise 4.3.1 Use the Kuhn-Tucker conditions to solve

$$\max \ln(1 + \xi_1) + \xi_2 \text{ subject to } 2\xi_1 + \xi_2 \leq 3 \, , \; \xi_1 \geq 0 \text{ and } \xi_2 \geq 0 \, .$$

Exercise 4.3.2 Use the Kuhn-Tucker conditions to find equations that determine the solution of

$$\max \alpha\xi_1^2 + \beta\xi_1\xi_2 \text{ subject to } \xi_1^2 + \gamma\xi_2^2 \leq 1 \, , \; \xi_1 \geq 0 \text{ and } \xi_2 \geq 0 \, .$$

For which values of α, β and γ does a solution for the unknowns ξ_1 and ξ_2 exist?

4.3.4 Lagrangian

Our next definition will introduce a functional that has an interesting relationship with the nonlinear programming problem in Definition 4.3.1. This functional will be used to describe the dual of this nonlinear programming problem in Sect. 4.3.5, and to develop interesting numerical methods in Sect. 4.5.3.

Definition 4.3.7 Suppose that $p < n$ and q are nonnegative integers. Let $\mathscr{D}$ be a closed subset of real n-vectors, and assume that the real functional ϕ is defined and continuous on $\mathscr{D}$. If $p > 0$, let the function $\mathbf{a}$ map $\mathscr{D}$ to real p-vectors. If $q > 0$, let the function $\mathbf{b}$ map $\mathscr{D}$ to real q-vectors. Assume that $\mathscr{D}$ contains the feasible region

$$\mathscr{F} = \{\mathbf{x} : \text{ if } p > 0 \text{ then } \mathbf{a}(\mathbf{x}) = \mathbf{0} \, , \text{ and if } q > 0 \text{ then } \mathbf{b}(\mathbf{x}) \leq \mathbf{0}\} \, .$$

Then for $\mathbf{x} \in \mathscr{D}$, real p-vectors $\boldsymbol{\ell}$ and real q-vectors $\mathbf{m}$ the **Lagrangian** of the associated nonlinear programming problem is

$$\lambda(\mathbf{x}, \boldsymbol{\ell}, \mathbf{m}) \equiv \phi(\mathbf{x}) + \boldsymbol{\ell} \cdot \mathbf{a}(\mathbf{x}) + \mathbf{m} \cdot \mathbf{b}(\mathbf{x}) . \tag{4.35}$$

The following lemma provides a fundamental relationship between the Lagrangian and the nonlinear programming problem.

Lemma 4.3.5 *Let $0 \leq p < n$ and q be nonnegative integers, and let $\mathscr{D}, \mathscr{A}$ and $\mathscr{B}$ be non-disjoint open subsets of real n-vectors. Suppose that* $\mathbf{a}$ *is a continuous mapping from $\mathscr{A}$ to real p-vectors, that* $\mathbf{b}$ *is a continuous mapping from $\mathscr{B}$ to real q-vectors, and that the real functional ϕ is continuous on $\mathscr{D}$. Denote the feasible set by*

$$\mathscr{F} = \{\mathbf{x} : \textit{ if } p > 0 \textit{ then } \mathbf{a}(\mathbf{x}) = \mathbf{0} \text{ , } \textit{and if } q > 0 \textit{ then } \mathbf{b}(\mathbf{x}) \leq \mathbf{0}\} \ .$$

Then the real n-vector $\mathbf{z}$ *minimizes ϕ over $\mathscr{F}$ with Lagrange multipliers $\boldsymbol{\ell}$ and $\mathbf{m} \geq \mathbf{0}$ if and only if for all $\widetilde{\mathbf{z}} \in \mathscr{F}$, all real p-vectors $\widetilde{\boldsymbol{\ell}}$ and all real q-vectors $\widetilde{\mathbf{m}} \geq \mathbf{0}$ we have*

$$\lambda(\mathbf{z}, \widetilde{\boldsymbol{\ell}}, \widetilde{\mathbf{m}}) \leq \lambda(\mathbf{z}, \boldsymbol{\ell}, \mathbf{m}) \leq \lambda(\widetilde{\mathbf{z}}, \boldsymbol{\ell}, \mathbf{m}) . \tag{4.36}$$

In other words, the minimizer of the general nonlinear programming problem corresponds to a saddle point of the Lagrangian.

Proof Suppose that $\mathbf{z}$ minimizes ϕ over $\mathscr{F}$. Then $\mathbf{a}(\mathbf{z}) = \mathbf{0}$, $\mathbf{b}(\mathbf{z}) \leq \mathbf{0}$, and the Kuhn-Tucker first order necessary conditions in Theorem 4.3.2 show that $\mathbf{m} \cdot \mathbf{b}(\mathbf{z})\ 0$. Since $\widetilde{\mathbf{m}} \geq \mathbf{0}$, it follows that $\widetilde{\mathbf{m}} \cdot \mathbf{b}(\mathbf{z}) \leq 0 = \mathbf{m} \cdot \mathbf{b}(\mathbf{z})$. Since $\widetilde{\boldsymbol{\ell}} \cdot \mathbf{a}(\mathbf{z}) = 0 = \boldsymbol{\ell} \cdot \mathbf{a}(\mathbf{z})$, we can add either of these to the opposite sides of the previous inequality to get

$$\widetilde{\boldsymbol{\ell}} \cdot \mathbf{a}(\mathbf{z}) + \widetilde{\mathbf{m}} \cdot \mathbf{b}(\mathbf{z}) \leq \boldsymbol{\ell} \cdot \mathbf{a}(\mathbf{z}) + \mathbf{m} \cdot \mathbf{b}(\mathbf{z}) .$$

Finally, we can add $\phi(\mathbf{z})$ to both sides of this inequality to get the left-hand inequality in (4.36). Next, since $\mathbf{z}$ and $\widetilde{\mathbf{z}}$ are both feasible and $\mathbf{z}$ is optimal, we must have $\phi(\mathbf{z}) \leq \phi(\widetilde{\mathbf{z}})$. We can add $\boldsymbol{\ell} \cdot \mathbf{a}(\mathbf{z}) + \mathbf{m} \cdot \mathbf{b}(\mathbf{z})$ to both sides of this inequality to obtain the right-hand inequality in (4.36).

Next, suppose that $\mathbf{z}$ is a real n-vector, and there is an p-vector $\boldsymbol{\ell}$ and a q-vector $\mathbf{m} \geq \mathbf{0}$ so that for all feasible vectors $\widetilde{\mathbf{z}}$, all p-vectors $\widetilde{\boldsymbol{\ell}}$ and all nonnegative q-vectors $\widetilde{\mathbf{m}}$ the inequalities (4.36) are satisfied. We can cancel $\phi(\mathbf{z})$ from both sides of the left-hand inequality in (4.36) to get

$$\widetilde{\boldsymbol{\ell}} \cdot \mathbf{a}(\mathbf{z}) + \widetilde{\mathbf{m}} \cdot \mathbf{b}(\mathbf{z}) \leq \boldsymbol{\ell} \cdot \mathbf{a}(\mathbf{z}) + \mathbf{m} \cdot \mathbf{b}(\mathbf{z}) . \tag{4.37}$$

Afterward, we can choose $\widetilde{\mathbf{m}} = \mathbf{m}$ and $\widetilde{\boldsymbol{\ell}} = \boldsymbol{\ell} \pm \mathbf{a}(\mathbf{z})$ to see that

$$\boldsymbol{\ell} \cdot \mathbf{a}(\mathbf{z}) \pm \|\mathbf{a}(\mathbf{z})\|_2^2 + \mathbf{m} \cdot \mathbf{b}(\mathbf{z}) \leq \boldsymbol{\ell} \cdot \mathbf{a}(\mathbf{z}) + \mathbf{m} \cdot \mathbf{b}(\mathbf{z}) .$$

Finally, we can cancel terms to get $\pm\|\mathbf{a}(\mathbf{z})\|_2^2 \leq 0$. We conclude that $\mathbf{a}(\mathbf{z}) = \mathbf{0}$. Inequality (4.37) now implies that

$$\widetilde{\mathbf{m}} \cdot \mathbf{b}(\mathbf{z}) \leq \mathbf{m} \cdot \mathbf{b}(\mathbf{z})$$

for all $\widetilde{\mathbf{m}} \geq \mathbf{0}$. For all $1 \leq j \leq q$ we can choose $\widetilde{\mathbf{m}} = \mathbf{m} + \mathbf{e}_j$ and find that

$$\mathbf{m} \cdot \mathbf{b}(\mathbf{z}) + \mathbf{e}_j \cdot \mathbf{b}(\mathbf{z}) \leq \mathbf{m} \cdot \mathbf{b}(\mathbf{z}) .$$

We can cancel terms to see that $\mathbf{e}_j \cdot \mathbf{b}(\mathbf{z}) \leq 0$. We conclude that $\mathbf{b}(\mathbf{z}) \leq \mathbf{0}$, and that $\mathbf{z} \in \mathscr{F}$. Finally, cancel terms in the right-hand inequality of (4.36) to see that

$$\phi(\mathbf{z}) \leq \phi(\widetilde{\mathbf{z}})$$

for all $\widetilde{\mathbf{z}} \in \mathscr{F}$. This show that $\mathbf{z}$ is the optimal solution to the nonlinear programming problem.

4.3.5 Duality

Sometimes, it is useful to understand duality for nonlinear programming problems. Discussions of this notion can be found in developments of numerical methods for nonlinear programming problems. Although we will not make much use of duality in nonlinear programming, we will provide a definition and a basic result.

Definition 4.3.8 Suppose that $p < n$ and q are nonnegative integers. Let $\mathscr{D}$ be a closed subset of real n-vectors, and assume that the real functional ϕ is defined on $\mathscr{D}$. If $p > 0$, let the function $\mathbf{a}$ map $\mathscr{D}$ to real p-vectors. If $q > 0$, let the function $\mathbf{b}$ map $\mathscr{D}$ to real q-vectors. Assume that $\mathscr{D}$ contains the feasible region

$$\mathscr{F} = \{\mathbf{x} : \text{ if } p > 0 \text{ then } \mathbf{a}(\mathbf{x}) = \mathbf{0} \text{ , and if } q > 0 \text{ then } \mathbf{b}(\mathbf{x}) \leq \mathbf{0}\} .$$

For any real p-vector $\boldsymbol{\ell}$ and any real q-vector $\mathbf{m}$, define

$$\theta(\boldsymbol{\ell}\ , \mathbf{m}) = \inf\{\phi(\mathbf{x}) + \boldsymbol{\ell}^\top \mathbf{a}(\mathbf{x}) + \mathbf{m}^\top \mathbf{b}(\mathbf{x}) : \mathbf{x} \in \mathscr{D}\} .$$

Then the **dual of the nonlinear programming problem**

$$\min \phi(\mathbf{x}) \text{ subject to } \mathbf{a}(\mathbf{x}) = \mathbf{0} \text{ and } \mathbf{b}(\mathbf{x}) \leq \mathbf{0}$$

is

$$\max \theta(\boldsymbol{\ell}\ , \mathbf{m}) \text{ subject to } \mathbf{m} \geq \mathbf{0} . \tag{4.38}$$

The following simple result explains the connection between the primal nonlinear programming problem and its dual.

Theorem 4.3.5 (Weak Duality for Nonlinear Programming) *Suppose that $p < n$ and q are nonnegative integers. Let $\mathscr{D}$ be a closed subset of real n vectors, and assume that the real functional ϕ is defined on $\mathscr{D}$. If $p > 0$, let the function $\mathbf{a}$ map $\mathscr{D}$ to real p-vectors. If $q > 0$, let the function $\mathbf{b}$ map $\mathscr{D}$ to real q-vectors. Assume that $\mathscr{D}$ contains the feasible region*

$$\mathscr{F} = \{\mathbf{x} : \text{ if } p > 0 \text{ then } \mathbf{a}(\mathbf{x}) = \mathbf{0} \ , \text{ and if } q > 0 \text{ then } \mathbf{b}(\mathbf{x}) \leq \mathbf{0}\} \ .$$

For any real p-vector $\boldsymbol{\ell}$ and any real q-vector $\mathbf{m}$, define

$$\theta(\boldsymbol{\ell} \ , \ \mathbf{m}) = \inf\{\phi(\mathbf{x}) + \boldsymbol{\ell}^\top \mathbf{a}(\mathbf{x}) + \mathbf{m}^\top \mathbf{b}(\mathbf{x}) : \mathbf{x} \in \mathscr{D}\} \ .$$

If $\mathbf{x} \in \mathscr{F}$, $\boldsymbol{\ell}$ is a real p-vector and $\mathbf{m}$ is a nonnegative q-vector, then

$$\phi(\mathbf{x}) \geq \theta(\boldsymbol{\ell} \ , \ \mathbf{m}) \equiv \inf\{\phi(\mathbf{x}) + \boldsymbol{\ell}^\top \mathbf{a}(\mathbf{x}) + \mathbf{m}^\top \mathbf{b}(\mathbf{x}) : \mathbf{x} \in \mathscr{D}\} \ .$$

Proof By definition,

$$\begin{aligned}\theta(\boldsymbol{\ell} \ , \ \mathbf{m}) &= \inf\{\phi(\mathbf{x}') + \boldsymbol{\ell}^\top \mathbf{a}(\mathbf{x}') + \mathbf{m}^\top \mathbf{b}(\mathbf{x}') : \mathbf{x}' \in \mathscr{D}\} \\ &\leq \phi(\mathbf{x}) + \boldsymbol{\ell}^\top \mathbf{a}(\mathbf{x}) + \mathbf{m}^\top \mathbf{b}(\mathbf{x})\end{aligned}$$

then since $\mathbf{a}(\mathbf{x}) = \mathbf{0}$ we get

$$\leq \phi(\mathbf{x}) + \mathbf{m}^\top \mathbf{b}(\mathbf{x})$$

and because $\mathbf{m} \geq \mathbf{0}$ and $\mathbf{b}(\mathbf{x}) \leq \mathbf{0}$ we obtain

$$\leq \phi(\mathbf{x}) \ .$$

This theorem easily implies that

$$\inf\{\phi(\mathbf{x}) : \mathbf{x} \in \mathscr{F}\} \geq \sup\{\theta(\boldsymbol{\ell} \ , \ \mathbf{m}) : \mathbf{m} \geq \mathbf{0}\} \ .$$

Furthermore, if $\mathbf{x} \in \mathscr{F}$, $\mathbf{m} \geq \mathbf{0}$ and $\phi(\mathbf{x}) = \theta(\boldsymbol{\ell} \ , \ \mathbf{m})$, then $\mathbf{x}$ is optimal for the primal nonlinear programming problem and $(\boldsymbol{\ell}, \mathbf{m})$ is optimal for the dual.

Example 4.3.2 Consider the nonlinear programming problem

$$\min \mathbf{c}^\top \mathbf{x} \text{ subject to } \mathbf{A}\mathbf{x} \geq \mathbf{b} \text{ and } \mathbf{x} \geq \mathbf{0} \ ,$$

which is actually the standard min linear programming problem. According to Definition 4.3.8, the dual objective for this nonlinear programming problem is

$$\theta(\mathbf{m_A}\ ,\ \mathbf{m_I}) = \inf_{\mathbf{x}} \left\{ \mathbf{c}^\top \mathbf{x} + \left[\mathbf{m_A}^\top, \mathbf{m_I}^\top\right] \begin{bmatrix} -\mathbf{A} \\ -\mathbf{I} \end{bmatrix} \mathbf{x} + \mathbf{m_A}^\top \mathbf{b} \right\}$$

$$= \begin{cases} -\infty, \ \mathbf{c}^\top - \mathbf{m_A}^\top \mathbf{A} - \mathbf{m_I}^\top \neq \mathbf{0}^\top \\ \mathbf{m_A}^\top \mathbf{b}, \ \mathbf{c}^\top - \mathbf{m_A}^\top \mathbf{A} - \mathbf{m_I}^\top = \mathbf{0}^\top \end{cases},$$

and is considered only when $\left[\mathbf{m_A}^\top, \mathbf{m_I}^\top\right] \geq \mathbf{0}^\top$. Thus at a maximum of θ, we must have

$$\mathbf{c}^\top - \mathbf{m_A}^\top \mathbf{A} - \mathbf{m_I}^\top = \mathbf{0}^\top .$$

Together with the nonnegative conditions on the Lagrange multipliers $\mathbf{m}$, we see that $\mathbf{m_A}$ is feasible for the standard max problem

$$\max \mathbf{y}^\top \mathbf{b} \text{ subject to } \mathbf{y}^\top \mathbf{A} \leq \mathbf{c}^\top \text{ and } \mathbf{y} \geq \mathbf{0} .$$

Lemma 4.2.5 shows that the maximum value of $\theta(\mathbf{m_A}\ ,\ \mathbf{m_I}) = \mathbf{m_A}^\top \mathbf{b}$ is $\mathbf{c}^\top \mathbf{x}$, where $\mathbf{x}$ is optimal for the original constrained minimization problem.

Unlike linear programs, it is not necessarily the case that $\inf\{\phi(\mathbf{x}) : \mathbf{x} \in \mathscr{F}\}$ is equal to $\sup\{\theta(\boldsymbol{\ell}\ ,\ \mathbf{m}) : \mathbf{m} \geq \mathbf{0}\}$ for nonlinear programs. We invite readers to examine Bazaraa et al. [14] for more detailed discussion of duality in nonlinear programming problems.

Exercise 4.3.3 Use Definition 4.3.8 to find the dual of the standard linear programming problem

$$\min \mathbf{c}^\top \mathbf{x} \text{ subject to } \mathbf{A}\mathbf{x} = \mathbf{b} \text{ and } \mathbf{x} \geq \mathbf{0} .$$

4.3.6 Perturbation Analysis

The Kuhn-Tucker first-order necessary conditions in Theorem 4.3.2 provide a system of nonlinear equations for the optimal solution $\mathbf{z}$ and the Lagrange multipliers $\boldsymbol{\ell}$ and $\mathbf{m}$. Let $\mathbf{P}$ be the permutation matrix described in Definition 4.3.6, which allows us to partition the inequality constraints into active and inactive constraints:

$$\mathbf{P}^\top \mathbf{b}(\mathbf{z}) = \begin{bmatrix} \mathbf{b}_{\mathcal{B}}(\mathbf{z}) \\ \mathbf{b}_{\mathcal{N}}(\mathbf{z}) \end{bmatrix}$$

where the first q_B components on the right can be written $\mathbf{b}_B(\mathbf{z}) = \mathbf{0}$. Then the Kuhn-Tucker first-order necessary conditions can be written

$$\mathbf{f}(\mathbf{z}, \boldsymbol{\ell}, \mathbf{m}_B) \equiv \begin{bmatrix} \mathbf{g}_\psi(\mathbf{z}) + \mathbf{J}_\mathbf{a}(\mathbf{z})^\top \boldsymbol{\ell} + \mathbf{J}_{\mathbf{b}_B}(\mathbf{z})^\top \mathbf{m}_B \\ \mathbf{a}(\mathbf{z}) \\ \mathbf{b}_B(\mathbf{z}) \end{bmatrix} = \begin{bmatrix} \mathbf{0} \\ \mathbf{0} \\ \mathbf{0} \end{bmatrix} .$$

Without loss of generality, we will assume that the active constraints are ordered first. We can use this system of nonlinear equations to study perturbations to the nonlinear programming problem, by applying Lemma 3.2.15 to the function $\mathbf{f}$ just above.

That lemma refers to the Jacobian

$$\mathbf{J}_\mathbf{f}(\mathbf{z}, \boldsymbol{\ell}, \mathbf{m}) = \begin{bmatrix} \mathbf{H}_\phi(\mathbf{z}) + \sum_{i=1}^{p} \mathbf{H}_{\mathbf{a}_i}(\mathbf{z})\boldsymbol{\ell}_i + \sum_{j=1}^{q_B} \mathbf{H}_{\mathbf{b}_j}(\mathbf{z})\mathbf{m}_j & \mathbf{J}_\mathbf{a}(\mathbf{z})^\top & \mathbf{J}_{\mathbf{b}_B}(\mathbf{z})^\top \\ \mathbf{J}_\mathbf{a}(\mathbf{z}) & \mathbf{0} & \mathbf{0} \\ \mathbf{J}_{\mathbf{b}_B}(\mathbf{z}) & \mathbf{0} & \mathbf{0} \end{bmatrix} ,$$

specifically to its inverse and Lipschitz continuity bound. Both of these quantities are difficult to estimate in general, so we will leave it to the reader to estimate these quantities in specific circumstances.

Exercise 4.3.4 Show that $\mathbf{z} = \mathbf{e}_1$ with Lagrange multiplier $\boldsymbol{\ell} = -1/2$ is a local minimum of the nonlinear programming problem

$$\min \mathbf{x}_1 + \mathbf{x}_2^2 + \mathbf{x}_2\mathbf{x}_3 + 2\mathbf{x}_3^2 \text{ subject to } \|\mathbf{x}\|_2^2 = 1\ ;.$$

Exercise 4.3.5 Show that

$$\mathbf{x} = \begin{bmatrix} 1 \\ 2 \end{bmatrix} \text{ and } \mathbf{m} = \begin{bmatrix} 1 \\ 0 \end{bmatrix}$$

are the solution and Lagrange multiplier for the nonlinear programming problem

$$\min 2\mathbf{x}_1^2 + 2\mathbf{x}_1\mathbf{x}_2 + \mathbf{x}_2^2 - 10\mathbf{x}_1 - 10\mathbf{x}_2 \text{ subject to } \mathbf{x}_1^2 + \mathbf{x}_2^2 \le 5 \text{ and } 3\mathbf{x}_1 + \mathbf{x}_2 \le 6 .$$

Exercise 4.3.6 Consider the nonlinear programming problem

$$\min \frac{1}{2}\mathbf{x}^\top \mathbf{F}\mathbf{x} \text{ subject to } \mathbf{a}^\top \mathbf{x} - 3 = 0$$

where

$$\mathbf{F} = \begin{bmatrix} 0 & 1 & 1 \\ 1 & 0 & 1 \\ 1 & 1 & 0 \end{bmatrix} \text{ and} \mathbf{a}^\top = \begin{bmatrix} 1.\ 1,\ 1 \end{bmatrix} .$$

1. Write down the system of equations that determine the optimal solution $\mathbf{z}$ for this nonlinear programming problem, and its associated Lagrange multiplier.
2. Use Lemma 3.2.15 to estimate the sensitivity of the optimal solution to errors in the nonlinear system determined in the previous step.

Exercise 4.3.7 Find the optimal solution to the constrained minimization problem

$$\min e^{-\mathbf{x}_1} + e^{-\mathbf{x}_2} \text{ subject to } \mathbf{e} \cdot \mathbf{x} \leq 1 \text{ and } \mathbf{x} \geq \mathbf{0} .$$

Here $\mathbf{e} = [1, 1]$.

Exercise 4.3.8 Show that $\mathbf{z} = \mathbf{e}_1$ is the solution of the constrained minimization problem

$$\min -\mathbf{x}_1 \text{ subject to } \mathbf{x}_2 - (1 - \mathbf{x}_1)^3 \leq 0 , \ -\mathbf{x}_1 \leq 0 \text{ and } -\mathbf{x}_2 \leq 0 .$$

Then verify that the first-order Kuhn-Tucker conditions (4.30) are not satisfied for this problem, and explain why this does not violate Theorem 4.3.2.

4.4 Quadratic Programming

A common form of the nonlinear programming problem involves a quadratic objective subject to linear constraints.

Definition 4.4.1 Suppose that $p < n$ and q are nonnegative integers. Let $\mathbf{F}$ be an $n \times n$ real symmetric matrix, $\mathbf{f}$ be a real n-vector, $\mathbf{A}$ be a real $p \times n$ matrix, $\mathbf{a}$ be a real p-vector, $\mathbf{B}$ be a real $q \times n$ matrix and $\mathbf{b}$ be a real q-vector. Then the associated **quadratic programming problem** is

$$\min \frac{1}{2}\mathbf{x}^\top \mathbf{F}\mathbf{x} + \mathbf{f}^\top \mathbf{x} \text{ subject to } \mathbf{A}\mathbf{x} - \mathbf{a} = \mathbf{0} \text{ and } \mathbf{B}\mathbf{x} - \mathbf{b} \leq \mathbf{0} . \tag{4.39}$$

Recall that we discussed quadratic programming problems with equality constraints in Sect. 6.12 in Chap. 6 of Volume I, in connection with least squares problems. The more general quadratic programming problem in (4.39) will be useful in numerical methods for general nonlinear programming problems, as we will see in Sect. 4.5.4.

Our goal in this section will be to develop an active constraint strategy for solving problem (4.39), by finding a sequence of feasible approximate solutions that reduces the objective. The resulting algorithm will make essential use of the symmetric indefinite matrix factorization that we previously discussed in Sect. 3.13.2 in Chap. 3 of Volume I.

4.4.1 Theory

Suppose that $\mathbf{z}$ is a constrained local minimum of the quadratic programming problem (4.39). We can find a $q \times q$ permutation matrix $\mathbf{P}$ that orders the active inequality constraints first:

$$\mathbf{P}^\top(\mathbf{B}\mathbf{z}-\mathbf{b}) = \begin{bmatrix}\mathbf{B}_\mathcal{B}\\ \mathbf{B}_\mathcal{N}\end{bmatrix}\mathbf{z} - \begin{bmatrix}\mathbf{b}_\mathcal{B}\\ \mathbf{b}_\mathcal{N}\end{bmatrix} \text{ and } \mathbf{m}^\top\mathbf{P} = \left[\mathbf{m}_\mathcal{B}{}^\top, \mathbf{m}_\mathcal{N}{}^\top\right] . \tag{4.40}$$

Here we have $\mathbf{B}_\mathcal{B}\mathbf{z} - \mathbf{b}_\mathcal{B} = \mathbf{0}$ and $\mathbf{B}_\mathcal{N}\mathbf{z} - \mathbf{b}_\mathcal{N} < \mathbf{0}$. Then the Kuhn-Tucker first-order necessary conditions in Theorem 4.3.2 for a solution of (4.39) imply that

$$\begin{bmatrix}\mathbf{F} & \mathbf{A}^\top & \mathbf{B}_\mathcal{B}{}^\top\\ \mathbf{A} & \mathbf{0} & \mathbf{0}\\ \mathbf{B}_\mathcal{B} & \mathbf{0} & \mathbf{0}\end{bmatrix}\begin{bmatrix}\mathbf{z}\\ \boldsymbol{\ell}\\ \mathbf{m}_\mathcal{B}\end{bmatrix} = \begin{bmatrix}-\mathbf{f}\\ \mathbf{a}\\ \mathbf{b}_\mathcal{B}\end{bmatrix} \tag{4.41}$$

with $\mathbf{m}_\mathcal{N}{}^\top = \mathbf{0}^\top$. Thus we have a system of linear equations to solve for the optimal solution $\mathbf{z}$ and the associated Lagrange multipliers $\boldsymbol{\ell}$ and $\mathbf{m}_\mathcal{B}$, provided that we know which inequality constraints are active.

Suppose that $\mathbf{z}$ is feasible for the quadratic programming problem (4.39), and that $\mathbf{z}$ is a regular point of its active constraints. In order to solve the first-order Kuhn-Tucker conditions, it will be helpful to factor

$$\left[\mathbf{A}^\top, \mathbf{B}_\mathcal{B}{}^\top\right] = \left[\mathbf{U}_\mathscr{R}, \mathbf{U}_\mathscr{N}\right]\begin{bmatrix}\mathbf{R}_\mathbf{A} & \mathbf{R}_{\mathbf{B}_\mathcal{B}}\\ \mathbf{0} & \mathbf{0}\end{bmatrix} \tag{4.42}$$

where $\left[\mathbf{U}_\mathscr{R}, \mathbf{U}_\mathscr{N}\right]$ is an orthogonal matrix, and $\left[\mathbf{R}_\mathbf{A}, \mathbf{R}_{\mathbf{B}_\mathcal{B}}\right]$ is right-triangular. Such a factorization could be computed by means of the Householder QR Factorization Algorithm 6.7.4 in Chap. 6 of Volume I. This factorization implies that

$$\begin{bmatrix}\mathbf{A}\\ \mathbf{B}_\mathcal{B}\end{bmatrix}\mathbf{U}_\mathscr{N} = \begin{bmatrix}\mathbf{R}_\mathbf{A}{}^\top\\ \mathbf{R}_{\mathbf{B}_\mathcal{B}}{}^\top\end{bmatrix}\mathbf{U}_\mathscr{R}{}^\top\mathbf{U}_\mathscr{N} = \mathbf{0} ,$$

so the columns of $\mathbf{U}_\mathscr{N}$ form a basis for the nullspace of $\begin{bmatrix}\mathbf{A}\\ \mathbf{B}_\mathcal{B}\end{bmatrix}$. We will use this factorization to reduce the number of unknowns in Sect. 4.4.2. This approach will be similar to the more reduced gradient method for general nonlinear programming problems, which we will discuss later in Sect. 4.5.2.3.

Since the quadratic programming problem (4.39) involves linear constraints, a feasible initial guess can be determined by using linear programming ideas discussed in Sect. 4.2.5.

Exercise 4.4.1 Use Definition 4.3.8 to find the dual of the quadratic programming problem (4.39) in the case where $\mathbf{F}$ is positive.

4.4.2 Constrained Objective

The algorithm we are about to described is due to Bunch and Kaufman [33], who were in turn motivated by an algorithm developed by Gill and Murray [79]. The computational technique we are about to develop is summarized in Algorithm 4.4.1. Readers may find it helpful to refer to this algorithm during the following discussion.

Suppose that we are given a feasible initial guess $\mathbf{x}$ for the quadratic programming problem (4.39), together with an identification of the active inequality constraints through a permutation matrix $\mathbf{P}$ as in Eq. (4.40). We assume that we have factored the active constraint matrices as in Eq. (4.42). Our first goal is to search for a better solution to this quadratic programming problem within the plane tangent to the currently active constraints. Thus we consider improved solutions of the form

$$\widetilde{\mathbf{z}} = \mathbf{x} + \mathbf{U}_{\mathscr{N}}\mathbf{u}\upsilon \; ,$$

where the vector $\mathbf{u}$ and scalar υ are yet to be determined.

The objective at a feasible point $\widetilde{\mathbf{z}}$ is

$$\phi(\widetilde{\mathbf{z}}) = \phi(\mathbf{x}) + \mathbf{g}_\phi(\mathbf{x})^\top \mathbf{U}_{\mathscr{N}}\mathbf{u} + \frac{1}{2}\mathbf{u}^\top {\mathbf{U}_{\mathscr{N}}}^\top \mathbf{F}\mathbf{U}_{\mathscr{N}}\mathbf{u} \, .$$

To see if ${\mathbf{U}_{\mathscr{N}}}^\top \mathbf{F}\mathbf{U}_{\mathscr{N}}$ is a positive matrix, we can factor

$$Q^\top {\mathbf{U}_{\mathscr{N}}}^\top \mathbf{F}\mathbf{U}_{\mathscr{N}} Q = M\mathbf{D}M^\top$$

as in Sect. 3.13.2 in Chap. 3 of Volume I. Recall that $\mathbf{F}$ is the matrix in the quadratic programming problem objective ϕ, and $\mathbf{U}_{\mathscr{N}}$ was determined by the orthogonal factorization (4.42). Also, $Q^\top$ is a permutation matrix, M is a unit left triangular matrix and $\mathbf{D}$ is a block-diagonal matrix with either 1×1 or 2×2 diagonal blocks. We should also recall from the discussion in Sect. 3.13.2.2 in Chap. 3 of Volume I that whenever a 2×2 diagonal block occurs, its determinant is negative. Thus, whenever $\mathbf{D}$ is nonnegative, its diagonal blocks are all 1×1.

Furthermore, we can define

$$\mathbf{w} = M^\top Q^\top \mathbf{u} \text{ and } \mathbf{h} = M^{-1} Q^\top {\mathbf{U}_{\mathscr{N}}}^\top \mathbf{g}_\phi(\mathbf{x}) \; ,$$

and see that the objective at

$$\widetilde{\mathbf{z}} = \mathbf{x} + \mathbf{U}_{\mathscr{N}}\mathbf{u}\upsilon = \mathbf{x} + \mathbf{U}_{\mathscr{N}} Q M^{-\top} \mathbf{w}\upsilon \tag{4.43}$$

is

$$\begin{aligned}\phi(\widetilde{\mathbf{z}}) &= \phi(\mathbf{x}) + \mathbf{g}_\phi(\mathbf{x})^\top \mathbf{U}_{\mathscr{N}} Q M^{-\top} \mathbf{w}\upsilon + \frac{\upsilon^2}{2}\mathbf{w}^\top M^{-1} Q^\top {\mathbf{U}_{\mathscr{N}}}^\top \mathbf{F}\mathbf{U}_{\mathscr{N}} Q M^{-\top} \mathbf{w} \\ &= \phi(\mathbf{x}) + \upsilon\mathbf{h}^\top \mathbf{w} + \frac{\upsilon^2}{2}\mathbf{w}^\top \mathbf{D}\mathbf{w} \, . \end{aligned} \tag{4.44}$$

At this point, we need to determine a vector **w** that will allow us to reduce the objective, and then we need to select the scalar υ to achieve the smallest possible objective value. The determination of the vector **w** will depend on whether the matrix **D** is positive, nonnegative, or has negative eigenvalues.

4.4.3 Positive Constrained Hessian

If **D** is positive, the objective value $\phi(\widetilde{\mathbf{z}})$ is minimized by choosing **w** to solve

$$\mathbf{D}\mathbf{w} = -\mathbf{h} \; . \tag{4.45}$$

Then Eq. (4.44) shows that

$$\phi(\widetilde{\mathbf{z}}) = \phi(\mathbf{x}) - \left(\upsilon - \frac{\upsilon^2}{2}\right)\mathbf{h}^\top\mathbf{D}^{-1}\mathbf{h} \; .$$

This proves that the objective is decreased whenever $\upsilon \in (0, 2)$ and $\mathbf{h} \neq \mathbf{0}$. With this choice of **w**, we also have

$$\mathbf{U}_{\mathscr{N}}{}^\top \mathbf{g}_\phi(\widetilde{\mathbf{z}}) = \mathbf{U}_{\mathscr{N}}{}^\top \left\{\mathbf{g}_\phi(\mathbf{x}) + \mathbf{F}\mathbf{U}_{\mathscr{N}}\mathbf{u}\right\} = \mathrm{QM}\left\{\mathbf{h} + \mathbf{D}\mathbf{w}\right\} = \mathbf{0} \; . \tag{4.46}$$

This implies that

$$\mathbf{g}_\phi(\widetilde{\mathbf{z}}) = \mathbf{g}_\phi(\mathbf{x}) + \mathbf{F}\mathbf{U}_{\mathscr{N}}\mathbf{u} \perp \mathscr{R}\left(\mathbf{U}_{\mathscr{N}}\right) \; ,$$

so we conclude that from the orthogonal factorization (4.42) that $\mathbf{g}_\phi(\widetilde{\mathbf{z}}) \in \mathscr{R}\left(\mathbf{U}_{\mathscr{R}}\right)$. This fact implies that the linear system (4.41) is solvable.

4.4.4 Nonnegative Constrained Hessian

Next, consider the case when **D** is nonnegative and each zero diagonal block of **D** corresponds to a zero component of **h**. In this situation, we will define the block components of **w** by

$$\mathbf{w}_j = \begin{cases} 0, & \mathbf{D}_j = \mathbf{0} \\ -\mathbf{D}_j^{-1}\mathbf{h}_j, & \mathbf{D}_j^{-1} \neq \mathbf{0} \end{cases} \tag{4.47}$$

Then Eq. (4.44) shows that

$$\phi(\widetilde{\mathbf{z}}) = \phi(\mathbf{x}) - \left(\upsilon - \frac{\upsilon^2}{2}\right) \sum_{\mathbf{D}_j^{-1} \neq \mathbf{0}} \mathbf{h}_j{}^\top \mathbf{D}_j^{-1} \mathbf{h}_j$$

If $\mathbf{h} \neq \mathbf{0}$, then for at least one index j with $\mathbf{D}_j \neq 0$ we will have $\mathbf{h}_j \neq \mathbf{0}$. As a result, the objective is decreased whenever $\upsilon \in (0, 2)$. With this choice of $\mathbf{w}$, Eq. (4.46) is satisfied as well, so again we have $\mathbf{g}_\phi(\widetilde{\mathbf{z}}) \in \mathscr{R}(\mathbf{U}_\mathscr{R})$, and again the linear system (4.41) is solvable.

4.4.5 Nonpositive Constrained Hessian

If $\mathbf{D}$ is not nonnegative, let us define the 2×2 rotation

$$\mathbf{W} = \begin{bmatrix} 0 & 1 \\ -1 & 0 \end{bmatrix} .$$

As we have remarked above, the pivoting strategy for the symmetric indefinite strategy guarantees that $\det \mathbf{D}_j < 0$ for any 2×2 diagonal block.

Let δ be a positive scalar that is yet to be determined, and let us choose the block components of $\mathbf{w}$ to be

$$\mathbf{w}_j = \begin{cases} -\mathbf{h}_j, & \mathbf{D}_j = \mathbf{0} \\ \mathbf{D}_j^{-1}\mathbf{j}_j, & \mathbf{h}_j{}^\top \mathbf{D}_j^{-1} \mathbf{h}_j < 0 \\ -\mathbf{h}_j, & \mathbf{h}_j{}^\top \mathbf{D}_j^{-1} \mathbf{h}_j \geq 0 \text{ and } \mathbf{h}_j{}^\top \mathbf{D}_j \mathbf{h}_j < 0 \\ -\mathbf{D}_j^{-1}\mathbf{h}_j\delta, & \mathbf{D}_j \; 1 \times 1 \text{ and } \mathbf{D}_j > 0 \\ -\mathbf{W}\mathbf{h}_j, & \mathbf{h}_j{}^\top \mathbf{D}_j^{-1} \mathbf{h}_j \geq 0 \text{ and } \mathbf{h}_j{}^\top \mathbf{D}_j \mathbf{h}_j \geq 0 \end{cases} \tag{4.48}$$

We also define the nonnegative scalars

$$\alpha_0 = \sum_{\mathbf{D}_j = \mathbf{0}} \mathbf{h}_j{}^\top \mathbf{h}_j + \sum_{\substack{\mathbf{h}_j{}^\top \mathbf{D}_j^{-1} \mathbf{h}_j \geq 0 \text{ and} \\ \mathbf{h}_j{}^\top \mathbf{D}_j \mathbf{h}_j < 0}} \mathbf{h}_j{}^\top \mathbf{h}_j$$

$$\alpha_1 = -\sum_{\substack{\mathbf{h}_j{}^\top \mathbf{D}_j^{-1} \mathbf{h}_j \geq 0 \text{ and} \\ \mathbf{h}_j{}^\top \mathbf{D}_j \mathbf{h}_j < 0}} \mathbf{h}_j{}^\top \mathbf{D}_j \mathbf{h}_j - \sum_{\substack{\mathbf{D}_j \; 2 \times 2 \\ \mathbf{h}_j{}^\top \mathbf{D}_j^{-1} \mathbf{h}_j \geq 0 \text{ and } \mathbf{h}_j{}^\top \mathbf{D}_j \mathbf{h}_j \geq 0}} \mathbf{h}_j{}^\top \mathbf{W}^\top \mathbf{D}_j \mathbf{W} \mathbf{h}_j$$

$$\alpha_{-1} = -\sum_{\mathbf{h}_j{}^\top \mathbf{D}_j^{-1} \mathbf{h}_j < 0} \mathbf{h}_j{}^\top \mathbf{D}_j^{-1} \mathbf{h}_j$$

$$\beta = \sum_{\substack{\mathbf{D}_j \; 1 \times 1 \\ \mathbf{D}_j > 0}} \mathbf{h}_j{}^\top \mathbf{D}_j^{-1} \mathbf{h}_j \; .$$

To see that the second sum in the definition of α_1 is negative, we recall that $\det \mathbf{D}_j < 0$ for any 2×2 diagonal block, and that

$$\mathbf{D}_j^{-1} = \mathbf{W}^\top \mathbf{D}_j \mathbf{W} \frac{1}{\det \mathbf{D}_j}$$

It is also helpful to note that $\mathbf{h}^\top \mathbf{W} \mathbf{h} = 0$ for all 2-vectors $\mathbf{h}$. Then Eq. (4.44) shows us that

$$\begin{aligned}
\phi(\widetilde{\mathbf{z}}) &= \phi(\mathbf{x}) + \mathbf{h}^\top \mathbf{w} \upsilon + \frac{\upsilon^2}{2} \mathbf{w}^\top \mathbf{D} \mathbf{w} \\
&= \phi(\mathbf{x}) - \sum_{\mathbf{D}_j = \mathbf{0}} \mathbf{h}_j{}^\top \mathbf{h}_j \upsilon + \left(\upsilon + \frac{\upsilon^2}{2} \right) \sum_{\mathbf{h}_j{}^\top \mathbf{D}_j^{-1} \mathbf{h}_j < 0} \mathbf{h}_j{}^\top \mathbf{D}_j^{-1} \mathbf{h}_j \\
&+ \sum_{\substack{\mathbf{h}_j{}^\top \mathbf{D}_j^{-1} \mathbf{h}_j \geq 0 \text{ and} \\ \mathbf{h}_j{}^\top \mathbf{D}_j \mathbf{h}_j < 0}} \left[-\mathbf{h}_j{}^\top \mathbf{h}_j \upsilon + \mathbf{h}_j{}^\top \mathbf{D}_j \mathbf{h}_j \frac{\upsilon^2}{2} \right] - \left(\upsilon - \frac{\upsilon^2}{2} \right) \sum_{\substack{\mathbf{D}_j \; 1\times 1 \\ \mathbf{D}_j > 0}} \mathbf{h}_j{}^\top \mathbf{D}_j^{-1} \mathbf{h}_j \\
&+ \frac{\upsilon^2}{2} \sum_{\substack{\mathbf{D}_j \; 2\times 2 \\ \mathbf{h}_j{}^\top \mathbf{D}_j^{-1} \mathbf{h}_j \geq 0 \text{ and } \mathbf{h}_j{}^\top \mathbf{D}_j \mathbf{h}_j \geq 0}} \mathbf{h}_j{}^\top \mathbf{W}^\top \mathbf{D}_j \mathbf{W} \mathbf{h}_j \delta \\
&= \phi(\mathbf{x}) - \alpha_0 \upsilon - \alpha_{-1} \left(\upsilon + \frac{\upsilon^2}{2} \right) - \alpha_1 \frac{\upsilon^2}{2} - \beta \left(\upsilon \delta - \frac{\upsilon^2 \delta^2}{2} \right) \\
&= \phi(\mathbf{x}) - (\alpha_0 + \alpha_{-1} + \beta \delta) \upsilon - \left(\alpha_{-1} + \alpha_1 - \beta \delta^2 \right) \frac{\upsilon^2}{2} \,.
\end{aligned}$$

This function decreases as $\upsilon > 0$ decreases provided that

$$\alpha_{-1} + \alpha_1 - \beta \delta^2 \geq 0 \,.$$

Accordingly, we choose

$$\delta = \begin{cases} 1, \; \beta \leq 0 \\ \sqrt{\frac{\alpha_{-1} + \alpha_1}{\beta}}, \; \beta > 0 \end{cases} . \tag{4.49}$$

Note that with this choice of $\mathbf{w}$, it is not necessarily the case that the linear system (4.41) is solvable.

4.4.6 New Active Constraint

At this point, we have used a previous guess $\mathbf{x}$ for the solution of the quadratic programming problem to compute a search direction $\mathbf{w}$. Our next goal is to determine a step length υ so that the new approximate solution $\widetilde{\mathbf{z}}$, defined by (4.43), has smaller objective value than $\mathbf{x}$.

If $\mathbf{D}$ is positive, or is nonnegative and the components of $\mathbf{h}$ corresponding to zero diagonal blocks are $\mathbf{D}$ are all zero, then our step direction $\mathbf{w}$ is given by either (4.45) or (4.47), and the step length $\upsilon = 1$ leads to the minimum objective value. However, it is possible that choosing $\upsilon = 1$ will violate one of the inactive constraints. Recall that

$$\mathbf{w} = \mathbf{M}^\top \mathbf{Q}^\top \mathbf{u} \; .$$

For each inactive constraint index i, let

$$\varrho_i = \mathbf{e}_i \cdot (\mathbf{B}_N \mathbf{x} - \mathbf{b}_N) \text{ and } \beta_i = \mathbf{e}_i \cdot \mathbf{B}_N (\mathbf{U}_{\mathscr{N}} \mathbf{u}) \; . \tag{4.50}$$

Since $\mathbf{x}$ is assumed to be feasible, we already have $\varrho_i < 0$ for each inactive constraint index i. We want to choose the largest possible value of $\upsilon \in (0, 1]$ so that for all inactive constraint indices i we have

$$0 \geq \varrho_i + \beta_i \upsilon \; .$$

Thus we choose

$$\upsilon = \min \left\{ 1 \, , \, \min \left\{ -\frac{\varrho_i}{\beta_i} \; : \; \beta_i > 0 \right\} \right\} \; . \tag{4.51}$$

If $\upsilon = 1$, then all constraints that are currently inactive at $\mathbf{x}$ remain inactive at $\widetilde{\mathbf{z}} = \mathbf{x} + \mathbf{U}\mathbf{u}$. If $\upsilon < 1$, then there is some inactive constraint index i_N that determines the value of υ. We use this value of υ to compute $\widetilde{\mathbf{z}}$ as in Eq. (4.43), add this constraint to the set of active constraints, and restart the minimization process at $\mathbf{x} = \widetilde{\mathbf{z}}$.

On the other hand, if $\mathbf{D}$ is not nonnegative, or is nonnegative and $\mathbf{h}$ has a nonzero component corresponding to a zero diagonal block of $\mathbf{D}$, then our step direction is given by (4.48). In this case, the decrease in the objective subject to the currently active constraints is unbounded as $\upsilon \to \infty$. For each inactive constraint index i, we can compute the residual ϱ_i and coefficient β_i as in (4.50). We choose

$$\upsilon = \min \left\{ -\frac{\varrho_i}{\beta_i} \; : \; \beta_i > 0 \right\} \; . \tag{4.52}$$

If $\beta_i \leq 0$ for all inactive constraints, then the quadratic programming problem is unbounded. Otherwise, there is some inactive constraint index i_N that determines the value of υ. We use this value of υ to compute $\widetilde{\mathbf{z}}$ as in Eq. (4.43), add this constraint to the set of active constraints, and restart the minimization process at $\mathbf{x} = \widetilde{\mathbf{z}}$.

4.4.7 New Inactive Constraint

We have almost completed the development of our active constraint algorithm for the quadratic programming problem. We have chosen a step direction $\mathbf{w}$ and a step length υ so that the new solution $\widetilde{\mathbf{z}}$ defined by (4.43) is feasible. It remains to determine if this vector is optimal. If it is not, then we need to update the set of active constraints.

If we choose the descent direction $\mathbf{w}$ by either Eq. (4.45) or (4.47), and if no inequality constraint was activated, then we need to examine the first-order Kuhn-Tucker conditions (4.30a) to see if the new solution $\widetilde{\mathbf{z}}$ is optimal. Since $\widetilde{\mathbf{z}}$ was computed to satisfy the current set of active constraints, all that remains is to solve

$$\mathbf{0} = \mathbf{F}\widetilde{\mathbf{z}} + \mathbf{f} + \mathbf{A}^\top \widetilde{\boldsymbol{\ell}} + \mathbf{B}_B{}^\top \widetilde{\mathbf{m}}_B = \mathbf{g}_\phi(\mathbf{x}) + \mathbf{F}\mathbf{U}_\mathscr{N}\mathbf{u} + \mathbf{U}_\mathscr{R}\left[\mathbf{R}_\mathbf{A}, \mathbf{R}_{\mathbf{B}_B}\right]\begin{bmatrix}\widetilde{\boldsymbol{\ell}} \\ \widetilde{\mathbf{m}}_B\end{bmatrix}$$

for the Lagrange multipliers $\boldsymbol{\ell}$ and $\widetilde{\mathbf{m}}_B$. Since we have shown above that $\mathbf{g}_\phi(\mathbf{x}) + \mathbf{F}\mathbf{U}_\mathscr{N}\mathbf{u} \in \mathscr{R}(\mathbf{U}_\mathscr{R})$, this equation for the Lagrange multipliers is solvable. We can compute the Lagrange multipliers by back-solving the right-triangular system

$$\left[\mathbf{R}_\mathbf{A}, \mathbf{R}_{\mathbf{B}_B}\right]\begin{bmatrix}\widetilde{\boldsymbol{\ell}} \\ \widetilde{\mathbf{m}}_B\end{bmatrix} = -\mathbf{U}_\mathscr{R}{}^\top\left\{\mathbf{g}_\phi(\mathbf{x}) + \mathbf{F}\mathbf{U}_\mathscr{N}\mathbf{u}\right\} . \tag{4.53}$$

If $\widetilde{\mathbf{m}}_B \geq \mathbf{0}$, then $\widetilde{\mathbf{z}}$ is optimal. Otherwise, there is an index i_B for which $\widetilde{\mathbf{m}}_{B,i_B} < 0$. We drop this constraint from the set of active constraints, and repeat the calculations above.

4.4.8 Active Constraint Algorithm

We can summarize our strategy for solving the quadratic programming problem by the following

Algorithm 4.4.1 (Active Constraint Strategy for Quadratic Programming)

given feasible $\widetilde{\mathbf{z}}$ /* see Sect. 4.2.5 */

$\mathbf{g} = \mathbf{F}\widetilde{\mathbf{z}} + \mathbf{f}$ /* gradient of ϕ */

while not converged

  factor $\begin{bmatrix}\mathbf{A}^\top, \mathbf{B}_B{}^\top\end{bmatrix} = \begin{bmatrix}\mathbf{U}_{\mathscr{R}}, \mathbf{U}_{\mathscr{N}}\end{bmatrix}\begin{bmatrix}\mathbf{R}_\mathbf{A} & \mathbf{R}_{\mathbf{B}_B} \\ \mathbf{0} & \mathbf{0}\end{bmatrix}$ /* $\mathbf{U}$ orthogonal, $\mathbf{R}$ right-triangular */

  factor $\mathrm{Q}^\top \mathbf{U}_{\mathscr{N}}{}^\top \mathbf{F} \mathbf{U}_{\mathscr{N}} \mathrm{Q} = \mathrm{M}\mathbf{D}\mathrm{M}^\top$ /* M left-triangular, $\mathbf{D}$ block diagonal */

  solve $\mathrm{M}\mathbf{h} = \mathrm{Q}^\top \mathbf{U}_{\mathscr{N}}{}^\top \mathbf{g}$ for $\mathbf{h}$

  if $\mathbf{D}$ is positive

    solve $\mathbf{D}\mathbf{w} = -\mathbf{h}$ and set $\upsilon = 1$

  else if $\mathbf{D}$ nonnegative and $\mathbf{D}_j = \mathbf{0} \Longrightarrow \mathbf{h}_j = \mathbf{0}$

    compute $\mathbf{w}$ by (4.47) and set $\upsilon = 1$

  else

    compute $\mathbf{w}$ by (4.48) and set $\upsilon = \infty$

  solve $\mathrm{M}^\top \mathrm{Q}^\top \mathbf{u} = \mathbf{w}$

  $i_N =$ undefined

  for each inactive constraint i

    $\varrho_i = \mathbf{e}_i \cdot (\mathbf{B}_N \widetilde{\mathbf{z}} - \mathbf{b}_N)$ and $\beta_i = \mathbf{e}_i \cdot (\mathbf{B}_N \widetilde{\mathbf{z}} - \mathbf{b}_N)$

    if $\beta_i > 0$ and $-\varrho_i < \upsilon\beta_i$

      $i_N = i$ and $\upsilon = -\varrho_i/\beta_i$

  if i_N is defined

    append constraint i_N to $\mathbf{B}_B$ and $\mathbf{b}_B$

    continue

  if $\upsilon = \infty$ break /* problem is unbounded */

  $\widetilde{\mathbf{z}} = \widetilde{\mathbf{z}} + \mathbf{U}_{\mathscr{N}}\mathbf{u}\upsilon$ and $\mathbf{g} = \mathbf{g} + \mathbf{F}\mathbf{U}_{\mathscr{N}}\mathbf{u}\upsilon$

  solve $\begin{bmatrix}\mathbf{R}_\mathbf{A}, \mathbf{R}_{\mathbf{B}_B}\end{bmatrix}\begin{bmatrix}\boldsymbol{\ell} \\ \mathbf{m}_B\end{bmatrix} = -\mathbf{U}_{\mathscr{R}}{}^\top \mathbf{g}$

  if $\mathbf{m}_B \geq \mathbf{0}$ break /* $\widetilde{\mathbf{z}}$ is optimal */

  for each active inequality constraint i

    if $\mathbf{m}_{B,i} < 0$

      drop active constraint i from $\mathbf{B}_B$ and $\mathbf{b}_B$

      continue

Bunch and Kaufman [33] discuss other alternatives for choosing a descent direction, and techniques for updating matrix factorizations when constraints are added or dropped. For details, see their paper.

4.4.9 *Software*

The GNU Octave project provides an interpreted programming language to solve quadratic programming problems by means of the qp function. Software for solving quadratic programming problems is available for a fee at Stanford Business Software Inc. MATLAB contains the quadprog command for solving quadratic programming problems.

Exercise 4.4.2 Solve the quadratic programming problem (4.39) with

$$\mathbf{F} = \begin{bmatrix} 2 & -2 \\ -2 & 4 \end{bmatrix} , \ \mathbf{f} = \begin{bmatrix} -15 \\ -30 \end{bmatrix} , \ \mathbf{B} = \begin{bmatrix} 1 & 2 \\ -1 & 0 \\ 0 & -1 \end{bmatrix} \text{ and } \mathbf{b} = \begin{bmatrix} 30 \\ 0 \\ 0 \end{bmatrix} .$$

There are no equality constraints in this problem.

4.5 General Constrained Optimization

For the remainder of this chapter, we will discuss minimization of general nonlinear functionals subject to nonlinear equality and inequality constraints. These problems suffer from the obvious difficulty that descent directions for the objective may not point into the feasible region. A varicty of strategies have been developed to deal with this problem. We will group many of these approaches into four broad categories: penalty methods, feasible search strategies, augmented Lagrangian methods and sequential quadratic programming. Here is a brief summary of these methods for solving constrained minimization problems.

Penalty methods modify the original objective by adding a term that is positive whenever a constraint is violated. In order to guarantee convergence to the true solution, the added term must be multiplied by a cocfficicnt that becomes infinite. The advantage of penalty methods is that they are easy to program, because they allow the user to use well-developed methods for solving constrained minimization problems. Two minor disadvantages are that the solutions of the modified problems are infeasible for the original constrained problem, and that the modified objective is continuous but not continuously differentiable at the boundary of the feasible region. The major disadvantage is that very large penalties cause serious conditioning problems.

Feasible search methods are designed to respect the constraints in the original minimization problem. For example, *feasible direction methods* choose descent directions for the original objective, and then use line searches to reduce the objective subject to *inequality constraints*. These methods are also fairly easy to program. However, nonlinear equality constraints make the use of fixed descent directions and line searches impossible in feasible direction methods. On the other hand, *gradient projection methods* use equality constraints to reduce the number of unknown parameters in the original minimization problem. These methods respect the constraints. However, gradient projection methods are a bit more difficult to program than penalty methods or feasible direction methods, especially if inequality constraints are involved. *Reduced gradient methods* are the more popular than the previous two feasible search methods. These methods use the active equality or inequality constraints to determine a descent direction for the constrained objective, and then search in that direction to reduce the original objective. When such a step violates a constraint, then the infeasible point is projected back to the feasible region. Reduced gradient methods involve minimization steps that are easier to solve than those in gradient projection methods, but involve an additional projection step to regain feasibility. Finally, *interior point methods* work by adding a barrier function to the original problem. The barrier function is designed to become infinite as the boundary of the feasible region is approach from inside. Like penalty methods, interior point methods are easy to program because they solve unconstrained minimization problems. Convergence to the true solution occurs if the coefficient of the barrier function approaches zero, and the modified objective is typically as smooth as the original objective and the constraint functions allow. However, some caution must be applied to deal with very large values of the barrier function, and an outer iteration over barrier function coefficients must surround the inner iteration to minimize the modified objective.

Augmented Lagrangian methods are similar to penalty methods, in that they modify the original objective with a term that involves the constraints. When properly designed with a sufficiently large coefficient for the modifications, the augmented Lagrangian objective has the same local minima as the original objective. Thus augmented Lagrangian methods can use unconstrained minimization techniques without necessarily encountering the conditioning problems of penalty methods. However, the magnitude of the coefficient of the modification cannot be known in advance, and must be discovered experimentally. Furthermore, augmented Lagrangian methods may still converge to the true solution of the original constrained problem along a sequence of infeasible points.

Finally, *sequential quadratic programming* methods approximate constrained minimization problems by a sequence of quadratic programming problems. These methods are closest in spirit to our development of Newton methods for unconstrained minimization, in that they use local quadratic models for the objective and local linear models for the constraints. As a result, we expect higher convergence rates with sequential quadratic programming methods than with the previous methods for constrained minimization. Furthermore, no large penalties are involved, and no barrier functions with embedded infinite values are used. However, sequential quadratic programming methods are probably the most difficult of all to program. Furthermore, the individual quadratic programming problems must be modified in order to guarantee feasibility.

4.5.1 *Penalty Methods*

As we have previously mentioned, the basic idea in penalty methods is to replace the constrained minimization problem by a sequence of unconstrained problems involving increasingly larger penalties for violating the constraints. This approach will use the following idea.

Definition 4.5.1 Given a set $\mathscr{F}$ of real n-vectors, a **penalty function** for $\mathscr{F}$ is a continuous real functional $\pi(\mathbf{x})$ such that $\pi(\mathbf{x}) \geq 0$ for all real n-vectors, and $\pi(\mathbf{x}) = 0$ if and only if $\mathbf{x} \in \mathscr{F}$.

If, for example, the feasible region is the set

$$\mathscr{F} = \{\mathbf{x} : \mathbf{a}(\mathbf{x}) = \mathbf{0} \text{ and } \mathbf{b}(\mathbf{x}) \leq \mathbf{0}\} \ ,$$

then we could choose our penalty function to be

$$\pi(\mathbf{x}) = \|\mathbf{a}(\mathbf{x})\|_2^2 + \|\max\{\mathbf{0}\ ,\ \mathbf{b}(\mathbf{x})\}\|_2^2 \ .$$

In this expression, the maximum of a vector is computed componentwise.

If we wish to minimize a continuous real functional ϕ defined on n-vectors $\mathbf{x}$ subject to the constraint that $\mathbf{x} \in \mathscr{F}$, we could design a penalty function π for $\mathscr{F}$ and execute the following

Algorithm 4.5.1 (Penalty Method)

given $\{\varrho_k\}_{k=0}^{\infty}$ such that $0 \leq \varrho_k \to \infty$ as $k \to \infty$

for $0 \leq k$

find $\mathbf{z}_k$ to minimize $\phi(\mathbf{x}) + \varrho_k \pi(\mathbf{x})$

As $\varrho_k \to \infty$, the penalty becomes more severe, and we expect that $\mathbf{z}_k$ should approach a local constrained minimum of ϕ.

In order to understand the properties of a penalty method, we will begin with the following result.

Lemma 4.5.1 *Suppose that ϕ is a real functional on real n-vectors, and $\pi(\mathbf{x})$ is a penalty function for some set $\mathscr{F}$ of real n-vectors. Let $\{\varrho_k\}_{k=0}^{\infty}$ be a strictly increasing sequence of nonnegative real scalars. Assume that for each $k \geq 0$, $\mathbf{z}_k$ minimizes $\phi(\mathbf{x}) + \varrho_k \pi(\mathbf{x})$. Also, assume that $\mathbf{z}$ minimizes $\phi(\mathbf{x})$ subject to the constraint that $\mathbf{x} \in \mathscr{F}$. Then for all $k \geq 0$*

$$\phi(\mathbf{z}_k) + \varrho_k \pi(\mathbf{z}_k) \leq \phi(\mathbf{z}_{k+1}) + \varrho_{k+1} \pi(\mathbf{z}_{k+1}) \ ,$$

$$\pi(\mathbf{z}_k) \geq \pi(\mathbf{z}_{k+1}) \ ,$$

$$\phi(\mathbf{z}_k) \leq \phi(\mathbf{z}_{k+1}) \text{ and}$$

$$\phi(\mathbf{z}_k) \leq \phi(\mathbf{z}_k) + \varrho_k \pi(\mathbf{z}_k) \leq \phi(\mathbf{z}) \ .$$

Proof Since $\mathbf{z}_k$ minimizes $\phi(\mathbf{x}) + \varrho_k \pi(\mathbf{x})$,

$$\phi(\mathbf{z}_k) + \varrho_k \pi(\mathbf{z}_k) \leq \phi(\mathbf{z}_{k+1}) + \varrho_k \pi(\mathbf{z}_{k+1})$$

and since π is nonnegative and $\varrho_k \leq \varrho_{k+1}$, we get

$$\leq \phi(\mathbf{z}_{k+1}) + \varrho_{k+1} \pi(\mathbf{z}_{k+1}) \,.$$

This proves the first claim.

We have just shown that

$$\phi(\mathbf{z}_k) + \varrho_k \pi(\mathbf{z}_k) \leq \phi(\mathbf{z}_{k+1}) + \varrho_k \pi(\mathbf{z}_{k+1}) \,.$$

Since $\mathbf{z}_{k+1}$ minimizes $\phi(\mathbf{x}) + \varrho_{k+1} \pi(\mathbf{x})$, we also have

$$\phi(\mathbf{z}_{k+1}) + \varrho_{k+1} \pi(\mathbf{z}_{k+1}) \leq \phi(\mathbf{z}_k) + \varrho_{k+1} \pi(\mathbf{z}_k) \,.$$

We can add these two inequalities, cancel terms involving ϕ and rearrange to get

$$(\varrho_{k+1} - \varrho_k) \pi(\mathbf{z}_{k+1}) \leq (\varrho_{k+1} - \varrho_k) \pi(\mathbf{z}_k) \,.$$

Since $\varrho_k < \varrho_{k+1}$, this inequality implies the second claim.

Again, since $\mathbf{z}_k$ minimizes $\phi(\mathbf{x}) + \varrho_k \pi(\mathbf{x})$, we have

$$\phi(\mathbf{z}_k) + \varrho_k \pi(\mathbf{z}_k) \leq \phi(\mathbf{z}_{k+1}) + \varrho_k \pi(\mathbf{z}_{k+1}) \,.$$

This implies that

$$\phi(\mathbf{z}_k) \leq \phi(\mathbf{z}_{k+1}) + \varrho_k \left[\pi(\mathbf{z}_{k+1}) - \pi(\mathbf{z}_k)\right]$$

then the second claim and the assumption that $\varrho_k \geq 0$ lead to

$$\leq \phi(\mathbf{z}_{k+1}) \,.$$

This proves the third claim.

Finally, since $\varrho_k \geq 0$ and $\pi(\mathbf{z}_k) \geq 0$

$$\phi(\mathbf{z}_k) \leq \phi(\mathbf{z}_k) + \varrho_k \pi(\mathbf{z}_k)$$

then since $\mathbf{z}_k$ minimizes $\phi(\mathbf{x}) + \varrho_k \pi(\mathbf{x})$, we have

$$\leq \phi(\mathbf{z}) + \varrho_k \pi(\mathbf{z})$$

and since $\mathbf{z} \in \mathscr{F}$ and $\pi(\mathbf{x}) = 0$ for all $\mathbf{x} \in \mathscr{F}$ we get

$$= \phi(\mathbf{z}) .$$

This proves the final claim

We can use the previous lemma to prove the convergence of the penalty method.

Theorem 4.5.1 *Suppose that ϕ is a continuous real functional on real n-vectors, and $\pi(\mathbf{x})$ is a penalty function for some set $\mathscr{F}$ of real n-vectors. Let $\{\varrho_k\}_{k=0}^{\infty}$ be a strictly increasing sequence of nonnegative real scalars that approaches infinity as k becomes large. Assume that for each $k \geq 0$, $\mathbf{z}_k$ minimizes $\phi(\mathbf{x}) + \varrho_k \pi(\mathbf{x})$. Also, assume that $\mathbf{z} \in \mathscr{F}$ minimizes $\phi(\mathbf{x})$ subject to the constraint that $\mathbf{x} \in \mathscr{F}$. Then any limit point of $\{\mathbf{z}_k\}_{k=0}^{\infty}$ minimizes $\phi(\mathbf{x})$ subject to the constraint that $\mathbf{x} \in \mathscr{F}$.*

Proof Suppose that $\{\mathbf{z}_k\}_{k=0}^{\infty}$ has a convergence subsequence. In order to simplify the notation, we will denote this convergent subsequence by $\{\mathbf{z}_k\}_{k=0}^{\infty}$, and its limit by $\mathbf{z}_\infty$. Let $\mathbf{z}$ minimize $\phi(\mathbf{x})$ subject to the constraint that $\mathbf{x} \in \mathscr{F}$. Then Lemma 4.5.1 shows that $\{\phi(\mathbf{z}_k) + \varrho_k \pi(\mathbf{z}_k)\}_{k=0}^{\infty}$ is nondecreasing and bounded above by $\phi(\mathbf{z})$. It follows that this sequence converges, and

$$\lim_{k \to \infty} [\phi(\mathbf{z}_k) + \varrho_k \pi(\mathbf{z}_k)] \leq \phi(\mathbf{z}) .$$

As a result,

$$\lim_{k \to \infty} \varrho_k \pi(\mathbf{z}_k) = \lim_{k \to \infty} [\phi(\mathbf{z}_k) + \varrho_k \pi(\mathbf{z}_k)] - \lim_{k \to \infty} \phi(\mathbf{z}_k) \leq \phi(\mathbf{z}) - \phi(\mathbf{z}_\infty) .$$

Since $\varrho_k \to \infty$ and $\pi(\mathbf{z}_k) \geq 0$, we conclude that

$$\lim_{k \to \infty} \pi(\mathbf{z}_k) = 0 .$$

Since penalty functions are continuous, we conclude that $\pi(\mathbf{z}_\infty) = 0$, which implies that $\mathbf{z}_\infty \in \mathscr{F}$. Thus $\mathbf{z}_\infty$ is feasible for the constrained minimization problem Also, $\mathbf{z}$ is optimal because Lemma 4.5.1 implies that

$$\phi(\mathbf{z}_\infty) = \lim_{k \to \infty} \phi(\mathbf{z}_k) \leq \phi(\mathbf{z}) .$$

In practice, it is difficult to choose a good value for the penalty function coefficient ϱ_k. Small values of ϱ_k lead to unconstrained minimization problems with minima that do not come close to satisfying the constraints, and large values of ϱ_k lead to conditioning problems in the gradient and Hessian of the unconstrained objective.

Avriel [5, pp. 371–412], Gill et al. [81, pp. 207–218] and Luenberber [120, pp. 278–300] all discuss penalty methods in much greater detail. Interested readers should consult these and other resources for more information about penalty methods.

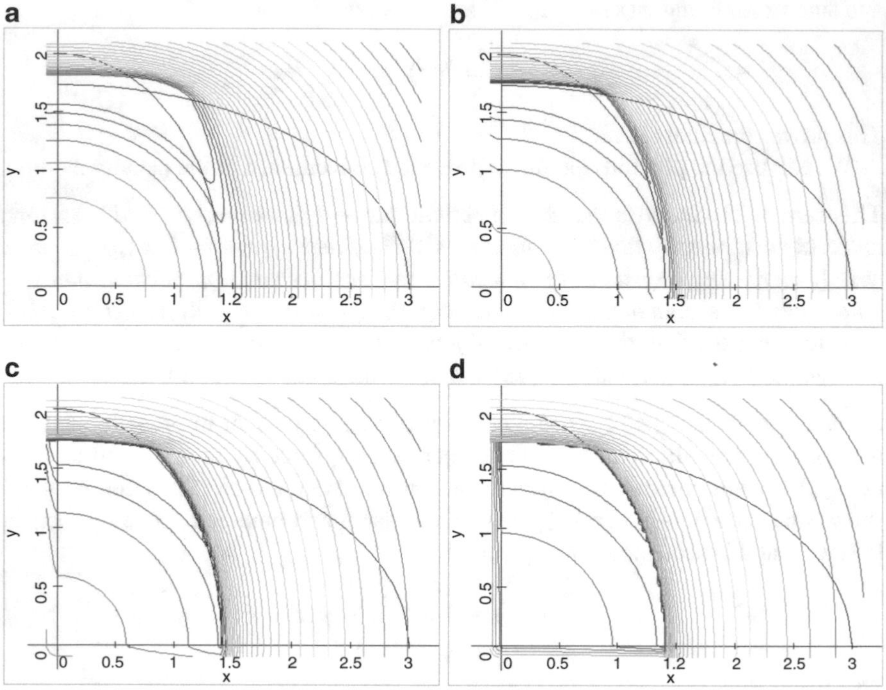

Fig. 4.4 Penalty method: contours of $\phi(\mathbf{x}) + \varrho\pi(\mathbf{x})$ for $\phi(\mathbf{x}) = -\mathbf{x}_1^2 - \mathbf{x}_2^2$, $\pi(\mathbf{x}) = \|\max\{\mathbf{0}\ ,\ \mathbf{b}(\mathbf{x})\}\|_2^2$ with $\mathbf{b}_1(\mathbf{x}) = 2\mathbf{x}_1^2 + \mathbf{x}_2^2 - 4$, $\mathbf{b}_2(\mathbf{x}) = \mathbf{x}_1^2 + 3\mathbf{x}_2^2 - 9$, $\mathbf{b}_3(\mathbf{x}) = -\mathbf{x}_1$ and $\mathbf{b}_4(\mathbf{x}) = -\mathbf{x}_2$. Constraints drawn in black. (**a**) $\varrho = 1$. (**b**) $\varrho = 10$. (**c**) $\varrho = 100$. (**d**) $\varrho = 1000$

In Fig. 4.4, we show contours of the function used in the penalty method for the constrained minimization problem displayed on the left in Fig. 4.3. The original constrained minimization problem is

$$\min -\mathbf{x}_1^2 - \mathbf{x}_2^2 \text{ subject to } 2\mathbf{x}_1^2 + \mathbf{x}_2^2 \le 4\ ,\ \mathbf{x}_1^2 + 3\mathbf{x}_2^2 \le 9 \text{ and } \mathbf{x} \ge \mathbf{0}\ .$$

We show logarithmically spaced contours of the penalty function for various values of the penalty parameter ϱ. Note that as the penalty coefficient ϱ increases, the modified function within the original feasible region becomes relatively flatter. This flatness, in combination with the lack of smoothness of the penalty function, can cause convergence problems for many unconstrained minimization algorithms. These figures were drawn by the C^{++} program penalty.C.

4.5.2 Feasible Search Strategies

Penalty methods converge to a constrained minimum along a sequence of points outside the feasible region. For problems in which constraint violation is unacceptable, we need a different approach.

4.5.2.1 Feasible Direction Methods

Feasible direction methods are easy to understand. Suppose that we are given a feasible initial guess $\mathbf{x}$ for a constrained minimization problem. Also assume that we can find some search direction $\mathbf{s}$ so that points $\mathbf{x} + \mathbf{s}\alpha$ are feasible for sufficiently small α and the objective decreases in the direction $\mathbf{s}$. We will see that we can determine such a search direction by solving a linear programming problem. Then we can perform a line search in this direction to improve the objective, and repeat the process.

To describe the ideas mathematically, we will employ the following definition.

Definition 4.5.2 Let $\mathscr{F}$ be a closed subset of real n-vectors, and let $\mathbf{x} \in \mathscr{F}$. Then the real n-vector $\mathbf{s}$ is a **feasible direction** for $\mathscr{F}$ at $\mathbf{x}$ if and only if there exists a real scalar $\overline{\alpha} > 0$ so that for all $\alpha \in (0, \overline{\alpha}]$ we have $\mathbf{x} + \mathbf{s}\alpha \in \mathscr{F}$.

Here is one possible implementation of a feasible direction method. Suppose that we want to solve the constrained minimization problem

$$\min \phi(\mathbf{x}) \text{ subject to } \mathbf{b}(\mathbf{x}) \leq \mathbf{0} \,.$$

Consider the following

Algorithm 4.5.2 (Feasible Direction)

given $\mathbf{z}_0$ such that $\mathbf{b}(\mathbf{z}_0) \leq \mathbf{0}$

for $0 \leq k$

solve the linear programming problem

$\min \delta$

subject to

$\mathbf{g_b}(\mathbf{z}_k) \cdot \mathbf{s}_k - \delta \leq 0$

$\mathbf{b}(\mathbf{z}_k) + \mathbf{J_b}(\mathbf{z}_k)\mathbf{s}_k - \mathbf{e}\delta \leq \mathbf{0}$

$\|\mathbf{s}_k\|_1 \leq 1$

if $\delta \geq 0$ then break

find $\alpha_k \in (0, 1]$ to minimize $\phi(\mathbf{z}_k + \mathbf{s}_k\alpha_k)$ subject to $\mathbf{b}(\mathbf{z}_k + \mathbf{s}_k\alpha_k) \leq \mathbf{0}$

$\mathbf{z}_{k+1} = \mathbf{z}_k + \mathbf{s}_k\alpha_k$

We remark that the constraint $\|\mathbf{s}_k\|_1 \leq 1$ in the linear programming problem within this algorithm can be converted to a standard linear programming problem constraint by writing

$$\mathbf{s}_k = \mathbf{p}_k - \mathbf{m}_k \text{ where } \mathbf{p}_k \,,\ \mathbf{m}_k \geq \mathbf{0}$$

and noting that

$$\|\mathbf{s}_k\|_1 = \mathbf{e}^\top(\mathbf{p}_k + \mathbf{m}_k) .$$

If $\delta < 0$ in the linear programming problem for the step direction $\mathbf{s}_k$, then $\mathbf{s}_k$ is a both a feasible direction for the constrained minimization problem, and a descent direction for the objective.

Otherwise we have $\delta \geq 0$. We will assume, in addition, that $\mathbf{z}_k$ is a regular point of the active constraints. Lemma 4.2.2 shows that the dual of this linear programming problem is

$$\max \mathbf{u}^\top \mathbf{b}(\mathbf{z}_k) - \omega$$

$$\text{subject to } \upsilon \mathbf{g}_\phi(\mathbf{z}_k)^\top + \mathbf{u}^\top \mathbf{J}_\mathbf{b}(\mathbf{z}_k)^\top = \omega \mathbf{e}^\top \;,\; \upsilon + \mathbf{u}^\top \mathbf{e} = 1 \text{ and } \upsilon \;,\; \mathbf{u} \;,\; \omega \geq \mathbf{0} .$$

The second Duality Theorem 4.2.8 implies that $\mathbf{u}^\top \mathbf{b}(\mathbf{z}_k) - \omega \geq 0$, so

$$\mathbf{u}^\top \mathbf{b}(\mathbf{z}_k) \geq \omega \geq 0 .$$

Since $\mathbf{u} \geq \mathbf{0}$ and $\mathbf{b}(\mathbf{z}_k) \leq \mathbf{0}$, we conclude that

$$\mathbf{u}^\top \mathbf{b}(\mathbf{z}_k) = 0 = \omega .$$

In particular, this implies that the non-basic components of $\mathbf{u}$ are all zero. Let us show by contradiction that $\upsilon > 0$. If not, then

$$\mathbf{J}_\mathbf{b}{}^\top \mathbf{u} = \mathbf{0} .$$

Then the assumption that $\mathbf{z}_k$ is a regular point of the active constraints implies that the basic components of $\mathbf{u}$ are also zero. This contradicts the dual linear programming problem constraint that

$$\upsilon + \mathbf{u}^\top \mathbf{e} = 1 .$$

Since $\upsilon > 0$, we can define

$$\mathbf{m} = \mathbf{u}/\upsilon \geq \mathbf{0} .$$

Then the constraints for the dual linear programming problem show that

$$\mathbf{g}_\phi(\mathbf{z}_k)^\top + \mathbf{m}^\top \mathbf{J}_\mathbf{b}(\mathbf{z}_k) = \mathbf{0}^\top \text{ and}$$

$$\mathbf{m}^\top \mathbf{b}(\mathbf{z}_k) = 0 .$$

These are the first-order Kuhn-Tucker necessary conditions (4.30) for a constrained minimum of ϕ. Thus when $\delta \geq 0$, the point $\mathbf{z}_k$ satisfies the first-order Kuhn-Tucker conditions for a local minimum of the original constrained minimization problem.

Feasible direction methods have lost popularity because they are unable to handle equality constraints easily, and because the linear programs being solved at each step are roughly as big as the original constrained optimization problem. For more information about feasible direction methods, see Zoutendijk [191].

4.5.2.2 Gradient Projection Methods

Suppose that we want to solve the nonlinear constrained optimization problem

$$\min \phi(\mathbf{x}) \text{ subject to } \mathbf{a}(\mathbf{x}) = \mathbf{0} \; .$$

The basic idea in the **gradient projection method** is to step in the direction of the projection of the steepest descent direction onto the tangent plane of the active constraints, and then move orthogonal to the tangent plane back to the feasible region. Let us develop some analysis to support this idea.

Given a guess $\widetilde{\mathbf{z}}$ for the optimal solution of the constrained minimization problem, we can find a permutation matrix $\mathbf{P}$ so that we can partition the active and inactive inequality constraints:

$$\mathbf{P}\mathbf{b}(\widetilde{\mathbf{z}}) = \begin{bmatrix} \mathbf{b}_B(\widetilde{\mathbf{z}}) \\ \mathbf{b}_N(\widetilde{\mathbf{z}}) \end{bmatrix} \text{ where } \mathbf{b}_B(\widetilde{\mathbf{z}}) = \mathbf{0} \text{ and } \mathbf{b}_N(\widetilde{\mathbf{z}}) < \mathbf{0} \; .$$

If $\widetilde{\mathbf{z}}$ is a regular point of the active constraints, then by using techniques from either Sect. 6.7 or 6.8 in Chap. 6 of Volume I, we can factor

$$\left[\mathbf{J}_{\mathbf{a}}(\widetilde{\mathbf{z}})^\top , \mathbf{J}_{\mathbf{b}_B}(\widetilde{\mathbf{z}})^\top\right] = \mathrm{Q}\left[\mathbf{R}_{\mathbf{a}}, \mathbf{R}_{\mathbf{b}_B}\right] ,$$

where Q has orthonormal columns and $\left[\mathbf{R}_{\mathbf{a}}, \mathbf{R}_{\mathbf{b}_B}\right]$ is right-triangular. The projection of the negative of the gradient of the objective onto the tangent plane of the active constraints is

$$\mathbf{s} = -\left[\mathbf{I} - \mathrm{Q}\mathrm{Q}^\top\right] \mathbf{g}_\phi(\widetilde{\mathbf{z}}) \; .$$

If Q was computed by the Gram-Schmidt process, then this projection should be computed by successive orthogonal projection, as in Algorithm 3.4.9 in Chap. 3 of Volume I. Note that

$$\mathbf{g}_\phi(\widetilde{\mathbf{z}}) + \mathbf{s} = \mathrm{Q}\mathrm{Q}^\top \mathbf{g}_\phi(\widetilde{\mathbf{z}}) \perp \mathbf{s} \; .$$

If $\mathbf{s}$ is nonzero, then

$$\mathbf{g}_\phi(\widetilde{\mathbf{z}}) \cdot \mathbf{s} = [\mathbf{g}_\phi(\widetilde{\mathbf{z}}) + \mathbf{s} - \mathbf{s}] \cdot \mathbf{s} = -\|\mathbf{s}\|_2^2 < 0 \; ,$$

so $\mathbf{s}$ is a descent direction.

Next, we search for $\alpha > 0$ to solve the line search

$$\min \phi(\widetilde{\mathbf{z}} + \mathbf{s}\alpha) \text{ subject to } \mathbf{b}_N(\widetilde{\mathbf{z}} + \mathbf{s}\alpha) \le \mathbf{0} \; .$$

At $\mathbf{x} = \widetilde{\mathbf{z}} + \mathbf{s}\alpha$, the constraints that were inactive at $\widetilde{\mathbf{z}}$ are still satisfied, but the constraints that were active at $\widetilde{\mathbf{z}}$ may be violated. At this point, we perform a fixed point iteration to move orthogonal to the tangent plane of the active constraints back to the feasible set. The iteration takes the form

$$\mathbf{x} \leftarrow \mathbf{x} - \mathbf{Q}\mathbf{c} \text{ where } \begin{bmatrix} \mathbf{R_a}^\top \\ \mathbf{R_{b_B}}^\top \end{bmatrix} \mathbf{c} = \begin{bmatrix} \mathbf{a}(\mathbf{x}) \\ \mathbf{b}_B(\mathbf{x}) \end{bmatrix}.$$

It is easy to see that if this fixed point iteration converges, then $\mathbf{c} = \mathbf{0}$ and the constraints that were previously active at $\widetilde{\mathbf{z}}$ will be active at $\mathbf{x}$.

In summary, we have the following

Algorithm 4.5.3 (Gradient Projection)

given $\widetilde{\mathbf{z}}$ feasible

until converged

partition $\mathbf{b}_B(\widetilde{\mathbf{z}}) = \mathbf{0}$ and $\mathbf{b}_N(\widetilde{\mathbf{z}}) < \mathbf{0}$

factor $\left[\mathbf{J_a}(\widetilde{\mathbf{z}})^\top, \mathbf{J_{b_B}}(\widetilde{\mathbf{z}})^\top\right] = \mathbf{Q}\left[\mathbf{R_a}, \mathbf{R_{b_B}}\right]$ where $\begin{cases} \mathbf{Q}^\top\mathbf{Q} = \mathbf{I} \text{ and} \\ \left[\mathbf{R_a}, \mathbf{R_{b_B}}\right] \text{ is right-triangular} \end{cases}$

$\mathbf{s} = -\left[\mathbf{I} - \mathbf{Q}\mathbf{Q}^\top\right]\mathbf{g}_\phi(\widetilde{\mathbf{z}})$

if $\|\mathbf{s}\|_2 = 0$ break

find $\alpha > 0$ to minimize $\phi(\widetilde{\mathbf{z}} + \mathbf{s}\alpha)$ subject to $\mathbf{b}_N(\widetilde{\mathbf{z}} + \mathbf{s}\alpha) \le \mathbf{0}$

$\mathbf{x} = \widetilde{\mathbf{z}} + \mathbf{s}\alpha$

until converged

solve $\begin{bmatrix} \mathbf{R_a}^\top \\ \mathbf{R_{b_B}}^\top \end{bmatrix} \mathbf{c} = \begin{bmatrix} \mathbf{a}(\mathbf{x}) \\ \mathbf{b}_B(\mathbf{x}) \end{bmatrix}$

$\widetilde{\mathbf{x}} = \mathbf{x} - \mathbf{Q}\mathbf{c}$

if $\phi(\widetilde{\mathbf{x}}) \ge \phi(\widetilde{\mathbf{z}})$ or $\mathbf{b}_N(\widetilde{\mathbf{x}}) \not\le \mathbf{0}$ then

$\alpha = \alpha/2$

$\mathbf{x} = \widetilde{\mathbf{z}} + \mathbf{s}\alpha$

else

$\mathbf{x} = \widetilde{\mathbf{x}}$

$\widetilde{\mathbf{z}} = \mathbf{x}$

Better gradient projection algorithms would implement a more sophisticated strategy for guaranteeing that the iteration reduces the objective and maintains feasibility. In particular, it would be useful to guarantee that each outer iteration achieves a sufficient decrease in the objective, and that each inner fixed-point iteration operates to enforce all inequality constraints. For more details regarding the successful implementation of gradient projection methods, see Rosen [151]. Luenberger [120, pp. 247–262] also discusses the gradient projection algorithm in greater detail, and includes an analysis of the convergence of the fixed point iteration to return to the feasible region.

4.5.2.3 Reduced Gradient Methods

The gradient projection algorithm worked with the full vector of unknowns. In contrast, **reduced gradient methods** use the active constraints to reduce the working number of unknowns. Let us see how this can be accomplished.

Suppose that we want to solve the nonlinear constrained optimization problem

$$\min \phi(\mathbf{x}) \text{ subject to } \mathbf{a}(\mathbf{x}) = \mathbf{0} \text{ and } \mathbf{b}(\mathbf{x}) \le \mathbf{0} ,$$

where $\mathbf{x}$ is a real n-vector. Given a feasible initial guess $\widetilde{\mathbf{z}}$ for the solution of this problem, we can find a permutation matrix $\mathbf{P}$ so that we can partition

$$\mathbf{P}^\top \mathbf{b}(\widetilde{\mathbf{z}}) = \begin{bmatrix} \mathbf{b}_B(\widetilde{\mathbf{z}}) \\ \mathbf{b}_N(\widetilde{\mathbf{z}}) \end{bmatrix} \text{ where } \mathbf{b}_B(\widetilde{\mathbf{z}}) = \mathbf{0} \text{ and } \mathbf{b}_N(\widetilde{\mathbf{z}}) < \mathbf{0} .$$

We will assume that $\widetilde{\mathbf{z}}$ is a regular point of the active constraints $\mathbf{a}(\widetilde{\mathbf{z}}) = \mathbf{0}$ and $\mathbf{b}_B(\widetilde{\mathbf{z}}) = \mathbf{0}$. Let $m_\mathbf{a}$ be the number of components of $\mathbf{a}$, and m_{b_B} be the number of components of $\mathbf{b}_B$. The LR Theorem 3.7.1 in Chap. 3 of Volume I shows that there is an $n \times n$ permutation matrix $\mathbf{Q}_L$ and an $(m_\mathbf{a} + m_{\mathbf{b}_B}) \times (m_\mathbf{a} + m_{\mathbf{b}_B})$ permutation matrix $\mathbf{Q}_R$ so that

$$\mathbf{Q}_L{}^\top \left[\mathbf{J}_\mathbf{a}(\widetilde{\mathbf{z}})^\top , \mathbf{J}_{\mathbf{b}_B}(\widetilde{\mathbf{z}})^\top \right] = \begin{bmatrix} \mathbf{L}_B \\ \mathbf{L}_N \end{bmatrix} \mathbf{R} \mathbf{Q}_R{}^\top ,$$

where $\mathbf{L}_B$ is unit left triangular and $\mathbf{R}$ is right-triangular with nonzero diagonal entries. We can also partition

$$\mathbf{Q}_L{}^\top \widetilde{\mathbf{z}} = \begin{bmatrix} \widetilde{\mathbf{z}}_B \\ \widetilde{\mathbf{z}}_N \end{bmatrix} \text{ and } \mathbf{Q}_L{}^\top \mathbf{g}_\phi(\widetilde{\mathbf{z}}) = \begin{bmatrix} \mathbf{g}_B(\widetilde{\mathbf{z}}) \\ \mathbf{g}_N(\widetilde{\mathbf{z}}) \end{bmatrix} ,$$

where both $\widetilde{\mathbf{z}}_B$ and $\mathbf{g}_B(\widetilde{\mathbf{z}})$ have $m_\mathbf{a} + m_{\mathbf{b}_B}$ components.

Since $\widetilde{\mathbf{z}}$ is a regular point of the active constraints, the Implicit Function Theorem 4.3.1 shows that there is a function $\mathbf{d}_B(\mathbf{x}_N)$ so that for all $\mathbf{x}_N$ in a neighborhood of $\widetilde{\mathbf{z}}_N$ we have

$$\begin{bmatrix} \mathbf{a}\left(\mathbf{Q}_L \begin{bmatrix} \mathbf{d}_B(\mathbf{x}_N) \\ \mathbf{x}_N \end{bmatrix}\right) \\ \mathbf{b}_B\left(\mathbf{Q}_L \begin{bmatrix} \mathbf{d}_B(\mathbf{x}_N) \\ \mathbf{x}_N \end{bmatrix}\right) \end{bmatrix} = \mathbf{0}\ ,\ \mathbf{d}_B(\widetilde{\mathbf{z}}_N) = \widetilde{\mathbf{z}}_B \text{ and } \mathbf{J}_{\mathbf{d}_B}(\widetilde{\mathbf{z}}_N) = -\mathbf{L}_B{}^{-\top}\mathbf{L}_N{}^{\top}\ . \tag{4.54}$$

Define the constrained objective function to be

$$\gamma(\mathbf{x}_N) = \phi\left(\mathbf{Q}_L \begin{bmatrix} \mathbf{d}_B(\mathbf{x}_N) \\ \mathbf{x}_N \end{bmatrix}\right)\ ,$$

and the inactive constraint function to be

$$\mathbf{c}_N(\mathbf{x}_N) = \mathbf{b}_N\left(\mathbf{Q}_L \begin{bmatrix} \mathbf{d}_B(\mathbf{x}_N) \\ \mathbf{x}_N \end{bmatrix}\right)\ .$$

Then our original constrained minimization problem can be written in the form

$$\min \gamma(\mathbf{x}_N) \text{ subject to } \mathbf{c}_N(\mathbf{x}_N) \le \mathbf{0}\ .$$

Of course, $\widetilde{\mathbf{z}}_N$ is feasible for this revised problem, and the constraints are all inactive at this point. Note that the gradient of the constrained objective is

$$\mathbf{g}_\gamma(\widetilde{\mathbf{z}}_N)^\top = \mathbf{g}_\phi(\widetilde{\mathbf{z}})^\top \left.\frac{\partial \mathbf{Q}_L \begin{bmatrix} \mathbf{d}_B(\mathbf{x}_N) \\ \mathbf{x}_N \end{bmatrix}}{\partial \mathbf{x}_N}\right|_{\mathbf{x}_N=\widetilde{\mathbf{z}}_N} = \left[\mathbf{g}_B(\widetilde{\mathbf{z}})^\top,\ \mathbf{g}_N(\widetilde{\mathbf{z}})^\top\right] \mathbf{Q}_L{}^\top \mathbf{Q}_L \begin{bmatrix} \mathbf{J}_{\mathbf{d}_B}(\widetilde{\mathbf{z}}_N) \\ \mathbf{I} \end{bmatrix}$$
$$= \mathbf{g}_N(\widetilde{\mathbf{z}})^\top - \mathbf{g}_B(\widetilde{\mathbf{z}})^\top \mathbf{L}_B{}^{-\top}\mathbf{L}_N{}^{\top}\ .$$

This is often called the **reduced gradient**.

If $\mathbf{g}_\gamma(\widetilde{\mathbf{z}}_N) \neq \mathbf{0}$, then we can find a descent direction $\mathbf{s}_N$ so that

$$0 > \mathbf{g}_\gamma(\widetilde{\mathbf{z}}_N)^\top \mathbf{s}_N\ .$$

For example, we could use the steepest descent direction

$$\mathbf{s}_N = -\mathbf{g}_\gamma(\widetilde{\mathbf{z}}_N)\ ,$$

or we could use another descent direction described in Sect. 3.5.3. Then we can perform a line search to find $\alpha > 0$ to solve

$$\min \gamma(\widetilde{\mathbf{z}}_N + \mathbf{s}_N\alpha) \text{ subject to } \mathbf{c}_N(\widetilde{\mathbf{z}}_N + \mathbf{s}_N\alpha) \leq \mathbf{0} . \tag{4.55}$$

Ideas for performing line searches were presented in Sect. 3.5.3.4. Note that for each value of α, the evaluation of either the objective or the constraint in this minimization problem requires that we evaluate $\mathbf{d}_B(\widetilde{\mathbf{z}}_N + \mathbf{s}_N\alpha)$, which is defined implicitly by the nonlinear system

$$\mathbf{0} = \begin{bmatrix} \mathbf{a}\left(\mathrm{Q}_L \begin{bmatrix} \mathbf{d}_B \\ \widetilde{\mathbf{z}}_N + \mathbf{s}_N\alpha \end{bmatrix}\right) \\ \mathbf{b}_B\left(\mathrm{Q}_L \begin{bmatrix} \mathbf{d}_B \\ \widetilde{\mathbf{z}}_N + \mathbf{s}_N\alpha \end{bmatrix}\right) \end{bmatrix} .$$

A Newton iteration to determine $\mathbf{d}_B$ would take the form

$$\mathbf{d}_B^{(k+1)} = \mathbf{d}_B^{(k)} - \mathrm{L}_B{}^{-\top}\mathbf{R}^{-\top}\mathbf{Q}_R{}^\top \begin{bmatrix} \mathbf{a}\left(\mathrm{Q}_L \begin{bmatrix} \mathbf{d}_B^{(k)} \\ \widetilde{\mathbf{z}}_N + \mathbf{s}_N\alpha \end{bmatrix}\right) \\ \mathbf{b}_B\left(\mathrm{Q}_L \begin{bmatrix} \mathbf{d}_B^{(k)} \\ \widetilde{\mathbf{z}}_N + \mathbf{s}_N\alpha \end{bmatrix}\right) \end{bmatrix} ,$$

where we have factored

$$\mathrm{Q}_L{}^\top \left[\mathbf{J}_\mathbf{a}\left(\mathrm{Q}_L \begin{bmatrix} \mathbf{d}_B^{(k)} \\ \widetilde{\mathbf{z}}_N + \mathbf{s}_N\alpha \end{bmatrix}\right)^\top , \mathbf{J}_{\mathbf{b}_B}\left(\mathrm{Q}_L \begin{bmatrix} \mathbf{d}_B^{(k)} \\ \widetilde{\mathbf{z}}_N + \mathbf{s}_N\alpha \end{bmatrix}\right)^\top \right] = \begin{bmatrix} \mathrm{L}_B \\ \mathrm{L}_N \end{bmatrix} \mathbf{R}\mathbf{Q}_R{}^\top .$$

In other word, this Newton iteration requires repeated evaluation of the Jacobians of the active constraints, as well as their factorization in preparation for solving linear systems.

At the end of the line search (4.55), we update the current guess at the solution:

$$\widetilde{\mathbf{z}} \leftarrow \widetilde{\mathbf{z}} + \mathbf{s}\alpha .$$

If the solution of the line search (4.55) occurs at a constraint, then there exists an index i_N such that

$$0 = \mathbf{e}_{i_N} \cdot \mathbf{b}_N(\widetilde{\mathbf{z}}) .$$

In such a case, constraint i_N becomes active, and we repeat the previous computations subject to the new set of active constraints.

If the reduced gradient is zero at some feasible guess, then we cannot perform a line search. In such a case,

$$\mathbf{0}^\top = \mathbf{g}_\gamma(\widetilde{\mathbf{z}}_N)^\top = \mathbf{g}_N(\widetilde{\mathbf{z}})^\top - \mathbf{g}_B(\widetilde{\mathbf{z}})^\top \mathrm{L}_B{}^{-\top} \mathrm{L}_N{}^\top ,$$

so

$$\begin{aligned}
\mathbf{g}_\phi(\widetilde{\mathbf{z}})^\top &= \left[\mathbf{g}_B(\widetilde{\mathbf{z}})^\top , \mathbf{g}_N(\widetilde{\mathbf{z}})^\top\right] \mathrm{Q}_L{}^\top \\
&= \mathbf{g}_B(\widetilde{\mathbf{z}})^\top \mathrm{L}_B{}^{-\top} \mathbf{R}^{-\top} \mathbf{Q}_R{}^\top \left[\mathbf{Q}_R \mathbf{R}^\top \mathrm{L}_B{}^\top , \mathbf{Q}_R \mathbf{R}^\top \mathrm{L}_N{}^\top\right] \mathrm{Q}_L{}^\top \\
&= \mathbf{g}_B(\widetilde{\mathbf{z}})^\top \mathrm{L}_B{}^{-\top} \mathbf{R}^{-\top} \mathbf{Q}_R{}^\top \begin{bmatrix} \mathbf{J_a}(\widetilde{\mathbf{z}}) \\ \mathbf{J}_{\mathbf{b}_B}(\widetilde{\mathbf{z}}) \end{bmatrix} \equiv -\left[\boldsymbol{\ell}^\top , \mathbf{m}_B\right] \begin{bmatrix} \mathbf{J_a}(\widetilde{\mathbf{z}}) \\ \mathbf{J}_{\mathbf{b}_B}(\widetilde{\mathbf{z}}) \end{bmatrix} .
\end{aligned}$$

If $\mathbf{m}_B \geq \mathbf{0}$, then the first-order Kuhn-Tucker necessary conditions for the original constrained minimization problem are satisfied at $\widetilde{\mathbf{z}}$. Otherwise, there is a component j_B of $\mathbf{m}_B$ with

$$0 > \mathbf{e}_{j_B} \cdot \mathbf{m}_B .$$

We can solve

$$\mathbf{Q}_R \mathbf{R}^\top \mathrm{L}_B{}^\top \mathbf{s}_B = \begin{bmatrix} \mathbf{0} \\ \mathbf{e}_{j_B} \end{bmatrix}$$

for the $(m_\mathbf{a} + m_{\mathbf{b}_B})$-vector $\mathbf{s}_B$. Then

$$\begin{aligned}
\mathbf{g}_\phi(\widetilde{\mathbf{z}})^\top \mathrm{Q}_L \begin{bmatrix} \mathbf{s}_B \\ \mathbf{0} \end{bmatrix} &= -\left[\boldsymbol{\ell}^\top , \mathbf{m}_B\right] \begin{bmatrix} \mathbf{J_a}(\widetilde{\mathbf{z}}) \\ \mathbf{J}_{\mathbf{b}_B}(\widetilde{\mathbf{z}}) \end{bmatrix} \mathrm{Q}_L \begin{bmatrix} \mathbf{s}_B \\ \mathbf{0} \end{bmatrix} \\
&= -\left[\boldsymbol{\ell}^\top , \mathbf{m}_B\right] \mathbf{Q}_R \mathbf{R}^\top \left[\mathrm{L}_B{}^\top , \mathrm{L}_N{}^\top\right] \mathrm{Q}_L{}^\top \mathrm{Q}_L \begin{bmatrix} \mathbf{s}_B \\ \mathbf{0} \end{bmatrix} = -\left[\boldsymbol{\ell}^\top , \mathbf{m}_B\right] \begin{bmatrix} \mathbf{0} \\ \mathbf{e}_{j_B} \end{bmatrix} < 0 ,
\end{aligned}$$

and

$$\begin{bmatrix} \mathbf{J_a}(\widetilde{\mathbf{z}}) \\ \mathbf{J}_{\mathbf{b}_B}(\widetilde{\mathbf{z}}) \end{bmatrix} \mathrm{Q}_L \begin{bmatrix} \mathbf{s}_B \\ \mathbf{0} \end{bmatrix} = \mathbf{Q}_R \mathbf{R}^\top \left[\mathrm{L}_B{}^\top , \mathrm{L}_N{}^\top\right] \begin{bmatrix} \mathbf{s}_B \\ \mathbf{0} \end{bmatrix} = \begin{bmatrix} \mathbf{0} \\ \mathbf{e}_{j_B} \end{bmatrix} .$$

Thus

$$\mathbf{s} = \mathrm{Q}_L \begin{bmatrix} \mathbf{s}_B \\ \mathbf{0} \end{bmatrix}$$

is a descent direction for ϕ, and $\mathbf{s}$ lies in the plane tangent to the hypersurface of all active constraints at $\widetilde{\mathbf{z}}$ except for inequality constraint with index $\mathbf{j}_B$. In this case, we drop active inequality constraint j_B and repeat the previous computations with the remaining active constraints.

In summary, we have the following

Algorithm 4.5.4 (Reduced Gradient)

given $\widetilde{\mathbf{z}}$ such that $\mathbf{a}(\widetilde{\mathbf{z}}) = \mathbf{0}$ and $\mathbf{b}(\widetilde{\mathbf{z}}) \leq \mathbf{0}$

while not converged

permute and partition $\mathbf{P}^\top \mathbf{b}(\widetilde{\mathbf{z}}) = \begin{bmatrix} \mathbf{b}_B(\widetilde{\mathbf{z}}) \\ \mathbf{b}_N(\widetilde{\mathbf{z}}) \end{bmatrix}$ where $\begin{cases} \mathbf{b}_B(\widetilde{\mathbf{z}}) = \mathbf{0} \text{ and} \\ \mathbf{b}_N(\widetilde{\mathbf{z}}) < \mathbf{0} \end{cases}$

factor $\mathbf{Q}_L{}^\top \left[\mathbf{J_a}\left(\mathbf{Q}_L \begin{bmatrix} \mathbf{d}_B^{(k)} \\ \widetilde{\mathbf{z}}_N + \mathbf{s}_N\alpha \end{bmatrix} \right)^\top , \mathbf{J}_{\mathbf{b}_B}\left(\mathbf{Q}_L \begin{bmatrix} \mathbf{d}_B^{(k)} \\ \widetilde{\mathbf{z}}_N + \mathbf{s}_N\alpha \end{bmatrix} \right)^\top \right] = \begin{bmatrix} \mathbf{L}_B \\ \mathbf{L}_N \end{bmatrix} \mathbf{R}\mathbf{Q}_R{}^\top$.

permute and partition $\mathbf{g}_\phi(\widetilde{\mathbf{z}})^\top \mathbf{Q}_L = \left[\mathbf{g}_B(\widetilde{\mathbf{z}})^\top , \mathbf{g}_N(\widetilde{\mathbf{z}})^\top \right]$

$\mathbf{g}_\gamma(\widetilde{\mathbf{z}}_N)^\top = \mathbf{g}_N(\widetilde{\mathbf{z}})^\top - \mathbf{g}_B(\widetilde{\mathbf{z}})^\top \mathbf{L}_B{}^{-\top} \mathbf{L}_N{}^\top$

if $\mathbf{g}_\gamma(\widetilde{\mathbf{z}}_N)^\top \neq \mathbf{0}$

choose a descent direction $\mathbf{s}_N$ so that $\mathbf{g}_\gamma(\widetilde{\mathbf{z}}_N)^\top \mathbf{s}_N < 0$

let $\mathbf{d}_B$ be defined implicitly by (4.54)

find $\alpha > 0$ to solve $\min \gamma(\widetilde{\mathbf{z}}_N + \mathbf{s}_N\alpha) \equiv \phi\left(\mathbf{Q}_L \begin{bmatrix} \mathbf{d}_B(\widetilde{\mathbf{z}}_N + \mathbf{s}_N\alpha) \\ \widetilde{\mathbf{z}}_N + \mathbf{s}_N\alpha \end{bmatrix} \right)$

subject to $\mathbf{0} \geq \mathbf{c}_N(\widetilde{\mathbf{z}}_N + \mathbf{s}_N\alpha) \equiv \mathbf{b}_N\left(\mathbf{Q}_L \begin{bmatrix} \mathbf{d}_B(\widetilde{\mathbf{z}}_N + \mathbf{s}_N\alpha) \\ \widetilde{\mathbf{z}}_N + \mathbf{s}_N\alpha \end{bmatrix} \right)$

$\widetilde{\mathbf{z}} = \mathbf{Q}_L \begin{bmatrix} \mathbf{d}_B(\widetilde{\mathbf{z}}_N + \mathbf{s}_N\alpha) \\ \widetilde{\mathbf{z}}_N + \mathbf{s}_N\alpha \end{bmatrix}$

if there exists i_N so that $\mathbf{e}_{i_B} \cdot \mathbf{b}_N(\widetilde{\mathbf{z}}) = 0$ then add constraint i_N to list of active constraints

else

solve $\mathbf{L}_B\mathbf{R}\mathbf{Q}_R \begin{bmatrix} \boldsymbol{\ell} \\ \mathbf{m} \end{bmatrix} = -\mathbf{g}_B(\widetilde{\mathbf{z}})$

if $\mathbf{m} \geq \mathbf{0}$ break

find j_B so that $\mathbf{m}^\top \mathbf{e}_{j_B} < 0$

remove inequality constraint j_B from list of active constraints

For more information about reduced gradient methods, see Abadie [1], Gill et al. [81, pp. 219–224] or Luenberger [120, pp. 240–272]. Lasdon et al. [119] discuss implementation details of a particular reduced gradient algorithm. Extensions to quasi-Newton methods are discussed by Sargent and Murtagh [156].

4.5.2.4 Interior Point Methods

Our next method for solving nonlinear programming problems was originally suggested by Carroll [35], and developed more fully by Fiacco and McCormick [66]. This approach uses the following idea.

Definition 4.5.3 Let β be a real-valued functional operating on positive q-vectors. Then β is a **barrier function** if and only if for all component indices i we have

$$\beta(\mathbf{b}) \uparrow \infty \text{ as } \mathbf{e}_i \cdot \mathbf{b} \downarrow 0 \; .$$

The basic idea is to replace the constrained minimization problem

$$\min \phi(\mathbf{x}) \text{ subject to } \mathbf{a}(\mathbf{x}) = \mathbf{0} \text{ and } \mathbf{b}(\mathbf{x}) \leq \mathbf{0}$$

by a sequence of nonlinear programming problems

$$\min \phi(\mathbf{x}) + \tau(\varrho_k)\beta(-\mathbf{b}(\mathbf{x})) \text{ subject to } \mathbf{a}(\mathbf{x}) = \mathbf{0} \text{ and } \mathbf{b}(\mathbf{x}) < \mathbf{0} \; ,$$

where β is a barrier function. The next lemma provides circumstances under which this approach is successful.

Lemma 4.5.2 *Suppose that $p < n$ and q are nonnegative integers. Let ϕ be a real functional on real n-vectors,* **a** *be a mapping from real n-vectors to real p-vectors, and* **b** *be a mapping from real n-vectors to real q-vectors. Let*

$$\mathscr{F} = \{\mathbf{x} : \mathbf{a}(\mathbf{x}) = \mathbf{0} \textit{ and } \mathbf{b}(\mathbf{x}) \leq \mathbf{0}\} \textit{ and } \mathscr{F}^o = \{\mathbf{x} : \mathbf{a}(\mathbf{x}) = \mathbf{0} \textit{ and } \mathbf{b}(\mathbf{x}) < \mathbf{0}\} \; .$$

Assume that there is a real n-vector **z** *that solves the nonlinear programming problem*

$$\min \phi(\mathbf{x}) \textit{ subject to } \mathbf{x} \in \mathscr{F} \; . \tag{4.56}$$

Next, suppose that the sequence $\{\varrho_k\}_{k=0}^{\infty}$ of real positive numbers strictly decreases to zero. Also assume that $\tau(\varrho)$ is a strictly increasing function on positive real numbers. Let β be a barrier function, and for each index $k \geq 0$ define the modified objective

$$\psi_k(\mathbf{x}) = \phi(\mathbf{x}) + \tau(\varrho_k)\beta(-\mathbf{b}(\mathbf{x})) \; . \tag{4.57}$$

Suppose that $\mathbf{z}_k$ *solves the nonlinear programming problem*

$$\min \psi_k(\mathbf{x}) \text{ subject to } \mathbf{x} \in \mathscr{F}^o \,. \tag{4.58}$$

Then

1. *the sequence* $\{\psi_k(\mathbf{z}_k)\}_{k=0}^{\infty}$ *is nonincreasing, and bounded below by* $\phi(\mathbf{z})$*;*
2. *if* ϕ, $\mathbf{a}$ *and* $\mathbf{b}$ *are continuous, then* $\{\psi_k(\mathbf{z}_k)\}_{k=0}^{\infty} \downarrow \phi(\mathbf{z})$*; and*
3. *if, in addition, there exists* $\mathbf{x}_L \in \mathscr{F}$ *so that* $\phi(\mathbf{x}_L) > \phi(\mathbf{z})$ *and*

$$\mathscr{L}(\mathbf{x}_L) = \{\mathbf{x} \in \mathscr{F} : \phi(\mathbf{x}) \leq \phi(\mathbf{x}_L)\} \tag{4.59}$$

   is closed and bounded, then there is a subsequence of $\{\mathbf{z}_k\}_{k=0}^{\infty}$ *that converges to an optimal solution of* (4.56).
4. *if* ϕ, $\mathbf{a}$ *and* $\mathbf{b}$ *are continuous, the feasible set* $\mathscr{F}$ *is bounded and* $\mathscr{F}^o$ *is nonempty, then* $\{\mathbf{z}_k\}_{k=0}^{\infty} \subset \mathscr{F}^o$.

Proof It is easy to prove the first claim. Since $\varrho_k > \varrho_{k+1}$ for all $k \geq 0$, τ is strictly increasing and β is positive, we have

$$\psi_k(\mathbf{z}_k) = \phi(\mathbf{z}_k) + \tau(\varrho_k)\beta(-\mathbf{b}(\mathbf{z}_k)) \geq \phi(\mathbf{z}_k) + \tau(\varrho_{k+1})\beta(-\mathbf{b}(\mathbf{z}_k))$$

then since $\mathbf{z}_{k+1}$ is optimal, we get

$$\geq \phi(\mathbf{z}_{k+1}) + \tau(\varrho_{k+1})\beta(-\mathbf{b}(\mathbf{z}_{k+1}))$$

and because both τ and β are positive, we have

$$> \phi(\mathbf{z}_{k+1})$$

and finally, because $\mathbf{z}_{k+1} \in \mathscr{F}$ and $\mathbf{z}$ is optimal, we conclude that

$$\geq \phi(\mathbf{z}) \,.$$

This proves the first claim.

Since $\{\psi_k(\mathbf{z}_k)\}_{k=0}^{\infty}$ is nonincreasing and bounded below by $\phi(\mathbf{z})$, there exists $\gamma \geq \phi(\mathbf{z})$ so that

$$\psi_k(\mathbf{z}_k) \downarrow \gamma \,.$$

We will now prove the second claim by contradiction. Suppose that $\gamma > \phi(\mathbf{z})$. Since $\mathbf{a}$ and $\mathbf{b}$ are continuous, the feasible set $\mathscr{F}$ is closed. Since ϕ is continuous, there exists $\delta > 0$ so that the feasible neighborhood

$$\mathscr{F}_\delta(\mathbf{z}) \equiv \{\mathbf{x} \in \mathscr{F} : \|\mathbf{x} - \mathbf{z}\|_2 \leq \varepsilon\}$$

is nonempty and

$$\phi(\mathbf{x}) \leq [\gamma + \phi(\mathbf{z})]/2$$

for all $\mathbf{x} \in \mathscr{F}_\delta(\mathbf{z})$. Choose $\widetilde{\mathbf{x}} \in \mathscr{F}_\delta(\mathbf{z})$. Since $\{\varrho_k\}_{k=0}^\infty$ is strictly decreasing to zero and τ is strictly increasing, there exists an integer $\underline{k} > 0$ so that for all $k \geq \underline{k}$ we have

$$\tau(\varrho_k)\beta(-\mathbf{b}(\widetilde{\mathbf{x}})) < [\gamma - \phi(\mathbf{z})]/4 \; .$$

Since $\mathbf{z}_k$ is optimal and $\widetilde{\mathbf{x}}$ is feasible, for all $k \geq \underline{k}$ we have

$$\psi_k(\mathbf{z}_k) = \phi(\mathbf{z}_k) + \tau(\varrho_k)\beta(-\mathbf{b}(\mathbf{z}_k)) \leq \phi(\widetilde{\mathbf{x}}) + \tau(\varrho_k)\beta(-\mathbf{b}(\widetilde{\mathbf{x}}))$$

then, since $\widetilde{\mathbf{x}} \in \mathscr{F}_\delta(\mathbf{z})$ and $k \geq \underline{k}$ we get

$$\leq [\gamma + \phi(\mathbf{z})]/2 + [\gamma - \phi(\mathbf{z})]/4 = \gamma - [\gamma - \phi(\mathbf{z})]/4 \; .$$

This contradicts the fact that $\psi_k(\mathbf{z}_k)$ converges to γ from above, and proves the second claim.

Next, let us prove the third claim. Since $\{\psi_k(\mathbf{z}_k)\}_{k=0}^\infty$ converges to $\phi(\mathbf{z})$, $\psi_k(\mathbf{z}_k) \geq \phi(\mathbf{z}_k)$ and $\phi(\mathbf{x}_L) > \phi(\mathbf{z})$, there exists $\underline{k}' \geq 0$ so that for all $k \geq \underline{k}'$ we have

$$\phi(\mathbf{z}_k) \leq \phi(\mathbf{x}_L) \; .$$

Since the feasible level set $\mathscr{L}(\mathbf{x}_L)$ is closed and bounded, there is a subsequence of $\{\mathbf{z}_k\}_{k=0}^\infty$ that converges to $\widetilde{\mathbf{z}}$. Since the feasible set $\mathscr{F}$ is closed, we must have $\widetilde{\mathbf{z}} \in \mathscr{F}$. Suppose that $\phi(\widetilde{\mathbf{z}}) > \phi(\mathbf{z})$. Then

$$\psi_k(\mathbf{z}_k) - \phi(\mathbf{z}) = \phi(\mathbf{z}_k) + \tau(\varrho_k)\beta(-\mathbf{b}(\mathbf{z}_k)) - \phi(\mathbf{z}) \rightarrow \phi(\widetilde{\mathbf{z}}) - \phi(\mathbf{z}) > 0 \; .$$

This contradicts the second claim, in which we proved that $\psi_k(\mathbf{z}_k) \downarrow \phi(\mathbf{z})$. We conclude that $\phi(\widetilde{\mathbf{z}}) = \phi(\mathbf{z})$, so $\widetilde{\mathbf{z}}$ is optimal.

Finally, we will prove the fourth claim. For any $\varrho > 0$, let

$$\alpha(\varrho) = \inf\{\phi(\mathbf{x}) + \tau(\varrho)\beta(-\mathbf{b}(\mathbf{x})) : \mathbf{x} \in \mathscr{F}^o\} \; .$$

Since $\mathscr{F}^o$ is nonempty, we conclude that $\alpha(\varrho)$ is finite. Then there exists a sequence $\{\mathbf{x}_k\}_{k=0}^\infty \subset \mathscr{F}$ such that

$$\phi(\mathbf{x}_k) + \tau(\varrho)\beta(-\mathbf{b}(\mathbf{x}_k)) \rightarrow \alpha(\varrho) \; .$$

Since $\mathscr{F}$ is closed and bounded, there is a subsequence that converges to $\widetilde{\mathbf{x}} \in \mathscr{F}$. Suppose that $\widetilde{\mathbf{x}} \notin \mathscr{F}^o$. Then $\mathbf{b}(\widetilde{\mathbf{x}}) \not> \mathbf{0}$, so $\beta(-\mathbf{b}(\widetilde{\mathbf{x}})) = \infty$ and

$$\alpha(\varrho) = \phi(\widetilde{\mathbf{x}}) + \lim_{\mathbf{x}_k \rightarrow \widetilde{\mathbf{x}}} [\tau(\varrho)\beta(-\mathbf{b}(\mathbf{x}_k))] = \infty \; .$$

This contradicts the fact that $\alpha(\varrho)$ is finite. Thus $\widetilde{\mathbf{x}} \in \mathscr{F}^o$, and the minimum of $\phi(\mathbf{x}) + \tau(\varrho)\beta(-\mathbf{b}(\mathbf{x}))$ is attained at $\widetilde{\mathbf{x}}$.

The fourth result in Lemma 4.5.2 explains why techniques based on minimizing the modified objective in (4.57) are called **interior point methods**.

An early paper using trust regions for interior point methods is due to Byrd et al. [34]. Readers may also be interested in interior point methods developed by Forsgren et al. [72] or Wächter and Biegler [179]. Our remaining discussion of interior point methods will follow the ideas in Waltz et al. [180].

In a loop over strictly decreasing values of some penalty parameter τ, we want to solve the nonlinear programming problem with barrier

$$\min \phi(\mathbf{x}) + \tau\beta(\mathbf{s}) \text{ subject to } \mathbf{a}(\mathbf{x}) = \mathbf{0} \text{ and } \mathbf{b}(\mathbf{x}) + \mathbf{s} = \mathbf{0} . \tag{4.60}$$

The first-order Kuhn-Tucker conditions for a solution of this problem generate the nonlinear system

$$\mathbf{0} = \mathbf{g}_\lambda(\mathbf{x}\,,\,\mathbf{s}\,,\,\boldsymbol{\ell}_\mathbf{a}\,,\,\boldsymbol{\ell}_\mathbf{b}) \equiv \begin{bmatrix} \mathbf{g}_\phi(\mathbf{x}) + \mathbf{J}_\mathbf{a}(\mathbf{x})^\top \boldsymbol{\ell}_\mathbf{a} + \mathbf{J}_\mathbf{b}(\mathbf{x})^\top \boldsymbol{\ell}_\mathbf{b} \\ \mathbf{g}_\beta(\mathbf{s})\tau + \boldsymbol{\ell}_\mathbf{b} \\ \mathbf{a}(\mathbf{x}) \\ \mathbf{b}(\mathbf{x}) + \mathbf{s} \end{bmatrix} . \tag{4.61}$$

Here λ is the Lagrangian function for the barrier problem, namely

$$\lambda(\mathbf{x}\,,\,\mathbf{s}\,,\,\boldsymbol{\ell}_\mathbf{a}\,,\,\boldsymbol{\ell}_\mathbf{b}) = \phi(\mathbf{x}) + \tau\beta(\mathbf{s}) + {\boldsymbol{\ell}_\mathbf{a}}^\top \mathbf{a}(\mathbf{x}) + {\boldsymbol{\ell}_\mathbf{b}}^\top [\mathbf{b}(\mathbf{x}) + \mathbf{s}] .$$

Newton's method for the solution of the nonlinear system (4.61) takes the form

$$\begin{bmatrix} \mathbf{H}(\mathbf{x}\,,\,\boldsymbol{\ell}_\mathbf{a}\,,\,\boldsymbol{\ell}_\mathbf{b}) & \mathbf{0} & \mathbf{J}_\mathbf{a}(\mathbf{x})^\top & \mathbf{J}_\mathbf{b}(\mathbf{x})^\top \\ \mathbf{0} & \mathbf{H}_\beta(\mathbf{s})\tau & \mathbf{0} & \mathbf{I} \\ \mathbf{J}_\mathbf{a}(\mathbf{x}) & \mathbf{0} & \mathbf{0} & \mathbf{0} \\ \mathbf{J}_\mathbf{b}(\mathbf{x}) & \mathbf{I} & \mathbf{0} & \mathbf{0} \end{bmatrix} \begin{bmatrix} \delta\mathbf{x} \\ \delta\mathbf{s} \\ \delta\boldsymbol{\ell}_\mathbf{a} \\ \delta\boldsymbol{\ell}_\mathbf{b} \end{bmatrix} = - \begin{bmatrix} \mathbf{g}_\phi(\mathbf{x}) + \mathbf{J}_\mathbf{a}(\mathbf{x})^\top \boldsymbol{\ell}_\mathbf{a} + \mathbf{J}_\mathbf{b}(\mathbf{x})^\top \boldsymbol{\ell}_\mathbf{b} \\ \mathbf{g}_\beta(\mathbf{s})\tau + \boldsymbol{\ell}_\mathbf{b} \\ \mathbf{a}(\mathbf{x}) \\ \mathbf{b}(\mathbf{x}) + \mathbf{s} \end{bmatrix} . \tag{4.62}$$

Here

$$\mathbf{H}(\mathbf{x}\,,\,\boldsymbol{\ell}_\mathbf{a}\,,\,\boldsymbol{\ell}_\mathbf{b}) \equiv \mathbf{H}_\phi(\mathbf{x}) + \sum_i {\boldsymbol{\ell}_\mathbf{a}}^\top \mathbf{e}_i \mathbf{H}_{\mathbf{a}_i}(\mathbf{x}) + \sum_j {\boldsymbol{\ell}_\mathbf{b}}^\top \mathbf{e}_j \mathbf{H}_{\mathbf{b}_j}(\mathbf{x}) .$$

In an outer loop over strictly decreasing values of some penalty parameter τ and an inner loop over guesses $\mathbf{x}$, $\mathbf{s}$, $\boldsymbol{\ell}_\mathbf{a}$ and $\boldsymbol{\ell}_\mathbf{b}$ for the minimization of (4.60), we want to choose a step length for the increments given by the Newton equations (4.62). This step length will be chosen by an approximate line search to find a sufficient decrease in the **merit function**

$$\psi_{\mu,\tau}(\mathbf{x},\mathbf{s}) \equiv \phi(\mathbf{x}) + \tau\beta(\mathbf{s}) + \mu\left\{\mathbf{a}(\mathbf{x})^\top\mathbf{a}(\mathbf{x}) + [\mathbf{b}(\mathbf{x})+\mathbf{s}]^\top[\mathbf{b}(\mathbf{x})+\mathbf{s}]\right\}^{1/2} . \tag{4.63}$$

Choose $\varrho \in (0,1)$. It is not hard to see that if μ is chosen to be positive and sufficiently large, namely

$$\mu \geq \frac{\mathbf{g}_\phi(\mathbf{x})^\top\delta\mathbf{x} + \tau\mathbf{g}_\beta(\mathbf{s})^\top\delta\mathbf{s} + \frac{1}{2}\max\left\{(\delta\mathbf{x})^\top\mathbf{H}(\mathbf{x}\,,\,\boldsymbol{\ell}_\mathbf{a}\,,\,\boldsymbol{\ell}_\mathbf{b})\delta\mathbf{x} + (\delta\mathbf{s})^\top\mathbf{H}_\beta(\mathbf{s})\delta\mathbf{s}\,,\,0\right\}}{(1-\varrho)\left\{\mathbf{a}(\mathbf{x})^\top\mathbf{a}(\mathbf{x}) + [\mathbf{b}(\mathbf{x})+\mathbf{s}]^\top[\mathbf{b}(\mathbf{x})+\mathbf{s}]\right\}^{1/2}} \tag{4.64}$$

then

$$\begin{aligned}
&\mathbf{g}_{\psi_{\mu,\tau}}(\mathbf{x}\,,\,\mathbf{s})^\top\begin{bmatrix}\delta\mathbf{s}\\ \delta\mathbf{s}\end{bmatrix}\\
&= \mathbf{g}_\phi(\mathbf{x})^\top\delta\mathbf{x} + \mu\frac{\mathbf{a}(\mathbf{x})^\top\mathbf{J}_\mathbf{a}(\mathbf{x})\delta\mathbf{x} + [\mathbf{b}(\mathbf{x})+\mathbf{s}]^\top\mathbf{J}_\mathbf{b}(\mathbf{x})\delta\mathbf{x}}{\left\{\mathbf{a}(\mathbf{x})^\top\mathbf{a}(\mathbf{x}) + [\mathbf{b}(\mathbf{x})+\mathbf{s}]^\top[\mathbf{b}(\mathbf{x})+\mathbf{s}]\right\}^{1/2}} + \tau\mathbf{g}_\beta(\mathbf{s})^\top\delta\mathbf{s}\\
&+ \mu\frac{[\mathbf{b}(\mathbf{x})+\mathbf{s}]^\top\delta\mathbf{s}}{\left\{\mathbf{a}(\mathbf{x})^\top\mathbf{a}(\mathbf{x}) + [\mathbf{b}(\mathbf{x})+\mathbf{s}]^\top[\mathbf{b}(\mathbf{x})+\mathbf{s}]\right\}^{1/2}}
\end{aligned}$$

then we use the last two equations in the Newton equations (4.62) to get

$$= \mathbf{g}_\phi(\mathbf{x})^\top\delta\mathbf{x} + \tau\mathbf{g}_\beta(\mathbf{s})^\top\delta\mathbf{s} - \mu\left\{\mathbf{a}(\mathbf{x})^\top\mathbf{a}(\mathbf{x}) + [\mathbf{b}(\mathbf{x})+\mathbf{s}]^\top[\mathbf{b}(\mathbf{x})+\mathbf{s}]\right\}^{1/2}$$

then the lower bound (4.64) on μ implies that

$$\begin{aligned}
&\leq -\mu\varrho\left\{\mathbf{a}(\mathbf{x})^\top\mathbf{a}(\mathbf{x}) + [\mathbf{b}(\mathbf{x})+\mathbf{s}]^\top[\mathbf{b}(\mathbf{x})+\mathbf{s}]\right\}^{1/2}\\
&-\frac{1}{2}\max\left\{(\delta\mathbf{x})^\top\mathbf{H}(\mathbf{x}\,,\,\boldsymbol{\ell}_\mathbf{a}\,,\,\boldsymbol{\ell}_\mathbf{b})\delta\mathbf{x} + (\delta\mathbf{s})^\top\mathbf{H}_\beta(\mathbf{s})\delta\mathbf{s}\,,\,0\right\} .
\end{aligned}$$

This means that in yet a third iteration inside the previous two, we can conduct a line search to reduce the merit function (4.63) along the search direction given by the Newton equations (4.62). This line search should be designed to avoid nonpositive components of the slack variables **s**. For more details, readers should read the paper by Waltz et al. [180].

Algorithm 4.5.5 (Interior Point)

given $\mathbf{z}_0$, $(\boldsymbol{\ell}_\mathbf{a})_0$, $(\boldsymbol{\ell}_\mathbf{b})_0$, $\varrho \in (0\,,\,1)$

[illegible]

for $\tau(\varrho_p) \downarrow 0$

$\widetilde{\mathbf{z}}_0 = \mathbf{z}_p$, $\widetilde{\mathbf{s}}_0 = \mathbf{s}_p$, $(\widetilde{\boldsymbol{\ell}}_\mathbf{a})_0 = (\boldsymbol{\ell}_\mathbf{a})_p$, $(\widetilde{\boldsymbol{\ell}}_\mathbf{b})_0 = (\boldsymbol{\ell}_\mathbf{b})_p$

for $k = 0, 1, \ldots$

solve (4.62) for $\delta\mathbf{x}$, $\delta\mathbf{s}$, $\delta\boldsymbol{\ell}_\mathbf{a}$, $\delta\boldsymbol{\ell}_\mathbf{b}$

choose μ to satisfy inequality (4.64)

choose $\lambda > 0$ to minimize the merit function $\psi_{\mu,\tau}(\widetilde{\mathbf{z}}_k + \delta\mathbf{x}\lambda\,,\,\widetilde{\mathbf{s}}_k + \delta\mathbf{s}\lambda)$ defined by (4.63)

$\widetilde{\mathbf{z}}_{k+1} = \widetilde{\mathbf{z}}_k + \delta\mathbf{x}\lambda$, $\widetilde{\mathbf{s}}_{k+1} = \widetilde{\mathbf{s}}_k + \delta\mathbf{s}\lambda$, $(\widetilde{\boldsymbol{\ell}}_\mathbf{a})_{k+1} = (\widetilde{\boldsymbol{\ell}}_\mathbf{a})_k + \delta\boldsymbol{\ell}_\mathbf{a}\lambda$, $(\widetilde{\boldsymbol{\ell}}_\mathbf{b})_{k+1} = (\widetilde{\boldsymbol{\ell}}_\mathbf{b})_k + \delta\boldsymbol{\ell}_\mathbf{b}\lambda$

$\mathbf{z}_{p+1} = \widetilde{\mathbf{z}}_\infty$, $\mathbf{s}_{p+1} = \widetilde{\mathbf{s}}_\infty$, $(\boldsymbol{\ell}_\mathbf{a})_{p+1} = (\widetilde{\boldsymbol{\ell}}_\mathbf{a})_\infty$, $(\boldsymbol{\ell}_\mathbf{b})_{p+1} = (\widetilde{\boldsymbol{\ell}}_\mathbf{b})_\infty$

In Fig. 4.5, we show contours of the function used in the interior point method for the constrained minimization problem displayed on the left in Fig. 4.3. The original constrained minimization problem is

$$\min -{\mathbf{x}_1}^2 - \mathbf{x}_2^2 \text{ subject to } 2\mathbf{x}_1^2 + \mathbf{x}_2^2 \le 4\,,\ \mathbf{x}_1^2 + 3\mathbf{x}_2^2 \le 9 \text{ and } \mathbf{x} \ge \mathbf{0}\,.$$

We show logarithmically spaced contours of the interior point function $\phi(\mathbf{x}) + \tau\beta(-\mathbf{b}(\mathbf{x}))$ where

$$\beta(\mathbf{s}) = \sum_{j=1}^{p} \frac{1}{\mathbf{s}_j} \text{ and } \mathbf{b}(\mathbf{x}) = \begin{bmatrix} 2\mathbf{x}_1^2 + \mathbf{x}_2^2 - 4 \\ \mathbf{x}_1^2 + 3\mathbf{x}_2^2 - 9 \\ -\mathbf{x}_1 \\ -\mathbf{x}_2 \end{bmatrix}\,.$$

for various values of the parameter τ. Note that as the barrier coefficient τ decreases, the minimizer of the modified objective approaches the minimizer of the original problem. At the same time, the modified objective becomes relatively flatter near the minimizer, indicating potential convergence difficulty for an unconstrained minimization algorithm. These figures were drawn by the C^{++} program interiorPoint.C

4.5.3 Augmented Lagrangian

In Sect. 4.5.1 we saw that penalty methods could approximate the solution of a constrained minimization problem by minimizing the original objective plus a penalty. The minimizer of the modified function approaches the true solution

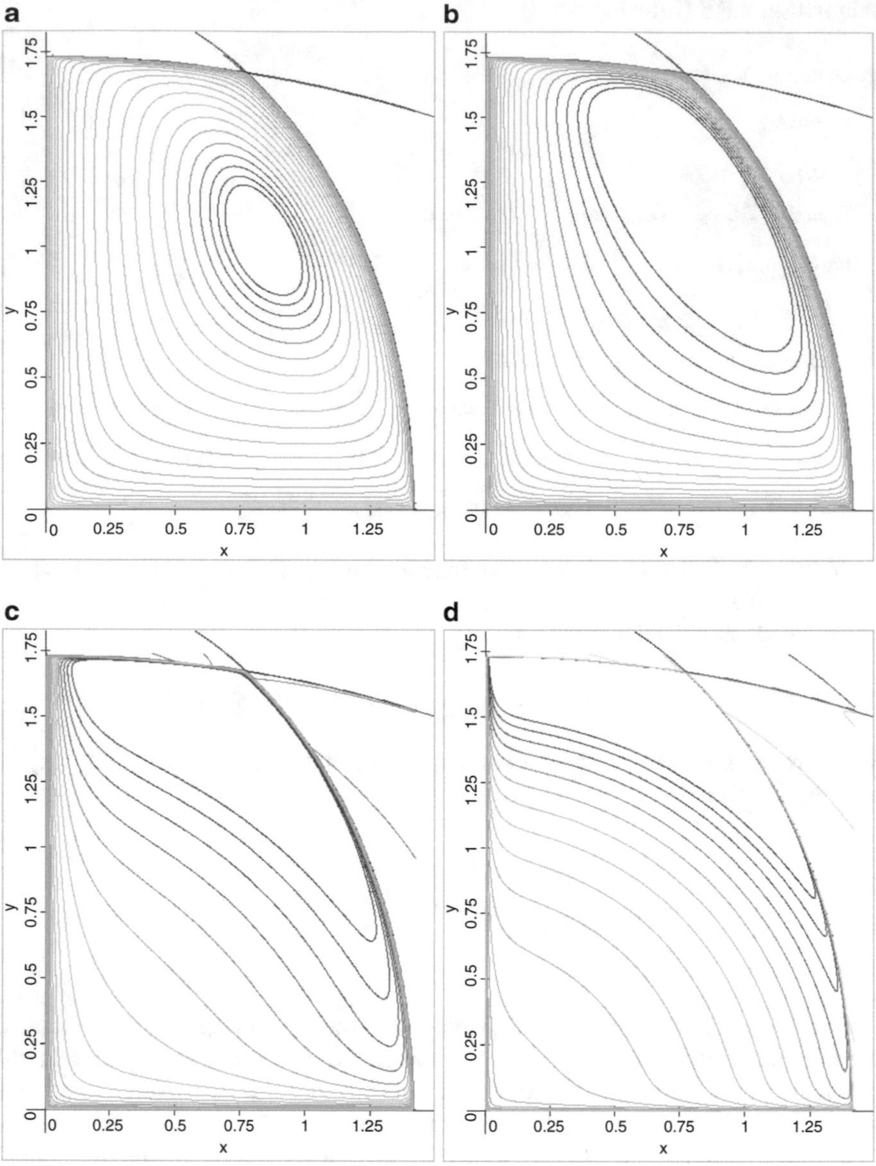

Fig. 4.5 Interior point method: contours of $\phi(\mathbf{x}) + \tau\beta(-\mathbf{b}(\mathbf{x}))$ for $\phi(\mathbf{x}) = -\mathbf{x}_1^2 - \mathbf{x}_2^2$, $\beta(\mathbf{s}) = \sum_{j=1}^{4} 1/bfs_j$ with $\mathbf{b}_1(\mathbf{x}) = 2\mathbf{x}_1^2 + \mathbf{x}_2^2 - 4$, $\mathbf{b}_2(\mathbf{x}) = \mathbf{x}_1^2 + 3\mathbf{x}_2^2 - 9$, $\mathbf{b}_3(\mathbf{x}) = -\mathbf{x}_1$ and $\mathbf{b}_4(\mathbf{x}) = -\mathbf{x}_2$. (**a**) $\tau = 10$. (**b**) $\tau = 1$. (**c**) $\tau = 0.1$. (**d**) $\tau = 0.01$

of the original constrained problem only in the limit as the coefficient of the penalty approaches infinity. We also saw in Sect. 4.5.2.4 that interior point methods approximate the solution of a constrained minimization problem by finding a point in the interior of the feasible region that minimizes a modified function involving a barrier. The minimizer of the modified function approaches the true solution only in the limit as the coefficient of the barrier approaches zero. In this section, we will discover a modified objective whose minimizer also minimizes the original constrained problem, provided only that the coefficient of the modification is chosen sufficiently large. Unlike the penalty or barrier function approaches, it is not necessary to use unbounded functions or parameters to obtain convergence.

We will begin in Sects. 4.5.3.1 and 4.5.3.2 by discussing two approaches that are applicable only to equality constraints. We will subsequently provide two approaches for more general constrained minimization problems in Sects. 4.5.3.3 and 4.5.3.4.

4.5.3.1 Powell's Algorithm

Let ϕ be a real linear functional on real n-vectors, and let $\mathbf{a}$ map real n-vectors to real p-vectors. Suppose that we want to solve the constrained minimization problem

$$\min \phi(\mathbf{x}) \text{ subject to } \mathbf{a}(\mathbf{x}) = \mathbf{0} \ .$$

Powell [142] noticed that for real p-vectors $\mathbf{d}$ and positive $p \times p$ diagonal matrices $\boldsymbol{\Gamma}$ the **augmented Lagrangian** objective

$$\lambda(\mathbf{x} \, ; \, \mathbf{d} \, , \, \boldsymbol{\Gamma}) = \phi(\mathbf{x}) + [\mathbf{a}(\mathbf{x}) + \mathbf{d}]^\top \boldsymbol{\Gamma} [\mathbf{a}(\mathbf{x}) + \mathbf{d}]$$

has its global minimum at a point $\mathbf{x}_*$, which also minimizes $\phi(\mathbf{x})$ subject to the constraint that $\mathbf{a}(\mathbf{x}) = \mathbf{a}(\mathbf{x}_*)$. The simple proof of this fact is by contradiction. If $\mathbf{x}_*$ does not solve

$$\min \phi(\mathbf{x}) \text{ subject to } \mathbf{a}(\mathbf{x}) = \mathbf{a}(\mathbf{x}_*) \ ,$$

then there exists a real n-vector $\mathbf{z}$ so that

$$\mathbf{a}(\mathbf{z}) = \mathbf{a}(\mathbf{x}_*) \text{ and } \phi(\mathbf{z}) < \phi(\mathbf{x}_*) \ .$$

But this inequality implies that

$$\begin{aligned} &\lambda(\mathbf{z} \, ; \, \mathbf{d} \, , \, \boldsymbol{\Gamma}) = \phi(\mathbf{z}) + [\mathbf{a}(\mathbf{z}) + \mathbf{d}]^\top \boldsymbol{\Gamma} [\mathbf{a}(\mathbf{z}) + \mathbf{d}] = \phi(\mathbf{z}) + [\mathbf{a}(\mathbf{x}_*) + \mathbf{d}]^\top \boldsymbol{\Gamma} [\mathbf{a}(\mathbf{x}_*) + \mathbf{d}] \\ &< \phi(\mathbf{x}_*) + [\mathbf{a}(\mathbf{x}_*) + \mathbf{d}]^\top \boldsymbol{\Gamma} [\mathbf{a}(\mathbf{x}_*) + \mathbf{d}] = \lambda(\mathbf{x}_* \, ; \, \mathbf{d} \, , \, \boldsymbol{\Gamma}) \ . \end{aligned}$$

This inequality contradicts the hypothesis that $\mathbf{x}_*$ is the global minimum of λ.

Powell then provided the following

Algorithm 4.5.6 (Powell's Augmented Lagrangian)

given $\mathbf{z}_0$ and τ

$\boldsymbol{\Gamma}_0 = \mathbf{I}$, $\mathbf{d}_0 = \mathbf{0}$, incremented_p $=$ false

choose $\alpha_0 > \|\mathbf{a}(\mathbf{z}_0)\|_\infty$

for $k = 1, 2, \ldots$

find $\mathbf{z}_k$ to minimize $\lambda(\mathbf{x} \,;\, \mathbf{d}_{k-1} \,,\, \boldsymbol{\Gamma}_{k-1})\phi(\mathbf{x}) + \{\mathbf{a}(\mathbf{x}) + \mathbf{d}_{k-1}\}^\top \boldsymbol{\Gamma}_{k-1}\{\mathbf{a}(\mathbf{x}) + \mathbf{d}_{k-1}\}$

$\alpha_k = \|\mathbf{a}(\mathbf{z}_k)\|_\infty$

if $\alpha_k \leq \tau$ then break

if ($\alpha_k < \alpha_{k-1}$ and (incremented_p $=$ false or $\alpha_k \leq \alpha_{k-1}/4$)) then

$\mathbf{d}_k = \mathbf{d}_{k-1} + \mathbf{a}(\mathbf{z}_k)$, $\boldsymbol{\Gamma}_k = \boldsymbol{\Gamma}_{k-1}$, incremented_p $=$ true

else

if ($\alpha_k \geq \alpha_{k-1}$ and incremented_p) then $j = k - 2$ else $j = k - 1$

incremented_p $=$ false , $\alpha_k = \min\{\alpha_k \,,\, \alpha_{k-1}\}$

for $1 \leq i \leq p$

if $|\mathbf{e}_i \cdot \mathbf{a}(\mathbf{z}_k)| > \alpha_{k-1}/4$ then

$\mathbf{e}_i \cdot \boldsymbol{\Gamma}_k \mathbf{e}_i = \mathbf{e}_i \cdot \boldsymbol{\Gamma}_j \mathbf{e}_i \, * \, 10$, $\mathbf{e}_i \cdot \mathbf{d}_k = \mathbf{e}_i \cdot \mathbf{d}_j/10$

else

$\mathbf{e}_i \cdot \boldsymbol{\Gamma}_k \mathbf{e}_i = \mathbf{e}_i \cdot \boldsymbol{\Gamma}_j \mathbf{e}_i$, $\mathbf{e}_i \cdot \mathbf{d}_k = \mathbf{e}_i \cdot \mathbf{d}_j$

Suppose that both ϕ and $\mathbf{a}$ are continuously differentiable. If this algorithm finds that $\{\mathbf{z}_k\}_{k=0}^\infty \to \mathbf{z}$, $\{\mathbf{d}_k\}_{k=0}^\infty \to \mathbf{d}$ and $\{\boldsymbol{\Gamma}_k\}_{k=0}^\infty \to \boldsymbol{\Gamma}$ then

$$\mathbf{0}^\top = \mathbf{g}_\phi(\mathbf{z} \,;\, \mathbf{d} \,,\, \boldsymbol{\Gamma})^\top + 2\{\mathbf{a}(\mathbf{z}) + \mathbf{d}\}^\top \boldsymbol{\Gamma} \mathbf{J}_\mathbf{a}(\mathbf{z}) \,.$$

If the algorithm also achieves $\mathbf{a}(\mathbf{z}) = \mathbf{0}$, then the Lagrange multipliers for the constrained problem are

$$\boldsymbol{\ell}^\top = 2\mathbf{d}^\top \boldsymbol{\Gamma} \,.$$

This algorithm increases $\boldsymbol{\Gamma}$ occasionally, in order to gain some of the advantages of a penalty method. Unlike a penalty method, however, it is not necessary to drive $\boldsymbol{\Gamma}$ to infinite values in order to achieve convergence to the solution of the constrained minimization problem.

Powell [142, pp. 289–293] proved the following two results.

Lemma 4.5.3 *Let $p < n$ be positive integers. Suppose that* $\mathbf{a}$ *maps real n-vectors to real p-vectors, and that the real functional ϕ is defined on real n-vectors* *and ϕ is bounded below by the scalar $\underline{\phi}$. Then the sequence $\{\alpha_k\}_{k=0}^{\infty}$ defined by Algorithm 4.5.6 converges to zero.*

Proof Note that $\{\alpha_k\}_{k=0}^{\infty}$ is nondecreasing and is bounded below by zero. We will identify three mutually exclusive cases that control the evaluation of $\mathbf{d}_k$ and $\boldsymbol{\Gamma}_k$:

$$\|\mathbf{a}(\mathbf{z}_k)\|_\infty \le \alpha_{k-1}/4 \tag{4.65a}$$

$$\alpha_{k-1}/4 < \|\mathbf{a}(\mathbf{z}_k)\|_\infty < \alpha_{k-1} \text{ and not increment\_p} \tag{4.65b}$$

$$\|\mathbf{a}(\mathbf{z}_k)\|_\infty \ge \alpha_{k-1} \text{ or (increment\_p and } \|\mathbf{a}(\mathbf{z}_k)\|_\infty > \alpha_{k-1}/4 \text{) .} \tag{4.65c}$$

The first two cases imply that incremented_p will be set to true, and the third case implies that incremented_p will be set to false. Thus the second case cannot occur for subsequent values of the iteration counter k. Every other step, either the first case or the third case must occur. In particular, if the step counter is

$$k = 4\ell - 1 \ ,$$

then the second case occurred at most 2ℓ times, leaving $2\ell - 1$ steps for the other two cases. By the pigeon hole principle, either the first case or the third case must have occurred at least ℓ times.

If case (4.65a) occurred at least ℓ times, then

$$\alpha_k \le \|\mathbf{a}(\mathbf{z}_0)\|_\infty / 4^\ell \ . \tag{4.66}$$

Otherwise, case (4.65c) occurred at least ℓ times. Next, let us assume that

$$\ell \ge (j-1)p + 1 \ .$$

Then the pigeon hole principle implies that at least one of the p diagonal entries of $\boldsymbol{\Gamma}$ was increased at least j times. In this situation, we will show that this increase in a particular diagonal entry of $\boldsymbol{\Gamma}$ will place an upper bound on α_k.

Since $\mathbf{z}_k$ minimizes $\lambda(\mathbf{x}\ ;\ \mathbf{d}_k\ ,\ \boldsymbol{\Gamma}_k)$, we see that

$$\begin{aligned}
&\phi(\mathbf{z}_k) + \{\mathbf{a}(\mathbf{z}_k) + \mathbf{d}_k\}^\top \boldsymbol{\Gamma}_k \{\mathbf{a}(\mathbf{z}_k) + \mathbf{d}_k\} = \lambda(\mathbf{z}_k\ ;\ \mathbf{d}_k\ ,\ \boldsymbol{\Gamma}_k) \\
&\le \lambda(\mathbf{z}\ ;\ \mathbf{d}_k\ ,\ \boldsymbol{\Gamma}_k) = \phi(\mathbf{z}) + \mathbf{d}_k{}^\top \boldsymbol{\Gamma}_k \mathbf{d}_k \ .
\end{aligned}$$

Using the fact that ϕ is bounded below, we can rewrite the previous inequality to get

$$\begin{aligned}
&\{\mathbf{a}(\mathbf{z}_k) + \mathbf{d}_k\}^\top \boldsymbol{\Gamma}_k \{\mathbf{a}(\mathbf{z}_k) + \mathbf{d}_k\} \le \phi(\mathbf{z}) - \phi(\mathbf{z}_k) + \mathbf{d}_k{}^\top \boldsymbol{\Gamma}_k \mathbf{d}_k \\
&\le \phi(\mathbf{z}) - \underline{\phi} + \mathbf{d}_k{}^\top \boldsymbol{\Gamma}_k \mathbf{d}_k \ .
\end{aligned} \tag{4.67}$$

We will use this inequality to prove inductively that

$$\mathbf{d}_k{}^\top \boldsymbol{\Gamma}_k \mathbf{d}_k \leq k\{\phi(\mathbf{z}) - \underline{\phi}\} \ . \tag{4.68}$$

Since $\mathbf{d}_0 = \mathbf{0}$, this inequality is satisfied for $k = 0$. Assume inductively that (4.68) is true for all step indices less than k. It is useful to note that Algorithm 4.5.6 implies that $\boldsymbol{\Gamma}_k \geq \boldsymbol{\Gamma}_{k'}$ for all $k' \leq k$. Also in Algorithm 4.5.6, there are three ways in which $\mathbf{d}_k$ is evaluated. If $\mathbf{d}_k$ is scaled componentwise from $\overline{\mathbf{d}}$ then we incremented $\mathbf{d}$ by $\mathbf{a}$ in step $k-1$, so $\overline{\mathbf{d}} = \mathbf{d}_{k-2}$ and

$$\begin{aligned}
&\mathbf{d}_k{}^\top \boldsymbol{\Gamma}_k \mathbf{d}_k \\
&= \sum_{\substack{1\leq i\leq p \\ |\mathbf{e}_i\cdot\mathbf{a}(\mathbf{z}_k)|\leq\alpha_{k-1}/4}} (\mathbf{e}_i \cdot \overline{\mathbf{d}})^2 \mathbf{e}_i \cdot \boldsymbol{\Gamma}_{k-1}\mathbf{e}_i + \frac{1}{10} \sum_{\substack{1\leq i\leq p \\ |\mathbf{e}_i\cdot\mathbf{a}(\mathbf{z}_k)|>\alpha_{k-1}/4}} (\mathbf{e}_i \cdot \overline{\mathbf{d}})^2 \mathbf{e}_i \cdot \boldsymbol{\Gamma}_{k-1}\mathbf{e}_i \\
&< \sum_{i=1}^p (\mathbf{e}_i \cdot \overline{\mathbf{d}})^2 \mathbf{e}_i \cdot \boldsymbol{\Gamma}_{k-1}\mathbf{e}_i = \mathbf{d}_{k-2}{}^\top \boldsymbol{\Gamma}_{k-1} \mathbf{d}_{k-2} = \mathbf{d}_{k-2}{}^\top \boldsymbol{\Gamma}_{k-2} \mathbf{d}_{k-2} \\
&\leq (k-2)\{\phi(\mathbf{z}) - \underline{\phi}\} \ .
\end{aligned}$$

If $\mathbf{d}_k$ is scaled componentwise from $\mathbf{d}_{k-1}$, then

$$\begin{aligned}
&\mathbf{d}_k{}^\top \boldsymbol{\Gamma}_k \mathbf{d}_k \\
&= \sum_{\substack{1\leq i\leq p \\ |\mathbf{e}_i\cdot\mathbf{a}(\mathbf{z}_k)|\leq\alpha_{k-1}/4}} (\mathbf{e}_i \cdot \mathbf{d}_{k-1})^2 \mathbf{e}_i \cdot \boldsymbol{\Gamma}_{k-1}\mathbf{e}_i + \frac{1}{10} \sum_{\substack{1\leq i\leq p \\ |\mathbf{e}_i\cdot\mathbf{a}(\mathbf{z}_k)|>\alpha_{k-1}/4}} (\mathbf{e}_i \cdot \mathbf{d}_{k-1})^2 \mathbf{e}_i \cdot \boldsymbol{\Gamma}_{k-1}\mathbf{e}_i \\
&< \sum_{i=1}^p (\mathbf{e}_i \cdot \mathbf{d}_{k-1})^2 \mathbf{e}_i \cdot \boldsymbol{\Gamma}_{k-1}\mathbf{e}_i = \mathbf{d}_{k-1}{}^\top \boldsymbol{\Gamma}_{k-1} \mathbf{d}_{k-1} \leq (k-1)\{\phi(\mathbf{z}) - \underline{\phi}\} \ .
\end{aligned}$$

Finally, if $\mathbf{d}_k$ is incremented from $\mathbf{d}_{k-1}$ then

$$\mathbf{d}_k{}^\top \boldsymbol{\Gamma}_k \mathbf{d}_k = \{\mathbf{d}_{k-1} + \mathbf{a}(\mathbf{z}_k)\}^\top \boldsymbol{\Gamma}_{k-1} \{\mathbf{d}_{k-1} + \mathbf{a}(\mathbf{z}_k)\}$$

then inequality (4.67) gives us

$$\leq \phi(\mathbf{z}) - \underline{\phi} + \mathbf{d}_{k-1}{}^\top \boldsymbol{\Gamma}_{k-1} \mathbf{d}_{k-1} \ .$$

then the inductive hypothesis (4.68) produces

$$\leq \phi(\mathbf{z}) - \underline{\phi} + (k-1)\{\phi(\mathbf{z}) - \underline{\phi}\} = k\{\phi(\mathbf{z}) - \underline{\phi}\} \ .$$

Thus, with any one of the three ways we evaluate $\mathbf{d}_k$, we have shown that (4.68) holds for k. This completes the inductive proof of (4.68).

There are some interesting implications from inequality (4.68). First, we should notice that

$$\left\| \boldsymbol{\Gamma}_k^{1/2} \mathbf{d}_k \right\|_\infty \leq \sqrt{(k+1)\{\phi(\mathbf{z}) - \underline{\phi}\}}$$

Similarly, inequality (4.67) and inequality (4.68). lead to

$$\left\| \boldsymbol{\Gamma}_k^{1/2} \{\mathbf{a}(\mathbf{z}_k) + \mathbf{d}_k\} \right\|_\infty \leq \sqrt{(k+1)\{\phi(\mathbf{z}) - \underline{\phi}\}} .$$

Combining these last two inequalities, we conclude that

$$\left\| \boldsymbol{\Gamma}_k^{1/2} \mathbf{a}(\mathbf{z}_k) \right\|_\infty \leq 2\sqrt{(k+1)\{\phi(\mathbf{z}) - \underline{\phi}\}} .. \tag{4.69}$$

Recall that $k = 4\ell - 1$ where $\ell \geq (j-1)p + 1$. If component i of $\boldsymbol{\Gamma}_k$ has been increased at least j times, then

$$\mathbf{e}_i \cdot \boldsymbol{\Gamma}_k \mathbf{e}_i \geq 10^j .$$

Suppose that the value for this component of $\boldsymbol{\Gamma}_k$ was increased from a previous value during step $q \leq k$. In that step, we must have had

$$\left| \mathbf{e}_i \cdot \mathbf{a}(\mathbf{z}_q) \right| \geq \alpha_{q-1}/4 .$$

We can reorder terms and combine with inequality (4.69) to get

$$\alpha_{q-1} \leq 4 \left| \mathbf{e}_i \cdot \mathbf{a}(\mathbf{z}_q) \right| \leq 8\sqrt{(k+1)\{\phi(\mathbf{z}) - \underline{\phi}\}/10^j} . \tag{4.70}$$

In summary, for $k = 4\ell - 1$ with $\ell \geq (j-1)p + 1$ either (4.66) or (4.70) is satisfied. Either of these inequalities implies that $\alpha_k \to 0$.

Theorem 4.5.2 *Let $p < n$ be positive integers. Suppose that $\mathbf{a}$ maps real n-vectors to real p-vectors, and that the real functional ϕ is defined on real n-vectors and ϕ is bounded below by the scalar $\underline{\phi}$. Assume that the sequence $\{\mathbf{z}_k\}_{k=0}^\infty$ is computed by Algorithm 4.5.6 and is contained in the closed and bounded set $\mathscr{S}$. Suppose that both ϕ and $\mathbf{a}$ are continuous on $\mathscr{S}$. Assume that the constrained minimization problem*

$$\min \phi(\mathbf{x}) \textit{ subject to } \mathbf{a}(\mathbf{x}) = \mathbf{0}$$

has a unique solution $\mathbf{z}$. For all $\varepsilon > 0$ and all $\mathbf{z} \in \mathscr{S}$ let

$$\mathscr{R}_\varepsilon(\mathbf{z}) = \{\mathbf{x} : |\phi(\mathbf{x}) - \phi(\mathbf{z})| < \varepsilon\} .$$

Assume that for all $\varepsilon > 0$ *there exists* $\delta > 0$ *so that for all n-vectors* $\mathbf{h}$ *with* $\|\mathbf{h}\|_\infty < \delta$ *there exists* $\mathbf{x} \in \mathscr{R}_\varepsilon$ *so that* $\mathbf{a}(\mathbf{x}) = \mathbf{h}$. *Then there is a subsequence of* $\{\mathbf{z}_k\}_{k=0}^\infty$ *that converges to* $\mathbf{z}$.

Proof First, consider the subsequence $\{\mathbf{z}_{k_j}\}_{j=0}^\infty$ for which Algorithm 4.5.6 finds that

$$\|\mathbf{a}(\mathbf{z}_{k_j})\|_\infty < \alpha_{k_j-1} \ .$$

Lemma 4.5.3 implies that

$$\lim_{j\to\infty} \|\mathbf{a}(\mathbf{z}_{k_j})\|_\infty = 0 \ .$$

Since $\{\mathbf{z}_{k_j}\}_{j=0}^\infty \subset \mathscr{S}$, which is closed and bounded, there is yet another subsequence that converges to some real n-vector $\widetilde{\mathbf{z}} \in \mathscr{S}$. To simplify notation, we will write this second subsequence as $\{\mathbf{z}_k\}_{k=0}^\infty$. Since $\mathbf{a}$ is continuous on $\mathscr{S}$,

$$\mathbf{a}(\widetilde{\mathbf{z}}) = \lim_{k\to\infty} \mathbf{a}(\mathbf{z}_k) = \mathbf{0} \ .$$

Given any $\varepsilon > 0$, there exists $\delta > 0$ and there exists a positive integer K so that for all $k \geq K$ we have

$$\|\mathbf{a}(\mathbf{z}_k)\|_\infty < \delta \ ,$$

so the hypotheses of this theorem imply that for each $k \geq K$ there exists $\mathbf{x}_k \in \mathscr{R}_\varepsilon(\mathbf{z})$ so that

$$\mathbf{a}(\mathbf{x}_k) = \mathbf{a}(\mathbf{z}_k) \ .$$

Since Algorithm 4.5.6 computes $\mathbf{z}_k$ in order to minimize

$$\lambda(\mathbf{x}\ ;\ \mathbf{d}_k\ ,\ \boldsymbol{\Gamma}_k) = \phi(\mathbf{x}) + \{\mathbf{a}(\mathbf{x}) + \mathbf{d}_k\}^\top \boldsymbol{\Gamma}_k \{\mathbf{a}(\mathbf{x}) + \mathbf{d}_k\} \ ,$$

we see that

$$\lambda(\mathbf{x}_k\ ;\ \mathbf{d}_k\ ,\ \boldsymbol{\Gamma}_k) \geq \lambda(\mathbf{z}_k\ ;\ \mathbf{d}_k\ ,\ \boldsymbol{\Gamma}_k) = \phi(\mathbf{z}_k) + \{\mathbf{a}(\mathbf{z}_k) + \mathbf{d}_k\}^\top \boldsymbol{\Gamma}_k \{\mathbf{a}(\mathbf{z}_k) + \mathbf{d}_k\}$$
$$= \phi(\mathbf{z}_k) + \{\mathbf{a}(\mathbf{x}_k) + \mathbf{d}_k\}^\top \boldsymbol{\Gamma}_k \{\mathbf{a}(\mathbf{x}_k) + \mathbf{d}_k\}\ , = \lambda(\mathbf{x}_k\ ;\ \mathbf{d}_k\ ,\ \boldsymbol{\Gamma}_k) + \phi(\mathbf{z}_k) - \phi(\mathbf{x}_k)$$

This inequality implies that

$$\phi(\mathbf{z}_k) \leq \phi(\mathbf{x}_k)$$

and since $\mathbf{x}_k \in \mathscr{R}_\varepsilon(\mathbf{z})$ we also have

$$\leq \phi(\mathbf{z}) + \varepsilon \ .$$

Since ϕ is continuous on $\mathscr{S}$, this implies that the limit $\widetilde{\mathbf{z}}$ of the second subsequence $\{\mathbf{z}_k\}_{k=0}^{\infty}$ is the solution $\mathbf{z}$ of the constrained minimization problem.
Powell also proved linear convergence for Algorithm 4.5.6.

4.5.3.2 Hestenes' Algorithm

For solving nonlinear minimization problems with equality constraints, Hestenes [95] suggested an **augmented Lagrangian** that differs a bit from Powell's approach. Hestenes' idea can be summarized by the following

Algorithm 4.5.7 (Hestenes' Augmented Lagrangian)

given $\mathbf{z}_0$ and τ

$\gamma_0 = 1\ ,\ \boldsymbol{\ell}_0 = \mathbf{0}$

choose $\alpha_0 > \|\mathbf{a}(\mathbf{z}_0)\|_2$

for $k = 1, 2, \ldots$

find $\mathbf{z}_k$ to minimize $\lambda(\mathbf{x}\ ;\ \boldsymbol{\ell}_{k-1}\ ,\ \gamma_{k-1}) = \phi(\mathbf{x}) + \boldsymbol{\ell}_{k-1}{}^\top\mathbf{a}(\mathbf{x}) + \frac{\gamma_k}{2}\|\mathbf{a}(\mathbf{x})\|_2^2$

$\alpha_k = \|\mathbf{a}(\mathbf{z}_k)\|_2$

if $\alpha_k \leq \tau$ then break

$\boldsymbol{\ell}_k = \boldsymbol{\ell}_{k-1} + \mathbf{a}(\mathbf{z}_k)\gamma_{k-1}\ ,$

if $\alpha_k \geq \alpha_{k-1}/4$ then $\gamma_k = \gamma_{k-1}\ *\ 10$ else $\gamma_k = \gamma_{k-1}$

If ϕ and $\mathbf{a}$ are continuously differentiable, then for each value of $k \geq 1$, the minimization problem for $\mathbf{z}_k$ implies that

$$\mathbf{0}^\top = \mathbf{g}_\lambda{}^\top(\mathbf{z}_k\ ;\ \boldsymbol{\ell}_{k-1}\ ,\ \gamma_{k-1}) = \mathbf{g}_\phi(\mathbf{z}_k)^\top + \left[\boldsymbol{\ell}_{k-1}{}^\top + \gamma_{k-1}\mathbf{a}(\mathbf{z}_k)^\top\right]\mathbf{J}_\mathbf{a}(\mathbf{z}_k)\ .$$

In comparison, the first-order Kuhn-Tucker necessary conditions for the constrained minimization problem imply that

$$\mathbf{0}^\top = \mathbf{g}_\phi(\mathbf{z})^\top + \boldsymbol{\ell}^\top\mathbf{J}_\mathbf{a}(\mathbf{z})\ .$$

If we can force $\mathbf{a}(\mathbf{z}_k) \rightarrow \mathbf{0}$, then whenever $\mathbf{J}_\mathbf{a}$ has full rank we expect $\boldsymbol{\ell}_k$ to tend to the Lagrange multipliers for the constrained minimization problem.

In support of this observation, we offer the following evidence.

Lemma 4.5.4 (Finsler's) *[46] Suppose that $\mathscr{D}$ is a closed and bounded set of real n-vectors. Let υ and ω be continuous real functionals on $\mathscr{D}$, and assume that $\omega(\mathbf{x}) \geq 0$ for all $\mathbf{x} \in \mathscr{D}$. Then*

$$\mathbf{x} \in \mathscr{D} \text{ and } \omega(\mathbf{x}) \leq 0 \text{ implies } \upsilon(\mathbf{x}) > 0 \tag{4.71}$$

if and only if

$$\text{there exists } \underline{\gamma} \geq 0 \text{ so that for all } \gamma \geq \underline{\gamma} \text{ and for all } \mathbf{x} \in \mathscr{D} \text{ we have } \upsilon(\mathbf{x}) + \gamma\omega(\mathbf{x}) > 0\,. \tag{4.72}$$

Proof It is easy to prove that (4.72) implies (4.71). If $\mathbf{x} \in \mathscr{D}$ and $\omega(\mathbf{x}) \leq 0$, then for all $\gamma \geq \underline{\gamma}$ we have

$$\upsilon(\mathbf{x}) > -\gamma\omega(\mathbf{x}) \geq 0\,.$$

All that remains is to prove that (4.71) implies (4.72). Let

$$\mathscr{D}_- = \{\mathbf{x} \in \mathscr{D} \,:\, \upsilon(\mathbf{x}) \leq 0\}\,.$$

If $\mathscr{D}_- = \emptyset$, then we are done. Otherwise, since $\mathscr{D}$ is bounded and υ is continuous, $\mathscr{D}_-$ is closed and bounded. Let $\mathbf{x}_- \in \mathscr{D}_-$ be such that for all $\mathbf{x} \in \mathscr{D}_-$

$$\omega(\mathbf{x}_-) \leq \omega(\mathbf{x})\,.$$

If $\omega(\mathbf{x}_-) = 0$, then (4.72) implies that $\upsilon(\mathbf{x}_-) > 0$, which contradicts the assumption that $\mathbf{x}_- \in \mathscr{D}_-$. We conclude that $\omega(\mathbf{x}_-) > 0$. We can choose

$$\underline{\gamma} > -\frac{\min_{\mathbf{x}\in\mathscr{D}_-} \upsilon(\mathbf{x})}{\omega(\mathbf{x}_-)}$$

and note that $\underline{\gamma} \geq 0$. Choose any $\gamma \geq \underline{\gamma}$. Then for all $\mathbf{x} \in \mathscr{D}_-$

$$\upsilon(\mathbf{x}) + \gamma\omega(\mathbf{x}) \geq \min_{\mathbf{x}\in\mathscr{D}_-} \upsilon(\mathbf{x}) + \gamma\omega(\mathbf{x}) > 0\,,$$

and for all $\mathbf{x} \in \mathscr{D} \setminus \mathscr{D}_-$

$$\upsilon(\mathbf{x}) + \gamma\omega(\mathbf{x}) \geq \upsilon(\mathbf{x}) > 0\,.$$

Avriel [5, p. 401] notes the following easy implication of Finsler's lemma.

Corollary 4.5.1 *Let p and n be positive integers. Suppose that $\mathbf{A}$ is a real $n \times n$ matrix and $\mathbf{B}$ is a real $p \times n$ matrix. Then*

$$\mathbf{B}\mathbf{x} = \mathbf{0} \text{ implies } \mathbf{x}^\top\mathbf{A}\mathbf{x} > 0$$

if and only if

$$\text{there exists } \underline{\gamma} \geq 0 \text{ so that for all } \gamma \geq \underline{\gamma} \text{ and for all } \mathbf{x} \neq \mathbf{0} \text{ we have } \mathbf{x}^\top \left(\mathbf{A} + \mathbf{B}^\top \mathbf{B}\right) \mathbf{x} > 0 \, .$$

Proof Let $\upsilon(\mathbf{x}) = \mathbf{x}^\top \mathbf{A} \mathbf{x}$, $\omega(\mathbf{x}) = \|\mathbf{B}\mathbf{x}\|_2^2$ and

$$\mathscr{D} = \{\mathbf{x} : \|\mathbf{x}\|_2 = 1\} \, .$$

Then Finsler's Lemma 4.5.4 shows that

$$\|\mathbf{x}\|_2 = 1 \text{ and } \|\mathbf{B}\mathbf{x}\|_2^2 \leq 0 \text{ implies } \mathbf{x}^\top \mathbf{A} \mathbf{x} > 0$$

if and only if

$$\begin{aligned} &\text{there exists } \underline{\gamma} \geq 0 \text{ so that for all } \gamma \geq \underline{\gamma} \text{ and for all } \|\mathbf{x}\|_2 = 1 \\ &\text{we have } \mathbf{x}^\top \left(\mathbf{A} + \mathbf{B}^\top \mathbf{B}\right) \mathbf{x} > 0 \, . \end{aligned}$$

We can scale $\mathbf{x}$ in this equivalence to get the claim in the corollary.

This corollary of Finsler's lemma allows us to prove the following result, which can be found in Avriel [5, p. 401] and is a simplification of a result in Arrow et al. [4, p. 228].

Theorem 4.5.3 *Let $p < n$ be positive integers. Assume that ϕ is a twice continuously differentiable real functional on real n-vectors. Also assume that* $\mathbf{a}$ *is a twice continuously differentiable mapping from real n-vectors to real p-vectors. Suppose that there exists a real n-vector* $\mathbf{a}$ *and a real p-vector* $\boldsymbol{\ell}$ *satisfying the second-order Kuhn-Tucker sufficient conditions in Theorem 4.3.4 for the constrained minimization problem*

$$\min \phi(\mathbf{x}) \text{ subject to } \mathbf{a}(\mathbf{x}) = \mathbf{0} \, ,$$

namely

$$\mathbf{g}_\phi(\mathbf{z})^\top + \boldsymbol{\ell}^\top \mathbf{J}_\mathbf{a}(\mathbf{z}) = \mathbf{0}^\top \, , \tag{4.73a}$$

$$\mathbf{a}(\mathbf{z}) = \mathbf{0} \text{ and} \tag{4.73b}$$

$$\text{if } \mathbf{J}_\mathbf{a}(\mathbf{z})\mathbf{t} = \mathbf{0} \text{ and } \mathbf{t} \neq \mathbf{0} \text{ then } \mathbf{t}^\top \left[\mathbf{H}_\phi(\mathbf{z}) + \sum_{i=1}^{p} \ell_i \mathbf{H}_{\mathbf{a}_i}(\mathbf{z})\right] \mathbf{t} > 0 \, . \tag{4.73c}$$

It follows that there exists $\underline{\gamma} \geq 0$ so that for all $\gamma \geq \underline{\gamma}$, the n-vector $\mathbf{z}$ *also minimizes*

$$\lambda(\mathbf{x}\,;\,\boldsymbol{\ell}\,,\,\gamma) \equiv \phi(\mathbf{x}) + \boldsymbol{\ell}^\top \mathbf{a}(\mathbf{x}) + \frac{\gamma}{2}\mathbf{a}(\mathbf{x})^\top \mathbf{a}(\mathbf{x})\,.$$

Furthermore, if $\mathbf{a}(\widetilde{\mathbf{z}}) = \mathbf{0}$ *and* $\widetilde{\mathbf{z}}$ *satisfies the second-order sufficient conditions in lemma for a minimum of* λ*, namely*

$$\mathbf{g}_\lambda(\widetilde{\mathbf{z}}\,;\,\boldsymbol{\ell}\,,\,\gamma) = \mathbf{0} \text{ and } \mathbf{H}_\lambda(\widetilde{\mathbf{z}}\,;\,\boldsymbol{\ell}\,,\,\gamma) \text{ positive}$$

then $\widetilde{\mathbf{z}}$ *is a strict local minimum of the constrained minimization problem.*

Proof Let us prove the first claim. We first note that the first-order Kuhn-Tucker conditions (4.73a) and (4.73b) for the constrained minimization problem imply the first-order condition

$$\mathbf{g}_\lambda(\mathbf{z}\,;\,\boldsymbol{\ell}\,,\,\gamma) = \mathbf{g}_\phi(\mathbf{z})^\top + \boldsymbol{\ell}^\top \mathbf{J}_\mathbf{a}(\mathbf{z}) + \gamma\mathbf{a}(\mathbf{z})^\top \mathbf{J}_\mathbf{a}(\mathbf{z}) = \mathbf{g}_\phi(\mathbf{z})^\top + \boldsymbol{\ell}^\top \mathbf{J}_\mathbf{a}(\mathbf{z}) = \mathbf{0}$$

for an unconstrained minimum of λ. Next, we apply Corollary 4.5.1 to the matrices

$$\mathbf{A} = \mathbf{H}_\phi(\mathbf{z}) + \sum_{i=1}^{p} \ell_i \mathbf{H}_{\mathbf{A}_i}(\mathbf{z}) \text{ and } \mathbf{B} = \mathbf{J}_\mathbf{a}(\mathbf{z})\,.$$

This corollary shows that $\mathbf{z}$ satisfies

$$\mathbf{A}\mathbf{t} = \mathbf{0} \text{ and } \mathbf{t} \neq \mathbf{0} \text{ implies } \mathbf{t}^\top \mathbf{A}\mathbf{t} > 0\,,$$

if and only if there exists $\underline{\gamma} \geq 0$ so that for all $\gamma \geq \underline{\gamma}$ and for all $\mathbf{t} \neq \mathbf{0}$ we have

$$\mathbf{t}^\top\left[\mathbf{A} + \gamma\mathbf{B}^\top\mathbf{B}\right]\mathbf{t} > 0\,.$$

The former of these conditions is equivalent to the second-order Kuhn-Tucker condition (4.73c), and the latter is equivalent to the second-order sufficient condition

$$\begin{aligned}
&\mathbf{t}^\top \mathbf{H}_\lambda(\mathbf{z}\,;\,\boldsymbol{\ell}\,,\,\gamma)\mathbf{t} \\
&= \mathbf{t}^\top\left[\mathbf{H}_\phi(\mathbf{z}) + \sum_{i=1}^{p} \ell_i \mathbf{H}_{\mathbf{a}_i}(\mathbf{z}) + \gamma\sum_{i=1}^{p} \mathbf{a}_i(\mathbf{z})\mathbf{H}_{\mathbf{a}_i}(\mathbf{z}) + \mathbf{J}_\mathbf{a}(\mathbf{z})^\top \mathbf{J}_\mathbf{a}(\mathbf{z})\right]\mathbf{t} \\
&= \mathbf{t}^\top\left[\mathbf{H}_\phi(\mathbf{z}) + \sum_{i=1}^{p} \ell_i \mathbf{H}_{\mathbf{a}_i}(\mathbf{z}) + \mathbf{J}_\mathbf{a}(\mathbf{z})^\top \mathbf{J}_\mathbf{a}(\mathbf{z})\right]\mathbf{t} > 0\,.
\end{aligned}$$

Lemma 3.2.12 now shows that $\mathbf{z}$ is a strict local minimizer of λ.

Next, let us prove the second claim. Suppose that $\mathbf{a}(\widetilde{\mathbf{z}}) = \mathbf{0}$ and that $\widetilde{\mathbf{z}}$ satisfies the second-order necessary conditions for an unconstrained minimum of λ. The first-order necessary conditions for this minimum are

$$\mathbf{0}^\top = \mathbf{g}_\lambda(\mathbf{z}\,,\,\boldsymbol{\ell}\,,\,\gamma)^\top = \mathbf{g}_\phi(\mathbf{z})^\top + \boldsymbol{\ell}^\top \mathbf{J}_\mathbf{a}(\mathbf{z}) + \mathbf{a}(\mathbf{z})^\top \mathbf{J}_\mathbf{a}(\mathbf{z}) = \mathbf{g}_\phi(\mathbf{z})^\top + \boldsymbol{\ell}^\top \mathbf{J}_\mathbf{a}(\mathbf{z})\,,$$

and the second-order sufficient conditions for this minimum are

$$\begin{aligned} 0 &< \mathbf{t}^\top \mathbf{H}_\lambda(\widetilde{\mathbf{z}}\,;\,\boldsymbol{\ell}\,,\,\gamma)\mathbf{t} \\ &= \mathbf{t}^\top \left[\mathbf{H}_\phi(\widetilde{\mathbf{z}} + \sum_{i=1}^p \ell_i \mathbf{H}_{\mathbf{a}_i}(\widetilde{\mathbf{z}}\,;\,\boldsymbol{\ell}\,,\,\gamma) + \gamma \sum_{i=1}^p \mathbf{a}_i(\widetilde{\mathbf{z}}) \mathbf{H}_{\mathbf{a}_i}(\widetilde{\mathbf{z}}\,;\,\boldsymbol{\ell}\,,\,\gamma) + \mathbf{J}_\mathbf{a}(\widetilde{\mathbf{z}})^\top \mathbf{J}_\mathbf{a}(\widetilde{\mathbf{z}}) \right] \mathbf{t} \\ &= \mathbf{t}^\top \left[\mathbf{H}_\phi(\widetilde{\mathbf{z}} + \sum_{i=1}^p \ell_i \mathbf{H}_{\mathbf{a}_i}(\widetilde{\mathbf{z}}\,;\,\boldsymbol{\ell}\,,\,\gamma) + \mathbf{J}_\mathbf{a}(\widetilde{\mathbf{z}})^\top \mathbf{J}_\mathbf{a}(\widetilde{\mathbf{z}}) \right] \mathbf{t} \end{aligned}$$

for all $\mathbf{t} \neq \mathbf{0}$. The first-order conditions for the unconstrained minimization problem are the same as the first-order condition (4.73a) for the constrained minimization problem. Corollary 4.5.1 shows that the second-order condition for the unconstrained minimization problem is equivalent to the second-order condition (4.73c) for the constrained minimization problem. Theorem 4.3.4 shows that $\widetilde{\mathbf{z}}$ is a strict local minimum of the constrained minimization problem.

4.5.3.3 General Constraints

A number of authors have suggested methods for dealing with inequality constraints in augmented Lagrangian methods. One approach can be found in Conn et al. [38], who discuss problems in the form

$$\min \phi(\mathbf{x}) \text{ subject to } \mathbf{a}(\mathbf{x}) = \mathbf{0} \text{ and } \underline{\mathbf{x}} \leq \mathbf{x} \leq \overline{\mathbf{x}}\,.$$

These authors note that inequality constraints of the form

$$\mathbf{b}(\mathbf{x}) \leq \mathbf{0}$$

can be written in the form

$$\mathbf{b}(\mathbf{x}) + \mathbf{s} = \mathbf{0} \text{ and } \mathbf{s} \geq \mathbf{0}\,.$$

In other words, inequality constraints can be converted to equality constraints by the introduction of slack variables subject to a lower bound.

Conn et al. [38, p. 551] suggested the following

Algorithm 4.5.8 (Conn-Gould-Toint Augmented Lagrangian)

given initial Lagrange multipliers $\boldsymbol{\ell}_0$

given initial penalty ϱ_0

given tolerances $\tau_0 < 1$, $\omega_0 < 1$, $\tau_\infty \ll 1$ and $\omega_\infty \ll 1$

given positive tolerance exponents α_τ , β_τ , α_ω and β_ω

given penalty scaling factor $\sigma \in (0, 1)$ and penalty bound $\overline{\varrho} \in (0, 1)$

for $k = 0, 1, \ldots$

find $\mathbf{z}_k$ to minimize $\lambda(\mathbf{x} \,;\, \boldsymbol{\ell}_k \,,\, \varrho_k) = \phi(\mathbf{x}) + \boldsymbol{\ell}_k{}^\top \mathbf{a}(\mathbf{x}) + \mathbf{a}(\mathbf{x})^\top \mathbf{a}(\mathbf{x})/2\varrho_k$

subject to $\underline{\mathbf{x}} \leq \mathbf{x} \leq \overline{\mathbf{x}}$ with error $\|\min\{\mathbf{x} - \underline{\mathbf{x}} \,,\, \max\{\mathbf{x} - \overline{\mathbf{x}} \,,\, \mathbf{g}_\lambda(\mathbf{x} \,;\, \boldsymbol{\ell}_k \,,\, \varrho_k)\}\}\|_\infty \leq \tau_k$

if $\|\mathbf{a}(\mathbf{z}_k)\|_\infty \leq \omega_k$ then

$\boldsymbol{\ell}_{k+1} = \boldsymbol{\ell}_k + \mathbf{a}(\mathbf{z}_k)/\varrho_k \,,\, \varrho_{k+1} = \varrho_k$

$\mu = \min\{\varrho_{k+1} \,,\, \overline{\varrho}\} \,,\, \tau_{k+1} = \tau_k \mu^{\alpha_\tau} \,,\, \omega_{k+1} = \omega_k \mu^{\alpha_\omega}$

else

$\boldsymbol{\ell}_{k+1} = \boldsymbol{\ell}_k \,,\, rho_{k+1} = \sigma \varrho_k$

$\mu = \min\{\varrho_{k+1} \,,\, \overline{\varrho}\} \,,\, \tau_{k+1} = \tau_0 \mu^{\beta_\tau} \,,\, \omega_{k+1} = \omega_0 \mu^{\beta_\omega}$

Conn et al. also suggested a second form of an augmented Lagrangian algorithm, and proved global linear convergence.

4.5.3.4 Diagonalized Methods

The previous augmented Lagrangian methods of Powell, Hestenes and Conn et al. involve an outer iteration in which the Lagrange multipliers and penalty are updated, and an inner iteration in which an unconstrained (or bound-constrained) minimization problem is solved. Using quasi-Newton methods, Tapia [170] developed and analyzed the following general

Algorithm 4.5.9 (Tapia Diagonalized Augmented Lagrangian)

given $\mathbf{z}_0 \,,\, \boldsymbol{\ell}_0 \,,\, \varrho_0$

choose approximate inverse Hessian $\mathbf{H}_0$

for $k = 1, 2, \ldots$

provisionally update $\widetilde{\boldsymbol{\ell}}_{k+1}$

perform quasi-Newton update $\mathbf{z}_{k+1} = \mathbf{z}_k - \mathbf{H}_k \mathbf{g}_\lambda(\mathbf{z}_k \,;\, \widetilde{\boldsymbol{\ell}}_{k+1} \,,\, \varrho_k)$

update $\boldsymbol{\ell}_{k+1} \,,\, \varrho_{k+1}$ and $\mathbf{H}_{k+1}$

Tapia considered a number of choices for the update formulas for each of the parameters in the algorithm, and proved convergence under fairly general circumstances.

There are several software packages that implement augmented Lagrangian methods. Conn et al. [39] have used augmented Lagrangian methods in the LANCELOT package to solve nonlinear optimization problems subject to nonlinear equality constraints and simple bounds. Birgin and Martinez [18] have used diagonalized quasi-Newton methods within augmented Lagrangian techniques to develop ALGENCAN within the TANGO project.

4.5.4 *Sequential Quadratic Programming*

We would like to describe yet another scheme for solving nonlinear programming problems. As a first approach, we could try to find the solution $\mathbf{z}$ of a constrained minimization problem

$$\min \phi(\mathbf{x}) \text{ subject to } \mathbf{a}(\mathbf{x}) = \mathbf{0} \text{ and } \mathbf{b}(\mathbf{x}) \leq \mathbf{0} \tag{4.74}$$

by finding an approximate constrained minimum $\mathbf{z}_{k+1}$ of the quadratic programming problem

$$\begin{aligned} &\min \phi(\mathbf{z}_k) + \mathbf{g}_\phi(\mathbf{z}_k)^\top(\mathbf{x} - \mathbf{z}_k) + \frac{1}{2}(\mathbf{x} - \mathbf{z}_k)^\top \mathbf{H}_k(\mathbf{x} - \mathbf{z}_k) \\ &\text{subject to } \mathbf{a}(\mathbf{z}_k) + \mathbf{J}_\mathbf{a}(\mathbf{z}_k)(\mathbf{x} - \mathbf{z}_k) = \mathbf{0} \text{ and } \mathbf{b}(\mathbf{z}_k) + \mathbf{J}_\mathbf{b}(\mathbf{z}_k)(\mathbf{x} - \mathbf{z}_k) \leq \mathbf{0}\,. \end{aligned} \tag{4.75}$$

Here $\mathbf{H}_k$ is an approximate Hessian, possibly generated by a quasi-Newton method as in Sect. 3.7.4. Ideally, we would choose an initial approximate solution $\mathbf{z}_0$, then in a loop over step indices k we would solve the quadratic programming problem (4.75). Hopefully, the iterates $\mathbf{z}_k$ converge to the solution $\mathbf{z}$ of the original problem (4.74).

In practice, such an approach is not always possible. One difficulty is that the sequential quadratic programming problem may not be feasible, even when the original problem is feasible. The next example illustrates this problem.

Example 4.5.1 ([171, p. 145]) Let

$$\mathbf{x} = \begin{bmatrix} \xi_1 \\ \xi_2 \end{bmatrix}, \ \phi(\mathbf{x}) = \xi_1^3 + \xi_2^2, \ \mathbf{a}(\mathbf{x}) = \begin{bmatrix} \xi_1^2 + \xi_2^2 - 10 \end{bmatrix} \text{ and } \mathbf{b}(\mathbf{x}) = \begin{bmatrix} 1 - \xi_1 \\ 1 - \xi_2 \end{bmatrix}.$$

It is easy to find the optimal solution of the nonlinear programming problem (4.74) with this choice of functions. If $\xi_1 = 1$, then $\mathbf{a}(\mathbf{x}) = \mathbf{0}$ and $\mathbf{b}(\mathbf{x}) \leq \mathbf{0}$ imply that $\xi_2 = 3$; we then have $\phi(\mathbf{x}) = 10$. On the other hand, if $1 < \xi_1 \leq 3$, then the constraints imply that $\xi_2 = \sqrt{10 - \xi_1^2}$; these imply that

$$\phi(\mathbf{x}) = \xi_1^3 + 10 - \xi_1^2 > 10 \; .$$

Since the constraints cannot be satisfied for $\xi_1 > 3$ or $\xi_1 < 1$, the optimal solution has $\xi_1 = 1$ and $\xi_2 = 3$.

Now, consider the sequential quadratic programming problem (4.75) with $\mathbf{z}_k = [-10, -10]$. We have

$$\mathbf{a}(\mathbf{z}_k) = [190] \; , \; \mathbf{J_a}(\mathbf{z}_k) = \begin{bmatrix} -20, \, -20 \end{bmatrix} \; , \; \mathbf{b}(\mathbf{z}_k) = \begin{bmatrix} 11 \\ 11 \end{bmatrix} \text{ and } \mathbf{J_b}(\mathbf{z}_k) = \begin{bmatrix} -1 & 0 \\ 0 & -1 \end{bmatrix} \; .$$

The equality constraint in (4.75) is

$$0 = \mathbf{a}(\mathbf{z}_k) + \mathbf{J_a}(\mathbf{z}_k)(\mathbf{x} - \mathbf{z}_k) = 190 - \begin{bmatrix} 20, \, 20 \end{bmatrix} \begin{bmatrix} \xi_1 + 10 \\ \xi_2 + 10 \end{bmatrix} = -210 - 20\xi_1 - 20\xi_2 \; .$$

The inequality constraints in (4.75) are

$$\mathbf{0} \geq \mathbf{b}(\mathbf{z}_k) + \mathbf{J_b}(\mathbf{z}_k)(\mathbf{x} - \mathbf{z}_k) = \begin{bmatrix} 11 \\ 11 \end{bmatrix} - \begin{bmatrix} \xi_1 + 10 \\ \xi_2 + 10 \end{bmatrix} = \begin{bmatrix} 1 - \xi_1 \\ 1 - \xi_2 \end{bmatrix} \; .$$

These constraints are inconsistent, and the corresponding sequential programming problem (4.75) is infeasible.

A successful approach for avoiding infeasible sequential quadratic programming problems has been provided by Tone [171, p. 150ff]. If the sequential programming problem (4.75) is infeasible, he suggests that we choose nonnegative vectors **c** and **d** in order to solve the modified problem

$$\begin{aligned} &\min \phi(\mathbf{z}_k) + \mathbf{g}_\phi(\mathbf{z}_k)^\top(\mathbf{x} - \mathbf{z}_k) + \frac{1}{2}(\mathbf{x} - \mathbf{z}_k)^\top \mathbf{H}_k(\mathbf{x} - \mathbf{z}_k) + \mathbf{c}^\top(\mathbf{u} + \mathbf{w}) + \mathbf{d}^\top \mathbf{v} \\ &\text{subject to } \begin{cases} \mathbf{a}(\mathbf{z}_k) + \mathbf{J_a}(\mathbf{z}_k)(\mathbf{x} - \mathbf{z}_k) - \mathbf{u} + \mathbf{w} = \mathbf{0} \\ \mathbf{b}(\mathbf{z}_k) + \mathbf{J_b}(\mathbf{z}_k)(\mathbf{x} - \mathbf{z}_k) - \mathbf{v} \leq \mathbf{0} \text{ and} \\ \mathbf{u} \geq \mathbf{0} \; , \; \mathbf{w} \geq \mathbf{0} \; , \; \mathbf{v} \geq \mathbf{0} \; . \end{cases} \; . \end{aligned} \tag{4.76}$$

This problem is always feasible. For example, the vectors

$$\mathbf{x} = \mathbf{z}_k \; , \; \mathbf{u} = \max\{\mathbf{a}(\mathbf{z}_k) \; , \; \mathbf{0}\} \; , \; \mathbf{w} = \min\{\mathbf{a}(\mathbf{z}_k) \; , \; \mathbf{0}\} \text{ and } \mathbf{v} = \max\{\mathbf{b}(\mathbf{z}_k) \; , \; \mathbf{0}\}$$

satisfy the constraints in (4.76). In these expressions, the maxima and minima are taken componentwise.

The modified sequential quadratic programming problem (4.76) is often associated with the modified objective

$$\psi(\mathbf{x} \; ; \; \mathbf{c} \; , \; \mathbf{d}) = \phi(\mathbf{x}) + \mathbf{c}^\top |\mathbf{a}(\mathbf{x})| + \mathbf{d}^\top \max\{\mathbf{b}(\mathbf{x}) \; , \; \mathbf{0}\} \; . \tag{4.77}$$

In this expression, the absolute value and maximum are again taken componentwise. The following lemma explains why this modified objective is relevant, and how to choose the penalties $\mathbf{c}$ and $\mathbf{d}$.

Lemma 4.5.5 [68, p. 28] *Let* 0 [illegible] *that* ϕ *is a twice continuously differentiable functional defined on real n-vectors, that* $\mathbf{a}$ *is a twice continuously differentiable function mapping real n-vectors to real p-vectors, and* $\mathbf{b}$ *is a twice continuously differentiable function mapping real n-vectors to real p-vectors. If the real n-vector* $\mathbf{z}$ *is a local minimum of the nonlinear programming problem*

$$\min \phi(\mathbf{x}) \text{ subject to } \mathbf{a}(\mathbf{x}) = \mathbf{0} \text{ and } \mathbf{b}(\mathbf{x}) \leq \mathbf{0} ,$$

$\mathbf{z}$ *is a regular point of the active constraints, and the constrained Hessian*

$$\mathbf{H}_\phi(\mathbf{z}) + \sum_{i=1}^{p} \boldsymbol{\ell}_i \mathbf{H}_{\mathbf{a}_i}(\mathbf{z}) + \sum_{j=1}^{p} \mathbf{m}_j \mathbf{H}_{\mathbf{b}_j}(\mathbf{z})$$

is positive on the tangent plane of the active constraints, then there is an p-vector $\underline{\mathbf{c}} \geq \mathbf{0}$ *and a p-vector* $\underline{\mathbf{d}} \geq \mathbf{0}$ *so that for all* $\mathbf{c} \geq \underline{\mathbf{c}}$ *and all* $\mathbf{d} \geq \underline{\mathbf{d}}$ *the n-vector* $\mathbf{z}$ *is also a local minimum of*

$$\psi(\mathbf{x}) = \phi(\mathbf{x}) + \mathbf{c}^\top |\mathbf{a}(\mathbf{x})| + \mathbf{d}^\top \max\{\mathbf{b}(\mathbf{x}) \, , \, \mathbf{0}\} \, .$$

Conversely, suppose that $\mathbf{z}$ *is a local minimum of* ψ *such that* $\mathbf{a}(\mathbf{z}) = \mathbf{0}$, $\mathbf{b}(\mathbf{z}) \leq \mathbf{0}$ *and assume that the limit of the constrained Hessian of* ψ *as we approach* $\mathbf{z}$ *from inside the constrained region* $\mathbf{a}(\mathbf{z}) \leq \mathbf{0}$ *and* $\mathbf{b}(\mathbf{z}) \leq \mathbf{0}$ *is positive on the tangent plane of the active constraints. Then* $\mathbf{z}$ *is a local minimum of* $\phi(\mathbf{x})$ *subject to* $\mathbf{a}(\mathbf{x}) = \mathbf{0}$ *and* $\mathbf{b}(\mathbf{x}) \leq \mathbf{0}$.

Proof Suppose that $\mathbf{z}$ minimizes $\phi(\mathbf{x})$ subject to $\mathbf{a}(\mathbf{x}) = \mathbf{0}$ and $\mathbf{b}(\mathbf{x}) \leq \mathbf{0}$. Then Theorem 4.3.3 implies that there is a real p-vector $\boldsymbol{\ell}$ and a real p-vector $\mathbf{m} \geq \mathbf{0}$ so that

$$\mathbf{g}_\phi(\mathbf{z})^\top + \boldsymbol{\ell}^\top \mathbf{J}_\mathbf{a}(\mathbf{z}) + \mathbf{m}^\top \mathbf{J}_\mathbf{b}(\mathbf{z}) = \mathbf{0}^\top \text{ with } \mathbf{m}^\top \mathbf{b}(\mathbf{z}) = 0 \, .$$

Furthermore, if the nonzero n-vector $\mathbf{t}$ satisfies

$$\mathbf{J}_\mathbf{a}(\mathbf{z})\mathbf{t} = \mathbf{0} \text{ and } \mathbf{m}_j > 0 \text{ implies } \mathbf{e}_j{}^\top \mathbf{J}_\mathbf{b}(\mathbf{z})\mathbf{t} = 0$$

then

$$\mathbf{t}^\top \left[\mathbf{H}_\phi(\mathbf{z}) + \sum_{i=1}^{p} \boldsymbol{\ell}_i \mathbf{H}_{\mathbf{a}_i}(\mathbf{z}) + \sum_{j=1}^{p} \mathbf{m}_j \mathbf{H}_{\mathbf{b}_j}(\mathbf{z}) \right] \mathbf{t} > 0 \, .$$

We choose

$$\underline{\mathbf{c}} = |\boldsymbol{\ell}| \text{ and } \underline{\mathbf{d}} = \mathbf{m} \; .$$

We also let $\mathbf{D_a}$ and $\mathbf{D_b}$ be real diagonal matrices with diagonal entries either one or minus one. Let $\mathbf{S_b}$ be the real diagonal matrix with diagonal entries

$$\mathbf{e}_j{}^\top \mathbf{S_b} \mathbf{e}_j = \begin{cases} 1, \; \mathbf{e}_j{}^\top \mathbf{D_b} \mathbf{e}_j = 1 \\ 0, \; \mathbf{e}_j{}^\top \mathbf{D_b} \mathbf{e}_j = -1 \end{cases} .$$

Consider the constrained nonlinear optimization problem

$$\min \psi(\mathbf{x}) \text{ subject to } -\mathbf{D_a a}(\mathbf{x}) \leq \mathbf{0} \text{ and } -\mathbf{D_b b}(\mathbf{x}) \leq \mathbf{0} \; .$$

Let

$$\mathbf{m_a} = \mathbf{c} - \mathbf{D_a}\boldsymbol{\ell} \text{ and } \mathbf{e}_j \cdot \mathbf{m_b} = \begin{cases} \mathbf{e}_j \cdot (\mathbf{d} - \mathbf{m}), \; \mathbf{e}_j \cdot \mathbf{D_b} \mathbf{e}_j = 1 \\ \mathbf{e}_j \cdot \mathbf{m}, \qquad\qquad \mathbf{e}_j \cdot \mathbf{D_b} \mathbf{e}_j = -1 \end{cases} .$$

Then $\mathbf{m_a} \geq \mathbf{0}$, $\mathbf{m_b} \geq \mathbf{0}$ and

$$\begin{aligned} &\mathbf{g}_\psi(\mathbf{z})^\top - \mathbf{m_a}{}^\top \mathbf{D_a J_a}(\mathbf{z}) - \mathbf{m_b}{}^\top \mathbf{D_b J_b}(\mathbf{z}) \\ &= \mathbf{g}_\phi(\mathbf{z})^\top + \mathbf{c}^\top \mathbf{D_a J_a}(\mathbf{z}) + \mathbf{d}^\top \mathbf{S_b J_b}(\mathbf{z}) - \mathbf{m_a}{}^\top \mathbf{D_a J_a}(\mathbf{z}) - \mathbf{m_b}{}^\top \mathbf{D_b J_b}(\mathbf{z}) \\ &= (\mathbf{D_a c} - \mathbf{D_a m_a} - \boldsymbol{\ell})^\top \mathbf{J_a}(\mathbf{z}) + (\mathbf{S_b d} - \mathbf{D_b m_b} - \mathbf{m})^\top \mathbf{J_b}(\mathbf{z}) = \mathbf{0}^\top \; . \end{aligned}$$

This shows that $\mathbf{z}$ satisfies the first-order Kuhn-Tucker conditions for a local minimum of ψ subject to $-\mathbf{D_a a}(\mathbf{x}) \leq \mathbf{0}$ and $-\mathbf{D_b b}(\mathbf{x}) \leq \mathbf{0}$. Furthermore, the constrained Hessian

$$\begin{aligned} &\mathbf{H}_\psi(\mathbf{z}) - \sum_{i=1}^{p} \mathbf{e}_i{}^\top \mathbf{D_a m_a H}_{\mathbf{a}_i}(\mathbf{z}) - \sum_{j=1}^{p} \mathbf{e}_j{}^\top \mathbf{D_b m_b H}_{\mathbf{b}_j}(\mathbf{z}) \\ &= \mathbf{H}_\phi(\mathbf{z}) + \sum_{i=1}^{p} \mathbf{e}_i{}^\top \mathbf{D_a}(\mathbf{c} - \mathbf{m_a}) \mathbf{H}_{\mathbf{a}_i}(\mathbf{z}) + \sum_{j=1}^{p} \mathbf{e}_j{}^\top (\mathbf{S_b d} - \mathbf{D_b m_b}) \mathbf{H}_{\mathbf{b}_j}(\mathbf{z}) \\ &= \mathbf{H}_\phi(\mathbf{z}) + \sum_{i=1}^{p} \mathbf{e}_i{}^\top \boldsymbol{\ell} \mathbf{H}_{\mathbf{a}_i}(\mathbf{z}) + \sum_{j=1}^{p} \mathbf{e}_j{}^\top \mathbf{m} \mathbf{H}_{\mathbf{b}_j}(\mathbf{z}) \end{aligned}$$

is assumed to be positive on the tangent plane of the active constraints. As a result, Theorem 4.3.4 implies that $\mathbf{z}$ is a local minimum of ψ subject to $-\mathbf{D_a a}(\mathbf{x}) \leq \mathbf{0}$ and $-\mathbf{D_b b}(\mathbf{x}) \leq \mathbf{0}$. The arbitrariness of the signs of the diagonal entries of $\mathbf{D_a}$ and $\mathbf{D_b}$ implies that $\mathbf{z}$ is a local minimum of ψ.

It remains to prove the converse. Suppose that $\mathbf{a}(\mathbf{z}) = \mathbf{0}$, $\mathbf{b}(\mathbf{z}) \leq \mathbf{0}$ and $\mathbf{z}$ is a local minimum of ψ. Then $\mathbf{z}$ is a local minimum of ψ subject to $\mathbf{a}(\mathbf{z}) \leq \mathbf{0}$ and $\mathbf{b}(\mathbf{z}) \leq \mathbf{0}$. Consequently, there exists $\mathbf{m_a} \geq \mathbf{0}$ and $\mathbf{m_b} \geq \mathbf{0}$ so that

$$\begin{aligned}\mathbf{0}^\top &= \mathbf{g}_\psi(\mathbf{z})^\top + \mathbf{m_a}^\top \mathbf{J_a}(\mathbf{z}) + \mathbf{m_b}^\top \mathbf{J_b}(\mathbf{z}) \\ &= \mathbf{g}_\phi(\mathbf{z})^\top + (\mathbf{m_a} - \mathbf{c})^\top \mathbf{J_a}(\mathbf{z}) + \mathbf{m_b}^\top \mathbf{J_b}(\mathbf{z}) \;.\end{aligned}$$

It follows that $\mathbf{z}$ satisfies the first-order Kuhn-Tucker conditions for a minimum of ϕ subject to $\mathbf{a}(\mathbf{z}) = \mathbf{0}$ and $\mathbf{b}(\mathbf{z}) \leq \mathbf{0}$. Since the constrained Hessian of ψ is assumed to be positive on the tangent plane of the active constraints, we conclude that $\mathbf{z}$ is a local minimum of ϕ subject to $\mathbf{a}(\mathbf{z}) = \mathbf{0}$ and $\mathbf{b}(\mathbf{z}) \leq \mathbf{0}$.

The next lemma explains the inter-relationship of the modified sequential programming problem (4.76) and the modified objective (4.77).

Lemma 4.5.6 *[171, p. 150] Let $0 \leq p < n$ and q be nonnegative integers. Suppose that $\mathbf{z}$ is a real n-vector, that $\mathbf{c}$ is a positive p-vector, that $\mathbf{d}$ is a positive q-vector and that $\mathbf{H}$ is a real $n \times n$ symmetric positive matrix. Assume that ϕ is a continuously differentiable functional defined on real n-vectors, that $\mathbf{a}$ is a continuously differentiable function mapping real n-vectors to real p-vectors, and $\mathbf{b}$ is a continuously differentiable function mapping real n-vectors to real q-vectors. Let the n-vector $\mathbf{x}_*$, nonnegative p-vectors $\mathbf{u}$ and $\mathbf{w}$, and nonnegative q-vector $\mathbf{v}$ solve the quadratic programming problem*

$$\min \phi(\mathbf{z}) + \mathbf{g}_\phi(\mathbf{z}) \cdot (\mathbf{x} - \mathbf{z}) + \frac{1}{2}(\mathbf{x} - \mathbf{z}) \cdot \mathbf{H}(\mathbf{x} - \mathbf{z}) + \mathbf{c} \cdot (\mathbf{u} + \mathbf{w}) + \mathbf{d} \cdot \mathbf{v}$$

$$\textit{subject to} \begin{array}{l} \mathbf{a}(\mathbf{z}) + \mathbf{J_a}(\mathbf{z})(\mathbf{x} - \mathbf{z}) - \mathbf{u} + \mathbf{w} = \mathbf{0} \;, \\ \mathbf{b}(\mathbf{z}) + \mathbf{J_b}(\mathbf{z})(\mathbf{x} - \mathbf{z}) - \mathbf{v} \leq \mathbf{0} \textit{ and} \\ \mathbf{u} \geq \mathbf{0} \;,\; \mathbf{w} \geq \mathbf{0} \;,\; \mathbf{v} \geq \mathbf{0} \end{array} \;. \tag{4.78}$$

If $\mathbf{z}$ is a regular point of (4.74) *correspond to the active constraints in* (4.78) *and $\mathbf{x}_* \neq \mathbf{z}$, then $\mathbf{x}_* - \mathbf{z}$ is a descent direction for*

$$\psi(\mathbf{x}) = \phi(\mathbf{x}) + \mathbf{c} \cdot |\mathbf{a}(\mathbf{x})| + \mathbf{d} \cdot \max\{\mathbf{b}(\mathbf{x}) \;,\; \mathbf{0}\} \tag{4.79}$$

at $\mathbf{x} = \mathbf{z}$.

Proof The Kuhn-Tucker first-order necessary conditions in Theorem 4.3.2 show that at a minimum of (4.78) we have

$$\mathbf{a}(\mathbf{z}) + \mathbf{J_a}(\mathbf{z})(\mathbf{x}_* - \mathbf{z}) - \mathbf{u} + \mathbf{w} = \mathbf{0} \tag{4.80a}$$

$$\mathbf{b}(\mathbf{z}) + \mathbf{J_b}(\mathbf{z})(\mathbf{x}_* - \mathbf{z}) - \mathbf{v} \leq \mathbf{0} \tag{4.80b}$$

$$\mathbf{u} \geq \mathbf{0} \;,\; \mathbf{w} \geq \mathbf{0} \;,\; \mathbf{v} \geq \mathbf{0} \tag{4.80c}$$

$$\mathbf{m_b} \geq \mathbf{0} \;,\; \mathbf{m_u} \geq \mathbf{0} \;,\; \mathbf{m_w} \geq \mathbf{0} \;,\; \mathbf{m_v} \geq \mathbf{0} \tag{4.80d}$$

$$\mathbf{m_b} \cdot [\mathbf{b}(\mathbf{z}) + \mathbf{J_b}(\mathbf{z})(\mathbf{x}_* - \mathbf{z}) - \mathbf{v}] - \mathbf{m_u} \cdot \mathbf{u} - \mathbf{m_w} \cdot \mathbf{w} - \mathbf{m_v} \cdot \mathbf{v} = \mathbf{0} \tag{4.80e}$$

$$\mathbf{g}_\phi(\mathbf{z}) + \mathbf{H}(\mathbf{x}_* - \mathbf{z}) + \mathbf{J_a}(\mathbf{z})^\top \boldsymbol{\ell}_\mathbf{a} + \mathbf{J_b}(\mathbf{z})^\top \mathbf{m_b} = \mathbf{0} \tag{4.80f}$$

$$\mathbf{c} - \boldsymbol{\ell}_\mathbf{a} - \mathbf{m_u} = \mathbf{0} \tag{4.80g}$$

$$\mathbf{c} + \boldsymbol{\ell}_\mathbf{a} - \mathbf{m_w} = \mathbf{0} \text{ and} \tag{4.80h}$$

$$\mathbf{d} - \mathbf{m_b} - \mathbf{m_v} = \mathbf{0} \; . \tag{4.80i}$$

Let the diagonal matrices $\mathbf{S_a}$ and $\mathbf{S_b}$ have diagonal entries

$$\mathbf{e}_i \cdot \mathbf{S_a}\mathbf{e}_i = \begin{cases} 1 & , \mathbf{a}_i(\mathbf{z}) > 0 \\ = 1 & , \mathbf{a}_i(\mathbf{z}) \le 0 \end{cases} \text{ and } \mathbf{e}_j \cdot \mathbf{S_b}\mathbf{e}_j = \begin{cases} 1, & \mathbf{b}_i(\mathbf{z}) > 0 \\ 0, & \mathbf{b}_i(\mathbf{z}) \le 0 \end{cases}$$

Then the directional derivative of ψ is

$$\mathbf{g}_\psi(\mathbf{z}) \cdot (\mathbf{x}_* - \mathbf{z}) = \mathbf{g}_\phi(\mathbf{z}) \cdot (\mathbf{x}_* - \mathbf{z}) + \mathbf{c}^\top \mathbf{S_a} \mathbf{J_a}(\mathbf{z})(\mathbf{x}_* - \mathbf{z}) + \mathbf{d}^\top \mathbf{s_b} \mathbf{J_b}(\mathbf{z})(\mathbf{x}_* - \mathbf{z})$$

and the first-order Kuhn-Tucker condition (4.80f) implies that

$$\begin{aligned} &= -(\mathbf{x}_* - \mathbf{z})^\top \mathbf{H}(\mathbf{x}_* - \mathbf{z}) + (\mathbf{S_a}\mathbf{c} - \boldsymbol{\ell}_\mathbf{a})^\top \mathbf{J_a}(\mathbf{z})(\mathbf{x}_* - \mathbf{z}) \\ &\quad + (\mathbf{S_b}\mathbf{d} - \mathbf{m_b})^\top \mathbf{J_b}(\mathbf{z})(\mathbf{x}_* - \mathbf{z}) \; . \end{aligned} \tag{4.81}$$

We would like to examine the componentwise contributions to the last two terms in this expression.

Suppose that $\mathbf{a}_i(\mathbf{z}) > 0$ and $\mathbf{e}_i \cdot \mathbf{J_a}(\mathbf{z})(\mathbf{x}_* - \mathbf{z}) > 0$. Then

$$\mathbf{e}_i \cdot \mathbf{u} = \mathbf{e}_i \cdot [\mathbf{a}(\mathbf{z}) + \mathbf{J_a}(\mathbf{z})(\mathbf{x}_* - \mathbf{z})]$$

and $\mathbf{e}_i \cdot \mathbf{w} = 0$. The complementarity condition (4.80e) then implies that $\mathbf{e}_i \cdot \mathbf{m_u} = 0$, and condition (4.80g) implies that

$$\mathbf{e}_i \cdot \boldsymbol{\ell}_\mathbf{a} = \mathbf{e}_i \cdot \mathbf{c} = \mathbf{e}_i \cdot \mathbf{S_a}\mathbf{c} \; .$$

Thus there is no contribution to the second term in (4.81) for this component index i.

Suppose that $\mathbf{a}_i(\mathbf{z}) \le 0$ and $\mathbf{e}_i \cdot \mathbf{J_a}(\mathbf{z})(\mathbf{x}_* - \mathbf{z}) < 0$. Then

$$\mathbf{e}_i \cdot \mathbf{w} = -\mathbf{e}_i \cdot [\mathbf{a}(\mathbf{z}) + \mathbf{J_a}(\mathbf{z})(\mathbf{x}_* - \mathbf{z})]$$

and $\mathbf{e}_i \cdot \mathbf{u} = 0$. The complementarity condition (4.80e) then implies that $\mathbf{e}_i \cdot \mathbf{m_w} = 0$, and condition (4.80h) implies that

$$\mathbf{e}_i \cdot \boldsymbol{\ell}_\mathbf{a} = -\mathbf{e}_i \cdot \mathbf{c} = \mathbf{e}_i \cdot \mathbf{S_a}\mathbf{c} \; .$$

Again, there is no contribution to the second term in (4.81) for this component index i.

Suppose that $\mathbf{b}_j(\mathbf{z}) > 0$ and $\mathbf{e}_j \cdot \mathbf{J}_{\mathbf{b}}(\mathbf{z})(\mathbf{x}_* - \mathbf{z}) > 0$. Then

$$\mathbf{d}_j = \mathbf{v}_j = \mathbf{e}_j \cdot [\mathbf{b}(\mathbf{z}) + \mathbf{J}_{\mathbf{b}}(\mathbf{z})(\mathbf{x}_* - \mathbf{z})] \; .$$

The complementarity condition (4.80e) then implies that $\mathbf{e}_j \cdot \mathbf{m}_{\mathbf{v}} = 0$, and condition (4.80i) implies that

$$\mathbf{e}_j \cdot \mathbf{m}_{\mathbf{b}} = \mathbf{e}_j \cdot \mathbf{d} = \mathbf{e}_j \cdot \mathbf{S}_{\mathbf{b}}\mathbf{d} \; .$$

Thus there is no contribution to the third term in (4.81) for this component index j.

Finally, suppose that $\mathbf{b}_j(\mathbf{z}) \le 0$ and $\mathbf{e}_j \cdot \mathbf{J}_{\mathbf{b}}(\mathbf{z})(\mathbf{x}_* - \mathbf{z}) < 0$. Then $\mathbf{e}_j \cdot \mathbf{v} = 0$. The complementarity condition (4.80e) then implies that $\mathbf{e}_j \cdot \mathbf{m}_{\mathbf{b}} = 0$, and the definition of $\mathbf{S}_{\mathbf{b}}$ implies that $\mathbf{e}_j \cdot \mathbf{S}_{\mathbf{b}}\mathbf{d} = 0$. Again, there is no contribution to the third term in (4.81) for this component index j.

The remaining contributions to the second term in (4.81) must be such that $\mathbf{e}_i \cdot \mathbf{a}(\mathbf{z})$ and $\mathbf{e}_i \cdot \mathbf{J}_{\mathbf{a}}(\mathbf{z})(\mathbf{x}_* - \mathbf{z})$ have opposite signs. If $\mathbf{e}_i \cdot \mathbf{a}(\mathbf{z}) > 0$ and $\mathbf{e}_i \cdot \mathbf{J}_{\mathbf{a}}(\mathbf{z})(\mathbf{x}_* - \mathbf{z}) < 0$, then

$$\mathbf{e}_i \cdot [\mathbf{S}_{\mathbf{a}}\mathbf{c} - \boldsymbol{\ell}_{\mathbf{a}}] = \mathbf{e}_i \cdot [\mathbf{c} - \boldsymbol{\ell}_{\mathbf{a}}] = \mathbf{e}_i \cdot \mathbf{m}_{\mathbf{u}} \ge 0 \; ,$$

and the contribution to the second term in (4.81) is non-positive. Otherwise, $\mathbf{e}_i \cdot \mathbf{a}(\mathbf{z}) \le 0$ and $\mathbf{e}_i \cdot \mathbf{J}_{\mathbf{a}}(\mathbf{z})(\mathbf{x}_* - \mathbf{z}) > 0$, so

$$\mathbf{e}_i \cdot [\mathbf{S}_{\mathbf{a}}\mathbf{c} - \boldsymbol{\ell}_{\mathbf{a}}] = \mathbf{e}_i \cdot [-\mathbf{c} - \boldsymbol{\ell}_{\mathbf{a}}] = -\mathbf{e}_i \cdot \mathbf{m}_{\mathbf{w}} \le 0 \; ,$$

and the contribution to the second term in (4.81) is also non-positive.

Furthermore, the remaining contributions to the third term in (4.81) must be such that $\mathbf{e}_i \cdot \mathbf{b}(\mathbf{z})$ and $\mathbf{e}_i \cdot \mathbf{J}_{\mathbf{b}}(\mathbf{z})(\mathbf{x}_* - \mathbf{z})$ have opposite signs. If $\mathbf{e}_i \cdot \mathbf{b}(\mathbf{z}) > 0$ and $\mathbf{e}_j \cdot \mathbf{J}_{\mathbf{b}}(\mathbf{z})(\mathbf{x}_* - \mathbf{z}) < 0$, then

$$\mathbf{e}_i \cdot [\mathbf{S}_{\mathbf{b}}\mathbf{d} - \mathbf{m}_{\mathbf{b}}] = \mathbf{e}_i \cdot [\mathbf{d} - \mathbf{m}_{\mathbf{b}}] = \mathbf{e}_i \cdot \mathbf{m}_{\mathbf{v}} \ge 0 \; ,$$

and the contribution to the third term in (4.81) is non-positive. Otherwise, $\mathbf{e}_i \cdot \mathbf{b}(\mathbf{z}) \le 0$ and $\mathbf{e}_j \cdot \mathbf{J}_{\mathbf{b}}(\mathbf{z})(\mathbf{x}_* - \mathbf{z}) > 0$, so

$$\mathbf{e}_i \cdot [\mathbf{S}_{\mathbf{b}}\mathbf{d} - \mathbf{m}_{\mathbf{b}}] = \mathbf{e}_i \cdot [-\mathbf{m}_{\mathbf{b}}] \le 0$$

and the contribution to the third term in (4.81) is also non-positive.

Inequality (4.81) and the discussion above has shown that

$$\begin{aligned}
&\mathbf{g}_{\psi}(\mathbf{z})\cdot(\mathbf{x}_*-\mathbf{z})\\
&=-(\mathbf{x}_*-\mathbf{z})^\top\mathbf{H}(\mathbf{x}_*-\mathbf{z})+(\mathbf{S}_\mathbf{a}\mathbf{c}-\boldsymbol{\ell}_\mathbf{a})^\top\mathbf{J}_\mathbf{a}(\mathbf{z})(\mathbf{x}_*-\mathbf{z})+(\mathbf{S}_\mathbf{b}\mathbf{d}-\mathbf{m}_\mathbf{b})^\top\mathbf{J}_\mathbf{b}(\mathbf{z})(\mathbf{x}_*-\mathbf{z})\\
&\le-(\mathbf{x}_*-\mathbf{z})^\top\mathbf{H}(\mathbf{x}_*-\mathbf{z})\;.
\end{aligned}$$

Since $\mathbf{H}$ is a symmetric positive matrix, this final expression is negative.

At this point, the design of a sequential quadratic programming algorithm should be somewhat straightforward. Given some guess $\mathbf{z}_k$ for the solution of the original nonlinear programming problem (4.74) and a symmetric positive matrix $\mathbf{H}_k$, we can choose a penalty $\gamma > 0$, let $\mathbf{c} = \mathbf{e}\gamma$ and $\mathbf{d} = \mathbf{e}\gamma$ and solve the modified quadratic programming problem (4.78). We use the solution $\mathbf{x}_k$ of this sequential quadratic programming problem to perform a line search on the modified objective (4.79). The result of the line search is taken to be $\mathbf{z}_{k+1}$, but the matrix $\mathbf{H}_{k+1}$ must be updated in a somewhat delicate fashion. Details for a careful implementation of such an algorithm can be found in Gill et al. [82]. The ideas in this paper have been implemented in the SNOPT software package.

In Fig. 4.6, we show contours of minimization objectives within their feasible regions. In this case, the original problem is

$$\min -\mathbf{x}_0^2-\mathbf{x}_1^2 \text{ subject to } 2\mathbf{x}_0^2+\mathbf{x}_1^2\le 4\;,\;\mathbf{x}_0^2+3\mathbf{x}_1^2\le 9 \text{ and } \mathbf{x}\ge\mathbf{0}\;.$$

The left-hand graph shows equally spaced contours of this objective within the feasible region. The right-hand graph shows contours of a sequential quadratic programming problem objective within the linearized constraints, as described by Eq. (4.75). This quadratic programming problem is interpolated from the at $\mathbf{z}_k$ equal to the solution $\mathbf{z} = [\sqrt{3/5}\;,\;\sqrt{14/5}]$ of the original problem. Note that the global minimum of the quadratic programming problem occurs along the horizontal

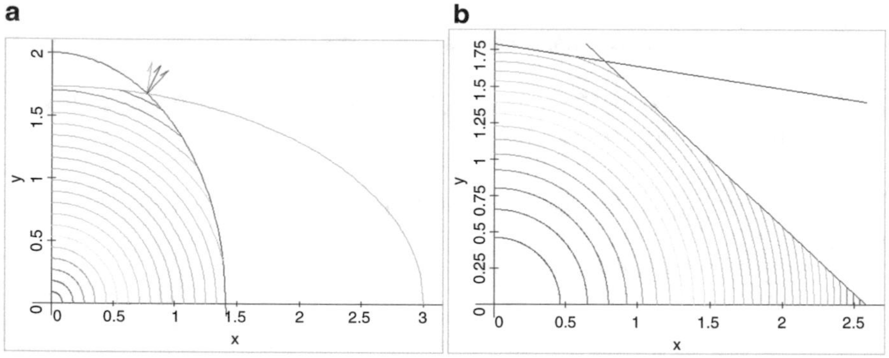

Fig. 4.6 Contours of $-\mathbf{x}_1^2-\mathbf{x}_2^2$ for $\mathbf{x}$ feasible. (**a**) Original problem. (**b**) Problem interpolating at $\mathbf{x} = [\sqrt{3/5}\;,\;\sqrt{14/5}]$

axis, and is very different from a local minimum at the solution of the original problem. Furthermore, the global minimum of the quadratic programming problem is infeasible for the original problem. This figure demonstrates why the sequential programming algorithm uses the solution of each individual quadratic programming problem to perform a line search that maintains feasibility of the original problem. This figure also provides motivation for the use of trust regions. The right-hand graph was drawn by the C^{++} program sqp.C, and the left-hand graph appeared previously in Fig. 4.3.

Alternative sequential quadratic programming algorithms for nonlinear constrained optimization include NLPQL (which is due to Schittkowski [158]), SLSQP (which is due to Kraft [112]), LSSQP (which is due to Eldersveld [60]), NPSOL (which is due to Gill et al. [80]), and TOMP (which is due to Kraft [113]). A number of other algorithms are available in NLopt.

For a survey paper on sequential quadratic programming algorithms, see Gould and Toint [86]. For an interesting paper regarding the use of trust regions with sequential quadratic programming, as well as an alternative to penalty functions, see Fletcher et al. [71].

4.5.5 Derivative-Free Methods

To conclude the chapter, we will merely mention that there are two interesting algorithms for solving constrained minimization problems without using derivatives. Powell's [145] **BOBYQA** (Bound Optimization BY Quadratic Approximation) algorithm and his earlier [144] **COBYLA** (Constrained Optimization BY Linear Approximations) algorithm are both available at either Powell Software or NLopt. Analysis supporting the convergence of COBYLA is due to Conn et al. [40].

References

1. J. Abadie, The GRG method for nonlinear programming, in *Design and Implementation of Optimization Software*, ed. by H.J. Greenberg (Sijthoff and Noordhoff, Alphen aan den Rijn, 1978), pp. 335–362 4.5.2.3
2. H. Akaike, On a successive transformation of probability distribution and its application to the analysis of the optimum gradient method. Ann. Inst. Stat. Math. Tokyo **11**, 1–17 (1959) 2.5.2
3. F. Aluffi-Pentini, V. Parisi, F. Zirilli, Algorithm 667: sigma–a stochastic-integration global minimimization algorithm. ACM TOMS **14**(4), 366–380 (1988) 3.10.1
4. K.J. Arrow, F.J. Gould, S.M. Howe, A general saddle point result for constrained optimization. Math. Program. **5**, 225–234 (1973) 4.5.3.2
5. M. Avriel, *Nonlinear Programming: Analysis and Methods* (Dover, New York, 2003) 4.1, 4.5.1, 4.5.3.2, 4.5.3.2
6. O. Axelsson, Incomplete block matrix factorization preconditioning methods. The ultimate answer? J. Comput. Appl. Math. **12**, 3–18 (1985) 2.5.4
7. O. Axelsson, *Iterative Solution Methods* (Cambridge University Press, New York, 1994) 2.1, 2.3.2, 2.4.2, 2.4.4, 2.5.3, 2.5.4, 2.6.3, 2.6.4, 2.6.1, 2.6.1, 2.6.2
8. K. Aziz, A. Settari, *Petroleum Reservoir Simulation* (Applied Science, London, 1979) 2.1
9. Z. Bai, J. Demmel, J. Dongarra, A. Ruhe, H. van der Vorst, *Templates for the Solution of Algebraic Eigenvalue Problems* (SIAM, Philadelphia, 2000) 1.1
10. R. Barrett, M. Berry, T.F. Chan, J. Demmel, J. Donato, J. Dongarra, V. Eijkhout, R. Pozo, C. Romine, H. Van der Vorst, *Templates for the Solution of Linear Systems: Building Blocks for Iterative Methods*, 2nd edn. (SIAM, Philadelphia, PA, 1994) 2.1, 2.6.3
11. R.H. Bartels, A stabilization of the simplex method. Numer. Math. **16**, 414–434 (1971) 4.2.6, 4.2.8
12. R.H. Bartels, G.H. Golub, The simplex method of linear programming using LU decomposition. Commun. Assoc. Comput. Mach. **12**, 266–268 (1969) 4.2.8
13. F.L. Bauer, C.T. Fike, Norms and exclusion theorems. Numer. Math. **2**, 123–144 (1960) 1.4.8
14. M.S. Bazaraa, H.D. Sherali, C.M. Shetty, *Nonlinear Programming: Theory and Algorithms* (Wiley, Hoboken, 2006) 4.3.5
15. A. Ben-Israel, D. Cohen, On iterative computation of generalized inverses and associated projections. SIAM J. Numer. Anal. **3**, 410–419 (1966) 1.5.3
16. G. Beylkin, R. Coifman, V. Rokhlin, Wavelets in Numerical Analysis, in *Wavelets and Their Applications* (Jones and Bartlett, Boston, 1992), pp. 181–210 1.5.3
17. R. Bhatia, *Perturbation Bounds for Matrix Eigenvalues* (SIAM, Philadelphia, PA, 2007) 1.4.2

J.A. Trangenstein, *Scientific Computing*, Texts in Computational Science and Engineering 19, https://doi.org/10.1007/978-3-319-69107-7

18. E.G. Birgin, J.M. Martínez, Improving ultimate convergence of an augmented Lagrangian method. Optim. Methods Softw. **23**(2), 177–195 (2008) 4.5.3.4
19. Å. Björck, S. Hammarling, A Schur method for the square root of a matrix. Linear Algebra Appl. **52/53**, 127–140 (1983) 1.7.3
20. Å. Björck, G.H. Golub, Numerical methods for computing angles between linear subspaces. Math. Comput. **27**(3), 579–594 (1973) 1.2.11
21. D. Braess, *Finite Elements* (Cambridge University Press, Cambridge, 2007) 2.7, 2.7.1, 2.7.3, 2.7.4, 2.7.1
22. K. Braman, R. Byers, R. Mathias, The multi-shift QR algorithm part I: maintaining well focused shifts, and level 3 performance. SIAM J. Matrix Anal. **23**, 929–947 (2002) 1.4.8.7
23. K. Braman, R. Byers, R. Mathias, The multi-shift QR algorithm part II: aggressive early deflation. SIAM J. Matrix Anal. **23**, 948–973 (2002) 1.4.8.7
24. J.H. Bramble, *Multigrid Methods*. Pitman Research Notes in Mathematics, vol. 294 (Longman Scientific and Technical, London, 1993) 2.4.3, 2.7
25. A. Brandt, Multi-level adaptive solutions to boundary-value problems. Math. Comput. **31**(193), 333–390 (1977) 2.7
26. S.C. Brenner, L.R. Scott, *The Mathematical Theory of Finite Element Methods* (Springer, Heidelberg, 2002) 2.7, 2.7.3
27. C. Brezinski, M.R. Zaglia, H. Sadok, Avoiding breakdown and near-breakdown on Lanczos type algorithms. Numer. Algorithms **1**, 199–206 (1991) 2.5.5.1
28. W. Briggs, *A Multigrid Tutorial* (SIAM, Philadelphia, 1987) 2.7.1
29. P.N. Brown, A local convergence theory for combined inexact-Newton/finite-difference projection methods. SIAM J. Numer. Anal. **24**, 407–434 (1987) 3.8.3, 3.8.1
30. P.N. Brown, Y. Saad, Convergence theory of nonlinear Newton-Krylov algorithms. SIAM J. Optim. **4**, 297–330 (1994) 3.8.3
31. C.G. Broyden, A class of methods for solving nonlinear simultaneous equations. Math. Comput. **19**, 577–593 (1965) 3.7.1
32. C.G. Broyden, J.E. Dennis Jr., J.J. Mor/'e, A class of methods for solving nonlinear simultaneous equations. Math. Comput. **19**, 577–593 (1965) 3.7.2, 3.7.2
33. J.R. Bunch, L. Kaufman, A computational method for the indefinite quadratic programming problem. Linear Algebra Appl. **34**, 341–370 (1980) 4.4.2, 4.4.8
34. R.H. Byrd, J.Ch. Gilbert, J. Nocedal, A trust region method based on interior point techniques for nonlinear programming. Math. Program. **89**, 149–185 (2000) 4.5.2.4
35. C.W. Carroll, The created response surface technique for optimizing nonlinear restrained systems. Oper. Res. **9**, 169–184 (1961) 4.5.2.4
36. V. Chvátal, *Linear Programming* (Freeman, New York, 1983) 4.1
37. E.A. Coddington, N. Levinson, *Theory of Ordinary Differential Equations* (McGraw-Hill, New York, 1955) 1.7
38. A.R. Conn, N.I.M. Gould, P.L. Toint, A globally convergent augmented lagrangian algorithm for optimization with general constraints and simple bounds. SIAM J. Numer. Anal. **28**(2), 545–572 (1991) 4.5.3.3, 4.5.3.3
39. A.R. Conn, N.I.M. Bould, P.L. Toint, *LANCELOT: A Fortran Package for Large Scale Nonlinear Optimization* (Springer, Berlin, 1992) 4.5.3.4
40. A.R. Conn, K. Scheinberg, P.L. Toint, A Globally convergent augmented lagrangian algorithm for optimization with general constraints and simple bounds, in *Approximation Theory and Optimization: Tributes to MJD Powell*, ed. by M.D. Buhmann, A. Iserles (Cambridge University Press, Cambridge, 1997) 4.5.5
41. R. Courant, D. Hilbert, *Methods of Mathematical Physics, Volume I* (Interscience, New York, 1953) 1.3.1.4
42. J.J.M. Cuppen, A divide and conquer method for the symmetric tridiagonal eigenproblem. Numer. Math. **36**(2), 177–195 (1981) 1.3.8, 1.3.8
43. J. Daniel, W.B. Gragg, L. Kaufman, G.W. Stewart, Reorthogonalization and stable algorithms for updating the Gram-Schmidt factorization. Math. Comput. **30**(136), 772–795 (1976) 2.6.2, 3.7.3.2, 4.2.8

44. G.B. Dantzig, *Linear Programming and Extensions* (Princeton University Press, Princeton, 1963) 4.1, 4.2.4, 4.2.10
45. W.C. Davidon, Variable metric methods for minimization. Technical Report ANL-5990, Argonne National Laboratory (1959) 3.7
46. G. Debreu, Definite and semidefinite quadratic forms. Econometrica **20**, 295–300 (1952) 4.1.4
47. R.S. Dembo, S.C. Eisenstat, T. Steihaug, Inexact Newton methods. SIAM J. Numer. Anal. **2**, 400–408 (1982) 3.4.6
48. J.W. Demmel, *Applied Numerical Linear Algebra* (SIAM, Philadelphia, 1997) 1.1
49. J.W. Demmel, O.A. Marques, B.N. Parlett, C. Vömel, Performance and accuracy of LAPACK's symmetric tridiagonal eigensolvers. Technical report, Lawrence Berkeley National Laboratory (2008) 1.3.10
50. E.D. Denman, A.N. Beavers, The matrix sign function and computations in systems. Appl. Math. Comput. **2**, 63–94 (1976) 1.7.3
51. J.E. Dennis Jr., J.J. Moré, A characterization of superlinear convergence and its application to quasi-Newton methods. Math. Comput. **28**, 549–560 (1974) 3.7.2, 3.7.2
52. J.E. Dennis Jr., R.B. Schnabel, *Numerical Methods for Unconstrained Optimization and Nonlinear Equations* (Prentice-Hall, Englewood Cliffs, 1983) 3, 3.1, 3.4.3, 3.4.5.2, 3.5.3.2, 3.5.3.2, 3.5.3.4, 3.6.1, 3.6.1.2, 3.6.1.3, 3.6.1.4, 3.7.2
53. I.S. Dhillon, B.N. Parlett, Multiple representations to compute orthogonal eigenvectors of symmetric tridiagonal matrices. Linear Algebra Appl. **387**(1), 1–28 (2004) 1.3.9
54. I.S. Dhillon, B.N. Parlett, Orthogonal eigenvectors and relative gaps. SIAM J. Matrix Anal. Appl. **257**(3), 858–899 (2004) 1.3.9, 1.3.9, 1.3.9
55. J. Dieudonné, *Foundations of Modern Analysis* (Academic, New York, 1967) 3.2.2, 3.2.3, 4.3.1
56. P.G. Drazin, *Nonlinear Systems* (Cambridge University Press, Cambridge, 1992) 1.1
57. Z. Drmac, K. Veselic, New fast and accurate Jacobi SVD algorithm II. SIAM J. Matrix Anal. Appl. **35**(2), 1343–1362 (2008) 1.5.10
58. T. Dupont, R.P. Kendal, H.H. Rachford Jr., An approximate factorization procedure for solving self-adjoint elliptic difference equations. SIAM J. Numer. Anal. **5**, 554–573 (1968) 2.5.4
59. C. Eckhart, G. Young, The approximation of one matrix by another of lower rank. Psychometrika **1**, 211–218 (1936) 1.5.6
60. S.K. Eldersveld, Large-scale sequential quadratic programming algorithms, Ph.D. thesis, Department of Operations Research, Stanford University, 1991 4.5.4
61. H.C. Elman, Iterative methods for large sparse nonsymmetric systems of linear equations, Ph.D. thesis, Computer Science Department, Yale University, 1982 2.6.4
62. J. Farkas, Über die Theorie der einfachen Ungleichungen. J. Reine Angew. Math. **214**, 1–24 (1902) 4.2.2
63. R.P. Fedorenko, The speed of convergence of one iterative process. USSR Comput. Math. Math. Phys. **4**, 1092–1096 (1964) 2.7
64. W. Feller, *An Introduction to Probability Theory and Its Applications*. Probability and Mathematical Statistics (Wiley, New York, 1968) 1.6
65. H.E. Fettis, J.C. Caslin, Eigenvalues and eigenvectors of Hilbert matrices of order 3 through 10. Math. Comput. **21**, 431–441 (1967) 1.3.17
66. A.V. Fiacco, G.P. McCormick, *Nonlinear Programming: Sequential Unconstrained Minimization Techniques* (Wiley, New York, 1968) 4.5.2.4
67. R. Fletcher, Conjugate gradient methods for indefinite systems, in *Proceedings of the Dundee Biennial Conference on Numerical Analysis 1974*, ed. by G.A. Watson (Springer, Heidelberg, 1975), pp. 73–89 2.5.5.2
68. R. Fletcher, An ℓ_1 penalty method for nonlinear constraints, in *Numerical Optimization 1984*, ed. by P.T. Boggs, R.H. Byrd, R.B. Schnabel (SIAM, Philadelphia, 1984), pp. 26–40 4.5.5
69. R. Fletcher, M.J.D. Powell, A rapidly convergent descent method for minimization. Comput. J. **6**, 163–168 (1963) 3.7

70. R. Fletcher, C.M. Reeves, Function minimization by conjugate gradients. Comput. J. **7**, 149–154 (1964) 3.8.2
71. R. Fletcher, N.I.M. Gould, S. Leyffer, Ph.L. Toint, A. Wächter, Global convergence of a trust-region SQP-filter algorithm for general nonlinear programming. SIAM J. Optim. **13**(3), 635–659 (2002) 4.5.4
72. A. Forsgren, P.E. Gill, M.H. Wright, Interior point methods for nonlinear optimization. SIAM Rev. **44**, 525–597 (2002) 4.5.2.4
73. J.G.F. Francis, The QR transformation, part I. Comput. J. **4**, 265–271 (1961) 1.3.7.7
74. R.W. Freund, N.M. Nachtigal, QMR: a quasi-minimum residual algorithm for non-Hermitian linear systems. Numer. Math. **60**, 315–339 (1991) 2.6.3, 2.6.3
75. J. Fried, H.D. Hansson, *Getting Real: The Smarter, Faster, Easier Way to Build a Successful Web Application* (37 Signals, Chicago, 2009) 4
76. Y.C. Fung, *Foundations of Solid Mechanics* (Prentice-Hall, Englewood Cliffs, 1965) 1.1
77. K. Gates, W.B. Gragg, Notes on TQR algorithms. J. Comput. Appl. Math. **86**, 195–203 (1997) 1.3.7.6
78. P.E. Gill, W. Murray, Newton-type methods for unconstrained and linearly constrained optimization. Math. Program. **28**, 311–350 (1974) 3.5.3.2
79. P.E. Gill, W. Murray, Numerically stable methods for quadratic programming. Math. Program. **14**, 349–372 (1978) 4.4.2
80. P.E. Gill, W. Murray, M.A. Saunders, M.H. Wright, Some theoretical properties of an augmented Lagrangian merit function, in *Advances in Optimization and Parallel Computing*, ed. by P.M. Pardalos (North-Holland, Amsterdam, 1992), pp. 101–128 4.5.4
81. P.E. Gill, W. Murray, M.H. Wright, *Practical Optimization* (Academic, New York, 1993). Tenth printing 3.1, 3.6.1, 4.5.1, 4.5.2.3
82. P.E. Gill, W. Murray, M.A. Saunders, SNOPT: an SQP algorithm for large-scale optimization. SIAM Rev. **47**(1), 99–131 (2005) 4.5.4
83. A.J. Goldman, A.W. Tucker, Theory of linear programming, in *Linear Inequalities and Related Systems*, ed. by H.W. Kuhn, A.W. Tucker. Annuals of Mathematical Studies, vol. 38 (Princeton University Press, Princeton, 1956) 4.2.10
84. G.H. Golub, C.F. van Loan, *Matrix Computations*, 4th edn. (Johns Hopkins, Baltimore, 2013) 1.2.11, 1.2.11, 1.4.1.3, 1.4.1.4, 1.4.2, 1.4.9, 1.4.7, 1.4.8.3, 1.4.8.6, 2.4.13, 2.5.4, 2.5.4, 3.7.3.2
85. G.H. Golub, J.H. Wilkinson, Ill-conditioned eigensystems and the computation of the Jordan canonical form. SIAM Rev. **18**, 578–619 (1976) 1.4.1.5
86. N.I.M. Gould, P.L. Toint, SQP methods for large-scale nonlinear programming (Cambridge, 1999), in *System Modelling and Optimization* (Kluwer Academic, Boston, MA, 2000) 4.5.4
87. V. Granville, M. Krivanek, J.-P. Rasson, Simulated annealing: a proof of convergence. IEEE Trans. Pattern Anal. Mach. Intell. **16**(6), 652–656 (1994) 3.10.2
88. A. Greenbaum, *Iterative Methods for Solving Linear Systems* (SIAM, Philadelphia, 1997) 2.1, 2.5.5.5
89. W. Hackbusch, *Multigrid Methods and Applications* (Springer, Berlin, 1985) 2.7
90. W. Hackbusch, *Iterative Solution of Large Sparse Systems of Equations*. Applied Mathematical Sciences, vol. 95 (Springer, New York, 1994) 2.1, 2.4.3, 2.4.6
91. L.A. Hageman, D.M. Young, *Applied Iterative Methods* (Academic, New York, 1981) 2.1
92. P. Halmos, *Finite-Dimensional Vector Spaces*. University Series in Higher Mathematics (Van Nostrand, Princeton, 1958) 1.4.1.5, 2.5.1
93. P. Henrici, *Error Propagation for Difference Methods* (Wiley, New York, 1963) 1.6
94. I.N. Herstein, *Topics in Algebra* (Blaisdell Publishing Company, Toronto, 1964) 1.4.1.5
95. M. Hestenes, Multiplier and gradient methods. J. Optim Theory Appl. **4**(5), 303–320 (1969) 4.5.3.2
96. M.R. Hestenes, E. Stiefel, Methods of conjugate gradients for solving linear systems. J. Res. Natl. Bur. Stand. **49**, 409–435 (1952) 2.5.3
97. N.J. Higham, Newton's method for the matrix square root. Math. Comput. **46**(174), 537–549 (1986) 1.7.3

98. N.J. Higham, Computing real square roots of a real matrix. Linear Algebra Appl. **88/89**, 405–430 (1987) 1.7.3
99. N.J. Higham, Stable iterations for the matrix square root. Numer. Algorithms **15**, 227–242 (1997) 1.7.3
100. D.S. Johnson, C.R. Argon, L.A. McGeoch, C. Schevon, Optimization by simulated annealing: an experimental evaluation, part I, graph partitioning. Oper. Res. **37**, 865–892 (1989) 3.10.2
101. M.T. Jones, P.E. Plassmann, An improved incomplete Cholesky factorization. ACM Trans. Math. Softw. **21**, 5–17 (1995) 2.5.4
102. L.V. Kantorovich, Functional analysis and applied mathematics. Uspechi Mat Nauk **3**, 89–185 (1948) 2.5.3
103. L.V. Kantorovoch, G.P. Akilov, *Functional Analysis in Normed Spaces*. International Series of Monographs in Pure and Applied Mathematics, vol. 46 (Pergaon Press, New York, 1964) 3.2.5.1, 3.2.5.1, 3.2.5.2
104. N. Karmarkar, A new polynomial time algorithm for linear programming. Combinatorica **4**(4), 373–394 (1984) 4.2.13, 4.2.13
105. R.M. Karp, George Dantzig's impact on the theory of computation. Discrete Optim. **5**(2), 174–185 (2008) 4
106. T. Kato, *Perturbation Theory for Linear Operators* (Springer, New York, 1966) 1.2.11
107. C.T. Kelley, *Iterative Methods for Linear and Nonlinear Equations* (SIAM, Philadelphia, 1995) 2.1, 3.1, 3.4.7, 3.8.3
108. C.T. Kelley, *Iterative Methods for Optimization* (SIAM, Philadelphia, 1999) 3.6.1, 3.9, 3.9.2
109. W. Kelley, A. Peterson, *The Theory of Differential Equations Classical and Qualitative* (Pearson Education, Upper Saddle River, 2004) 1.1, 1.7
110. S. Kirkpatrick, C.D. Gelatt, M.P. Vecchi, Optimization by simulated annealing. Science **220**, 671–680 (1983) 3.10.2
111. V. Klee, G.J. Minty, How good is the simplex algorithm? in *Inequalities-III*, ed. by O. Sisha (Academic, New York, 1972) 4.2.1
112. D. Kraft, A software package for sequential quadratic programming. Technical Report DFVLR-FB 88-28, DLR German Aerospace Center – Institute for Flight Mechanics, Koln, Germany (1988) 4.5.4
113. D. Kraft, Algorithm 733: Tomp-fortran modules for optimal control calculations. ACM Trans. Math. Softw. **20**(3), 262–281 (1994) 4.5.4
114. E. Kreyszig, *Introductory Functional Analysis with Applications* (Wiley, New York, 1978) 3.2.5.1, 3.2.5.1
115. V.N. Kublanovskaya, On some algorithms for the solution of the complete eigenvalue problem. Zh. Vychisl. Mat. **1**, 555–570 (1961) 1.3.7.7
116. J.D. Lambert, *Numerical Methods for Ordinary Differential Systems* (Wiley, New York, 1991) 1.6
117. C. Lanczos, An iteration method for the solution of the eigenvalue problem of linear differential and integral operators. J. Res. Natl. Bur. Stand. **45**, 255–282 (1950) 1.3.12
118. C. Lanczos, Solution of systems of linear equations by minimized iterations. J. Res. Natl. Bur. Stand. **49**, 33–53 (1952) 2.5.3
119. L.S. Lasdon, A.D. Waren, M. Ratner, Design and testing of a GRG code for nonlinear optimization. ACM Trans. Math. Softw. **4**, 34–50 (1978) 4.5.2.3
120. D.G. Luenberger (ed.), *Introduction to Linear and Nonlinear Programming* (Addison-Wesley, Reading, 1973) 2.5.2, 2.5.2, 2.5.3, 2.5.4, 3.1, 3.8.1, 4.1, 4.5.1, 4.5.2.2, 4.5.2.3
121. L.E. Malvern, *Introduction to the Mechanics of a Continuous Medium* (Prentice-Hall, Englewood Cliffs, 1969) 1.1, 1.7
122. J. Mandel, S. McCormick, R. Bank, Variational multigrid theory, in *Multigrid Methods*, ed. by S.F. McCormick (SIAM, Philadelphia, 1987) pp. 131–178 2.7
123. O.L. Mangasarian, *Nonlinear Programming* (McGraw-Hill, New York, 1969) 4.1, 4.2.2, 4.2.9.2
124. J.E. Marsden, T.J.R. Hughes, *Mathematical Foundations of Elasticity* (Prentice-Hall, Englewood Cliffs, 1983) 1.2.6, 1.7

125. S. Mehrotra, On the implementation of a primal-dual interior point method. SIAM J. Optim. **2**, 575–601 (1992) 4.2.13
126. C. Moler, C. Van Loan, Nineteen dubious ways to compute the exponential of a matrix, twenty-five years later. SIAM Rev. **45**(1), 3–49 (2003) 1.7.2
127. J.J. Mor/'e, D.C. Sorensen, On the use of directions of negative curvature in a modified Newton method. Math. Program. **16**, 1–20 (1979) 3.5.3.3
128. J.J. Mor/'e, J.A. Trangenstein, On the global convergence of Broyden's method. Math. Comput. **30**(135), 523–540 (1976) 3.6.2
129. J.J. Mor/'e, B.S. Garbow, K.E. Hillstrom, Testing unconstrained optimization software. ACM TOMS **7**(1), 17–41 (1981) 3.11
130. W. Murray, *Numerical Methods for Unconstrained Optimization* (Academic, New York, 1972) 3.1
131. N.M. Nachtigal, S. Reddy, L.N. Trefethen, How fast are nonsymmetric matrix iterations? SIAM J. Matrix Anal. Appl. **13**, 778–795 (1992) 2.5.5.5
132. J.A. Nelder, R. Mead, A simplex method for function minimization. Comput. J. **7**, 308–313 (1965) 3.9.2
133. J.M. Ortega, W.C. Rheinboldt, *Iterative Solution of Nonlinear Equations in Several Variables* (Academic, New York, 1970) 3.1, 3.2.2, 4.3.1
134. E.E. Osborne, On pre-conditioning of matrices. J. Assoc. Comput. Mach. **7**, 338–345 (1960) 1.4.8.1
135. A.M. Ostrowski, On the linear iteration procedures for symmetric matrices. Rend. Math. Appl. **14**, 140–163 (1954) 2.4.5
136. B.N. Parlett, Global convergence of the basic QR algorithm on Hessenberg matrices. Math. Comput. **22**(4), 803–817 (1968) 1.4.8.6
137. B.N. Parlett, *The Symmetric Eigenvalue Problem*. Series in Computational Mathematics (Prentice-Hall, Englewood Cliffs, 1980) 1.1, 1.3.2.4, 1.3.2.4, 1.3.6.4, 1.3.7, 1.3.7.4, 1.3.7.6
138. B.N. Parlett, W.G. Poole Jr., A geometric theory for the QR, LU and power iterations. SIAM J. Numer. Anal. **10**(2), 389–412 (1973) 1.4.7
139. B.N. Parlett, C. Reinsch, Balancing a matrix for calculation of eigenvalues and eigenvectors, in *Handbook for Automatic Computation Volume II Linear Algebra*, ed. by J.H. Wilkinson, C. Reinsch. Die Grundlehren der mathematischen Wissenschaften in Einzeldarstellungen, vol. 186 (Springer, Berlin, 1971), pp. 315–326 1.4.8.1
140. B.N. Parlett, D.R. Taylor, Z.A. Liu, A look-ahead Lanczos algorithm for unsymmetric matrices. Math. Comput. **44**, 105–124 (1985) 2.5.5.1
141. F.A. Potra, S.J. Wright, Interior-point methods? J. Comput. Appl. Math. **124**, 281–302 (1985) 4.2.13
142. M.J.D. Powell, A method for nonlinear constraints in minimization problems, in *Optimization*, ed. by R. Fletcher (Academic, London/New York, 1969), pp. 283–298 4.5.3.1, 4.5.3.1
143. M.J.D. Powell, A hybrid method for nonlinear equations, in *Numerical Methods for Nonlinear Algebraic Equations*, ed. by P. Rabinowitz (Gordon and Breach, London, 1970), pp. 87–114 3.6.1, 3.6.1.4
144. M.J.D. Powell, A direct search optimization method that models the objective and constraint functions by linear interpolation, in *Advances in Optimization and Numerical Analysis*, ed. by S. Gomez, J.-P. Hennart (Kluwer Academic, Dordrecht, 1994), pp. 51–67 4.5.5
145. M.J.D. Powell, Least Frobenius norm updating of quadratic models that satisfy interpolation conditions. Math. Program. **100**, 183–215 (2004) 4.5.5
146. G. Quintana-Orti, R. van de Geijn, Improving the performance of reduction to Hessenberg form. ACM Trans. Math. Softw. **32**(2), 180–194 (2006) 1.4.8.1
147. F.M. Rabinowitz, Algorithm 744: a stochastic algorithm for global optimization with constraints. ACM TOMS **21**(2), 194–213 (1995) 3.10.1
148. J.K. Reid, On the method of conjugate gradients for the solution of large sparse systems of linear equations, in *Large Sparse Sets of Linear Equations*, ed. by J.K. Reid (Academic, New York, 1971), pp. 231–254 2.5.3

149. J.K. Reid, A sparsity-exploiting variant of the Bartels-Golub decomposition for linear programming bases. Math. Program. **24**, 55–69 (1982) 4.2.8
150. F.S. Roberts, *Applied Combinatorics* (Prentice-Hall, Upper Saddle River, 1984) 1.1, 1.6
151. J.B. Rosen, The gradient projection method for nonlinear programming. part II. nonlinear constraints. J. Soc. Ind. Appl. Math. **9**(4), 514–532 (1961) 4.5.2.2
152. H.L. Royden, *Real Analysis*, 2nd edn. (Macmillan, New York, 1968) 3.2.5.1, 3.2.5.1
153. J.W. Ruge, K. Stüben, Algebraic multigrid, in *Multigrid Methods*, ed. by S.F. McCormick (SIAM, Philadelphia, 1987), pp. 73–130 2.7
154. Y. Saad, *Iterative Methods for Sparse Linear Systems* (PWS Publishing Co., Boston, 1996) 2.1, 2.5.4, 2.6.2, 2.6.2
155. Y. Saad, M. Schultz, GMRES: a generalized minimal residual algorithm for solving nonsymmetric linear equations. SIAM J. Sci. Stat. Comput. **7**(3), 856–869 (1986) 2.6.2, 2.6.2, 2.6.8, 2.6.1, 2.6.2
156. R.W.H. Sargent, B.A. Murtagh, Projection methods for nonlinear programming. Math. Program. **4**, 245–268 (1973) 4.5.2.3
157. M.A. Saunders, The complexity of LU updating in the simplex method, in *The Complexity of Computational Problem Solving*, ed. by R.S. Anderssen and R.P. Brent (Queensland University Press, Queensland, 1976), pp. 214–230 4.2.8
158. K. Schittkowski, NLPQL: a Fortran subroutine for solving constrained nonlinear programming problems. Ann. Oper. Res. **5**, 485–500 (1985/1986) 4.5.4
159. A. Schrijver, *Theory of Linear and Integer Programming* (Wiley, New York, 1986) 4.1, 4.2.10
160. G. Sierksma, Linear and Integer Programming: Theory and Practice (CRC Press, Boca Raton, 2001) 4.1, 4.2.10
161. P. Sonneveld, CGS, a fast Lanczos-type solver for nonsymmetric linear systems. SIAM J. Sci. Stat. Comput. **10**, 35–52 (1989) 2.5.5.3
162. W. Spendley, G.R. Hext, F.R. Himsworth, Sequential application of simplex designs in optimization and evolutionary operation. Technometrics **4**, 441–461 (1962) 3.9.2
163. G.W. Stewart, Error and perturbation bounds for subspaces associated with certain eigenvalue problems. SIAM Rev. **15**(4), 727–764 (1973) 1.2.11, 1.2.11, 1.2.11, 1.2.11, 1.3.2.2
164. G.W. Stewart, *Introduction to Matrix Computations* (Academic, New York, 1973) 1.4.1.3, 1.4.5, 1.4.6.3, 1.4.8.3
165. G.W. Stewart, On the perturbation of pseudo-inverses, projections and linear least squares problems. SIAM Rev. **19**(4), 634–662 (1977) 1.2.11
166. G.W. Stewart, *Matrix Algorithms Volume II: Eigensystems* (SIAM, Philadelphia, 2001) 1.1
167. K. Stüben, Appendix A: an introduction to algebraic multigrid, in *Multigrid*, ed. by U. Trottenberg, C. Oosterlee, A. Schüller (Academic, New York, 2001), pp. 413–532 2.7, 2.7.5
168. W.H. Swann, Direct search methods, in *Numerical Methods for Unconstrained Optimization*, ed. by W. Murray (Academic, 1972), pp. 13–28 3.9
169. D.B. Szyld, L. Moledo, B. Sauber, Positive solutions for the Leontief dynamic input-output model, in *Input-Output Analysis: Current Developments*, ed. by M. Ciaschini (Chapman and Hall, London, 1988) 1.1
170. R.A. Tapia, Diagonalized multiplier methods and quasi-Newton methods for constrained optimization. J. Optim. Theory Appl. **22**, 135–194 (1977) 4.5.3.4
171. K. Tone, Revisions of constraint approximations in the successive QP method for nonlinear programming problems. Math. Program. **26**, 144–152 (1983) 4.5.1, 4.5.4, 4.5.6
172. J.A. Trangenstein, *Numerical Solution of Hyperbolic Partial Differential Equations* (Cambridge University Press, Cambridge, 2009) (document)
173. J.A. Trangenstein, *Numerical Solution of Elliptic and Parabolic Partial Differential Equations* (Cambridge University Press, Cambridge, 2013) (document), 2.1, 2.2, 2.7.4, 2.7.5, 3.8.3
174. C. Truesdell, *The Elements of Continuum Mechanics* (Springer, New York, 1966) 1.1, 1.2.6
175. A. Tucker, *Applied Combinatorics* (Wiley, New York, 1995) 1.1, 1.6
176. H.A. van der Voorst, Bi-CGSTAB: a fast and smoothly converging variant of BI-CG for the solution of non-symmetric linear systems. SIAM J. Sci. Stat. Comput. **12**, 631–644 (1992) 2.5.5.4

177. R.S. Varga, *Matrix Iterative Analysis* (Springer, Berlin, 2000) 2.1, 2.4.5, 2.4.6, 2.4.4, 2.4.5
178. E.L. Wachspress (ed.), *Iterative Solution of Elliptic Systems and Applications to the Neutron Diffusion Equations of Reactor Physics* (Prentice-Hall, Englewood Cliffs, 1966) 2.1, 2.4.5, 2.4.12, 2.4.13
179. A. Wächter, L.T. Biegler, On the implementation of an interior-point filter line-search algorithm for large-scale nonlinear programming. Math. Program. **106**, 25–57 (2006) 4.5.2.4
180. R.A. Waltz, J.L. Morales, J. Nocedal, D. Orban, An interior algorithm for nonlinear optimization that combines line search and trust region steps. Math. Program. **107**, 391–408 (2006) 4.5.2.4, 4.5.2.4
181. D.S. Watkins, Understanding the QR algorithm. SIAM Rev. **24**(4), 427–440 (1982) 1.4.8.5
182. D.S. Watkins, Some perspectives on the eigenvalue problem. SIAM Rev. **35**(3), 430–471 (1993)
183. D.S. Watkins, The QR algorithm revisited. SIAM Rev. **50**(1), 133–145 (2008)
184. D.S. Watkins, Francis's algorithm. Am. Math. Mon. **118**, 387–403 (2011) 1.4.8.5
185. D.S. Watkins, L. Elsner, Convergence of algorithms of decomposition type for the eigenvalue problem. Linear Algebra Appl. **143**, 19–47 (1991) 1.4.8.6
186. J.H. Wilkinson, *The Algebraic Eigenvalue Problem* (Oxford University Press, Oxford, 1965) 1.1, 1.4.2, 1.4.6, 1.4.8.1, 1.4.8.2, 1.4.8.6, 1.4.8.6
187. J.H. Wilkinson, C. Reinsch, *Handbook for Automatic Computation Volume II Linear Algebra*, Die Grundlehren der mathematischen Wissenschaften in Einzeldarstellungen, vol. 186 (Springer, Berlin, 1971) 1.1
188. W.L. Winston, *Introduction to Mathematical Programming* (Duxbury Press, Belmont, 1995) 4.2.11
189. D.M. Young, *Iterative Solution of Large Linear Systems* (Academic, New York, 1971) 2.1, 2.4.1, 2.4.3, 2.4.3
190. A.A. Zhigljavsky, *Theory of Global Random Search* (Kluwer Academic, Dordrecht, 1991) 3.10.1
191. G. Zoutendijk, *Methods of Feasible Directions* (Elsevier, Amsterdam, 1960) 4.5.2.1

Notation Index

J.A. Trangenstein, *Scientific Computing*, Texts in Computational Science and Engineering 19, https://doi.org/10.1007/978-3-319-69107-7

Author Index

J.A. Trangenstein, *Scientific Computing*, Texts in Computational Science and Engineering 19, https://doi.org/10.1007/978-3-319-69107-7

Subject Index

J.A. Trangenstein, *Scientific Computing*, Texts in Computational Science and Engineering 19, https://doi.org/10.1007/978-3-319-69107-7

Editorial Policy

1. Textbooks on topics in the field of computational science and engineering will be considered. They should be written for courses in CSE education. Both graduate and undergraduate textbooks will be published in TCSE. Multidisciplinary topics and multidisciplinary teams of authors are especially welcome.

2. Format: Only works in English will be considered. For evaluation purposes, manuscripts may be submitted in print or electronic form, in the latter case, preferably as pdf- or zipped ps-files. Authors are requested to use the LaTeX style files available from Springer at: http://www.springer.com/authors/book+authors/helpdesk?SGWID=0-1723113-12-971304-0 (Click on ⟶ Templates ⟶ LaTeX ⟶ monographs)
Electronic material can be included if appropriate. Please contact the publisher.

3. Those considering a book which might be suitable for the series are strongly advised to contact the publisher or the series editors at an early stage.

General Remarks

Careful preparation of manuscripts will help keep production time short and ensure a satisfactory appearance of the finished book.

The following terms and conditions hold:

Regarding free copies and royalties, the standard terms for Springer mathematics textbooks hold. Please write to martin.peters@springer.com for details.

Authors are entitled to purchase further copies of their book and other Springer books for their personal use, at a discount of 33.3% directly from Springer-Verlag.

Series Editors

Timothy J. Barth
NASA Ames Research Center
NAS Division
Moffett Field, CA 94035, USA
barth@nas.nasa.gov

Michael Griebel
Institut für Numerische Simulation
der Universität Bonn
Wegelerstr. 6
53115 Bonn, Germany
griebel@ins.uni-bonn.de

David E. Keyes
Mathematical and Computer Sciences
and Engineering
King Abdullah University of Science
and Technology
P.O. Box 55455
Jeddah 21534, Saudi Arabia
david.keyes@kaust.edu.sa

and

Department of Applied Physics
and Applied Mathematics
Columbia University
500 W. 120 th Street
New York, NY 10027, USA
kd2112@columbia.edu

Risto M. Nieminen
Department of Applied Physics
Aalto University School of Science
and Technology
00076 Aalto, Finland
risto.nieminen@tkk.fi

Dirk Roose
Department of Computer Science
Katholieke Universiteit Leuven
Celestijnenlaan 200A
3001 Leuven-Heverlee, Belgium
dirk.roose@cs.kuleuven.be

Tamar Schlick
Department of Chemistry
and Courant Institute
of Mathematical Sciences
New York University
251 Mercer Street
New York, NY 10012, USA
schlick@nyu.edu

Editor for Computational Science
and Engineering at Springer:
Martin Peters
Springer-Verlag
Mathematics Editorial IV
Tiergartenstrasse 17
69121 Heidelberg, Germany
martin.peters@springer.com

Texts in Computational Science and Engineering

1. H. P. Langtangen, *Computational Partial Differential Equations*. Numerical Methods and Diffpack Programming. 2nd Edition
2. A. Quarteroni, F. Saleri, P. Gervasio, *Scientific Computing with MATLAB and Octave*. 4th Edition
3. H. P. Langtangen, *Python Scripting for Computational Science*. 3rd Edition
4. H. Gardner, G. Manduchi, *Design Patterns for e-Science*.
5. M. Griebel, S. Knapek, G. Zumbusch, *Numerical Simulation in Molecular Dynamics*.
6. H. P. Langtangen, *A Primer on Scientific Programming with Python*. 5th Edition
7. A. Tveito, H. P. Langtangen, B. F. Nielsen, X. Cai, *Elements of Scientific Computing*.
8. B. Gustafsson, *Fundamentals of Scientific Computing*.
9. M. Bader, *Space-Filling Curves*.
10. M. Larson, F. Bengzon, *The Finite Element Method: Theory, Implementation and Applications*.
11. W. Gander, M. Gander, F. Kwok, *Scientific Computing: An Introduction using Maple and MATLAB*.
12. P. Deuflhard, S. Röblitz, *A Guide to Numerical Modelling in Systems Biology*.
13. M. H. Holmes, *Introduction to Scientific Computing and Data Analysis*.
14. S. Linge, H. P. Langtangen, *Programming for Computations* - A Gentle Introduction to Numerical Simulations with MATLAB/Octave.
15. S. Linge, H. P. Langtangen, *Programming for Computations* - A Gentle Introduction to Numerical Simulations with Python.
16. H.P. Langtangen, S. Linge, *Finite Difference Computing with PDEs* - A Modern Software Approach.
17. B. Gustafsson, *Scientific Computing from a Historical Perspective*
18. J.A. Trangenstein, *Scientific Computing* - Vol. I. – Linear and Nonlinear Equations
19. J.A. Trangenstein, *Scientific Computing* - Vol. II. – Eigenvalues and Optimization
20. J.A. Trangenstein, *Scientific Computing* - Vol. III. – Approximation and Integration

For further information on these books please have a look at our mathematics catalogue at the following URL: `www.springer.com/series/5151`

Monographs in Computational Science and Engineering

1. J. Sundnes, G.T. Lines, X. Cai, B.F. Nielsen, K.-A. Mardal, A. Tveito, *Computing the Electrical Activity in the Heart*.

For further information on this book, please have a look at our mathematics catalogue at the following URL: `www.springer.com/series/7417`

Lecture Notes in Computational Science and Engineering

1. D. Funaro, *Spectral Elements for Transport-Dominated Equations.*
2. H.P. Langtangen, *Computational Partial Differential Equations.* Numerical Methods and Diffpack Programming.
3. W. Hackbusch, G. Wittum (eds.), *Multigrid Methods V.*
4. P. Deuflhard, J. Hermans, B. Leimkuhler, A.E. Mark, S. Reich, R.D. Skeel (eds.), *Computational Molecular Dynamics: Challenges, Methods, Ideas.*
5. D. Kröner, M. Ohlberger, C. Rohde (eds.), *An Introduction to Recent Developments in Theory and Numerics for Conservation Laws.*
6. S. Turek, *Efficient Solvers for Incompressible Flow Problems.* An Algorithmic and Computational Approach.
7. R. von Schwerin, ***M**ulti **B**ody **S**ystem **SIM**ulation.* Numerical Methods, Algorithms, and Software.
8. H.-J. Bungartz, F. Durst, C. Zenger (eds.), *High Performance Scientific and Engineering Computing.*
9. T.J. Barth, H. Deconinck (eds.), *High-Order Methods for Computational Physics.*
10. H.P. Langtangen, A.M. Bruaset, E. Quak (eds.), *Advances in Software Tools for Scientific Computing.*
11. B. Cockburn, G.E. Karniadakis, C.-W. Shu (eds.), *Discontinuous Galerkin Methods.* Theory, Computation and Applications.
12. U. van Rienen, *Numerical Methods in Computational Electrodynamics.* Linear Systems in Practical Applications.
13. B. Engquist, L. Johnsson, M. Hammill, F. Short (eds.), *Simulation and Visualization on the Grid.*
14. E. Dick, K. Riemslagh, J. Vierendeels (eds.), *Multigrid Methods VI.*
15. A. Frommer, T. Lippert, B. Medeke, K. Schilling (eds.), *Numerical Challenges in Lattice Quantum Chromodynamics.*
16. J. Lang, *Adaptive Multilevel Solution of Nonlinear Parabolic PDE Systems.* Theory, Algorithm, and Applications.
17. B.I. Wohlmuth, *Discretization Methods and Iterative Solvers Based on Domain Decomposition.*
18. U. van Rienen, M. Günther, D. Hecht (eds.), *Scientific Computing in Electrical Engineering.*
19. I. Babuška, P.G. Ciarlet, T. Miyoshi (eds.), *Mathematical Modeling and Numerical Simulation in Continuum Mechanics.*
20. T.J. Barth, T. Chan, R. Haimes (eds.), *Multiscale and Multiresolution Methods.* Theory and Applications.
21. M. Breuer, F. Durst, C. Zenger (eds.), *High Performance Scientific and Engineering Computing.*
22. K. Urban, *Wavelets in Numerical Simulation.* Problem Adapted Construction and Applications.
23. L.F. Pavarino, A. Toselli (eds.), *Recent Developments in Domain Decomposition Methods.*

24. T. Schlick, H.H. Gan (eds.), *Computational Methods for Macromolecules: Challenges and Applications.*

25. T.J. Barth, H. Deconinck (eds.), *Error Estimation and Adaptive Discretization Methods in Computational Fluid Dynamics.*

26. M. Griebel, M.A. Schweitzer (eds.), *Meshfree Methods for Partial Differential Equations.*

27. S. Müller, *Adaptive Multiscale Schemes for Conservation Laws.*

28. C. Carstensen, S. Funken, W. Hackbusch, R.H.W. Hoppe, P. Monk (eds.), *Computational Electromagnetics.*

29. M.A. Schweitzer, *A Parallel Multilevel Partition of Unity Method for Elliptic Partial Differential Equations.*

30. T. Biegler, O. Ghattas, M. Heinkenschloss, B. van Bloemen Waanders (eds.), *Large-Scale PDE-Constrained Optimization.*

31. M. Ainsworth, P. Davies, D. Duncan, P. Martin, B. Rynne (eds.), *Topics in Computational Wave Propagation.* Direct and Inverse Problems.

32. H. Emmerich, B. Nestler, M. Schreckenberg (eds.), *Interface and Transport Dynamics.* Computational Modelling.

33. H.P. Langtangen, A. Tveito (eds.), *Advanced Topics in Computational Partial Differential Equations.* Numerical Methods and Diffpack Programming.

34. V. John, *Large Eddy Simulation of Turbulent Incompressible Flows.* Analytical and Numerical Results for a Class of LES Models.

35. E. Bänsch (ed.), *Challenges in Scientific Computing - CISC 2002.*

36. B.N. Khoromskij, G. Wittum, *Numerical Solution of Elliptic Differential Equations by Reduction to the Interface.*

37. A. Iske, *Multiresolution Methods in Scattered Data Modelling.*

38. S.-I. Niculescu, K. Gu (eds.), *Advances in Time-Delay Systems.*

39. S. Attinger, P. Koumoutsakos (eds.), *Multiscale Modelling and Simulation.*

40. R. Kornhuber, R. Hoppe, J. Périaux, O. Pironneau, O. Wildlund, J. Xu (eds.), *Domain Decomposition Methods in Science and Engineering.*

41. T. Plewa, T. Linde, V.G. Weirs (eds.), *Adaptive Mesh Refinement – Theory and Applications.*

42. A. Schmidt, K.G. Siebert, *Design of Adaptive Finite Element Software.* The Finite Element Toolbox ALBERTA.

43. M. Griebel, M.A. Schweitzer (eds.), *Meshfree Methods for Partial Differential Equations II.*

44. B. Engquist, P. Lötstedt, O. Runborg (eds.), *Multiscale Methods in Science and Engineering.*

45. P. Benner, V. Mehrmann, D.C. Sorensen (eds.), *Dimension Reduction of Large-Scale Systems.*

46. D. Kressner, *Numerical Methods for General and Structured Eigenvalue Problems.*

47. A. Boriçi, A. Frommer, B. Joó, A. Kennedy, B. Pendleton (eds.), *QCD and Numerical Analysis III.*

48. F. Graziani (ed.), *Computational Methods in Transport.*

49. B. Leimkuhler, C. Chipot, R. Elber, A. Laaksonen, A. Mark, T. Schlick, C. Schütte, R. Skeel (eds.), *New Algorithms for Macromolecular Simulation.*

50. M. Bücker, G. Corliss, P. Hovland, U. Naumann, B. Norris (eds.), *Automatic Differentiation: Applications, Theory, and Implementations.*

51. A.M. Bruaset, A. Tveito (eds.), *Numerical Solution of Partial Differential Equations on Parallel Computers.*

52. K.H. Hoffmann, A. Meyer (eds.), *Parallel Algorithms and Cluster Computing.*

53. H.-J. Bungartz, M. Schäfer (eds.), *Fluid-Structure Interaction.*

54. J. Behrens, *Adaptive Atmospheric Modeling.*

55. O. Widlund, D. Keyes (eds.), *Domain Decomposition Methods in Science and Engineering XVI.*

56. S. Kassinos, C. Langer, G. Iaccarino, P. Moin (eds.), *Complex Effects in Large Eddy Simulations.*

57. M. Griebel, M.A Schweitzer (eds.), *Meshfree Methods for Partial Differential Equations III.*

58. A.N. Gorban, B. Kégl, D.C. Wunsch, A. Zinovyev (eds.), *Principal Manifolds for Data Visualization and Dimension Reduction.*

59. H. Ammari (ed.), *Modeling and Computations in Electromagnetics: A Volume Dedicated to Jean-Claude Nédélec.*

60. U. Langer, M. Discacciati, D. Keyes, O. Widlund, W. Zulehner (eds.), *Domain Decomposition Methods in Science and Engineering XVII.*

61. T. Mathew, *Domain Decomposition Methods for the Numerical Solution of Partial Differential Equations.*

62. F. Graziani (ed.), *Computational Methods in Transport: Verification and Validation.*

63. M. Bebendorf, *Hierarchical Matrices.* A Means to Efficiently Solve Elliptic Boundary Value Problems.

64. C.H. Bischof, H.M. Bücker, P. Hovland, U. Naumann, J. Utke (eds.), *Advances in Automatic Differentiation.*

65. M. Griebel, M.A. Schweitzer (eds.), *Meshfree Methods for Partial Differential Equations IV.*

66. B. Engquist, P. Lötstedt, O. Runborg (eds.), *Multiscale Modeling and Simulation in Science.*

67. I.H. Tuncer, Ü. Gülcat, D.R. Emerson, K. Matsuno (eds.), *Parallel Computational Fluid Dynamics 2007.*

68. S. Yip, T. Diaz de la Rubia (eds.), *Scientific Modeling and Simulations.*

69. A. Hegarty, N. Kopteva, E. O'Riordan, M. Stynes (eds.), *BAIL* 2008 – *Boundary and Interior Layers.*

70. M. Bercovier, M.J. Gander, R. Kornhuber, O. Widlund (eds.), *Domain Decomposition Methods in Science and Engineering XVIII.*

71. B. Koren, C. Vuik (eds.), *Advanced Computational Methods in Science and Engineering.*

72. M. Peters (ed.), *Computational Fluid Dynamics for Sport Simulation.*

73. H.-J. Bungartz, M. Mehl, M. Schäfer (eds.), *Fluid Structure Interaction II - Modelling, Simulation, Optimization.*

74. D. Tromeur-Dervout, G. Brenner, D.R. Emerson, J. Erhel (eds.), *Parallel Computational Fluid Dynamics 2008.*

75. A.N. Gorban, D. Roose (eds.), *Coping with Complexity: Model Reduction and Data Analysis.*

76. J.S. Hesthaven, E.M. Rønquist (eds.), *Spectral and High Order Methods for Partial Differential Equations.*

77. M. Holtz, *Sparse Grid Quadrature in High Dimensions with Applications in Finance and Insurance.*

78. Y. Huang, R. Kornhuber, O. Widlund, J. Xu (eds.), *Domain Decomposition Methods in Science and Engineering XIX.*

79. M. Griebel, M.A. Schweitzer (eds.), *Meshfree Methods for Partial Differential Equations V.*

80. P.H. Lauritzen, C. Jablonowski, M.A. Taylor, R.D. Nair (eds.), *Numerical Techniques for Global Atmospheric Models.*

81. C. Clavero, J.L. Gracia, F.J. Lisbona (eds.), *BAIL 2010 – Boundary and Interior Layers, Computational and Asymptotic Methods.*

82. B. Engquist, O. Runborg, Y.R. Tsai (eds.), *Numerical Analysis and Multiscale Computations.*

83. I.G. Graham, T.Y. Hou, O. Lakkis, R. Scheichl (eds.), *Numerical Analysis of Multiscale Problems.*

84. A. Logg, K.-A. Mardal, G. Wells (eds.), *Automated Solution of Differential Equations by the Finite Element Method.*

85. J. Blowey, M. Jensen (eds.), *Frontiers in Numerical Analysis - Durham 2010.*

86. O. Kolditz, U.-J. Gorke, H. Shao, W. Wang (eds.), *Thermo-Hydro-Mechanical-Chemical Processes in Fractured Porous Media - Benchmarks and Examples.*

87. S. Forth, P. Hovland, E. Phipps, J. Utke, A. Walther (eds.), *Recent Advances in Algorithmic Differentiation.*

88. J. Garcke, M. Griebel (eds.), *Sparse Grids and Applications.*

89. M. Griebel, M.A. Schweitzer (eds.), *Meshfree Methods for Partial Differential Equations VI.*

90. C. Pechstein, *Finite and Boundary Element Tearing and Interconnecting Solvers for Multiscale Problems.*

91. R. Bank, M. Holst, O. Widlund, J. Xu (eds.), *Domain Decomposition Methods in Science and Engineering XX.*

92. H. Bijl, D. Lucor, S. Mishra, C. Schwab (eds.), *Uncertainty Quantification in Computational Fluid Dynamics.*

93. M. Bader, H.-J. Bungartz, T. Weinzierl (eds.), *Advanced Computing.*

94. M. Ehrhardt, T. Koprucki (eds.), *Advanced Mathematical Models and Numerical Techniques for Multi-Band Effective Mass Approximations.*

95. M. Azaïez, H. El Fekih, J.S. Hesthaven (eds.), *Spectral and High Order Methods for Partial Differential Equations ICOSAHOM 2012.*

96. F. Graziani, M.P. Desjarlais, R. Redmer, S.B. Trickey (eds.), *Frontiers and Challenges in Warm Dense Matter.*

97. J. Garcke, D. Pflüger (eds.), *Sparse Grids and Applications – Munich 2012.*

98. J. Erhel, M. Gander, L. Halpern, G. Pichot, T. Sassi, O. Widlund (eds.), *Domain Decomposition Methods in Science and Engineering XXI.*

99. R. Abgrall, H. Beaugendre, P.M. Congedo, C. Dobrzynski, V. Perrier, M. Ricchiuto (eds.), *High Order Nonlinear Numerical Methods for Evolutionary PDEs - HONOM 2013.*

100. M. Griebel, M.A. Schweitzer (eds.), *Meshfree Methods for Partial Differential Equations VII.*

101. R. Hoppe (ed.), *Optimization with PDE Constraints - OPTPDE 2014.*

102. S. Dahlke, W. Dahmen, M. Griebel, W. Hackbusch, K. Ritter, R. Schneider, C. Schwab, H. Yserentant (eds.), *Extraction of Quantifiable Information from Complex Systems.*

103. A. Abdulle, S. Deparis, D. Kressner, F. Nobile, M. Picasso (eds.), *Numerical Mathematics and Advanced Applications - ENUMATH 2013.*

104. T. Dickopf, M.J. Gander, L. Halpern, R. Krause, L.F. Pavarino (eds.), *Domain Decomposition Methods in Science and Engineering XXII.*

105. M. Mehl, M. Bischoff, M. Schäfer (eds.), *Recent Trends in Computational Engineering - CE2014.* Optimization, Uncertainty, Parallel Algorithms, Coupled and Complex Problems.

106. R.M. Kirby, M. Berzins, J.S. Hesthaven (eds.), *Spectral and High Order Methods for Partial Differential Equations - ICOSAHOM'14.*

107. B. Jüttler, B. Simeon (eds.), *Isogeometric Analysis and Applications 2014.*

108. P. Knobloch (ed.), *Boundary and Interior Layers, Computational and Asymptotic Methods – BAIL 2014.*

109. J. Garcke, D. Pflüger (eds.), *Sparse Grids and Applications – Stuttgart 2014.*

110. H. P. Langtangen, *Finite Difference Computing with Exponential Decay Models.*

111. A. Tveito, G.T. Lines, *Computing Characterizations of Drugs for Ion Channels and Receptors Using Markov Models.*

112. B. Karazösen, M. Manguoğlu, M. Tezer-Sezgin, S. Göktepe, Ö. Uğur (eds.), *Numerical Mathematics and Advanced Applications - ENUMATH 2015.*

113. H.-J. Bungartz, P. Neumann, W.E. Nagel (eds.), *Software for Exascale Computing - SPPEXA 2013-2015.*

114. G.R. Barrenechea, F. Brezzi, A. Cangiani, E.H. Georgoulis (eds.), *Building Bridges: Connections and Challenges in Modern Approaches to Numerical Partial Differential Equations.*

115. M. Griebel, M.A. Schweitzer (eds.), *Meshfree Methods for Partial Differential Equations VIII.*

116. C.-O. Lee, X.-C. Cai, D.E. Keyes, H.H. Kim, A. Klawonn, E.-J. Park, O.B. Widlund (eds.), *Domain Decomposition Methods in Science and Engineering XXIII.*

117. T. Sakurai, S. Zhang, T. Imamura, Y. Yusaku, K. Yoshinobu, H. Takeo (eds.), *Eigenvalue Problems: Algorithms, Software and Applications, in Petascale Computing.* EPASA 2015, Tsukuba, Japan, September 2015.

118. T. Richter (ed.), *Fluid-structure Interactions.* Models, Analysis and Finite Elements.

119. M.L. Bittencourt, N.A. Dumont, J.S. Hesthaven (eds.), *Spectral and High Order Methods for Partial Differential Equations ICOSAHOM 2016.*

120. Z. Huang, M. Stynes, Z. Zhang (eds.), *Boundary and Interior Layers, Computational and Asymptotic Methods BAIL 2016.*

121. S.P.A. Bordas, E.N. Burman, M.G. Larson, M.A. Olshanskii (eds.), *Geometrically Unfitted Finite Element Methods and Applications.* Proceedings of the UCL Workshop 2016.

122. A. Gerisch, R. Penta, J. Lang (eds.), *Multiscale Models in Mechano and Tumor Biology.* Modeling, Homogenization, and Applications.

For further information on these books please have a look at our mathematics catalogue at the following URL: www.springer.com/series/3527